MAPPING PARADISE

MAPPING PARADISE
A History of Heaven on Earth

ALESSANDRO SCAFI

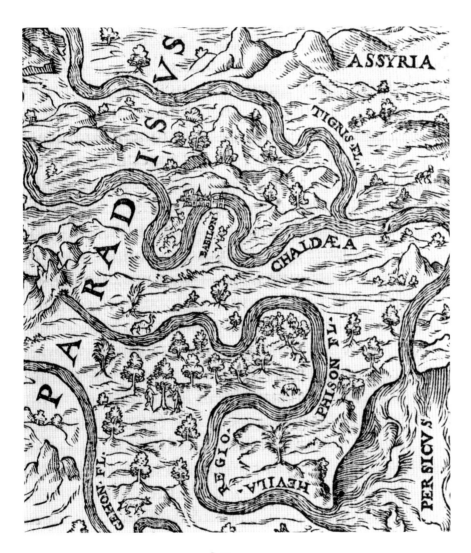

The University of Chicago Press

The University of Chicago Press, Chicago 60637
The British Library, London NW1 2DB

Text and drawings © 2006 Alessandro Scafi
Illustrations © 2006 The British Library Board
and other named copyright holders.
All rights reserved. Published 2006
Printed in Hong Kong by South Sea International Press

15 14 13 12 11 10 09 08 07 06   1 2 3 4 5

ISBN: 0-226-73559-1 (cloth)

Endpapers: Athanasius Kircher's map of paradise from his *Arca Noe*
(Amsterdam: Ioannes Ianssonius, 1675), between pp. 196 and 197.
Size of the original 28.5 × 43 cm. London, British Library, 460.c.9.
Detail of the plan of the Garden of Eden in Mesopotamia.

Frontispiece: Map of the vast region of paradise
(labeled *Paradisus*). From Matthaeus Beroaldus,
Chronicum, Scripturae Sacrae autoritate constitutum
(Geneva: A. Chuppinus, 1575), p. 88. Size of the original 12.5 × 15 cm. Detail.

Library of Congress Cataloging-in-Publication Data

Scafi, Alessandro.
Mapping Paradise : a history of heaven on earth / Alessandro Scafi.
p. cm.
Includes bibliographical references (p. 375) and index.
ISBN 0-226-73559-1 (cloth : alk. paper)
1. Paradise—Maps. 2. Geographical myths. 3. Paradise. I. Title.
BL540.S23 2006
202'.3—dc22
2005055973

∞ The paper used in this publication meets
the minimum requirements of the American
National Standard for Information Sciences—Permanence
of Paper for Printed Library Materials, ANSI Z39.48-1992.

Typeset by Hope Services Ltd
Designed by Andrew Shoolbred

The British Library would like to thank all copyright holders
for permission to reproduce material and illustrations.
While every effort has been made to trace and acknowledge copyright holders,
we would like to apologise for any errors or omissions.

Contents

Colour section between pages 192–3

Preface	6	
List of Abbreviations	9	

Prologue: Journeying to Paradise   10

1 Changing Views on Paradise on Maps   16
Nineteenth-Century Pioneers   19
Twentieth-Century Followers   23
A New Road to the Past   27

2 Paradise in the Bible   32
Rendering and Reading the Text   32
Early Debate   36
    Philosophical Interpretation: the Spiritual Eden   36
    Literal Interpretation: the Physical Eden   39

3 Locating Paradise in Space   44
Paradise as a Real Place on Earth   44
Naming the Place   47
    Isidore of Seville and Bede   47
    The Glossa ordinaria and Peter Lombard   49
Searching for Paradise   51
Finding Salvation   52

4 Locating Paradise in Time   62
The Relevance of Paradise Lost   63
Naming the Time   64
    How Long Ago? The Past Linked to the Present   64
    For How Long? The Past Linked to the Future   66
    Paradise and the Pattern of History   68
    Merging Eden, Church and Heaven   69
    Displaying History from Eden to Heaven   76

5 Mapping Paradise in Space and Time   84
Mapping the Land   84
Paradise on Christianized Maps   88
Mapping with Theology   94
    The Pseudo-Isidorean Vatican Map of the World   95
    Mapping Time   98
    Mapping Biblical Space   99
    Mapping Universal History   103
Mapping Heaven on Earth   104
    Beatus of Liébana's Map of the World   104
    Apocalypse Now   105
    Mapping the Church   108
    Eden and the Sixth Age   113

6 The Heyday of Paradise on Maps   125
The Ordering of Space and Time   125
The Dominance of History over Geography   128
The Prime Epochal Zone: Paradise   131
The Invisible Paradise on Maps   138
Mapping God's Creation and Redemption   141
The Utopian Search for a Place   152

7 Where Is Nowhere?   160
Paradise between East and West   160
A Garden between Heaven and Earth   163
Paradise and the Climatic Zones   165
Grasping the Mystery: Paradise and the Geographical Renaissance   170
    Paradise and the Terrestrial Globe   172
    Paradise On or Beyond the Equator   173
    The Mathematics of Paradise   176
An Unresolved Debate   179
A Poetic Flight to Paradise   182

8 The Twilight of Paradise on Maps   191
The End of the Honeymoon   191
Paradise East of Anywhere   193
A Humanist Search for Paradise   195
Paradise and Cartography in Flux   198
    Conventional Paradise in the Far East   202
    Conventional Paradise on Disorientated Maps   211
Mapping Paradise in Africa   218
    Paradise in the South   219
    Paradise in Equatorial Africa   226
The Coordinates of the Earthly Paradise   230
Paradise Cornered   235
Paradise and the Geographical Discoveries   240

9 Paradise Lost and Found   254
Changing Cartography   254
Changing Theology   258
    The Waning of the Idea of Corporeal Perfection   259
Changing Exegesis: Paradise Lost in Mesopotamia   261
Paradise as the Whole Earth   264
Paradise Flooded   266
Paradise Mapped in the Middle East   270

10 The Afterlife of Paradise on Maps   284
Everything Changes, Nothing Changes   285
A New Genre: Sacred Geography   288
Confluent Streams: Paradise in Mesopotamia   291
    The Problem of the Single River   295
    Variations on the Mesopotamian Theme   303
    An Underground Solution   313
Return to the Sources: Paradise in Armenia   317
Back to the Future: Paradise in the Holy Land   322

11 The Eclipse of the Theological Eden   342
The Theologian Gives up   343
A Babylonian Paradise   347
Ways of Thinking: Mainstream and Fringe   352

Epilogue: Paradise Then and Now   365

Bibliography   375
Index   391

Preface

Place is the beginning of our existence, just as a father.
Roger Bacon, *Opus maius*, I.1.5 (1265–7).

Back in the Summer of 1994, I paid a visit to the Vatican Library to order some of the images for this book. On the application form for the various items, I specified that in each case I wanted to have a reproduction not just of the map as a whole, but also of the earthly paradise as a detail. Assuming that my requests would be promptly processed and the images ready for dispatch within two to three weeks, I hastily returned to London to attend to a number of other pressing commitments. It was only after some considerable time that the bad news eventually reached me: one of my relatives at home in Rome had received an urgent call from a member of staff at the Vatican Library to let me know that, although they had not experienced any problems tracking down the maps I requested, they were unable to locate the earthly paradise... Now if, in the Vatican, I thought, they did not know where paradise was, then we really were in trouble.

Actually, the difficulties involved in finding paradise – whether on a map or on the very face of the earth itself – are such that the inability to do so is quite excusable for anyone in any circumstances. To find paradise, after all, would be equivalent to answering the paradoxical question: *where* is *nowhere*? Indeed, I myself have been engaged in the almost Sisyphean task of studying the various attempts made throughout history to find this *nowhere*. My starting point was the study of the utopian imagination in early modern Europe, but after repeated encounters with the cliché about the readiness of medieval people to believe straightforwardly in monsters and marvels, in contrast to the apparently critical circumspection of modern scientific thinkers, I was keen to look more closely at the notion, regularly ridiculed, that in the Middle Ages the earthly paradise was located on maps of the world, and to give it a fairer hearing. After all, Sir Thomas More, who is generally regarded as an open-minded thinker – credited with reviving the classical notion of utopia after centuries of religious obscurantism – firmly believed in the existence of a terrestrial paradise and even branded as 'heretike' anybody who thought otherwise.

Encouraged by the beauty of the cartographical documents and the depth and complexity of the issues involved, I was led well beyond the chronological boundaries of the Middle Ages and the Renaissance, drawn ever further by the irresistible temptation to trace as far as possible – up to the start of the third millennium – the seductive history of the mapping of paradise.

Very soon, though, I came across someone else who had also attempted to chart some regions of the scholarly territory on the topic. At the end of the nineteenth century an Italian, Edoardo Coli, had written about the appearance of paradise on medieval maps in *Il paradiso terrestre dantesco* (1897), which is still a valuable source on the subject. When, a century later, I discovered that Coli had taught my grandfather at school in a town near the Adriatic, I was happy to think that I was destined to write this book. I embraced destiny, and destiny transformed my solitary scholarly quest for paradise into a cooperative adventure, by placing along my path many people who have helped enormously. This book would not have been written without the extremely generous assistance of Catherine Delano-Smith, who gave invaluable and much appreciated guidance, encouragement and advice from the very outset of the project. Not only did she prove to be a safe guide in the field of the history of cartography, but she also edited my English with extreme rigour, dedication and competence. Her help in structuring and polishing my prose also enabled me to clarify my thoughts. Moreover, Catherine has put me in touch over the years with an international network of map scholars, who have assisted me in many different ways: Peter Barber, Anne-Dorothee von den Brincken, Tony Campbell, Angelo Cattaneo, Daniel Connolly, Denis Cosgrove, Evelyn Edson, Francis Herbert, Sumathy Ramaswamy, Zur Shalev, the late David Woodward and many others. In particular, I have benefited from the constructive comments made on my text by Michael Coogan, Patrick Gautier Dalché, Paul D. A. Harvey, Elizabeth M. Ingram and Robert A. Markus. Moreover, the book has been greatly improved by the thoughtful comments and detailed criticism of Scott D. Westrem, who proved to be an extremely sharp and careful reader.

The intellectual environment of the Warburg Institute undoubtedly enhanced my research a great deal, not only with the outstanding material resources of its library, but also because of its stimulating scholarly community and the enriching exchange of knowledge I have enjoyed there for so long. I owe a debt of gratitude to all staff, students and fellows who helped me in various ways with inspiring conversation, encouragement and practical help. Most of the research on the medieval exegesis of the earthly paradise was carried out for my doctoral thesis at the Warburg, and I am deeply grateful to my supervisor, Jill Kraye, for the extremely competent and friendly way in which she helped me to develop scholarly skills and has continued to support me ever since. Her sound help contributed greatly to the development of this book. Moreover, the alert criticism and rigorous advice of Christopher Ligota helped me focus and organize my arguments on the Christian Middle Ages. He read, discussed and edited earlier drafts of the book, asking me the right questions, pricking my curiosity to reach for the right answers and providing me with abundant bibliographical information. His help was crucial as I struggled to organize massive amounts of seemingly incoherent material. Likewise, I owe a special debt of gratitude to Jonathan Rolls, who provided invaluable support as well as essential bibliographical guidance and generous advice on the text as a whole. I am also grateful to Magnus Ryan, Mariana Giovino and Darin Hayton, who have read different parts of my work and made useful comments, and to Ronit Yoeli-Tlalim, who helped me with the Hebrew. Ian Bavington Jones also gave me the benefit of his expertise. In addition, the Warburg offered me the chance to profit from the advice and friendship of many people, including Julie Boch, Jenny Boyle, Charles Burnett, Imogen Cornwall-Jones, Ann Giletti, Ulrike Ilg, David Juste, Dorothea McEwan, Lucy McGuinness, the late Nicolai Rubinstein, Yuri Stoyanov, Michael Sylwanowicz, Hanna Vorholt and Claudia Wedepohl. Shortage of space prevents me from acknowledging everyone who has helped.

I have greatly benefited also from the possibility of discussing my work during various seminars and conferences at the Warburg Institute, Royal Holloway, the recently re-named Institute of Germanic and Romance Studies (University of London), at Hereford Cathedral, at the 18th and 21st International Conferences in the History of Cartography (Lisbon and Budapest), at the Musée du Louvre, at the Mishkenoth Sha'ananim of Jerusalem (where Tilo Schabert drew my attention to the photograph reproduced on page 15 below), at the Università Cattolica of Milan and at the Universities of Limoges and Bologna.

I would like to thank the libraries which provided me with images and with their kind permission to reproduce them: Bischofszell (Canton of Thurgovia, Switzerland), Ortsmuseum, Dr.-Albert-Knoepfli-Stiftung; Brussels, Bibliothèque Royale de Belgique/Koninklijke Bibliotheek van België; Burgo de Osma, Archivo de la Catedral; Cambridge, Parker Library, Corpus Christi College; Ebstorf, Kloster Ebstorf; Einsiedeln, Benediktinerabtei, Stiftsbibliothek; Florence, Biblioteca Medicea Laurenziana; Biblioteca Nazionale Centrale; Genoa, Biblioteca Durazzo Giustiniani; Hereford Cathedral; Konstanz, Heinrich-Suso-Gymnasium, Bibliothek; Lisbon, Arquivo Nacional da Torre do Tombo; London, College of Arms; Lambeth Palace Library; The British Library; The Warburg Institute; The Weisskopf Institute; Macon, Bibliothèque Municipale; Modena, Biblioteca Estense; New York, Metropolitan Museum; New York, Pierpont Morgan Library; Paris, Bibliothèque Nationale de France; Parma, Biblioteca Palatina; Philadelphia, Library of the American Philosophical Society; Stuttgart, Württembergische Landesbibliothek; Turin, Biblioteca Nazionale Universitaria; Utrecht, Rijksuniversiteit Bibliotheek; Vatican City, Biblioteca Apostolica Vaticana; Venice, Biblioteca Nazionale Marciana; Vercelli, Archivio e Biblioteca Capitolare; Verona, Biblioteca Comunale; Vicenza, Biblioteca Civica Bertoliana; Vienna, Österreichische Akademie der Wissenschaften. I am also grateful to the artists Ilya and Émilia Kabakov, Hendrikje Kühne and Beat Klein, and to Trish Boardman of Air Mauritius.

I am grateful to all the British Library staff who supervised the production of this book with such professionalism and commitment: David Way, who believed in it from the outset, as well as Bernard Dod, Kathy Houghton, Andrew Shoolbred, Lara Speicher, Belinda Wilkinson and Paul Wilson. Laura Nuvoloni helped me in the Department of Manuscripts. My stay in London during the writing of this book was made possible and enjoyable through the intervention of David Ward and Jemma Street. Finally, and above all, I am deeply grateful to my family for their constant support and for sharing – together with my friends, in England, Italy and elsewhere – my ongoing search for paradise.

Alessandro Scafi
February 2006

List of Abbreviations

CCCM	*Corpus Christianorum, Continuatio Medievalis*
CCSL	*Corpus Christianorum, Series Latina*
GCS	*Griechische christliche Schriftsteller*
CISC SL	*Corpus Islamo-Christianorum, Series Latina*
CSCO	*Corpus Scriptorum Christianorum Orientalium*
CSEL	*Corpus Scriptorum Ecclesiasticorum Latinorum*
MGH AA	*Monumenta Germaniae Historica, Auctores Antiquissimi*
PG	*Migne: Patrologia Graeca*
PL	*Migne: Patrologia Latina*
RBMAS	*Rerum Britannicarum Medii Aevi Scriptores*
RED SM	*Rerum Ecclesiasticarum Documenta, Series Maior*
SC	*Sources chrétiennes*

Author's Note

Classical texts are cited by author and title, with references given according to the standard division of the texts. As for other works, full bibliographical references are given the first time they are quoted in each chapter and in the Bibliography. Punctuation and spelling of Latin have been normalized; accents have been introduced or changed according to modern usage; abbreviations have been expanded. All translations into English are my own, unless otherwise indicated. The drawings accompanying many of the maps are my own; but they are not facsimiles, as they include only certain features, selected to help the reader understand the original maps or follow a particular line of argument.

Paradise begins in Air Mauritius

The smoothness of a trip, the warmth of a service and the promise of a dreamland... Air Mauritius

AIR MAURITIUS

Air Mauritius Centre, President John Kennedy Street, Port Louis, Mauritius.

Prologue: Journeying to Paradise

It is not down in any map; true places never are.
Herman Melville, *Moby Dick, or The Whale* (1851)

For many people today, the promise of paradise is just a commercial gimmick enticing jaded urban citizens of the Western world to a distant dreamland. The tourist industry proclaims that this or that paradise is only a short journey or a simple flight away (Figure 0.1). Once the price of initiation has been paid at the travel agency, the airline assures those who feel oppressed by their complicated and busy lives that they may lie back in their seats cocooned in the comfort of a smooth trip and soothed by warm, friendly service. For the paradise-seeker, the flight of the imagination begins the moment the aeroplane leaves the earth to spear the clouds. The stressed-out businessman dreams of a utopian island – in our case Mauritius – promising escape, if only for a week or two, from the cares of modern life and the claustrophobia of the city, guaranteeing renewal through contact with nature, and assuring total relaxation and complete happiness. Exploiting this theme, in 2002 two Swiss artists, Hendrikje Kühne and Beat Klein, created a *Map of Paradise.* Their huge cartographical pastiche, a collage of maps cut from brochures they had collected from travel agents along the high streets of Western Europe, shows a world composed entirely of holiday resorts (Figures 0.2 and 0.3). Kühne and Klein have created a world where places have become destinations and destinations have become places. Each dreamland has been built up from the relevant brochures of the manipulative geography of tourism that targets the imagination of the potential traveller. Their world is a world composed of palm trees, endless empty sandy beaches and brightly coloured swimming pools, all promising paradise.

Look more closely, though, and you will find that the only map-making rule observed by Kühne and Klein is their aesthetic use of colour. There is no other cartographical apparatus. Their map lacks a key, a scale bar, any system of coordinates and consistent orientation. The dreamlands are elusive, without firm borders. The *Map of Paradise* presents a mirage; it is impossible to put the finger on paradise itself. The world of Kühne and Klein is higgledy-piggledy, the places all jumbled up, and the map does not function as we think a map should. What, anyway, is paradise and for whom is it an attraction? For the Mauritian islander, the global crossroads of a big city may offer a paradise of unfamiliar thrills and new opportunities. Yet this is just the New York, the London, or the Paris that the businessman of Air Mauritius's advertisement has been enticed into escaping. Paradise, we discover, changes according to time and place. With their map, the artists Kühne and Klein seem to be inviting us to search for paradise within ourselves, rather than to heed the siren call of the tourist brochures.[1] After all, no sane person would buy a return ticket from a true and proper paradise.

The nineteenth-century German philosopher Arthur Schopenhauer observed that the enchantment of distance reveals paradises that later vanish like optical illusions. For Schopenhauer, happiness was always in the future or in the past; the present is a small dark cloud driven by the wind over a sunny plain; all before and all behind is

Fig. 0.1 (opposite). *Paradise begins in Air Mauritius.* Advertisement, 2001. It is no longer so easy to plough the Seas of the South in the same spirit of adventure as that which drove Herman Melville's heroes, or to arrive in the Caribbean expecting to be close to the earthly paradise, as Christopher Columbus is alleged to have thought in 1498. Nonetheless, the tourist industry assures us all that it is possible for every one of us to book a holiday in paradise. The prospect of a luxurious and welcoming natural environment suggests a return to the primordial sources of life itself. Today, though, it is cash, not divine assistance, which is needed to be able to cross the threshold of paradise. Mauritius was described as a paradise in the eighteenth century by the French botanist Pierre Poivre.

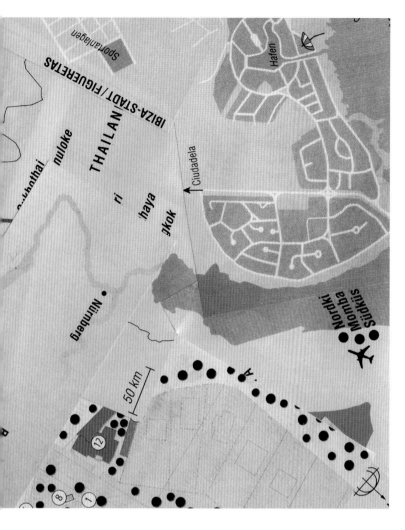

Fig. 0.2. Hendrikje Kühne and Beat Klein, *Map of Paradise*, exhibited at The Gallery, Guernsey, 2002. 200 × 360 cm. Collection of the artists. Detail of work in progress. Photo: Fiona Adams, Guernsey. The artists Kühne and Klein collected tourist brochures from which they cut out and glued together the maps found therein to create a cartographical collage to show a reformulated map of the world on which the world's tourist destinations are represented. The word 'paradise' is frequently adopted by those who write the advertisements for the tourist industry as a means of firing the imagination of potential travellers. A far-away idyllic spot may be isolated by geographical distance but, thanks to the travel agency, it can be reached. Of course the encouragement to 'book a holiday in paradise' is a persuasive advertising ploy. Nevertheless it echoes the old religious invitation to think in terms beyond those of ordinary space and time.

bright light, only the cloud itself casts a deep shadow.[2] Looking back at history, however, we find that, however gloomy the place and however cloudy the season, mankind has always searched for that light beyond the horizon and yearned for the brightness that went before and that is to come. Elusive it may be, but throughout history paradise has appeared everywhere in a variety of secular and religious guises, always thought of as 'elsewhere' and 'out of time'. Visions of the perfect happiness of past and future times and of present but distant places are common to all humanity. The world's literature is rich in stories about times of peace, plenty and justice, and packed with accounts of secluded realms where there are fountains of immortality and where it is always spring. All ancient peoples, such as Sumerians and Babylonians, and civilizations such as those of the ancient Chinese, Indians and Egyptians, had their paradises. The Greeks and the Romans also conceived of a golden age and blessed islands. The Garden of Eden was a Hebrew tradition before Christians adopted the Hebrew Bible, and with it the Genesis story. Through the Christian conquest of Rome, the history and culture of virtually the whole of Europe and the Mediterranean basin came to be deeply influenced by Christian teaching. For two millennia the word 'paradise' has indicated the paradise described in the Christian Bible, the paradise with which we are concerned in this book. The paradises of other religious traditions (such as the Jewish, Muslim and Hindu) do not feature in the history of heaven on earth that will be told in the following pages, which are concerned with the Western Christian tradition; the Orthodox Christian and the Judaic traditions only hover in the background.

For Jews and Christians, the biblical narrative opens with the description of one paradise, the Garden of Eden (Genesis 2.8–14) and for Christians it closes with the description of another paradise, that of the Heavenly Jerusalem (Revelation 21.1–3; 22.1–2). The Garden of Eden is the enchanting place where the first human couple, Adam and Eve, lived in a blissful state of perfection. The Heavenly Jerusalem is the final perfection that is to be established by God at the end of human history. In the Holy Scriptures, and consequently in Christian theology, the symmetrical images of the beginning and the end – the garden and the city – tended to overlap, but there is a distinction, which should always be borne in mind. For Christian believers, the Heavenly Jerusalem of Revelation is the total and definitive redemption in heaven, while the Garden of Eden of Genesis is the place on earth of original human innocence. Long before travel agencies promised the modern urban masses of the West a paradise on earth, generations of Christians believed in the earthly existence of a pristine spot where God had placed Adam and Eve at the dawn of time, and which was supposed to have been the ideal and perfect habitat for mankind.

Beside the maps of the tourist brochures, there are modern maps showing the biblical earthly paradise or the Garden of Eden (the terms are interchangeable). In 1998, David Rohl claimed that he had found, once and for all and after nearly two millennia of debate, the location of the original Garden of Eden (Figures 0.4 and 0.5).[3] According to Rohl, Adam and Eve's paradise was an agriculturally rich plain some 60 miles wide and 200 long, enclosed by mountain ranges, in north-western Iran, not far from the city of Tabriz. Rohl has also identified, and plotted on a map, the four rivers of Eden that Genesis tells us flowed out of paradise: the Tigris, Euphrates, Gihon, and Pishon (2.10–14). From the time of Augustine (fifth century) to the Renaissance, the most learned scholars in all Europe, Africa and Asia, agreed that the Gihon and the Pishon were the Nile and the Ganges, an idea put forward by the first-century Jewish historian Flavius Josephus.[4] According to Rohl, however, the Gihon and the Pishon are two rivers in Azerbaijan and Kurdistan, which flow into the Caspian Sea: the Araxes, known before the Islamic invasion of the eighth century as the Gaihun (the linguistic equivalent of the Hebrew Gihon), and the Uizhun (a variation on the Hebrew Pishon). Rohl also identifies the biblical land of Cush (across which, according to Genesis, the Gihon flows) as the range of the Kusheh Dagh, sited about 100 miles east along the valley in which his Garden of Eden is situated, and claims to have discovered the single river (unnamed in the Bible) that watered the Garden, identifying it with the Adji Chay. A village called Noqdi is a relic of the land of Nod, the land to the east of Eden to which, according to Genesis, Cain was exiled after murdering his brother Abel (4.16). Havilah, described in Genesis as a place rich in gold and surrounded by the Pishon, is in Rohl's opinion in Kurdistan, where there used to be, it is true, a couple of gold mines. The case of Rohl confirms that the story of the biblical paradise has lost none of its appeal over the centuries. The fertile paradise of Genesis remains in the memory in other ways too: Figure 0.6 shows a group of modern actors using their bodies to form the word 'paradise'. But it is not only in the theatre or in the churches that the words 'paradise' and 'Eden' are still rehearsed. The two words are common currency in a diversity of contexts, ranging from hotel and restaurant management, catering, novel writing, film making to tourism, as we have seen above. Whichever way we turn, we come across echoes of the ancient story of Adam and Eve in paradise.

Fig. 0.3. Hendrikje Kühne and Beat Klein, *Map of Paradise*. Detail of work in progress. Photo: Fiona Adams, Guernsey. Nobody today claims that the Garden of Eden is still somewhere on the globe, but many still long for 'Eden-like' places on earth. Geographical discovery has always pushed paradise one step further beyond the boundary of the known world. But now that these boundaries are pushed *ad infinitum* to outer space – in 1961 Yuri Gagarin, the first man into orbit, declared that he did not encounter God during his flight – humanity's longing for a 'beyond' has been replaced by attempts to rebuild it within the confines of our own world.

The history of Christian belief in an earthly paradise fully merits exploration. Paradise has appeared in almost every form of cultural expression: poetry, visual art, literary and philosophical writing. It has also appeared on maps, and Rohl's map is only one of the latest in a long series of attempts to locate paradise cartographically. Maps offer the historian of ideas a peculiarly valuable point of entry for rediscovering a forgotten world where the notion of an earthly paradise was never irrational, absurd or trivial, but always predicated on belief and knowledge. The enterprise of depicting paradise on a map of the world, or part of the world, has for centuries involved a major intellectual challenge. The mapping of the earthly paradise, the history of heaven on earth as depicted on maps from Late Antiquity to the twenty-first century, is the subject of this book.

Interpreting the text of the Bible has never been easy or straightforward. Two thousand years of Christian exegesis have seen generations of scholars, from medieval monks to humanist writers and Reformers and to contemporary historians, struggling

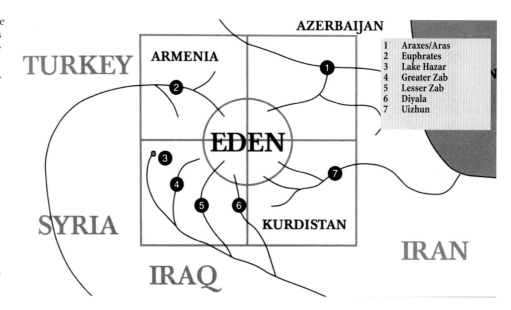

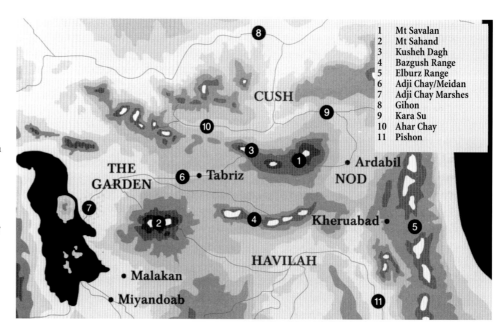

Fig. 0.4. The four rivers of paradise flowing out from the four quarters of Eden, from David Rohl's *Legend: The Genesis of Civilisation* (London: Century, 1998), p. 55. 10 × 17 cm. Rohl announces his intention to venture 'where others fear to tread' in order to explain 'one of the most difficult puzzles handed down to us from the ancient world'. After acknowledging his debt to the scholar Reginald Arthur Walker (1917–89), Rohl establishes the identities of the four rivers of paradise, showing that they all flow in Armenia. Both the Pishon (which he identi-fies with the Uizhun, in the south-east, no. 7 on the map) and the Gihon (identified with the Araxes, in the north-east, no. 1) flow into the Caspian Sea. The Tigris (whose main source rises in Lake Hazar, no. 3, and which receives the waters of the Greater and the Lesser Zab and the Diyala, nos 4, 5 and 6 respectively, in the south-western sector) and the Euphrates (in the north-western sector, no. 2) empty into the Persian Gulf.

Fig. 0.5. Simplified relief map showing the location of the Garden of Eden and the lands of Cush and Havilah, from David Rohl, *Legend: The Genesis of Civilisation* (London: Century, 1998), p. 67. 10 × 17 cm. Rohl identifies the biblical Eden with the Adji Chay valley (formerly known as the Meidan valley) in north-west Iran, at the heart of which is the regional capital of Tabriz. The Garden is 'in the east of Eden', protected on its north, east and south sides by the Savalan (no. 1) and Sahand (no. 2) mountains, and on its west side by the marshy delta of the Adji Chay (no. 7), which stretches out into Lake Urmia. To the north of the Garden lies the Kusheh Dagh ('Mountain of Kush', no. 3). Havilah is south of the Bazgush range (no. 4), in the Iranian district of Anguran, renowned for its mineral wealth. The town of Kheruabad (the 'settlement of the Kheru') possibly owes its name to the Cherubim, the angels guarding Eden (Genesis 3.24). The land of Nod, 'east of Eden' (Genesis 4.16), the destination of Cain's exile, is in the plain west of the Elburz mountains (no. 5) around the city of Ardabil.

to understand the scriptural account of a paradise on earth and to comprehend what the Garden of Eden was and, above all, where it was. Their learned arguments have left an enormous heritage of books and commentaries, treatises, pamphlets and letters – a veritable mountain of parchment and paper – to be deciphered and distilled. A fifth river of paradise would be the ink spilt over the years in the attempt to penetrate the mystery of the written word. In their turn, the compilers of the maps that will be discussed in this book also had to persevere to find ways of translating the theological monochrome into the often colourful spectrum of the cartographers's visual language. We cannot but admire the careful judgement of these map makers in their displays of the complex and paradoxical relationship between paradise and earth that the exegetes had debated with erudition and ingenuity. In a number of cases, it was the same hand that wrote the commentary and drew the map.

The history of the mapping of the earthly paradise thus involves confronting profound theological matters as well as yielding to the seductive imagery of the maps.

Fig. 0.6. Actors forming the word 'paradise' with their bodies. Production staged in the 1970s by the *Living Theatre*, an experimental group created by Julien Beck and Judith Malina. Photo: Gianfranco Mantegna. The actors were expressing with their bodies the nostalgia for paradise, the ideal and perfect state of mankind. Ancient words such as 'paradise' and 'Eden' still have meaning today, and not only within the realm of religion. They are extremely popular in the most diverse fields of human activities, from hotels and restaurants to novels and movies. 'Eden Garden' is the name of the biggest cricket ground in the world, built in Calcutta; the 'Eden Project' aims to re-create in Cornwall a microcosm of the planet with its different environments inside huge glass spheres.

A world of difference, of course, lies between our modern map – mathematically constructed for use in a post-Enlightenment society – and the map compiled by those for whom Christian belief governed the description of the natural world. The mathematical precision of modern satellite- and computer-generated representations of the earth may seem to have little in common with the 'picturesque' medieval maps of the world, or even with the maps that were reconstructed towards the end of the Middle Ages and early in the Renaissance from the text of Ptolemy's *Geography*. In the rush to praise the technological progress in mapmaking, it might be tempting to dismiss the mapping of paradise as a naïve fairy tale. The tendency to shame the simple-mindedness of earlier times is not uncommon in modern scholarship as well as popular books. The aim in this book is to visit the foreign country of the past with as open a mind as possible and in the belief that, as Shakespeare's Fluellen expressed it, 'there is [sic] occasions and causes why and wherefore in all things'.[5] There have been reasons, we learn, for putting paradise on a map, and the task we have set ourselves is to discover them.

1   The artists Kühne and Klein produced their cartographical collage at the International Artist in Residence Programme in Guernsey, Channel Islands, in November 2002. I am grateful to the artists and to Joanna Littlejohns, curator of The Gallery, Guernsey, for providing me with images of the map and the catalogue of the installation. For examples of how mapping has been reconsidered by artists see Denis Cosgrove, 'Maps, Mapping, Modernity: Art and Cartography in the Twentieth Century', *Imago Mundi*, 57/1 (2005), pp. 35–54 and Arlette Lemonnier, ed., *Le Dessus des cartes: Art et cartographie* (Brussels: ISELP, 2004).
2   Arthur Schopenhauer, *Die Welt als Wille und Vorstellung*, 3rd edn, 2 vols (Leipzig: Brochaus, 1859), II, p. 655; *The World as Will and Idea*, transl. by Richard B. Haldane and John Kemp, 3 vols (London: Trübner, 1883–6), III (1886), p. 383.
3   David Rohl, *Legend: The Genesis of Civilisation* (London: Century, 1998), pp. 46–68.
4   Flavius Josephus, *Jewish Antiquities* I.37–9, Loeb Classical Library, transl. by Henry St. John Thackeray, Ralph Marcus, and Louis H. Feldman, 9 vols (London: Heinemann; Cambridge, MA: Harvard University Press, 1926–65), IV, transl. by Thackeray (1930), pp. 18–21. On Augustine see Chapter 3, p. 46. See also Reinhold R. Grimm, *Paradisus coelestis, paradisus terrestris: Zur Auslegungsgeschichte des Paradieses im Abendland bis um 1200* (Munich: Fink, 1977), pp. 121–8.
5   William Shakespeare, *The Life of King Henry V*, Act V, Scene I.

Changing Views on Paradise on Maps

A map of the world that does not include utopia is not even worth glancing at.
Oscar Wilde, The Soul of Man under Socialism (1892)

One day in 1442 the Venetian cartographer Giovanni Leardo laid down his pen, having signed and dated the map he had just completed. The map was a masterpiece, blending text and image into a portrait of the earth (Figure 1.1). In a text below the map proper, he identified the three concentric circles that surround the earth. The innermost indicated the twelve months of the year, the second contained the twelve signs of the zodiac, and the outermost gave the dates for Easter for the next hundred years or so.[1] Leardo knew that by computing the date of Easter – which had to be the first Sunday following the first full moon after the spring equinox – human time, religious truths, and heavenly rhythms would be reconciled.[2] On the circle with the zodiac, the top of the map coincides with the passage from Pisces to Aries, the constellation believed to have been in the ascendant when God created the world, Mary conceived the Saviour, and Christ was resurrected from the dead.[3] Once it is realized that east, marked by a decorative cross, is at the top of the map, the outlines of the Mediterranean Sea and Europe are readily recognized, for Leardo delineated them with great accuracy. As was not unusual in late medieval and early Renaissance times, he must have taken his outlines from a nautical chart, a quite different type of map used in contemporary navigation.[4] For other aspects of his geographical description, he followed the biblical text. For example, he emphasized the centrality of Jerusalem by placing the city at the intersection of the two lines which divide the map into four quarters.[5]

On Leardo's map, the east–west (vertical) line cuts across the Indian Ocean, crosses the Persian Gulf, reaches the Mediterranean just south of Cyprus, and touches the coast of Africa at Tunis. Passing through Spain and Portugal, it terminates in the *Mar de Spagnia*, the Sea of Spain. The horizontal line links north with south. It passes first through the inscription *Dixerto dexabitado per fredo* ('the desert uninhabited because of the cold'), then cuts across the river Don and the Black Sea, the Red Sea, Ethiopia and the 'land of the Pharaohs' before reaching another inscription which announces *Dixerto dexabitado per chaldo e per serpenti* ('the desert uninhabited because of heat and snakes'). The two deserts on the map – one in the north and one in the south, both uninhabitable because of extremes of temperature – coincide with the two solstices (December and June) indicated on the encircling calendar.

Leardo's map depicts the whole of the habitable world as it was known in the mid fifteenth century. Fully aware that the earth is a sphere, but faced with the problem of

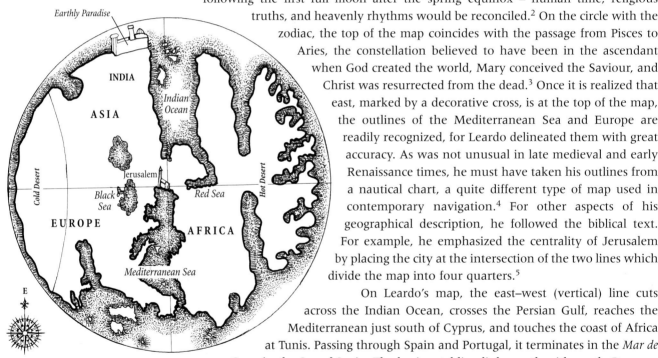

Fig. 1.1a. Diagram of Leardo's map (see Fig. 1.1b).

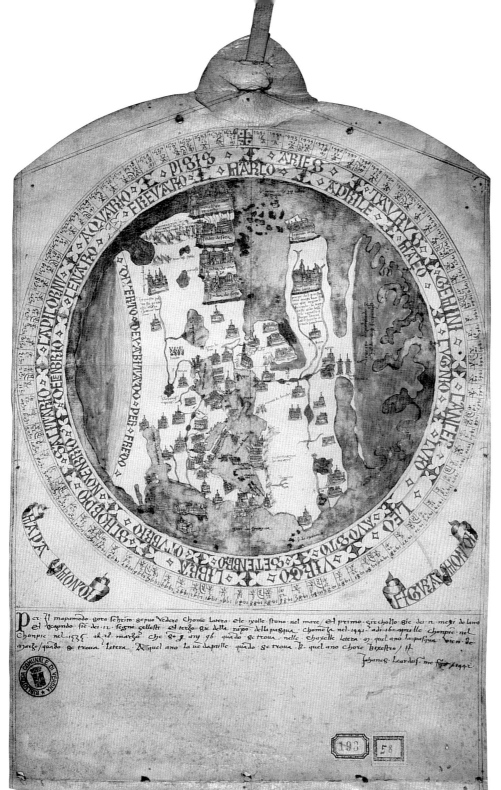

Fig. 1.1b. Giovanni Leardo, *Mapa Mondi. Figura Mondi.* Venice, 1442. 350 × 280 mm. Verona, Biblioteca Civica, MS 3119, unfoliated (see also Plate 2a). In modern mapping, measurement matters and a 'good' map shows outlines and contours accurately. Accuracy of measurement, however, has not always been the paramount cartographical condition. Five hundred years ago, a Western world map reflected what was written in the Bible. In a text just below the map, Leardo presented his cartographical masterpiece both as a portrait of the earth and as a device for the computation of the dates of Easter.

representing a globe on a flat surface, Leardo could show on his circular map only the known and inhabited portion of the earth, that is the northern hemisphere.[6] Accordingly, his world stretches from the Far East to the Atlantic, and from the dark and cold northern deserts, corresponding to the Arctic Circle, to the torrid regions of Africa. It is still a world of Asia, Africa and Europe; it would be another fifty years before Christopher Columbus added America to the map of the world. Leardo's map is highly detailed. Mountain ranges, islands, rivers and lakes are shown. Vignettes of castles, walled towns and churches represent regions as well as towns and cities. In Europe, places are located accurately, but in Asia the entire Indian subcontinent is missing. Rather unexpectedly, Africa is divided into two parts, a northern and a southern part. The seas are coloured blue, except for the Red Sea, which is suitably red. Here and there little notes tell us that, for example, on some islands in the Indian Ocean pepper and other spices are found, and that in the middle of Asia there are people who eat human flesh.[7]

By the standards of his time, Leardo was an accurate map maker. He had evidently taken considerable pains over the layout of the Mediterranean and Europe. Several inscriptions in Asia include information taken from travel accounts of relatively recent and trustworthy explorers, such as Marco Polo. Place names in Africa and the Far East came from Ptolemy's *Geography*, which had been translated from Greek into Latin earlier in the fifteenth century. Care had also been taken over the calendar. So we may be surprised to find the earthly paradise also depicted on Leardo's map of the world. A vignette placed on the map close to India portrays an exceptionally beautiful city, with a tall column in the centre surrounded by splendid buildings, clearly labelled in red ink, and in fifteenth-century Venetian vernacular, *paradixo teresto* (Plate 2a).

To the modern mind the presence of the Garden of Eden on a map of the world throws into doubt the seriousness of Leardo's cartographical enterprise. How could a careful cartographer, who used up-to-date navigational charts as his base map, include on a modern map of the world something as intangible as the earthly paradise, a place described in the Bible as inaccessible to man? Why, if he had to show it, did he choose to place it near India? Were his contemporaries expected to believe that Leardo's *paradixo teresto* was really in that particular place?

As it happened, Leardo was one of the last to produce a map of the world showing the Garden of Eden. After him, geographical exploration and developments in cartographical practice changed the appearance of world maps, and the Garden of Eden was shown only on regional, historical maps. By the end of the sixteenth century, Leardo's 'desert uninhabited because of the cold' had been defined as the Arctic Circle. His 'desert uninhabited because of heat and snakes' and his 'land of the Pharaohs' had begun to be explored, the Sahara desert and the forests of equatorial Africa were being slowly differentiated, and the Indian peninsula was receiving its shape. World maps were beginning to show earthly space with an unprecedented mathematical precision as map makers focused on lines of latitude and longitude. New places were being added to the map as Europeans reached distant lands in their voyages of exploration.[8] So, for example, the world map by Jeronimo de Girava, made in Milan in 1556, shows the Moluccas and Madagascar and takes account of the discovery of the Americas and the Pacific Ocean.[9] And as new regions were included on the maps, so some of the old places, such as the earthly paradise, disappeared entirely from 'modern' world maps such as Jeronimo's.

Today, in the twenty-first century, maps are made with the help of satellite and radar electronics. In February 2000, for example, the Shuttle Radar Topography Mission (SRTM) sent the space shuttle Endeavour into space to complete an ambitious, eleven-day mission to gather radar images of the whole earth (Figure 1.2).[10] Six astronauts brought back 300 digital tapes containing radar data to be processed for the most detailed map of the planet ever made. The astronauts collected sufficient data to fill 20,600 compact discs. The radar aboard Endeavour mapped 112 million square kilometres at least twice for the double imaging that is considered essential for the creation of really accurate three-dimensional maps of the earth. Such maps show not only the location of each geographical detail, but also indicate altitude. The impossible has been achieved: modern world maps contain data on places that are difficult if not impossible to reach from ground level. Access to any part of the globe – the most impenetrable mountain chains, the most inhospitable deserts or the densest and most dangerous tropical rainforests – is no longer a problem, for radar can reach everywhere.

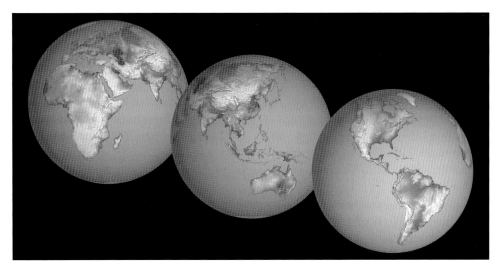

Fig. 1.2. Three globes showing the coverage of the Shuttle Radar Topography Mission (2000), NASA/JPL/NIMA. The shuttle Endeavour carried radar instruments for the collection of geographical data. The United States Defence Department declared that the function of Endeavour's mapping mission was to improve the aim of missiles and aid the deployment of troops. Usually modern world maps eschew theological meaning and do not feature any paradise on earth. There is a place known as Paradise, at about 600 metres above sea-level, not far from Chico in California, but it is not what Leardo and his contemporaries were thinking of. In fact, there are at least eight different towns called 'Paradise', all, significantly, in the New World. Paradise Island lies south of the coast of Florida. It is not, however, the original abode of Adam and Eve, but only a tourist resort offering tropical beaches and a vibrant nightlife.

Yet, no radar sensor has even hinted at, much less disclosed, the existence of a paradise on earth, neither near Leardo's site in eastern Asia nor anywhere else. Of course modern world maps have no room for representations of paradise: they take no notice of the Bible. The assumptions, as well as procedures, that govern modern mapping are different from those that held sway throughout the thousand-year span of medieval Europe. In modern mapping it is measurement that matters, and a 'good' map is expected to depict outlines and contours as they 'really are'. Geodetic measurement, however, has not always been the paramount cartographical feature. A millennium ago, a world map was expected to reflect what was written in the Bible and not to dispute fundamental Christian doctrine.

If a map is considered a form of graphic representation that 'facilitates a spatial understanding of things, concepts, conditions, processes, or events in the human world' – and not only a 'representation of the earth's surface, or a part of it, [...] delineated on a flat surface of paper, etc., according to a definite scale of projection' – the appearance on early world maps of a location recorded in the Bible presents no problem.[11] All mapping involves much more than the drawing of lines on a flat surface. No less than the writing of text, the creation of a map is in the first instance a social act, unavoidably involving the thoughts and beliefs of the maker of the map as well as those of the readers.[12] Leardo's placing of the earthly paradise on his map of the world expressed not only his belief in its existence but also a conception of cartography different from our own. Unlike modern mapping, which takes its point of reference from accurate measurement, medieval map makers depended not only on their own observations, and on the authority of classical geographers, but also on the Bible. Paradise was accepted as a place no less real than Rome or Paris or any of the towns and villages surrounding the map maker's home – it was just a better place than any other. We begin to understand that only a special kind of cartography can accommodate a place such as paradise, a cartography very different from our own. This difference needs to be borne in mind if we are to visit the 'foreign country' of the past with any understanding.

Nineteenth-Century Pioneers

The nineteenth century saw the beginning not only of the history of cartography in general, but also of the history of the mapping of paradise in particular. At the start of the century, the vast majority of surviving medieval maps were to all intents and purposes out of sight, hidden and closeted in dusty archives and libraries. Nobody bothered to search out and study medieval maps in monastic libraries. In the course of the century, however, historians of antiquarian bent and historians of cartography began to show an interest in old maps.[13] By the 1840s an international scattering of scholars were laying the foundations of the history of cartography as a field of study – the Portuguese Manuel Francisco de Barros e Sousa, 2nd Viscount of Santarém, the Frenchman Edmé-François Jomard, followed by the Pole Joachim Lelewel and the German Konrad Miller – and for the first time, the rich cartographical archive dating back to the Middle Ages was being systematically explored and meticulously studied. The remarkable efforts of these scholars in discovering, documenting, reproducing and comparing maps from all over Europe has especial value today in view of the loss of so much of that material during subsequent European wars. Nor are the comments of these talented pioneers without interest, for their disparaging opinions coloured the way the theological content of medieval maps came to be regarded in the twentieth century. To such scholars, medieval maps were little more than the primitive beginnings of a progressive development that was reaching its peak in their own time. In short, medieval maps were judged according to nineteenth-century criteria, and not in the light of their own cultural and technological context.

Viscount Santarém's *Essai sur l'histoire de la cosmographie et de la cartographie* (written between 1849 and 1852) opens by praising geography: 'Of all the sciences, geography shows best by what a long and difficult route the human spirit emerged from the darkness of uncertainty, and achieved wide and positive knowledge.'[14] For ten centuries (from the fifth century to the fifteenth), Santarém asserted, knowledge of the globe made no progress and medieval maps represent 'a formless and often barbarous continuation of the maps of the Ancients'.[15] Only in the fifteenth century was science restored, and the erroneous placing of regions and cities on medieval maps corrected, as a result of geographical discoveries and astronomical calculations.[16] Edmé-François Jomard also stressed the significance of geographical discovery. The preface to *Les Monuments de la géographie* (1854) states that medieval maps help us understand the history of geographical discovery and appreciate the merit of seafaring nations.[17] Konrad Miller considered that the medieval *mappae mundi* ('maps of the world') were no more than copies of Roman maps dressed up in Christian symbolism, that their sole value was the light they shed on their classical prototypes, and that all medieval maps show scant cartographical progress over the medieval period.[18]

These nineteenth-century historians of cartography, preoccupied as they were with notions of scientific progress and with assessing cartographical advancement through accuracy of delineation and location, had no time for the representation of paradise on medieval maps. It was left to a different kind of historian, one interested in myth and legend, to comment on the presence of paradise on early maps of the world. In 1843, Alfred Maury introduced the subject by explaining in his *Essai sur les légendes pieuses du Moyen-Âge* that if medieval people believed in the existence of mythical places, it was because the Middle Ages was an age 'forgetful' of science and ruled by imagination.[19] In similar vein, in 1878, Arturo Graf delivered a lecture at the University of Turin in which he remarked that the scientists of his day had washed their hands of the bizarre muddle of medieval theories about the location of Eden. He himself thought that the cradle of mankind lay in a region somewhere in the neighbourhood of the Tibetan plateau, and that the memory of this primordial place, coloured by religious sentiment, could be held responsible for the paradise tradition, a legendary and popular fantasy common, Graf also noted, to both Arian and Semitic mythology. It was also Graf who suggested that during disturbed and turbulent times, the vision of Eden quenched the Christian West's thirst for poetry, tempering the rigour of a religion which had a cross as its symbol and which constantly promoted penance and contemplation.[20] Graf, a firm believer in the unstoppable march of reason and yet interested in the most irrational aspects of medieval culture, displayed remarkable erudition in his studies of the medieval legend of the earthly paradise. He eventually embodied his views in a publication tellingly entitled *Miti, leggende e superstizioni del Medio Evo* (1892–3).[21] In this book, he concluded that the paradise myth was a universal notion that preserved to a certain extent the memory of the original birthplace of mankind – and perhaps also of 'a golden age', before society was encumbered by property – although its origins really lay in 'the projection into time and space of an inner fantasy'.[22] He argued that it is inevitable for humankind to imagine a state of perfect happiness and to project this onto a remote space and a distant moment in time. In the Middle Ages, when large parts of the globe remained unexplored by Europeans, it was easy to let the imagination conjure up nostalgic images of distant regions of delight, a tendency much favoured, according to Graf, by Christian doctrine because of the need to highlight the gravity of the Fall by emphasizing the beauty of the lost Garden of Eden. Although Graf's primary interest did not lie in maps, he noted in passing that medieval maps incorporated the imaginary notion of paradise and that they placed the Garden of Eden in the east.[23]

What must be the first historical study devoted specifically to the question of paradise on maps was by another Italian scholar, Pompeo Durazzo. His book is entitled *Il paradiso terrestre nelle carte medievali* (1886). Durazzo agreed with the point made by Graf in his lecture of 1878, that the legend of paradise had grown out of the historical memory of a real place of origin of the peoples of the world. Finally free from religious prejudice and mystical rhetoric, and supported by comparative philology pointing to the Tibetan region, nineteenth-century science in Durazzo's view was at last in a position to solve the problem of that earthly paradise. Durazzo dismissed medieval belief in an earthly paradise as a monstrosity born of the encounter of pagan fantasies and Christian reveries. This 'geographical legend', as he called it, 'preoccupied so many writers and turned so many minds away from the truth' because in times of distress only thoughts about distant and unknown places could feed the need for poetry and fantasy lands of the Latin world. Distant Asia, Africa and the ocean could shelter all the realms of happiness and islands of delight conceived by the human imagination; everybody, 'the learned as well as the ignorant, saints and unbelievers, all contributed to the development of this great mass of oddities, which appear to a serious analysis only as an assortment of fantasies, produced by sick minds'.[24] The search for the location of paradise was unrelenting because Christian believers were real people who needed the vision of an earthly place of physical delight. In the confusion of medieval geographical knowledge the most extravagant theories and arbitrary opinions had gained ground. Even the noble mind of Christopher Columbus, said Durazzo, was entangled in a net of prejudices, scruples and naïve beliefs that the most fervent Catholic of modern times would never share.[25] Durazzo went on to observe that 'the utmost confusion' reigned in the more diagrammatic maps. Medieval *mappae mundi*, he said, could 'sometimes be artistic and were always original and interesting', but they always 'lacked all the necessary elements to be called true geographical maps'.[26] Medieval legends about paradise filled a lacuna left by a complete lack of proper geographical information, and they impeded the advancement of knowledge about the globe. Eventually, after a long struggle, truth prevailed. With the exploration of all the remote corners of the earth, Durazzo concluded, all the enchanted places of the medieval imagination vanished.[27]

The seminal work on paradise on maps appeared three years before the end of the nineteenth century when Edoardo Coli published his clearly structured synthesis, *Il paradiso terrestre dantesco* (1897). In order to explain Dante's ideas, Coli dealt systematically with medieval theology and cartography, explaining that Dante's vision of the Garden of Eden grew – like a sudden and luxuriant flower – out of an undigested muddle of theological quibbles, absurd and grotesque cosmic visions, picturesque descriptions and naïve legends.[28] Following Graf and Durazzo, Coli also presented Eden as a

Fig. 1.3. Edoardo Coli's redrawing of the earthly paradise, from the map of the world in a manuscript of Beatus of Liébana's commentary on the Apocalypse (Saint-Sever, eleventh century, Paris, Bibliothèque Nationale de France, MS Lat. 8878, fols 45v–46r), in his book on *Il paradiso terrestre dantesco* (Florence: Carnesecchi, 1897), p. 103. Coli describes the attempt to unravel medieval ideas about paradise as 'arduous', and refers to 'a most ardent love, in the name of Dante, for the cultural tradition of Italy' as the motivation behind his book.

generous reverie, an escape from the rigour of the Christian faith to 'the diaphanous meadows of those dreams that swarm from the flesh'.[29] He did add, however, that Dante had enhanced the ideal qualities of the paradise tradition at the expense of the marvellous and the exotic. The Garden of Eden of Italy's greatest poet was the place where mankind could be restored to perfect harmony between heaven and earth and to full communion between the realm of the spirit and natural beauty. Relying on his faith in the progress announced by Dante's poetry, Coli celebrated the rebirth of both mankind and art and foresaw a day when the entire world will be one paradise.[30]

Coli failed to see the Middle Ages as anything but an era of 'fantastic exuberance', but he noticed that, whereas other medieval legends fell by the wayside with the passing of time, 'the empire of the ultramundane legend' survived.[31] He explained that although theologians pointed out that paradise had been concealed by divine providence 'there was no poor mortal who, leaving home for distant lands, did not think that somewhere on the way, one day or another, he might come across the earthly paradise'.[32] Hence the inevitable, in his view, interest in the whereabouts of Eden. Coli listed all the places in which medieval and Renaissance scholars had placed Eden – India, Ceylon, China, Armenia, Syria, Persia, Arabia, Africa, the Holy Land, the Alps, the Caspian Sea, America, the South Pole, underground, even the third heaven or the moon.[33] For Coli, the mapping of Eden was a bizarre transference of a spiritual idea to a physical dimension, and medieval *mappae mundi* were 'extravagantly complicated'.[34] Nevertheless, his book provides reproductions of the most significant medieval depictions of paradise on maps. It is not surprising that, despite its florid prose and the limitations of its nineteenth-century viewpoint, Coli's work has been deeply quarried – not always with acknowledgement – by later scholars for the impressive amount of coherently organised material it contains and the consistent picture it presents of the cartography of paradise (Figure 1.3).

Coli's research into the earthly paradise was informed by his belief in the progress of humankind. He complained, when analysing his cartographical material, that he could not see in the variety of maps he was studying the clear concatenation and succession of phases he was looking for as a reflection of the linear development of cartography, as befitted the prevalent model of scientific progress.[35] In fact, though, Coli showed a greater awareness than many of his contemporaries of the distance that separated his age from the time when the Bible constituted the horizon of scientific knowledge. After denying that there could be any scientific value in the fantastical ideas on paradise that had flourished in the Middle Ages, he qualified his rejection by adding that even the most foolish aberration should be studied to help penetrate the mysteries of human thinking. Nobody any longer gets excited over the 'pseudo-geographical' problem of paradise, said Coli, but it would be unfair to ridicule the work of more than fifty generations of scholars who had attempted to solve it. His love of and respect for Dante encouraged him, like others after him, to take seriously the belief in the earthly paradise that such a genius had seen fit to celebrate in his poetry. At the very least, the debate stimulated explorers, including Christopher Columbus, to open up new horizons to mankind.[36]

Twentieth-Century Followers

In the twentieth century, the assumption that paradise on maps was the expression of a bizarre superstition that was eventually wiped out by the march of progress continued to be held by historians of cartography, geographers and non-geographers alike. The tendency was to dismiss the medieval belief in the earthly paradise out of hand, as if it were nothing more than a picturesque legend and an example of the period's many superstitions, and to avoid seeking reasons for its presence on maps. In Raymond Beazley's *The Dawn of Modern Geography* (1897–1906), the Middle Ages are seen as a period of decadence, worthy of study only because 'a true view of history will not ignore the weakness, or the degradation, or even the lifelessness of the past'.[37] Beazley adhered to the notion of progress, saying that the advance of mankind was better understood by examining those centuries 'when the tide of life seemed ebbing', as well as 'those other and brighter times, which, taken at the flood, led on to fortune'. In the course of tracing the gradual developments that eventually led to the radiant dawn of modern scientific geographical knowledge, Beazley discussed the medieval geography of paradise in terms of a conflict between the literal meaning of Scripture and the tacit opposition of geography to, as he put it, 'the undying vigour of the oldest and most poetic of physical myths'.[38] Similarly, in *The Geographical Lore at the Time of the Crusades* (1925), John Kirtland Wright was convinced that, however 'irrational' they might have been, medieval scientific theories deserved some attention from modern scholars. Although Wright was interested in medieval geographical learning mainly as an antecedent to the Great Age of Discovery, his treatment of the terrestrial paradise was well grounded, not least because he was building on the earlier work of Graf and Coli. Both Beazley and Wright acknowledged the dangers of anachronism, Wright warning that 'cartographic accuracy was not the aim of the map maker of the time, and we are not justified in criticizing his maps in the light of modern standards'.[39]

It is hard not to suspect even Wright of a patronising attitude to past times. He labelled the Middle Ages as 'credulous', in the sense that 'credulity is an inevitable concomitant of the undue respect for authority'. He asserted that the acceptance of a geographical lore that was based on authority and tradition, that conformed to the Bible, and that gave rise to the mapping of fabulous monsters reflected the 'normal intellectual habit' of a period in love with the marvellous and the bizarre. He did have praise, though, for those pioneering medieval writers who managed to follow a 'geography of observation' and to place their work on a sound and scientific footing. William of

Conches, Robert Grosseteste, Roger Bacon, and even Albert the Great, were all held up by Wright as 'oases of fresh observation and clear reason in the midst of the arid deserts of plagiarism that constitute so much of medieval literature', oases that marked out 'the pathway of the history of science'.[40]

The writers of the 1930s came from the same mould. In England, George Kimble (*Geography in the Middle Ages*, 1938) insisted no less than had his predecessors on the model of linear progress from the medieval 'dark ages' to the establishment of scientific geography in the sixteenth century. Like Beazley and Wright, Kimble admitted that medieval maps could be useful for understanding the culture of the time, but he regarded the majority of *mappae mundi* as works of art, not as sources of geographical information. They could not have been used for measuring distances, for instance. If medieval maps document myths and marvels rather than geographical knowledge, that was what was wanted at the time. His explanation for the presence of paradise on such maps was that it reflected 'the strength of the [period's] religious bias' and 'the clerical hold on scholarship', two typically medieval attitudes that were later discarded.[41] In Italy, Leonardo Olschki also acknowledged that medieval *mappae mundi* should not be dismissed simply because of their religious content or geographical inaccuracies, for they portray an 'ideological', pre-scientific geography. But he too tended to see the post-medieval shift to the 'scientific method' as part of the linear sequence of progress rather than as a separate option.[42] In America, soon after the Second World War, Lloyd Arnold Brown published his *The Story of Maps* (1949), in which he presented medieval map makers as superstitious monks making rough sketches to entertain the common people or to support some theological doctrine regardless of experience and knowledge to the contrary. So fables, monsters and marvels – which included paradise – were mapped indiscriminately. The 'lamp of scientific knowledge, a tremulous flame at best,' said Brown, 'was obscured for a time by the blinding light of religious ecstasy'. Eventually and – in Brown's scenario – in spite of threats and the fear of eternal damnation, man shook the cobwebs out of his head and began to exercise his critical faculties. No less than the others, Brown was a progressivist even though, like Coli before him, he was well aware of the difficulty of tracing a line of steady improvement in cartographical procedures and in the presentation of facts and ideas.[43]

Despite its entrenched views, Brown's *Story of Maps* is still the first port of call for students seeking a world history of cartography in a single handy volume. The earthly paradise was usually dealt with by students from a different field of study. In *The Other World* (1950), a synthesis of the medieval visionary literature, Howard Patch included a chapter on journeys to the earthly paradise, citing the presence of Eden on medieval maps as an example of popular imagination.[44] In Gérard de Champeaux and Sébastien Sterckx's *Introduction au monde des symboles* (1966), an underlying theme is the long-term struggle between the progress of cartography towards scientific rigour and what they called the 'insuperable resistance of the symbol', this time, however, with the authors on the side of the symbol.[45] In *Milton's Earthly Paradise* (1972), Joseph E. Duncan applied Coli's approach to Dante to his own analysis of the treatment of the paradise issue in Milton's great poetic epic. For Duncan, ideas in the Middle Ages about 'the location of the earthly paradise as reflected in map or legend were usually imaginative, romantic, and unsupported'.[46] Despite offering a clear and generally well-grounded account of some crucial shifts in the history of the paradise tradition, Duncan, too, fell back into the received pattern of a medieval paradise coloured by myth and legend that disappeared when mankind raised the banner of reason.[47]

Duncan's approach provided modern historians with an interpretative model of the cartography of Eden. On the one hand, he saw the medieval mapping of paradise as an irrational and old-fashioned attempt to give cartographical legitimacy to religious imagery. On the other, he paid attention to the ways paradise was mapped after the Renaissance, seeing these in a more favourable light in view of the fact that they took into account the new horizons opened up by geographical discovery and by the advance of modern science, which included biblical philology and comparative history. To modern historians such as Duncan, the mapping of the Garden of Eden during the Renaissance and the Reformation was more approachable than the medieval. In fact many twentieth-century writers fully embraced the attitude of those Renaissance thinkers who condemned as fantasies medieval ideas about paradise. Renaissance treatises on the Garden of Eden with their synthesis of medieval thought became the main sources for twentieth-century historical analysis. Duncan's detailed discussion of post-Renaissance scholarly efforts to locate paradise was followed by other studies. In some, the focus was expressly on the location of the earthly paradise, whereas in others the paradise question was only part of a wider theme. An exhibition in Antwerp in 1982 on the earthly paradise in sixteenth- and seventeenth-century Flemish art, for example, contained a section on cartography, in which maps of paradise produced by authors such as Gerard Mercator, Iodocus Hondius and Peter Plancius were displayed.[48] Catherine Delano-Smith and Elizabeth M. Ingram included Lutheran and Calvinistic maps of paradise in their illustrated catalogue of maps in Protestant sixteenth-century Bibles (1991).[49] In contrast, the late seventeenth-century treatise on the location of paradise by the Catholic bishop Pierre-Daniel Huet, who also compiled a map, was the subject in 1988 of an essay by Jean-Robert Massimi.[50] Specialists in philosophy and biblical exegesis also explored the implications of the cartographical and geographical approach to the Garden of Eden by sixteenth- and seventeenth-century authors such as John Calvin, Jacques Lapeyre d'Auzoles, Samuel Bochart and Pierre-Daniel Huet.[51] The eighteenth- and nineteenth-century debate on the location of paradise has also attracted interest.[52] Although the whole picture of the mapping of paradise from the Renaissance to the twentieth century has yet to be explored, some modern cartographical artefacts featuring paradise have received sound scholarly attention, while interest in the medieval debate about the location of paradise has been kept alive by Dante specialists.[53]

Such scholars apart, though, the medieval mapping of paradise remained the object of a condescending smirk. Compared with the march of cartographic progress and with the scientific 'objectivity' of modern maps – seen as a 'truthful' mirror of nature – medieval maps with their depictions of the earthly paradise continued to be considered as naïve representations of an imaginary geography. Claude Kappler (*Monstres, démons et merveilles*, 1980) expressed admiration for the way medieval and early Renaissance travellers – Christopher Columbus not excepted – journeyed far and wide in their search for the earthly paradise.[54] Jacques Le Goff set a new trend, elaborating the progressivist line of argument. In his *L'Imaginaire médiéval* (1985), Le Goff described the Middle Ages as a fascinating period of blurred frontiers between the visible and the invisible and between the natural and the supernatural. For him, paradise was to be included in his inventory of the 'medieval marvellous'.[55] Le Goff explained that for medieval man the visible was the trace of the invisible and that 'there was no dividing line, let alone a barrier, between this world and the next' as the supernatural overflowed into daily life.[56] In similar vein, Giuseppe Tardiola (*Atlante fantastico del Medioevo*, 1990) invited his readers to take an 'imaginary trip among the marvellous and surprising coordinates of a plausible "Atlas of fantasy"' and discover the mythical and legendary lands of the past, 'from the earthly paradise to the *India mirabilis*, from the disturbing kingdom of Prester John to the gloomy domains of Satan and the Antichrist'.[57] Quoting Graf and Olschki, Tardiola explained that the geographical myth of paradise bears witness to the tenacious and naïve faith of medieval Europeans given to fantasy and to visionary exoticism. Like Le Goff, Tardiola emphasized the idea that all medieval society was firmly persuaded of the continuity between the physical senses and the spiritual realm that made all kinds of marvels and monsters possible until the amazing atlas of medieval fantasy vanished at the dawn of the modern era.[58] Old ideas die hard, though, and not least the idea of linear progress. Like Duncan, Jean Delumeau (*Une Histoire du paradis*, I, *Le Jardin des délices*, 1992) offered a detailed discussion of Renaissance views on paradise, but apparently thought it appropriate to discuss medieval mapping of paradise in the same terms as those employed in nineteenth-century scholarship; namely, as a struggle between the progress of geography and cartography on the one hand, and the tenacious resistance of traditional belief on the other.[59] Delumeau considered that the religious illusion, as he put it, of Eden as a concrete and geographical place somewhere on earth evaporated once European explorers failed to find it, thus allowing a new reading of the story of the Fall in which the West could banish the image of a jealous God and reappraise mankind's feeling of guilt.[60]

If Delumeau saw the Western myth of the earthly paradise as conveying a dark image of the relationship between man and God, the German-born psychologist Fred Plaut explored the psychological implications of the cartography of paradise in his essay on 'General Gordon's Map of Paradise' (1982). Plaut's analysis of Gordon's map of the location of Eden gave him the opportunity to offer a brief survey of the cartographical depiction of paradise throughout the centuries.[61] In Plaut's view, paradise was included on the cosmological and encyclopedic maps of the Middle Ages, which he saw as expression of a largely mythological and legendary world, and then banished from cartography with the gradual expansion of scientific knowledge during and after the Renaissance. In modern times, what Plaut considered as more plausible hypotheses were drawn in when attempts were made to reconstruct the original site of the prediluvian paradise in a world not yet altered by the Flood. Plaut's search as a psychologist for Gordon's innermost motives in recognizing the site of paradise in the Seychelles in the late nineteenth century, by when most scholars and theologians had abandoned the question entirely, encouraged him to publish another short article addressing the fundamental reason for the representation of paradise on any map, either medieval or modern ('Where is Paradise? The Mapping of a Myth', 1984). Plaut saw the idea of the earthly paradise as coming from the need for an orientating myth that would give meaning to life.[62] Explaining that some people are not able to see symbolic meaning without the support of literal facts, he pointed to the Christian defenders of faith and of the authority of the Bible who insisted on knowing where the Garden of Eden really was and who looked to a map for concrete evidence. Still affected by the idea of progress, Plaut highlighted the single-mindedness and fanaticism of those literalists 'who desperately wanted to reconstruct the site of paradise'.[63] Finally he offered a cartographical model of 'an alternating progress and regress and development from literal to symbolic thought' that lasted until modern cartography made room for a new world picture, although even then, he admitted, 'the vagrant paradise did not give up that easily before becoming relegated to a merely decorative position'.[64] In a chapter on the role of fantasy maps in psychoanalysis (*Analysis Analysed: When the Map Becomes the Territory*, 1993), Plaut later returned to Gordon's maps for the location of paradise. Starting with a brief overview of earlier maps showing the earthly paradise, Plaut again painted his progressivist picture, crediting Renaissance and Enlightenment science with the removal of the Garden of Eden from maps of the world. In the nineteenth century, 'variously motivated persons' did return to attempt to reconstruct the site.[65] One of these was Gordon, whose 'bizarre' thesis Plaut diagnosed as 'a remarkable regress', an expression of 'literal-mindedness', an 'anachronism', and as 'pseudo-scientific and … absurd' as the maps drawn in the sixth century by the Alexandrian monk Cosmas Indicopleustes.[66] By drawing paradise on a map in the late nineteenth century, Plaut reckoned, Gordon was 'turning the clock back'.[67]

As recently as 2004 – thus already in the twenty-first century – Corin Braga (*Le Paradis interdit au Moyen Âge: La Quête manquée de l'Eden oriental*) pointed to the 'enchanted way of thinking' (*pensée enchantée*) of the Middle Ages as the key to the presence of paradise on maps of the world and as an explanation for the medieval bending of empirical geography to fit the biblical text, which continued until philosophical empiricism and the new Cartesian science triumphed in the seventeenth century.[68] Braga saw the eastern legend of human immortality in a divine garden as having been first censored by the Judaeo-Christian tradition and then definitively expelled from the European imagination by modern rationalism. A consequence of the Christianization of the pagan myth was that the geographical search for the garden was condemned to inevitable failure, and the idea of an earthly paradise had to be replaced by the promise of beatitude in heaven. This is why, for Braga, medieval maps became a depositary of the unconscious dreams and nightmares of medieval man, a 'gallery of monsters and marvels', which he saw as 'imaginary workshops, with specimens in the test-tube'.[69] In his sight, it was 'a positive glance' on Asia during the Renaissance that eventually destroyed the mirage of the eastern paradise.[70] Subsequent attempts to locate paradise during the Renaissance were for him no more than 'the reaction of a more rationalist and pragmatic age to the fantastical constructions of the "enchanted way of thinking" of the Middle Ages'.[71] The Renaissance maps, however, still lacked empirical proof and precision and still made history and geography subservient to theology. Only in the Enlightenment was the phantom of paradise finally destroyed.[72] Braga, keen to demonstrate how medieval travellers always saw the 'extra-European *other*' by means of their cultural and mythical stereotypes, appears to be imposing his own cultural clichés on what may be for us (and him) 'the medieval *other*'.

A New Road to the Past

As the twentieth century turned into the twenty-first century, the presence of paradise on a map of the world was still being labelled as a medieval superstition or an imaginative mirage. The conditions that had allowed a theological notion to be mapped remained unexplored. Now, though, the decline of the idea of progress, the arrival of new cartographical theories, and a revisionist history of cartography have made medieval mapping more approachable. General acceptance of the idea that the mathematically scaled map is not the only kind of map, and that mapping may involve representation of the wider cosmos as much as of the terrestrial globe, means that it is a good time to make the uncondescending conceptual journey into the past that is needed in order to explore the mapping of paradise.

Modern cartographical theorists recognize the difference between maps that represent aspects of the physical world, and of entities that can be experienced ('phenomenon-representations'), and maps that depict an abstract idea about reality, or even about any alternative reality that the human mind can devise ('concept-representations').[73] The referent for a 'concept-representation' can be a hypothesis or an abstraction that does not exist outside a particular theoretical framework. The writings of modern practising cartographers leave room for the medieval idea of showing paradise on a map like Leardo's, a place that is inaccessible to foot-weary travellers in the physical world and yet reachable through spiritual pilgrimage. From a lengthy sojourn with maps seen exclusively as vehicles for spatial relationships, cartographers have come to see the relevance of time, sequence and process, and of the pairing of time and space, to an understanding of maps. The animated cartography and visual representations produced by the computer technology of practitioners of Geographical Information Systems depict dynamic processes and interact with the map viewer to reveal relationships in the data displayed.[75] They offer a glimpse of the three-dimensional, and dynamic, environment in which time and space are blended. For example, animated weather maps predict temperature and precipitation and illustrate the succession of climate and vegetation, while dynamic and narrative maps simulate earthquake epicentres, depict global patterns of natural occurrences, and describe the diffusion across space and over time of an epidemic or of a nation's demographic characteristics. In a perhaps weird and certainly an unexpected manner, most modern cartographical technology is making the medieval conception of a map that can accommodate not only geographical space but also historical time appear less exotic and alien. It has become easier to comprehend the medieval view of the Garden of Eden – a place which had existed in the distant past and still existed somehow and somewhere – as an essential element in a complex theological framework that involved the city of Christ's death and resurrection, Jerusalem, and that drew the map observer into the wider divine plan for the salvation of all humanity.

For thoughtful cartographers and historians of cartography today, the idea of progress – the notion that mankind has reached full maturity after passing a medieval childhood and a Renaissance adolescence – is no longer as seductive as it was once. The epistemological myth that cartographical methods reflect, in Brian Harley's words, the 'cumulative progress of an objective science always producing better delineations of reality' has been denounced.[76] Since Harley wrote, other scholars have attacked the pretence that cartography can attain a scientific and objective *non plus ultra* in the representation of the world, leaving behind mythical and religious mapping as quaint irrelevances.[77] Modern historians of cartography have abandoned the simplistic vision of a linear progression to recognize that the corpus of medieval maps, which until yesterday they rejected as absurd and irrational, represents an alternative cartographical system.[78]

The premise that maps are always a reflection of the culture in which they are produced and that they always transform reality, highlighting some phenomena at the expense of others, is becoming widely recognized as fundamental to the understanding of maps, their image and their role. Even the images provided by contemporary satellite and computer-generated representation, despite their apparent objectivity, are no more than the result of sophisticated cartographical techniques. The 'scientific' mapping of the Enlightenment and post-Enlightenment is no longer regarded as a neutral transfer of information from an 'objective' reality, but as an act of human imagination that allows the cartographer to disclose, or the map user to discern, patterns in perceived reality. It is the representations that result from the mapping process that constitute human knowledge. So a map of the world with a portrayal of paradise is not necessarily any less objective or more manipulative or more fanciful than a modern map without it. Both deal with areas that cannot possibly be seen in the same way as they are represented.[79] To dismiss paradise on maps as a naïve representation of an imaginary land that in due course was wiped out by the radiant light of progress is itself a naïvety. Maps created in medieval and Renaissance times need to be judged in terms of the world outlook of the time. To understand every age on its own terms does not mean that one has to adopt that world view oneself, only that the anachronistic arrogance of judging it from one's personal viewpoint is to be avoided.

The collapse of an unwavering belief in the inevitability of scientific advance is liberating. New horizons are opened. Medieval mapping no longer has to be considered an inferior stage in the development of the history of cartography. We are free to see it as a foreign country that can be visited and understood, and as a different system with its own intrinsic values and its own internal consistency: a cartographical and calendrical system with a logic that worked for people who measured space and time differently from ourselves. The comparison between modern, Renaissance and medieval cartography has become a comparison between equals and between different systems of representation, each valid on its own terms. The moment has arrived for a fresh look at the mapping of paradise, for an interrogation of the reasons for the presence of the Garden of Eden on medieval world maps, and for an exploration of the intellectual conditions that made possible the early modern mapping of paradise. The gate guarded by the biblical Cherubim is ajar; behind it lies the Garden of Eden that is to be reviewed in the following pages.

1   Giuseppe Crivellari, *Alcuni cimeli di cartografia medievale* (Florence: Seeber, 1903), p. 11: 'Per il mapamondo sora schrito sepuo vedere chome latera e le ixolle stano nel mare; el primo çircollo si e dei 12 mesi de lano el segondo sie dei 12 segni çellesti; el terzo sie della raxon della pasqua; comenza nel 1441 adi 16 aprille chonpie nel chonpie nel (sic) 1535 adi 28 marzo, che so per ani 46. Quando se trova nelle caxelle letera M quel ano la pasqua vien de marzo; quando se trova letera A quel ano la ve daprille; quando se trova B quel ano chore Bixestro. Johanes Leardus me fecit 1442.' 'The world map delineated above makes it possible to see the position of the earth and the islands in the sea; the first circle is of the twelve months of the year; the second circle indicates the twelve celestial signs; the third shows the rationale of Easter; it starts in 1441, on 16 April and finishes in 1535 on the 28 March, which is 46 years. When the letter *M* is in the square, Easter comes in March; when we find the letter *A*, Easter comes in April; when the letter *B* is found, the year is a leap year. Giovanni Leardo made me in 1442.' On Leardo's 1442 map (now in the Biblioteca Comunale, Verona) see Crivellari, op. cit., pp. 5–28; Pompeo Durazzo, *Il planisfero di Giovanni Leardo* (Mantua: Eredi Segna, 1885); Tullia Gasparrini Leporace, ed., *L'Asia nella cartografia degli Occidentali: Catalogo descrittivo della mostra* (Venice: Biblioteca Nazionale Marciana, 1954), p. 17; Agostino Contò, 'Giovanni Leardo, *Mapa Mondi/Figura Mondi*', in Guglielmo Cavallo, ed., *Due mondi a confronto 1492–1728: Cristoforo Colombo e l'apertura degli spazi*, 2 vols (Rome: Istituto Poligrafico e Zecca dello Stato–Libreria dello Stato, 1992), I, p. 157.

2   On maps and the calculation of the date of Easter see Evelyn Edson, *Mapping Time and Space: How Medieval Mapmakers Viewed their World* (London: British Library, 1997), pp. 52–71, and 'World Maps and Easter Tables: Medieval Maps in Context', *Imago Mundi*, 48 (1996), pp. 25–42. The method for calculating Easter was decreed by the Council of Nicaea (325).

3   Up to the end of the Republic (1797), the Venetian year began in March: John K. Wright, *The Leardo Map of the World, 1452 or 1453, in the Collections of the American Geographical Society* (New York: American Geographical Society, 1928), p. 3 n. 8.

4   On nautical charts see Tony Campbell, 'Portolan Charts from the Late Thirteenth Century to 1500', in Brian Harley and David Woodward, eds, *The History of Cartography*, I, *Cartography in Prehistoric, Ancient and Medieval Europe and the Mediterranean* (Chicago: University of Chicago Press, 1987), pp. 371–463; on the use of nautical charts in medieval world maps see, for example, Peter Barber, 'Old Encounters New: The Aslake World Map', in Monique Pelletier, ed., *Géographie du monde au Moyen Âge et à la Renaissance* (Paris: Comité des Travaux Historiques et Scientifiques, 1989), pp. 69–88, esp. 84–6, and David Woodward, 'Medieval *Mappaemundi*', in Harley and Woodward, eds, *The History of Cartography*, I (1987), pp. 314–15, 357–8. The representation of paradise on fifteenth-century world maps is discussed below, in Chapter 8.

5   Jerusalem is described by the prophet Ezekiel (5.5) as being 'in the midst of all nations' (*in medio gentium*). See below, Chapter 5, p. 119 n. 51.

6   On the erroneous belief that in the Middle Ages the world was thought to be flat, see Jeffrey B. Russell, *Inventing the Flat Earth: Columbus and Modern Historians* (New York: Praeger, 1991); Woodward, 'Medieval *Mappaemundi*' (1987), pp. 318–21 and works cited in n. 152; Woodward, 'Reality, Symbolism, Time, and Space in Medieval World Maps', *Annals of American Geographers*, 75/4 (1985), pp. 517–19.

7   The information is from Herodotus, *Historiae*, III.99.

8   A theme explored in Jerry Brotton, *Trading Territories: Mapping the Early Modern World* (Ithaca, NY: Cornell University Press, 1997).

9   See Rodney W. Shirley, *The Mapping of the World: Early Printed World Maps, 1472–1700* (London: Early World Press, 2001), pp. 114–15.

10   See the website: http://www2.jpl.nasa.gov/srtm [last accessed: 15 February 2006].

11   The revisionist definition comes from Harley and Woodward, 'Preface', in *The History of Cartography*, I (1987), p. XVI; the second, the traditional, definition is from the Shorter Oxford English Dictionary.

12   Donald F. McKenzie, *Bibliography and the Sociology of Text*, The Panizzi Lectures 1985 (London: British Library, 1986), p. 34.

13   For the beginning of the history of cartography, see Harley and Woodward, 'Preface' (1987), p. XVII; Brian Harley, 'The Map and the Development of the History of Cartography', in Harley and Woodward, eds, *The History of Cartography*, I (1987), pp. 12–13.

14   Manuel Francisco de Barros e Sousa, Viscount of Santarém, *Essai sur l'histoire de la cosmographie et de la cartographie pendant le Moyen-Âge et sur les progrès de la géographie après les grandes découvertes du XVe siècle*, 3 vols (Paris: Maulde et Renou, 1849–52), I, (1849), p. XIII: 'La géographie est de toutes les sciences celle qui fait le mieux voir par quelle route longue et pénible l'esprit humain sortit des ténèbres de l'incertitude, et parvint à des connaissances étendues et positives.'

15   Ibid., I, p. 178: 'une continuation informe et souvent barbare de celles des anciens'.

16   Ibid., pp. XV, LXIX. Santarém praises the Portuguese contribution to the progress of geographical science: ibid., p. LXI.

17   Edmé-François Jomard, *Les Monuments de la géographie; ou, Recueil d'anciennes cartes européennes et orientales* (Paris: Duprat, 1842–62), 'Note préliminaire'.

18   Konrad Miller, *Mappaemundi: Die ältesten Weltkarten*, 6 vols (Stuttgart: J. Roth, 1895–8), IV, *Die Herefordkarte* (1896), p. 1. According to Miller, the early thirteenth-century Ebstorf map, for example, contained nothing, apart from certain details in the depiction of Germany, that can be seen as progress (*Die Erbstorfkarte: Eine Weltkarte aus dem 13. Jahrhundert* (Stuttgart: J. Roth, 1900), p. 79. In contrast, he considered that the Hereford map (*c*.1300) was more

accurate in its geographical features than the Ebstorf map (1235–40): *Mappaemundi*, IV, *Die Herefordkarte* (1896), pp. 1–4. Instead of the modern compound form *mappaemundi* (plural, *mappaemundi*), the original Latin expression *mappa mundi* (plural, *mappae mundi*) is the form preferred throughout the present book. For a discussion of the Latin term *mappa(e) mundi* see below, Chapter 5, pp. 85–6.

19 Alfred Maury, *Essai sur les légendes pieuses du Moyen-Âge* (Paris: Ladrange, 1843), p. 84. Maury claimed that in the Middle Ages the figures of poetry and the metaphors of language were equated with reality, producing coarse and irrational conceptions.

20 Arturo Graf, *La leggenda del paradiso terrestre* (Turin: Loescher, 1878), pp. 8, 10–13, 20–1.

21 Arturo Graf, *Miti, leggende e superstizioni del Medio Evo*, 2 vols (Turin: (Loescher, 1892–3), I (1892), pp. XI–XXIII, 1–238. On Graf's cultural background, personal interests, and ambivalence towards the medieval world, see Gianfranco De Turris, 'Introduzione: Trasformazioni del "mito edenico"', in Arturo Graf, *Il mito del paradiso terrestre* (Rome: Manilo Basaia, 1982), pp. 16–23 and Luigi Firpo, 'Introduzione', in Arturo Graf, *Il Diavolo* (Rome: Salerno Editrice, 1980), pp. 7–9.

22 Graf, *Miti, leggende e superstizioni del Medio Evo*, I (1892), I, p. XVII: 'Più lo scruto e lo sviscero, e più mi sembra che il mito, s'è, per qualche picciola parte, un ricordo, sia per la massima parte una visione ideale, nasca dalla proiezione di un fantasma interiore nel tempo e nello spazio.'

23 Ibid., pp. XVII–XXIII, 3–4, 7, 19.

24 Pompeo Durazzo, *Il paradiso terrestre nelle carte medievali* (Mantua: Arnaldo Forni, 1886), pp. 6, 7–8: 'Dotti ed ignoranti, santi e miscredenti, tutti portarono l'opera loro, tutti influirono allo sviluppo di questo grande ammasso di stranezze, le quali non appariscono più di un'accozzaglia di fantasmi creati da menti malate, quando vengano seriamente studiate.'

25 Ibid., pp. 10–25.

26 Ibid., p. 34: 'mappamondi, che se sono qualche volta artistici e sempre originali ed interessanti, mancano di tutti gli elementi necessari perché possano essere chiamati vere carte geografiche'.

27 Ibid., pp. 8–9, 42–4.

28 Edoardo Coli, *Il paradiso terrestre dantesco* (Florence: Carnesecchi, 1897), p. 185. Coli, a respected Dantist, taught Italian Literature at the Regio Liceo G. B. Vico in Chieti (Italy).

29 Ibid., p. 130: 'regione campata nella diafana landa dei sogni che pullulan dalla carne'.

30 Ibid., pp. 250–2.

31 Ibid., pp. 127–9.

32 Ibid., p. 93: 'non v'era meschino mortale che si movesse da casa sua per regioni lontane, che non si figurasse di trovare per via, un dì o l'altro, il paradiso terrestre'.

33 Ibid., pp. 120–1. The fate of the list of the medieval and Renaissance locations of paradise is discussed below, in the Epilogue.

34 Ibid., p. 110: 'I mappamondi … sono per solito … stravagantemente complicati.'

35 Ibid., p. 101.

36 Ibid., pp. 121–2, 185. Coli's scholarship and tolerance of earlier thinkers are worthy of respect.

37 Charles Raymond Beazley, *The Dawn of Modern Geography*, 3 vols (London: John Murray, 1897–1901; repr. New York: Peter Smith, 1949), I (1949), p. VI.

38 Ibid., pp. 332–4.

39 John K. Wright, *The Geographical Lore of the Time of the Crusades: A Study in the History of Medieval Science and Tradition in Western Europe* (New York: American Geographical Society, 1925; repr. New York: Dover, 1965), pp. 357–61.

40 Ibid., p. 361.

41 George H. T. Kimble, *Geography in the Middle Ages* (London: Methuen, 1938), pp. 181–204, esp. 181–5.

42 Leonardo Olschki, *Storia letteraria delle scoperte geografiche: Studi e ricerche* (Florence: Olschki, 1937), pp. 139–48.

43 Lloyd Arnold Brown, *The Story of Maps* (Boston: Little, Brown, 1949), pp. 81–112.

44 Howard R. Patch, *The Other World, According to Descriptions in Medieval Literature* (Cambridge, MA: Harvard University Press, 1950), pp. 134–74.

45 Gérard de Champeaux and Sébastien Sterckx, *Introduction au monde des symboles* (Yonne: Atelier Monastique de l'Abbaye Ste-Marie de la Pierre-qui-vire, 1966), p. 209: 'Les progrès de la cartographie et son évolution vers la rigueur des relevés scientifiques se heurteront longtemps à l'incoercible résistance du symbole.'

46 Joseph E. Duncan, *Milton's Earthly Paradise* (Minneapolis: University of Minnesota Press, 1972), p. 79.

47 Ibid., pp. 67–101.

48 *Het aards paradijs: Dierenvoorstellingen in de Nederlanden van de 16de en 17de eeuw* (Antwerp: Koninklijke Maatschappij voor Dierkunde van Antwerpen, 1982), pp. 137–52. For a more recent discussion on the cartography of paradise included in an exhibition catalogue on Flemish art see Ute Kleinmann, 'Wo lag das Paradies? Beobachtungen zu den Paradieslandschaften des 16. und 17. Jahrhunderts', in *Die flämische Landschaft 1520–1700* (Lingen: Luca Verlag, 2003), pp. 279–303.

49 Catherine Delano-Smith and Elizabeth M. Ingram, *Maps in Bibles, 1500–1600: An Illustrated Catalogue* (Geneva: Droz, 1991). See also Catherine Delano-Smith, 'Maps in Bibles in the Sixteenth Century', *The Map Collector*, 39 (1987), pp. 9–10; Wilco C. Poortman and Joost Augusteijn, *Karten in Bijbels (16e–18e eeuw)* (Zoetmeer, Uitgeverij Boekencentrum, 1995), pp. 78–82; Elizabeth M. Ingram, 'Maps as Readers' Aids: Maps and Plans in Geneva Bibles', *Imago Mundi*, 45 (1993), pp. 34–5.

50 Jean-Robert Massimi, 'Montrer et démontrer: Autour du *Traité de la situation du paradis terrestre* de P. D. Huet (1691)', in Alain Desrumeaux and Francis Schmidt, eds, *Moïse géographe: Recherches sur les représentations juives et chrétiennes de l'espace* (Paris: Vrin, 1988), pp. 203–25.

51 Claudine Poulouin, *Le Temps des origines: L'Eden, le Déluge et 'les temps reculés'. De Pascal à l'Encyclopédie* (Paris: Honoré Champion, 1998); Franco Motta, '"Geographia sacra": Il luogo del paradiso nella teologia francese del tardo Seicento', *Annali di Storia dell'Esegesi*, 14/2 (1997), pp. 477–506; Max Engammare, 'Portrait de l'exégète en géographe: La Carte du paradis comme instrument herméneutique chez Calvin et ses contemporains', *Annali di Storia dell'Esegesi*, 13/2 (1996), pp. 565–81.

52 Charles W. J. Withers, 'Geography, Enlightenment, and the Paradise Question', in David N. Livingstone and Charles W. J. Withers, eds, *Geography and Enlightenment* (Chicago–London: University of Chicago Press, 1999), pp. 67–92. Walter Goffart, *Historical Atlases: The First Three Hundred Years, 1570–1870* (Chicago–London: University of Chicago Press, 2003), pp. 156, 226, referred to late eighteenth-century maps of paradise.

53 See, e.g., Peter S. Hawkins, 'Out upon Circumference: Discovery in Dante', in Scott Westrem, ed., *Discovering New Worlds: Essays on Medieval Exploration and Imagination* (New York–London: Garland, 1991), pp. 193–220 (repr. in Peter S. Hawkins, ed., *Dante's Testaments: Essays in Scriptural Imagination* (Stanford, CA: Stanford University Press, 1999), pp. 265–83); Patrick Boyde, *Dante Philomythes and Philosopher: Man in the Cosmos* (Cambridge: Cambridge University Press, 1981), pp. 96–111; Bruno Nardi, 'Il mito dell'Eden', in his *Saggi di filosofia dantesca*, 2nd edn (Florence: La Nuova Italia, 1967), pp. 311–40; Charles S. Singleton, *Dante's "Commedia": Elements of Structure* (Baltimore:

Johns Hopkins University Press, 1954) ('The Return to Eden'); Edward Moore, *Studies in Dante, Third Series: Miscellaneous Essays* (Oxford: Oxford University Press, 1903; repr. 1968), pp. 109–43, esp. 134–9.

54 Claude Kappler, *Monstres, démons et merveilles à la fin du Moyen Âge* (Paris: Payot, 1980), pp. 85–95.

55 Jacques Le Goff, *L'Imaginaire médiéval: Essai* (Paris: Gallimard, 1985), pp. 30, 111.

56 Jacques Le Goff, 'Introduction: Medieval Man', in *The Medieval World*, ed. by Jacques Le Goff, transl. by Lydia G. Cochrane (London: Collins and Brown, 1990), p. 29. Le Goff often quotes Graf.

57 Giuseppe Tardiola, *Atlante fantastico del Medioevo* (Anzio: De Rubeis, 1990), p. 7: 'Dal paradiso terrestre all'*India mirabilis*, dall'inquietante regno del Prete Gianni ai tenebrosi domini di Satana e dell'Anticristo, avremo modo di compiere un immaginario viaggio tra le coordinate meravigliose e inattese di un plausibile "Atlante fantastico".'

58 Ibid., pp. 7–46.

59 Jean Delumeau, *Une histoire du paradis*, I, *Le Jardin des délices* (Paris: Fayard, 1992), pp. 59–97; *History of Paradise: The Garden of Eden in Myth and Tradition*, transl. by Matthew O'Connell (New York: Continuum, 1995), pp. 39–70.

60 Delumeau, *Une histoire du paradis* (1992), pp. 301–7 (pp. 229–33 in the English edition, 1995).

61 Fred Plaut, 'General Gordon's Map of Paradise', *Encounter* (June/July 1982), pp. 20–32. On Gordon's map of paradise see below, Chapter 11, pp. 353–6.

62 Fred Plaut, 'Where is Paradise? The Mapping of a Myth', *The Map Collector*, 29 (1984), pp. 2–7.

63 Ibid., p. 7.

64 Ibid., p. 4.

65 Fred Plaut, *Analysis Analysed: When the Map Becomes the Territory* (London–New York: Routledge, 1993), p. 160. The classification of maps such as those of the location of the Garden of Eden as fantasy maps as opposed to religious maps is discussed elsewhere in the book, e.g. on p. 215.

66 Cosmas described the world as shaped like a Tabernacle and Eden as an oblong land in the east: see below, Chapter 7, pp. 354–6.

67 Plaut, *Analysis Analysed* (1993), pp. 159–61.

68 Corin Braga, *Le Paradis interdit au Moyen Âge: La Quête manquée de l'Eden oriental* (Paris: L'Harmattan, 2004), for example, pp. 135–7, 149, 234, 329–30.

69 Ibid., p. 233: 'Semblables à des laboratoires imaginaires, avec des spécimens conservés dans des éprouvettes, les mappemonde de l'époque …'; p. 14: 'galerie de monstres et de merveilles'. See also pp. 13, 137–43, 150–1, 205, 215, 237, 255, 304, 314, 334.

70 Ibid., p. 377: 'Ce regard positif jeté sur des zones jusqu'alors floues de la mappemonde détruit le mirage du paradis.' See also pp. 146, 384.

71 Ibid., p. 133: 'Ces nouveaux positionnements reflètent en général la réaction d'une époque plus rationaliste et pragmatique aux constructions fantastiques de la "pensée enchantée" du Moyen Âge.' See also p. 382.

72 Ibid., p. 384.

73 Alan M. MacEachren, *How Maps Work: Representation, Visualization, and Design* (New York–London: Guilford Press, 1995), pp. 255–6.

74 Ibid., pp. 28, 190–93, 252, 278–90, 361.

75 Brian Harley, 'Deconstructing the Map', *Cartographica*, 26/2 (1989), p. 15, quoted in McEachren, *How Maps Work* (1995), p. 10.

76 See David Woodward, 'The "Two Cultures" of Map History – Scientific and Humanistic Traditions: A Plea for Reintegration', in David Woodward, Catherine Delano-Smith and Cordell D. K. Yee, *Plantejaments i objectius d'una història universal de la cartografia/Approaches and Challenges in a Worldwide History of Cartography* (Barcelona: Institut Cartogràfic de Catalunya, 2001), pp. 49–67; Denis Cosgrove, 'Introduction: Mapping Meaning', in Denis Cosgrove, ed., *Mappings* (London: Reaktion, 1999), pp. 1, 7–8; David Woodward and Brian Harley, 'Preface', in Brian Harley and David Woodward, eds, *The History of Cartography*, II/1, *Cartography in the Traditional Islamic and South Asian Societies* (Chicago–London: University of Chicago Press, 1992), esp. pp. XIX–XXIV.

77 See, for example, Patrick Gautier Dalché's penetrating analyses: 'Décrire le monde et situer les lieux au XIIe siècle: L'*Expositio mappe mundi* et la généalogie de la mappemonde de Hereford', in *Mélanges de l'École Française de Rome*, 113 (2001), pp. 376–7; 'Principes et modes de la représentation de l'espace géographique durant le haut Moyen Âge', in *Uomo e spazio nell'alto Medioevo* (Spoleto: Centro Italiano di Studi sull'Alto Medioevo, 2003), pp. 119–50.

78 See, for example, Brian Harley, *The New Nature of Maps: Essays in the History of Cartography*, ed. by Paul Laxton (Baltimore and London: Johns Hopkins University Press, 2001); MacEachren, *How Maps Work* (1995), pp. 355–60; Denis Wood, *The Power of Maps* (New York: Guilford Press, 1992); Mark S. Monmonier, *How to Lie with Maps* (Chicago–London: University of Chicago Press, 1991); John Keates, *Understanding Maps* (New York: John Wiley, 1982), pp. 52–3. A number of writers have commented on the inherent subjectivity of maps and the dangers of viewing cartography as an objective activity.

2

Paradise in the Bible

The Holy Spirit ... in any one expression of Scripture meant far more than any expositor can expound or discover.
Thomas Aquinas, *Quaestiones quodlibetales*, VII.14, (1256)

The story of paradise and the Garden of Eden comes from the Book of Genesis in the Hebrew Bible. No more than a dozen or so biblical verses lie at the origin of the age-old debate on the existence and the location of the earthly paradise. Few as they are, however, the interpretation of these verses is crucial for understanding the history of the representation of the Garden of Eden in art, poetry and literature, and, from the eighth century onwards, on the maps with which this book is concerned.

The Bible opens with the story of the creation of the world by God (see Plate 1). The first chapter of the Book of Genesis records how God completed his work in six days. On the first day, God created light, separating it from darkness. On the second day he formed the firmament of the sky. On the third day he created the land, with its plants and trees, distinguishing it from the sea. On the fourth day, the stars, the sun and the moon were created, and on the fifth day, the fish and the birds. Man was fashioned on the sixth day, together with the animals. Finally, on the seventh day, as described at the beginning of the second chapter, God finished his work and rested.

Confusingly, however, later in the second chapter of Genesis, God is once again described fashioning a human being, a man into whose nostrils God breathes life. The presence of a second account gave rise to a textual contradiction that, despite all the interpretative difficulties involved, in the end only enriches the paradise motif. God's new being needed a place in which to lay his head, and it is in this second account that we read how God created a garden for him (verse 8), and how he ordained that all kinds of trees should grow there, trees that were 'pleasant to the sight and good for food' (verse 9). In the middle of the garden were two special trees, the Tree of Life and the Tree of the Knowledge of Good and Evil, and four rivers sprang forth to water the garden. Then God noticed that of all the creatures he had created, the man alone lacked a mate. Causing him to fall into a deep sleep, and taking one of his ribs, God shaped a woman. When Adam – the man – awoke, he found himself, much to his liking, with a companion, Eve. Entrusting to Adam the care of a garden that provided the perfect conditions for human life, God ordered the pair never to touch the fruit growing on the Tree of the Knowledge of Good and Evil. Eve, however, yielded to the serpent's temptation, plucked a forbidden fruit, ate of it, and gave some to her husband. For their disobedience, man and woman were expelled forever from the Garden of Eden.

So much for the story of Adam and Eve in paradise, in simple outline.

Rendering and Reading the Text

Not only has the duplication of the account of the Creation in the second chapter of the Book of Genesis complicated the interpretation of the biblical text, but there is no extant 'original' version to turn to for a definitive account of the origin of the world and for a coherent description of the garden into which the newly created man was placed.[1] The words of the original author, whoever he was (Moses, as Judaeo-Christian tradition claimed, or an anonymous compiler of material from different sources, as modern Old Testament specialists hold) have long been lost. All that remains is a miscellany of texts that were translated into different languages at different times from a prototype text that was almost certainly based on ancient oral tradition and that was probably written down in Hebrew in the ninth or tenth century BC. The earliest allusions to a written version are no older, however, than about 300 BC, and the earliest surviving biblical texts (discovered in the 1950s among the Dead Sea Scrolls) date from some time between the second century BC and the beginning of the Christian era. What has been taken by Jewish scholars as the authoritative Hebrew version – the so-called Masoretic text (from מָסוֹרָה (*masorah*), meaning 'transmission') – is the work of Jewish grammarians from the sixth to the eighth centuries AD.

A Greek translation of the first five books of the lost original Hebrew text (the Pentateuch) became available in the third century BC in Egypt for Greek-speaking Jews. This is the version today known as the Septuagint, in recognition of the 72 elders of Israel who prepared the translation for the library in Alexandria.[2] It was the Septuagint that was used in the first two centuries AD by the early Christians.[3] In the third century, the Christian scholar Origen assembled six different texts into the *Hexapla*: four Greek translations, a Hebrew text, and a transliteration of the Hebrew text into Greek characters. Early in the fifth century AD, Jerome used the *Vetus Latina* (Old Latin), a version derived from numerous early but fragmentary Latin translations from the Greek, together with the earliest Hebrew text and some other Greek texts, for a thorough revision of the whole Bible that he wrote down in Latin. Jerome's translation, now known as the Vulgate, became the standard and authoritative Latin version for the Western Christians throughout the Middle Ages.

Apart from Jerome's contemporary, Augustine, who used the *Vetus Latina*, almost every Christian commentator up to the Reformation read the description of the Garden of Eden in the Vulgate, Jerome's Latin version:

> *Plantaverat autem Dominus Deus paradisum voluptatis a principio in quo posuit hominem quem formaverat / produxitque Dominus Deus de humo omne lignum pulchrum visu et ad vescendum suave / lignum etiam vitae in medio paradisi / lignumque scientiae boni et mali / et fluvius egrediebatur de loco voluptatis ad inrigandum paradisum qui inde dividitur in quattuor capita / nomen uni Phison / ipse est quod circuit omnem terram Evilat / ubi nascitur aurum / et aurum terrae illius optimum est / ibique invenitur bdellium et lapis onychinus / et nomen fluvio secundo Geon / ipse est qui circuit omnem terram Aethiopiae / nomen vero fluminis tertii Tigris / ipse vadit contra Assyrios / fluvius autem quartus ipse est Eufrates / tulit ergo Dominus Deus hominem et posuit eum in paradiso voluptatis ut operaretur ibi et custodiret illum.* (Genesis 2.8–15)

From 1611, readers in England could also appreciate the description of the Garden of Eden in one of the most famous English-language texts, the version authorized by King James I:[4]

> *And the Lord God planted a garden eastward in Eden; and there he put the man whom he had formed. And out of the ground made the Lord God to grow every tree that is pleasant to the sight and good for food; the tree of life also in the midst of the garden, and the tree of knowledge of good and evil. And a river went out of Eden to water the garden; and from thence it was parted, and became into four heads. The name of the first is Pison: that is it which compasseth the whole land of Havilah, where there is gold; And the gold of that land is good: there is bdellium and the onyx stone. And the name of the second river*

is Gihon: the same is it that compasseth the whole land of Ethiopia. And the name of the third river is Hiddekel: that is which goeth toward the east of Assyria. And the fourth river is Euphrates. And the Lord God took the man and put him into the Garden of Eden to dress it and to keep it. (Genesis 2.8–15)

Irrespective of language, the interpretation of these eight verses has led to an immense tangle of assumptions about the nature and location of the earthly paradise – the Garden of Eden – because of problems of translation and interpretation in the absence of an original standard version. Even today different versions of the Bible are read in different places in the Christian world.

Variations in the rendering of the text are liable to entrain quite different meanings. For example, the word 'paradise' appeared first in the Septuagint of the third century BC. When, in the Hebrew version of the second chapter of Genesis, the earthly paradise is first mentioned it is described with the words גַּן־בְּעֵדֶן (*gan-beEden*), meaning 'a garden in Eden' (2.8) and thus defining Eden as the place in which God had placed a garden. Later on, in Genesis, we encounter other forms of the words; a river is described as going מֵעֵדֶן (*meEden*) 'out of Eden' (2.10), Adam as being put בְּגַן־עֵדֶן (*began-Eden*) 'in the Garden of Eden' (2.15), or, eventually, expelled מִגַּן־עֵדֶן (*megan-Eden*) 'from the Garden of Eden' (3.23). The translators of the Septuagint rendered the Hebrew גַּן *gan* ('garden') as παράδεισος (*paradeisos*), a word which in ancient Greek

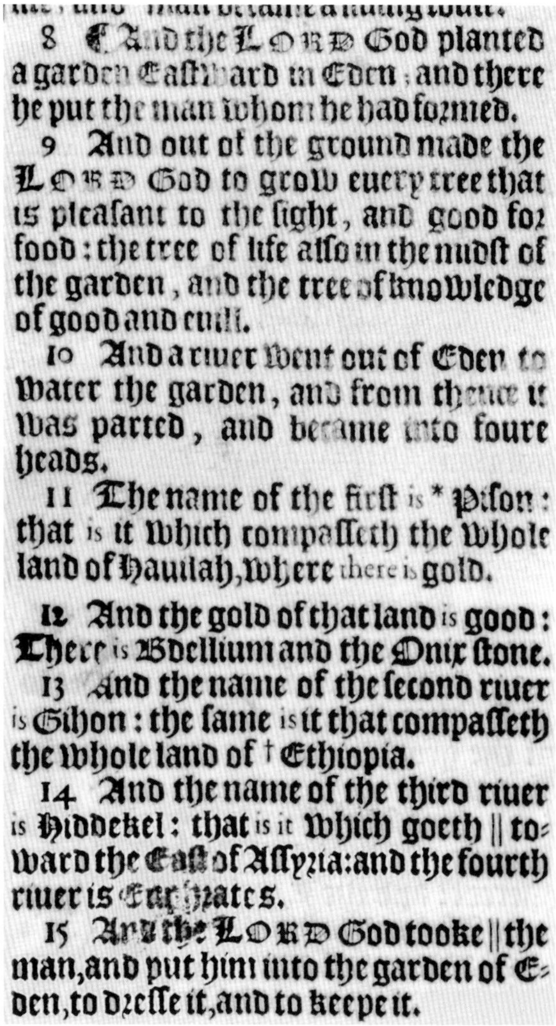

Fig. 2.1 (left). The text of Genesis. Detail showing the description of the Garden of Eden (2.8–15) in the Latin Vulgate, from the *Biblia sacra latina*. Pre-Monstratensian monastery of St Maria de Parco, Louvain, 1148. London, British Library, Add. MS 14788, fol. 7v.

Fig. 2.2 (right). The text of Genesis, 2.8–15 in English, from King James's (Authorized) version of *The Holy Bible* (London: Robert Barker, 1611), sig. A1v–A2r. London, British Library, L13.f.2.

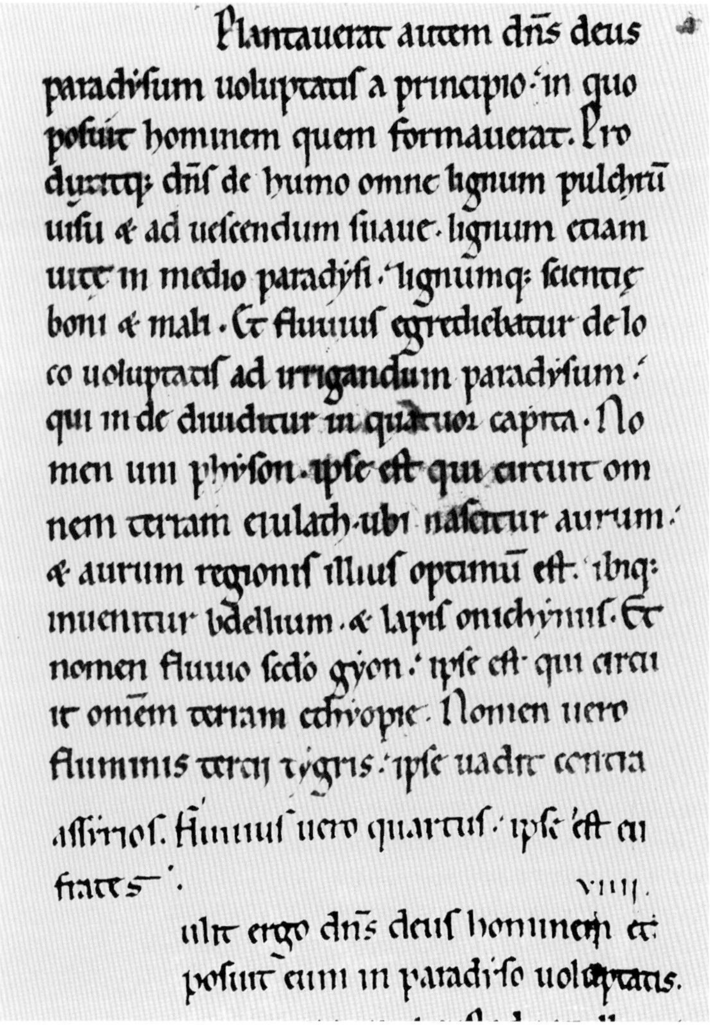

meant 'enclosed park' or 'pleasure ground'.[5] The term παράδεισος (paradeisos) was first used in Western literature by the Greek historian Xenophon to indicate a large well-watered field containing trees, flowers and animals, and surrounded by a wall. The original Median word, paridaeza, meant an enclosure: pari corresponds to the Greek περί (meaning 'around') and daeza indicated a wall made of a sticky substance like clay (from diz, 'to mould' or 'to form'). Thus, the Septuagint's παράδεισος (paradeisos) denoted much more specifically than did the Hebrew גן (gan) an *enclosed* park.[6] Further qualifications appeared in early Christian translations. Whereas in the *Vetus Latina* the word *paradisus* on its own had sufficed, in Jerome's Vulgate the qualification *voluptatis* ('of delight') was added in recognition of the fact that *Eden* in Hebrew means 'delight', as many later biblical scholars liked to point out. Not all translators, however, used the word *Eden* as a common noun to indicate the pleasure of the park. In the Septuagint, in the *Vetus Latina*, and in the Authorized Version of King James, the word was deployed as a proper noun to designate a specific place called Eden.

The translation of the word Eden as a place inevitably invoked in due course the question: where was this garden? Further confusion resulted from the translation of other words in the text that could be interpreted as relating to its geographical location. The Hebrew qualifies גַּן־בְּעֵדֶן (gan-beEden), 'a garden in Eden', with the term מִקֶּדֶם (miqedem) (2.8), a word that has two quite different meanings, one referring to space, the other to time. Faced with the ambiguity of the words, translators had to choose between rendering the Hebrew spatially – 'away to the east' – or temporally – 'from before the beginning'. Jerome adopted the latter for the Vulgate, translating מִקֶּדֶם (miqedem) as *a principio*, to convey the idea that the earthly paradise had been an integral part of God's primordial creation.[7] In contrast, the translators of the Septuagint, the *Vetus Latina* and the English Authorized Version all chose to give the expression a spatial meaning: κατὰ ἀνατολὰς (kata anatolas) in the Greek, *in oriente* in the Latin, and 'eastward' in the English versions respectively.

Yet another geographical problem arose in the process of translation. This concerned the identity of the four rivers that watered the Garden of Eden.[8] Two were easily recognized. The Tigris – Hebrew חִדֶּקֶל, *Hiddekel* – is described in Genesis as running along the eastern side of Assyria (2.14). The Euphrates – Hebrew פְּרָת (*Prat*) (2.14) – was known in Antiquity as the river by which Babylon stood. The other two rivers, the Pishon (פִּישׁוֹן, 2.11) and the Gihon (גִּיחוֹן, 2.13), posed greater problems of identification, and from the start translators and exegetes struggled to find suitable rivers to match the text. The most common interpretation, originally put forward by the first-century Jewish historian Flavius Josephus, was to identify the Pishon and the Gihon with the Ganges and the Nile respectively.[9] Other authors identified the Gihon with the Indus (for instance, Hippolitus of Rome in the third century) and the Pishon with the Danube (for instance, Ephrem the Syrian in the fourth century, and Severian of Gabala in the fifth century).[10] In the event, the problem of the four rivers of paradise was to occupy some of the best minds in exegetical mapping well into modern times.

The names of key places also posed problems. In all the early translations, 'Havilah' (הַחֲוִילָה, 2.11) appears to indicate the region surrounded by the river Pishon, but its exact geographical location proved elusive and was never made absolutely clear. To judge from references elsewhere in the Bible, the region lay somewhere between Asia and Africa. Josephus and most of the Church Fathers identified Havilah with the Ganges valley, but Havilah has been also located in Abyssinia, Syria and Arabia.[11] Another region in the vicinity of the Garden of Eden was the land of Cush, described in Genesis as surrounded by the river Gihon (2.13). Jerome followed Josephus, who thought Gihon meant the Nile, and rendered the Hebrew word כּוּשׁ (*Kush*) as 'Ethiopia', meaning the land of Nubia, south of Egypt. Other translators, though, thought Cush was the Mesopotamian kingdom of the Kassites.[12] Not only words, but entire phrases posed problems of translation. For example, Genesis 3.24 was variously rendered to mean either that God allowed Adam and his progeny to live on the opposite side (i.e. to the east) of the Garden of Eden or that he placed Cherubim on the opposite side (i.e. to the east) of the Garden of Eden.[13] What was lost in translation was the original Hebrew pun in the words גַּן־בְּעֵדֶן מִקֶּדֶם (gan-beEden miqedem), found at the beginning of the paradise narrative (2.8), and מִקֶּדֶם לְגַן־עֵדֶן (miqedem legan-Eden), found at the end of the story (3.24). Throughout the ages, in short, each act of translation has been in effect an act of textual redefinition.

Early Debate

Quite apart from the hazards of translation, the description in Genesis of the Garden of Eden has always left biblical commentators with puzzling questions. Did Adam and Eve's garden really exist on earth? What exactly was it – a real park somewhere in the east? Were the Trees of Life and of the Knowledge of Good and Evil actual plants? And, if the four paradisiacal rivers that were said to emerge from the Garden of Eden are to be identified with the Ganges, the Nile, the Tigris and the Euphrates, how could they all come from a single source? From the beginning of Christianity, biblical exegetes had to face the challenge of the meaning of the Garden of Eden described in Genesis, and the difficulties involved meant that the idea of an earthly paradise was not clearly defined until relatively late. Until the fifth century, when Augustine of Hippo cut through the Gordian knot of accumulated controversy by stressing that the Book of Genesis was referring to a real place on earth, ideas on Eden remained unsettled. Early Christian theologians oscillated between different notions in a fluid process of interpretation and reinterpretation, debating every element until each had crystallized and all had come together in a standard formulation.

From these early debates came two main lines of approach, one allegorical and one literal. In the former, the elements of the scriptural narrative were reduced to figurative expressions; in the latter, the description of the Garden of Eden was taken as a report about a real place somewhere on earth. Of course, centuries of theological exegesis can hardly be reduced to a simple opposition between two such clear-cut positions, and the full complexity of the theological issues involved is grasped only after detailed study of all the writings concerned, since there were many possibilities for compromise between radical realism and multi-layered symbolical reading. Another difficulty inherent in any attempt to present such unstable material coherently yet without oversimplification is that the biblical text could be read on different levels that were not necessarily exclusive. The Church Fathers were themselves aware of additional interpretative strategies that they themselves omitted to explore.

For the purpose of reconstructing the conditions that ultimately led to the representation of paradise on medieval maps, only a few of the most influential thinkers from the first four centuries of Christianity need be introduced here in order to illustrate how the two approaches were applied to the problem of the nature and location of the earthly paradise. More often than not, the different interpretations expressed fundamentally different lines of inquiry: thus, whereas the allegorical elevated Eden from the realm of geography into a spiritual dimension, the literal produced a geography of paradise tied closely to the physical world.

Philosophical Interpretation: the Spiritual Eden

Major difficulties were bound to emerge as soon as the paradise narrative was interpreted as a series of real events in a real place. God's creation of an earthly paradise is described in Genesis in a strikingly anthropomorphic way, and the physical existence of the Garden of Eden became a geographical conundrum. But the initial problem lay in the duplication of the story of Adam's creation in the first two chapters of Genesis. In the first chapter, Adam is described as having been shaped as an ideal being made in the image of God. In the second chapter the creation of Adam is immediately followed by an account of the Temptation and the resultant Expulsion of Adam and Eve from the Garden of Eden. One of the first tasks in the early centuries of Christian exegesis was to reconcile these textual inconsistencies.[14] A way out of the difficulty was to bring philosophical principles to bear on the biblical account and to transfer its interpretation to an allegorical plane. This meant considering all the elements of the narrative as merely figurative expressions.

Two Alexandrian writers, one a Jew and the other a Christian, can be taken as examples of the allegorical approach. In fact, Western exegesis as a whole owes its conception of a spiritual Eden to Philo, the Jewish thinker whose writings in the first half of the first century AD came in due course to influence heavily those of the Christian exegete Origen. Although separated by nearly two centuries, both Philo and Origen came from a milieu in which Hellenistic culture and Judaeo-Christian tradition were intricately interwoven. Both read Genesis as Platonic philosophers, and both attempted to reveal deeper meanings hidden in the text. Philo pointed out that the garden of Adam and Eve was not a park of material plants, but rather a place of heavenly virtues, and associated Eden, defined as 'the wisdom of God', with the heavenly Adam, the divine archetype of man.[15] To explain the duplication of the creation story, Philo distinguished between the creation of a heavenly, immortal human being made in God's image, and the creation of an earthly and corruptible man, intrinsically unstable and naturally inclined to sin as a consequence of the Fall.[16] In Philo's philosophical interpretation, the second – the earthly – Adam is seen as only a shadow of his true self, the first and perfect man, who was created in the image of God and whose task is to return to his original, spiritual perfection in the true Eden.[17]

Philo also accepted an allegorical level of interpretation, in which the events described in Genesis were taken as referring to processes within the soul.[18] To imagine the Lord of the universe tilling the soil and planting a garden with vines, olive, apple and pomegranate trees, still more the existence of a Tree of Life and a Tree of the Knowledge of Good and Evil, was absurd.[19] Instead, Philo urged the initiated reader to seek the true meaning of the story of Adam and Eve in 'what lies beneath the surface'.[20] The individual elements of the biblical account were merely ways of making deeper notions visible and were to be understood figuratively.[21] For Philo, the paradise narrative referred to the moral struggle over the acquisition of virtue. The Garden of Eden stood for the rational human soul, the trees in the garden that bore the fruit of immortal life and were 'beautiful to look upon' and 'good to eat' were the virtues planted by God in man, and the four rivers were the four cardinal virtues of Greek philosophy.[22] This was why the garden was called Eden, which – Philo reminded his readers – means 'delight'. It symbolized the ecstasy that a soul finds in the service of the Lord. The Garden was said to be in the east, where the sun rises, because virtue, like the sun, disperses darkness with its radiant light.[23] The river Pishon, which encircled the land of Havilah – which represented speciousness or the land of foolish aspirations – was Prudence, who gleamed like gold and prevented the soul from wrongdoing. The Gihon, which surrounded the land of Cush, represented courage encircling cowardice. The Tigris, which flowed towards Assyria, stood for the virtue of self-mastery, the control of desire, and a bulwark against over-indulgence. Euphrates meant 'fruitfulness' and was the symbolic name for justice, a fruitful virtue indeed, for justice appears when the three parts of the Platonic soul – the rational, the spirited and the appetitive, corresponding to the first three virtues, the first three rivers of Eden, and the head, breast and abdomen – are in harmony; that is, when reason rules.[24] Paradise thus represented for Philo the imitation within the human soul of an archetypal heavenly excellence. The man made in God's image (the heavenly Adam) and the Tree of Life (which refers to the heavenly Adam) stand together for the divine guiding power by which the human mind (represented by the earthly Adam) is enabled to return to original perfection.

In being allotted the task of tending the plants in the Garden of Eden, the earthly Adam – figuratively, the human mind – had been instructed to pursue the virtues of the rational soul (represented by the Tree of Life and the other trees).[25] The Tree of the Knowledge of Good and Evil signified the soul's choice between virtue and sin and between life and death.[26] If the mind chose evil, the Tree of Life became the Tree of Knowledge. Arguing that it was ridiculous, on a literal level, to imagine a woman being created from a man's rib, Philo suggested that Eve – who was born when Adam fell asleep, that is, when the mind was off-guard – symbolized the physical senses. This would explain how it was that Eve was corrupted by pleasure (represented by the serpent) and seduced Adam, and that both were expelled from the Garden. Interpreted by Philo, the Genesis story discloses the pattern of all human sin and the submission of the mind to the physical senses. However, exclusion from the Garden of Eden (i.e. from the delight of divine communion) need be only temporary for those who repent and return to the path of virtue.[27]

Philo's Jewish reading of Genesis had a marked influence on the Christian school in Alexandria, the most important representative of which was Origen.[28] Following closely in Philo's footsteps, and drawing on philosophical speculations and on other biblical texts, Origen also expounded the mysteries of mankind's beginnings. Before creating the visible world, according to Origen, God had created 'intellects' and had given them free will. One of these creatures, however, rebelled and turned into the devil, thus initiating a process of degradation of the intellects from their originally fervent worship of God towards a cool neglect of their creator. The process brought about the emergence of the material world and, for man, the burden of a body and a perpetual internal struggle between rational dictates and irrational impulses.[29]

Like Philo, Origen dealt with the creation of Adam by differentiating the man that had been made in God's image – spiritual and immortal – from the earthly and corporeal man remoulded as a consequence of his Fall.[30] Also like Philo, Origen insisted that true human nature is that of the man originally made in God's image (that is, with mankind's highest spiritual attainments, which are related to the divine), and not that of man's post-lapsarian confinement within a physical body.[31] Again like Philo, he urged the reader of the Bible to look for the spiritual treasures hidden in its text.[32] He recognized, however, that the Scriptures contained a complex mixture of historical narrative and spiritual truth, yet contended that the biblical description of paradise had a purely spiritual and not a historical significance.[33] Adopting Philo's words, Origen argued that it would be foolish to imagine God as a peasant farmer planting trees somewhere in the east, and to regard the Trees of Life and of the Knowledge of Good and Evil as visible and tangible trees whose fruit would bestow immortal life or the knowledge of good and evil. Origen also took over Philo's idea that the Garden of Eden was the paradise of the soul and its virtues and that the paradise story was a parable about the nature of man and the life of every Christian.[34] If the heart (Eve), married to the spirit (Adam), succumbs to corporeal pleasures, the Christian soul loses its condition of divine bliss, having been carried away by the gratification of the flesh.[35] Going beyond Philo, and as a Christian, Origen developed an interpretation of the Garden of Eden as an image of the Church, and of the Tree of Life as an allusion to Christ, creating an allegorical reading of the paradise narrative that was to endure for centuries.[36]

Like Philo's, Origen's thought was strongly influenced by Platonic philosophy and was consequently unable to accommodate the idea of a condition of human perfection involving physical bodies and, above all, a physical location. At the same time,

Origen speculated about a 'good land', or 'pure earth', different from our own. He equated it with both the original dwelling place of Adam and the ultimate place of beatitude, where human souls could purge themselves on their way back to heaven. We find no mention, however, in Origen's writings, of any paradise on earth.[37]

Literal Interpretation: the Physical Eden

Philo and Origen were both anxious to ensure consistency between the Bible and Platonic philosophy. Neither considered that the story of paradise could be placed on the plane of earthly history, so the Garden of Eden could not be an earthly place.[38] Had their reading of Genesis prevailed in the Latin Middle Ages, Eden would never have been perceived as a mappable feature. Other Christian exegetes, however, saw things differently. The early Christian Church of Antioch, for example, in contrast to the Platonic and mystical tradition of Alexandria, advocated a historical exegesis of Scripture.[39]

The Antiochenes emphasized the need to adhere closely to the sacred text. In their view, the description of Eden was a report about a real place on earth. The differences of interpretation between the Antiochenes and the Alexandrians initiated an enduring opposition of, on the one hand, the biblical account of historical events and, on the other hand, philosophical principles as expounded in Greek Platonism. Until Augustine provided an authoritative solution to this struggle, the concept of the Garden of Eden as a specific region on earth locatable in space and time had to compete with the notion of a spiritual paradise of divine perfection beyond geography and history and known only through allegory.[40] An exegesis orientated by philosophy emphasized the need to bring philosophical principles to bear on the biblical narrative if Scripture was to be defended from the mockery of the pagans. Whereas allegory purported to unveil the deep spiritual truths that lay beneath the surface of the biblical text, those who favoured a literal reading of Genesis argued that an exclusively allegorical interpretation challenged the essential beliefs of the Christian faith. To doubt the historicity of the events narrated in the Bible implied that the word of God could be questioned by human reason; and to deny the historical existence of Adam and Eve and the story of their Fall was to put at risk the whole system of Christian dogma, for it was the Fall that necessitated the Incarnation of Christ, his redeeming sacrifice on the Cross, and the sacramental life of the Church.

There were, then, some early exegetes who stressed the importance of a literal reading of Genesis and the geographical aspects of paradise. One of these was the Greek 'golden-mouthed' John Chrysostom, bishop of Constantinople from 398 to 404.[41] Chrysostom warned that anyone who wanted to penetrate divine mysteries 'should know that man cannot behold all the works of God relying on his own reasoning' and condemned those who claimed that paradise was not on earth.[42] Ignoring the geographical details given in the Genesis text, which, he pointed out, included even the name (Eden) of the location of paradise, some so-called wise men had dreamed up foolish fables, attributing to heaven what belonged to earth. Trusting in their own eloquence, or in some exotic philosophy, they wanted to deceive simple people. The author of Genesis (Moses, according to Chrysostom) had been inspired by the Holy Spirit, whereas those people heeded only what amused or attracted them. They contradicted the word of God by asserting that paradise was not on earth and that the water of the rivers of paradise was not in fact water. Chrysostom urged his readers to plug their ears so as not to hear those who so dangerously imposed their thoughts on Scripture; it was far better to follow the text of the Bible literally. Chrysostom never doubted that God had created paradise in the east, in the place described in the Bible, and that, in honour of man, he had planted therein trees, which were pleasing to the eye and which would provide food.[43]

Another late antique writer, Theodore of Mopsuestia, also stated unhesitatingly that paradise was an earthly region somewhere in the east, that the Tree of the Knowledge of Good and Evil was a real tree (a fig tree, in fact), and that Adam and Eve had plucked real fruit from it.[44] Severian, bishop of Gabala, Syria, at the beginning of the fifth century, explained why God had placed paradise in the Orient, saying that the Creator wanted to prefigure the future of mankind by linking human existence with the rising and setting of heavenly bodies. As the sun rises in the east, so the human race also had its beginning in the east, and as the sun sets in the west so, with the Fall of Adam, mankind was plunged into darkness. Just as the sun rises again, however, man with the advent of Christ would be given the opportunity of another dawn. Thus, as with all heavenly bodies, man has to pass from life to death in order to rise again at the time of the resurrection of the dead. Severian also described the topography of Eden. He identified the four rivers of paradise as the Tigris, the Euphrates, the Nile and the Danube. These four streams, he said, have the same water and the same characteristics as their source, but as they flow across lands lying outside Eden they become polluted, and because they pass underground to reach the inhabited earth, the exact location of paradise will remain forever a secret. Since the waters of paradise turn into the great rivers of the world, paradise itself is no miniature garden, but a vast region that had been prepared not only for Adam and Eve, but also for the whole of mankind.[45]

Epiphanius of Salamis is perhaps the best-known of the early opponents of an allegorical reading of the biblical text.[46] In a letter to John, bishop of Jerusalem (394), Epiphanius listed the doctrinal errors commonly attributed to Origen. Waxing indignant at Origen's ideas, he said, he disagreed in particular with Origen's allegorical interpretation of the paradise story.[47] Our first ancestors were not ghosts, he protested, nor were the trees of paradise abstract angelic powers. It had to be acknowledged that man was made in God's image, as is frequently stated in the Scriptures. Adam had a physical body, Eve was formed from his rib, the fig leaves covered real flesh, Adam lay with Eve, who gave birth to Cain and later to Abel. When Adam was 130 years old, he had another son, Seth, and yet other sons and daughters, and eventually died at the age of 930 (Genesis 5.3–5). Thus, from Adam came all subsequent generations. Contradicting Origen's speculations, Epiphanius claimed that the Bible unambiguously locates the Garden of Eden on earth and not in heaven; Adam and Eve were expelled from an earthly garden, and not 'from heaven to earth'; after the Expulsion, they did not live 'under paradise', but east of Eden; the Cherubim guard the 'way' to paradise, not the 'ascent' to it; and, finally, the rivers 'flow out of' Eden and do not 'descend from' it. Indeed, Epiphanius had seen with his own eyes the Gihon – the Nile – and he had drunk real, not spiritual, water from the Euphrates.[48]

Epiphanius made the same points in two major works. In the *Panarion*, he asserted that there was no doubt that the paradise of Adam and Eve was an extraordinary place in the eastern regions of the earth, remote in space and separate from the rest of the world. The Tigris, the Euphrates and the other two rivers flowed out of the garden to follow circuitous courses across the known regions of the earth. These rivers could not have fallen from heaven, for the earth would never have withstood the impact.[49] These arguments are repeated in the *Ancoratus*, where Epiphanius added further geographical information about the rivers. The Pishon, which in India and Ethiopia is called the Ganges, and is known as the Indus to the Greeks, first encircles the land of Havilah (the region of the Elymaeans) then, having entered Ethiopia, flows away towards the south.[50] The Gihon is also no shadowy allegory, but can be identified as one of the rivers of the world, taking a course through Ethiopia and Egypt to reach eventually the Mediterranean. The Tigris and Euphrates travel underground before springing up in Armenia to flow on to Assyria and Persia respectively. If there were no visible paradise, Epiphanius insisted, there could be no source and hence no rivers. The Pishon, the Gihon, the Tigris and the Euphrates would disappear, there would be neither fig tree nor leaves. Nor would the forbidden fruit have been tasted, for there would have been no Eve to eat from the tree. Adam would never have existed. Without him, there would be no mankind. By turning everything into an allegory, warned Epiphanius, truth turns to fable.[51]

1. For the problems of textual transmission, I have made use of several biblical encyclopedias, including Wilfrid R. F. Browning, *A Dictionary of the Bible* (Oxford: Oxford University Press, 1996); *Encyclopaedia Judaica*, ed. by Cecil Roth, 17 vols (Jerusalem: Encyclopaedia Judaica, 1972); *Encyclopedia of Biblical Theology*, ed. by Johannes B. Bauer, 3 vols (London–Sydney: Sheed and Ward, 1970); *Peake's Commentary on the Bible*, ed. by Matthew Black and Harold H. Rowley (London: Nelson, 1962); *The Oxford Companion to the Bible*, ed. by Bruce M. Metzger and Michael D. Coogan (New York–Oxford: Oxford University Press, 1993).
2. *Septuaginta* in Latin means seventy. According to the apocryphal *Letter of Aristeas* (variously dated between 200 BC and 33 AD), six members from each of the twelve tribes of Israel (72 scholars in all) were summoned from Jerusalem by the Egyptian ruler Ptolemy II Philadelphus (285–246 BC). They took 72 days to achieve a perfect Greek translation, which was deposited in the Alexandrian library. Each scholar had worked independently, yet their finished versions were identical and, as a result of divine inspiration, superior to the original text: see *Encyclopaedia Judaica* (1972), IV, cols 851–5.
3. The communities of the Jewish diaspora abandoned the Septuagint, in favour of other revised Greek translations, after the Christians adopted it. Further Greek translations were made in the second century AD by Aquila, Theodotion and Symmachus.
4. The Authorized Version (1611) was translated by a team of 54 scholars drawn from the universities of Oxford and Cambridge and the clerics of Westminster Abbey.
5. The Hebrew פַּרְדֵּס (*pardes*, 'grove', 'orchard') that, like the Greek παράδεισος (*paradeisos*), derives from the Median *paridaeza*, does not appear in the Book of Genesis but is found elsewhere in the Old Testament (e.g. Songs of Songs 4.13; Ecclesiastes 2.5; Nehemiah 2.8). The word פַּרְדֵּס (*pardes*) was also used in medieval talmudic literature as a mnemonic for the four types of biblical exegesis, being an acronym of פְּשָׁט (*peshat*, 'literal meaning'), רֶמֶז (*remez*, 'implied meaning'), דְּרוּשׁ (*derash*, homiletic interpretation), and סוֹד (*sod*, 'mystical meaning'). We shall see in Chapter 4, pp. 69–70, that a fourfold reading of Scripture was also adopted by Christian exegetes in the Middle Ages.
6. On the history of the term 'paradise' see Jan N. Bremmer, 'Paradise: From Persia, via Greece, into the Septuagint', in Gerard P. Luttikhuizen, ed., *Paradise Interpreted: Interpretations of Biblical Paradise: Judaism and Christianity* (Leiden–Boston, MA: Brill, 1999), pp. 1–20.
7. On the Hebrew word מִקֶּדֶם (*miqedem*) and the biblical Garden of Eden see Ed Noort, 'Gan-Eden in the Context of the Mythology of the Hebrew Bible', in Luttikhuizen, ed., *Paradise Interpreted* (1999), p. 22. The earliest attestation of a temporal reading of the Hebrew term מִקֶּדֶם (*miqedem*), promoting the idea that God created paradise before the creation of the world, is found in IV Ezra, III.6: see Michael E. Stone, *Fourth Ezra: A Commentary on the Book of Fourth Ezra* (Minneapolis: Fortress, 1968), p. 68.
8. See, for example, Noort, op. cit., pp. 28–32.
9. Flavius Josephus, *Jewish Antiquities*, I.37–9 (see above, Prologue, p. 15 n. 4).
10. Hippolytus, *Die Chronik*, 236, ed. by Adolf Bauer and Rudolf Helm, GCS XLVI (Berlin: Akademie Verlag, 1955), p. 41; Ephrem the Syrian, *The Commentary on Genesis*, 6, in *Hymns on Paradise*, ed. by Sebastian Brock (Crestwood, NY: St Vladimir's Seminary Press, 1990), p. 201; Severian, *De mundi creatione*, V.5, PG LVI, col. 478. See also Monique Alexandre, *Le Commencement du livre: Genèse I–V: La Version grecque de la Septante et sa réception* (Paris: Beauchesne, 1988), pp. 259–60; Alexandre, 'Entre Ciel et terre: Les Premiers Débats sur le site du Paradis (Gen. 2, 8–15 et ses réceptions)', in François Jouan and Bernard Deforge, eds, *Peuples et pays mythiques* (Paris: Belles Lettres, 1988), pp. 206–7. In the late sixteenth century, Abraham Ortelius alluded to the Pishon as the Danube: see below, Chapter 10, p. 317.
11. Havilah is found in the Bible both as a personal and a place name. At Genesis 10.7, Havilah refers to a son of Cush, the son of Ham who inherited Africa from Noah. The Ishmaelite tribes are described in Genesis 25.18 as extending from Havilah into Shur on the eastern frontier of Egypt. See also 1 Samuel 15.7. In Genesis 10.29 and 1 Chronicles 1.23 another Havilah is mentioned, the son of Joktan, the personal name apparently standing for a locality in southern Arabia.
12. A Mesopotamian land of Cush was the favourite reading of Renaissance exegetes: see below, Chapters 9 and 10.
13. In the original Hebrew, הַכְּרֻבִים (*Cherubim*) is plural and refers to more than one angel, whereas in the Christian tradition only one angel, armed with a flaming sword, guards the entrance to Eden.
14. For the implications for the geography of paradise of the so-called documentary hypothesis (according to which the two creation narratives represent the fusion of two quite distinct sources, the Priestly and the Jahwist), see below, Chapter 11, pp. 343–6. On the problem of the six days of creation see Johannes Zahlten, *Creatio mundi: Darstellungen des sechs Schöpfungstage und naturwissenschaftliches Weltbild im Mittelalter* (Stuttgart: Klett-Cotta, 1979); André Caquot and others, eds, *In principio: Interprétations des premiers versets de la Genèse* (Paris: Études Augustiniennes, 1973); Yves-M. J. Congar, 'Le Thème de *Dieu-Créateur* et les explications de l'hexameron dans la tradition chrétienne', in *L'Homme devant Dieu: Mélanges offerts au père Henri de Lubac*, 3 vols (Paris: Aubier, 1963–4), I, (1963), pp. 189–222.
15. Philo, *Legum allegoriae*, I.I.65, [Works], Loeb Classical Library, transl. by Francis H. Colson and George H. Whitaker, 10 vols and 2 supplementary vols (London: Heinemann; New York: Putnam's Sons, 1929–62), I (1929), p. 189; ibid., p. 188: 'τῆς τοῦ θεοῦ σοφίας'. See also *Questions and Answers on Genesis*, I.4, Supplement 1, ed. by Levi A. Post and others, transl. by Ralph Marcus (London: Heinemann; Cambridge, MA: Harvard University Press, 1953), p. 3; *De confusione linguarum*, XI.41, [Works], IV, (1932), p. 33. On Philo and the letter of Scripture see David M. Hay, ed., *Literal and Allegorical: Studies in Philo of Alexandria's Questions and Answers on Genesis and Exodus* (Atlanta, GA: Scholars Press, 1991).
16. Philo, *De opificio mundi*, XLVI.134, [Works], I (1929), p. 106. See also *Legum allegoriae*, I.XII.31–2, I.XXIX.90, II.II.4 (1929), pp. 166–7,

17 See, e.g., Philo, *De opificio mundi*, XLVI.135, LI.145–7 (1929), pp. 106–7, 114–17; *De plantatione*, XI.44, [Works], III (1930), pp. 234–5. See David T. Runia, *Philo of Alexandria and the Timaeus of Plato* (Leiden: Brill, 1986), pp. 337, 467–75, and Thomas Tobin, *The Creation of Man: Philo and the History of Interpretation* (Washington, DC: Catholic Biblical Association of America, 1983), pp. 36–55. See also Pieter W. Van der Horst, 'Philo and the Rabbis on Genesis: Similar Questions, Different Answers', in *Eratopokriseis: Early Christian Question-and-Answer Literature in Context* (Louvain: Peeters, 2004), pp. 55–70.

18 Tobin, who identifies earlier material in Philo's writings, sees the allegory of the soul as Philo's own contribution: *Creation of Man* (1983), pp. 135–76. On the importance of the allegory of the soul in Philo's exegesis see also Pierre Boyancé, 'Études philoniennes', *Revue des Études Grecques*, 76 (1963), p. 68, and Jean Daniélou, *Philon d'Alexandrie* (Paris: Fayard, 1958), pp. 135–7. Certainly, Philo was the first to develop from Genesis a spiritual allegory referring to internal processes. On Alexandrian predecessors of Philo, see also Fearghus Ó Fearghail, 'Philo and the Fathers: The Letter and the Spirit', in Thomas Finan and Vincent Twomey, eds, *Scriptural Interpretations in the Fathers: Letter and Spirit* (Blackrock, Co. Dublin–Portland, OR: Four Courts Press, 1995), pp. 43–6.

19 Philo, *De plantatione*, VIII.32 (1930), pp. 228–9.

20 Philo, *De opificio mundi*, LVI.157 (1929), p. 125; ibid., p. 124: 'δι' ὑπονοιῶν'.

21 Ibid.; see also *De opificio mundi*, LIV.154 (1929), pp. 122–3.

22 Genesis 2.9. Philo, *Legum allegoriae*, I.XVII.56–8 (1929), pp. 182–5.

23 Philo, *Legum allegoriae*, I.XIV.45–6 (1929), pp. 174–7; *De plantatione*, IX.36–40 (1930), pp. 230–3. On the relationship between Philo, Plato and Alexandrian Middle Platonists in connection with the allegory of the soul see Tobin, *Creation of Man* (1983), pp. 149–54.

24 Philo, *Legum allegoriae*, I.XIX–XXIV.63–76 (1929), pp. 186–97.

25 Ibid., I.XVI.53–5, XXVIII.88–XXX.96 (1929), pp. 180–3, 204–11; see also *De opificio mundi*, LIV.153–LV.156 (1929), pp. 120–5; *De plantatione*, XI.44–XII.46 (1930), pp. 234–7.

26 Philo, *Legum allegoriae*, I.XVIII.59–61, XXXI.97–104 (1929), pp. 184–5, 210–17.

27 Philo, *De opificio mundi*, LVI.157–LIX.166; *Legum allegoriae*, II.VII.19–IX.34 (1929), pp. 124–33, 236–47.

28 For Origen's life see Henri Crouzel, *Origène* (Paris–Namur: Lethielleux–Culture et Vérité, 1985), pp. 17–61. For a bibliography see John Clark Smith, *The Ancient Wisdom of Origen* (Lewisburg–London–Cranbury, NJ: Bucknell University Press–Associated University Presses, 1992), pp. 337–57.

29 See Smith, *Ancient Wisdom of Origen* (1992), pp. 36–41, 61–2; Henri Crouzel, *Origène* (1985), pp. 267–342; Crouzel, 'L'Anthropologie d'Origène: De L'Arche au telos', in Ugo Bianchi, ed., *Arché e telos: L'antropologia di Origene e di Gregorio di Nissa: Analisi storico-religiosa* (Milan: Vita e Pensiero, 1981), pp. 37–42.

30 Origen, *Homélies sur la Genèse*, I.13, ed. by Louis Doutreleau (Paris: Cerf, 1976), pp. 56–65; see also *Traité des principes*, I.2.6, ed. by Henri Crouzel and Manlio Simonetti, 5 vols (Paris: Cerf, 1978–84), I (1978), pp. 120–5; *In epistolam Pauli ad Romanos*, II.13, *Commento alla lettera ai Romani*, ed. and transl. by Francesca Cocchini, 2 vols (Casale Monferrato: Marietti, 1985–6), I (1985), pp. 174–5. For a detailed discussion of Origen's views on this subject see, e.g., Adalbert G. Hamman, *L'Homme image de Dieu: Essai d'une anthropologie chrétienne dans l'Église des cinq premiers siècles* (Paris: Desclée, 1987), pp. 128–52; Crouzel, *Origène* (1985), pp. 130–7; Crouzel, *Théologie de l'image de Dieu chez Origène* (Paris: Aubier, 1956), esp. pp. 148–53.

31 Origen, *Commentaire sur Saint Jean*, XX.3, ed. by Cécile Blanc, 5 vols, SC CXX, CLVII, CCXXII, CCXC, CCCLXXXV (Paris: Cerf, 1966–92), IV, SC CCXC (1982), p. 248. See also ibid., I.17, I, SC CXX (1966), pp. 112–13. See Crouzel, *Origène* (1985), pp. 132–4. Origen may have intended to interpret the formation of Adam from the dust of the ground (Genesis 2.7) as the creation of the ethereal body of the intellect, which in his view had occurred before the Fall. With the exception of the members of the Trinity, every creature, although ethereal and incorruptible, had a body, and the Fall did not cause the emergence of the body, but its thickening and a qualitative change into a corruptible and gross state, as indicated in the report that after the Fall God made for Adam and Eve 'coats of skins' (Genesis 3.21): Crouzel, *Origène* (1985), pp. 128–9, 283–4; Crouzel, 'Le Thème platonicien du "véhicule de l'âme" chez Origène', *Didaskalia*, 7 (1977), pp. 225–37; Manlio Simonetti, 'Alcune osservazioni sull'interpretazione origeniana di *Genesi* 2,7 e 3,21', *Aevum*, 36 (1962), pp. 370–81.

32 On Origen's reading of Scripture see Ronald E. Heine, 'Reading the Bible with Origen', in Paul M. Blowers, ed., *The Bible in Greek Christian Antiquity* (Notre Dame, IN: University of Notre Dame Press, 1997), pp. 131–48; Ó Fearghail, 'Philo and the Fathers'; Gerard Watson, 'Origen and the Literal Interpretation of Scripture', in Finan and Twomey, eds, *Scriptural Interpretation* (1995), pp. 55–7 and 75–84; Charles J. Scalise, 'Origen and the *sensus literalis*' and Joseph W. Trigg, 'Divine Deception and the Truthfulness of Scripture', in Charles Kannengiesser and William L. Petersen, eds, *Origen of Alexandria: His World and his Legacy* (Notre Dame, IN: University of Notre Dame Press, 1988), pp. 117–29 and 147–64; Richard P. C. Hanson, *Allegory and Event: A Study of the Sources and Significance of Origen's Interpretation of Scripture* (London: SCM Press, 1959), esp. pp. 235–88; Henri de Lubac, *Histoire et esprit: L'Intelligence de l'Écriture d'après Origène* (Paris: Aubier, 1950).

33 See Origen, *Traité des principes*, IV.2.5; 3.4, III (1980), pp. 316, 358. On those cases in the Bible where the literal sense is unacceptable and demands an allegorical interpretation, see Hanson, *Allegory and Event* (1959), pp. 237–42. Origen claims that impossible features have been included in the sacred text precisely in order to push the reader towards allegory: Origen, *Traité des principes*, IV.2.9, III (1980), p. 337.

34 Origen, *Homélies sur le Lévitique*, XVI.4, ed. by Marcel Borret, SC CCLXXXVII (Paris: Cerf, 1981), pp. 276–81; *Homiliae in Librum Jesu Nave*, XIII.4, in *Homilien zum Hexateuch*, ed. by Wilhelm A. Baehrens, *Origenes Werke*, VII, GCS (Leipzig: Hinrichs, 1921), p. 374; *In Matthaeum*, XVII.8, ed. by Ernst Benz and others, *Origenes Werke*, X, GCS (Leipzig: Hinrichs, 1935), pp. 605–7; *Traité des principes*, IV.3.1, III (1980), p. 344: see Manlio Simonetti, 'Origene e i vignaioli perfidi', *Orphaeus*, 17 (1996), p. 39.

35 Origen, *Contre Celse*, IV.39–40, ed. by Marcel Borret, 5 vols, SC CXXXII, CXXXVI, CXLVII, CL, CCXXVII (Paris: Cerf, 1967–76), II, SC CCXXVII (1968), pp. 282–91; *Homélies sur la Genèse*, I.15 (1976), pp. 66–9.

36 Origen, *In Genesim*, II.16–17, PG XII, cols 99–100. See Reinhold R. Grimm, *Paradisus coelestis, paradisus terrestris: Zur Auslegungsgeschichte des Paradieses im Abendland bis um 1200* (Munich: Fink, 1977), pp. 38–9. Origen's reading of Genesis, like Philo's, is by no means straightforward, and it is unclear how he regarded the Fall. The issue has been widely debated since in some passages Origen seems to understand Adam as a historical figure whereas in others he appears to

37 treat him as a symbolic figure, standing for all mankind. On balance, Origen seems not to have believed in the historicity of the story of Adam and Eve. Such a conclusion fits in well with his doctrine of the pre-existence of souls and the occurrence of a pre-cosmic Fall. On the issue of Origen's view of the historicity of Adam and Eve, see Hanson, *Allegory and Event* (1959), pp. 269–70.

37 Origen distinguishes between two worlds (*mundi*): one, visible and transient, containing our earth and planetary system; and another that is beyond ours and related to it as the sun is to the earth. This other world was made on the first day – in the beginning – with its own heaven and its own earth, seat of the primordial and final unity. In contrast, our world is composed of the visible firmament and the 'dry land' that was created on the second day, the land to which Adam was exiled after the Fall: *Traité des principes*, II.3.6, I (1978), pp. 264–71; *Homiliae in Numeros*, XXVI.5, ed. by Wilhelm A. Baehrens and Louis Doutreleau, 3 vols (Paris: Cerf, 2001), III, *SC* CCCCLXI, pp. 251–7. Origen also seems to have conceived of an intermediate paradise for the souls, corresponding to the 'bosom of Abraham' that is mentioned in the Lazarus story (Luke 16.19–24) or to the 'paradise' promised by Jesus to the good thief (Luke 23.42–3): *Homiliae in Numeros*, XXVI.4 (2001), pp. 249–51; *Traité des principes*, II.11.6–7, I (1978), pp. 407–13; *Homélies sur Ézéchiel*, XIII.2, ed. by Marcel Borret, *SC* CCCLII (Paris: Cerf, 1989), pp. 422–3 and n. 3. See also Grimm, *Paradisus* (1977), pp. 34–8.

38 Not all the expositions on Genesis in either the Judaic or the Christian tradition were purely allegorical or reflected such a strong influence of Greek philosophy. The narrative aspects of the paradise story were already developed in Jewish writings such as the Targums, in the Aramaic versions of the Bible that included additional explanatory material, and in classical rabbinical literature: for a bibliography see John Westerdale Bowker, *The Targums and Rabbinic Literature: An Introduction to Jewish Interpretations of Scripture* (London: Cambridge University Press, 1969).

39 On the Antiochene school of exegesis see Lucas Van Rompay, 'Antiochene Biblical Interpretation: Greek and Syriac', in Judith Frishman and Lucas Van Rompay, eds, *The Book of Genesis in Jewish and Oriental Christian Interpretation: A Collection of Essays* (Louvain: Peeters, 1997), pp. 103–23; Dimitri Z. Zaharopoulos, *Theodore of Mopsuestia on the Bible: A Study of his Old Testament Exegesis* (New York: Paulist Press, 1989); David S. Wallace-Hadrill, *Christian Antioch: A Study of Early Christian Thought* (Cambridge–New York: Cambridge University Press, 1982); Christoph Schäublin, *Untersuchungen zu Methode und Herkunft der Antiochenischen Exegese* (Cologne–Bonn: [n.publ.], 1974); Jacques Guillet, 'Les Exégèses d'Alexandrie et d'Antioche, conflit ou malentendu?' *Recherches de Sciences Religieuses*, 34 (1947), pp. 257–302; Alberto Vaccari, 'La "teoria" esegetica antiochena', *Biblica*, 15 (1934), pp. 94–101; see also Alexandre, 'Entre Ciel et terre' (1988), pp. 187–224, and Grimm, *Paradisus* (1977), pp. 40–3.

40 It is impossible to do justice here to the many authors who contributed in different ways to this tension in the early centuries of Christianity: e.g. Tertullian (*c*.160–225), who assumed the existence of an earthly paradise surrounded by a zone of fire and containing the souls of the martyrs; Cyprian (d. 258), who believed that Enoch and the patriarchs had resided in Eden and who made, for the first time, the association between the four rivers and the four gospels (see Grimm, *Paradisus* (1977), pp. 44–8); Basil (330–79), who wrote the *Hexaemeron*, a treatise that established the tradition of commenting on the six days of creation; Gregory of Nyssa (*c*.330–*c*.395), who maintained that the Garden of Eden was a metaphor of the world to come, the spiritual place into which only the chosen would enter: *De beatitudinibus*, II, in *Opera*, VII/2, ed. by Werner Jäger, Hermann Langerbeck, Heinrich Dörrie, John F. Callahan (Leiden: Brill, 1992), p. 92, and *De hominis opificio*, XIX, *PG* XLIV, cols 196–7. Ephrem the Syrian (*c*.306–73), whose symbolic and imaginative reading of the Bible presents a vivid picture of the complex 'mosaic' of paradise, will be discussed below in Chapter 7, pp. 162–3.

41 For a study of John Chrysostom, with a good bibliography, see John N. D. Kelly, *Golden Mouth: The Story of John Chrysostom, Ascetic, Preacher, Bishop* (London: Duckworth, 1995). See also Charles Kannengiesser, ed., *Jean Chrysostome et Augustin* (Paris: Beauchesne, 1975).

42 John Chrysostom, *In Genesim*, III.18, *PG* LIII, col. 151: 'μανθανέτω μὴ πάντα μετὰ ἀκριβείας τοῖς οἰκείοις λογισμοῖς ἀκολουθοῦντα δύνασθαι τὸν ἄνθρωπον καὶ τὰ τοῦ θεοῦ ἔργα κατοπτεύειν'.

43 Ibid., II.3, cols 107–9.

44 Theodore of Mopsuesta, *In Genesim*, II.8, III.8, *PG* LXVI, cols 638–40.

45 Severian, *De mundi creatione*, V.5, *PG* LVI, cols 477–80.

46 For a study of Epiphanius and his contradiction of Origen see John F. Dechow, *Dogma and Mysticism in Early Christianity: Epiphanius of Cyprus and the Legacy of Origen* (Louvain–Macon, GA: Mercer University Press, 1988). A series of controversies involving Origen's opinions dominated early theological debate. Origen was eventually condemned by various Church councils on a number of counts: at Alexandria in 400, and Constantinople in 543 and 553. See Elizabeth A. Clark, *The Origenist Controversy: The Cultural Construction of an Early Christian Debate* (Princeton, NJ: Princeton University Press, 1992), and Pierre Lardet, 'Introduction', in Jerome, *Apologie contre Rufin*, *SC* CCCIII, ed. by Lardet (Paris: Cerf, 1983), pp. 1–75.

47 Epiphanius, 'Epistula … ad Iohannem episcopum a Sancto Hieronymo translata', in Jerome, *Epistulae*, LI.4–5, ed. by Isidorus Hilberg, *CSEL* LIV (Vienna–Leipzig: Tempsky, 1910); repr. Vienna: Österreichische Akademie der Wissenschaften, 1996), pp. 400–3.

48 Ibid., 5–6, pp. 403–9.

49 Epiphanius, *Panarion*, II.64.47–8, *PG* XLI, cols 1147–50.

50 The region of the Elymaeans corresponds to the west of modern Iran: see Pliny the Elder, *Naturalis historia*, VI.28.111, and Livy, *Ab urbe condita*, XXXVII.40.

51 Epiphanius, *Ancoratus*, LVIII, *PG* XLIII, cols 117–20.

3

Locating Paradise in Space

The only true paradise is always the paradise we have lost.
Marcel Proust, Le Temps retrouvé (1922, publ. 1927)

The early debate on paradise shows that the Garden of Eden presented the Church Fathers with a tangled problem. It was Augustine who exercised the greatest single influence on Latin Christianity and who worked out an exegetical framework that put the notion of an earthly paradise beyond controversy. Augustine's literal reading of Genesis provided a new synthesis that welded four centuries of Christian interpretation into an exegesis that would dominate the Church's thinking for generations to come, creating the conditions for understanding the Garden of Eden as a specific place and sanctioning the belief in an earthly paradise that led later scholars to place the Garden of Eden on maps of the world.[1] Ironically, the medieval geography of the earthly paradise was only a by-product of Augustine's main theological concerns. Augustine was not himself particularly interested in geographical matters, but his novel emphasis on history fostered a keen interest in geography. His attempt to explain the creation of the world and the shift from eternity into time produced the geographical assumptions about the location of Eden that preoccupied theologians and map makers for centuries thereafter. Space proved to be a product of time and geography an outcome of history: 'here time becomes space', as Wagner puts it in *Parsifal*.[2]

Paradise as a Real Place on Earth

Augustine of Hippo was born in 354 in Tagaste, North Africa, of a pagan father and a Christian mother. He studied rhetoric in Carthage, where he was soon attracted by literature and philosophy and came to share the views of the heretical Manichaeans. He then moved to Italy where he became interested in Neo-Platonism. He was converted to Christianity in the summer of 386. In 388, he returned to Africa to become first a monk, then a priest, and eventually bishop of Hippo. He died in 430. The focus in and after the Middle Ages on biblical geography and the mapping of paradise can be traced back directly to the focus on time and history found in the writings of this restless man of letters turned theologian.

    The theology that Augustine worked out and that laid the foundation for the Church's teaching down to the thirteenth century was formulated in an atmosphere dominated by threats from various heresies in which the question of the earthly paradise was of particular significance. Gnostics, for instance, propounded that visible creation was not made by a good God, but rather by an evil principle, and was therefore imperfect and opposed to the spiritual realm.[3] Manichaeans believed in a cosmic conflict between light and dark in which a divine and spiritual spark had become trapped within man and needed to be rescued from its evil and material environment.[4] A particular threat came from the Neo-Platonic concept of emanationism, a theory that postulated a hierarchy of intelligences between the divine being and the physical world and argued for the pre-existence of the soul and its equality with God.[5]

Having been a Manichaean himself, Augustine was keen to demonstrate that Manichaean views were heterodox and to prove the goodness of the visible creation.[6] With this in mind, he argued that God had created man perfect in mind and body and that Adam had lived in a state of perfection in a physically perfect earthly paradise. Convinced of the literal truth of the Bible, he was anxious to defend the relevance of the Old Testament in Christianity against those who wanted to distance themselves from a Jewish heritage. And it was his deep sense of history that supported his reading of the biblical account of creation. Augustine's objective was to give a consistent picture of the crucial shift from divine eternity into human history. Interestingly, however, his medieval successors found this temporal aspect much less engaging than the spatial and geographical consequences of his teaching with their implication of the earthly location of Eden.[7]

Augustine refined Philo and Origen's reading of Genesis and their idea of a double creation, jettisoning in the process some of the embarrassing legacies of Platonism. His new interpretative model of the first two chapters of Genesis harmonized the two creation accounts.[8] Aware that the notion of 'beginning' implies the prior existence of time, and that the human mind was inadequate to imagine how historical time could have sprung from eternity, Augustine tried to resolve the mystery by suggesting that a single, instantaneous, creation out of time (described in the first chapter of Genesis) was followed by a creation within time (described in the second chapter). In this way, he eliminated the implication, conveyed by the concept of a double creation, of an opposition between soul and body and between an original – heavenly and godlike – perfection and the lower world of materiality.[9] He acknowledged that the biblical account of an extra-temporal and instantaneous creation had been adapted to the capacity of the human mind and that the linear process of writing things down had made the first chapter of Genesis appear to relate a sequence of events spread over six days. This, he said, was a literary device to convey the mystery of the world's creation.[10] Thus the first account of the week of creation was to be taken figuratively, and the second account, from God's rest on the seventh day (Genesis 2.2–3) onwards, was to be taken literally as a report of true historical events and a description of how God's creation had taken place in time.

In the context of representing the earthly paradise on maps, then, the outstanding significance of Augustine's reading of Genesis lies first in his rejection of Platonic dualism and second in his insistence on the transcendent importance of Scripture. By insisting on the uniqueness of the first human being, Augustine was able to reject the idea that human nature was intrinsically imperfect.[11] God did not create two beings, a heavenly as opposed to an earthly Adam. The man formed from the dust of the ground (Genesis 2.7) was God's fulfilment in time of the being in his own image that he had been envisaging in the extra-temporal, instantaneous creation (Genesis 1.26–7).[12] He wanted to create the earthly Adam. History was not the result of some cosmic fall, but a process set in motion by God himself. The visible world was no mere reflection of a higher and eternal reality, but the work of an omnipotent creator and the stage on which God's plan for the salvation of mankind was to be enacted.[13] By breaking with the Platonic nostalgia for divine perfection outside time, and by insisting on the dynamic of salvation history, Augustine also produced a sharper focus on the exegesis of the biblical text instead of dwelling on a particular philosophical framework. For Augustine, Scripture was absolutely transcendent, the place of divine revelation and a 'firmament of authority', itself a kind of heaven extended by God over mankind to assist its pilgrimage on earth.[14]

The authority of the biblical text was enhanced by Augustine's combination of two approaches, the literal and the allegorical. In this way, he was able to claim that the

Garden of Eden described in Genesis was simultaneously a real place on earth and had figurative importance. By espousing a theory of signs in a way that has led some modern writers to consider him the father of semiology, Augustine could argue that the words written in Scripture were *signa translata (*figurative signs). They alluded to things that were in turn used to signify something else, for God spoke to mankind through the events recorded in the Bible.[15] Thus the Genesis text might refer to the Garden of Eden and the story of the Fall, but this real garden, and the historical events that took place there, referred in turn – as signs – to something else. This is why, in his literal commentary on the Book of Genesis, Augustine declared his preference for the opinion of those exegetes who interpreted the Garden of Eden as both a material fact and a spiritual allegory:

> *I am aware that many authors have written at great length on paradise, but their theories on the subject in general can be reduced to three. There is, first, the opinion of those who interpret the word 'paradise' in an exclusively corporeal sense. Then there are those who prefer to give an exclusively spiritual meaning to the word. Finally, there are those who accept the word 'paradise' in both senses, sometimes corporeally and at other times spiritually. Briefly, then, I admit that the third interpretation appeals to me. According to it I have now undertaken to treat of paradise, if God will make this possible. Man was made from the slime of the earth – and that certainly means a human body – and was placed in a corporeal paradise. Adam, of course, signifies something else where Saint Paul speaks of him as a type of the One who was to come. But he is understood here as a man constituted in his own proper nature, who lived a certain number of years, was the father of numerous children, and died as other men die, although he was not born of parents as other men are but made from the earth, as was proper for the first man. Consequently, paradise, in which God placed him, should be understood as simply a place, that is, a land, where an earthly man would live.*[16]

In short, Adam was for Augustine a corporeal creature, from which it followed that he had been placed in a corporeal paradise.

The link between body and place established – almost incidentally – the Garden of Eden as a geographical location within the physical world. Seen as a historical occurrence, Adam and Eve's original sin called for a spatial framework. Although the site of the Garden of Eden might not be known, or could not be accurately described, Augustine did not doubt that 'those first human beings lived in a land that abounded in woods and fruit and that received the name "paradise"'.[17] He also reasserted the physical reality of the Garden of Eden by speaking of its extent: 'For it was no small place that was watered by such a mighty river.'[18] Equally real were the four rivers, which flowed from a single source in the middle of the Garden and emerged into the known regions of the earth having travelled a considerable distance underground. The biblical Gihon was the Nile that watered Egypt, the Pishon was the Ganges that crossed India, and the Tigris and Euphrates (given their usual names in Genesis) flowed through Mesopotamia. Augustine's authoritative reading hastened the demise of other theories as to the identity of the Gihon and the Pishon.[19] He commented:

> *In discussing these rivers, need I make any further effort to establish the fact that they are true rivers, not just figurative expressions without a corresponding reality in the literal sense, as if the names would signify something else and not rivers at all, in spite of the fact that they are well known in the lands through which they flow and are spoken of in nearly all the world?*[20]

The issue of Eden's spatial dimension was an outcome of Augustine's discourse on time and history. In his affirmation of the earthly reality of paradise, Augustine neither saw a need nor made an attempt to point to its location. But the rivers described in the text – the Tigris, Euphrates, Nile and Ganges – implicitly linked the textual Garden of Eden with places that could be situated geographically, whether in Assyria, Mesopotamia, Egypt, India or elsewhere. Sooner or later, the question of the location of paradise would be raised. The geography of Eden, initially a marginal aspect of Augustine's vision of history, became an issue that snowballed through the Middle Ages and bequeathed to the modern period a conundrum which, in one form or another, has continued to haunt Western thought to the present day.

Naming the Place

The new geographical notion of Eden was born out of Augustine's speculations on eternity and time and out of his reading of the paradise narrative as a real historical event. The question of the whereabouts of the Garden, of secondary importance for Augustine, caught the imagination of later biblical exegetes who drew out of Augustine's writings the latent geographical discourse and who named the place where map makers could put paradise. The grounds for identifying the site of the earthly paradise rested on the *Vetus Latina* (the old Latin translation of the Hebrew text of Genesis), which had explained that paradise had been planted not 'from the beginning' – as Jerome rendered it in the Vulgate – but 'in the east'.[21] From among the most influential of the early medieval scholars who discussed paradise and who distilled Augustine's theories for posterity, we need select only a few in order to illustrate the direction taken by the debate: Isidore of Seville, Bede the Northumbrian scholar, the unnamed compilers of the *Glossa ordinaria* (the standard medieval commentary on the Bible), and the commentator Peter Lombard. It was these clerics' naming of the place that in due course guided the compilers of maps where the sign for the Garden of Eden should be placed on their representations of the world.

Isidore of Seville and Bede

Isidore, bishop of Seville, and Bede, the scholar monk of Tyneside, Northumbria, were two profoundly learned men whose works, written mainly in the seventh and eighth centuries respectively, are still widely read. In their different ways, both carried forward Augustine's views on the paradise narrative. But whereas Isidore was a compiler of encyclopedias, seeking to record for posterity the best of classical and biblical learning, Bede was a highly original thinker, didactic in his exegetical simplifications. Ignoring the finer points of Augustine's complex thinking, both Isidore and Bede expanded Augustine's notion about the physicality of the Garden of Eden and its geographical location, and it is to these two scholars above all that we owe the popularity of the geographical notion of an earthly paradise and, from later in the eighth century, its representation on maps.

Isidore included a discussion of the geography of Eden in his *Etymologiae* (*c.*635) without dwelling on Augustine's philosophical and exegetical concerns. His much-quoted remarks contributed greatly to medieval belief in a distant paradise:

> *Asia includes many provinces and regions. I shall briefly list their names and locations, starting with paradise. Paradise is a place in the east, whose name translated from Greek into Latin is* hortus *[garden]. Moreover, in Hebrew it is called* Eden, *which means in our language* deliciae *[delights]. The combination of the two words produces* hortus deliciarum *[garden of delights]. This garden is planted with every kind of tree and of fruit tree, and it also has the Tree of Life. There cold and heat are unknown, the air is always temperate. In the middle there is a spring watering the entire grove, which*

separates into four rivers. After original sin, this place was inaccessible to man; for it is surrounded on all sides by a flaming sword, that is, by a wall of fire, reaching almost to heaven. Cherubim, that is, an angelic guard, is also arrayed in addition to the glowing sword to keep away evil spirits, so that the fire and the angels banish evil men and angels respectively, preventing the threshold of paradise from being crossed either by flesh or by spirit.[22]

A number of points in Isidore's description should be noted, first of all his inclusion of the earthly paradise amongst the regions of Asia and his depiction of paradise as walled in by a ring of flames soaring almost to heaven. He also reminded his readers that the Garden of Eden was not to be confused with the *Fortunatae Insulae*, the Fortunate or Blessed Islands situated in the western ocean, whose fertility had led classical authors to mistake them for paradise.[23] All these features would in due course be shown on maps.

Moreover, Isidore's remarks bear witness to a significant shift in the way the world was described that had been taking place since Augustine's influential work on the literal reading of Genesis. Earlier classical accounts of world geography – particularly those by Pliny the Elder and Solinus – had been organized to read from west to east. They start with Europe, then Africa is described and finally Asia.[24] By beginning with Asia, Isidore called attention to the fact that his starting point was *paradisus*, beyond which there was no east. Orosius had already prepared the ground for the transition. His survey of the whole of human history, *Historia adversus paganos*, which is thought to have been instigated by Augustine himself, opens with a short gazetteer. This starts with a brief overview of Asia, Europe and Africa (in that order) then returns to Asia with a much more detailed account of that continent, as Orosius moved from the Far East to the Middle East and Egypt (the Nile being the border between Asia and Africa) to conclude with the land of the Amazons, near the Caspian Sea.[25] No specific reference to paradise is made, but the length of Orosius's description of Asia in relation to either Europe or Africa conveys something of the east–west movement that can be detected in his understanding of Christian historiography. Isidore's more overtly east–west description provided the pattern followed by later medieval encyclopedists.[26]

Bede's treatment of the earthly paradise is even more revealing than Isidore's of the way later exegetes tended to simplify Augustine's complex vision of history. Taking a theological stance, Bede deliberately rejected Augustine's elaborate reading of the Genesis account of creation and his understanding of the week of creation as the figurative rendering of a single creative act. Instead, Bede accepted that the creation really did take place over six days.[27] The second chapter of Genesis (2.8–16) merely reiterated, in his view, what had happened on the third day of creation, when the land was first clothed in vegetation (1.11–12), and gave further details on, for instance, the Trees of Life and of Knowledge.[28] Bede's interpretation was closer to the letter of the biblical text than Augustine's complicated idea of an instantaneous creation, which never fully convinced medieval scholars, and it was Bede's version that was to prove the more acceptable throughout the Middle Ages.

Bede had intended his commentary on Genesis to be no more than a handy summary of earlier traditions, but in the event he had created an accessible and enduring model for the interpretation and transmission of Augustinian exegesis.[29] He placed special emphasis on the harmonization of the biblical story with contemporary teaching of geography and the natural sciences. It goes without saying that he accepted the existence of an earthly paradise: 'we are not allowed to doubt that the place was and is on earth'.[30] He saw the garden of Eden as representing the Church on earth or as an allusion to the Church in heaven, but also as simultaneously a vast, pleasant and wholly delightful place, copiously watered and shaded by trees heavy with fruit.[31] In creating a place on earth imprinted with heavenly characteristics, God wanted to remind mankind

– through the rivers that flowed from this earthly paradise and through the proximity of the garden to the inhabited regions of the earth – of mankind's ultimate destiny, the celestial kingdom.[32] As to the location of this wonderful place, Bede agreed with the general view, taken from the *Vetus Latina*, that paradise was planted 'in the east'. According to Bede, this eastern Eden was kept separate from the rest of the world by a vast expanse of land or sea, and was at such an altitude that it had survived untouched by the Flood that covered the rest of the earth (Genesis 7, 8.1–19). He appears to have been less certain, however, about the exact site of Eden, saying that only God knows its place.[33]

Notwithstanding the divine veiling of the precise location of paradise, Bede did go so far as to suggest that the land of Havilah described in the Bible as surrounded by the river Pishon (Genesis 2.11) was a region in India that took its name from its holder, Havilah, son of Joktan and grandson of the patriarch Eber (Genesis 10.29). In naming a place bordering paradise, in an allusion to classical geography, Bede was the first medieval commentator to quote extensively from Pliny the Elder's *Naturalis historia* in the exegesis of Genesis. In Pliny's description of the regions of India as 'plentiful in veins of gold, much more than other regions', Bede saw a clear match with the biblical qualification of Havilah, the land 'where there is gold' (Genesis 2.11).[34] His description of the courses of the rivers of paradise also accorded with classical geography – the Pishon or Ganges has its source in the Caucasus, the Gihon or Nile, as Orosius had reported, rises in west Africa near the Atlas mountains, and the Tigris and Euphrates descend from Armenia – and with Augustine's idea of their underground passage from paradise to the inhabited world. The ultimate source of all four rivers, in paradise, was beyond human knowledge.[35]

The Glossa ordinaria and Peter Lombard

Two highly influential twelfth-century texts offer the modern historian a shortcut through the vast number of medieval theological works concerning the location of the earthly paradise to the essential elements of the exegetical tradition on which later medieval scholastic thought was founded. The anonymous *Glossa ordinaria* and Peter Lombard's *Sententiae* were both widely used in the Middle Ages as standard reference works, the *Glossa ordinaria* as a commentary on the Bible and the *Sententiae* as a theological textbook.[36] The two texts were also systematically quarried by later medieval clerics eager to take advantage of the *Glossa* as a single source for quotations from the Church Fathers for any verse of Scripture, and of the *Sententiae* as a thematic digest of Christian theology drawn from the most important patristic and ecclesiastical works.[37]

The *Glossa ordinaria* was compiled in the middle of the twelfth century in the monastic school at Laon, France, and completed through the collaborative effort of various masters from Auxerre and Paris as well as Laon.[38] The work was intended as a guide to the Bible, providing explanatory comments on the same page as the scriptural text, either as interlinear insertions or as marginal commentary, and from the twelfth to the fifteenth century was regarded as an undisputed authority. Overall, the *Glossa* reflects the overwhelming importance of Augustine. The compilers were not troubled by differences of opinion. In their comments on Genesis, they presented the reader with both Augustine's challenging idea of the instantaneous creation of the world and Bede's simpler alternative of a creation spread over six days. They were happy to accept, as Augustine had been and Thomas Aquinas would be, that Scripture could be explained in different ways without denying its truth.[39] Their readers certainly would have learnt from the *Glossa* that the Bible asserted that paradise was an earthly garden, and that Eden was a region in the east, at an exceedingly high altitude and thus untouched by the Flood.[40] As confirmation of this easterly and earthly location, the compilers cited Walafrid Strabo, the most modern authority invoked in the *Glossa* on the subject of paradise:

> *Some manuscripts have Eden 'in the sunrise'. We can conclude from this that paradise is in the east. However, wherever it is, we know that it is on earth; and that there is the ocean in between, and that there are mountains situated so as to form a barrier. [We also know] that it is very far from our world, located on high, and reaches the sphere of the moon. This is why the waters of the Flood did not touch it at all.*[41]

The expression 'in the sunrise' (*ad ortum* in the Latin original) had been referred to by Strabo as an alternative rendering of the controversial Hebrew term מִקֶּדֶם (*miqedem*), which, as already noted, described both 'where' and 'when' God had planted paradise. Strabo's term seemed an effective way of conveying in Latin the spatial and temporal complexity of the original Hebrew, even while emphasizing the spatial rather than the temporal aspects. Other important features about paradise are defined in this particular gloss: the reality of the earthly Eden, its eastern location, and its insulation from the inhabited world by a barrier of ocean, mountains and altitude. Strabo's statement that the Garden of Eden was located on a very high mountain matched the description found in Ezekiel's prophetical lament for the Prince of Tyre of Eden as 'the holy mountain of God' (Ezekiel 28.13–14).[42] Especially crucial was Strabo's explanation of a mountain-island reaching as high as the sphere of the moon for paradise's survival of the Flood that covered the rest of the earth. In the Aristotelian and Ptolemaic visions of the universe, the sphere of the moon marked the frontier between the changing and corruptible sublunary world (made up of the four elements of water, earth, air, and fire) and the stable and eternal dimension of the heavens.[43] The symbolical significance of the extreme altitude of the *Glossa*'s earthly paradise was powerful. The image of a paradisiacal mountain reaching from earth to the very heavens suggested the continuing existence of Eden on earth as a separate and alternative reality despite God's punishment of human wickedness by the Expulsion and the Flood. At the same time, the compilers of the *Glossa* made it clear, through their quotations from Augustine and Bede, that the Garden of Eden was anchored to the inhabited earth by the four rivers and that, by implication, they supported the theory of the rivers' underground course. They also repeated Bede's remarks about Havilah and India.[44]

Peter Lombard's *Sententiae* are thought to have been written between 1155 and 1158. Their immediate popularity lay in the brilliance of his synthesis of what had by then become established Christian theological teaching.[45] Peter Lombard confirmed the view, inherited from Augustine and mediated through the *Glossa*, that Adam and Eve were real human beings with physical bodies who had dwelt a short while in a delightful place on earth called Eden. Commenting on the week of creation, Peter Lombard referred to the two different theories, noting that while Augustine held that everything was created instantaneously, in both matter and form, theologians such as Gregory the Great and Bede thought that original matter was first created unformed – with the four elements intermingled – and that only later were the different kinds of physical bodies given their shape, in turn and at real temporal intervals. The latter view, Peter Lombard concluded, seemed to fit the biblical account most satisfactorily.[46] He also referred to Bede's comment that paradise was created *a principio*, that is, on the third day, when God created the plants. Taking this to mean that the seven days mentioned in Genesis indicated seven actual days, each of twenty-four hours, he said that the idea of a week of creation was the Church's preferred interpretation.[47] Siding on this matter with Bede, Peter Lombard then compromised by finding room for Augustine's idea of an instantaneous creation, if only to confine it to the first day, saying that God created the angels (heaven) and matter (earth) on this day, and that the differentiation of matter into genera and species took place over the following five days.[48]

Peter Lombard's *Sentences* were compiled more than four hundred years after Bede. They show that medieval theology had by then settled on a notion of an earthly paradise without the encumbering ballast of Augustine's complex explanation of the origin of the universe. The emergent paradigm embraced the literal and historical truth of the Genesis description of the Garden of Eden as Augustine had first worked it out. Peter Lombard confirmed the allegorical interpretation of paradise as an image of the Church on earth and in heaven, but also emphasized that the Garden of Eden was 'to be understood as a corporeal place, where man was placed';[49] an Eden that was distant in the east, separated from the inhabited world by a great desert or sea (as Bede had said), and situated on a mountain so high that it reached the sphere of the moon so that the paradise on its peak remained untouched by the Flood (as Strabo and the compilers of the *Glossa* had said).[50] Certain details of the paradise narrative would continue to challenge theologians and exegetes, but from now on the critical aspects would remain firm: paradise was a real place on earth, inaccessible because of original sin, yet connected to the inhabited world through the four rivers. The Garden of Eden was out of sight, but very probably situated somewhere in the remotest eastern corner of Asia.

Searching for Paradise

The ambiguous status of the Garden of Eden, a region existing on earth yet beyond the reach of mankind, fuelled the growth of myths and legends about the earthly paradise. Endless stories were fabricated about all those heroes, adventurers and monks who attempted to reach the rainbow's end. The elusiveness of paradise was stressed in different ways. Some stories told of journeys of great daring, involving both the greatest, such as the ambitious and undaunted Alexander the Great, and the humblest and most pious amongst pilgrims. Notwithstanding their status or piety, all alike were refused access to Eden itself, a reminder that re-entry into paradise could not be achieved through human effort alone. Other stories concerned more fortunate pilgrims, usually courageous monks, whose success in crossing the threshold of Eden and penetrating an inaccessible place served to highlight the supernatural dimension of the Garden of Eden.[51]

According to most of the legends, those who travelled in search of Eden found their way through the gate of the Garden of Eden blocked. Their adventures served to underscore the point that access to paradise was granted by divine grace only. One twelfth-century version of the Alexander legend told how, after conquering India, the hero embarked on the Ganges with an elite force of five hundred young warriors in the hope that, by following the river upstream, he would reach its source in the Garden of Eden. For a month the group struggled onward, against the powerful current, until Alexander noticed that the houses of the inhabitants of the land they were passing through were roofed with huge leaves taken from the river that, once dry, emitted a wonderful fragrance. Thus encouraged that he was approaching his objective, Alexander pressed on until he found himself below the smooth and high walls of a wondrous city, which, he realized, was the earthly paradise wherein the righteous were awaiting the Last Judgement. From the sole window in an otherwise featureless wall a single inhabitant, an old man with a long white beard, passed down to the proud king a miraculous stone weighing more than all the gold on earth. The old man warned Alexander, as a lesson against the vanity of human ambition, that if dust were allowed to cover this wonder stone, it would weigh no more than a single gold piece.[52] The widely popular epic poem *Alexandreis*, composed in the late twelfth century by Walter of Châtillon and extensively copied throughout the Middle Ages, also reported Alexander's intention to track down the source of another of the four rivers of Eden, this time the Nile, and to 'lay siege to paradise'.[53] Paradise was thought to be extant and yet forbidden to all humans, and thus served to highlight Alexander's impious ambition.

In the oldest version of another legend, that of Saint Macarius of Rome, it was related how three monks set out from their monastery, situated between the Tigris and the Euphrates, in search of the Garden of Eden. After visiting Jerusalem and Bethlehem, the monks – Theophilus, Sergius and Hyginus – journeyed on towards Persia and India. They wandered through various lands and faced numerous adventures. They passed through the lands of dog-headed men and pygmies, they climbed high, dark mountains, and they met all sorts of animals and giants before finally arriving at the cave of the hermit saint, Macarius. Macarius informed them that paradise was only twenty miles further on, but that no mortal was ever permitted to enter Eden. He had himself once tried to approach the Garden, but had been stopped by a guardian angel. Here again, legend insisted on the inapproachability of the earthly paradise.[54]

One group of legends suggested that evidence of the existence of the earthly paradise was to be found in the inhabited world. Jean, sire de Joinville, the French author of the *History of Saint Louis* (1305–9), described how Egyptian fishermen would cast their nets into the Nile and find precious spices and plants caught in them – ginger, rhubarb, aloes and cinnamon – blown from the trees within paradise and transported downstream by the current. Hearing of this, the Sultan of Cairo had organized a number of expeditions in an attempt to reach the source of the river and the earthly paradise, only to find that his every effort was blocked by an unassailable mountain.[55] According to other legendary journeys, certain individuals seemed to have been more successful. A lucky pilgrim might experience paradise for what he thought was a brief instant, or at most a few hours, but which for the rest of humanity was a period of many decades or even generations. A fourteenth-century Italian legend told how a trio of monks travelled upstream and arrived at the earthly paradise, into which they were admitted by a guardian angel who showed them the splendours of the garden. The monks believed that they had sojourned in Eden for three days, but when they returned to the world they discovered that three centuries had gone by.[56] In his *Pantheon* (1187–91), Godfrey of Viterbo told of a group of monks who set sail for paradise from the shores of Britain. They, too, enjoyed the wonders of the Garden of Eden for a brief moment, but once back home they found that they had aged, did not even understand the language now spoken in their native country, and could no longer recognize the world around them.[57] The ineffable beauty of the garden had overwhelmed them, as it did all such privileged visitors, but in paradise time moves at a different pace. Whether the legends spoke of insurmountable physical obstacles barring the pilgrim's path or described the effects of time suspended, all made the point that Eden belonged to a different dimension.

Finding Salvation

One of the most popular legends of all medieval literature told of the voyages of Saint Brendan. This was the story of a Christian odyssey that at the same time articulated the ideals of monastic life, since the ocean journeys of a group of adventurous monks searching for the earthly paradise in the form of a delightful and distant island also stood for man's vocation to reach the Kingdom of Heaven.[58] The *Navigatio Sancti Brendani* belongs to the medieval genre of Irish maritime tales known as *immrama*. The anonymous Latin text, composed between the eighth and the tenth century, possibly by an Irish writer, survives in more than 120 Latin manuscripts and in numerous vernacular adaptations and translations.[59] It was read throughout Europe, and has given rise to speculation that the legend was based on a real sea voyage and that the historical Brendan, a sixth-century Irish monk, had reached the New World before the Vikings and long before Christopher Columbus.[60] The legend related how the pilgrims set out westwards in their boat, but then shipped their oars, tied the rudder, left the sails spread, and waited for God to pilot the boat. Their willingness to put their trust in God, instead of pursuing a specific route, allowed the divine hand to turn the prow of their boat towards the island of paradise. The godly wind that filled their sails marks a shift in the legend from the ordinary to the marvellous: 'When they got a wind, they did not know from what direction it came or in which direction the boat was heading.'[61]

The point of the story is that it is useless to speculate exactly where Brendan and his fellow monks made their fantastical discoveries since they had by then ceased sailing to the west and were no longer in the ordinary world. In the course of their seven-year journey, the voyagers encountered many extraordinary sights. They found themselves in a sea that at times was coagulated, at times as clear as glass, and at yet other times was transformed into a pillar of bright crystal. They disembarked on fantastical islands: on an island of giant sheep, an island of fallen angels with the bodies of birds, an island of monks perpetually singing praises to God, an island populated by three choirs chanting Psalms, an island of grapes as big as apples, and an island of demon smiths on the confines of hell. They met Judas Iscariot sitting unhappily on a rock in the midst of waves, freed each week from Saturday night to Sunday morning from his torment in hell. They met beasts of immense size and mighty monsters, including a great fish called Iasconius, which they mistook for a piece of land and on which they celebrated Easter. Brendan described many of the islands as Eden-like, rich in herbs, fruit and flowers, exempt from corruption and unmarked by the passing of the seasons. The rhythm of each island visit was set by liturgies, fast and prayers, and was measured in a symbolic number of days.

It should be noted that when Saint Brendan and his monks had set sail from their Irish monastery they were searching not for the Garden of Eden of Adam and Eve, but for what is described in the *Navigatio* as the *terra repromissionis sanctorum*, the 'Land of Promise of the Saints'. In medieval vernacular versions of the legend the statement that Brendan set sail for the Garden of Eden of Adam is followed by the qualification that his destination was the 'Land of Promise of the Saints', an implication of the apocalyptic dimension of a heavenly abode.[62] The delightful island that Brendan reached at the end of his odyssey is described in the *Navigatio* as densely wooded with trees laden with fruit throughout the year, as a place that had remained uncontaminated since the world began, and as a heavenly land of bliss and eternal light – the light being Christ – promised by God to his saints at the end of time.[63] As always with the earthly paradise, however, there was a border to the land beyond. Brendan was permitted to set foot on the *terra repromissionis sanctorum*, but was prevented by an angel from crossing the river that ran across the island. The angel told Brendan that the part beyond the river was the place destined by God as a refuge for Christians in times of persecution, an allusion to the persecutions of the Antichrist at the end of the world.[64] So, after dwelling on this side of the river on that delightful island for forty days, Brendan was sent back into the world, where he died fortified by the holy sacraments.

The legend of Saint Brendan contains other hints that his quest for the earthly paradise was in fact an eschatological search. In the *Navigatio*, the *terra repromissionis sanctorum* takes on the characteristics of the Promised Land, as described in the Bible (in the books of Exodus and Numbers) and prefigures the New Jerusalem of the Book of Revelation.[65] Thus it was not by accident that on the penultimate island the monks encountered Paul the Hermit waiting for the Last Judgement in a state of angelic perfection, for the geographical distance between the hermit's island and the *terra repromissionis sanctorum* in the legend corresponded to the immeasurable time-span the faithful had to wait before the coming of the Lord.

The adventures of Brendan and his monks in their search for a paradise on earth, and their arrival at a sort of Promised Land, highlights the point of the legends

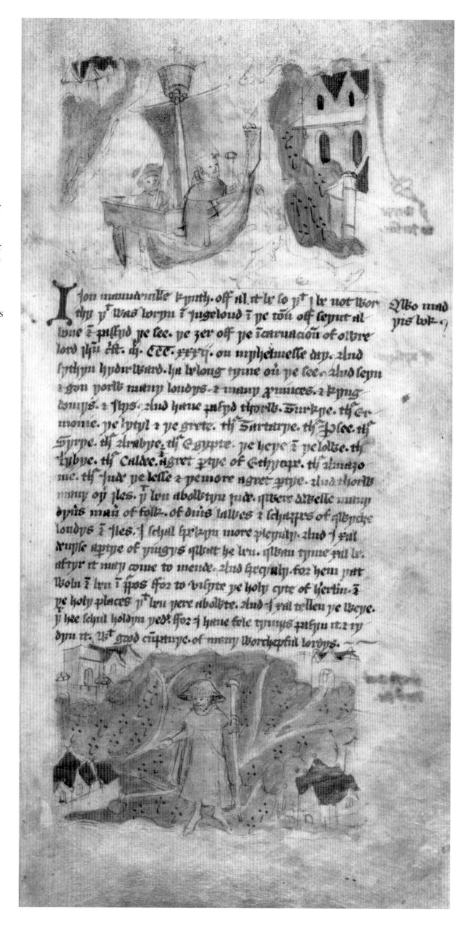

Fig. 3.1. Page from an English manuscript of *The Book of Sir John Mandeville*, originally compiled around 1360. East Anglia, *c.*1430. 28 × 14 cm. London, British Library, Harley MS 3954, fol. 2r. At the top, Sir John Mandeville is seen setting out on his journey. At the bottom, he is portrayed as a pilgrim in the countryside. The format of this fifteenth-century manuscript implies that it was produced to be carried while travelling, and the prologue introduces the work as a guidebook for pilgrims journeying to Jerusalem. It is a fictitious account, however, although the Mandeville character (who was almost certainly not the author) claims to base his report on a real journey to the East. He gives an elaborate account of pilgrimage itineraries and locations in the Middle East and describes regions and peoples in the Orient. He begins at Constantinople and proceeds east through India and Cathay. He admitted that he failed to get close to the earthly paradise, not only because the sea between Cathay and paradise was impassable but also because he was 'unworthy' to approach paradise, still less to enter it.

Fig. 3.2. Seth asking the angel at the gates of paradise for the oil of mercy. Illumination from *The Book of Sir John Mandeville* (see Fig. 3.1). 8 × 9.5 cm. London, British Library, Harley MS 3954, fol. 4r. Medieval belief traced the history of the Cross of Christ back to the Garden of Eden, and thence down to the early centuries of Christianity. According to the Legend of the Cross, Seth asked the angel guarding paradise for the oil of mercy, but was given instead some seeds (or a branch) from the Tree of Life or the Tree of Knowledge. The tree that grew from the seed was cut down and became the pole on which Moses raised the brazen serpent (Numbers 21.4–9). The pole served as a bridge in Jerusalem and was worshipped by the Queen of Sheba at the time of her visit to King Solomon. It floated in the pool of Bethesda, granting its waters miraculous curative powers. From there it was taken to be made into the Cross of Christ. Helena, mother of the Emperor Constantine I, discovered the True Cross when she was founding churches in the Holy Land. Part of it was recovered in the seventh century by the Emperor Eraclius when he was fighting against the Persian king Chosroes II. According to the text illustrated here (Chapter 2), Seth gets 'treiz grainz' from the Tree of Life. He returns home to find his father dying and he puts the seeds in Adam's mouth. From those three seeds grew the three trees that were used for the Cross on which Christ was crucified.

about the few who did manage to enter the earthly paradise, people such as Seth, the son of Adam, and Enoch and Elijah, the Old Testament patriarchs, which was to stress the importance of Christian redemption. The story of Seth emphasized the connection between Adam and Eve's Fall and Christ's Crucifixion. The legend was of Jewish origin, but passed into Christian literature through an apocryphal text, the *Life of Adam and Eve*, possibly written between the second and the fourth century AD, which expanded the rather concise account in Genesis of the Expulsion. It told of Seth's journey back to the Garden of Eden, together with Eve, on behalf of his old and dying father, Adam, to ask for the oil of mercy promised by God as he banished Adam from paradise.[66] The legend was transmitted in various forms, and in another rather more widely disseminated version, the legend of Seth was Christianized by amalgamating it with the Legend of the Cross to record how Seth succeeded in his search.[67] He travelled eastwards, following the barren footsteps left by Adam and Eve in their flight from the Garden of Eden. When he had reached the gate of paradise, described as a place of shining beauty, the angel that guarded the way allowed him three tantalizing glimpses of the marvels within. First, Seth saw the beauty of the garden, with its fruits and flowers of all kinds, the fountain with the four rivers flowing from it, and the Tree of the Knowledge of Good and Evil, now leafless. Second, he saw the tree with the snake encircling it. Finally, in a vision that announced the coming of the Redeemer, he saw that the tree appeared to reach to heaven, that its roots penetrated deep into hell, and that, on the highest branches, there was a child. Before turning back, Seth was given by the angel three seeds from the Tree of the Knowledge of Good and Evil that he was to place under the tongue of Adam at the time of burial, and from which would grow the wood for Christ's Cross.[68] In this apocryphal account of Seth's journey back to Eden, the link between the Fall of Adam, who had tasted the forbidden fruit of the Tree of the Knowledge of Good and Evil, and Christ's sacrifice on the Cross, by which mankind was to be redeemed, was reiterated yet again.

While Seth was allowed only three brief glances at Eden, Enoch and Elijah were far more successful in actually reaching the earthly paradise. A later chapter in Genesis (Genesis 5.24) describes how Enoch, having lived for 365 years, 'walked with God: and he was not; for God took him'. The words suggest that Enoch did not die, but was

Locating Paradise in Space   55

transported by God to some divine abode, an idea confirmed in the New Testament: 'By faith Enoch was translated that he should not see death; and was not found, because God had translated him: for before his translation he had this testimony, that he pleased God' (Hebrews 11.5). According to another apocryphal text, the *Book of Enoch*, the place to which the patriarch was taken has never been found.[69] Enoch's companion in paradise, the prophet Elijah, was – according to the Book of Kings – taken up by God 'into heaven by a whirlwind' (4 Kings, 2.1, 11).[70] In an early Christian rendering of the story, the heaven into which the two Old Testament patriarchs escaped death in this way was a lower heaven, that is to say, the earthly paradise. Enoch and Elijah thus became the first inhabitants of the Garden of Eden after Adam and Eve's banishment. Commenting on Enoch's translation to Eden, Augustine pointed out that Enoch was the seventh man in line of descent from Adam, the significance of seven being that it was the number that made the Sabbath a consecrated day, and the sixth in descent from Seth, the significance of six being that it was on the sixth day that man was created and that God had completed his work.[71] Augustine's pairing of the admission of Enoch and Elijah into paradise with the expulsion of Adam and Eve from it was a way of saying that, in allowing the two Old Testament patriarchs into Eden, God was showing mankind what Adam and Eve would have enjoyed had they not sinned; namely, a period of fleshly immortality on earth (granted by the Tree of Life) before being transferred to heaven. Now, all Adam and Eve's descendants are obliged to experience death before being admitted to celestial bliss. However, Enoch and Elijah were allowed to remain in the flesh in the earthly paradise.[72] Other Church Fathers identified Enoch and Elijah with the two witnesses whose appearance at the end of time to fight the Antichrist is announced in the Book of Revelation (Chapter 11).[73] Having been slain by the Antichrist, Enoch and Elijah would be resurrected after the Last Judgement and received into the eternal glory of heaven. The two patriarchs were allowed to enter the original garden of delights, but were also called to witness the final destruction and ultimate renewal of the universe. In this interpretation, the presence of Enoch and Elijah in Eden was linked with the wider Christian scheme of salvation history: their waiting in the Garden of Eden for the Apocalypse served to associate the beginning and the end of time.

The idea that the first things are linked to the last also emerged in a number of Christian apocryphal texts, depository of earlier Jewish traditions, where paradise is revealed in a vision. Whereas, for example, the Bible (at 2 Corinthians 12.2) contains a somewhat mysterious reference to a visit by Saint Paul to paradise, in the late fourth-century *Apocalypse of Paul*, Paul's visit is described in detail. Paul is said to go to a variety of places: the third heaven, the Land of Promise, the City of Christ, and finally to the paradise of Adam and Eve, where he sees the rivers Pishon, Gihon, Tigris and Euphrates. Paul also meets the patriarchs and the blessed, an indication that he is now in the heavenly abode of the just.[74] In the *Greek Apocalypse of Ezra*, Ezra (the main character in the dialogue) asks the Lord to allow him to see paradise. He is then taken somewhere towards the east, where he sees not only the Tree of Life, Enoch and Elijah, but also Moses, Peter and Paul, Luke and all the righteous and patriarchs.[75] The association of the primordial Garden of Eden with the paradise of the future is clearly reinforced in another apocryphal work, the *Fourth Ezra*, a Jewish apocalyptic text composed in Palestine about the end of the first century AD, which inspired the *Greek Apocalypse of Ezra* and was widely read in various Latin versions throughout the Christian Middle Ages.[76] Here, the terms 'Garden of Eden' or 'paradise' are used to describe both Adam's garden, created by God before the creation of the world, and the eschatological reward that awaits the righteous and that is viewed both as a garden – with incorruptible fruit and flowers – and as a walled city.[77] In Ezra's prophetic vision, at the end of time, the Messiah and the heavenly city, presently invisible, shall appear and 'the land which now is hidden shall be disclosed'.[78] What did Ezra mean? Was it the location of Eden that was to be revealed in the 'last days' of human history? A common feature of all these apocryphal texts seems to be the intimate association of the earthly paradise with the heavenly city.

Underlying the ambiguity of the Hebrew term מִקֶּדֶם (*miqedem*), already discussed (Chapter 2) is the Hebrew root ק.ד.ם. (*qdm*), which means 'in front' or 'before', 'east' or 'ancient'. The Hebrew word for past קֶדֶם (*qedem*) derives from the same root and means 'that which is in front', and therefore already known, insofar as it lies ahead. In contrast, the Hebrew word אַחַר (*ahar*) means both 'behind' and 'after', and the adjective אַחֲרוֹן (*aharon*) derived from it indicates not only 'that which is behind', and which thus remains unknown, but also that which comes after and which therefore may be subsequent in time ('the last'). It seems that the Semitic view is that one walks backward into the future.[79] The subtleties of the Hebrew language and the insights of apocryphal literature appear to combine to point to the eschatological significance of the terrestrial paradise in the drama of salvation history. The remote antiquity of the Garden of Eden, created before mankind and before the whole world, is projected towards the ultimate end of the universe. The world to come was created by God in the beginning.

The medieval legends are also an aid to the appreciation of the significance of the temporal dimension of the Garden of Eden. The adventures of Brendan in reaching the Land of Promise of the Saints, the journey of Seth back to Eden, and Enoch and Elijah's wait in Eden for the end of the world, all bore witness to the Christian vision of human history, in which the beginning and the end of time are intimately linked. In this respect it is significant that the Alexander legend described the earthly paradise as a city in which the righteous await the Last Judgement. The corollary of the geographical place of the earthly paradise is its place in human history and its inextricable association with the entire scheme of Christian salvation. Understanding the interconnection of three pivotal events – original sin in Eden, the sacrifice of Christ on Golgotha, and his Second Coming at the end of time – is of paramount importance for understanding the role of the representation of the earthly paradise on medieval maps. In the next chapter, medieval thinking about the location of paradise in time is explored.

1. Augustine's *De Genesi ad litteram* was sold at the University of Paris in the late thirteenth century as a textbook: see Lynn Thorndike, *University Records and Life in the Middle Ages* (New York: Columbia University Press, 1944), p. 113.
2. 'Zum Raum wird hier die Zeit', says Gurnemanz leading Parsifal towards the Temple of the Graal, a variation on the theme of paradise.
3. The Garden of Eden in Gnostic literature merits fuller discussion than there is space for here, but see Gerard P. Luttikhuizen, 'A Resistant Interpretation of the Paradise Story in the Gnostic *Testimony of Truth* (Nag Hamm. Cod. IX.3) 45–50', in Luttikhuizen, ed., *Paradise Interpreted: Interpretations of Biblical Paradise: Judaism and Christianity* (Leiden–Boston, MA: Brill, 1999), pp. 140–52; Roelof Van den Broeck, *Studies in Gnosticism and Alexandrian Christianity* (Leiden–New York–Cologne: Brill, 1996), pp. 67–85; Jean Magné, *From Christianity to Gnosis and from Gnosis to Christianity: An Itinerary through the Texts to and from the Tree of Paradise* (Atlanta, GA: Scholars Press, 1993); Philip S. Alexander, 'The Fall into Knowledge: The Garden of Eden/Paradise in Gnostic Literature', in Paul Morris and Deborah Sawyer, eds, *A Walk in the Garden: Biblical, Iconographical and Literary Images of Eden* (Sheffield: Journal for the Study of the Old Testament Press, 1992), pp. 91–104.
4. See Samuel N. C. Lieu, *Manichaeism in the Later Roman Empire and Medieval China*, 2nd edn (Tübingen: J. C. B. Mohr, 1992), pp. 8–32.
5. See, e.g., Arthur H. Armstrong, 'Man in the Cosmos: A Study of Some Differences between Pagan Neoplatonism and Christianity', in Willem den Boer, and others, eds, *Romanitas et Christianitas* (Amsterdam–London: North-Holland, 1973), pp. 5–14; Armstrong, *Saint Augustine and Christian Platonism* (Villanova, PA: Villanova University Press, 1967).
6. See Johannes Van Oort, 'Augustinus und der Manichäismus', in Alois van Tongerloo, ed., *The Manichaean ΝΟΥΣ* (Louvain: International Association of Manichaean Studies, 1995), pp. 289–315; Lieu, *Manichaeism* (1992), pp. 151–91; Robert A. Markus, 'Augustine's *Confessions* and the Controversy with Julian of Eclanum: Manichaeism Revisited', in Bernard Bruning, and others, eds, *Collectanea Augustiniana: Mélanges T. J. Van Bavel* (Louvain: Louvain University Press, 1990), pp. 913–25; Markus, *Conversion and Disenchantment in Augustine's Spiritual Development* (Villanova, PA: Villanova University Press, 1989); Prosper Alfaric, *L'Évolution intellectuelle de Saint Augustin*, I, *Du Manichéisme au Néoplatonisme* (Paris: E. Nourry, 1918).
7. Augustine asked God for the grace to 'understand how in the beginning you made heaven and earth' (Genesis 1.1): *Confessiones*, XI.3.5, ed.

by Lucas Verheijen, *CCSL* XXVII (Turnhout, Brepols, 1981), p. 196: 'Audiam et intelligam, quomodo in principio fecisti caelum et terram.' Transl. by Henry Chadwick (Oxford–New York: Oxford University Press, 1992), p. 223. Augustine's sense of history has been discussed by Robert A. Markus, *Saeculum: History and Society in the Theology of Saint Augustine*, 2nd edn (Cambridge–New York: Cambridge University Press, 1988), and more recently by Christopher Ligota, 'La Foi historienne: Histoire et connaissance de l'histoire chez S. Augustin', *Revue des Études Augustiniennes*, 43 (1997), pp. 111–71.

8 On Philo and Origens see above, Chapter 2, pp. 36–9.

9 The idea had been ambiguously anticipated by Philo and Origen within the context of their Platonizing interpretation of Genesis. In their view, the week of creation referred to the simultaneous creation of the intelligible world: Philo, *De opificio mundi*, IX.35; *Legum allegoriae*, I.2, [Works], Loeb Classical Library, transl. by Francis H. Colson and George H. Whitaker, 12 vols (London: Heinemann; New York: Putnam's Sons, 1929–62), I (1929), pp. 26–7, 146–9; Origen, *Homélies sur la Genèse*, I.1, ed. by Louis Doutreleau (Paris: Cerf, 1976), pp. 24–7. See also William Adler, *Time Immemorial: Archaic History and its Sources in Christian Chronography from Julius Africanus to George Syncellus* (Washington, DC: Dumbarton Oaks Research Library and Collection, 1989), pp. 43–6; Richard Sorabji, *Time, Creation and the Continuum* (London: Duckworth, 1983), pp. 208–9.

10 Augustine developed his concept of an instantaneous creation in the first eight books of *De Genesi ad litteram libri duodecim*, ed. by Joseph Zycha, *CSEL* XXVIII/1 (Prague–Vienna– Leipzig: F. Tempsky–G. Freytag, 1894), pp. 3–267. See Ligota, 'La Foi historienne' (1997), pp. 144–8 and n. 220; see also Marie-Anne Vannier, '*Creatio*', '*conversio*', '*formatio*', *chez S. Augustin*, 2nd edn (Fribourg, Switzerland: Éditions Universitaires, 1997); Reinhold R. Grimm, *Paradisus coelestis, paradisus terrestris: Zur Auslegungsgeschichte des Paradieses im Abenland bis um 1200* (Munich: Fink, 1977), pp. 61–2, provides a useful synthesis. Augustine's changing views on creation are discussed in Gilles Pelland, *Cinq Études d'Augustin sur le début de la Genèse* (Tournai: Desclée, 1972). For a discussion of Augustine's approach to the paradoxes entailed by the passage from eternity to the temporal existence of creatures see Roland J. Teske, *Paradoxes of Time in Saint Augustine* (Milwaukee: Marquette University Press, 1996).

11 Augustine, *De civitate Dei*, XIII.14, ed. by Bernard Dombart and Alfons Kalb, *CCSL* XLVIII (Turnhout: Brepols, 1955), pp. 384, 395–6. On Augustine's view of the unity of body and soul in man, see John M. Rist, *Augustine: Ancient Thought Baptized* (Cambridge: Cambridge University Press, 1994), pp. 92–147; Ludger Hölscher, *The Reality of the Mind: Augustine's Philosophical Argument for the Human Soul as a Spiritual Substance* (London–New York: Routledge & Kegan Paul, 1986) pp. 213–20.

12 Augustine, *De Genesi ad litteram*, III.19–22 (1894), pp. 84–90.

13 Augustine often emphasized the novelty of the world. See, for example, *De Genesi ad litteram*, VIII.1, VIII.3 (1894), pp. 229–30, 234; for the notion that God allowed the free development of history even while ruling over it, see Ligota, 'La Foi historienne' (1997), pp. 143–4.

14 Augustine, *Confessiones*, XIII.14.15–15.18 (1981), pp. 250–2. See also Ligota, 'La Foi historienne' (1997), pp. 143, 165–6.

15 Augustine put forward the principles of his exegetical practice in *De doctrina christiana*, ed. by Klaus D. Daur, *CCSL* XXXII (Turnhout: Brepols, 1962). On Augustine's theory of signs, see Robert A. Markus, *Signs and Meanings: World and Text in Ancient Christianity* (Liverpool: Liverpool University Press, 1996), pp. 3–8, 22–31, 71–104 and bibliography p. 1 nn. 1 and 2, and pp. 120–4; Cornelius P. Mayer, *Die Zeichen in der geistigen Entwicklung und in der Theologie des jungen Augustinus* (Würzburg: Augustinus Verlag, 1969). On Augustine and semiology, see, for example, Umberto Eco, *Semiotics and the Philosophy of Language* (London: Macmillan, 1984), p. 33.

16 Augustine, *The Literal Meaning of Genesis*, VIII.1.1, transl. and annotated by John H. Taylor, 2 vols (New York–Ramsey, NJ: Newman Press, 1982), II, pp. 32–3; *De Genesi ad litteram*, VIII.1 (1894), p. 229: 'Non ignoro de paradiso multos multa dixisse; tres tamen de hac re quasi generales sunt sententiae. Una eorum, qui tantummodo corporaliter paradisum intellegi volunt, alia eorum, qui spiritaliter tantum, tertia eorum, qui utroque modo paradisum accipiunt, alias corporaliter, alias autem spiritaliter. Breviter ergo ut dicam, tertiam mihi fateor placere sententiam. Secundum hanc suscepi nunc loqui de paradiso, quod dominus donare dignabitur, ut homo factus e limo – quod utique corpus humanum est – in paradiso corporali conlocatus intellegatur, ut, quemadmodum ipse Adam, etsi aliquid aliud significant secundum id, quod eum formam futuri esse dixit apostolus, homo tamen in natura propria expressus accipitur, qui vixit certo numero annorum et propagata numerosa prole mortuus est, sicut moriuntur et ceteri homines, etsi non sicut ceteri ex parentibus natus, sed sicut primitus oportebat ex terra factus est, ita et paradisus, in quo cum conlocavit deus, nihil aliud quam locus quidam intellegatur terrae scilicet, ubi habitaret homo terrenus.'

17 Augustine, *De Genesi ad litteram*, XIII.18 (1894), pp. 198–201: 'Primi quidem illi homines in terra erant nemorosa atque fructuosa, quae paradisi nomen obtinuit.'

18 Augustine, *The Literal Meaning of Genesis*, VIII.10.21 (1982), II, p. 48; *De Genesi ad litteram*, VIII.10 (1894), p. 246: 'Neque enim exiguus locus erat, quem tantus fons irrigabat.'

19 See above Chapter 2, p. 35.

20 Augustine, *The Literal Meaning of Genesis*, VIII.7.13 (1982), II, p. 43; *De Genesi ad litteram*, VIII.7 (1894), pp. 240–1: 'De his fluminibus quid amplius satagam confirmare, quod vera sint flumina nec figurate dicta, quae non sint, quasi tantummodo aliquid nomina ipsa significent, cum et regionibus, per quas fluunt, notissima sint et omnibus fere gentibus diffamata?'

21 See above Chapter 2, p. 35.

22 Isidore, *Etymologiae*, XIV.3.2–4, ed. by Wallace Martin Lindsay, 2 vols (Oxford: Clarendon Press, 1911): 'Habet [Asia] autem provincias multas et regiones, quarum breviter nomina et situs expediam, sumpto initio a Paradiso. Paradisus est locus in orientis partibus constitutus, cuius vocabulum ex Graeco in Latinum vertitur hortus: porro Hebraice Eden dicitur, quod in nostra lingua deliciae interpretatur. Quod utrumque iunctum facit hortum deliciarum; est enim omni genere ligni et pomiferarum arborum consitus, habens etiam et lignum vitae: non ibi frigus, non aestus, sed perpetua aeris temperies. E cuius medio fons prorumpens totum nemus irrigat, dividiturque in quattuor nascentia flumina. Cuius loci post peccatum hominis aditus interclusus est; septus est enim undique romphea flammea, id est muro igneo accinctus, ita ut eius cum caelo pene iungat incendium. Cherubin quoque, id est angelorum praesidium, arcendis spiritibus malis super rompheae flagrantiam ordinatum est, ut homines flammae angelos vero malos angeli submoveant, ne cui carni vel spiritui trangressionis aditus Paradisi pateat.'

23 Isidore, *Etymologiae*, XIV.6.8. The wall of fire is found in Tertullian and Lactantius. As Howard R. Patch notes, Isidore's suggestion that paradise may be confused with the Fortunate Isles implies that the garden is situated on an

island: *The Other World according to Descriptions in Medieval Literature* (Cambridge, MA: Harvard University Press, 1950), p. 145. See also Franco Cardini, 'Alla cerca del paradiso', in *Columbeis V* (Genoa: Università di Genova, 1993), pp. 67–88, esp. 77–9; and Christiane Deluz, 'Le Paradis terrestre, image de l'Orient lointain dans quelques documents géographiques médiévaux', in *Images et signes de l'Orient lointain dans l'Occident médiéval: Littérature et civilisation* (Aix-en-Provence: Université de Provence; Marseille: Laffitte, 1982), pp. 143–61.

24 The same is true in Martianus Capella, *De nuptiis Philologiae et Mercurii* (fifth century).

25 Orosius, *Historiae adversos paganos*, I.2.1–106, ed. by Marie-Pierre Arnaud-Lindet, 3 vols (Paris: Belles Lettres, 1990–1), I (1990), pp. 13–42.

26 Hugh of St Victor, for example, was closely echoing (as will be seen below in Chapter 6, pp. 125–7), Isidore's line of thought when he started his geography (and history) with paradise itself.

27 Bede, *Libri quatuor in principium Genesis usque ad nativitatem Isaac et electionem Ismahelis adnotationum*, I.1.1; I.1.31–2.1; I.2.4–5, ed. by Charles W. Jones, CCSL CXVIIIA (Turnhout: Brepols, 1967), pp. 3, 32, 39–41; see Grimm, *Paradisus* (1977), p. 81.

28 Bede, *In Genesim*, I.2.8–9 (1967), pp. 45–6.

29 For the following discussion of Bede I have mainly followed Grimm, *Paradisus* (1977), pp. 80–2. Rabanus Maurus (*c*.776/784–856) quoted extracts from the commentaries of Ambrose, Isidore, Bede and Gregory the Great; he copied his sources so accurately that his edition can be used for their textual criticism. He comments on the paradise narrative in *In Genesim*, I.12, PL CVII, cols 476–80 and *De universo*, XII.3 PL CXI, cols 334–5. On the relationship between Augustine's and Bede's views on creation, see Paolo Siniscalco, 'Due opere a confronto sulla creazione dell'uomo: Il *De Genesi ad litteram libri XII* di Agostino e i *Libri IV in principium Genesis* di Beda', in *Miscellanea di studi agostiniani in onore di P. Agostino Trapè, OSA* (Rome: Istituto Patristico Augustinianum, 1985), pp. 435–52.

30 Bede, *In Genesim*, I.2.8 (1967), p. 46: 'nos tamen locum hunc fuisse et esse terrenum dubitare non licet.'

31 Ibid., I.2.8, p. 46.

32 Ibid., I.2.10, p. 48.

33 Ibid., I.2.8, p. 46.

34 Ibid., I.2.11–12, p. 49; Pliny the Elder describes the riches of India in his *Naturalis historia*, VI.21–4. See also *Biblia latina cum Glossa ordinaria*, facsimile of the *editio princeps* ... (Strasbourg: Adolph Rusch, 1480/1), ed. by Margaret T. Gibson and Karlfried Fröhlich, 2 vols (Turnhout: Brepols, 1992), I, p. 22.

35 Bede, *In Genesim*, I.10 (1967), pp. 48–9; Orosius, *Historiae adversos paganos*, I.2.27–33 (1990), I, pp. 19–20. See Scott D. Westrem, *The Hereford Map: A Transcription and Translation of the Legends with Commentary* (Turnhout: Brepols, 2001), pp. 90, 374. See also below, Chapter 5, p. 121 n. 81.

36 See Jenny Swanson, 'The Glossa Ordinaria', in Gillian R. Evans, ed., *The Medieval Theologians* (Oxford: Blackwell, 2001), pp. 156–67; Margaret T. Gibson, 'The Place of the *Glossa ordinaria* in Medieval Exegesis', in Mark D. Jordan and Kent Emery Jr., eds, *Ad Litteram: Authoritative Texts and their Medieval Readers* (Notre Dame, IN: University of Notre Dame Press, 1992), pp. 21–2; Beryl Smalley, *The Study of the Bible in the Middle Ages*, 3rd edn (Oxford: Blackwell, 1984), pp. 56–62.

37 Commentaries on the *Glossa* were still being written in the seventeenth century. On the history of commentaries on the *Sentences* see the works cited by Marcia L. Colish, *Peter Lombard*, 2 vols (Leiden–New York–Cologne: Brill, 1994), I, p. 2.

38 Gibson, 'The Place of the *Glossa* in Medieval Exegesis' (1992), pp. 5–6; Gibson, 'The Twelfth-Century Glossed Bible', *Studia Patristica*, 23 (1990), pp. 232–44; Christopher F. R. de Hamel, *Glossed Books of the Bible and the Beginnings of the Paris Booktrade* (Woodbridge: Brewer, 1984); Smalley, *Study of the Bible* (1984), pp. 46–66.

39 Augustine, *De Genesi ad litteram*, I.19–21 (1894), pp. 27–31; Thomas Aquinas, *Summa theologiae*, 1a, q.68, a.1, ed. by Pietro Caramello, Leonine edn, 3 vols (Turin–Rome: Marietti, 1948–50), I (1950), p. 331.

40 *Biblia latina cum Glossa ordinaria* (1480/1; 1992), I, p. 21; see also Bede, *In Genesim*, I.2.8 (1967), p. 46.

41 *Biblia latina cum Glossa ordinaria* (1480/1; 1992), I, p. 21: 'Quidam codices habent eden ad ortum. Ex quo possumus coniicere paradisum in oriente situm. ubicumque autem sit. scimus eum terrenum esse: et interiecto oceano. et montibus oppositis. remotissimum a nostro orbe. in alto situm. pertingentem usque ad lunarem circulum: Unde aquae diluvii illuc minime pervenerunt.' See Strabo (Remigius of Auxerre), *In Genesim*, II.8, PL CXXXI, col. 60. As Jean de Blic, 'L'Oeuvre exégétique de Walafrid Strabon et la *Glossa ordinaria*', *Recherches de Théologie Ancienne et Médiévale*, 16 (1949), pp. 5–28, has pointed out, the fact that some glosses are ascribed to Walafrid Strabo does not mean that the *Glossa* is the work of Walafrid Strabo, who added his own comments to the patristic extracts. It is more plausible to assume that the compilers of the *Glossa* made extracts from Strabo's commentaries, adding them to other sources.

42 In order to illustrate the fall of the Phoenician king, Ezekiel compared the prince's destiny to the fall of the first man, who lapsed into corruption having previously enjoyed a blissful and privileged existence. The expulsion from Eden thus became a metaphor for the impending judgement against the nations of the earth, whereas the Garden of Eden was described by Ezekiel as a place of perfection and happiness that man lost through his pride. Ezekiel also used the image of Eden against Egypt: Ezekiel 31.8, 9, 16, 18. Eden as a metaphor for the coming judgement of the day of the Lord is also found in Joel 2.3.

43 The medieval use of Aristotelian visions of the universe is discussed in greater detail below in Chapter 7, pp. 170–6.

44 *Biblia latina cum Glossa ordinaria* (1480/1; 1992), I, p. 22; see also Bede, *In Genesim*, I.2.10–11 (1967), p. 49. Similar words are found in Andrew of St Victor, *Opera*, I, *Expositio super Heptateuchum*, 2.11–12, ed. by Charles H. Lohr and Rainer Berndt, CCCM LIII (Turnhout: Brepols, 1986), pp. 31–2, 880–95.

45 See Gillian R. Evans, ed., *Medieval Commentaries on the Sentences of Peter Lombard: Current Research*, 2 vols (Leiden–Boston: Brill, 2002–), I. On the life of Peter Lombard, see Colish, *Peter Lombard* (1994), I, pp. 15–23; see also II, pp. 779–818 for a thorough bibliography.

46 Peter Lombard, *Sententiae in IV libris distinctae*, II, d.12, ed. by Ignatius C. Brady, 3rd edn, 2 vols (Grottaferrata, Rome: Editiones Collegii S. Bonaventura Ad Claras Aquas, 1971), I, pp. 384–9. See Gregory the Great, *Moralia in Iob*, XXXII.12.16, ed. by Marcus Adriaen, CCSL CXLIIIB (Turnhout: Brepols, 1985), pp. 1640–1; Bede, *In Genesim*, I.2.8–9 (1967), pp. 3, 32, 40–2.

47 Peter Lombard, *Sententiae*, II, d.2, c.1–3, d.12, c.1–2, d.15, c.5–6, d.17, c.5.1–4 (1971), I, pp. 336–9, 384–5, 402–14. See Colish, *Peter Lombard* (1994), I, pp. 340–1.

48 Peter Lombard, *Sententiae*, II, d.12, c.1 (1971), I, p. 384. In Peter Lombard's view, God initially planned a creation for all eternity, but then chose to execute it in stages: d.12, c.6, pp. 388–9. Peter Lombard appears to be reintroducing the Augustinian notion of seminal reasons when he specifies that the forms used by God in the

second phase of his creation were already created: see Ermenegildo Bertola, 'La dottrina della creazione nel *Liber Sententiarum* di Pier Lombardo', *Pier Lombardo*, 1/1 (1957), pp. 35–40. To clarify this new hexaemeral model, he quotes Alcuin, who maintained that there were four stages in which God performed his creative work: *Sententiae*, II, d.12, c.6 (1971), I, pp. 388–9; see Alcuin, *Super Genesim*, Interr. 19, *PL* C, col. 519.

49 Peter Lombard, *Sententiae*, II, d.17, c.5 (1971), I, p. 413: 'Intelligitur autem paradisus localis et corporalis, in quo homo locatus est.'

50 Ibid., c.5.1–4, pp. 413–14. For Bede and the *Glossa*, see above, pp. 49–50 and nn. 33 and 41.

51 General accounts on legendary journeys to paradise in the Middle Ages are found in Peter Dronke, *Imagination in the Late Pagan and Early Christian World: The First Nine Centuries* AD (Florence: SISMEL, 2003), pp. 132–42; Gerhardus Johannes Marinus Bartelink, 'De terugkeer naar het paradijs: Paradijsverhalen uit de Oudheid', *Hermeneus: Tijdschrift voor antieke cultuur*, 62 (1990), pp. 203–8; Patch, *The Other World* (1950), pp. 134–74; Arturo Graf, *Miti, leggende e superstizioni del Medio Evo*, 2 vols (Turin: Loescher, 1892–3), I (1892), pp. 73–126 (nn. pp. 175–93).

52 Lawton P. G. Peckham and Milan S. La Du, *La Prise de Defur and le voyage d'Alexandre au paradis terrestre* (Princeton: Princeton University Press, 1935; repr. New York: Kraus Reprint, 1965), pp. 33–40, 48–52; *Alexandri Magni iter ad paradisum*, ed. by Julius Zacher (Königsberg: Th. Theile, 1859), pp. 19–32. See also Catherine Gaullier-Bougassas, *Les Romans d'Alexandre: Aux frontiers de l'épique et du Romanesque* (Paris: Honoré Champion, 1998), pp. 432–3, 478–84; David J. A. Ross, *Alexander Historiatus: A Guide to Medieval Illustrated Alexander Literature* (Frankfurt am Main: Athenäum, 1963; repr. 1988); George Cary, *The Medieval Alexander* (Cambridge: Cambridge University Press, 1956); George Cary, 'Alexander the Great in Mediaeval Theology', *Journal of the Warburg and Courtauld Institutes*, 17 (1954), pp. 98–114; Mary M. Lascelles, 'Alexander and the Earthly Paradise in Mediaeval English Writings', *Medium Aevum*, 5 (1936), pp. 31–104, 173–88. The story of the journey of Alexander to paradise derives from a Jewish tradition found in the treatise *Tamid* of the Babylonean *Talmud*.

53 Walter of Châtillon, *Alexandreis*, X.108–10, in *The Alexandreis of Walter of Châtillon: A Twelfth-Century Epic*, transl. by David Townsend (Philadelphia: University of Pennsylvania Press, 1996), p. 172: 'And if the Fates / should lend his sails kind winds, he plans to seek / the Nile's source, and lay siege to Paradise'; Latin ed. by Marvin L. Colker (Padua: Antenore, 1978), X.95–8, p. 257: 'cuius si fata secundis / Vela regant ventis, caput indagare remotum / A mundo Nyli et Paradysum cingere facta /Obsidione parat'. The poem was an immediate success: over 200 manuscripts survive; it was used in the schools and replaced the classical poets as a tool for learning grammar; it was translated into Old Norse, Czech, Dutch and Spanish. Nine world maps are known to have been included in manuscripts of the *Alexandreis*: see Marcel Destombes, *Mappemondes AD 1200–1500: Catalogue preparé par la Commission des Cartes Anciennes de l'Union Géographique Internationale* (Amsterdam: N. Israel, 1964), pp. 167–72. See also Evelyn Edson, *Mapping Time and Space: How Medieval Mapmakers Viewed their World* (London: British Library, 1997), pp. 102–5, where a world map inserted in a thirteenth-century English manuscript of Walter of Châtillon's text and featuring paradise is reproduced.

54 A Latin version of the legend, *Vita fabulosa sancti Macarii romani, servi Dei, qui inventus est iuxta paradisum*, was published in *Acta sanctorum*, LVIII, entry for 23 October, pp. 566–71. For this and other versions of the legend, where the monks are described as reaching the earthly paradise, see Giuliana Ravaschietto, *Il viaggio dei tre monaci al paradiso terrestre* (Alessandria: Edizioni dell'Orso, 1997).

55 Jean, sire de Joinville, *The History of Saint Louis*, English translation of *Histoire de Saint Louis* (ed. by Natalis de Wailly) by Joan Evans (London–New York: Oxford University Press, 1938), pp. 55–6.

56 See Graf, *Miti, leggende e superstizioni del Medio Evo*, I (1892), pp. 87–8, and Patch, *The Other World* (1950), pp. 165–6.

57 Godfrey of Viterbo, *Pantheon*, in *Illustres veteres scriptores qui rerum a Germanis per multas aetates gestarum historias vel annales posteris reliquerunt*, ed. by Johann Pistorius (Frankfurt am Main: [n.p.], 1613), II, pp. 58–60.

58 Dorothy A. Bray, 'Allegory in the *Navigatio Sancti Brendani*', *Viator*, 26 (1995), pp. 1–10.

59 *Navigatio Sancti Brendani abbatis: From Early Latin Manuscripts*, ed. by Carl Selmer (Notre Dame, IN: University of Notre Dame Press, 1959); *The Voyage of Saint Brendan: Journey to the Promised Land*, transl. by John J. O'Meara (Portlaoise: Dolmen, 1985). A *Voyage of Saint Brendan*, probably composed around 1150 and surviving in Middle-Dutch and German versions, presents some differences compared to the older Latin text. For a recent bibliography see Glyn S. Burgess and Clara Strijbosch, *The Legend of Saint Brendan: A Critical Bibliography* (Dublin: Royal Irish Academy, 2000). See also Christoph Fasbender and Reinhard Hahn, eds, *Brandan, Die mitteldeutsche 'Reise'-Fassung* (Heidelberg: Universitätsverlag C. Winter, 2002); Clara Strijbosch, *The Seafaring Saint: Sources and Analogues of the Twelfth-Century Voyage of Saint Brendan*, transl. by Thea Summerfield (Dublin: Four Courts, 2000); Dominik Pietrzik, *Die Brandan-Legende: Ausgewählte Motive in der frühneuhochdeutschen sogenannten 'Reise'-Version* (Frankfurt am Main: P. Lang, 1999); Pierre Bouet, *Le Fantastique dans la littérature latine du Moyen Âge: La Navigation de Saint Brendan* (Caen: Presses Universitaires de Caen, 1986); Graf, *Miti, leggende e superstizioni del Medio Evo*, I (1892), pp. 97–110.

60 See, for example, Louis Kervran, *Brandan: Le Grand Navigateur du Ve siècle* (Paris: Laffont, 1977). See also Renata A. Bartoli, *La Navigatio Sancti Brendani e la sua fortuna nella cultura romanza dell'età di mezzo* (Fasano, Brindisi: Schena, 1993), p. 103 n. 166, and *The Voyage of Saint Brendan* (1985), pp. XII–XIV.

61 *Navigatio Sancti Brendani abbatis*, VI.10–12 (1959), p. 12: 'Et aliquando ventum habebant, sed tamen ignorabant ex qua parte veniret aut in quam partem ferebatur navis.' English translation from *The Voyage of Saint Brendan* (1985), p. 10.

62 See, for example, the twelfth-century Anglo-Norman text of Benedeit, *The Anglo-Norman Voyage of St Brendan*, ed. by Ian Short and Brian Merrilees (Manchester: Manchester University Press, 1979), pp. 74–5. See Marie-Louise Rotsaert, *San Brandano: Un antitipo germanico* (Rome: Bulzoni, 1996), p. 20. In the Middle-Dutch version, *De reis van Sint Brandaan: Een reisverhall uit de twaalfde eeuw*, ed. by Willem Wilmink, intro. by Willem P. Gerritsen (Amsterdam: Uitgeverij Prometheus–Bert Bakker, 1994), pp. 67–73, the monks spend some time on the 'eylant', where they find wonderful weather, a lovely 'Kasteel', angels all named 'Cherubin', Saint Michael and great peace. They know from the Bible that that place is the earthly paradise (p. 73).

63 *Navigatio Sancti Brendani abbatis*, I.38–9, 42, 60; XXVIII.12–15, 33–4 (1959), pp. 5–7, 79–80.

64 Ibid., XXVIII.30–2, p. 80. See David N. Dumville, 'Two Approaches to the Dating of "Navigatio

65 Grimm, *Paradisus* (1977), p. 109 n. 40.
66 There are extant Greek, Latin, Armenian, Georgian and Slavonic versions, and a fragmentary Coptic version of the *Life of Adam and Eve*. The original language, provenance and date of the work have been variously debated by scholars. See, for instance, Michael D. Eldridge, *Dying Adam with his Multiethnic Family* (Leiden–Boston–Cologne: Brill, 2001); Gary A. Anderson, 'The Original Form of the *Life of Adam and Eve*: A Proposal', and Marinus De Jonge, 'The Christian Origin of the *Greek Life of Adam and Eve*', in Gary A. Anderson, Michael E. Stone and Johannes Tromp, eds, *Literature on Adam and Eve* (Leiden–Boston–Cologne: Brill, 2000), pp. 215–31 and 347–63; John R. Levison, *Texts in Transition: The Greek Life of Adam and Eve* (Atlanta, GA: Society of Biblical Literature, 2000); Gary A. Anderson and Michael E. Stone, eds, *A Synopsis of the Books of Adam and Eve*, 2nd edn (Atlanta, GA: Scholars Press, 1999); Marinus De Jonge and Johannes Tromp, *The Life of Adam and Eve and Related Literature* (Sheffield: Sheffield Academic Press, 1997); Michael E. Stone, *A History of the Literature of Adam and Eve* (Atlanta, GA: Scholars Press, 1992); *La Vie grecque d'Adam et Ève*, ed. by Daniel A. Bertrand (Paris: Maisonneuve, 1987); Esther C. Quinn, *The Quest of Seth for the Oil of Life* (Chicago: University of Chicago Press, 1962), pp. 1–32; 'The Books of Adam and Eve', ed. by L. S. A. Wells, in *The Apocrypha and Pseudepigrapha of the Old Testament*, ed. by Robert H. Charles, 2 vols (Oxford: Clarendon Press, 1913), II, pp. 123–54; Wilhelm Meyer, 'Die Geschichte des Kreuzholzes vor Christus', *Abhandlungen der Königlichen Bayerischen Akademie der Wissenschaften, Philosoph.-Philologische Klasse*, 16/2 (1881), pp. 103–60; 'Vita Adae et Evae', ibid., 14/3 (1879), pp. 187–250.
67 An early Christianized version of Seth's journey is found in a second- or third-century account, *Christ's Descent into Hell*, incorporated in the fifth-century *Gospel of Nicodemus*: James K. Elliott, *The Apocryphal New Testament: A Collection of Apocryphal Christian Literature in an English Translation* (Oxford: Clarendon Press, 1993), pp. 186–7, 191–2, 201. The story was later amalgamated into a complex of legends centred on the Cross of Christ, whose earliest extant versions date to the eleventh century and which were reported, for example, in Godfrey of Viterbo's *Pantheon* (1187–91), Honorius Augustodunensis's *Imago mundi* (first completed around 1110) and the thirteenth-century *Legenda aurea* by Jacobus de Voragine. See, for example, Iacobus de Voragine, *The Golden Legend*, transl. by William Granger Ryan, 2 vols (Princeton, NJ: Princeton University Press, 1993), I, p. 277. The complete fusion of the two stories was achieved in an anonymous thirteenth-century Latin text: Meyer, 'Die Geschichte des Kreuzholzes' (1881), pp. 128–30. On Seth and the Legend of the Cross see Barbara Baert, *Een erfenis van heilig hout: De neerslag van het teruggevonden kruis in tekst en beeld tijdens de Middeleeuwen* (Louvain: Universitaire Pers Leuven, 2001), esp. pp. 215–54; Baert, 'Seth of de terugkeer naar het paradijs: Bijdragen tot het kruishoutmotief in de middeleeuwen', *Bijdragen, tijdschrift voor filosofie en theologie*, 56 (1995), pp. 313–39; Quinn, *The Quest of Seth* (1962), pp. 34–136; Albert Frederick J. Klijn, *Seth in Jewish, Christian and Gnostic Literature* (Leiden: Brill, 1977). A good summary and a rich bibliography on the Legend of the Cross are found in Hans Martin von Erffa, *Ikonologie der Genesis: Die Christlichen Bildthemen aus dem Alten Testament und ihre Quellen*, 2 vols (Munich: Deutscher Kunstverlag, 1989–1995), I (1989), pp. 114–19.
68 Meyer, 'Die Geschichte des Kreuzholzes' (1881), pp. 131–48. See also Quinn, *The Quest of Seth* (1962), pp. 105–6; Patch, *The Other World* (1950), pp. 155–7; Graf, *Miti, leggende e superstizioni del Medio Evo*, I (1892), pp. 76–83.
69 *Book of Enoch*, XII.1, ed. by Matthew Black (Leiden: Brill, 1985), p. 31.
70 In the modern numbering, 2 Kings 2.1, 11.
71 Augustine, *De civitate Dei*, XV.19 (1955), pp. 481–2. See also below, Chapter 4, p. 66.
72 Augustine, *Contra Iulianum*, VI.30, *PL* XLV, cols 1580–2.
73 The identification of Enoch and Elijah as the two witnesses was widespread. See, for example, Tertullian, *De anima*, L.1, ed. by Jan Hendrik Waszink (Amsterdam: J. M. Meulenhoff, 1947), p. 68; Jerome, *Epistula* LIX.3, ed. by Isidorus Hilberg, *CSEL* LIV (Vienna–Leipzig: Tempsky, 1910; repr. Vienna: Österreichische Akademie der Wissenschaften, 1996), I, pp. 543–4; Augustine, *De Genesi ad litteram*, IX.6 (1894), pp. 273–5; Hippolytus, *De Antichristo*, 43, ed. by Enrico Norelli (Florence: Nardini, 1987), pp. 114–15. See Éliane Poirot, *Les Prophètes Élie et Élisée dans la littérature chrétienne ancienne* (Turnhout: Brepols, 1997); Maria Magdalena Witte, *Elias und Henoch als Exempel, typologische Figuren und apokalyptischen Zeugen: Zu Verbindungen von Literatur und Theologie im Mittelalter* (Frankfurt am Main–New York: P. Lang, 1987); James C. VanderKam, *Enoch and the Growth of an Apocalyptic Tradition* (Washington, DC: Catholic Biblical Association of America, 1984).
74 *Visio Pauli*, esp. 45–51, in Theodore Silverstein and Anthony Hilhorst, eds, *Apocalypse of Paul: A New Critical Edition of Three Long Latin Versions* (Geneva: Cramer, 1997), pp. 201–7; see also Anthony Hilhorst, 'A Visit to Paradise: *Apocalypse of Paul* 45 and its Background', in Luttikhuizen, ed., *Paradise Interpreted* (1999), pp. 128–39, esp. 134, and Elliott, *The Apocryphal New Testament* (1993), pp. 616–44, esp. 639–44.
75 Ezras's account is in fact part of an eschatological vision of the Last Judgement: *Apocalypse of Ezra*, V.20–2, in *Apocalypsis Esdrae, Apocalypsis Sedrach, Visio Beati Esdrae*, ed. by Otto Wahl, *Pseudepigrapha Veteris Testamenti*, ed. by A. M. Denis and M. de Jonge, IV (Leiden: Brill, 1977), pp. 31–2; *Revelation of Ezra*, in *Apocryphal Gospels, Acts and Revelations*, transl. by Alexander Walker, *Ante-Nicene Christian Library*, ed. by Alexander Roberts and James Donaldson, XVI (Edinburgh: Clark, 1870), p. 474.
76 See John J. Collins, *The Apocalyptic Imagination: An Introduction to Jewish Apocalyptic Literature*, 2nd edn (Grand Rapids, MI–Cambridge: Eerdmans, 1998), pp. 195–212; Tom W. Willett, *Eschatology in the Theodicies of 2 Baruch and 4 Ezra* (Sheffield: Sheffield Academic Press, 1989), pp. 51–75; Michael E. Stone, 'The Metamorphosis of Ezra: Jewish Apocalypse and Medieval Vision', *Journal of Theological Studies*, 33/1 (1982), pp. 1–18; Stone, *Fourth Ezra: A Commentary on the Book of Fourth Ezra* (Minneapolis: Fortress, 1968), esp. pp. 43–7 on Ezra in Christian tradition. See also the bibliography quoted in *Die Esra-Apokalypse (IV.Esra)*, ed. by Albert Frederik J. Klijn (Berlin: Akademie Verlag, 1992), pp. XXXI–XXXV.
77 See, for example, *Fourth Ezra*, III.6, VI.2, VII.6, VIII.52, *Der lateinische Text der Apokalypse des Esra*, ed. by Albert Frederik J. Klijn (Berlin: Akademie Verlag, 1983), pp. 25, 38, 43, 63; Stone, *Fourth Ezra* (1968), pp. 58, 68–9, 142, 156, 190, 196–8, 277, 286–7.
78 *Fourth Ezra*, VII.26, Stone, *Fourth Ezra* (1968), p. 202; see also pp. 213–14; *Der lateinische Text der Apokalypse des Esra*, ed. Klijn (1983), p. 45: 'et ostendetur quae nunc subducitur terra'.
79 See *The Oxford Companion to the Bible* (1993), pp. 743–4; Francis Brown, and others, *A Hebrew and English Lexicon of the Old Testament* (Oxford: Clarendon Press, 1972), pp. 30–1.

4

Locating Paradise in Time

Time present and time past
Are both perhaps present in time future,
And time future contained in time past.
T. S. Eliot, 'Burnt Norton' (1941)

Time is a difficult concept, not tangible like a geographical landscape. There is always a danger of reading the past with the eyes of the present, and one of the problems about our understanding of the concept of paradise and its representation on medieval maps is the difference in the way time and history are perceived today. For a modern historian, history is the story of human actions. In the Middle Ages it was understood that the past was part of a historical process which did not run an arbitrary course – the result of wilful human actions – but had an otherworldly dimension. For a medieval Christian writer, all history was orchestrated by a divine plan. Past, present and future were linked through a network of meaning that at every point referred to God and that referred all events to an overall scheme. When we come to examine medieval *mappae mundi* later in this book we shall find that they depict the world in both its temporal and spatial dimensions, and that the compilers had not only portrayed an orderly physical and human geography, but they had also made visible the invisible order that guided the course of human events. First, however, we need to consider time, for time is as important as space in explaining the presence of paradise on maps. The earthly paradise was not only a place, a distant island or remote land; it was also an 'event/place', that is, a historical occurrence at a specific place. In the case of paradise, the event/place was the original sin of Adam and Eve being committed in the Garden of Eden.

Crucial to medieval thinking about paradise was not only Augustine's historical interpretation, which led to the geography of paradise, but also his 'typological' approach, which was based on the assumption that the true meaning of the Old Testament is revealed in the New. Augustine understood Old Testament features as 'types' or prefigurations of Christian realities. So the Garden of Eden described in Genesis was seen as forecasting the Christian realities of the Church and the final paradise in heaven and Adam was interpreted as a 'type' of Christ. The typological approach prevented the event/place of paradise from being confined to a remote and irrelevant past. Although Eden was inaccessible and forbidden to mankind, the Garden remained present in its very loss, since the whole history of salvation – the historical process set in motion by God – was designed to repair the damages of original sin. Thus, the lost paradise was in some way still present and had a place in the future. Typology linked the historical events that occurred in Eden both with the saving sacrifice of Christ in the earthly Jerusalem that led to the establishment of the Church on earth and with the eternal reality of the Heavenly Jerusalem. It was precisely the connection between past, present and future suggested by typological thinking that helped to bring about the temporal dimension in medieval cartography and that favoured the depiction of the earthly paradise on maps of the world.

The Relevance of Paradise Lost

Once again it was Augustine who paved the way to an understanding of the interrelationship between past, present and future. Augustine insisted that the events in the Garden of Eden had actually taken place, that paradise was not a distant historical event locked away in a remote past, but that, on the contrary, the paradise story took on enduring meaning because it extended beyond its own time to connect with the future. Adam prefigured Christ, Eve foreshadowed the Virgin Mary, and the mishaps in Eden were part of God's *carmen universitatis* ('poem of the universe').[1] In his typological reading of the first chapters of Genesis, Augustine was following Paul when he proclaimed that 'no Christian will dare say that the narrative must not be taken in a figurative sense' and, again, when he adopted Paul's dictum that 'now all these things that happened to them were types'.[2] For Paul, a 'type' was a thing, a person or an action from the Old Testament that had its own historical existence, but that at the same time was designed by God to anticipate a future thing, person or action relating to Christian redemption.[3] As an example of a type, Augustine referred to Paul's explanation that the union between Adam and Eve described in Genesis 2.24 ('they shall be two in one flesh') was 'a great mystery in reference to Christ and to the Church'.[4] In other words, Adam and Eve's physical love prefigured the loving communion between Christ and his Church. Christ, for Augustine, was everywhere in the Bible, giving it full meaning.[5] The prophetic significance of the historical facts recorded in the Old Testament lay in their function as *signa*, deeds planted by God in history as signs of a reality that went beyond the deeds themselves and carried a meaning relating to Christ. These facts take their value from the eventual fulfilment of what they prefigured, that is, the establishment of the Church through the Incarnation and Crucifixion, and the final beatitude of the blessed after Christ's Second Coming. Hence, for Augustine, the episodes of the *sacra historia* recorded in the Old Testament were to be seen as ways by which God spoke to man about Christian salvation.[6] The earthly garden inhabited by Adam – a real man in the flesh – was a double reference: 'And yet a more thoughtful consideration of the matter might possibly suggest that the corporeal paradise in which Adam lived his corporeal life was a sign both of this life of the saints now existing in the Church and of that eternal life which will be when this life is done.'[7] The Garden of Eden thus signified far more than a mythical state of innocence or a fanciful dreamland, forever lost. The earthly paradise pointed to a present and future reality, that of Christian redemption.

The prophetic dimension of the paradise story was to be fulfilled in two stages, marked by the two advents of Christ. Adam prefigured the first coming of Christ. An earthly and worldly history, expressed in specific ages and places, separated the First Adam in Eden, man made by God (described in Genesis) from the Second Adam in Judaea, God made man (Jesus Christ). The fulfilling of the prophecies of the Old Testament by the first advent of the Son of God, however, initiated the expectation of a Second Coming, that of Christ in glory. In this context, the Garden of Eden was also a type of the heavenly paradise, the second and final stage of Christian salvation history. The delay between the *already* and the *not yet* – Christ's First and Second Coming – created a spiritual tension. Time continued to pass, but, unlike the time that structured the salvation history of the Bible down to the first advent of Christ, the time of waiting for his Second Coming has been unstructured time, lacking any chronological landmark.[8] The thirteenth-century Ebstorf *mappa mundi* (Figures 6.17–9 and Plate 9) perfectly reflects this temporal disjuncture. On the map, the depiction of the Garden of Eden and the portrayal of the earthly city of Jerusalem at the time of Christ's resurrection together refer to the vicissitudes of salvation history. Christ had *already* come and risen from the dead. The waiting for his triumph *yet* to come is expressed through the superimposition and iconographic assimilation of the earthly Jerusalem, the city of Judaea, and the heavenly Jerusalem of the end of time.[9]

Other medieval map makers, as we shall see, found other ways to signify the complex relationship between the different stages of Christian salvation history. The theology behind their diverse imaginative skill, however, was one and the same, and went back to Augustine's settlement of earlier exegetical and theological debate. In Augustine's exegesis, Adam's Fall in Eden, the sacrifice of Christ on Golgotha, and the Second Coming of Christ were intimately interrelated as the three pivotal events that structure the whole history of human salvation. Although Adam and Eve's banishment from Eden was irreversible, and the garden of delights was forever lost to mankind, the human race was saved from the consequences of Adam's sin by the Second Adam. The biblical history of salvation is the journey between Eden, the stage of original sin, and the earthly Jerusalem, the place of the Passion of Christ. The tension between the earthly Jerusalem and the Heavenly Jerusalem that has yet to come endows that journey with its ultimate meaning, giving the celestial city its duality as simultaneously a vision of the future and, in its representation of the Church on earth, an image of the present. Eden was a type of the Christian community in history and an anticipation of God's mysteries in heaven. While Augustine's historical interpretation of Genesis made clear the difference between the earthly Eden of the beginning of time and the heavenly paradise of the end of history, his typological reading linked them intimately.[10]

Naming the Time

The event/place of paradise was thus both anchored to the beginning of time and linked to the history of salvation. Augustine's emphasis on history encouraged him, and later exegetes, to attempt to place paradise on the historical time span. If, the argument effectively ran, the creation of man and original sin were historical occurrences, a certain number of years and a succession of epochs must have separated these events from, and connected them to, mankind's subsequent history since, from Adam onwards, the human race had multiplied itself. However mysterious the shift from eternity to history may have been, God had given time a precise beginning, setting mankind on a linear historical course. Just as the legendary theme of the search for paradise implied a reference to salvation, so the exegetical dating of the creation of Adam and his sojourn in paradise implied a reference to the later events of salvation history. The specific question 'when?' as applied to the Garden of Eden would open up the wider question of Christian redemption.

How long ago? The Past Linked to the Present
Locating paradise in time was an enterprise that fascinated many Christian historians. Exactly when were Adam and Eve created? Exactly when did they pick the forbidden fruit? And from how far back in the world's time should the years of human history be calculated? From the early centuries of Christianity onwards, historical chronology was reckoned from Adam and not from the first day of creation. Few medieval Christians doubted that the intangible and mysterious moment of the origin of the universe had taken place when God created heaven and earth. Understandably, however, historians let human history start from the day of the creation of the first human five days later. Sextus Iulius Africanus's *Chronography*, compiled in the late third century, presented a computation of the individual life spans of all the patriarchs named in the Bible and a biblical chronology that is compared with those in Near Eastern, Greek and Roman sources. The conclusion arrived at was that 5,500 years had elapsed between Adam and the birth of Christ.[11]

The question of the beginning of history was of particular urgency to those exegetes concerned with the date of the end of the world. Looking forward eagerly to the thousand-year reign of peace promised in the Book of Revelation (20.1–6), they sought every hint in the Scriptures that might pinpoint the beginning of that period. Combining the Genesis account of the creation of the universe in six working days, followed by one day of divine rest, with the notion that for God a single day is like a thousand years (and a thousand years like a single day: Psalm 90.4; 2 Peter, 3.8), they calculated that the world was meant to last for six thousand years before final sabbatical rest. How much time was left depended on the calculations of the individual exegete and the version of the Bible used.[12] For example, in about 200, Hippolytus of Rome assumed, as had Sextus Iulius Africanus, that Christ had been born 5,500 years after Adam, and he accordingly estimated that 300 years remained.[13] Hippolytus was commenting on the Book of Daniel, which, like the Book of Revelation, was a source of apocalyptic prophecy. When Nebuchadnezzar, king of Babylon, dreamt of a statue made of gold, silver, bronze and iron that was broken into pieces by a stone, which then became a great mountain filling the whole world, Daniel explained that the dream foretold that after a succession of four empires (symbolized by the four metals) the eternal kingdom of God would be eventually established (Daniel 2.31–45). Daniel also described other prophetic visions, such as the appearance of four beasts rising up out of the sea, one after the other (7.1–14). Christian writers used the four empires and the four beasts to structure the course of time up to the final restoration of divine sovereignty.

Even for those who, like Augustine, did not attribute a literal value to the concept of the millennium and who warned against vain efforts to calculate the exact timing of the Second Coming since Christ himself had declared that nobody may know that date (Matthew 24.36; Mark 13.32; Acts 1.7), the problem of locating paradise in time was still important.[14] With or without precise apocalyptic expectations, Christian historians needed a chronological scheme to give the past, present, and future a Christian framework. In the early eighth century, Bede estimated the amount of time that had elapsed since the creation of Adam and his Fall by comparing Jerome's Vulgate with the *Vetus Latina* and with Jewish computations and concluded that the week of creation had taken place 3,952 years before the birth of Christ.[15] In the eleventh century, a different date was reached by Marianus Scotus, who placed paradise 4,216 years before the birth of Christ.[16]

When, as was invariable, Adam in paradise was cited as the starting point of universal history, the critical issue then became the meaning of the expression 'from Adam'.[17] At the turn of the third into the fourth century, the historically minded bishop of Caesarea, Eusebius, argued that it was impossible to calculate the exact length of Adam's sojourn in paradise, not least because human life within the Garden of Eden was of a different order from post-lapsarian human existence. Eusebius's calculations for the beginning of true history started from the moment of the Expulsion. Using a different version of the Bible from Sextus Iulius Africanus, Eusebius dated this event to exactly 5,198 years before the birth of Christ.[18]

Eusebius was metaphorically washing his hands of the problem. Genesis makes no reference at all to the date of original sin, and throughout the early Middle Ages the period that Adam and Eve spent in the Garden of Eden remained a topic of speculation. In Eusebius's day, it was thought that Adam and Eve's stay in the Garden of Eden was not subject to the normal passage of time. From different sources, Judaic or Christian – various apocryphal writings on the life of Adam and Eve, for example – came different ideas, ranging from a few hours or a few years to a century or two.[19] Unlike Eusebius, Augustine held that human history began within the Garden of Eden and that every event reported in the second chapter of Genesis had taken place in a real moment of time. Adam's creation was followed by his introduction into the Garden, his instruction by divine commandment, the naming of the animals, the creation of Eve, the temptation by the serpent, the Fall, and the Expulsion of the guilty pair. In Augustine's view, the counting of years had to start from the creation of Adam. At the same time, he avoided committing himself to any precise figure, concluding simply that fewer than six thousand years had passed since the creation of man.[20] Acknowledging that there was no hint in Scripture of how much time had elapsed between the creation of Adam and Eve and the birth of Cain, Augustine merely observed that man had dwelt in the Garden 'for as long as he desired what God commanded'. Augustine's view seems to have been that Adam's bliss in Eden was of short duration, an idea supported by his remark that Adam and Eve had been encouraged by God to procreate, but simply did not have the time to do so (or, alternatively, that they were waiting for an explicit commandment from God).[21] After Augustine, the door remained open for later writers to speculate in greater detail on the question of time.

For How Long? The Past Linked to the Future

Early Christian exegetes were well aware that the Genesis account of the creation of Adam and his original sin in paradise was related to the most profound mysteries concerning the nature and destiny of mankind. Augustine's literal reading of the biblical narrative encouraged them to think that a precise timetable of the primordial events was possible. Their efforts to record the succession in time of the crucial incidents that took place in Eden reflect above all a profound wish to draw a typological parallel between Adam's Fall and Christ's Redemption. The idea that Adam, having sinned on the first day of his creation, was permitted only a few hours in paradise was soon established doctrine in Western Christian tradition and consistently presented in Christological terms.

The issue of the time spent by Adam and Eve in paradise became tied up with the question of the day on which they sinned. Irenaeus, a second-century bishop of Lyons, insisted that since the Redeemer was crucified on a Friday, to save mankind from the consequences of original sin, Adam, irrespective of whether he sinned on the day of his creation or had stayed longer in Eden, had definitely been created on a Friday, ate of the forbidden fruit on a Friday, and died on a Friday.[22] Victorinus, bishop of Pettau, Pannonia, who died about 304, wrote that the devil had deceived Eve on the day that the archangel Gabriel brought to the Virgin Mary – the new Eve – the good news of her miraculous pregnancy, that Christ was born on the same day as Adam's creation, and that Christ died on the Cross on the same day as the original sin was committed.[23] Even Augustine, who did not take the account of the six days of creation literally, and for whom the Genesis account of the creation of man on the sixth day made no reference to a historical Friday, pondered on the significance of the symbolism of the number six. Six, the result of multiplying one, two and three (the numbers of the Trinity, and a number suggestive of the perfection of the divine work of creation) is also the first number which is the sum of its own parts: one (a sixth), two (a third) and three (a half). Augustine pointed out that Genesis relates that the First Adam was created on the sixth day, and that the number three and its multiples played a crucial role in the Passion of the Second Adam: it was in the third hour (9 a.m.) that the mob shouted for Christ to be crucified; it was in the sixth hour that Christ was crucified (noon); and it was in the ninth hour that he died (3 p.m.) (Matthew 27.45–6).[24] Augustine also likened the creation of Eve from the rib of the sleeping Adam to the origin of the Church (the spouse of Christ) from the wounds of the Second Adam as he hung dying on the Cross: the blood and water that issued from the side of the crucified Christ initiated the two main Christian sacraments of the Eucharist and baptism.[25]

Other exegetes, believing (unlike Augustine) literally in a creation week of seven days, sought to identify the exact date of the beginning of history to demonstrate parallels between the date and the hour of the Fall and of the Redemption. Throughout the Middle Ages, the traditional date (accepted by Augustine) of 25 March for the Annunciation and the Passion was taken for granted.[26] Classical authors, like later Christian writers, had also settled on that date for the world's creation because of its coincidence with the spring equinox.[27] Irrespective of any differences in their computations, all medieval Christian authors saw the coincidence between the most important events in human history – the formation of the universe at the spring equinox, when nature was in full bloom, the succession of the six days of creation, the Fall of Adam, the Incarnation and Passion of Christ – as underscoring the consistency of God's salvation plan for mankind. Bede stated that the creation of heaven and earth occurred on Sunday, 18 March, that dry land was created on Monday, 19 March (the second day), that the land was clothed with vegetation on the same day that paradise was planted (Tuesday, 20 March, the third day), that God created the stars, sun, and moon on the day of the spring equinox (Wednesday, 21 March, the fourth day), and that on Thursday, 22 March (the fifth day) all the animals were created. Finally, on Friday, 23 March (the sixth day) man and woman were fashioned. According to Bede's sources, Christ was crucified on 23 March. For Bede the parallels were clear: it was on Friday, 23 March that the first woman was created from the rib taken from Adam's side, and another Friday 23 March when the Church, the 'Bride of the Lord', was born from the wound of the crucified Christ.[28] Bede also underlined the fact that Adam had sinned at noon, and was banished from Eden at the same moment – noon – that Christ was nailed to the Cross and re-opened the gate of paradise to the good thief.[29] Even when different dates were suggested, it was always the same occurrences that were compared. As things turned out, the date of 25 March prevailed for the three fundamental events: the beginning of the universe (which was the first day for some scholars and the third day for others, on the basis that time could not be calculated until the celestial bodies had come into existence), the Annunciation, and the Crucifixion.

Once a day and date had been decided upon for the creation of Adam and for his Fall, the way was open for the parallelism between Adam's sin and Christ's Crucifixion to be elaborated. The time of the day was also held to be revealing. According to a widespread tradition, reported in the twelfth century by Honorius Augustodunensis in his influential geographical, astronomical and historical compendium (*Imago mundi*), and also by the French theologian Peter Comestor in his biblical commentary (*Historia scholastica*), Adam and Eve had remained in paradise for no more than seven hours, an idea later taken up by the Italian poet, Dante Alighieri.[30] The problem was to determine the hour from which to start the calculation of Adam and Eve's stay in the Garden of Eden. Scholars counted either from the first (7 a.m.) to the seventh hour (1 p.m.), or from the third (9 a.m.) to the ninth (3 p.m.). According to the thirteenth-century encyclopedist, Vincent de Beauvais, Adam arrived in paradise at the third hour (9 a.m.) of Friday, 25 March, ate the forbidden fruit at the sixth hour (noon). Whatever the details, the object was always to connect in some way the two pivotal events in human history, the Fall and the Redemption, and to show that Christ's Passion exactly matched Adam's stay in Eden. Many centuries later, at the same time on the same day, Christ was crucified, and just as the First Adam, guilty and fearful of his having sinned, spent three hours in Eden, so Christ, the New Adam, hung on the Cross for the same length of time. Just as Adam was banished from paradise at the ninth hour (3 p.m.), so Christ promised the good thief access to paradise and died at the identical hour (Luke 23.43).[31]

Knowing the exact hour and day of Adam's sin was important. The linking of this remote event with other key events from the past, notably the Annunciation and the Crucifixion, structured the Church's liturgy through the constant – i.e. present – re-enactment of Christ's Passion. For some exegetes, the reckoning concealed the future.

The 25th of March, freighted with parallels, was anticipated as the last day of human history, the day on which Christ would return in glory (Matthew 24.42). The ninth-century Carolingian grammarian, Christian of Stavelot, noted that Adam sinned on Friday, 25 March – the same day as the creation of the world, the Incarnation, and the Crucifixion – and, aware of an ancient tradition according to which the Second Coming of Christ would take place during an Easter vigil, suggested that the world would end on an Easter Sunday of that date.[32] Other exegetes associated the correspondence between the beginning, the apex, and the end of human history with a broader perspective. For them, the primordial events recounted in Genesis set the pattern for the unfolding of entire human history, in this way tying the past to the future in the widest possible sense.

Paradise and the Pattern of History

Agreement that the beginning of human history dated from that crucial Friday on which God had created man and on which, only a few hours later, Adam and Eve were banished from Eden forever, meant that paradise now had a precise location in time. The historical process thereafter did not follow an arbitrary course, but was endowed with a meaningful direction established by the paradise narrative. According to the prevailing view amongst Christian historians in the Middle Ages (who were all following Augustine) the historical process was a divinely planned sequence of six epochs modelled on the six days of creation. History had begun in Eden, had passed through crucial tests such as the Flood and Abraham's sacrifice, had culminated in the Incarnation of Christ and was now in the last age. The imminent end of the world would reveal the full glory and power of God, but it was the beginning that provided the whole historical process with a structure.[33]

The conception of history as a field of divine activity articulated as six epochs was not Augustine's invention, but much older.[34] What Augustine had found important was the way the various epochs could be used to signpost the progression of human history towards a preordained conclusion and to reject heretical ideas either postulating the eternity of the world or promoting an endless repetition of historical cycles. History, for Augustine, took a linear course. He saw it as a dynamic of events set in motion and controlled by God, beginning uniquely in Eden, punctuated by the events recorded in the Bible, and ending in the final disclosure of God to mankind. Consequently, Augustine held that historical writing should not dwell on past deeds, but rather point to the divine order behind the course of time.[35] Just as the world was created in six days, with God resting on the seventh day, the six ages of human history would be followed by eternal rest in heaven.

The number six, we have seen, symbolized for Augustine the perfection of the Creation.[36] His six ages were each marked by a crucial event. The first age ran from the creation of Adam to Noah's building of the Ark. The second age went from the Flood to the sacrifice of Abraham. The third extended from Abraham to the kingdom of David. The fourth ran from David to captivity in Babylon and the fifth from Babylon to the advent of Christ.[37] The sixth and last age, with the Incarnation and Passion of the Son of God, was the culmination of history. As Augustine appreciated, Christ had died on the sixth hour of a sixth day of the week in the sixth age of the world.[38] The Second Coming of Christ at the end of the sixth age would validate the whole process. Augustine also associated the six world epochs with the classical periodization of world history according to the ages of man: the infancy of mankind had taken place between Adam and Noah, childhood between Noah and Abraham, adolescence between Abraham and David, young adulthood between David and Babylonian captivity, mature adulthood between Babylon and Christ, and old age between Christ's First and Second Coming. Augustine, who dismissed as vain and impious speculation about the exact date of the end of the world, embraced instead a kind of reckoning of time that increased wonder at the harmony of God's plan.[39]

However defined, and irrespective of the details of the temporal boundaries, the division of human history into six world ages became a commonplace.[40] Later writers, such as Isidore of Seville and Bede, used the model to structure historical events in order to emphasize the unity of mankind and the universality of the faith. Isidore referred to the six ages both in his chronicles, which he began with the Creation, and in his allegorical writing, where he endorsed the typological link between Adam and Christ.[41] As the first man was created on the sixth day in the image of God, so the Son of God became incarnate in the sixth age in order to renew human nature in the divine likeness. In the *Etymologiae*, Isidore provided a precise summary of world history in which he gave an exact number of years for each of the five ages that had already gone by and fixed the beginning of the last age at 5,155 years from the Creation.[42] Bede also adopted the model of the six ages, and his chronological manuals, notably the *De temporibus liber* and *De temporum ratione*, began with the Creation and took Christ's birth as the starting point of the sixth and final era.[43]

Merging Eden, Church and Heaven

In Christian art as well as exegesis, the three fundamental features of salvation history that were linked with each other in epochal time – the Garden of Eden of Adam and Eve; the Church, established on earth through the sacrifice of Christ; and paradise in heaven – overlapped.[44] Paradisiacal notions associated with images of Eden, the Church, and heaven were indifferently combined and interwoven by medieval artists and writers. Whether found in canonical, apocryphal or exegetical texts or in liturgy, the merging of the three concepts did not necessarily imply any confusion. For example, the Tree of Life, which according to Genesis is found in the Garden of Eden, was also seen in various apocalyptic and visionary texts as the central feature of the celestial paradise, notably in the Book of Revelation (22.2). Similarly, the desire to return to Eden, the native land of mankind, was an established *topos* in theological writing. The sacramental life of the Church, like the consecrated life of monks and priests, offered a readmission to Eden and an anticipation of heaven.[45] It is significant that in the legends the typological link between the Fall and the Crucifixion acquired a temporal dimension – the Cross was made from the wood of the Tree of Knowledge – and a spatial dimension – Adam was buried at the site of the Crucifixion.[46] The Cross, which in this manner was associated with the beginning of time, also pointed to the end of history through the belief that it would reappear before Christ at his Second Coming, destined to take place on Sion, the mountain of Jerusalem identified as the mountain of paradise of Ezekiel (28.13–16).[47]

The tendency to link past, present and future is especially clear in the medieval approach to the Bible. To early readers of Genesis, always concerned with the ultimate meaning of all things, paradise – the beginning – was bound up with both the first and the second advents of Christ. Augustine and the other Church Fathers had taught that Scripture was 'sacred' not only because it was a text inspired by the Holy Spirit, but also because the events reported in it were part of a sacred history, the purpose of which was to bring about the salvation of man. The events of the past recorded in the Bible were endowed with meaning for the present and the future. Bible readers were continuously invited to look *through* the text rather than just at it, but not, as had been the case with Hellenistic allegorical reading, to the detriment of the literal meaning.[48] The frame of the mirror was as important as the deep reflections it revealed. The teaching that the literal sense, the *sensus historicus* or *sensus literalis*, of Scripture was the indispensable foundation upon which spiritual interpretation rests, was repeated throughout the Middle Ages by, for example, Gregory the Great in the sixth century, Peter Comestor

Fig. 4.1 (opposite). The opening of Genesis from the first page of the Stavelot Bible. Stavelot, Mosan Valley (modern Belgium), late twelfth century. 58.5 × 37.5 cm. London, The British Library, Add. MS 28106, fol. 6r. Six major events in the life of Christ (in the centre, portrayed in a column, starting at the bottom, the Annunciation, Nativity, Baptism, Crucifixion, Deposition, Resurrection, Second Coming); important episodes in biblical history (on the left, scenes from the lives of Adam, Noah, Abraham, Moses and Christ); and the parable of the Labourers in the Vineyard (Matthew 20.1–16) (on the right, with the various times of the day in which the labourers were hired by the owner of the vineyard, an image of the various stages of human history). The paralleling of these three sets of scenes reiterated the inter-relationship of events in the story of Christian salvation while maintaining the focus on the life of Christ, so that *in principio* (in the beginning) could be interpreted to mean *in Christo* (in Christ).

and Hugh of St Victor both in the twelfth century, and Nicholas of Lyre in the early fourteenth century.[49] The literal sense of Scripture was strengthened by typology, which interpreted the Old Testament as the foreshadowing of the New.[50] Even so, the fruit of the spirit, hidden among the leaves of the text, as Jerome put it, had to be sought.[51]

In the search for deeper and less visible meanings of the biblical text beyond the letter, exegetes found multiple and coexisting levels of significance in their reading of Scripture, to which they applied different terminologies.[52] A fourfold model, proposed by John Cassian in the early fifth century and adopted by Augustine in his literal commentary on Genesis, became the standard medieval classification. Cassian's four senses of scriptural exegesis – literal, moral, allegorical, and anagogical – were accepted by the compilers of the *Glossa ordinaria*, by Thomas Aquinas in the thirteenth century and by Nicholas of Lyre in the fourteenth century. That the biblical account could be read on multiple levels was taken as evidence of the profundity of the text. The terms referred to history, the soul and its virtues, the Church and her sacraments, and the heavenly realities.[53] Searching for the spiritual treasures hidden in the biblical description of the earthly paradise, medieval commentators, like the Church Fathers before them, saw the Garden of Eden literally as the earthly garden where God put Adam, morally as the human soul, allegorically as the Church and anagogically as the heavenly paradise. A literal reading of the description of the Garden of Eden allowed the reader to accept that the paradise that had sheltered Adam and Eve was a real, if out of the ordinary, place on earth. On the moral level, the Genesis story became a parable about the nature of man, instructing the way every Christian should lead his or her life (the Garden of Eden standing for the human soul). An allegorical reading understood the Garden of Eden to be an image of the Church, the *hortus conclusus* planted by the Lord through Christ. Finally, paradise could also be seen as a promise for the future, as the description of the place of original delight was raised to convey higher things, an anagogical reference to the next world.

Seen through the lens of the four senses of Scripture, the paradise narrative encapsulated the history of mankind from Eden to heaven. Particularly interesting in such a context is Jerome's rendering of and comment on the critical Hebrew term מִקֶּדֶם (*miqedem*). In the Vulgate, paradise was said to have been not *in oriente* (in the east) but *a principio* (from the beginning), meaning from Christ, the foundation of all.[54] Significantly, the beginning of the first sentence of the Bible (*In principio*, followed by *creavit Deus caelum et terram*) was illuminated in the late twelfth-century Bible produced at the Abbey of Stavelot, modern Belgium, with a series of medallions illustrating six major events in Christ's life (Figure 4.1). The series culminates with Christ's Second Coming at the end of time, and is paired with the succession of the ages of world history (shown below) associated with the various times of the day at which the labourers in Jesus's parable of the kingdom of God were hired by the owner of the vineyard (Matthew 20.1–16).[55] The illumination in the Stavelot Bible demonstrates how the events of the story of salvation were seen as interrelated and centring on the life of Christ: God creating heaven and earth *in principio*, in the beginning, could be interpreted to mean *in Christo*, in Christ, especially given the parallel text of greatest significance: *In principio erat verbum*, 'In the beginning was the Word' (John 1.1). The graphic symbol that all the stages of human history were combined was also a visual confirmation of John Chrysostom's point that 'Sacred Scripture offers nothing in vain' and that 'even the smallest dot contains hidden treasures'.[56]

The typological association of the Garden of Eden, the Church, and heaven was also manifested in contemporary art. Palaeo-Christian images – bas-reliefs on sarcophagi and frescoes in catacombs – depicted the Garden of Eden not as an empty landscape, but as one that included Adam and Eve, the perpetrators of the event that made salvation necessary.[57] The transition in funerary art from pagan to Christian was often

creauit deus celu et terram. Terra aut erat ina
nis et vacua: et tenebre erant sup facie abissi:
et spiritus dni ferebatur sup aquas. Dixitq; de
fiat lux. Et facta e lux. Et vidit deus luce qp
esset bona: et diuisit luce a tenebris: appella
uitq; lucem die. et tenebras nocti. factumq;
est vesper et mane dies vnus. Dixit quoq; de
us. fiat firmamentu in medio aquarum: et
diuidat aquas ab aquis. Et fecit deus firma
mentu: diuisitq; aquas que erant sub fir
mamento ab hijs que erant sup firmamen
ti. Et factu est ita. Vocauitq; deus firma
mentu celu: et factu e vesper et mane dies
secundus. Dixit vero deus. Congregetur
aque que sub celo sunt in locu vnu: et appa
reat arida. Et factu est ita. Et vocauit deus
aridam terra: ꝯgregationesq; aquaru appella
uit maria. Et vidit deus qp esset bonu: ⁊ ait.
Germinet terra herba virente et faciente se
men: et lignu pomiferum faciens fructum
iuxta genus suu: cuius semen i semetipso
sit super terra. Et factu est ita. Et protulit
terra herbam virente et faciente semen iuxta
genus suu: lignumq; faciens fructum: et
habens vnumquodq; sementem secundum
speciem suam. Et vidit deus qp esset bonu:
et factum est vesper et mane dies tercius.
Dixit aut deus. fiant luminaria in fir
mamento celi: et diuidant diem ac nocte:
et sint in signa et tpa et dies et annos: vt
luceant in firmamento celi. et illuminent
terra. Et factu est ita. fecitq; deus duo lu
minaria magna: luminare mai' vt præs
set diei. et luminare min' vt præsset nocti:
et stellas. Et posuit eas in firmamento celi
vt lucerent super terra: et præessent diei ac no
cti: et diuideret luce ac tenebras. Et vidit
deus qp esset bonum: et factum est vesper
et mane dies quartus. Dixit etiam deus.
Producant aque reptile anime viuentis
et volatile super terra: sub firmamento ce
li. Creauitq; deus cete grandia: et omnem
anima viuentem atq; motabile quam p
duxerunt aque in species suas: et omne
volatile secundum genus suu. Et vidit
deus qp esset bonu: benedixitq; eis dicens. Cres
cite et multiplicamini: et replete aquas ma
ris: auesq; multiplicentur super terra. Et
factum est vesper et mane dies quintus. Di
xit quoq; deus. Producat terra animam
viuentem in genere suo: iumenta ⁊ repti
lia. et bestias terre secundum species suas.

Locating Paradise in Time 71

72 Mapping Paradise: A History of Heaven on Earth

marked by portrayals of Adam and Eve in paradise on the sarcophagi of the converted. These portraits were intended less to signify Adam and Eve as historical figures than to highlight their (and our) redemption through Christ.[58] In ambitious compositions such as the sarcophagus of Iunius Bassus (359 AD), for example, the presence of the figures of Adam and Eve demonstrated belief in the Christian scheme of salvation.[59] Even when the early artists emphasized the guilty posture and shameful nudity of Adam and Eve, the promise of Christian salvation was always conveyed. As Wladimir Weidlé remarks, the scenes of Adam and Eve in early Christian art refer to both 'Paradise Lost and Paradise Regained'.[60]

This is not the place to attempt to do justice to the many ways in which depictions of paradise in Christian art in general pointed to the relationship between the Fall of Adam and Eve and the history of the salvation of mankind, and to the complex entanglement of Eden, the Church on earth and the Church in heaven.[61] To understand paradise on maps, however, it is useful to bear in mind some of the traditional typological motifs associating Adam and Christ, Eve and Mary, and Eden and the Church. The permutations were limitless. A representation of the Garden of Eden could include an ecclesiastical feature such as a baptismal font or a temple to indicate the source of

Fig. 4.2 (opposite). The paradise narrative in Genesis, from the Bedford Book of Hours, France, Paris, c. 1423. 26 × 18 cm. London, British Library, Add. MS 18850, fol. 14r. At the bottom of the picture, the Trinitarian God is shown creating Adam outside the enclosed Garden of Eden and then carrying him into it. Above (upper left) Adam is naming the animals, God creates Eve (right of centre), and warns the couple not to eat from the Tree of Knowledge (left of centre). But they pick the forbidden fruit (right of centre, middle) and have to cover their nakedness with their hands (lower centre) as God reproaches them before they are expelled from the garden by an angel (bottom right). Outside the towered walls of paradise, Adam digs with a hoe and Eve holds a distaff. At the top of the picture, on the right, we see their children, Cain and Abel, making their sacrifices to God and, below that, Cain killing his brother. In the middle of paradise is the fountain of life, protected by an ornate Gothic structure.

Fig. 4.3. Central part of the apsidal mosaic in San Clemente, Rome. Photo: Alinari, Florence. On the arch is the symbol with the first two letters of the name of Christ in Greek and the first and last letters of the Greek alphabet. Below the hand of the Father plants a triumphal crown on a Crucifix decorated by twelve white doves, symbols of the apostles, on either side of which stand the Virgin Mary and Saint John, the embryonic Church. The Cross is the new Tree of Life. From its base issue the four Edenic rivers, which irrigate a luxuriant acanthus, symbol of the pastures of the faithful. In the inscription, the explanation of the meaning of the mosaic is interspersed with a reference to relics of the Cross and of Saint James and Saint Ignatius of Antioch incorporated inside the body of Christ in the mosaic itself when it was made. Below the inscription, Christ and the apostles are represented in the form of lambs.

Locating Paradise in Time 73

Fig. 4.4a. Paradise in the *Speculum virginum*. Southern Germany, probably after 1140. 27 × 18 cm. London, British Library, Arundel MS 44, fol. 13r. The *Speculum virginum* (*The Mirror of Virgins*) is a twelfth-century devotional book in the form of a dialogue. The protagonists are a nun and a Benedictine monk, who explains that Scripture is a mirror from which virgins may learn what to do in order to please their eternal spouse. In the central medallion, Christ holds an open book. Flowing from Christ are the four rivers of paradise, arranged in the form of a cross, with representations of the four virtues, the eight beatitudes (small figures below the virtues), the four evangelists, represented by their symbols (eagle, lion, angel, ox), and the four Doctors of the Church.

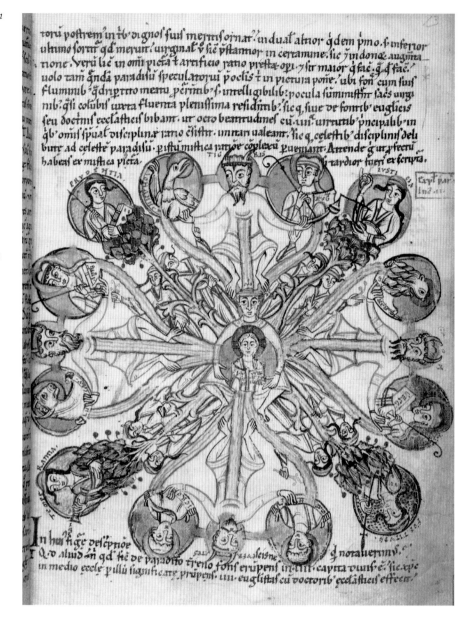

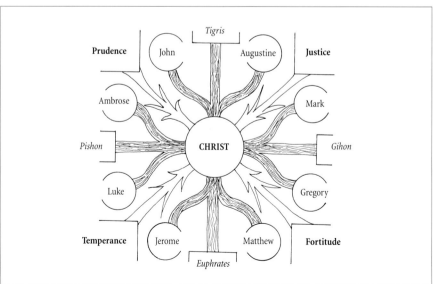

Fig. 4.4b. Diagram of Fig. 4.4a.

the four rivers, or a detail of church architecture such as a portal (Figure 4.2.). Or the Church could be portrayed metaphorically as a vineyard watered by the four rivers of Eden, as in the twelfth-century apsidal mosaic in San Clemente, Rome (Figure 4.3). Against a dense tapestry of tendrils, the mosaic shows how God allowed paradise to regerminate in human history through the sacrifice of Christ and the life of his Church.[62] Scenes of the Expulsion from Eden could contain references to Christ by including a cross, as in a mosaic in San Marco, Venice, where it appears on the flaming sword of the Cherubim that guard Eden.[63] A scene of the Crucifixion could contain a symbol of mankind's delivery from Adam's original sin. The placing of Adam's skull, spattered with the blood of Christ, at the foot of the Cross on Golgotha (itself 'the place of the skull', as well as Adam's burial place) signified the washing away of human sin. The proximity to the Cross of a serpent with some sort of fruit in its mouth was a direct allusion to the Fall. In some Crucifixion scenes the figure of Adam is shown emerging from his tomb below the Cross; in others, the Cross is portrayed as the new Tree of Life, from the base of which flow the four rivers of Eden. Particularly common from the Palaeo-Christian to the Gothic period was the way the layout of a church building and its architectural details were made to demonstrate the association of the Church in heaven with the Church on earth, an association confirmed by theological and liturgical sources.[64] Also common, in depictions of the celestial paradise inspired by the Book of Revelation, was the addition to the representation of the four rivers, to point to the link between heaven and Eden, as on the front of the so-called Sarcophagus of Constantius III in the Mausoleum of Galla Placidia, Ravenna.[65]

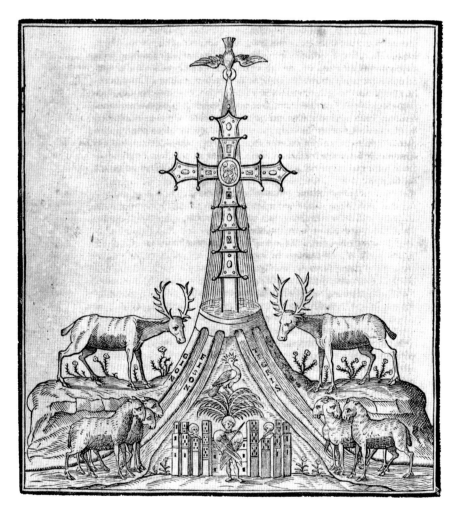

Fig. 4.5. Seventeenth-century engraving of the central part of a late thirteenth-century mosaic in San Giovanni in Laterano, Rome, from Giacomo Bosio, *Crux triumphans et gloriosa* (Antwerp: ex officina Plantiniana apud Balthasarem et Ioannem Moretos: 1617), p. 622. 21.5 × 19.5 cm. London, British Library, 471.g.2. The mosaic was modelled on an earlier one, dating from the fifth century, in the apse, which was rebuilt in 1884. The Holy Spirit, in the form of a dove, descends upon the Cross of Christ, out of which flow the four rivers of Eden. Below the Cross, in the midst of the heavenly city (or the earthly paradise), guarded by the Archangel Michael (or the Cherub with flaming sword), is the Tree of Life, surmounted by a phoenix, a symbol of Christ's Resurrection. The two deer, yearning for running streams, represent the faithful thirsting for the living God (Psalm 42). The six lambs stand for the twelve apostles.

The complex intermingling of the various motifs is demonstrated in Figures 4.4 and 4.5. Figure 4.4a, which comes from a twelfth-century Latin manuscript entitled *Speculum virginum* (*The Mirror of Virgins*), represents the internal and spiritual paradise.[66] A figure in the centre, which could be the Virgin Mary, the virginal nun of the book, or the Church personified, embraces a medallion from which flow the four named rivers of paradise arranged in the form of a cross. Each river is represented by a horned figure holding a pair of discs on which are depicted the evangelists and the Church Fathers. Four other medallions, each on a floral stem and flanked by two beatitudes, contain a representation of a cardinal virtue. The figure in the central medallion is Christ, who holds an open book on which is written *Si quis sitit, veniat et bibat* ('If any man thirst, let him come unto me, and drink', John 7.37) (see also Figure 4.4b). The message of the composition is that Christ pours forth – like the source of the four rivers in Eden – the four evangelists and the four major Church Fathers to irrigate the world. By drinking from the spring of the gospels and the teachings of the Church, and by following the cardinal virtues and the beatitudes, the human heart will bear the fruit of spiritual happiness.[67] Here, the paradise of the allegorical meaning (the Church) overlaps with the moral paradise (enjoyed by the faithful) just as the literally geographical rivers become the cardinal virtues of the moral sense. Figure 4.5 shows a detail from a seventeenth-century re-drawing, by the antiquarian Giacomo Bosio, of the cross in the thirteenth-century apsidal mosaic in San Giovanni in Laterano, Rome.[68] Richly bejewelled, the cross (*crux gemmata*) symbolizes the Second Coming of Christ. From the base of the cross flow the four rivers of Eden, each named, in recognition of the association in medieval exegesis of the four rivers of paradise with the waters of baptism and the four gospels. The Baptism of Christ is portrayed on the intersection of the arms of the cross. The dove that appeared at the moment of Christ's baptism is shown at the head of the cross.[69] Two deer drink from the rivers and six lambs approach a fortified city lying at the base of the cross. The city, guarded by an angel and inhabited by two figures and a phoenix on a palm tree, refers either to the Garden of Eden or to the Heavenly Jerusalem, or to both. The two unnamed figures could be Enoch and Elijah or Peter and Paul. The whole composition can be seen as a symbol of the Church ruled by Christ, and as the 'true image of the heavenly city and the celestial Jerusalem, vivid spiritual portrait of the earthly paradise'.[70]

Displaying History from Eden to Heaven

Chrysostom's smallest dot was indeed seen as containing many hidden treasures, and a single image could point to any of several paradises. Nevertheless, the temporal relationship between Eden, the Church in history and the Church in heaven – notions closely connected in art and exegesis – had to be explained to the faithful in the clearest possible way. Peter of Poitiers was a devoted pupil of Peter Lombard (and later chancellor of the chapter of Notre Dame of Paris) and the compiler some time between 1167 and 1170 of a successful abridgement of biblical history in which the linear unfolding of time from its origin in the Garden of Eden to the establishment of the Christian Church was portrayed as a genealogical tree reaching from Adam to Christ and covering the entire span of biblical events (Figures 4.6, 4.7, 4.8).[71] Peter of Poitiers's arrangement was probably inspired by the genealogy listed in the Gospel of Matthew (1.1–17). To Matthew's genealogy, which proceeded from Abraham to Christ in three successive sets each of fourteen generations, Peter of Poitiers inserted a period before Abraham to include the original paradise, with Adam in Eden, and another period following the birth of Jesus of Nazareth to include the paradise of the Church, with the mission of the Holy Spirit announcing the coming of Christ at the end of time.

Peter had been born at Poitiers, or in the Poitou, France, about 1130. His theological teaching career began in 1167. Two years later, he succeeded to Peter Comestor's chair of theology at the University of Paris at a time when Paris was the leading theological centre in Europe. He had warmly supported the introduction into the curriculum of Aristotle's logical works, which favoured a new dialectical and rational approach to theological questions that went back to the Church Fathers. Peter was also a keen student and teacher of biblical history. Like other medieval theologians and exegetes, he appreciated that history had to be the foundation of scriptural study, and his genealogical teaching device, painted on vellum, hung on the wall of his classroom.[72] The continuing importance of his abridgement of biblical history, the *Compendium historiae in genealogia Christi*, is shown by its widespread diffusion, evidenced today in the large number of surviving copies that date from the twelfth to the fifteenth century. In most of these copies, Peter's straightforward presentation contains numerous interpolations from major historical works, such as Peter Comestor's *Historia scholastica*, and from older texts such as those of Flavius Josephus, the Church Fathers and later exegetes. In turn, Peter's genealogy of Christ was incorporated into various universal chronicles, such as the summary of biblical history compiled in the mid fourteenth century by John of Udine, and the anonymous *Rudimentum noviciorum* (*Elementary Book for Beginners*),

Fig. 4.6. Adam and Eve, from a roll-version of Peter of Poitier's *Compendium historiae in genealogia Christi*. ?St Albans, early fourteenth century. 22 × 15 cm. London, British Library, Royal Roll 14 B IX. At the top, Adam and Eve, depicted at the moment of the Fall, are shown eating the forbidden fruit as the snake, with the head of a female, stares at them with satisfaction. Lower down, large vignettes feature the most important figures and events of biblical history, such as Noah's Ark and the Flood, Abraham's sacrifice of his son Isaac, and the Babylonian captivity.

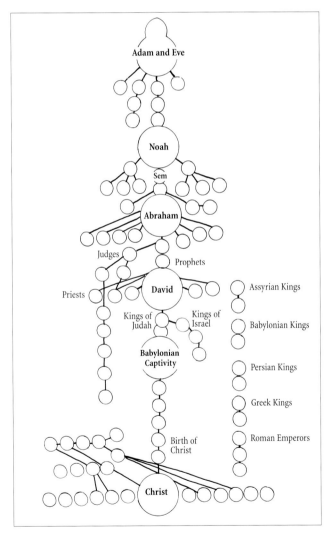

Fig. 4.7. Diagram showing the time scheme of Peter of Poitier's *Compendium historiae in genealogia Christi*. The genealogical tree starts with Adam, continues with Noah, and culminates with Christ. In the scroll the biblical and secular sequences run parallel.

Locating Paradise in Time 77

printed in 1475 in Lübeck, Northern Germany, by Lucas Brandis.[73] At the end of the sixteenth century, the Protestant Ulrich Zwingli had a version printed in Basle that ended with Saint Paul's martyrdom.[74]

In the prologue, Peter of Poitiers announced that he had accepted the task of compiling a single-pamphlet summary of biblical history as an aide-mémoire to the long and complex narrative contained in the Bible. The genealogical tree of Christ began with Adam and accounted for the lives and deeds of the patriarchs, the judges, and the kings of Israel and Judah, together with their prophets and priests.[75] All historical figures were enumerated in chronological order and a short biographical note for each was included in the text, together with their life span. From the note on Adam, for example, we learn that Adam was created in the Damascene Fields (*Ager Damascenus*), near Hebron, and only later placed in the Garden of Eden (from which he was expelled back to the Damascene Fields as soon as he had sinned), and that he had many children and lived 930 years.[76] After Adam, Peter continued the genealogy to Noah, his sons, and the patriarchs, thus providing a framework that later historians would embellish. Peter's *Compendium* also provided biographical notes on certain individuals mentioned in the Bible who were not amongst the ancestors of Christ, and gave an account of each of the ancient nations surrounding the Jewish people. Like his predecessors, Peter followed the medieval Christianization of Jewish ethnography, which was structured according

Fig. 4.8. The last stage of the genealogical tree of Christ, from a roll-version of Peter of Poitier's *Compendium historiae in genealogia Christi*. ?St Albans, early fourteenth century. 13.5 × 33 cm. London, British Library, Royal Roll 14 B IX. Scenes of the Nativity, the Crucifixion, and the Resurrection of Christ are followed along the genealogical tree by a scene showing the apostles, whose names are indicated in small circles either sides of the central circle with a portrait of Christ. The twelve are: Andrew, Philip, Bartholomew, James the Great, John the Evangelist, Peter, Thomas, James the Less, Simon, Thaddaeus (also called Jude), Matthew and Matthias (elected in the place of Judas Iscariot). Also shown are Paul and Barnabas, who preached together in Cyprus and Asia Minor (Acts 13–14) and later were both called apostles. The genealogical tree shows that James the Great and John the Evangelist were sons of Zebedee (Matthew 10.2) and Salome (as can be gathered from Matthew 27.55 and Mark 15.40), and that James the Less, Simon and Taddaeus (or Jude) were sons of Alphaeus and Mary of Cleophas, as related in Jacobus de Voragine's *Legenda aurea* (*Golden Legend*). They appear to be all cousins of Jesus. The genealogical tree that had begun with Adam ends with Christ and all his early disciples (not only his cousins), to indicate the Church itself as the final point of biblical history. Below the scene reproduced here, at the end of the roll, a rubric presents the reader with a fourfold structure of world history: the age of deviation, from Adam to Moses; the age of the call, from Moses to Christ; the age of reconciliation, from the birth of the Redeemer to the mission of the Holy Spirit; and the age of pilgrimage, from the sending of the Holy Spirit to the Day of Judgement. The final age, which comprised the Church's continuing pilgrimage on earth, was always a problem for medieval historians, who had to find ways of accounting for the lapse of time between the first coming of Christ and the world that is yet to come.

to genealogies, and conceived of 'nations' as groups of descendants of a single biblical progenitor.[77] Also listed were the descendants of each of Noah's sons – Sem, Cham and Japhet – and the countries in which they lived. Peter's main focus was on the descendants of Sem, since it was from these that came the progenitors of Christ, but he also provided information about Cham's and Japhet's gentile descendants.[78] In some versions of the *Compendium*, a reference is found to the division of world history into six ages, together with the explanation that the division was not mathematical, but structured according to important characters and events: the Creation, the Flood, Abraham, King David, the exile to Babylon, the Incarnation of Christ, and the Final Judgement.[79]

Although Peter's genealogy ran from Adam to Christ, Christ was seen as the beginning and the end of all history. It is significant that in an early fourteenth-century roll of the *Compendium*, pictures of Christ and of a seven-branched candelabrum precede the image of Adam and Eve and are accompanied by a text in which the cosmic role of the Incarnate God is emphasized.[80] The rubric explains that the seven branches of the candelabra symbolize the seven gifts of the Holy Spirit, the seven spirits before the Throne of God, the seven stars held by Christ and the seven messengers sent to the seven churches of Asia as described in Revelation (1.4–20). The rubric also explains that the candelabra provides an image of various groups of people within the Church – the religious, the celibate, and the married – and confirms that the Son of God

Locating Paradise in Time 79

is the Alpha and the Omega of the universe, the one who was, is, and will be. The linear path of human history, as traced by the generations presented in Peter's diagram, stretched *ab initio mundi usque ad diem iudicii* ('from the beginning of the world up to the Day of Judgement'). At the end of the roll, where the genealogical tree arrives at the life of Christ and his apostles, which marks the birth of the Church through the descent of the Holy Spirit, it is explained that Christ has been God since the beginning.

Peter of Poitiers chose to display the unfolding of human history in diagrammatic form. He had already attached to his historical scheme explanatory plans and maps, for example on the division of the earth amongst Noah's sons, the division of Canaan amongst the Twelve Tribes of Israel, the structure of the Ark and the tabernacle, and the city of Jerusalem. The entire succession of events that made up world history could equally well have been exhibited geographically on the stage of the whole inhabited earth. Medieval world maps, in which spatial and temporal dimensions were combined, provided precisely such a geographical framework. It was the unique ability of the map to accommodate both dimensions that allowed their compilers to situate the earthly paradise in time and space, and it was not by accident that a world map featuring paradise was included in a number of chronicles that obviously owe much to Peter of Poitier's *Compendium*.[81] As the reader may already have surmised, far from being the naïve depiction of a picturesque fantasy land, the incorporation of paradise on medieval world maps epitomized a vital element of Christian doctrine. Paradise was located spatially by means of the literal and historical reading of Genesis, and situated temporally by means of the typological approach that framed its role within salvation history. Armed with an understanding of the significance of both space and time in the Christian view of the paradise narrative, this is the point at which to embark on an exploration of the medieval mapping of paradise.

---

1   Augustine, *De musica*, VI.11.29, ed. by Giovanni Marzi (Florence: Sansoni, 1969), pp. 574–7; see also Christopher Ligota, 'La Foi historienne: Histoire et connaissance de l'histoire chez S. Augustin', *Revue des Études Augustiniennes*, 43 (1997), pp. 133–4.

2   Translation slightly modified from Augustine, *The Literal Meaning of Genesis*, I.1.1, transl. and annotated by John H. Taylor, 2 vols (New York–Ramsey, New Jersey: Newman Press, 1982), I, p. 19; *De Genesi ad litteram libri duodecim*, I.1, ed. by Joseph Zycha, *CSEL* XXVIII/1 (Prague–Vienna–Leipzig: F. Tempsky–G. Freytag, 1894), p. 3: 'Nam non esse accipienda figuraliter nullus christianus dicere audebit, adtendens apostolum dicentem: *omnia autem haec in figura contingebant in illis*.' See 1 Corinthians 10.11: 'Now all these things happened unto them for ensamples: and they are written for our admonition, upon whom the ends of the world are come.'

3   By way of an introduction to the extensive literature on Christian typology see Anthony T. Hanson, *The Living Utterances of God: The New Testament Exegesis of the Old* (London: Darton, Longman and Todd, 1983); Leonhard Goppelt, *Typos: Die typologische Deutung des Alten Testaments im Neuen* (Darmstadt: Wissenschaftliche Buchgesellschaft, 1969); Henri de Lubac, *Exégèse médiévale: Les Quatre Sens de l'Écriture*, 4 vols (Paris: Aubier, 1959–64); Pierre Grelot, *Sens chrétien de l'Ancien Testament: Esquisse d'un traité dogmatique*, 3rd edn (Tournai: Desclée, 1962); Geoffrey W. H. Lampe and Kenneth J. Woollcombe, *Essays on Typology* (London: SCM Press, 1957); Jean Daniélou, *Sacramentum futuri: Études sur les origines de la typologie biblique* (Paris: Études de Théologie historique, 1950), English transl. by Wulstan Hibberd, *From Shadows to Reality: Studies in the Biblical Typology of the Fathers* (London: Burns & Oates, 1960).

4   Augustine, *The Literal Meaning of Genesis*, I.1.1 (1982), I, p. 19; *De Genesi ad litteram*, I.1 (1894), pp. 3–4: 'et illud, quod in Genesi scriptum est: *et erunt duo in carne una*, magnum sacramentum commendantem in Christo et in ecclesia.' Exegetes from the Antioch school had developed a typological reading before Augustine: see Jean-Noël Guinot, 'La Typologie comme technique herméneutique', in *Figures de l'Ancien Testament chez les Pères* (Strasbourg: Centre d'Analyse et de Documentation Patristiques, 1989), pp. 1–34.

5   Augustine, *Contra Faustum*, XII.27, ed. by Joseph Zycha, *CSEL* VI/1 (Prague–Vienna–Leipzig: F. Tempsky–G. Freytag, 1891), p. 356: 'Christus mihi ubique illorum librorum.' 'I find Christ everywhere in those books.'

6   Augustine, *De civitate Dei*, XVI.2, ed. by Bernard Dombart and Alfons Kalb, *CCSL* XLVIII (Turnhout: Brepols, 1955), p. 500; see also Robert A. Markus, *Saeculum: History and Society in the Theology of Saint Augustine*, 2nd edn (Cambridge–New York: Cambridge University Press, 1988), pp. 187–96.

7   Augustine, *The Literal Meaning of Genesis*, XII.28.56 (1982), II, p. 220; *De Genesi ad litteram*, XII.28 (1894), p. 423: 'quamquam diligentius considerantibus fortassis occurrat illo paradiso corporali, in quo Adam corporaliter fuit, et istam vitam sanctorum significatam, quae nunc agitur in ecclesia, et illam, quae post hanc erit in aeternum'.

8   Christian millenarians, anticipating a thousand-

9 year period of blessedness on earth, have a different opinion, but in the context of the present book, only Augustine's vision of history, which became the accepted view in the Latin Church, is discussed.
9 See below, Chapter 6, pp. 151–2.
10 Augustine, *De Genesi ad litteram*, XII.1–14 (1894), pp. 379–400. See also Reinhold R. Grimm, *Paradisus coelestis, paradisus terrestris: Zur Auslegungsgeschichte des Paradieses im Abenland bis um 1200* (Munich: Fink, 1977), pp. 67–71.
11 The surviving fragments of Sextus Iulius Africanus's *Chronographia* are published in Martin J. Routh, *Reliquiae sacrae*, 2nd edn, 3 vols (Oxford: Oxford University Press, 1846), II, pp. 238–309, and *PL* X, cols 65–93. See Osvalda Andrei, 'L'esamerone cosmico e le Cronografie di Giulio Africano', in *La narrativa cristiana antica: Codici narrativi, strutture formali, schemi retorici* (Rome: Institutum Patristicum Augustinianum, 1995), pp. 169–83; Andrei, 'La formazione di un modulo storiografico cristiano: Dall'esamerone cosmico alle *Chronographiae* di Giulio Africano', *Aevum*, 69 (1995), pp. 147–70; Heinrich K. G. Gelzer, *Sextus Julius Africanus und die byzantinische Chronographie*, 2 vols (Leipzig: Hinrichs and Teubner, 1880–5, 1898; repr. Hildesheim: Gerstenberg, 1978). On Christian chronography in general, see William Adler, *Time Immemorial: Archaic History and its Sources in Christian Chronography from Julius Africanus to George Syncellus* (Washington, DC: Dumbarton Oaks Research Library and Collection, 1989), pp. 15–71.
12 The tradition of the duration of the universe as six thousand years had already been attested by the second-century *Epistle of Barnabas*: see *Épitre de Barnabé*, ed. by Pierre Prigent and Robert A. Kraft, *SC* CLXXII (Paris: Cerf, 1971), XV.4–5, p. 185 and n. 3; *Der Barnabasbrief*, ed. by Ferdinand R. Prostmeier (Göttingen: Vandenhoeck und Ruprecht, 1999), pp. 486–92.
13 Hippolytus, *Commentary on Daniel*, IV.23–4 (Daniel II.3–7), ed. by Georg Nathanael Bonwetsch, *GCS*, Neue Folge, VII (Berlin: Akademie Verlag, 2000), pp. 244–9; see also Auguste Luneau, *L'Histoire du salut chez les Pères de l'Église: La Doctrine des âges du monde* (Paris: Beauchesne, 1964), pp. 209–17; Jean Sirinelli, *Les Vues historiques d'Eusèbe de Césarée durant la période prénicéenne* (Dakar: Université de Dakar, 1961), p. 39.
14 Augustine, *De civitate Dei*, XVIII.52–3; XX.7 (1955), pp. 650–3, 708–12.
15 Bede, *De temporum ratione*, LXVI.9, ed. by Charles W. Jones, *CCSL* CXXIIIB (Turnhout: Brepols, 1977), pp. 464–5, 495. See also Bernard Guenée, *Histoire et culture historique dans l'Occident médiéval* (Paris: Aubier Montaigne, 1980; repr. 1991), pp. 148–54, and Ernst Breisach, *Historiography: Ancient, Medieval and Modern* (Chicago–London: University of Chicago Press, 1983), p. 92.
16 Breisach, op. cit., pp. 131–2.
17 Sextus Iulius Africanus had used the expression 'from Adam': *Chronographia*, *PL* X, col. 67. Gelzer, *Sextus Julius Africanus* (1880–5), I, p. 35; Sirinelli, *Les Vues historiques d'Eusèbe* (1961), p. 42.
18 Eusebius, *Chronicorum libri duo*, I.16.3–6, II.5–7, *PG* XIX, cols 146, 323–4; Adler, *Time Immemorial* (1989), pp. 41–50; Sirinelli, *Les Vues historiques d'Eusèbe* (1961), pp. 44–6.
19 On the Judaic tradition see Louis Ginzberg, *The Legends of the Jews*, transl. by Henrietta Szold, 7 vols (Philadelphia: Jewish Publication Society of America, 1946–7), I (1947), p. 82 and V (1947), pp. 106–7 and n. 97. For an overview of the apocryphal material, see Michael E. Stone, *Armenian Apocrypha Relating to Adam and Eve* (Leiden–New York–Cologne: Brill, 1996); Brian O. Murdoch, 'Commentary', in David Greene and Fergus Kelly, eds, *The Irish Adam and Eve Story from Saltair na Rann*, 2 vols (Dublin: Institute for Advanced Studies, 1976), II, pp. 16–25, 70–1; Arturo Graf, *Miti, leggende e superstizioni del Medio Evo*, 2 vols (Turin: Loescher, 1892–3), I (1892), pp. 53–5.
20 Augustine, *De civitate Dei*, XII.11, 13 (1955), pp. 365–8.
21 Ibid., XIV.26 (1955), pp. 449–50; Augustine, *De Genesi ad litteram*, IX.4 (1894), pp. 272–3.
22 Irenaeus, *Adversus haereses*, V.23.2, ed. by Adelin Rousseau, Louis Doutreleau, and others, 5 vols, *SC* CCLXIII, CCLXIV, CCXCIII, CCX and CCXI (2nd edn), C, CLII, CLIII (Paris: Cerf, 1952–74), V/2, *SC* CLIII, ed. by Rousseau, Doutreleau, and Charles Mercier (Paris: Cerf, 1969), pp. 290–5.
23 Victorinus of Pettau, *De fabrica mundi*, IX, in *Sur l'Apocalypse et autres écrits*, ed. by Martine Dulaey, *SC* CCCXXIII (Paris: Cerf, 1997), pp. 146–9.
24 Augustine, *De Trinitate*, IV.7, 10, ed. by William J. Mountain and François Glorie, *CCSL* L (Turnhout: Brepols, 1968), pp. 169, 173–4; Augustine, *De civitate Dei*, XI.30 (1955), p. 350.
25 Augustine, *In Iohannis Evangelium tractatus*, XV.8, ed. by Augustinus Mayer, *CCSL* XXXVI (Turnhout: Brepols, 1954), p. 153.
26 Augustine, *De Trinitate*, V.9 (1968), pp. 172–3.
27 Columella, *De re rustica*, IX.14.1 and Pliny, *Naturalis historia*, XVIII.246 put the spring equinox on 25 March.
28 Bede, *De temporum ratione*, LXVI.9 (1977), pp. 464–5.
29 Bede, *In Lucam*, VI.23.44–5, ed. by David Hurst, *CCSL* CXX (Turnhout: Brepols, 1960), p. 406.
30 Honorius Augustodunensis, *Imago mundi*, III.1, ed. by Valerie I. J. Flint, *Archives d'Histoire Doctrinale et Littéraire du Moyen Âge*, 49 (1982), p. 124; Peter Comestor, *Historia scholastica*, *Liber Genesis* XXIV, *PL* CXCVIII, col. 1075; Dante, *Paradiso* XXVI.139–42. On Dante and the Garden of Eden see below, Chapter 7, pp. 182–3.
31 Vincent of Beauvais, *Speculum maius*, I.56, 4 vols (Douai: Baltazar Bellerus, 1624; repr. Graz: Akademische Drück und Verlagsanstalt, 1965), IV, p. 22.
32 Christian of Stavelot, *Expositio in Matthaeum*, 56, *PL* CVI, cols 1460–2. See David C. Van Meter, 'Christian of Stavelot on Matthew 24:42, and the Tradition that the World will End on a March 25th', *Recherches de Théologie Ancienne et Médiévale*, 63 (1996), pp. 68–92.
33 Augustine, *De Genesis contra Manichaeos*, I.23.35–24.42, ed by Dorothea Weber, *CSEL* XCI (Vienna: Verlag der Österreichischen Akademie der Wissenschaften, 1998), pp. 104–12. See Markus, *Saeculum*, pp. 17–21. Also Revelation 5.1 suggests six periods between the opening of the first and the seventh seal.
34 Luneau, *L'Histoire du salut* (1964), pp. 37–52; 285–331.
35 Augustine, *De doctrina christiana*, II.27.41, 42, 44, ed. by Roger P. H. Green (Oxford: Clarendon Press, 1995), pp. 104–7.
36 Augustine, *De civitate Dei*, XX.7 (1955), pp. 708–12. On numbers and Augustine see Luneau, *L'Histoire du salut* (1964), pp. 333–56.
37 Augustine used different starting points in his attempts to work out the Second Coming of Christ, such as the date of Christ's birth, preaching, and baptism but these depended on the context and were only minor differences. See Luneau, *L'Histoire du salut* (1964), p. 314.
38 Augustine, *In Iohannis Evangelium tractatus*, XV.9 (1954), pp. 153–4.
39 Augustine also created a tripartite division of human history: the age before the Law, under the Law and under God's glory. See Breisach, *Historiography* (1983), pp. 84–6.
40 Moses was occasionally substituted for David to mark the passage from the third to the fourth age.
41 Isidore, *Chronica*, in *Chronica minora saec. IV.V.VI.VII*, ed. by Theodor Mommsen, *MGH AA*, XI (Berlin: Weidmann, 1893), pp. 391–497; *Chronicon*, *PL* LXXXIII, cols 1019–53; *Allegoriae quaedam Scripturae Sacrae*, III, *PL* LXXXIII, col. 99.

42 Isidore, *Etymologiae*, V.38–9, ed. by Wallace Martin Lindsay, 2 vols (Oxford: Clarendon Press, 1911).
43 Bede, *De temporibus liber*, XVI–XXII, ed. by Charles W. Jones, CCSL CXXIIIC (Turnhout: Brepols, 1980), pp. 600–11; Bede, *De temporum ratione*, LXVI, 1–8, 268 (1977), pp. 463–4, 495. See also Jan Davidse, 'On Bede as Christian Historian', in L. A. J. R. Houwen and Alasdair A. MacDonald, eds, *Beda Venerabilis: Historian, Monk and Northumbrian* (Groningen: Egbert Forsten, 1996), pp. 1–15; Peter Hunter Blair, *The World of Bede* (London: Secker and Warburg, 1970), pp. 259–71.
44 Already in Judaism the Garden of Eden was associated with the Temple of Solomon and the Garden of God in Jerusalem: Lawrence E. Stager, 'Jerusalem ad the Garden of Eden', *Eretz-Israel: Archaeological, Historical and Geographical Studies*, 26 (1999), pp. 183–94.
45 See, for example, Willemien Otten, *From Paradise to Paradigm: A Study of Twelfth-Century Humanism* (Leiden: Brill, 2004), pp. 15–26, 222–31, 256–9, 274–7; H. S. Benjamins, 'Paradisiacal Life: The Story of Paradise in the Early Church' and Christoph Auffarth, 'Paradise Now – But for the Wall Between: Some Remarks on Paradise in the Middle Ages', in Gerard P. Luttikhuizen, ed., *Paradise Interpreted: Interpretations of Biblical Paradise: Judaism and Christianity* (Leiden-Boston, MA: Brill, 1999), pp. 153–79; Jean Daniélou, 'Terre et paradis chez les Pères de l'Église', *Eranos-Jahrbuch*, 23 (1953), pp. 433–72.
46 For the tradition of Adam's burial at Calvary see, for example, Honorius Augustodunensis, *Imago mundi*, III.1 (1982), p. 124; Ambrose, *Expositio in Lucam* X.114, ed. by Mark Adriaen, CCSL XIV (Turnhout: Brepols, 1957), p. 37. See also Hans Martin von Erffa, *Ikonologie der Genesis: Die Christlichen Bildthemen aus dem Alten Testament und ihre Quellen*, 2 vols (Munich: Deutscher Kunstverlag, 1989–95), I (1989), pp. 408–13, and bibliography quoted above, Chapter 3, p. 61 nn. 66–8.
47 See Bernard McGinn, *Apocalyptic Spirituality* (New York: Paulist Press, 1979), p. 76.
48 Beryl Smalley, *The Study of the Bible in the Middle Ages*, 3rd edn (Oxford: Blackwell, 1984), p. 2.
49 Gregory the Great, *Epistola ad Leandrum*, 4, in *Moralia in Iob*, ed. by Mark Adriaen, CCSL CXLIIIA (Turnhout: Brepols, 1979), pp. 5–6; Peter Comestor, *Historia scholastica*, Introd., PL CXCVIII, cols 1053–4; Hugh of St Victor, *Didascalicon*, VI.2–3, ed. by Charles H. Buttimer (Washington, DC: Catholic University Press, 1939), p. 114 (PL CLXXVI, cols 799–801); Nicholas of Lyre, *Postilla super totam Bibliam*, 'Prologus secundus', in *Biblia sacra cum glossa ordinaria*, ed. by Leander de Sancto Martino, Ioannes Gallemart, and others, 6 vols (Douai: Baltazar Bellerus; Antwerp: Ioannes Keerbergius, 1617), I, sig. +4r F- +4v B.
50 See Rowan A. Greer and James L. Kugel, *Early Biblical Interpretation* (Philadelphia: Westminster Press, 1986), pp. 178–84.
51 Jerome, *Commentarium in Hiezechielem*, XLVII.12, ed. by François Glorie, CCSL LXXV (Turnhout: Brepols, 1964), p. 718. See Lubac, *Exégèse médiévale* (1959–64), II (1959), p. 440 n. 8.
52 Jean Pépin, *La Tradition de l'allégorie de Philon d'Alexandrie à Dante: Études historiques* (Paris: Études Augustiniennes, 1987), pp. 94–5; Ceslaus Spicq, *Esquisse d'une histoire de l'exégèse latine au Moyen Âge* (Paris: Vrin, 1944), p. 19.
53 On John Cassian see Lubac, *Exégèse médiévale* (1959–64), I (1959), pp. 190–8; Augustine, *De Genesi ad litteram*, 1.1 (1894), p. 3; *Biblia latina cum Glossa ordinaria*, facsimile of the *editio princeps* ... (Strasbourg: Adolph Rusch 1480/1), ed. by Margaret T. Gibson and Karlfried Fröhlich, 2 vols (Turnhout: Brepols, 1992), I, sig. a2v; Thomas Aquinas, *Summa theologiae*, 1a, q.1, a.10, Leonine edn, ed. by Pietro Caramello, 3 vols (Turin–Rome: Marietti, 1948–50), I (1950), p. 9; Nicholas of Lyre, *Postilla super totam Bibliam*, 'Prologus primus' in *Biblia sacra cum glossa ordinaria* (1617), I, sig. +4r C–E. See Alastair J. Minnis and A. Brian Scott, eds, *Medieval Literary Theory and Criticism, c.1100–c.1375: The Commentary Tradition* (Oxford–New York: Oxford University Press, 1988), p. 203. The term 'anagogical' derives from the Greek, ἀναγωγή (*anagoge*), meaning 'elevation'.
54 See Jerome, *Hebraicae quaestiones in libro Geneseos*, I.1, ed. by Paul de Lagarde, in *Opera*, I/1, CCSL LXXII (Turnhout: Brepols, 1959), p. 3. See also Rabanus Maurus, *De universo*, XII.3 PL CXI, cols 334–5; Isidore, *Mysticorum expositiones sacramentorum seu quaestiones in Vetus Testamentum*, III.2, PL LXXXIII, col. 216; *Biblia latina cum Glossa ordinaria* (1480/1; 1992), I, p. 21.
55 London, British Library, Add. MS 28106, fol. 6r. An established reading of the gospel of Matthew associated the parable of the kingdom of God with the various stages of human history: see, for example, Jerome, *In Matthaeum*, XX.1–2, ed. by Émile Bonnard, 2 vols SC CCXLII, CCLIX (Paris: Cerf, 1977–9), II, SC CCLIX (1979), pp. 84–9.
56 John Chrysostom, *In Genesim*, XVIII.4, PG LIII, col. 154: 'Scriptura sacra nihil frustra dicit'; 'at apiculus unicus reconditum habet thesaurum'.
57 The earliest extant images of Adam and Eve in Eden date from the third century: Helga Kaiser-Minn, *Die Erschaffung des Menschen auf den spätantiken Monumenten des 3. und 4. Jahrhunderts* (Münster in Westphalia: Aschendorff, 1981), pp. 2–31; *Lexikon der christlichen Ikonographie*, ed. by Egelbert Kirschbaum, 8 vols (Rome–Freiburg im Breisgau–Basle-Vienna: Herder, 1968), I, cols 42–70.
58 Walter Lowrie, *Art in the Early Church* (New York: Pantheon Books, 1947), p. 78.
59 André Grabar, *Christian Iconography: A Study of its Origin* (Princeton, NJ: Princeton University Press, 1968), pp. 12–13. See also Elizabeth Struthers Malbon, *The Iconography of the Sarcophagus of Junius Bassus* (Princeton, NJ: Princeton University Press, 1990).
60 Wladimir Weidlé, *The Baptism of Art: Notes on the Religion of the Catacomb Paintings* (Westminster: Dacre Press, 1950), p. 18 n. 1.
61 See, for instance, Jonathan J. G. Alexander, '"Jerusalem the Golden": Image and Myth in the Middle Ages in Western Europe'; Peter Low, 'The City Refigured: A Pentecostal Jerusalem in the San Paolo Bible'; Wendy Pullan, 'Jerusalem from Alpha to Omega in the Santa Pudenziana Mosaic', in Bianca Kühnel, ed., *The Real and Ideal Jerusalem in Jewish, Christian and Islamic Art: Studies in Honour of Bezalel Narkiss on the Occasion of his Seventieth Birthday* (Jerusalem: Center for Jewish Art, The Hebrew University of Jerusalem, 1998), pp. 255–64; 265–74; 405–17; Bianca Kühnel, *From the Earthly to the Heavenly Jerusalem: Representations of the Holy City in Christian Art of the First Millennium* (Rome–Freiburg im Breisgau–Vienna: Herder, 1987); William A. McClung, *The Architecture of Paradise: Survivals of Eden and Jerusalem* (Berkeley: University of California Press, 1983); on images of Eden in art: von Erffa, *Ikonologie der Genesis* (1989–95), I, pp. 150–248, esp. 161–71; Jennifer O'Reilly, 'The Trees of Eden in Medieval Iconography', in Paul Morris and Deborah Sawyer, eds, *A Walk in the Garden: Biblical, Iconographical and Literary Images of Eden*, (Sheffield: Journal for the Study of the Old Testament Press, 1992), pp. 167–204; Joseph B. Trapp, 'The Iconography of the Fall of Man', in C. A. Patrides, ed., *Approaches to Paradise Lost* (London: Edward Arnold, 1968), pp. 223–65.
62 The Latin inscription reads: 'We liken the Church of Christ to this vine..., which the Law causes to wither, but the Cross makes it flourish with green.' See Mary Stroll, 'The Twelfth-Century Apse Mosaic in San Clemente in Rome and its Enigmatic Inscription', *Storia e Civiltà*, 4/1–2

(1988), p. 6: 'Ecclesiam Christi viti similabimus isti ... quam lex arentem set crus facit ee virente.'

63 Otto Demus, *The Mosaics of San Marco*, 2 vols (Chicago: University of Chicago Press, 1984), Text II, pp. 144–7; Plates II, no. 32.

64 The relationship between the church as a building and the notion of the Church as the Heavenly Jerusalem had been explored by a number of scholars, for example: Jean Daniélou, *Le Signe du temple ou de la présence de Dieu* (Paris: Desclée, 1942; repr. 1975, 1990); Lawrence H. Stookey, 'The Gothic Cathedral as the Heavenly Jerusalem: Liturgical and Theological Sources', *Gesta*, 8 (1969), pp. 35–41; Alfred Stange, *Basiliken, Kuppelkirchen, Kathedralen: Das himmlische Jerusalem in der Sicht der Jahrhunderte* (Regensburg: F. Pustet, 1964); Stange, *Das frühchristliche Kirchengebäude als Bild des Himmels* (Köln: Comel, 1950); Lothar Kitschelt, *Die frühchristliche Basilika als Darstellung des himmlischen Jerusalem* (Munich: Neuer Filser-Verlag, 1938).

65 The bibliography on the representation of the Heavenly Jerusalem in art is rich. See, for example, Maria Luisa Gatti Perer, ed., *'La dimora di Dio con gli uomini' (Ap 21,3): Immagini della Gerusalemme celeste dal III al XIV secolo* (Milan: Vita e Pensiero, 1983).

66 London, British Library, Arundel MS 44, fol. 13r. Eleonor S. Greenhill, *Die Stellung der Handschrift British Museum Arundel 44 in der Überlieferung des Speculum Virginum* (Munich: Hueber, 1966); Martha Strube, *Die Illustrationen des Speculum Virginum* (Düsseldorf: Nolte, 1937), pp. 11–14; Arthur Watson, 'The *Speculum virginum* with Special Reference to the Tree of Jesse', *Speculum*, 3/4 (1928), pp. 447–9.

67 See also *Speculum virginum*, ed. by Jutta Seyfarth, CCCM V (Turnhout: Brepols, 1990), pp. 41–2.

68 Giacomo Bosio, *Crux triumphans et gloriosa* (Antwerp: ex officina Plantiniana apud Balthasarem et Ioannem Moretos, 1617), p. 622. See also Gatti Perer, ed., *'La dimora di Dio con gli uomini'* (1983), pp. 193–4.

69 Matthew 3.13–17; Mark 1.9–11; Luke 3.21–2; John 1.29–34.

70 Giacomo Bosio, *Crux triumphans et gloriosa* (1617), p. 623: 'Ideoque sub ipsa Cruce, ibi opere musivo in civitatis forma efficta conspicitur sancta Ecclesia, vera supernae civitatis et caelestis Ierosolymae imago, vivusque terrestris Paradisi spiritualis typus.'

71 In 1193, the chancellor was responsible on behalf of the bishop of Paris for the supervision of the official activities of the cathedral's chapter and for the administration of all its higher schools. For the life of Peter of Poitiers, see Philip S. Moore, *The Works of Peter of Poitiers, Master in Theology and Chancellor of Paris (1193–1205)* (Notre Dame, IN: University of Notre Dame Press, 1936), pp. 1–24. Peter of Poitiers, chancellor of Paris, is not to be confused with two other twelfth-century persons of the same name, a monk at Cluny and a regular canon at St Victor, Paris. Besides the *Sentences*, Peter of Poitiers wrote a historical treatise and two works on the allegorical interpretation of Scripture. For the *Sentences*, see *Sententiae*, I, ed. by Philip S. Moore and Marthe Dulong (Notre Dame, IN: University of Notre Dame Press, 1943), pp. VI–VII; see also *PL* CCXI, cols 789–1280. I am grateful to Daniel Connolly for having drawn my attention to Peter of Poitiers's genealogical tree.

72 Peter wrote that without the foundation of history the entire edifice of spiritual interpretation of the Bible is unstable; Alberic of Trois Fontaines attributed to Peter of Poitiers the invention of genealogical trees painted on skins as aids to teach biblical history: see Moore, *The Works of Peter of Poitiers* (1936), p. 7.

73 *Die Historienbible des Johannes von Udine (Ms 1000 Vad)*, ed. by Renate Frohne (Berne: P. Lang, 1992); *Rudimentum noviciorum* (Lübeck: Lucas Brandis, 1475).

74 *Die Historienbibel des Johannes von Udine (Ms 1000 Vad)* (1992); Moore, *The Works of Peter of Poitiers*(1936), pp. 97–117.

75 Hans Vollmer, ed., *Deutsche Bibelauszüge des Mittelalters zum Stammbaum Christi* (Postdam: Akademische Verlagsgesellschaft Athenaion, 1931), pp. 127–8. Vollmer edited the German translation of the *Compendium* from MS Cgm. 564, fol. 99v–128r, of the Bayerische Staatsbibliothek in Munich, and accompanied it with the Latin text of MS Theol. 2029, fols 1r–18v, in the Staats- und Universitätsbibliothek in Hamburg (fourteenth century). Another transcription, from Paris, Bibliothèque Nationale de France, MS Lat. 14435, fol. 136r, is found in Moore, *The Works of Peter of Poitiers*(1936), p. 99.

76 Vollmer, ed., *Deutsche Bibelauszüge* (1931), p. 128. The belief that Adam was created by God in the Damascene Fields was widespread in the Middle Ages and provided another link between the paradise narrative and biblical geography. As Jerome, and after him Peter Comestor and Roger Bacon, pointed out, the Damascene Fields where Adam was created were not in Syria but close to Hebron. According to Honorius Augustodunensis, *Imago mundi*, III.1 (1982), p. 124, Adam was created near Hebron; died, after his expulsion from paradise, in Jerusalem; and was buried on Golgotha. Later his remains were moved back to Hebron to fulfil Genesis 3.19: 'In the sweat of thy face shalt thou eat bread, till thou return unto the ground; for out of it wast thou taken.' See von Erffa, *Ikonologie der Genesis* (1989–95), pp. 81–2.

77 See Hervé Inglebert, *Interpretatio christiana: Les Mutations des savoirs (cosmographie, géographie, ethnographie, histoire) dans l'antiquité chrétienne: 30–630 après J.-C.* (Paris: Institut d'Études Augustiniennes, 2001), pp. 109–92.

78 Vollmer, ed., *Deutsche Bibelauszüge* (1931), p. 128.

79 Ibid., p. 25.

80 London, British Library, Royal MS 14.B.IX.

81 For example, the *Summa de etatibus* by John of Udine (died 1363) and the *Rudimentum noviciorum*, first published by Lucas Brandis at Lübek in 1475; see Vollmer, ed., *Deutsche Bibelauszüge* (1931), pp. 18–31. The maps found in John of Udine's historical chronicle and in the *Rudimentum noviciorum* are discussed in Chapter 8, pp. 204–7, 212–4.

5

Mapping Paradise in Space and Time

Out of bliss these beings are born, in bliss they are sustained,
and to bliss they go and merge again.
Taittiriya Upanishad, III.6.1, (?fifteenth–seventh century BC)

A map that shows paradise as part of the world involves a leap of the imagination. Mapping paradise is simultaneously a confession of belief in a God who operates within terrestrial space and of the limits of human reasoning. Most medieval maps featuring paradise were made in monastic and ecclesiastic scriptoria, and the people who gave visual expression to the Garden of Eden were mostly monks and clerics. This is not to say that their maps of the world should be seen as the result of naïve piety, any more than that they should be idealized as products of a spiritual age. The aim in this book is neither to attack nor to defend the relevant religious world view, but to make a conceptual journey into medieval Christianity in order to answer a specific historical question: what were the conditions that made it possible to represent paradise on maps? In the previous chapters it has been shown how the Book of Genesis was read literally and typologically and how Augustine's emphasis on history stimulated the belief that the Garden of Eden was a real place on earth and the stage of the primordial event in human history. In this and the following chapters it will be seen how paradise was mapped in space as well as in time. The challenge for the compilers of the maps in question was to render visible a place that was geographically inaccessible – yet linked to the inhabited earth by the four rivers – and remote in time – yet still relevant as the scene of an essential episode of salvation history. As the ambiguous Hebrew word מִקֶּדֶם (*miqedem*) intimates, through its references to space and time, Eden came to be placed both at the beginning of time and in the farthest east.[1] On maps, Eden thus represented the prime event of human history and the easternmost region of the earth. First, however, we need to establish the context of medieval mapping in general.

Mapping the Land

Maps were drawn in the Middle Ages for many reasons. Those people who were sufficiently literate were also likely to find, as many still do, that their own thinking, reading, and communicating were considerably enlightened when the relevant spatial relationships were represented visually. It is impossible today to have any real idea of what proportion of the medieval corpus is represented by the maps that have survived to the present day, or that have not survived but about which something is known through references in medieval or later sources. The assumption is that the majority have perished. It is also impossible to assess how many world maps of Roman origin had been transmitted, indirectly if not directly, to influence medieval mapping. None has survived. That Roman traditions had not percolated through the intervening centuries in one way or another is scarcely imaginable, although firm evidence is hard to come by.[2] It has been convincingly argued, however, on the grounds of the presence on this type of maps of rare place names found only in texts compiled between the fourth and the sixth centuries, that the original model for medieval world maps dates back to Late Antiquity.[3]

What is clear is that a whole range of different types of maps were produced at various times in medieval Europe. Plans and charts as well as topographical maps were used to clarify an argument and served as didactic aids, in legal affairs and in estate management, in celebration of achievements, as propaganda, and even as aide-mémoire for the positioning of actors in plays and priests in liturgical ceremonies, for example.[4] Some local maps were produced by religious communities in connection with their activities as intellectual and economic centres. Other maps were made by monks and clerics, sometimes with the help of lay brothers, in their capacity as book producers in every sense of the word – from creating new texts to replicating old ones. Such maps may relate to contemporary practical problems, to the study of the natural world and how it works, to biblical exegesis, or to human history.

Two kinds of maps familiar to modern map users, charts for navigation and maps for land travel, do not feature prominently or – in the case of the latter – even at all in the medieval map corpus. From time immemorial, the standard aid for travel by land or sea was the itinerary, a simple written list of the names of point of departure to destination with intervening places, possibly with the distances between them included.[5] Quite why or how charts for navigation began to appear in the Mediterranean basin by, at the latest, the early thirteenth century remains unclear, but from then up to the seventeenth century the nautical chart was in regular usage by Italian and Iberian navigators.[6] The appearance of charts in northern waters, however, was later still, and it was not until the sixteenth century that the Dutch and the English, for example, began to make and use charts rather than rutters (the seaman's itinerary). It may even have been that compilers of maps of the world were more eager than local navigators to acquire charts in order to copy the accurate outlines they contained for the Mediterranean and other areas well known to Mediterranean traders and explorers, such as the Atlantic coasts of southern Europe and northern Africa. Nor were maps used for land travel in medieval Europe. Again, the standard way-finding aid was the itinerary. Only when a potential traveller had the occasion to see a regional map or a map of the world might such a map have served in the preliminary stages of planning a journey.[7] Most of those treading the busy roads of medieval Europe, however, had no use for a map. They knew the names of the places they were heading for, day by day, and knew that so long as they kept to the highway they could always ask for directions. The English Benedictine monk Matthew Paris's remarkable and beautiful rendering of an itinerary between London and southern Italy was created for those making the spiritual, not a physical, journey, and yet offered a 'realistic' approach to spatial relationships.[8] The viewer was encouraged to travel mentally through the major political and religious centres of Britain, France and Italy to find the points of embarkation for the East and realize an imagined pilgrimage, the ultimate destination of which, Jerusalem, was indicated by a map of Acre, port of disembarkation, and the Holy Land, on the following folio in the codex.[9] A map featuring paradise, thus, was not necessarily indicating the physical route to paradise but it could express a highly sophisticated view of the cosmos.

Surviving medieval maps show anything from the entire universe to a tiny part of the local community. The genre of medieval maps we are concerned with in this book are maps of the world (in Latin, *mappae mundi*) for it was on this kind of map that paradise was featured. The literal translation of the Latin term is 'cloth' (*mappa*) and 'of the world' (*mundi*). The earliest reference of the term dates to the ninth century, when it was used to indicate maps of the world painted on some sort of surface; later in the

Middle Ages it was used, with other terms, to refer to any description of the world, written or drawn.[10] The element *mappa* passed into English giving the word 'map', although it has become standard practice amongst medieval historians and historians of cartography to reserve the term *mappa mundi* for medieval maps of the world.[11] More than a thousand surviving or known examples of *mappae mundi* have been recorded.[12] The enterprise of defining and classifying medieval *mappae mundi* has preoccupied historians of cartography since the end of the nineteenth century, but has increasingly proved to be a difficult exercise.[13] All sorts of categories have been suggested, the most basic of which attempt to classify the maps according to content and to form. Thus, for example, tripartite maps (those showing the distribution of the three known parts of the world) are distinguished from zonal maps (those depicting the sequence of climatic zones on the whole globe).[14] More recently, scholars have also privileged context and purpose. *Mappae mundi* produced as independent artefacts can be distinguished from *mappae mundi* in books. The former may have been a tiny minority, but they include some of the largest and some of the most impressive maps ever made to hang on a wall or lie flat on a table. *Mappae mundi* drawn as book illustrations may be further grouped according to the nature of the texts they illustrate.[15] World maps may be found, for example, in the geographical sections of late antique treatises on the nature of the universe, such as Macrobius's fifth-century commentary on Cicero's *Somnium Scipionis* (*The Dream of Scipio*). Others relate to events narrated in classical histories and were intended to illustrate the geographical digressions found in these texts. In medieval manuscripts of Sallust's history of the Jugurthine war (first century BC), for example, schematic world maps illustrate an ethnogeographical discussion on north Africa. Likewise, in a number of medieval manuscripts of Lucan's *Pharsalia* (first century AD), an account of the civil war in which the chief protagonists were Julius Caesar and Pompey, the geographical descriptions are illustrated with marginal sketch maps designed to serve as readers' aids.[16] Maps of the world were also inserted in the geographical chapters of Christian encyclopedic works that recycled classical learning, such as Orosius's *Historiae adversus paganos* (c. 418), Isidore's *De natura rerum* (c. 620) and *Etymologiae* (c. 635), Lambert of St-Omer's *Liber floridus*, Honorius Augustodunensis's *Imago mundi*, and Guido of Pisa's *Liber historiarum* (all dating from the twelfth century). Maps illustrated yet another type of medieval book, that of the chronicle or universal history, in which the events of human history from the world's creation to the chronicler's own day were detailed, as in Ranulf Higden's *Polychronicon* (begun in 1320 and regularly updated until Higden's death in 1363). Maps and plans were used to illuminate a range of particularly complex topics in exegetical treatises and biblical commentaries. In the Book of Genesis, world maps accompany the text to illustrate the creation of the universe and the repopulation of the world by Noah's sons.[17] Sometimes the maps were provided by the author, as in the eighth-century *Commentary on the Apocalypse of Saint John* by Beatus of Liébana, sometimes added to the margins by a glossator or reader.[18] In fact, many medieval exegetes and theologians, as well as chroniclers and encyclopedists, seem to have regarded maps as an essential component of their work, although it is not always easy to decide the extent to which the maps in surviving manuscripts represent the ideas of a late antique or early medieval author rather than those of a scribal intermediary. Far too few authors observed the helpful practice of Macrobius (c. 400), who tied each drawing to his text with words such as 'as this figure shows ...'

To the modern reader, however, possibly the most familiar type of world map is the kind that can be shown to date back to the Greek astronomers and geographers of the last millennium BC, namely maps that portray the world on a framework of measured lines constructed from astronomical data. This is the kind of world map to which the second-century AD Alexandrian geographer Claudius Ptolemaeus (Ptolemy) devoted his *Geography* and on which Renaissance cartography is based.[19] The type of world map known as *mappa mundi* is rather different. Definitions of the genre are usually far from clear-cut, but it may be suggested that the term should apply exclusively to maps of the world that are *not* ruled by astronomically defined coordinates and mathematical measurement, or on the type of geometry in which distance and direction are prime factors. In this respect, medieval *mappae mundi* are profoundly different from zonal maps (although these were also called *mappae mundi* in the Middle Ages), navigational charts and the maps associated with Ptolemy's *Geography*.[20] On the kind of world map in question here – medieval *mappae mundi* – places are located according not to a universal system of reference expressed in mathematical terms as dictated by the rules of Euclidean geometry, but according to principles generally referred to as 'topological'. In such an arrangement, the size, shape and positioning of each map sign (the mark on the map used to represent a place or the site of an event) is a function of contiguity. Neighbouring features and sites are depicted next to each other irrespective of the exact distance or direction between them. Individual features are singled out by the size of the map sign for their cultural or social importance, not for their physical dimension.[21] Another characteristic of a *mappa mundi* concerns the map's orientation. Whereas we have become accustomed since the late fifteenth century, when Ptolemy's *Geography* began to influence map making in the Latin West, to find north at the top of maps of the world, a range of orientations can be found on medieval maps in general. But, while zonal maps might have south, north or, more occasionally, west at the top, the compilers of *mappae mundi* were consistent in orienting their maps to the east, the direction of the rising sun.[22]

Topological mapping, based as it is on internal reciprocal rather than external relationships between units (temporal events as well as spatial features) can reach a precision, a rigour and a rationality of its own.[23] The topological characteristics of medieval *mappae mundi* (and of many other kinds of map, modern as well as medieval) have to be grasped if they – and, above all, the presence of paradise on them – are to be properly understood. Thus, the first thing to bear in mind when looking at paradise on a medieval *mappa mundi* is that this kind of map was not created to inform the observer of the precise latitude, longitude and size of the Garden of Eden, but to demonstrate its contiguity to the inhabited earth.

In structuring their maps, then, all medieval map makers (except those responsible for nautical charts) were working to fundamentally different principles from those of their modern counterparts. In other respects, however, *mappae mundi* are not all that different from modern maps of the world. Their content, for example, was no more selective than is that of a modern map. It is obviously impossible to represent every single landscape feature on a map or to show every intricacy of coastal outline, mountain contour or town or village layout. No less than any other map maker, the medieval compiler of *mappae mundi* had to decide what to include and what to omit in view of the map's purpose and context. As a general rule, the greater the variety of potential map users, the greater the quantity and variety of information that has to be given to ensure that the map will serve all comers, but in the case of a map carrying a specific message, or destined for a specific map-using group, information extraneous to the message or to the interests of that group is irrelevant. Elaboration is not only unnecessary, but can be confusing. Some medieval maps of the world inevitably contain a greater amount of information than others. Some are styled more formally than others.[24] Whether the *mappa mundi* in question is a visually spare and economical diagram or an intricate and visually rich representation of the world packed with scenic vignettes and figures in lively poses, paradise is often indicated in one way or another. How and why paradise was included on medieval maps of the world is the focus of the rest of this chapter and the following one.

Paradise on Christianized Maps

The mapping of paradise on medieval world maps was made possible by two main factors. The first was the structure of *mappae mundi.* Only on a map of the world showing places according to the principle of contiguity rather than in an astronomically defined and mathematically exact location was it possible to 'locate' the inherently unlocatable earthly paradise. The reason for wanting to show it at all was a consequence of the process of Christianization of classical geography that had been taking place since the early centuries of the Christian era. The incorporation of classical learning into Christian discourse favoured the development of the temporal dimension in medieval cartography, the second factor that allowed the representation of paradise on a map of the world. Paradise as an 'event/place' could be depicted only on a type of map that showed how the different layers of history were piled up on geographical space. Medieval maps flexible enough to contain simultaneously the two dimensions of time and space, and whose structure ignored the mathematical measurement of distances, could also be made to show the inaccessible earthly paradise that played such an important role in the Christian scheme of salvation.

The presence of the Garden of Eden on world maps thus bears witness to the complex process by which early European society gradually became desecularized or, in other words, Christianized.[25] As the practitioners of the new religion of Christianity filtered classical learning to fit their beliefs, aspects of pagan scholarship were rejected, adopted or adapted into a new synthesis to incorporate Jewish and Graeco-Roman geographical traditions into Christian discourse.[26] At first, Christians, seeing their calling as a pilgrimage or a passing through the earthly world on the way to the longed-for heavenly home, did not share the traditional classical enthusiasm for amassing knowledge about that world. Christian focus on geographical learning arose only about the fourth century, prompted, according to some modern historians, by Emperor Constantine's conversion to Christianity and his building programme in Jerusalem, where churches and other monuments were being constructed to mark places associated with the life and death of Jesus.[27] Other historians prefer to stress the gradual eclipse of apocalyptic expectations, the development of holy places, or the emergence of the cult of martyrs as factors in the rise of Christians' interest in geography.[28] However it came about, by the fifth century Christianity had shifted fundamentally from an initial indifference regarding earthly places to a keen interest in geography and a strong sense of sacred space. The increasing attention to the geography of the Holy Land, for instance, was reflected in the growing number of pilgrims to the eastern Mediterranean.[29] Exegetes too began to take an equally keen interest in geographical space. Eusebius's *Onomasticon,* an alphabetical list of biblical place names, was written in *c.*303, and Jerome's Latin version of the *Onomasticon, De situ et nominibus locorum Hebraicorum* was produced about 390.[30] Thus the ground was prepared for Augustine's interest in an earthly paradise and the question of its location to take root in biblical exegesis and flourish on maps of the world.

Even with all this renewed attention to geography, what mattered most of all to Christians was the ultimate meaning of human life and its relationship with God. As far as the structure of the world was concerned, the absence of an explicit biblical model meant that classical learning was accepted as perfectly compatible with Christianity. Graeco-Roman geographical literature provided the basis of the Christian description of the world, its regions, and its inhabitants, and classical geography slipped easily into the new religious discourse. Writers such as Eusebius, Jerome, Orosius, Isidore of Seville and Bede recycled the works of classical authors such as Pliny the Elder. Traditional geographical ideas about the dimensions of the globe, its division into parts, and the listing of the peoples and provinces of the inhabited world were adopted and refined to accommodate Christian themes such as the geography of the Holy Land, the apostles' mission throughout the three parts of the earth, and the conversion of the nations, and to emphasize Christ's sovereignty over the entire world. In this process of the Christianization of secular geographical lore, the Garden of Eden featured prominently, as is seen in one of the earliest available Christian versions of a pagan text, the sixth-century adaptation of a fourth-century description of the world called the *Expositio* [*Descriptio*] *totius mundi*.[31] The *Expositio* was originally compiled in Roman Syria. It contains a detailed account of all the territories under Roman rule, including a description of some regions in Asia well beyond the boundaries of the Empire. The Christian editor, who may have been one of Cassiodorus's associates, copied the Roman description, but substituted Christian allusions for the original pagan ones. One of his additions is a reference to the Garden of Eden, which is described as in eastern Asia, and to the four rivers that flow from it.[32] A later geographical work, now known as the *Ravenna Cosmography* (it was compiled in Ravenna, the centre of Byzantine power in Italy, by an anonymous cleric about 700) comprises a list of cities, rivers and islands probably derived from Roman and Greek geographical sources, but even this had room for a mention of the earthly paradise. The Ravenna cosmographer described the garden as situated at the very eastern limits of the earth, east of India. He also described the course of the four rivers, and reminded his readers that the aromatic fragrance of trees growing in eastern Asia was due to the vicinity of paradise.[33]

The process of adapting Graeco-Roman science to Christianity was also applied to maps. The major features of the inhabited world described in the classical geographies provided the basis for the display of a specifically Christian world. The degree to which elements from the Old Testament shown on the maps came from Jewish sources as opposed to the Christian scribes' and copyists' reading of the Bible is debatable. Specifically biblical features such as the division of the world amongst the three sons of Noah, Jerusalem, Sinai, and the Red Sea Crossing – even the Garden of Eden – appear to have been added to the map by the scribes as they copied classical works. Sometimes a cross was placed at the top of the map. The typical early Christianized map is the east-orientated diagrammatic map of the world that has survived only in medieval copies (dating mostly from after the early tenth century) of late antique histories and encyclopedias. This kind of map is commonly referred to as the T-O (or Y-O) map from the way the circle representing the inhabited earth is divided by straight lines to show the three parts of the world (Figure 5.1). While a similar T-O structure may be discerned in a detailed *mappa mundi*, it is in the diagrammatic map form that the cartographical genre presents its features most clearly. The lines that separate Asia, Europe and Africa and that form the T (or Y) represent the river Tanais (the Don, dividing Europe from Asia), the river Nile (dividing Africa from Asia), and the Mediterranean Sea (dividing Europe from Africa).[34] The origin of these maps is uncertain. Most scholars suppose a direct ancestry from the diagrams illustrating Roman geographies.[35] Others have argued for a Jewish influence, suggesting the importance of the second-century BC *Book of Jubilees* for the assimilation of the classical tripartite division into biblical ethnography, with Noah's sons Sem and his descendants in Asia, Cham and his descendants in Africa, and Japhet and his descendants in Europe.[36] Without an unambiguously original exemplar, Roman or Jewish, and in view of our general ignorance about the extent to which surviving Christian medieval world maps represent the original corpus, it is difficult to trace the ancestry of the T-O maps or to identify with any certainty their original function. Irrespective of the question of their origin, however, it can be said that the

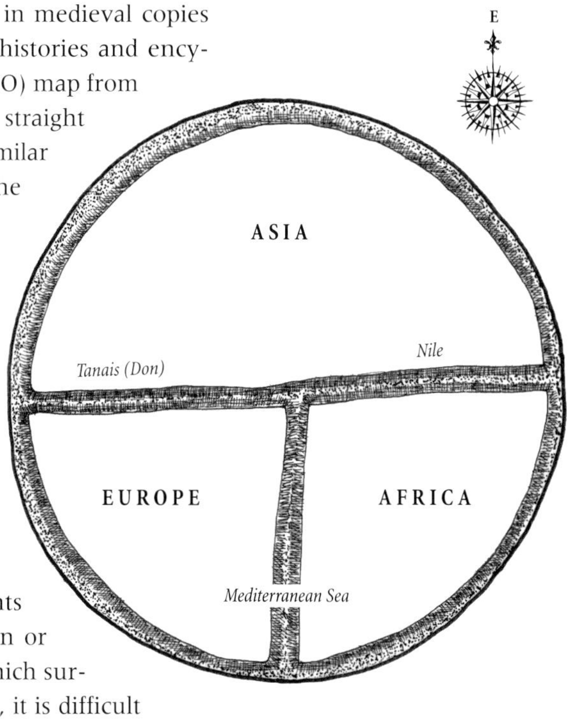

Fig. 5.1. Schematic diagram of the T-O map of the world.

Fig. 5.2. T-O map from a miscellaneous manuscript. Einsiedeln, Tenth century. 18 × 14 cm. Einsiedeln, Benediktinerabtei, Stiftsbibliothek, MS 263 (973), fol. 182v. The manuscript contains the stories of the lives of saints and martyrs including Paula, Paphnutius, Pelagius, and Petronilla (fols 1–124), Jerome's *Sermo de die Paschae* (fols 167–72), and computus material, where the map is inserted (fols 125–66; 180–5). The world map depicts the three parts of the world in a more elaborate structure than in the standard tripartite diagrams: in addition to the Nile, the Mediterranean and the river Don, the Meotides Paludes (Sea of Azov), mentioned by Isidore, *Etymologiae*, XIV.4, is also marked. The inscription *Orbis Paradisus* appears in a semicircle attached below to the world map. Extracts from Isidore's geographical chapter (*Etymologiae*, XIV.4) have been inserted by the copyist within the diagram itself (in Asia and in the semi-circle) and continue on the following page of the manuscript.

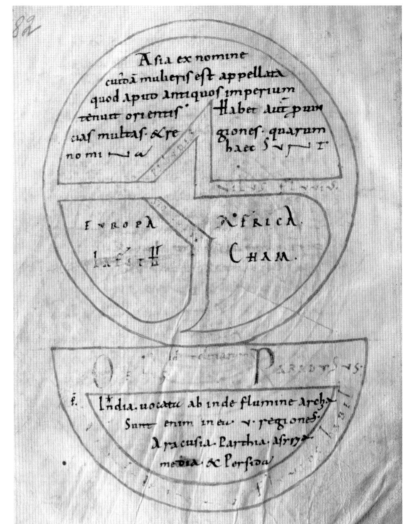

Fig. 5.3. T-O map (Y-O variant) from a manuscript of the fifth-century commentary by Macrobius on Cicero's *Somnium Scipionis*. France, the Loire Valley, ?Fleury, twelfth-century. Diameter, 8.7 cm. Paris, Bibliothèque Nationale de France, MS Lat. 16679, fol. 33v. Macrobius provided information on the size of the earth and on its climatic zones. Unlike the north- or south-orientated zonal maps of the earth that usually accompanied his commentary, this diagrammatic map shows only paradise, with its four rivers, and the inhabited part of the earth. The standard T-O model has been slightly modified by the medieval copyist who inserted the map into the manuscript, in that the rivers marking the boundaries between the three parts of the world take the shape of a Y.

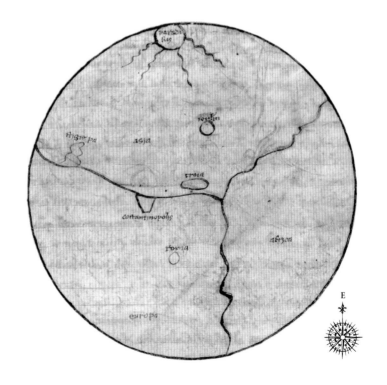

90   Mapping Paradise: A History of Heaven on Earth

T-O maps, like other surviving early cartographical artefacts, record the impact of the Bible on classical geographical notions.

However it came about, the fact remains that the earthly paradise is marked on many medieval T-O maps. A diagram in a tenth-century miscellany shows admirably how the addition of the biblical paradise to a map describing the three parts of the world – Asia, Europe, and Africa – implied that paradise was contiguous to and yet separate from the inhabited earth (Figure 5.2). In this manuscript, paradise is shown as a semicircle set at a tangent to the known terrestrial regions. The majority of T-O maps, however, featured the Garden of Eden as the first region of Asia and placed it within the circle containing the inhabited earth. This kind of map is commonly found in medieval copies of classical or late antique works that, together with the medieval encyclopedic compilations, continued to be heavily used throughout the Middle Ages. So paradise and the four rivers are found on a map in a twelfth-century copy of the fifth-century commentary by Macrobius on Cicero's *Somnium Scipionis* (Figure 5.3).[37] In some manuscripts of Sallust's *De bello Iugurthino*, the schematic T O map shows not only Carthage, Troy and Rome, as might be expected, but also Jerusalem and paradise (Figure 5.4).[38] Medieval manuscripts of Lucan's *Pharsalia* often include a T-O map on which paradise is marked.[39] The manner in which paradise is indicated varies. In both the Sallust and the Lucan manuscripts, paradise is indicated by the word *paradisus*. On T-O maps illustrating the geographical sections of Isidore's *Etymologiae*, the word *paradisus* usually labels a graphic representation of the four rivers that flow from their source in the Garden of Eden (Figures 5.5 and 5.6).[40] The merging of classical geographical learning with biblical lore on maps continued throughout the Middle Ages. Particularly striking is the arrangement found on a T-O map in a thirteenth-century manuscript of *L'image du monde* – an encyclopedic poem dealing with geography and astronomy – written about 1245 in France by Gautier (or Gossouin) de Metz, but based largely on Honorius Augustodunensis's early twelfth-century *Imago mundi* (Figure 5.7).[41] On each of the three parts of the world represented on Gautier's map is inscribed a long list of their principal regions, nations and kingdoms. Framing the circle of the world are verses from Ovid's *Metamorphoses* (I.52–68) describing the four principal winds: Eurus (east), Zephyrus (west), Boreas (north), and Auster (south). There is no sign for paradise, but a passage inscribed along the circle near the top (east) reads: 'assemble and it will become Adam'.[42] The inscription, inspired by Honorius Augustodunensis's text, is pointing out that the Greek initials of the four cardinal points – 'A' for ἀνατολή (*anatole*, east), 'D' for δύσις (*dusis*, west), 'A' for ἄρκτος (*arctos*, north), 'M' for μεσήμβριος (*mesembrios*, south) – spell out the name 'Adam'.[43]

From these examples, as from all Christianized T-O maps, we can see that the story of Adam and the Garden of Eden was still considered relevant in the Middle Ages no fewer than six thousand years or so after the Expulsion (as the clerics reckoned time) – and that classical geographical knowledge had been thoroughly infused with the tincture of Christianity. In fact, the typological reading of the Bible provided a way for situating Eden within a spatial network of peoples, places and histories. Exegetes and artists alike explained in their commentaries and through their works of art that the death and resurrection of Christ were prefigured in the Old Testament and that paradise had been both lost and regained. The scribes and illuminators responsible for the

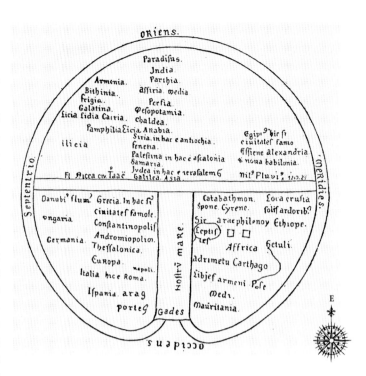

Fig. 5.4. T-O map from a manuscript of Sallust's *De bello Iugurthino* (first century BC). Uknown provenance, thirteenth century. Diameter 12cm. Deventer, Athenaeum bibliotheek, MS I, 81 (old 1791), fol. 1R. Thirteenth century. T-O maps are found in some 60 of the 200 extant manuscripts of Sallust's history of the Jugurthine war (first century BC), where they illustrate the geography of North Africa. On this thirteenth-century exemplar of the map, key places in the war mentioned in the text, such as Carthage and Rome, are also marked, but so too are two essentially Christian features, Jerusalem and the earthly paradise (*paradisus*).

Fig. 5.5. T-O map from Isidore of Seville, *Etymologiae* (*c.*620). Spain, *c.*1300, Diameter 11.5 cm. Florence, Biblioteca Medicea Laurenziana, Plut. 27 Sin. 8, fol. 64v. Isidore paid great attention to the physical features and the geographical location of the Garden of Eden. He listed paradise among the regions of Asia, describing it as surrounded by a wall of fire. On the map, paradise is indicated by the word *paradisus* and represented graphically by the four rivers flowing from a single source (*fons paradisi*) in the Garden of Eden. The Nile is shown crossing the boundary of paradise to flow through western Asia and to extend in an African branch, following a course that takes it through Egypt. Outside the map proper the four cardinal points are indicated. The T-O structure includes here the Meotides Paludes (Sea of Azov), mentioned by Isidore, *Etymologiae*, XIV.4, as in Fig. 5.2.

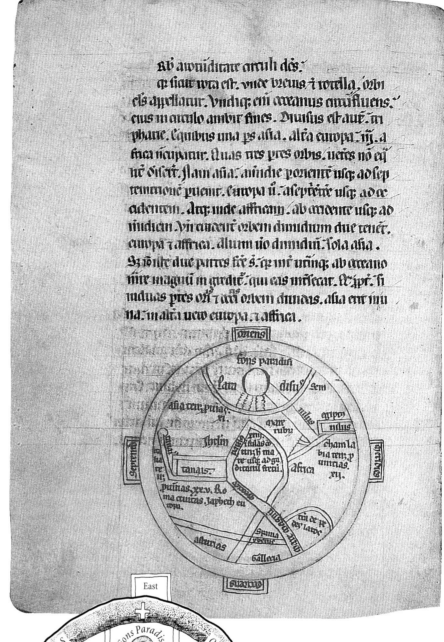

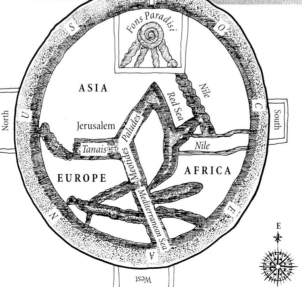

Fig. 5.6. Diagram of paradise on a T-O map from a manuscript of Isidore's *Etymologiae* (*c.*620). San Millán de la Cogolla, Spain, 946. Diameter of the original 10.4 cm. Madrid, Biblioteca de la Real Academia de la Historia, MS 76, fol. 108. The schematic Isidorean map has a similar structure to the map reproduced in Fig. 5.5, with the addition of the Meotides Paludes and the depiction of paradise and its four rivers at the top. A cross marks the east, however, and paradise is here bounded by a rectangle and not a curved line.

92   Mapping Paradise: A History of Heaven on Earth

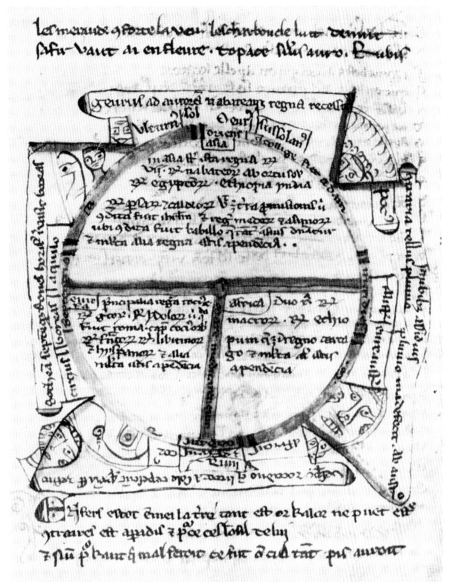

Fig. 5.7. T-O map of the world from a manuscript of Gautier (or Gossouin) de Metz's *L'image du monde*. France, thirteenth century. Diameter 6.6 cm. The original manuscript, formerly in the Bibliothèque Municipale, Verdun, is now lost. From Marcel Destombes, *Mappemondes* AD *1200–1500: Catalogue préparée par la Commission des Cartes Anciennes de l'Union Géographique Internationale* (Amsterdam: N. Israel, 1964), Plate VIIIa. East is at the top and an inscription near the east refers to the fact that the initial letters in Greek of the cardinal points spell the name of Adam. Windheads are depicted between Ovid's quotations on the four winds.

maps shared that understanding and were encouraged by typological art and exegesis to map more than the visible surface of the earth. For them, the history of salvation brought about a geography of salvation. Most of the biblical places – specific points on earth where, according to the Bible, God's intervention had taken place – were locatable in well-known regions and were seen as part of the same space–time network within which the space–time reality of Eden played a crucial role. All could be recorded on the map. There was no problem: a *mappa mundi* that included the holy city of Jerusalem – the place of Christ's redeeming sacrifice – could equally well include the Garden of Eden – the place of Adam's Fall. The two places were related through the events that had occurred in each. In fact, any place on earth could hint at the paradise of Adam and Eve: Rome, for instance, the city of the papacy at the heart of the Mediterranean world, or the centres of pilgrimage all over Europe, other approximations to Eden.

Classical learning and the Christian typological approach to the Bible together led to the introduction of a temporal dimension in the medieval mapping of the world. The continued influence of Graeco-Roman geographies and histories meant that the names of places whose day was long over remained common currency. The presence on medieval maps of towns such as ancient Troy in Asia Minor, and Leptis Magna and Carthage in North Africa, deepened a sense of the past and taught the observer that mapped space can contain different layers of time. In some cases, maps were decidedly non-modern, omitting most contemporary towns and cities but including a liberal distribution of mainly Roman settlements. In other cases, the names of classical peoples and tribes, their regions and their cities, jostled on the map with the names of new cities and newly formed nations. The boundary lines of Roman provinces such as Gallia, Germania, Achaea and Macedonia delineated a vanished Empire. The names of medieval emporia such as Genoa, Venice, Bologna and Barcelona, together with those of the great cities of Rome and Constantinople, also carried with them something of the glory of their past. Places in the Middle East evoking the life of Christ and his apostles – Jerusalem, Bethlehem, Nazareth, the Sea of Galilee, Damascus, Ephesus, Antioch, Tarsus – pointed to a different glory. From yet earlier in biblical history came the names of the Twelve Tribes of Israel, the Tower of Babel, Babylon, Mount Sinai, the Red Sea, the sons of Noah, the Ark, and paradise.

The linking of history with geography was no invention of the Middle Ages. Ancient authors such as Strabo, Herodotus, Polybius and Sallust had already blended geography and history. Their essentially historical texts included detailed descriptions of the geographical stage on which events took place. Conversely, their geographies were not geographies of an abstract space so much as geographies of peoples and places defined by the history of the territory. Human geography was privileged over physical geography.[44] Jewish as well as Greek and Roman writers shared a strong sense of the past and of the relationship between people and place. Medieval Christianity's awareness of the interrelationship of time and space was inherited from both classical and Judaic traditions.[45]

Mapping with Theology

The depiction of the earthly paradise came to play an important role in the mapping of a biblically based world view. The maps of the world produced in the monasteries and cathedrals of Western Europe, however, were not devotional, pastoral or theological documents, as opposed to our modern scientific representations of the earth, nor were they tools of religious propaganda or sermons in visual form. Rather, they were representations of the world according to a particular conception, one that took into account the scriptural text and the teachings of the Christian faith. Assessed on their own terms, the medieval maps of the world were in fact no less 'scientific' than any other type of map.

Today, over two centuries after the Enlightenment, it is sometimes difficult to understand how the Bible was read in earlier times and to grasp the degree to which medieval map making was related to theology and biblical exegesis. The conviction reflected in much of the scholarship of the twentieth century is that medieval world maps were in essence pictorial Bible stories attended by pious texts and gave 'a theological turn to geography'.[46] The situation, though, is more complex, not least because medieval maps included much more than biblical lore.[47] Since the rise of modern science, the tendency has been to see the Bible as a specifically religious book. At the same time, the physical world has come increasingly under the rational scrutiny of scholars who, preoccupied with natural laws, eschew the authority of divine revelation in a scientific context. The history of the rise and fall of paradise on maps has been too easily reconstructed in the naïve terms of a conflict between the scientist and the monk. It is an oversimplification to explain the inclusion of paradise on world maps – and the shift that eventually brought about its disappearance – in terms of a long drawn out struggle between science and religion, with the final triumph of the former. As we saw in Chapter 1, the representation of the earthly paradise on maps was explained in the nineteenth century, and by some scholars in the twentieth century, as a reflection of the medieval monks' narrow belief in the letter of the Bible and of their ignorance of the laws of nature and the principles of science as transmitted from Antiquity. According to that post-Enlightenment paradigm, once religion no longer pervaded every thought, action and interpretation, maps could be, and were, thoroughly cleansed of what could not be scientifically justified. It is important to realize, however, that for Christian scholars in Late Antiquity and in the Middle Ages the Bible provided the key to *all* forms of knowledge. Religious guidance was indeed found in the Holy Scriptures, but so also was the authoritative account of the world's creation and the history of the human race. The Bible may not provide a great deal of geographical information, but it was the fundamental point of reference in the study of cosmology, natural philosophy and history.[48] Specific geographical learning could be taken from classical texts, but it had had to be adapted to fit a biblical world view.

Scripture contributed several notions to the medieval understanding of geography, not only the terrestrial reality of the Garden of Eden and the life-giving role of the four great rivers of the earth, the Nile, the Tigris, the Euphrates and the Ganges. The reading of the Old Testament, for example, favoured the idea that the inhabited world was divided into three parts, Asia, Europe and Africa, which were inherited after the Flood by the three sons of Noah, Sem, Japhet and Ham, and peopled by their descendants (Genesis 9.18–19).[49] The existence of an inhabited fourth part (the Antipodes) was seen by some theologians, Augustine for example, as contradicting Scripture.[50] The biblical statement that God brought salvation 'in the midst of the earth' (*in medio terrae*) (Psalm 73[74].12), and that he placed Jerusalem 'in the midst of all nations' (*in medio gentium*) (Ezekiel 5.5), suggested that the holy city was found at the centre of the tripartite earth.[51] Another biblical feature that contributed a geographical notion was the reference, in the Book of Revelation, to the apocalyptic tribes of Gog and Magog who, it was predicted, would eventually storm the earth at the time of the Antichrist (Revelation 20.7–8). The way in which Gog and Magog were understood in geographical terms throughout the Middle Ages was neither consistent nor stable. Some map makers located the dwelling place of these cruel peoples at the north-eastern margin of Asia where, according to the *Revelations* of Pseudo-Methodius, Alexander the Great had walled them in behind two mountains.[52] The Bible also describes episodes of salvation history that took place in the Holy Land, such as the Exodus of the Israelites from Egypt to the Promised Land, the division of the land of Canaan amongst the Twelve Tribes of Israel, and the preaching of Christ, and that demanded a geographical framework. The notion that the apostles preached the good news throughout the three parts of the known world also encouraged the mapping of their burial places.

The rest of this chapter is devoted to a close examination of two maps that bear witness to the thoroughness of the process of Christianizing classical geographical learning. The maps have been selected not only because they may be relics of a much larger former corpus of highly detailed early maps of the world, but also – indeed chiefly – to show how they reflect the early medieval Christian view of the world and how the earthly paradise was integrated into a much more detailed geographical portrayal than was either possible or necessary on a diagrammatic T-O map.

The Pseudo-Isidorean Vatican Map of the World

The first map to be considered, the 'Pseudo-Isidorean Vatican map', is so called from its current location in the Biblioteca Apostolica Vaticana and from its former attribution (now disproved) to Isidore of Seville. Its description of the earth has a distinctly classi-

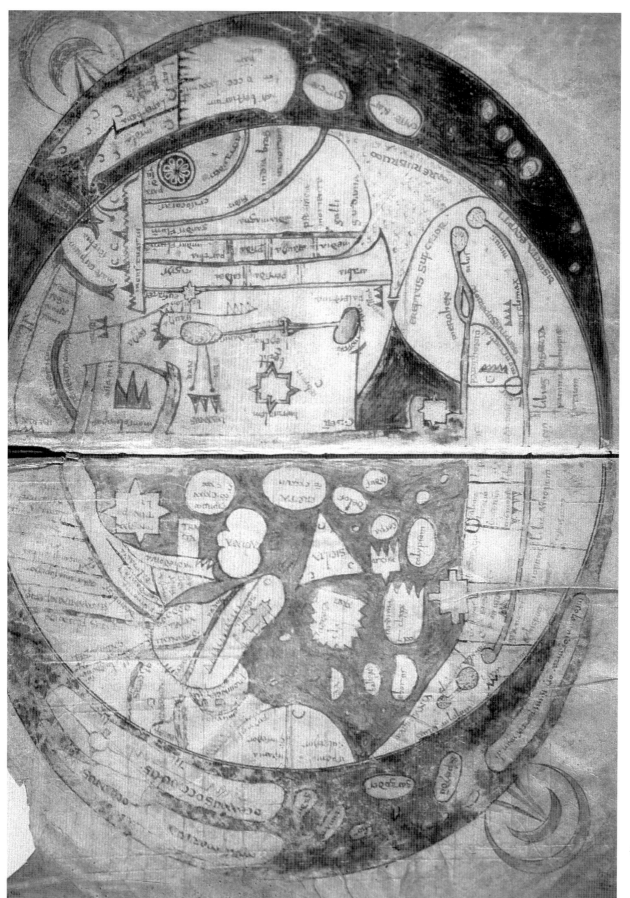

96 Mapping Paradise: A History of Heaven on Earth

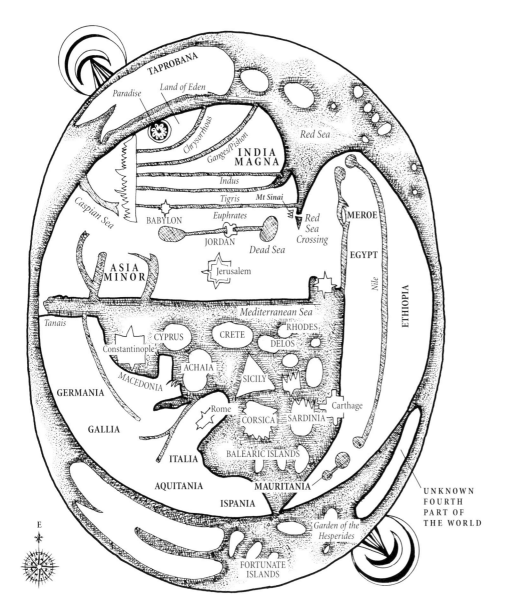

Fig. 5.8a (opposite). The Vatican (or Pseudo-Isidorean) map. Italy or southern France, late eighth century. 29 × 22 cm. Vatican City, Biblioteca Apostolica Vaticana, MS Vat. Lat. 6018, fols 63v–64r (63v Asia, 64r Europe). The Christian world view has been superimposed on to a classical geographical description. The Holy Land has been given great emphasis, and paradise is indicated in the farthest east. Following Orosius, the map maker has shown a lake in Mauritania at the head of the Nile; a second lake in Ethiopia, not far from the shore of the Red Sea, before the river sinks into the sands; and a third lake around the island of Meroe, where the Nile reappeared to continue its course to terminate in the Mediterranean Sea. The colours of the map and the ink of the writing have faded but they would have been brilliant when fresh.

Fig. 5.8b. Diagram of Fig.5.8a.

cal imprint that has been modified by the inclusion of biblical material, which makes it an important early medieval cartographical document. It was drawn probably in Italy or southern France, and probably in the late eighth century, and is thus the earliest known non-diagrammatic world map from the Middle Ages (Figure 5.8).[53] For a long time it was thought to be part of a manuscript of Isidore's *Etymologiae*, and some historians even went so far as to suggest that it was a copy of a map drawn by Isidore himself to illustrate his description of the world.[54] That idea has gone out of fashion and the map is now referred to as 'the Pseudo-Isidorean Vatican map' or simply as 'the Vatican map'. It is a small map (29 × 22 cm) yet in comparison with most T-O maps it is remarkably rich in geographical information, containing more than 130 inscriptions. The map is bound into a codex of miscellaneous content, where it occupies folios 63v and 64r, but the composition of the codex shows that it is part of a separate gathering made up of folios 55 to 75. Nevertheless, no conclusive evidence has so far been advanced to suggest that the various components of the codex should not be considered as a single whole.[55] The gathering with the map contains an assortment of matter that falls under the general heading of computus. This was material for the measurement of time in indictions or fifteen-year cycles, the calculation of the date of Easter, of the age of the moon, and of the years since the creation of Adam or the Crucifixion of Christ. The

Mapping Paradise in Space and Time

indiction tables, which form part of the same gathering and which come immediately after the map (on folios 68v–71v), point to the date of their compilation, which has been established as between the years 763 and 778.[56] Since map and tables are in the same hand, those dates can also be taken as the probable date of the map.

Mapping Time

Further confirmation of the link between the map of the world in Vat. Lat. 6018 and the computus material amongst which it is embedded it is found in the diagram on the verso of the map (fol. 64v), also seemingly in the same hand (Figure 5.9). At the centre of this diagram is a floral sign similar to the one used on the map on the other side of the folio to indicate paradise (Figure 5.10a), a point that could strengthen the case for a common author.[57] The diagram refers to the *Dies Aegyptiaci*, the days of the year thought to be unlucky. Above and below it, a few lines of text suggest ways of minimizing risk by, for example, avoiding bloodletting, taking medicine, marrying, harvesting grapes, building houses or moving house on any of those ill-omened days.[58] The whole diagram was an attempt to Christianize the zodiac by linking the twelve astrological signs with biblical characters and events. Thus, Aquarius is paired with the Baptism of Christ in the Jordan, Pisces with Jonah inside the whale, Aries with the sacrifice of Abraham, Gemini with Adam and Eve, Leo with Daniel in the lions' den, and Virgo with the Virgin Mary. The fact that the winter solstice coincides with the birth of Christ, and the summer solstice with the birth of John the Baptist is noted a few folios on (fol. 68r). In short, both diagram and map in the Vatican computus manuscript testify to the importance in medieval scholarship of ensuring a Christian gloss on pagan learning as regards heavenly as well as earthly matters.[59]

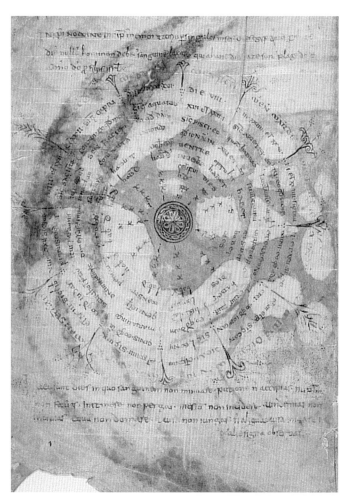

Fig. 5.9. Diagram of the *Dies Aegyptiaci* (the ill-omened days). Italy or southern France, late eighth century. 22 × 14 cm. Vatican City, Biblioteca Apostolica Vaticana, MS Vat. Lat. 6018, fol. 64v. The diagram shows that astrological divination, the attempt to prognosticate the future by interpreting the motions of celestial bodies, was widespread in the Middle Ages despite Ambrose's and Augustine's disapproval.

The chronological and computistical material that follows the world map (folios 65r–71r) was intended to show how time was regulated by the movement of the sun and the moon in relation to the earth. It was important in the Middle Ages to adjust the lunar and solar years to work out the correct date for Easter, the most important of the Christian feasts.[60] The reader of the Vatican computus manuscript, with its pages of lunar and solar calendars and horological diagrams, could discover not only the position of the sun at different hours of the day in any month of the year (using the *horologium*), and the beginning of each season, but also the days of a full moon in any given year and the reoccurrences of a specific date on a specific day of the week within each nineteen- or twenty-eight-year cycle. Especially significant for the Church were the tables of fifteen-year cycles that showed the date of Easter and the age of the moon on Easter for the next 229 years. The compiler of the Vatican computus manuscript calculated time from the moment when God created Adam and placed him in the Garden of Eden, 5,200 years before the birth of Christ.

The world map in the Vatican computus manuscript highlights the earth's relationship with the heavenly bodies. It offers a view of the geographical stage over which time passes according to a rhythm reckoned by the moon–sun system. There is, though, an explicit link between the map and the computus tables in the astral signs depicted in the map's north-eastern and south-western corners. These signs have long been a puzzle, and scholars have either ignored them or interpreted them in general terms as 'the sun and the moon' or just 'lunar symbols'.[61] Leonard Chekin is less vague in seeing these 'shining crescents' as representations of, respectively, the rising sun at the summer solstice and the setting sun at the winter solstice, functioning in much the same way as similar signs on the hemispherical map in Lambert of St-Omer's *Liber floridus* (1112–21).[62] We may go further and suggest that the symbols indicate the shadow produced by the earth on the moon, a phenomenon represented on other illustrations of Lambert's work.[63] The different colours used in the two signs indicate the status of the heavenly bodies, with orange red for the rising sun and the setting moon, and pale green for the setting sun and the rising moon.

The placing of the two astral signs on the Vatican computus map at the point of the solstices would have been a potent reminder of the link between the heavenly bodies and earthly phenomena. Their position in the north-east and south-west would have also served to indicate the cardinal points and to inform the reader that the map depicted the inhabited portion of the earth in the northern part of the globe, surrounded by the ocean. The unknown fourth part of the world, which was not destined for the children of Adam and that was conceived of as a landmass situated in the southern hemisphere, was represented on the map as a long, thin island to the south-west of Africa.[64] The Garden of Eden is clearly seen in the northern hemisphere, contiguous to the inhabited earth. In contemporary terms, and in its relationship with the text to which it belongs, the Vatican map was no less 'scientific' in its preoccupation with the location of paradise, and in its wealth of allusions to the Bible, than in its highlighting of the position of sun and moon and the depiction of other aspects of the earth.

Mapping Biblical Space
The emphasis the Vatican map gives to biblical geography is consistent with the strong interest in the Bible and the Holy Land manifested in other writings and glossaries in the codex. One of these is the apocryphal letter of Damasus to Jerome, in which it is pointed out that the priest-king Melchisedek was a real historical figure and a prefiguration of Christ (fols 89v–91v).[65] Another text is the *Decretum Gelasianum*, an early sixth-century document attributed to Pope Gelasius (402–6) that contains a list of apocryphal and canonical writings (fols 126r–126v).[66] There are dictionaries and lexicons (one particularly long glossary occupies fols 3r–50v); a discussion of the meaning of various letters of the alphabet (fols 51r–53v); and a comparative discussion of the Hebrew, Greek and Latin alphabets (fol. 54v).

All this material was potentially useful for understanding the literal and the allegorical meanings of the biblical text. Most significant of all, in relation specifically to the map, however, was the work of Eucherius of Lyons, the fifth-century bishop whose allegorical interpretations and interest in the Holy Land provided a popular and long-lasting exegetical tool. Extracts from Eucherius's *Formulae spiritalis intelligentiae* and *Instructiones ad Salonium*, which contain comments on Jewish customs and religion, Hebrew etymology and pagan idolatry, are found just before the map (fols 55v–62v).[67] What is missing from the extracts taken from the *Instructiones* is Eucherius's description of the world, its regions and its peoples, and his discussion of the four rivers of paradise, and it may well have been that the map was regarded as a sufficient substitute.[68] Following the map are further extracts from the *Instructiones*, including one section on the Jewish months (fol. 72r) and another on measures in general (fols 120r–120v).[69] Probably some time just before or after compiling the *Instructiones*, Eucherius wrote a scholarly letter describing the geography of the Holy Land, and other sections of the Vatican codex, which deal with biblical places, may be in part a reflection of those geographical interests.[70] The commentary on biblical sites (fols 117r–118v), for example,

Fig. 5.10a. The Vatican (or Pseudo-Isidorean) map (see Fig. 5.8). Detail of eastern Asia with paradise. The location of the Garden of Eden is marked by a circle containing a floret. Paradise is represented contiguous to the inhabited earth as if within an inset, giving the impression of an independent level of reality, separate and different from that of the ordinary world.

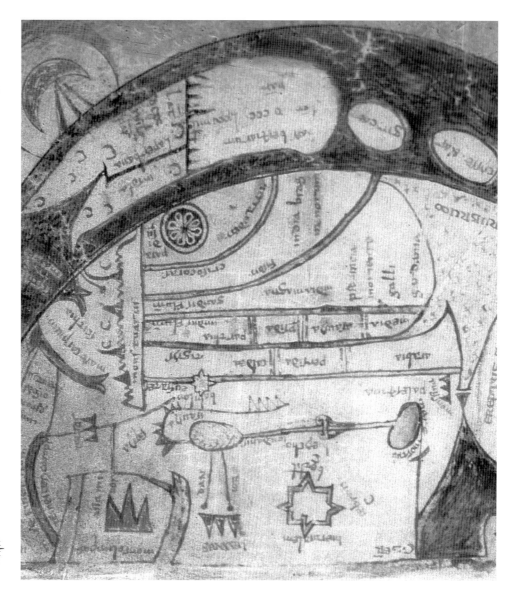

that begins with Jerusalem and ends with Babylon, would almost certainly have been compiled from Eucherius's works as well as those of Jerome. The description of Jerusalem and the Holy Land elsewhere in the codex (fols 121r–123v), though, came from a sixth-century pilgrimage account written by Theodosius.[71]

The fact that the Vatican map emphasizes the holy sites of Jerusalem and Bethlehem as well as the Old Testament towns of Babylon and Jericho is thus consistent with all aspects of the interest shown by the compiler of the codex in the geography of the Holy Land.[72] Geography, Eucherius pointed out, had a spiritual dimension: the earth stood for mankind; the mountains signified the Lord, the Church, or the apostles; and the valleys pointed to the contrition of humble saints (fol. 56v).[73] Biblical geography provided a still more potent reason for recognizing the spiritual significance of the world's physical features. Eucherius's discussion of the allegorical meaning of various biblical sites, all of which are indicated on the map, is found on fol. 58r. Jerusalem, he wrote, stood for the Church and for the human soul; Egypt was an image of the world and of the Gentiles; and Babylon was a symbol of the mundane and of pagan Rome.[74] The Vatican map could have suggested allegorical significances, but its main purpose would have been to facilitate an understanding of the literal and histori-

100   Mapping Paradise: A History of Heaven on Earth

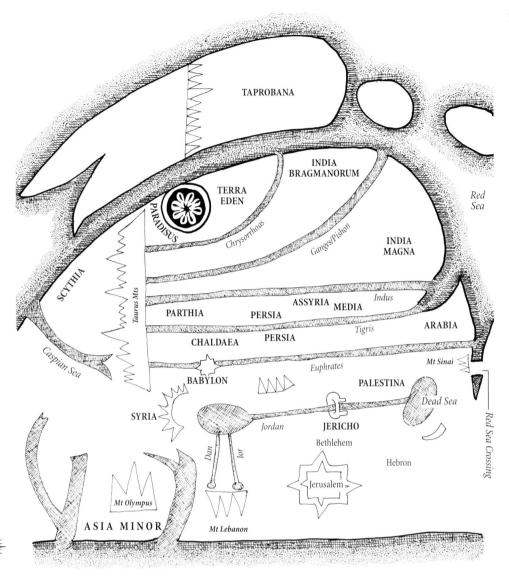

Fig. 5.10b. Diagram of Fig. 5.10a.

cal meaning of Scripture by showing the lands in which the events reported in the Bible had taken place. After reading about the series of plagues unleashed by God on Egypt (Exodus 8 and 9), and listed on fol. 55r, the reader finds, a few folios later, the land of the Pharaohs marked on the map. Likewise, Mount Sinai, where Moses received from God the ten commandments (Exodus 19; 20) that are written out on fol. 54v, is portrayed near the Red Sea crossing (Exodus 14.19–31). The map also depicts other places associated with biblical events such as Hebron, the mountains of Lebanon, and the river Jordan with its two sources (*Ior* and *Dan*) (see Figure 5.10).

On the map, Old Testament geography merges naturally with New Testament geography. The significance of the river Jordan is emphasized by the indication of the place where the Israelites crossed it to enter the land of Canaan (Joshua 3). At the same time, indicating the river would in itself have served as a reminder of the Baptism of Christ (Matthew 3.13–17; Mark 1.9–11; Luke 3.21–2; John 1.29–34). A brief exposition attributed to Jerome about the dates and places of crucial events in salvation history is reported on fol. 54v, just before the gathering containing the map. The account gives the names of the Roman consuls at various dates that were important for Christians: Christ's conception (25 March), and birth in Bethlehem (25 December); his baptism in

Mapping Paradise in Space and Time 101

the river Jordan (6 January) at the age of 30. Also noted are Christ's arrest (24 March), his Passion (25 March), Resurrection (27 March), and Ascension (forty days later), when Pontius Pilate was Roman governor of Judaea and Tiberius emperor of the world.[75] All these events needed to be situated in space as well as in time, and their sites were marked on the map.

Also marked on the Vatican map is paradise. The indiction tables in the codex indicated that the world was created 5,200 years before the birth of Christ. On fol. 102v, in the course of a didactic dialogue on various topics (*Interrogationes de diversis floralibus*, fols 100r–103r), it is specified that the first day of creation was Sunday, 25 March. In a theological summary, in question and answer format and attributed to Augustine (fols 103v–116v), it is noted that Adam was created on the following Friday, and put in the Garden of Eden (fol. 112r).[76] Paradise is shown on the map close to the eastern ocean in Asia (Figure 5.10). It is marked by a rosette labelled *Paradisus* and is contained within a land named *Terra Eden*. The cartographical representation is consistent with the interpretation of the Genesis account of paradise as a garden placed eastward in Eden (Genesis 2.8). Eucherius seems to make a similar distinction between paradise and Eden in the *Instructiones*, where he explained that, since Adam was an earthly creature, paradise must be an earthly region, although inaccessible since the Fall, and that Eden was 'the most holy place of paradise, its location situated in the east'.[77]

On the Vatican map, *Terra Eden* is bordered on the west by an India inhabited by Brahmins (*India Bragmanorum*) and distinct from *India magna*, which is shown between the Ganges and the Indus.[78] Beyond Eden, in the ocean, is the island of *Laperbana* (Taprobana, traditionally understood to be modern Sri Lanka, but also sometimes identified with other locations in the East Indies), with a note about its size, uninhabited parts, and inhabited areas with their cities.[79] The four rivers of paradise are not marked as if springing out of the Garden of Eden, but they are shown flowing across the inhabited earth: the medieval reader would have been well aware that the rivers reached the inhabited earth from paradise only after a mysterious underground journey. The Tigris, the Euphrates and the Ganges (which the map expressly identifies as the biblical Pishon) are shown rising from a long range of mountains stretching across Asia to the eastern ocean and labelled *Mons Taurus*.[80] The Nile, the river commonly understood as the biblical Gihon, flows the length of Africa in two separate streams. The complexity of its course illustrates Orosius's report of the ancient theory that the river had its source in western Africa, in the Atlas mountains, and that its waters disappeared on several occasions into the sand before finally reappearing to flow towards Egypt.[81] The region of paradise is surrounded by a river labelled *Crisacoras*, a corruption of the Greek *Chrysorrhoas*, meaning 'stream of gold'. In Antiquity various rivers were given this title,[82] but it is likely that the author of the Vatican map was simply drawing on Orosius's description of eastern Asia, where a river *Chrysorrhoas* – 'golden stream' – is noted as meeting the eastern ocean.[83] In any case, the presence of a 'golden stream' marking the border between Eden and India confirmed the idea of an abundance of gold both in India (as Pliny the Elder and, later, Bede noted) and in the land of Havilah (as described in Genesis).[84]

The pagan paradises of Antiquity also have their place on the map, in the form of the Fortunate Islands (spelt as *Furtunate*) and the Garden of the Hesperides (spelt as *Etperidum*) in the western ocean. The simultaneous presence of the classical and the biblical paradises, destined to become one of the hallmarks of medieval mapping, accents the interweaving of the different strands of human history in the process of Christianization. In the view of the compiler of the Vatican codex, however, the biblical Eden was of paramount importance, as it stood for the beginning of world history, which by his day had reached the sixth age.

Mapping Universal History
The depiction of paradise on the Vatican map may be best understood by taking into consideration the content of the codex as a whole. Its former identification as a copy of Isidore's *Etymologiae* was, scholars are now agreed, an error. However, there are grounds for associating it with Isidore. Some excerpts come straight from the *Etymologiae*, such as the chapters on orthography (fols 74v–75v) and weights and measures (fols 119r–120v).[85] Of special interest is a copy of the abbreviated version of a single universal chronicle written by Isidore probably about 615. This *Chronicon*, which Isidore had included in the *Etymologiae* as a summary of his longer chronicle, documents in a remarkable fashion the Christianization of the classical past.[86] In the Vatican codex, the *Chronicon* is found five folios after the gathering containing the map (i.e. between folios 80v and 89v), and after a collection of synonyms attributed to Cicero.[87]

What is noteworthy about Isidore's chronicles is the way in which the seventh-century bishop of Seville blended histories worked out by ancient Greeks and Romans with the history of the Jewish people as recorded in the Old Testament into a single narrative of global history viewed from the Christian perspective. Christian chroniclers, both in the Greek East and the Latin West, had already been trying to integrate the history of the Persians, Greeks, Romans and other nations with biblical chronology to create a framework into which salvation history could be placed. The aim was to show that human history was global, and that all peoples and nations shared the single chronology of the Christian six ages and were inexorably moving towards a common world end. Isidore himself, in his preface to the *Chronicon*, acknowledged the work of earlier authors such as Sextus Iulius Africanus, Eusebius of Caesarea and Jerome. Isidore's intention was to provide the reader of his chronicles with a fully up-to-date synthetic view of universal history from the beginning of the world to the reigns of the Visigoth king Sisebut in Toledo and the Emperor Heraclius in Constantinople. Instead, however, of setting down the chronologies of the individual nations in parallel columns, as Sextus Iulius Africanus and Eusebius, for example, had done, Isidore amalgamated them into a series of comprehensive paragraphs alongside which the six ages were indicated in the margins. Thus the Christian six ages became peopled not only with biblical patriarchs and Jewish rulers, such as Methuselah, Noah, Isaac and King David, but also with pagan and non-biblical characters, such as Zoroaster, Euripides, Sophocles, Romulus and Aeneas. Events reported in the Bible such as the Flood and the Jewish captivity in Babylon were accounted for, together with non-biblical events such as the capture of Troy and the establishment of the Olympiads. In Isidore's chronology, the entire population of the world was descended from Noah's sons: 36 nations coming from Sem, 15 from Japhet, and 30 from Cham. Christ was crucified for all nations in the eighteenth year of Tiberius's reign. The various streams of history had come together in Roman and Church history.

The great advantage of Isidore's concise synthesis and marginal indications of the six ages was the clarity and simplicity with which they communicated the totality of human history and conveyed a sense of the time that had elapsed since Creation. All mankind was proceeding under God's care from era to era in a single flow of time towards a climax in the sixth age, which would comprise the Incarnation of the Son of God, the diffusion of Christianity over the entire earth, and the final renewal of the world with the Second Coming of Christ. In portraying a 'globalized' world, the map that was placed before Isidore's chronicle in the Vatican codex offered a geographical perspective on the same universal history outlined so succinctly by Isidore. Just as Isidore blended biblical with non-biblical features to include all the world's peoples in one universal history, so the map depicted the interconnected locations of classical, Hebrew and Christian geography. The map enabled the entire geographical stage on which human history had been and was being played out to be visually summarized, and to represent at a glance mankind's dwelling-place. The left half of the Vatican map (top in the reproduction), which shows Asia, Egypt and the land of Canaan and the other lands of biblical history, has been given so much emphasis that the map has been seen as a regional map of the Holy Land inserted into a map of the world.[88] The right half of the map (bottom in the reproduction) shows the heartlands of the Graeco-Roman civilization, dominated by an island-studded Mediterranean and containing a somewhat compressed Europe and the rest of North Africa. To make his point, the eighth-century author of the Vatican map used knowledge about the world's human and physical features derived from Isidore and older writers such as Pliny the Elder, Solinus and Orosius. Thus biblical features coexist on the map with the regions of ancient civilizations, such as Media, Chaldaea and Persia. Classical sites, such as Achaia, Mount Olympus, Carthage, the region of the Amazons, and great imperial centres such as Rome and Constantinople are marked. The boundaries between the provinces of *Gallia Belgica*, *Gallia Lugdunensis* and *Aquitania* correspond to those decreed by Augustus and reported by Pliny the Elder and Orosius.[89] Likewise, the boundaries in Africa are those of the Roman provinces in the north and the Ethiopian countries to the south.[90]

The Pseudo-Isidorean Vatican map corroborates the point that mankind's history and geography had been unified by the universal message of Christianity. In an excerpt from a biblical commentary traditionally attributed to Isidore, included in the same codex (fols 124r–125v), each of the four evangelists is described as having written his gospel in a different part of the world: Matthew in Judaea, Mark in Italy, Luke in Achaia, and John in Asia Minor. The commentator then goes on to make the analogy that, just as the four rivers of paradise sprang from a single source in the Garden of Eden to re-emerge in different places to water the whole world, so too had the words of the evangelists, inspired by God and written in a particular region, come to be disseminated throughout the entire world.[91] At the time the map was being drawn, however, the Christianizing of the entire inhabited world, which is shown on the map as stretching from the island of Taprobana in the Far East to the western shores of Europe and Africa, had not yet been fully achieved.[92] The delay of Christ's Second Coming was to be ascribed to divine mercy, since before the end of the world could come about, 'the good news must be preached to all the heathen' (Mark 13.10) and the process of Christianization had to have reached 'the ends of the earth' (Acts 1.8). Although the indiction tables contained in the Vatican codex accounted for time only down to the year AD 1000, it was by no means certain, as Isidore had pointed out at the end of his chronicle, exactly when the world would end. Isidore's closing words were copied out again on folio 89r as a reminder to the reader that Christ himself had warned his disciples that it was not for humanity to speculate about the timing of God's final revelation (Matthew 24.36; Mark 13.32; Acts 1.7) and that, in the final analysis, each individual suffers the end of the world at the moment of his or her own death.[93] The complexity of the medieval conception of Christian global history as it stretched from the paradise of the time of the world's creation to the impending apocalyptic destruction of the world at the end of the sixth age was far greater than it might appear to the modern mind, as is revealed by another map from the eighth century.

Mapping Heaven on Earth

Beatus of Liébana's Map of the World

The second map of the world examined here was being drawn in northern Spain at more or less the same time as the anonymous scribe in Italy or France was sketching the Vatican map. This Spanish map was for a commentary on the Book of Revelation. It was the work of a monk in the Benedictine monastery of San Martín de Turieno

(Santo Toribio), Beatus of Liébana, in Asturia. Beatus is thought to have lived from about 711 to 800, a time when much of the rest of the Iberian Peninsula was being conquered by the Arabs. No less than the Vatican map, Beatus's map – on which paradise is featured much more prominently – was also a summary of the Christian vision of history. It underlined the theme of Beatus's *Commentary on the Apocalypse of Saint John,* in which he was urging Christians to keep their minds fixed on the heavenly kingdom.[94]

Beatus's commentary is richly illustrated throughout. Together with the map, some 68 illuminations offer a glimpse of a world that both belonged to and yet transcended history. Beatus's aim was to illustrate the Christian perspective on human history, and the significance of his map, with its depiction of the Garden of Eden, lies in the way he understood the Book of Revelation. Beatus's map was not a map of the world that happened to feature the earthly paradise, but a map of the world that depicted heaven on earth. For Beatus, Revelation presented the faithful with the vision of the universal Church. As the apocalyptic clock ticked away towards the approaching end, the message of Revelation was that the end had in fact already come and that through the Church the faithful could make the leap here and now from this world to the next.

Apocalypse Now

Scholars have not always resisted the temptation to detect a 'dark-age' element in Beatus's commentary. Not only does his world map distort the outlines of countries, seas, rivers and mountains, and include an imaginary garden (the paradise of Adam and Eve), but the accompanying text also appears to reflect the author's terror at the thought of the imminent end of the world.[95] The extant illuminated copies of Beatus's *Commentary* date mostly from the tenth and eleventh centuries, and include illustrations depicting the rise of the Antichrist and the Last Judgement. Together with Beatus's statement that the world was due to end in a few years' time, the illustrations have tended to be taken as confirming the stereotype of the perennial medieval fear of the impending Apocalypse. As a matter of fact, Beatus's *Commentary* – and his map – testify to the complex ways in which the relationship between history and eternity, and the struggle between good and evil, were viewed by Christians in the Middle Ages.[96] Beatus, like other Christian exegetes, held that Christ's final victory over evil, prophesied in Revelation, had already taken place with Christ's sacrifice on the Cross and the establishment of the Christian Church at Pentecost, and that since then, Christianity had been rejoicing in the Incarnation, the first descent to earth of the Heavenly Jerusalem (Revelation 21.2) and the fulfilment of the Old Testament prophecies. The Church was still awaiting Christ's Second Coming and the final and full establishment of heaven on earth. Beatus's point, though, was that it was not for humanity to speculate on the interval that separates the present from that future. What was important, for him, was that Christians would choose Jerusalem and not Babylon. As Augustine had taught, the contrasting images of the two cities, found in Revelation, referred to a mystical struggle. The faithful should not inhabit the earthly city (Babylon), devoting their love, loyalty and longing to purely secular matters, but should become citizens, by accepting that they are only temporary residents in this world, of the heavenly city (Jerusalem). To be a true Christian meant to be a pilgrim in this world, an earthly citizen of the Heavenly Jerusalem.

Beatus's *Commentary on the Apocalypse of Saint John* sheds light on the eschatological tension between the Church *post Christum* and the world to come, the Church in heaven. In his preface, Beatus acknowledged his debt to Ambrose, Jerome, Augustine, Tyconius, Gregory and Isidore.[97] Of these, it was above all Tyconius's fourth-century allegorical exegesis that had inspired Beatus with the idea that the Book of Revelation referred to the struggles and triumphs of the Church in the interval between the First and the Second Coming of Christ.[98] Tyconius's reading of the Heavenly Jerusalem as both the Church on earth and the community of the blessed in heaven had been adopted by Jerome and by Augustine, who pointed out that the Church, the house of God, was already built and was awaiting the end of time to be consecrated.[99] Beatus explained the meaning of the scriptural account of apocalyptic events: 'All these things which are rightly believed to take place at the Day of Judgement, they also occur every day in the spiritual life of the Church.'[100] He stressed the double application of apocalyptic prophecies and the overlapping of past, present and future: 'Whenever the Spirit promises future things, he relates the past at the same time, warning that what took place in the Church will occur in the future.'[101] Thus, according to Beatus, the 'battle of that great day of God Almighty' (Revelation 16.14) referred to the cosmic struggle between good and evil as taking place both in the present and the future. The 'great day' meant the time from the Passion of Christ up to end of the world, as well as the conclusive and most ferocious battle that would take place immediately before the Last Judgement.[102] The prophesied coming of Christ in glory would take place at the end of time, even as it took place daily for the saints.[103] Thanks to the Incarnation of the Son of God, the Heavenly City, prepared as a bride for her husband (Revelation 21.2), came down to earth each day through faith and the observance of a Christian life.[104] Likewise, the Antichrist would appear at the end of time, although he was a spiritual reality already operating against the Church; in fact, all sinners had already joined the apocalyptic tribes of Gog and Magog (Revelation 20.8). Here Beatus was following the allegorical interpretation of Gog and Magog that Augustine had already adopted. Augustine had pointed out that Gog and Magog were not a specific people in a remote corner of the earth, but the citizenry of the devil threatening the Church everywhere.[105] In fact, in Beatus's view, anyone opposing the Church was the Antichrist who kills the two witnesses, the Old and the New Testaments (Revelation 11.3–12). In this sense, Antichrists were already everywhere.[106]

In keeping with mainstream exegesis, Beatus was thus relating all the features of the apocalyptic description of the celestial city to Christ and the Church. So, where the Book of Revelation describes the seventh trumpet that announces a vision of God's heavenly temple and is followed by an earthquake and a storm (11.19), Beatus saw a reference to the birth of Christ.[107] Similarly, in the account of the appearance of a woman 'clothed with the sun, and the moon under her feet, and upon her head a crown of twelve stars' (12.1–6), the woman stood for the Church, which radiated with the virtues of the prophets, the patriarchs, the apostles, the martyrs and the saints, and which was robed with the light of Christ and was shining, like the moon, in the darkness of the night.[108] The Lamb standing on Mount Sion (14.1) was Christ present in his Church, in which his sacrifice was re-enacted every day.[109] The river of life (22.1) was identified with the redeeming effect of baptism or with Christ himself.[110] The angel measuring the city with a golden reed (21.15) was Christ distributing the gift of divine grace, the reed symbolizing faith in the Incarnation.[111] The exact dimensions of the square celestial city (21.16–17) conveyed the perfection and solidity of the Church and her saints.[112] The metals and precious stones adorning the city (21.18–21) were associated with the moral qualities of individual saints.[113] The fact that the gates of the Heavenly Jerusalem were never closed (21.25) indicated the solidity of the Apostolic doctrine.[114] Any number used to describe the Heavenly Jerusalem in the Book of Revelation was associated in some way with the four evangelists, the four virtues or the Trinity.[115]

Beatus differentiated the Heavenly Jerusalem already present in history as the Church from the Church that would eventually be established in heaven inasmuch as the Church in history had to confront evil, both within and without.[116] The City of God, symbolized by the woman robed in the sun (12.1), was opposed to the city of the devil, typified by the great prostitute riding the beast (17.1–7), an opposition that emerged on earth within the Church herself between saints, on the one hand, and schismatics and false Christians, on the other.[117] Outside the Church, were the enemies of the faith – pagans and unbelievers – who openly opposed the Christian faith.[118] Only at the end of the world would God judge mankind and consign the righteous to heaven and the reprobates to hell.[119] On that day, having suffered persecutions in the course of history, Christendom would finally receive the victor's crown.[120]

The unanswered question concerned the date when the triumph of the Church under Christ was to take place. Commenting on the symbolism of the number six, Beatus cited the theory of the six days of creation as a figure of the six ages and 6,000 years as the age of the universe. Allowing for the lapse of 5,227 years between the creation of Adam and the Incarnation of Christ and 786 years between the Nativity and his own day (the year 824 of the Spanish era), Beatus concluded that the world was 5,986 years old and that it was due to end in fourteen years' time.[121] These extraordinary remarks have been taken as evidence of Beatus's belief that the world was soon to end and confirmation of his gloomy apocalyptic tone. Beatus's foresight should not, however, be oversimplified. After reporting on the age of the world, Beatus quoted Isidore's warning that it was not for humanity to know how much time was left, and went on to point out that Christ himself had told his disciples that only God the Father knew the exact date of Last Judgement. Christ was well aware of the date of the Apocalypse, commented Beatus, but avoided disclosing it to his followers so that they would not put too much trust in this world and would be kept in a state of continuous tension as they waited for the end of the world. Beatus was convinced of the need for humanity to remain alert, hence his insistence that the world might end before the completion of 6,000 years. Equally, it might be that God would defer the end, as he did in the case of the city of Nineveh, whose destruction was delayed through the penance and supplications of its inhabitants (Jonah 3).[122] Beatus's hints at the impending destruction of the world were intended to inspire the necessary fear of God. They were not pronouncements giving a timetable for the ending of the world, which anyway would have been as illegitimate as impossible for, as he said, 'we do not know in what year, at what hour, on what day, or according to what means of calculation the day will come'.[123] The true Christian, insisted Beatus, 'ought to ponder, wait and fear, and to consider these fourteen years [that remain before the End] as if they were no more than an hour, and should weep day and night in sackcloth and ashes for him- or herself and for the destruction of the world, but not strive to calculate time'.[124] Underlying Beatus's argument here is the biblical idea that for God one day is like a thousand years and a thousand years like one day (Psalm 90.4; 2 Peter, 3.8). Quoting Isidore again, he warned that since each individual sooner or later suffers the end through death, the point was always to avoid sin.[125] In the tenth century, long after the year 800 and the supposed end of the sixth millennium, the scribe Magius felt free to write, as he copied Beatus's Commentary, that he had illuminated the account of the Apocalypse 'so that those who know may fear the coming of the future judgement of the world's end'.[126] Beatus's warning remained efficacious indefinitely.

References in Revelation to years, months, days or hours were, Beatus reminded his readers, to be understood as references to the entire span of time between the Incarnation of Christ and his Second Coming.[127] He read figuratively the description of the thousand years during which Christ will reign on earth at his Second Coming (Revelation 20.1–6), an account that had been taken literally by some earlier unorthodox exegetes. Beatus insisted that the Book of Revelation was referring here to the establishment on earth of the sovereignty of Christ and the Church, and that 'one thousand years' was a symbolic expression indicating the full span of the sixth age. To think that the phrase meant that the world would end a thousand years after the birth of Christ was pure heresy.[128]

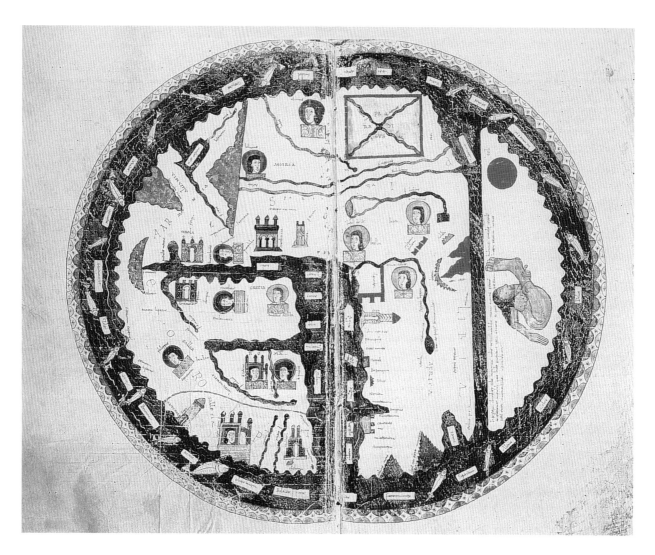

Fig. 5.11a. Map of the world from a manuscript of Beatus of Liébana's *In Apocalypsin*. Sahagún, Navarre, 1086. 36 × 45 cm. Burgo de Osma, Archivo de la Catedral, MS 1, fols 34v–35r (see also Plate 3a). The expansion of the Church of Christ over the whole earth is clearly illustrated on this map by the portrayal of the heads of the apostles who evangelized different parts of the known world, a characteristic feature of Beatus maps belonging to the Osma group of manuscripts. In the text, Beatus lists the places where the apostles preached the good news.

Mapping the Church

Beatus's world map illustrated the establishment on earth of the Christian Church (the Heavenly Jerusalem) in the sixth age. The map placed the struggle between good and evil, as narrated in Revelation, on to its geographical stage, and portrayed the Fall in Eden, and the transformation of the whole earth into the paradise of the universal Church. The spread of the faith over earthly space, through the preaching of the twelve apostles, had allowed, with the foundation of the Church and the diffusion of the Holy Spirit, an initial superimposition of heaven on earth. However, it had also initiated mankind's wait for the final 'new heaven and new earth' (Revelation 21.1), when the Antichrist would be defeated and the kingdom of God established forever.

The map that accompanied Beatus's original manuscript has not survived. Instead, we know of fourteen copies of his map in manuscripts dating from the tenth to the thirteenth centuries. The transmission of the *Commentary* has been reconstructed in various ways in attempts to identify the codex closest to Beatus's original, but the sequence is complicated by the fact that the *Commentary*, with its wealth of illuminations, was hugely successful, was widely copied (at least in Iberia), and each copy was updated as it was made.[129] Although all the maps depict the earthly paradise, historians have wondered about the map's original content and form and about Beatus's model and his sources. All the extant maps have been classified into two main groups that may represent two branches of transmission of the eighth-century prototype. The map in a

108   Mapping Paradise: A History of Heaven on Earth

manuscript of 1086, now in the library of the cathedral in Burgo de Osma and characterized by its oval shape and portraits of the apostles in their respective fields of mission, has hitherto been thought to be the closest to the original (Figure 5.11). Recently, though, John Williams has suggested that the tenth-century map in the oldest surviving manuscript, now in the Pierpont Morgan Library in New York and characterized by its rectangular shape and the absence of the apostles, may be a more faithful witness of the original (Figure 5.12).[130]

When Beatus's allegorical reading of Revelation and his Christian vision of history are taken into account, the presence of a world map in the *Commentary* cannot be regarded as either accidental or strange.[131] Beatus referred explicitly in the text to his illustrations. Furthermore, in the preface he declared his intention to provide the reader with an easily memorizable summary of Scripture, and a map would have been a powerful mnemonic device for such a summary.[132] Speculation is not necessary, though, since Beatus made the purpose of the map perfectly clear in his text. In the prologue to Book Two, the section dealing with the universality of the Christian Church, he invites his reader to contemplate the map to see how the apostles evangelized the earth and how they performed signs and miracles throughout the world, including Persia and India: 'And to be more clear, the following picture [*subiectae formulae pictura*] shows these grains of seed [i.e. the preaching of the apostles] throughout the field of this world, which the prophets prepared and the apostles reaped.'[133] Beatus quotes Saint Paul, who gave thanks to God for the spread of the faith throughout the world (Romans 1.8).[134] Citing a passage from a work attributed to Isidore, the *De ortu et obitu patrum* (*The Life and Death of the Fathers*), he lists the places where the apostles preached the Gospel: Peter in Rome, Andrew in Achaia, Thomas in India, James the Greater in Spain, John in Asia, Matthew in Macedonia, Philip in Gaul, Bartholomew in Licaonia, Simon Zelotes in Egypt, James the Less in Jerusalem. He omitted Jude (Thaddeus), who preached in Mesopotamia, and Matthias, who had replaced Judas Iscariot (Acts 1.23–6) and who preached in Judaea. The list concludes with Paul, whose activities were not limited to a particular region, but who preached to the Gentile community.[135] Thus, both map stemmata record the apostles' missions although only in the Osma group are the apostles actually portrayed, either in their field of mission or where they were buried. The other maps indicate only the mission fields.[136]

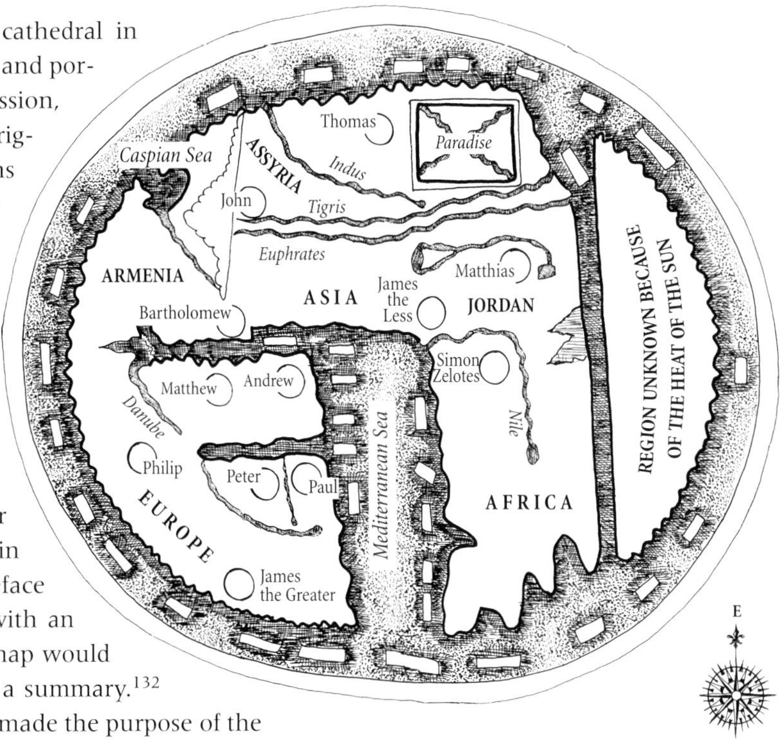

It is significant that Beatus was applying the imagery of Revelation to the earthly missions of the apostles and that in so doing he defined them as the twelve gates and the twelve foundations of the Heavenly Jerusalem (Revelation 21.12). Through the mouths of the apostles it was Christ, the only true gate and 'the foundation of foundations' of the Heavenly Jerusalem, who spoke:[137]

> *[The apostles] are the twelve hours of the day, lit up by Christ the sun. They are the twelve gates of the Heavenly Jerusalem, through which we enter the blessed life. They constitute the first apostolic Church, which we believe to be the foundation stone firmly built on Christ. They are the twelve thrones that are to judge the Twelve Tribes of Israel. This is the Church spread all over the earth. This is the holy and chosen seed, the regal priesthood sown throughout the whole world. They were few, but they were the chosen ones. And out of these little seeds arose a great harvest.*[138]

Fig. 5.11b. Diagram of Fig. 5.11a.

Mapping Paradise in Space and Time 109

Fig. 5.12. Map of the world from a manuscript of Beatus of Liébana's *In Apocalypsin*. Tábara, *c.* 940–5. 30 × 56 cm. New York, Pierpont Morgan Library, MS, 644, fols 33v–34r.

The apocalyptic ripeness of the harvest of the earth (14.15) was a reference, for Beatus, to the whole of the sixth age and not only to its ending; the end had already begun with the Incarnation and the descent of the Holy Spirit with his gifts. The map of the world shows the ground that had been prepared by the prophets and the harvest reaped by the apostles. All the regions and lands of the earth together now constituted the holy 'field' in which the history of salvation took place. Beatus's map could even be called a map of the sixth age for the way it represents the period of fulfilment of the divine promise to set the world free from the pains of evil. It portrays the renewed earth that followed the Incarnation and on which the Church had been established as the Heavenly Jerusalem.[139] Simultaneously, the map conveys, through its portrayal of the preaching of the gospel all over the earth that was supposed to precede the end, the expectation of the Second Coming of Christ.[140]

The cartographical depiction of the various regions into which the apostles were sent pointed to the final harvest – the Heavenly Jerusalem on earth. Beatus's map gave his theological message geographical form and context. The outer ocean, the Mediterranean, and the Black and Red seas are identified on most of the maps, as are islands such as the British Isles and the Fortunate Islands. Rivers (such as the Jordan, Nile, and Tanais/Don) and mountains (such as the Caucasus, the Lebanon and Sinai) are marked, together with cities such as Jerusalem, Babylon, Alexandria, Constantinople and Rome.

All the maps associated with Beatus's *Commentary* include a puzzling area of land in the extreme south that appears to be divided from Asia and Africa by a remarkably straight body of water.¹⁴¹ At first sight, it is unclear what this southern extension represents. Some maps carry a legend, taken from Isidore's *Etymologiae*, identifying the land as the fourth part of the earth, unknown to mankind because of the excessive heat and believed to be inhabited by the Antipodes.¹⁴² Most scholars have concluded that the area stood for the land mass in the southern hemisphere suggested by the unknown land in the southern hemisphere shown on the zonal model of the world described in Macrobius's commentary on Cicero's *Somnium Scipionis*.¹⁴³ In this case, Beatus's map would be portraying much more of the world than the inhabited portion that was the apostles' designated mission field.¹⁴⁴ There is, however, another interpretation. John Williams has pointed out, in the first place, that an analysis of the text in the Beatus manuscripts shows that there is no reference to a 'fourth part of the world' and to the Antipodes in the the oldest manuscripts, which would suggest that the words may have been added to the map in the tenth century. Williams also points out that the passage from Isidore, certainly a more relevant author to the Beatus map tradition than Macrobius, refers to a fourth part of the world lying beyond an inner ocean, not beyond the equatorial ocean, as the Macrobian zonal tradition would have it. Some scribes labelled this trans-African body of water as the Red Sea or coloured it red (as on the map reproduced in Figure 5.14), showing that they did not think of it as equatorial. Williams further noted that on other versions of the Beatus map the southernmost land is described as a *neighbouring* desert unknown because of the excessive heat, implying that it was a land contiguous to the habitable part of the earth – even though it appeared divided by a body of water – and not a land mass beyond the equatorial ocean.¹⁴⁵ Finally, Williams maintains, in the light of Isidore's reference in the *Etymologiae* to two races of Antipodes, that the legend on the map (and, on some maps, the vignette of a one-legged man, the Sciopod, shading himself from intense solar heat with his single gigantic foot, as in Figure 5.11a) is more likely to designate the monstrous races living in the southern and excessively hot part of Africa than to refer to Antipodeans living on the other side of the earth. As Augustine had taught, African races, however deformed or fabulous, were descendants of Adam.¹⁴⁶ They also needed to be evangelized, and their land could not, therefore, be completely inaccessible.¹⁴⁷ We may conclude, then, that an inhabited earth divided into four parts, with an unknown fourth land added to the three known parts of the world, provided a cartographical model that suited better than any other to Beatus's requirements for a map illustrating the spread of Christianity to the four corners of the earth.¹⁴⁸

The point that Beatus was making with his map was that Christianity was still in the process of spreading over the whole of the inhabited earth. An important feature of the map is the portrayal of the ocean that surrounds the earth. On most exemplars, this is shown as full of fishes, a possible allusion to the apostolic task of fishing men (Matthew 4.19; Mark 1.17; Luke 5.10) intended to emphasize the

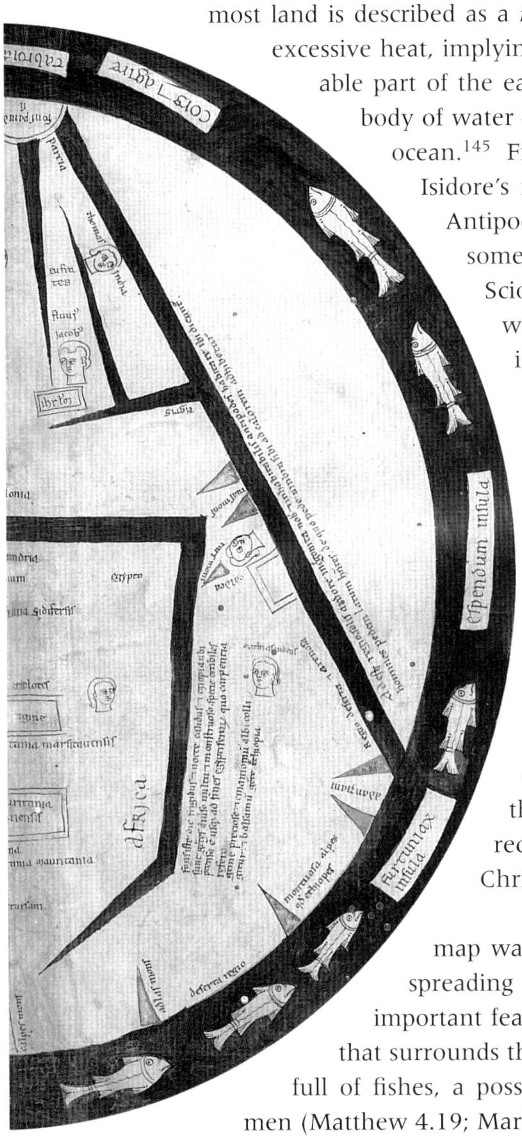

Fig. 5.13. The fragmentary Lorvão map from a manuscript of Beatus of Liébana's *In Apocalypsin*. Monastery of St Mammas, Lorvão, 1189. 34.5 × 24.5 cm. Lisbon, Arquivo Nacional da Torre do Tombo, MS 160, fol. 34 bis v. © IAN/TT/Jose Antonio Silva. The four rivers flow outside what appears to be a circular paradise. The regular updating of the extant manuscripts of Beatus's *Commentary* between the tenth and the thirteenth centuries shows that the work was hugely successful, at least on the Iberian peninsula.

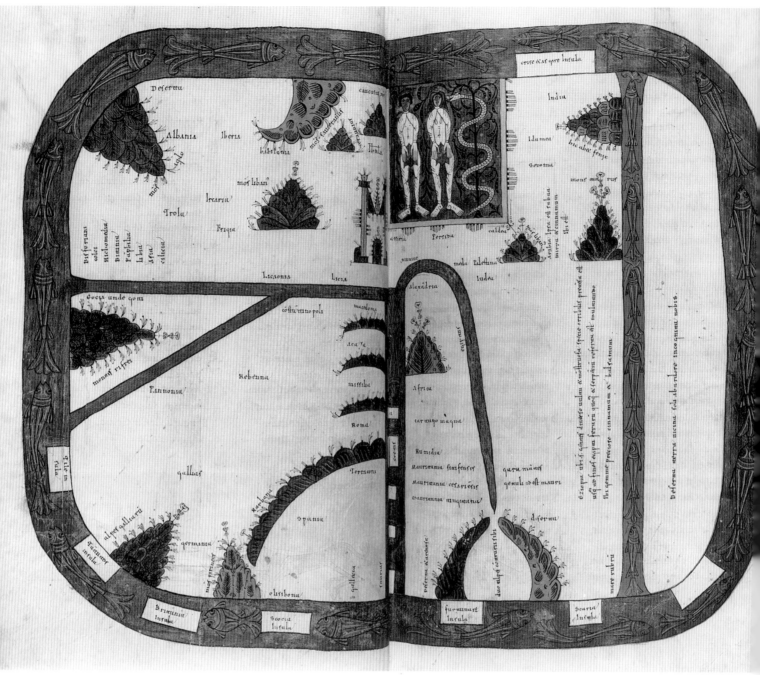

dissemination of Christianity (see, for example, Figures 5.11a, 5.12, 5.13, 5.14a). The fish was from earliest times a symbol of Christian baptism and thus of Christian believers. Those who followed Christ – the ἰχθύς (fish) – were thus called *pisciculi*, the 'little fishes'.[149]

Beatus's cartographical portrait of the earth was an integral part of his theological argument. In depicting the universal Church on his map, he was reiterating his point that the Church 'is not confined in some particular place, as in the case of heretical sects, but it is spread and extended all over the earth'.[150] His map also echoes his comparison of God's interventions in the various regions of the earth with the working of divine inspiration on the human heart.[151] The key to Beatus's use of the map lies in his conviction that the sixth age was the apocalyptic period in which the entire earth

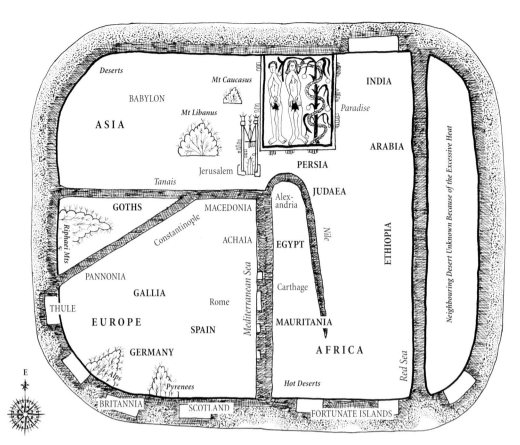

Fig. 5.14a (opposite). Map of the world from a manuscript of Beatus of Liébana's *In Apocalypsin*. Monastery of Santo Domingo de Silos, 1109. 32 × 43 cm. London, British Library, Add. MS 11695, fols 39v–40r (see also Plate 3b). Beatus, who commented on the apocalyptic vision experienced by the author of the Book of Revelation, mapped the establishment on earth of the universal Church in anticipation of the coming of the Kingdom of Heaven. The map from Santo Domingo de Silos lacks the portraits of the apostles but indicates their mission fields.

Fig. 5.14b. Diagram of Fig. 5.14a.

was the field of divine action. His emphasis on the allegorical reading of the Book of Revelation and on the universal spread of the Church would explain the absence on the map of the apocalyptic people of Gog and Magog.[152] For Beatus, as for Augustine before him, Gog and Magog were not confined to a remote valley, but were present everywhere. They represented the ubiquitous threat of evil against the Church of Christ, also ubiquitous, and could not be confined to one particular place on a world map.[153] In the text, Beatus also quoted the words uttered by the dying Isaac, as he blessed his youngest son Jacob in mistake for Esau, to underline the replacement of an essentially parochial Judaism by the universal Christian Church: 'See, the smell of my son is as the smell of a field which the Lord hath blessed' (Genesis 27.27). Quoting from the New Testament (Matthew 13.38) Beatus explained that the field meant the world, over which the aroma of Christian virtues had spread.[154] He also explained Isaiah's verse glorifying God – 'the whole earth is full of his glory' (Isaiah 6.3) – as a celebration of the Incarnation of Christ and the universal diffusion of the Church. In portraying a world in the process of being evangelized by the apostles, Beatus's map most effectively communicates the message that 'the preaching [of Christ] is spread over the whole earth and the voice of the apostles penetrates to the very edges of the world.'[155]

Eden and the Sixth Age

The representation of the Garden of Eden on Beatus's world map can be understood only by considering the map's broader purpose, which, as shown above, relates to his allegorical reading of Revelation and his wish to show the spread of the Church over all the earth. The map shows the same earth that had been cursed as a result of original sin but that eventually had been blessed by the mission of the apostles during the sixth age. It was essential to portray the paradise of Adam and Eve on such a map as a reminder of the Fall and the consequent necessity for redemption. The representation of Eden pointed to the saving reality of the universal Church. In detail, however, differences in the manner by which the earthly paradise is depicted on the maps in the two groups of the Beatus manuscript tradition add a different nuance to the basic message. On some manuscripts, paradise is marked by a geometric design within which a dot marks the point from which the four rivers flow. On other manuscripts, paradise is marked by a narrative vignette that includes the figures of Adam and Eve, but not the rivers.

The geometric diagram with the four rivers reinforces the idea of the diffusion of Christianity over the earth. Beatus himself compared the four rivers issuing from a single source to the four gospels issuing from the single mouth of Christ; both rivers and gospels give life to the world.[156] On the Osma map, paradise forms a rectangle with a point in the centre from which a line (a river) reaches each of the four corners (Figure 5.11 and Plate 3a). The image would have been seen as a direct reference to the preaching of the four gospels in the four corners of the world, often alluded to by Beatus and anticipated by Augustine in his interpretation of the fourfold division of Christ's clothes by the soldiers at the foot of the Cross (John 19.23–4) as a foretelling of the future diffusion of the Church into the four corners of the world.[157] Beatus explained that in the heavenly city there were three gates at each cardinal point (Revelation 21.13) and that these referred to the diffusion over the whole world of the four gospels, which were preached by three apostles in each of the four cardinal directions.[158] As Beatus reminded his readers, the Garden of Eden stood for the Church: 'Paradise is a figure of the Church, and the first man, Adam, is a shadow of the future, while the Second Adam, Christ, is a sun of justice, bringing light to the shadow of our blindness.'[159] Similar typological references are found in other maps of the same group. On a twelfth-century map

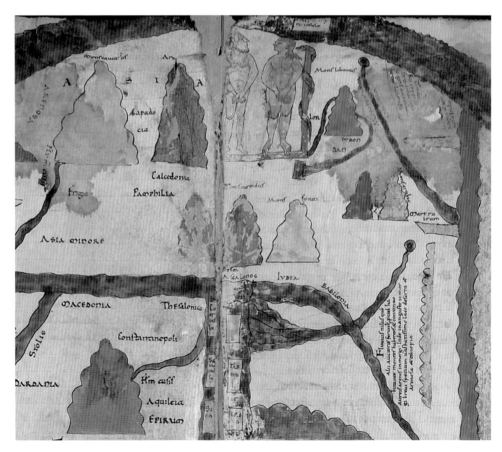

Fig. 5.15a. Map of the world from a manuscript of Beatus of Liébana's *In Apocalypsin*. Catalonia, probably Ripoll, early twelfth century (see Plate 2b). Size of the whole map 37 × 54 cm. Ministero per i Beni e le Attività Culturali. Turin, Biblioteca Nazionale Universitaria, MS I.II.1 (old D.V.39), fols 45v–46r. Detail of paradise in Asia. The river Jordan, where Christ was baptized, and the mountains of Lebanon (in Beatus's view a symbol of baptismal purity) are shown nearby, presumably to make the point that the waters of baptism flowing into the new paradise of the Church could purify the corruption of original sin.

114   Mapping Paradise: A History of Heaven on Earth

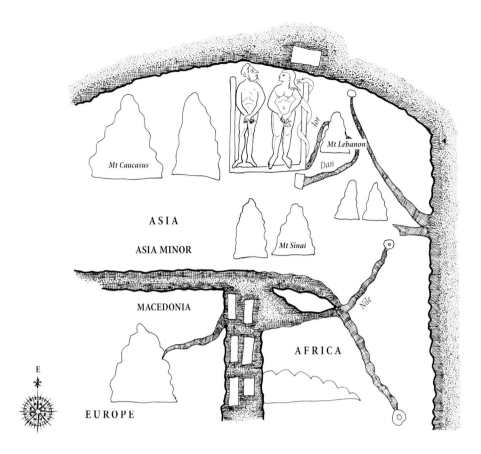

Fig. 5.15b. Diagram of Fig. 5.15a.

from the monastery of Oña, now in the Biblioteca Ambrosiana, Milan, the same rectangular sign for paradise as on the Osma map is found.[160] On a later map, thought to belong to the same group, paradise is depicted by a slight variation of the geometric form, a circle within which the four rivers issue from a central point to flow out in four different directions. On another (Figure 5.13) only the Euphrates is visible.[161]

The pictorial representations of paradise on the second and more common group of maps are rather more detailed. On the Pierpont Morgan Library map, for example, a rectangular Garden of Eden contains Adam and Eve, portrayed immediately after the Fall covering their shame with fig leaves and standing guiltily beside the Tree of the Knowledge of Good and Evil, around the trunk of which is coiled the snake (Figure 5.12). Later maps of the same group have a similar sign for paradise. A twelfth-century map from Santo Domingo de Silos, rectangular in outline and lacking the apostles, has Adam and Eve in a rectangular paradise (see Figure 5.14 and Plate 3b). The circular early twelfth-century map from a manuscript copied in Catalonia depicts all the main features of the paradise narrative – Adam and Eve, the Tree of Knowledge and the snake – without any enclosing rectangle or circle (Figure 5.15 and Plate 2b).[162] Interestingly, on this map – as well as on the tenth-century map now in Gerona and on other maps in the same group (for example, the map in a late twelfth-century manuscript now in Manchester, England)[163] – the river Jordan, where Christ was baptized, and the mountains of Lebanon, which Beatus called a symbol of baptismal purity, are shown in the immediate vicinity of the site of the Fall, as a token of the purification of corrupted man by the new paradise of the Church.[164]

The portrayal of Adam and Eve on the various copies of Beatus's map is also significant in making the link between the origin of the world's population in Adam and its conversion to Christianity through the work of the apostles.[165] The vignette marks the first age, while the map as a whole describes the earth at the sixth age. Beyond these two extremes – that is, before and after human history – there is, as Beatus explained, a mystery that is not for humans to explore: Holy Scripture deals only with the time span of the six ages and so, accordingly, does the map.[166] Beatus drew what he perceived to be a map of the Church, that is, a map of heaven on earth. As mankind waited for the Second Coming of Christ, the Church was already organized in a hierarchy of bishops and priests that united mankind with Christ. In Beatus's words, 'through these ranks the city of Jerusalem descends to earth, and every day, through the same ranks, it ascends to heaven'.[167] Beatus mapped the Garden of Eden as the starting point of human history, a type of the Church, and an anticipation of heaven.

Paradise was included on T-O maps, on the Vatican map and on Beatus's map, in order to give visual and geographical form to the typological and literal reading of the Holy Scriptures. Biblical history implied that the Garden of Eden had a geographical location and that it was the stage of a crucial event in salvation history. Later in the Middle Ages, the capacity of maps to include a conception of history in their geographical representation was exploited to the fullest. On later medieval *mappae mundi*, time was made to move across space. In the next chapter, we shall see how twelfth- and thirteenth-century *mappae mundi* offered an encyclopedic vision of the world as structured by mankind's historical pilgrimage from Eden to heaven.

1. For the dual meaning of the Hebrew term מִקֶּדֶם (*miqedem*) see Chapter 2, p. 35.
2. A number of modern scholars believe that sufficient memory of Roman cartographical methods may have survived well into medieval times to account for certain particulars on medieval regional and world maps. See, for example, Evelyn Edson, *Mapping Time and Space: How Medieval Mapmakers Viewed their World* (London: British Library, 1997), pp. 10–11, 18–25, 108–111; Paul D. A. Harvey, *Medieval Maps* (London: British Library, 1991), p. 21; Peter Barber, 'Old Encounters New: The Aslake World Map', in Monique Pelletier, ed., *Géographie du monde au Moyen Âge et à la Renaissance* (Paris: Comité des Travaux Historiques et Scientifiques, 1989), pp. 69–88, esp. 77, where he cites Anna-Dorothee von den Brincken, 'Mappa mundi und Chronographia: Studien zur *imago mundi* des abendländischen Mittelalters', *Deutsches Archiv für Erforschung des Mittelalters*, 24 (1968), pp. 118–86, esp. 146–50. See also the discussion on the survival of manuscripts of, or derived from, Ptolemy's *Geography* in Late Antiquity and the early Middle Ages in J. Lennart Berggren and Alexander Jones, *Ptolemy's Geography: An Annotated Translation of the Theoretical Chapters* (Princeton, NJ–Oxford: Princeton University Press, 2000), pp. 41–52. The issue of Roman prototypes for medieval world and regional representation should not be confused with the separate issue of whether Roman maps were drawn to mathematical scale.
3. Patrick Gautier Dalché, *La 'Descriptio Mappe Mundi' de Hugues de Saint-Victor* (Paris: Études Augustiniennes, 1988), pp. 62–77.
4. For examples of most of the types mentioned here, see Paul D. A. Harvey and Raleigh A. Skelton, eds, *Local Maps and Plans from Medieval England* (Oxford: Clarendon Press, 1986). For others, see Catherine Delano-Smith and Roger J. P. Kain, *English Maps: A History* (London: British Library, 1999), pp. 7–27, and 181–4; Harvey, *Medieval Maps* (1991), pp. 19–37, and *The History of Topographical Maps: Symbols, Pictures and Surveys* (London: Thames and Hudson, 1980).
5. For a general survey of the use of itineraries in pre-modern times, see Catherine Delano-Smith, 'Milieus of Mobility: Itineraries, Route Maps and Road Maps', in James R. Akerman, ed., *Cartographies of Travel and Navigation* (Chicago: University of Chicago Press, in press), and Delano-Smith and Kain, *English Maps* (1999), pp. 142–61. For Roman land travel, see, for example, Benet Salway, 'Travel, *Itineraria* and *Tabellaria*', in Colin Adams and Ray Lawrence, eds, *Travel and Geography in the Roman Empire* (London–New York: Routledge, 2001), pp. 22–66.
6. Patrick Gautier Dalché, in his *Carte marine et Portulan au XIIe siècle: Le 'Liber de existencia riveriarum et forma maris nostri Mediterranei' (Pise, circa 1200)* (Rome: École Française de Rome, 1995), challenges the idea that nautical charts first appeared later in the thirteenth century, as in, for example, Tony Campbell, 'Portolan Charts from the Late Thirteenth Century to 1500', in Brian Harley and David Woodward, eds, *The History of Cartography*, I, *Cartography in Prehistoric, Ancient and Medieval Europe and the Mediterranean* (Chicago: University of Chicago Press, 1987), pp. 371–463, esp. 380–1. See also Patrick Gautier Dalché, 'Portolans and the Byzantine World', in Ruth Macrides, ed., *Travel in the Byzantine World* (Aldershot, Hants: Ashgate, 2002), pp. 59–71, and Delano-Smith and Kain, *English Maps* (1999), pp. 153–7.
7. The compiler of the Ebstorf map pointed out that readers of the map might find it 'of no small usefulness to see the direction to be taken' ... and to chose for themselves which monuments they will contemplate and which routes to travel: Konrad Miller, *Die Ebstorfkarte: Eine Weltkarte aus dem 13. Jahrhundert* (Stuttgart: J. Roth, 1900), p. 8: 'que scilicet non parvam prestat legentibus utilitatem, viantibus directionem rerumque viarum gratissime speculationis directionem [or: dilectionem]'.
8. Patrick Gautier Dalché has convincingly argued against the dangers of anachronistic clichés that too easily oppose a 'symbolic and spiritual' cartography to 'practical' maps: see his 'Cartes de Terre Sainte, cartes de pèlerins', in Massimo Oldoni, ed., *Fra Roma e Gerusalemme nel Medioevo: Paesaggi umani ed ambientali del pellegrinaggio meridionale*, 3 vols (Salerno: Laveglia, 2005), I, pp. 573– 612, and, on Matthew Paris, 592–7.
9. Daniel Connolly, 'Imagined Pilgrimage in

Itinerary Maps of Matthew Paris', *Art Bulletin,* 81 (1999), pp. 598–622. On Matthew Paris in general see Suzanne Lewis, *The Art of Matthew Paris in the 'Chronica majora'* (Berkeley: University of California Press; Aldershot, Hants: Scholar Press with Corpus Christi College, Cambridge, 1987) and Richard Vaughan, *Matthew Paris* (Cambridge: Cambridge University Press, 1958). For reproductions of several of Matthew's maps see Harvey, *Medieval Maps* (1991), pp. 74–5, 91.

10 Beatus of Liébana referred to a *formulae pictura*: Beatus of Liébana, *Commentarius in Apocalypsin*, ed. by Eugenio Romero Pose, 2 vols (Rome: Istituto Poligrafico e Zecca dello Stato, 1985), I, p. 193. Other words used to designate a map include *orbis pictus, orbis terrarum descriptio, forma, figura, pictura, tabula, imago mundi,* and even wholly verbal geographical descriptions were sometimes called *mappae mundi*: Patrick Gautier Dalché, 'Le Sens de *mappa* (*mundi*): IVe–XIVe siècle', *Archivum Latinitatis Medii Aevi,* 62 (2004), pp. 187–202; Paul D. A. Harvey, 'The Sawley Map (Henry of Mainz) and Other World Maps in Twelfth-Century England', *Imago Mundi,* 49 (1997), p. 38; Gautier Dalché, *La 'Descriptio Mappe Mundi'* (1988), pp. 89–95; David Woodward, 'Medieval *Mappaemundi*', in Harley and Woodward, eds, *The History of Cartography,* I (1987), pp. 287–8.

11 The term *mappae mundi* is used in a broad sense as referring to medieval maps of the world by David Woodward as well as Edson, *Mapping Time and Space* (1997); Paul D. A. Harvey, *Mappa Mundi: The Hereford World Map* (London: Hereford Cathedral–British Library; Toronto: University of Toronto Press, 1996); Harvey, *Medieval Maps* (1991), pp. 19–37; Anna-Dorothee von den Brincken, *Fines terrae: Die Enden der Erde und der vierte Kontinent auf mittelalterlichen Weltkarten* (Hanover: Hahnsche Buchhandlung, 1992); von den Brincken, *Cartographische Quellen: Welt-, See- und Regionalkarten* (Turnhout: Brepols, 1988), pp. 23–38; von den Brincken, 'Mappa mundi und Chronographia' (1968), pp. 118–86; Marcel Destombes, *Mappemondes AD 1200–1500: Catalogue preparé par la Commission des Cartes Anciennes de l'Union Géographique Internationale* (Amsterdam: N. Israel, 1964).

12 A comprehensive inventory of *mappae mundi* is found in Destombes, *Mappemondes* (1964), and in Woodward, 'Medieval *Mappaemundi*' (1987), pp. 298, 359–68. Further examples are constantly being brought to the attention of historians of cartography.

13 See the detailed critique of the classifying obsession by Patrick Gautier Dalché, '*Mappae mundi* antérieures au XIIIe siècle dans les manuscrits latins de la Bibliothèque Nationale de France', *Scriptorium,* 52/1 (1998), pp. 102–61, esp. 102–9.

14 A useful survey of early attempts at classification is given in Woodward, 'Medieval *Mappaemundi*' (1987), pp. 294–6. Woodward proposes his own typology, pp. 296–9. On the problems of map classification see also Delano-Smith and Kain, *English Maps* (1999), pp. 30–40, and Edson, *Mapping Time and Space* (1997), pp. 11–16. The opening up of the definition of 'map' in the last two decades of the twentieth century led not only to lively scholarly debate but also to a more catholic attitude to early maps, and to the understanding of medieval *mappae mundi* on their own terms: see the editors' 'Preface' in Harley and Woodward, eds, *History of Cartography,* I (1987), p. XVI, and Woodward, 'Medieval *Mappaemundi*' (1987), pp. 291–9 and bibliography therein. On 'the liberalization' of the definition of a map see also Delano-Smith and Kain, *English Maps* (1999), p. 1 n. 1. For a cartographer's reaction, see Alan M. MacEachren, *How Maps Work: Representation, Visualization, and Design* (New York–London: Guilford Press, 1995), pp. 96–101, 251; Barbara Petchenik and Arthur Robinson, *The Nature of Maps* (Chicago–London: University of Chicago Press, 1976), pp. 115–19. The term 'part of the world' is adopted here and elsewhere, instead of 'continent', as *pars mundi*, 'part of the world', was the term used in the Latin Middle Ages.

15 On the relationship between map and text, Patrick Gautier Dalché, 'De la glose à la contemplation: Place et fonction de la carte dans les manuscrits du haut Moyen Âge', in *Testo e immagine nell'alto Medioevo*, 2 vols (Spoleto: Centro Italiano di Studi sull'Alto Medioevo, 1994), II, pp. 693–771.

16 The practice of annotating text with marginal diagrams was a late antique procedure: see Patrick Gautier Dalché, 'Les Diagrammes topographiques dans les manuscrits des classiques latins (Lucain, Solin, Salluste)', in Pierre Lardet, ed., *La Tradition vive: Mélanges d'histoire des textes en l'honneur de Louis Holtz* (Turnhout: Brepols, 2003), pp. 291–306.

17 For a list of the main exegetical map subjects see Catherine Delano-Smith, 'Maps and Religion in Medieval and Early Modern Europe', in David Woodward, Catherine Delano-Smith, and Cordell D. K. Yee, *Plantejaments i objectius d'una història universal de la cartografia/Approaches and Challenges in a Worldwide History of Cartography* (Barcelona: Institut Cartogràfic de Catalunya, 2001), p. 183.

18 On Beatus's map see this chapter, pp. 104–16.

19 On Ptolemy's *Geography* see Oswald A. W. Dilke, 'The Culmination of Greek Cartography in Ptolemy', in Harley and Woodward, eds, *History of Cartography,* I (1987), pp. 177–200. For a critical introduction to the three books containing instructions for the construction of maps on a projection, with an English translation, see Lennart Berggren and Jones, *Ptolemy's Geography.* (2000). See also below, Chapter 9, pp. 254–8.

20 See my 'Defining *Mappaemundi*', in Paul D. A. Harvey, ed, *The Hereford World Map: Medieval World Maps and their Context* (London: British Library, 2006), pp. 345–54.

21 For topology in the earliest maps, see Catherine Delano-Smith, 'Cartography in the Prehistoric Period in the Old World: Europe, the Middle East, and North Africa', in Harley and Woodward, eds, *The History of Cartography,* I (1987), pp. 54–101. Patrick Gautier Dalché discusses the importance of the principle of 'topographical contiguity' in structuring space on medieval maps: 'Principes et modes de la représentation de l'espace géographique durant le haut Moyen Âge', in *Uomo e spazio nell'alto Medioevo* (Spoleto: Centro Italiano di Studi sull'Alto Medioevo, 2003), pp. 119–50, esp. 136; Gautier Dalché, 'Décrire le monde et situer les lieux au XIIe siècle: l'*Expositio mappe mundi* et la généalogie de la mappemonde de Hereford', in *Mélanges de l'École Française de Rome,* 113 (2001), pp. 343–409, esp. 374–7.

22 On the orientation of medieval maps see Burton L. Gordon, 'Sacred Directions, Orientation, and the Top of the Map', *History of Religions,* 10 (1971), pp. 211–27, and von den Brincken, 'Mappa mundi und Chronographia' (1968), pp. 175–85.

23 Brigitte Englisch, *Ordo orbis terrae* (Berlin: Akademie Verlag, 2002), suggests that maps of the world from the eighth to the thirteenth century were structured on some divinely inspired geometrical order.

24 See, for instance, Edson, *Mapping Time and Space* (1997), pp. 4–9. The principles of diagrammatic maps are briefly spelt out in Catherine Delano-Smith, 'Smoothed Lines and Empty Spaces: The Changing Face of the Exegetical Map before 1600', in Isabelle Laboulais-Lesage, ed., *Combler les blancs de la carte: Modalités et enjeux de la construction des savoirs géographiques (XVIIe– XXe siècle)* (Strasbourg, Presses Universitaires de France, 2004), pp. 17–34, esp. 31.

25 The term desecularization is used by Robert A.

26 For a thorough overview, which takes into account the Christianization of classical geographical knowledge, see Hervé Inglebert, *Interpretatio christiana: Les Mutations des savoirs (cosmographie, géographie, ethnographie, histoire) dans l'antiquité chrétienne: 30–630 après J.-C.* (Paris: Institut d'Études Augustiniennes, 2001).

27 Bianca Kühnel, *From the Earthly to the Heavenly Jerusalem: Representations of the Holy City in Christian Art of the First Millennium* (Rome–Freiburg im Breisgau–Vienna: Herder, 1987), pp. 72–89. Woodward, 'Medieval *Mappaemundi*' (1987), p. 299, quotes the contrasting examples of Lactantius (early fourth century), suspicious of scientific inspections, and Jerome (later in the same century), devoted to pagan learning.

28 See Robert A. Markus, 'How on Earth Could Places Become Holy? Origin of the Christian Idea of Holy Places', *Journal of Early Christian Studies*, 2/3 (1994), pp. 257–71. The later phenomenon of the Crusades in the twelfth century is also mentioned in the context of a new Christian interest in geography.

29 For a list of early pilgrims, see John Wilkinson, *Jerusalem Pilgrims before the Crusades* (Warminster: Aris and Phillips, 1977).

30 Jerome's *Liber locorum* has been published in *Onomastica sacra*, ed. by Paul de Lagarde (Göttingen: Horstmann, 1887; repr. Hildesheim: Olms, 1966), pp. 118–90. See also Inglebert, *Interpretatio christiana* (2001), pp. 81–2; Delano-Smith, 'Maps and Religion' (2001), pp. 180–1; Markus, 'How on Earth Could Places Become Holy?' (1994), p. 262; Pierre Maraval, *Lieux saints et pèlerinages d'Orient: Histoire et géographie des origins à la conquête arabe* (Paris: Cerf, 1985), pp. 27–8.

31 *Expositio totius mundi et gentium*, ed. by Jean Rougé, SC CXXIV (Paris: Cerf, 1966). The *Expositio* is the original pagan text, compiled about 359; the *Descriptio*, of uncertain date and authorship, is the Christian adaptation. See also Inglebert, *Interpretatio christiana* (2001), p. 97.

32 *Descriptio totius mundi*, IV, in *Expositio totius mundi et gentium* (1966), pp. 142–3.

33 *Ravennatis anonymi Cosmographia et Guidonis Geographica*, I.6–8, ed. by Moritz E. Pinder and Gustav Parthey (Berlin: Fridericus Nicolaus, 1860; repr. Aalen: Zeller, 1962), pp. 13–21. See also Youssuf Kamal, *Monumenta cartographica Africae et Aegypti*, 5 vols (Cairo: 1926–51), II/3 (1932), pp. 411–12. On the Ravenna cosmographer see Oswald A. W. Dilke, 'Cartography in the Byzantine Empire', in Harley and Woodward, eds, *The History of Cartography*, I (1987), p. 260.

34 On T-O maps see Inglebert, *Interpretatio christiana* (2001), pp. 100–4; Edson, *Mapping Time and Space* (1997), pp. 4–5; Woodward, 'Medieval *Mappaemundi*' (1987), pp. 296–8.

35 For example, Uwe Ruberg, '*Mappae mundi* des Mittelalters in Zusammenwirken von Text und Bild', in Christel Meier and Uwe Ruberg, eds, *Text und Bild: Aspekte des Zusammenwirkens zweier Künste in Mittelalter und früher Neuzeit* (Wiesbaden: Reichert, 1980), p. 555, thinks that the T-O maps are Roman in origin but have been Christianized by labelling the three parts of the world with the sons of Noah.

36 See, for example, James M. Scott, *Geography in Early Judaism and Christianity: The Book of Jubilees* (Cambridge: Cambridge University Press, 2002), pp. 23–43, 59–170. The *Book of Jubilees* was a retelling of the Old Testament story from the creation of the world to Israel's arrival at Mount Sinai, written in the middle of the second century BC and combining Hellenistic learning with biblical, religious conceptions. Mention is made, for example, of a paradise in eastern Asia: *Book of Jubilees*, IV.24. See also Inglebert, *Interpretatio christiana* (2001), pp. 73–81; Francis Schmidt, 'Naissance d'une géographie juive', in Alain Desrumaux and Francis Schmidt, eds, *Moïse géographe: Recherches sur les représentations juives et chrétiennes de l'espace* (Paris: Vrin, 1988), pp. 13–30, with a reconstruction of a hypothetical map, p. 23.

37 Paris, Bibliothèque Nationale, MS Lat. 16679, fol. 33v.

38 On Sallust maps see Edson, *Mapping Time and Space* (1997), pp. 18–21.

39 Ibid., pp. 21–4. Christianized T-O maps illustrate the description of the parts of the world in Lucan's history of the civil war between Caesar and Pompey in the first century AD. See, for example, the map from a thirteenth-century manuscript of Lucan's *Pharsalia* in Copenhagen, Det Kongelige Bibliotek, G.K.S. 2020–4to, fol. 102r.

40 Isidore, *Etymologiae*, XIV, ed. by Wallace Martin Lindsay, 2 vols (Oxford: Clarendon Press, 1911). On Isidore and paradise see above, Chapter 3, pp. 47–8.

41 Verdun, Bibliothèque Municipale, MS 28 (Belles Lettres 5), fol. 22v. Destombes, *Mappemondes* (1964), pp. 132–3.

42 'Collige fiet Adam'. Destombes, *Mappemondes* (1964), pp. 132–3, Figure VIIIa. Destombes reads 'Collige licet Adam'.

43 See Honorius Augustodunensis, *Imago mundi*, I.92, ed. by Valerie I. J. Flint, *Archives d'Histoire Doctrinale et Littéraire du Moyen Âge*, 49 (1982), p. 81. The association between the name Adam and the four cardinal points had been put forward by Augustine, *In Iohannis Evangelium tractatus CXXIV*, IX.10, ed. by Augustinus Mayer, *CCSL* XXXVI (Turnhout: Brepols, 1954), p. 98; Augustine, *Enarrationes in Psalmos*, XCV.15, ed. by Eliquis Dekkers and Johannes Fraipont, 3 vols, *CCSL* XXXVIII–LX (Turnhout: Brepols, 1956), XXXIX, pp. 1352–3 and Isidore, *Etymologiae*, III, XLII.2, LXXI.7.

44 Inglebert, *Interpretatio christiana* (2001), pp. 21, 22 n. 15; Katherine Clarke, *Between Geography and History: Hellenistic Constructions of the Roman World* (Oxford: Clarendon Press, 1999), pp. 72–6; Javier Teixidor, 'Géographie du voyageur au Proche-Orient ancien', *Aula Orientalis*, 7 (1989), p. 110.

45 Von den Brincken, 'Mappa mundi und Chronographia' (1968), p. 169; Woodward, 'Medieval *Mappaemundi*' (1987), p. 328.

46 See, for example, John B. Friedman, *The Monstrous Races in Medieval Art and Thought* (Cambridge, MA: Harvard University Press, 1981; repr. Syracuse, NY: Syracuse University Press, 2000), pp. 37–8. Patrick Gautier Dalché, 'Sur l'"originalité" de la "géographie" médiévale', in Michel Zimmermann, ed., *Auctor et Auctoritas: Invention et conformisme dans l'écriture médiévale* (Paris: École de Chartes, 2001), pp. 131–43, has offered a sharp criticism of modern definitions of medieval cartography as 'symbolic' or 'theological'.

47 That medieval *mappae mundi* show a history limited to the Bible is a half-truth recently questioned by Scott D. Westrem, *The Hereford Map: A Transcription and Translation of the Legends with Commentary* (Turnhout: Brepols, 2001). Westrem has shown that references relating directly to the Bible account for fewer than twenty out of the total of almost 1100 legends on the Hereford map (which will be discussed in the following chapter). In his words, although the Hereford map 'certainly portrays the world from a medieval Latin Christian point of view – literally under the Judgment Seat of Jesus Christ – it is by no means a giant illustrated Bible story' (p. XXVIII).

48 The historical relationship between Christianity and science has given rise to a lively debate. See, for example, David C. Lindberg and Ronald L. Numbers, eds, *God and Nature: Historical Essays*

on the Encounter between Christianity and Science (Berkeley–Los Angeles–London: University of California Press, 1986), esp. the essays by David C. Lindberg, 'Science and the Early Church', pp. 19–48, and Edward Grant, 'Science and Theology in the Middle Ages', pp. 49–75.

49 The association between the three sons of Noah and the three parts of the *orbis terrarum* became common after Isidore of Seville: Gautier Dalché, 'De la glose à la contemplation' (1994), pp. 712–13.

50 Augustine, *De civitate Dei*, XVI.9 (1955), p. 510, had objected that the existence of a completely independent antipodean race was just a conjecture and that there was no mention of the Antipodes in Scripture. In the twelfth century Manegold attacked the heretical belief in a human race cut off from the rest of mankind: Pierre Duhem, *Le Système du monde: Histoire des doctrines cosmologiques de Platon à Copernic*, 10 vols (Paris: Hermann, 1913–59), III (1915), pp. 64–5; John K. Wright, *The Geographical Lore of the Time of the Crusades: A Study in the History of Medieval Science and Tradition in Western Europe* (New York: American Geographical Society, 1925; repr. New York: Dover, 1965), p. 161. On the problem of the antipodes in the Middle Ages see Patrick Gautier Dalché, 'Entre le folklore et la science: La Légende des antipodes chez Giraud de Cambrie et Gervais de Tilbury', in *La leyenda, antropología, historia, literatura* (Madrid: Casa de Velázquez– Universidad Complutense, 1989), pp. 103–14, esp. 112–13; see also Gautier Dalché, 'L'Oeuvre géographique du cardinal Fillastre († 1428): Représentation du monde et perception de la carte à l'aube des découvertes', in Didier Marcotte, ed., *Humanisme et culture géographique à l'époque du Concile de Constance, autour de Guillaume Fillastre* (Turnhout: Brepols, 2002), p. 313 n. 69.

51 Jerome, *Commentarium in Hiezechielem*, II.5.5, ed. by François Glorie, CCSL LXXV (Turnhout: Brepols, 1964), p. 56: 'Hierusalem in medio mundi sitam, hic idem propheta testatur, umbilicum terrae eam esse demonstrans.' See also Isidore, *Etymologiae*, XIV.3.21. See Ingrid Baumgärtner, 'Die Wahrnehmung Jerusalems auf mittelalterlichen Weltkarten', in Dieter Bauer, Klaus Herbers, and Nikolas Jaspert, eds, *Jerusalem im Hoch- und Spätmittelalter: Konflikte und Konfliktbewältigung-Vorstellungen und Vergegenwärtigungen* (Frankfurt am Main: Campus, 2001), pp. 271–334; Anna-Dorothee von den Brincken, 'Mappe del Medio Evo: Mappe del cielo e della terra', in *Cieli e terre nei secoli XI–XII: Orizzonti, percezioni, rapporti* (Milan: Vita e Pensiero, 1998), pp. 38–40.

52 Scott D. Westrem has highlighted the complexity of the medieval tradition of Gog and Magog, usually defined too glibly in modern research and misleadingly generalized: 'Against Gog and Magog', in Sylvia Tomasch and Sealy Gilles, eds, *Text and Territory* (Philadelphia: University of Pennsylvania Press, 1998), pp. 54–75. The literature on the legend of Gog and Magog is a rich one, and includes some studies on the lengthy career of these tribes on maps: see, for example, Sverre Bøe, *Gog and Magog* (Tübingen: Mohr Siebeck, 2001); Andrew Gow, 'Gog and Magog on *Mappaemundi* and Early Printed World Maps: Orientalizing Ethnography in the Apocalyptic Tradition', *Journal of Early Modern History*, 2/1 (1998), pp. 61–88; Anna-Dorothee von den Brincken, 'Gog und Magog', in Walther Heissig and Claudius C. Müller, eds, *Die Mongolen* (Innsbruck–Frankfurt am Main: Pinguin–Umschau, 1989), pp. 27–9; David J. A. Ross, *Alexander Historiatus: A Guide to Medieval Illustrated Alexander Literature* (Frankfurt am Main: Athenäum, 1963; repr. 1988), pp. 34–5; George Cary, *The Medieval Alexander* (Cambridge: Cambridge Univesity Press, 1956), pp. 130–1; Andrew R. Anderson, *Alexander's Gate, Gog and Magog and the Inclosed Nations* (Cambridge, MA: Mediaeval Academy of America, 1932).

53 Vatican City, Biblioteca Apostolica Vaticana, MS Lat. 6018, fols 63v–64r. Transcriptions of the legends are found in 'Mappa mundi e codice Vatic. Lat. 6018', ed. by François Glorie, in *Itineraria et alia geographica*, CCSL CLXXV (Turnhout: Brepols, 1965), pp. 455–66 and Richard Uhden, 'Die Weltkarte des Isidorus von Sevilla', *Mnemosyne: Bibliotheca Classica Batavia*, 3rd series, 3 (1935–6), pp. 1–28. See also Peter Barber, 'Mito, religione e conoscenza: La mappa del mondo medievale', in *Segni e sogni della terra: Il disegno del mondo dal mito di Atlante alla geografia delle reti* (Novara: De Agostini, 2001), p. 52; Leonid S. Chekin, 'Easter Tables and the Pseudo-Isidorean Vatican Map', *Imago Mundi*, 51 (1999), pp. 13–23; Edson, *Mapping Time and Space* (1997), pp. 61–4; Edson, 'World Maps and Easter Tables: Medieval Maps in Context', *Imago Mundi*, 48 (1996), pp. 30–2; Gautier Dalché, 'De la glose à la contemplation' (1994), pp. 759–61; Jörg-Geerd Arentzen, *Imago mundi cartographica: Studien zur Bildlichkeit mittelalterlicher Welt- und Ökumenekarten unter besonderer Berücksichtigung des Zusammenwirkens von Text und Bild* (Munich: Fink, 1984), pp. 108–11.

54 See, for example, Uhden, 'Die Weltkarte des Isidorus' (1935–6), p. 22.

55 For a brief description of the codex see Birger Munk Olsen, *L'Étude des auteurs classiques latins aux XI et XII siècles*, 3 vols (Paris: Centre National de la Recherche Scientifique, 1982–9), I (1985), p. 349, and Ludwig Bethmann, 'Nachrichten über die von ihm für die *Monumenta Germaniae Historica* benutzten Sammlungen von Handschriften und Urkunden Italiens aus dem Jahre 1854', *Archiv der Gesellschaft für Ältere Deutsche Geschichtskunde*, 12 (1872), pp. 253–5. Gautier Dalché, 'De la glose à la contemplation' (1994), p. 761, and Jeanne Bignami-Odier, 'Une lettre apocryphe de saint Damase à saint Jérôme sur la question de Melchisedech', *Mélanges d'Archéologie et d'Histoire de l'École Française de Rome*, 63 (1951), pp. 183–4 n. 2, though noting the irregular pagination and the variety of texts, have argued in favour of the unity of the manuscript. I am grateful to Evelyn Edson and Patrick Gautier Dalché for having shared with me the unpublished results of their research on this manuscript.

56 Several folios in the manuscript (93r, 98r–98v, 127r–128v) are palimpsests dating back to the seventh century: Elias Avery Lowe, ed., *Codices latini antiquiores*, 12 vols (Oxford: Clarendon Press, 1934–66), I (1934), p. 16. Michel Andrieu, *Les Ordines romani du haut moyen âge*, 5 vols (Louvain: Spicilegium sacrum lovaniense, 1931–48; 1961–74), II (1971), pp. 469–88, editing the collection of liturgical instructions for performing the Roman rite found at folios 129r–129v, dated this text to the early eighth century on the basis of liturgical history. Lowe, as well as Pierre Salmon, *Les Manuscrits liturgiques latins de la Bibliothèque Vaticane*, 5 vols (Vatican City: Biblioteca Apostolica Vaticana, 1968–72), III (1970), p. 8, assigned the manuscript to the ninth century. Bethmann, in 'Nachrichten' (1872), dated most of the manuscript to the ninth century, but put the gathering with the map earlier; Chekin, however, 'Easter Tables' (1999), noticing that time-reckoning in the indiction tables was influenced by the Byzantine chronological style, dated the Vatican map to between 762 (the year before the first year indicated on the tables) and 777 (the last year of the first indiction cycle). Chekin acknowledged that the gathering with the map is not necessarily contemporaneous with the rest of the manuscript; on the basis of traces of Greek learning in the manuscript, he also suggested that the map bears evidence of Byzantine influence.

57 A similarity also noted by Chekin, 'Easter Tables' (1999), p. 15.
58 The ill-omened days were dubbed 'Egyptian' as the practice of identifying them through the study of astral influence was attributed to Egyptian astrologers. Bede's writings confirm that blood-letting according to the cycles of the moon was common medical practice around the eighth century: Bede, *Ecclesiastical History of the English People*, V.3, ed. by Bertram Colgrave and Roger A. B. Mynors (Oxford: Clarendon Press, 1969), pp. 458–63; on the Egyptian days, see Gundolf Keil, 'Die verworfenen Tage', *Sudhoffs Archiv für Geschichte der Medizin und der Naturwissenschaften*, 41 (1957), pp. 27–58, and Lynn Thorndike, *A History of Magic and Experimental Science*, 8 vols (New York: Columbia University Press, 1934–58), I (1943), pp. 685–9. I am grateful to David Juste for having discussed with me the astrological aspects of this manuscript. See his *L'Astrologie latine du VIe au Xe siècle*, BA Dissertation, Université Libre de Bruxelles (1997), pp. 134–9.
59 Wolfgang Hübner, *Zodiacus Christianus: Jüdisch-christliche Adaptationen des Tierkreises von der Antike bis zur Gegenwart* (Königstein: Anton Hain, 1983).
60 Stephen C. McCluskey, *Astronomies and Cultures in Early Medieval Europe* (Cambridge: Cambridge University Press, 1998), pp. 77–96.
61 Edson, *Mapping Time and Space* (1997), p. 62; Edson, 'World Maps and Easter Tables' (1996), p. 32; Gautier Dalché, 'De la glose à la contemplation' (1994), p. 760; Uhden, 'Die Weltkarte des Isidorus' (1935–6), p. 3.
62 Similar symbols are found on other maps, see Chekin, 'Easter Tables' (1999), pp. 13–14; Uhden, 'Die Weltkarte des Isidorus' (1935–6), p. 8. See *Lamberti S. Audomari canonici liber floridus codex autographus Bibliothecae Universitatis Gandavensis*, ed. by Albert Derolez (Ghent: Story-Scientia, 1968), p. 50 (facsimile edition and transcription of Ghent, Rijksuniversiteit Bibliotheek, MS 92, fol. 24v).
63 See *Lamberti S. Audomari canonici liber floridus* (1968), pp. 52, 187, 190, 449, 450 (fols 25v, 92r, 93v, 225r, 225v).
64 Chekin, 'Easter Tables' (1999), p. 15; Edson, *Mapping Time and Space* (1997), pp. 62–3; 'Mappa mundi e codice Vatic. Lat. 6018', ed. by Glorie (1965), p. 463 nn. 184–5; Uhden, 'Die Weltkarte des Isidorus' (1935–6), pp. 4–8; Gautier Dalché, 'De la glose à la contemplation' (1994), p. 760, associates the representation of an unknown austral land on the Vatican map with a similar feature on the Beatus map.
65 Bignami-Odier, 'Une lettre apocryphe de saint Damase' (1951), pp. 183–90.
66 Ernst von Dobschütz, ed., *Das Decretum Gelasianum de libris recipiendis et non recipiendis* (Leipzig: Hinrichs, 1912), with a description of the manuscript at pp. 169–70.
67 Eucherius, *Formulae spiritalis intellegentiae, Instructionum libri duo*, ed. by Carmela Mandolfo *CCSL* LXVI (Turnhout: Brepols, 2004). See Gilbert Dahan, *L'Exégèse chrétienne de la Bible en Occident médiéval* (Paris: Cerf, 1999), pp. 329–30.
68 Eucherius, *Instructionum libri duo*, II, in *Formulae spiritalis intellegentiae* (2004), pp. 199–203.
69 Ibid., II, pp. 203–4, 212–13.
70 Eucherius, *De situ hierusolimitane urbis atque ipsius Iudaeae epistola ad Faustum presbyterum*, in *Itinera Hierosolymitana saeculi IIII–VIII*, ed. by Paul Geyer, *CSEL* XXXIX (Prague–Vienna–Leipzig: Tempsky, 1898), pp. 123–34. See Herbert Donner, *Pilgerfahrt ins Heilige Land: Die ältesten Berichte christlicher Palästinapilger (4.–7. Jahrhundert)* (Stuttgart: Katholisches Bibelwerk, 1979), pp. 171–89. The *Instructiones* are thought to have been written between 428 and 434: see William G. Rusch, *The Later Latin Fathers* (London: Duckworth, 1977), pp. 159–60. Doubt has been expressed whether the author of the *Epistola ad Faustum* (*Letter to Faustus*) was Eucherius of Lyons (in which case the letter would date from between 414 and 449) or another, later, Eucherius. See Wilkinson, *Jerusalem Pilgrims* (1977), pp. 3–4. An English translation of the *Letter* is on pp. 53–5.
71 Theodosius, *De situ Terrae Sanctae*, ed. by Paul Geyer, in *Itineraria et alia geographica* (1965), pp. 113–25; see Donner, *Pilgerfahrt ins Heilige Land* (1979), pp. 190–225.
72 The earliest partially surviving topographical representation of the Holy Land is the mosaic map on the floor of the church at Madaba, Jordan (542/562). For a summary of medieval maps relating to the Holy Land see Catherine Delano-Smith, 'The Intelligent Pilgrim: Maps and Medieval Pilgrimage to the Holy Land', in Rosamund Allen, ed., *Eastward Bound: Medieval Travel and Travellers 1050–1500* (Manchester: Manchester University Press, 2004), pp. 107–30. On the map that may derive from Burchard of Sion's ten-year sojourn in the Holy Land see Paul D. A. Harvey, 'The Biblical Content of Medieval Maps of the Holy land', in Dagmar Unverhau, ed., *Geschichtsdeutung auf alten Karten: Archäologie und Geschichte* (Wiesbaden: Harrassowitz, 2003), pp. 55–63. For reproductions of William Wey's and Gabriele Capodilista's maps see Kenneth Nebenzahl, *Maps of the Bible Lands: Images of Terra Sancta through Two Millennia* (London: Times Books; New York: Abbeville Press, 1986), Plates 17 and 18. Marino Sanudo's early fourteenth-century map of the Holy Land served a more specific propaganda purpose: see Evelyn Edson, 'Reviving the Crusade in the Fourteenth Century: Sanudo's Schemes and Vesconte's maps', in Allen, ed., *Eastward Bound* (2004), pp. 131–55.
73 Eucherius, *Instructionum libri duo*, III (2004), pp. 13–14.
74 Ibid., IX, pp. 61, 67–8.
75 *Ex dictis sancti Hieronymi*, in *Scriptores Hiberniae minores*, ed. by Robert E. McNally, I, *CCSL* CVIIIB (Turnhout: Brepols, 1973), pp. 225–30.
76 Pseudo-Augustine, *Dialogus questionum LXV*, *PL* XL, col. 744.
77 Eucherius, *Instructionum libri duo*, I (2004), p. 82: 'Paradisus ubi esse definitur? Diversa super hac re sententia plurimorum est. Aliqui tamen hunc in terra esse definiunt, ut non potuerit nisi in terra habitare terrenus, quem tamen homini inaccessibilem effici divina virtute confirmant. ... Eden in Genesi quid accipiemus? Sacratissimum ipsum ad orientem paradisi locum.' Here Eucherius slightly modifies the wording of Eusebius as translated in Jerome's *Liber locorum*, ed. by Lagarde (1887; 1966), p. 150: 'Eden sacri paradisi locus ad orientem, quod in voluptatem deliciasque transfertur.' Significantly, Eucherius omits Jerome's point that 'Eden' in Hebrew means 'delight', and takes 'Eden', more straightforwardly than Jerome, as a proper noun of a specific location.
78 'India magna', mentioned in Pliny, *Naturalis historia*, VI.64, is found on the Ripoll map (1056) and Pietro Vesconte's map (c.1320); 'India Bragmanorum' on the Ripoll, Ebstorf, and Higden maps. Also in the *Expositio totius mundi*, VIII (1966), p. 148, the *Braxmani* are mentioned.
79 The island of Taprobana was mentioned by Solinus, *Collectanea rerum memorabilium*, 53.1, ed. by Theodor Mommsen, 2nd edn (Berlin: Weidmann, 1958), pp. 195–6; Isidore, *Etymologiae*, XIV.6.12; Orosius, *Historiae adversos paganos*, I.2.16, ed. by Marie-Pierre Arnaud-Lindet, 3 vols (Paris: Belles Lettres, 1990–1), I (1990), p. 16. See Ananda Abeydeera, 'Aspects mythiques de la cartographie de Ceylan de l'Antiquité à la Renaissance', in François Moureau, ed., *L'Île, territoire mythique* (Paris: Aux Amateurs de Livres, 1989), pp. 9–15.
80 On the controversial identification of various mountain ranges in Asia, see Yves Janvier,

La Géographie d'Orose (Paris: Belles Lettres, 1982), pp. 85–115.

81 Orosius, *Historiae adversos paganos*, I.2.27–33 (1990), I, pp. 19–20. See above, Chapter 3, p. 49 and p. 59 n. 35. See also Janvier, *La Géographie d'Orose* (1982), pp. 206–12. The theory of the origin of the Nile in Mauritania is found in Solinus, *Collectanea rerum memorabilium*, XXXII.2–8; Pliny, *Naturalis historia*, V.9.10, and Servius, *In Vergilii carmina commentarii, Aeneis* VIII.713, ed. by Georg Thilo and Hermann Hagen, 3 vols (Leipzig: Teubner, 1881–1902), II (1883–4) p. 304. See also Danielle Bonneau, *La Crue du Nil, divinité égyptienne* (Paris: C. Klincksieck, 1964), pp. 147–9.

82 For example, a 'stream of gold' was the Lydian river Pactolus, mentioned by Virgil and Isidore as containing in its mud a large quantity of gold dust: Isidore, *Etymologiae*, XIII.21.21; Virgil, *Aeneid*, X.142. 'Golden streams' were also said to water the towns of Mastaura in northern Caria, Troezen in the Peloponnesus, and Damascus in Syria: see William Smith, ed., *Dictionary of Greek and Roman Geography*, 2 vols (London: John Murray, 1856–7), I (1856), p. 643, II (1857), pp. 295, 1235. The Nile itself, the biblical Gihon, was called in Greek 'the golden stream', as pointed out by the sixth-century Byzantine historian John Malalas, author of a widely circulated text, which chronicled world history from Adam to his own day: John Malalas, *Chronographia*, I.6, transl. by Elizabeth Jeffreys, Michael Jeffreys, Roger Scott, and others (Melbourne: Australian Association for Byzantine Studies, 1986) p. 5. It is unlikely, however, that '*Crisocoras*' on the Vatican map stands for the Gihon, as the river is shown rising from the Taurus mountains.

83 Orosius, *Historiae adversus paganos*, I.2.46 (1990), p. 23. The context of Orosius's description of eastern Asia makes it unlikely that he was referring to any of the rivers flowing in Syria, Lydia or Bitinia. Various distortions from Orosius's *Chrysorrhoas* are found on later maps: '*Grisogaris*', on the Ripoll map; '*Crisoroas*', on the Sawley map; '*Cristoas*' on the Hereford map. See Uhden, 'Die Weltkarte des Isidorus' (1935–6), p. 16 n. 1.

84 Pliny, *Naturalis historia*, VI.21–4; Genesis 2.11–12; Bede, *Libri quatuor in principium Genesis usque ad nativitatem Isaac et electionem Ismahelis adnotationum*, I.2.11–12, ed. by Charles W. Jones, *CCSL* CXVIIIA (Turnhout: Brepols, 1967), p. 49. See above, Chapter 3, pp. 49 and 59 n. 34.

85 Isidore, *Etymologiae*, I.27; XVI.25. Edson, 'The Oldest World Maps: Classical Sources of Three Eighth-Century *Mappaemundi*', *Exploration and Colonization in the Ancient World*, 24 (1993), pp. 169–84, esp. 172–3.

86 Isidore, *Etymologiae*, V.38–9; the best edition of Isidore's short chronicle is found in *Chronica minora saec. IV.V.VI.VII*, ed. by Theodor Mommsen, *MGH AA*, XI (Berlin: Weidmann, 1893), pp. 482–8. See above, Chapter 4, p. 81 n. 41 and Jacques Fontaine, *Isidore de Séville: Genèse et originalité de la culture hispanique au temps des Wisigoths* (Turnhout: Brepols, 2000), pp. 220–4; Pierre Cazier, *Isidore de Séville et la naissance de l'Espagne catholique* (Paris: Beauchesne, 1994), p. 51, and Paul Merritt Bassett, 'The Use of History in the *Chronicon* of Isidore of Seville', *History and Theory: Studies in the Philosophy of History*, 15/3 (1976), pp. 278–92.

87 The text in Vat. Lat. 6018 lacks Isidore's calculation of the years elapsed since Adam. On the *Synonima* attributed in Late Antiquity and early Middle Ages to Cicero see Paolo Gatti, *Synonima Ciceronis: La raccolta Accusat, lacescit* (Trento: Università degli Studi, 1994).

88 Catherine Delano-Smith, 'Geography or Christianity? Maps of the Holy Land Before AD 1000', *Journal of Theological Studies*, 42/1 (1991), pp. 150–2.

89 Pliny, *Naturalis historia*, III.31, IV.105; Orosius, *Historiae adversus paganos*, I.2.63–5 and 67 (1990), pp. 28–9. See Uhden, 'Die Weltkarte des Isidorus' (1935–6), p. 18.

90 Pliny, *Naturalis historia*, V.51–4. See Uhden, 'Die Weltkarte des Isidorus ' (1935–6), pp. 19–20.

91 Pseudo-Isidore, *In libros Veteris et Novi Testamenti*, *PL* LXXXIII, cols 175–6. The four evangelists with the regions where they wrote their gospels are represented in the upper basilica of St Francis in Assisi.

92 Gautier Dalché, 'De la glose à la contemplation' (1994), p. 760. The British Isles can be identified, despite the obscure names ('mare mortun ... oceanus', 'oceanus occiduus'), by comparing the Vatican map with other medieval maps: see Chekin, 'Easter Tables' (1999), p. 15; also Edson, *Mapping Time and Space* (1997), p. 62.

93 Isidore of Seville, *Chronicon*, in *Chronica minora* (1893), pp. 480–1. Another case in point are the writings of the Burgundian monk, Rodulfus Glaber (*c*.980–*c*.1046). His *Histories* are often described as an anxious warning of the apocalyptic destruction threatened for the year 1000. Glaber, however, only described the wonders and portents that accompanied the year 1000 (a millennium after the Incarnation) and the year 1033 (a millennium after the Passion). He certainly believed that he was living in the sixth, and last, age before the end and was particularly aware that the world around him was dramatically changing, but he gave no explicit hint that he expected the world to finish at either 1000 or 1033. See John France, 'Introduction' to Rodulfus Glaber, *The Five Books of History*, ed. and transl. by John France (Oxford: Clarendon Press, 1989), pp. LXIII–LXV.

94 We shall refer here to *Commentarius in Apocalypsin*, ed. Romero Pose (1985). See also Beatus of Liébana, *In Apocalipsin*, ed. by Henry A. Sanders, 2 vols (Rome: American Academy, 1930). See John Williams, *The Illustrated Beatus: A Corpus of the Illustrations in the Commentary on the Apocalypse*, 5 vols (London: Harvey Miller, 1994–2003); Williams, 'Isidore, Orosius and the Beatus Map', *Imago Mundi*, 49 (1997), pp. 7–32; Claudio Sánchez Albornoz, 'El Asturorum Regnum en los días de Beato de Liébana', and Luis Vázquez de Parga, 'Beato y el ambiente cultural de su época', in *Actas del Simposio para el estudio de los codices del 'Comentario al Apocalipsis' de Beato de Liébana*, 3 vols (Madrid: Joyas Bibliográficas, 1978–80), I (1978), pp. 19–32, 33–46.

95 See, for example, Juan Gil, 'Los terrores del año 800', in *Actas del Simposio ... Beato de Liébana*, I (1978), pp. 215–47. Giuseppe Tardiola talks about a 'naïve illumination featuring Adam and Eve', *Atlante fantastico del Medioevo* (Anzio: De Rubeis, 1990), p. 29: 'un'ingenua miniatura raffigurante Adamo ed Eva'. Discarding Beatus's remarks on the unknowability of the time of the end as a routine homage to orthodoxy, Richard Landes has emphasized the grip of apocalyptic concerns on even Beatus's imagination: 'Lest the Millennium Be Fulfilled: Apocalyptic Expectations and the Pattern of Western Chronography, 100–800 CE', in *The Use and Abuse of Eschatology in the Middle Ages*, ed. by Werner Werbeke, Daniel Verhelst, and Andries Welkenhuysen (Louvain: Louvain University Press, 1988), pp. 192–4.

96 The question of apocalyptic expectations in the Middle Ages is complex and the object of a vast literature. See, for instance, the discussions on apocalyptic speculations around the year 1000 in Richard Landes, Andrew Gow, and David C. Van Meter, eds, *The Apocalyptic Year 1000: Religious Expectation and Social Change, 950–1050* (Oxford: Oxford University Press, 2003); Johannes Fried, 'Endzeiterwartung um die Jahrthausendwende', *Deutsches Archiv für Erforschung des Mittelalters*, 45/2 (1989), pp. 381–473; and the critique of misleading generalizations about medieval

apocalyptic concerns in Sylvain Gouguenheim, *Les Fausses Terreurs de l'an mil* (Paris: Picard, 1999). On the discussion of the apocalyptic character of Beatus's Commentary see John Williams, 'Purpose and Imagery in the Apocalypse Commentary of Beatus of Liébana', in Richard K. Emmerson and Bernard McGinn, eds, *The Apocalypse in the Middle Ages* (Ithaca–London: Cornell University Press, 1992), pp. 217–33. On the apocalyptic character of Beatus manuscripts see Bianca Kühnel, *The End of Time in the Order of Things: Science and Eschatology in Early Medieval Art* (Regensburg: Schnell–Steiner, 2003), pp. 209–21.

97   Beatus of Liébana, *Commentarius in Apocalypsin* (1985), I, p. 4. See Adolfo Baloira Bértolo, 'El Prefacio del Comentario al Apocalipsis de Beato de Liébana', *Archivos Leoneses*, 71 (1982), pp. 7–25.

98   Williams, 'Isidore, Orosius and the Beatus Map' (1997), p. 15; Williams, 'Purpose and Imagery' (1992), pp. 218, 226; Kenneth B. Steinhauser, *The Apocalypse Commentary of Tyconius: A History of its Reception and Influence* (Frankfurt am Main–New York: P. Lang, 1987), pp. 141–96. Williams suggests that an illustrated version of Tyconius's commentary influenced both text and images of Beatus's commentary on the Apocalypse.

99   See, for example, Augustine, *Sermones de Vetere Testamento*, ed. by Cyrillus Lambot, CCSL XLI (Turnhout: Brepols, 1961), p. 361; and Clementina Mazzucco, 'La Gerusalemme celeste dell'"Apocalisse" nei Padri', in Maria Luisa Gatti Perer, ed., *La dimora di Dio con gli uomini (Ap. 21,3): Immagini della Gerusalemme celeste dal III al XIV secolo* (Milan: Vita e Pensiero, 1983), pp. 61–3.

100 Beatus of Liébana, *Commentarius in Apocalypsin* (1985), II. p. 207: 'Haec omnia et hodie spiritaliter in ecclesia geruntur quae adhuc perspicue in diem iudicii facta creduntur.'

101 Ibid., II, p. 105: 'Quotiens spiritus futura promittit, et praeterita narrat, et id futurum in ecclesia quod factum est praemonet.' See also II, pp. 280, 342, 414.

102 Ibid., II, pp. 249–58.

103 Ibid., II, pp. 105–6.

104 Ibid., I, p. 117: 'Ipsi enim sunt Iherusalem, quae cotidie de caelo descendit, id est, de populo sancto nascuntur sancti imitando sanctos, sicut bestia de abisso, quod est populus malus ex populo malo.' See also pp. 408, 540, 543; II, p. 387. See Tyconius, *Fragmenta Commentarii in Apocalypsin*, 93, in *The Turin Fragments of Tyconius' Commentary on Revelation*, ed. by Francesco Lo Bue (Cambridge: Cambridge University Press, 1963), pp. 72–3.

105 Augustine, *De civitate Dei*, XX.11 (1955), pp. 729, 730. Westrem, 'Against Gog and Magog' (1998), pp. 67–8.

106 Beatus of Liébana, *Commentarius in Apocalypsin* (1985), I, pp. 399–402; II, pp. 75–6, 358.

107 Ibid., II, pp. 93–4.

108 Ibid., II, pp. 98–100.

109 Ibid., II, pp. 180–8.

110 Ibid., II, p. 425.

111 Ibid., II, pp. 393–4. Beatus also viewed the angel that descends to earth (Revelation 21.17) as referring to the incarnated Christ: ibid., II, p. 304.

112 Ibid., II, pp. 393–410.

113 Ibid., II, pp. 405–11.

114 Ibid., II, pp. 411–12.

115 Ibid., II, pp. 395, 403–4.

116 Ibid., II, pp. 186–8.

117 Ibid., I, pp. 561–8; II, pp. 5–6, 295–8. According to Beatus, the red dragon in the sky (Revelation 12.3–4) indicated the presence of the devil within the Church herself: II, p. 101.

118 Ibid., II, pp. 258–62, 272.

119 Ibid., I, pp. 253–7.

120 See, for example, ibid., II, pp. 368, 380.

121 Ibid., I, pp. 608–09. The Spanish 'era' was a system of dating, popular in the Iberian peninsula from the fifth century down to the late Middle Ages, in which 38 years were added to the period elapsed since Christ's birth. The year 824 of the Spanish era thus corresponded to the year 786 elsewhere in the Latin West. Beatus's calculations were altered in different editions of his Commentary: see Williams, 'Purpose and Imagery' (1992), p. 224 n. 28.

122 Beatus of Liébana, *Commentarius in Apocalypsin* (1985), I, pp. 610–14. Beatus mentions other biblical accounts of the unpredictable mercy and justice of God, including Isaiah 38.1–5; Jeremiah 2–3, 7.16; 1 Kings 16.1.

123 Beatus of Liébana, *Commentarius in Apocalypsin* (1985), I, p. 612: 'quo anno, qua hora, qua die, qua era resurrectio sit nescimus'.

124 Ibid., I, p. 614: 'Ita … intellegere debet et expectare et timere omnis catholicus et hos XIIII annos tamquam unam horam putare et die noctuque in cinere et cilicio tam se quam mundi ruinam plangere et de supputatione annorum supra non quaerere.' As John Williams, 'Purpose and Imagery' (1992), pp. 223–5, notes, spiritual enrichment was the aim of Beatus's allusions to the imminent end of the world.

125 Beatus of Liébana, *Commentarius in Apocalypsin* (1985), I, p. 614. Isidore of Seville, *Chronicon*, in *Chronica minora* (1893), pp. 480–1.

126 New York, Pierpont Morgan Library, MS 644, fol. 293r: 'ut scientibus terreant iudicii futuri adventui peracturi seculi'. See Williams, *The Illustrated Beatus*, II (1994), p. 21; Williams, 'Purpose and Imagery' (1992), pp. 226–7.

127 In Beatus's view, references to the sixth age included the description of the peoples of the world mourning Babylon's downfall within a single hour (Revelation 18.19), and the day in which God would establish his house 'in the top of the mountains' (Isaiah 2.2), Beatus of Liébana, *Commentarius in Apocalypsin* (1985), II, pp. 314, 386. See also II, pp. 306–12, 624–36.

128 Ibid., II, pp. 344–54; see also II, p. 412. See Williams, 'Purpose and Imagery' (1992), p. 223. It is rather odd, then, that some modern scholars attribute to Beatus the very idea that he himself so clearly considered heretical. Beatus adhered to the biblical notion that, for God, days and millennia were identical (Psalm 90.4; 2 Peter 3.8). Augustine, too, had taken a broader reading of the 'thousand years' of the Book of Revelation as a reference either to the sixth millennium or to all of the world's history: *De civitate Dei*, XX.7 (1955), pp. 710–11.

129 Of the 26 extant illustrated manuscripts of Beatus's Commentary, only two are known to have been made outside Spain (the Saint-Sever Beatus, Paris, Bibliothèque Nationale de France, MS Lat. 8878, and the Berlin Beatus, Berlin, Staatsbibliothek Preussischer Kulturbesitz, MS Theol. Lat. Fol. 561), which is presumably a pointer of some sort to the original area of diffusion. See Williams, *The Illustrated Beatus* (1994–2003), and for a synthetic list: I (1994), pp. 10–11; Hermenegildo García-Aráez Ferrer, *La miniatura en los codices de Beato de Liébana (su tradición pictórica)* (Madrid: Enero, 1992), pp. 82–92 on the map; Betty A. Watson Al-Hamdani, 'Beatus of Liebana versus Elipandus of Toledo and Beatus's Illuminated Commentary on the Apocalypse', in *Andalucia medieval* (Cordoba: Monte de Piedad y Caja de Ahorros de Cordoba, 1978), pp. 153–63; Anscario M. Mundo and Manuel Sanchez Mariana, *El Comentario de Beato al Apocalipsis: Catálogo de los códices* (Madrid: Biblioteca Nacional, 1976); Peter K. Klein, *Der ältere Beatus-Kodex Vitr. 14–1 der Biblioteca Nacional zu Madrid: Studien zur Beatus-Illustration und der spanischen Buchmalerie des 10. Jahrhunderts*, 2 vols (Hildesheim–New York: Georg Olms, 1976); Wilhelm Neuss, *Die Apokalypse des Hl. Johannes in der altspanischen und altchristlichen Bibel-Illustration (das Problem der Beatus-Handschriften)* (Münster in Westphalia: Aschendorff, 1931).

130 Williams, 'Isidore, Orosius and the Beatus Map' (1997), pp. 9–26.
131 Patrick Gautier Dalché has already recognized the importance of Beatus's allegorical reading of Revelation for understanding the map, 'De la Glose à la contemplation' (1994), pp. 750–53.
132 Beatus of Liébana, *Commentarius in Apocalypsin* (1985), I, p. 3. Further on (II, p. 55), Beatus remarks on the need of preachers to clarify the difficult language of Scripture.
133 Ibid., I, p. 193: 'Et quo facilius haec seminis grana per agrum huius mundi, quem profetae laboraverunt et hi metent, subiectae formulae pictura demonstrat.' Mireille Mentré, *Contribucion al estudio de la miniatura en Leon y Castilla en la alta Edad Media (problemas de la forma y del espacio en la ilustración de los Beatos)* (León: Institución 'Fray Bernardino de Sahagún', 1976), pp. 67, 80–1 suggests that the map might have been used in teaching.
134 Beatus, *Commentarius in Apocalypsin* (1985), I, p. 169.
135 Ibid., I, pp. 186, 191–2. Beatus's source was the *De ortu et orbitu patrum*, attributed to Isidore, ed. by César Chaparro Gómez (Paris: Belles Lettres, 1985), pp. 215–17. Denise Cosgrove sees the allocation of specific regions to each apostle and of the world to Paul as an expression of the conflict in Christianity between 'local autonomy and imperial centralism': *Apollo's Eye: A Cartographic Genealogy of the Earth in the Western Imagination* (Baltimore–London: Johns Hopkins University Press, 2001), p. 57.
136 Edson, *Mapping Time and Space* (1997), p. 157. Williams, 'Isidore, Orosius and the Beatus Map' (1997), p. 24, suggests the possibility that the heads of the apostles were a later insertion.
137 Beatus, *Commentarius in Apocalypsin* (1985), II, pp. 390–2. In Beatus's view, the twelve angels and the twelve tribes of the children of Israel (Revelation 21.12) were a reference to the prophets and the patriarchs.
138 Beatus, *Commentarius in Apocalypsin* (1985), I, p. 192: 'Hi sunt duodecim horae diei, quae per Christum solem inluminantur. Hi sunt duodecim portae caelestis Iherusalem, per quas ad vitam beatam ingredimur. Hi sunt prima apostolica ecclesia, quam credimus fortissime supra Christum petram fondatam. Hi sunt duodecim throni iudicantes duodecim tribus Israhel. Haec est ecclesia per universum mundum seminatum. Rari fuerunt, sed electi. Et de his parvis granis multa seges surrexit.'
139 Beatus says that in the sixth age mankind has been renewed in the image and likeness of God: ibid., I, p. 636.
140 Beatus makes the point explicitly that the preaching of the gospel throughout the earth precedes the end of the world: ibid., I, pp. 554–61. See also Matthew 24.14.
141 Edson, *Mapping Time and Space* (1997), p. 157, defines the depiction of the land to the south of Africa 'the most striking common feature' of Beatus maps.
142 Isidore, *Etymologiae*, XIV.5.17 (1911): 'Extra tres autem partes orbis quarta pars trans oceanum interior est in meridie, quae solis ardore incognita nobis est. In cuius finibus antipodas fabulose inhabitare produntur.' 'In addition to the three parts of the earth, there is in the south, beyond the inner ocean, a fourth part, unknown to us because of the heat of the sun. The fabulous Antipodeans are said to live within its confines.' The passage is found, with some variations, on the map in Paris, Bibliothèque Nationale de France, MS Lat. 8878, fols 45bisv–45ter, and in Girona, Museu de la Catedral, Num. Inv. 7 (11), fols 54v–55r, for example.
143 For example, von den Brincken, *Fines terrae* (1992), pp. 56–8, and Arentzen, *Imago mundi cartographica* (1984), pp. 108–10; for a later example of Macrobian zonal map see below, Figure 8.17a.
144 Von den Brincken, *Fines terrae* (1992), p. 58.
145 The inscription on Pierpont Morgan Library 644 reads: 'Deserta terra vicina solida ardore incognita nobis', in Williams's translation ('Isidore, Orosius and the Beatus Map' (1997) p. 30 n. 52): 'The neighbouring desolate land is unknown to us due to its harshness.' It is a variation of 'Deserta terra vicina soli[s] ab ardore incognita est', in Williams's translation (p. 18): 'a neighbouring desert land unknown to us because of the heat of the sun'. The inscriptions on the Osma and Lorvão maps begin: 'Haec regio ab ardore solis incognita nobis', 'This region because of the heat of the sun is not known to us.'
146 Augustine, *De civitate Dei*, XVI.8 (1955), pp. 508–10.
147 Williams, 'Isidore, Orosius and the Beatus Map' (1997), pp. 17–23.
148 Beatus hints at the fourfold division of the earth, referrring it to the four cardinal points, throughout his *Commentary*: see, for instance, *Commentarius in Apocalypsin* (1985), I, pp. 221, 466, 519, 525; II, p. 210. See also below, nn. 150–51.
149 See Tertullian, *De baptismo*, I.3, ed. by Raymond F. Refoulé (Paris: Cerf, 2002), p. 65. The fish, an early symbol of Christian batpism, was also a symbol of Christ. The letters of the Greek word for fish ($\imath\chi\theta\acute{u}\varsigma$) formed the initials of the Greek words for Jesus ($\mathrm{'}I\eta\sigma o\tilde{u}\varsigma$) Christ ($X\rho\iota\sigma\tau\acute{o}\varsigma$), of God ($\theta\varepsilon o\tilde{u}$), the Son ($\upsilon\acute{\iota}\acute{o}\varsigma$), the Saviour ($\sigma\omega\tau\acute{\eta}\rho$).
150 Beatus, *Commentarius in Apocalypsin* (1985), I, p. 169: 'Non enim, sicut conventicula haereticorum, in aliquibus regionum partibus coartatur, sed per totum terrarum orbem dilatata diffunditur.' See also I, pp. 27, 127, 246; II, pp. 64, 192, 338, 655–9.
151 Ibid., I, p. 497.
152 The absence of Gog and Magog from Beatus's map has been noticed by Westrem, 'Against Gog and Magog' (1998), p. 59, and von den Brincken, 'Die Ebstorfer Weltkarte im Verhältnis zur spanischen und angelsächsischen Weltkartentradition', in Hartmut Kugler and Eckhard Michael, eds, *Ein Weltbild vor Columbus: Die Ebstorfer Weltkarte: Interdisziplinäre Colloquium 1988* (Weinheim: VCH, Acta Humaniora, 1991), p. 139.
153 See above nn. 105–6.
154 Beatus, *Commentarius in Apocalypsin* (1985), I, pp. 510–11.
155 Ibid., I, p. 650: '… in omnem terram praedicatio illius porrigatur et apostolorum sonus mundi limites penetret'. See also I, pp. 643–4.
156 Ibid., I, p. 501.
157 Augustine, *In Iohannis Evangelium tractatus CXXIV*, CXVIII.4, ed. by Rabdodus Willems, CCSL XXXVI (Turnhout: Brepols, 1954), p. 656: 'Quadripartita vestis Domini Iesu Christi, quadripartitam figuravit eius ecclesiam, toto scilicet, qui quatuor partibus constat, terrarum orbem diffusam, et omnibus eisdem partibus aequaliter, id est concorditer distributam. Propter quod alibi dicit missurum se angelos suos, ut colligant electos eius a quatuor ventis; quod quid est, nisi a quatuor partibus mundi, oriente, occidente, aquilone et meridie?'
158 Beatus, *Commentarius in Apocalypsin* (1985), II, p. 393. The tripartite structure of the Heavenly Jerusalem, Beatus pointed out, also refers to the Trinity and to the three theological virtues practised by the apostles: II, pp. 393, 401–3, 411.
159 Ibid., I, p. 279: 'Paradisus enim ecclesiae figura est: et primus homo Adam *umbra futuri* est: et secundus Adam Christus *sol iustitiae* est, qui umbram caecitatis nostrae inluminat.' See also I, pp. 280–5.
160 Milan, Biblioteca Ambrosiana, MS F. sup. 150, fols 71v–72r. See Luis Vázquez de Parga, 'Un mapa desconocido de la serie de los "Beatos"', in *Actas del Simposio … Beato de Liébana*, I (1978), pp. 272–8; García-Aráez Ferrer, *La miniatura en los códices de Beato de Liébana* (1992), pp. 91–2.

161 Paris, Bibliothèque Nationale de France, MS Lat. 1366, fols 24v–25r (late twelfth century); Lisbon, Arquivo Nacional da Torre do Tombo, Cod. 160, fol. 34bisv (1189). See Soledad de Silva y Verástegui, 'Le "Beatus" navarrais de Paris (Bibl. Nat., Nouv. Acq. Lat. 1366)', *Cahiers de Civilisation Médiévale*, 40, (1997), pp. 215–32, and Anne de Egry, *Um estudo de o Apocalipse do Lorvão e a sua relação com as ilustrações medievais do Apocalipse* (Lisbon: Fundação Calouste Gulbenkian, 1972).

162 Turin, Biblioteca Nazionale Universitaria, MS I.II.1, fols 45v–46r. See Destombes, *Mappemondes* (1964), p. 41.

163 Girona, Museu de la Catedral, Num. Inv. 7 (11), fols 54v–55r (975); Manchester, The John Rylands University Library, MS Lat. 8, fols 43v–44r (c.1175).

164 Beatus, *Commentarius in Apocalypsin* (1985), I, pp. 120–1.

165 Beatus explains that just as the whole of the world's population was descended from Adam, so God became incarnate for the salvation of whole mankind: ibid., II, p. 649.

166 For Beatus, Isaiah's vision of the divine glory represented the two Testaments. Isaiah describes (6.2) the two Seraphim who conceal with their wings God's face and feet or, as the Hebrew text puts it where the Vulgate is ambiguous, their own faces and feet. Beatus observes that it is impossible to know what went before and what would go on after the history of mankind as it is known from the Bible: ibid., I, pp. 179, 647–9. From the tenth century onwards, some manuscripts of the *Commentarius* include a lengthy genealogy detailing the intermediate stages of salvation history from Adam to Christ, with small T-O maps showing the division of the three parts of the world amongst Noah's sons, and Jerome's commentary on Daniel, which also underlined that human history was now in its sixth and final age. Beatus mentions the genealogy of Christ: ibid., I, pp. 465–6. Jerome and other Church Fathers expected the appearance of the Antichrist after the fall of pagan Rome, the last of the four empires of the world, as prophesied in Daniel 7. Christian messianic expectations in the Middle Ages, however, were complex, and Daniel's prophecy was used against millennarianism, on the basis of the argument that the ancient Roman Empire was considered to have continued in what historians now call the Holy Roman Empire. The progression of the four empires is discussed in many passages of Beatus commentary; see Williams, 'Purpose and Imagery' (1992), p. 226.

167 Beatus, *Commentarius in Apocalypsin* (1985), I, p. 452: 'Per hos gradus civitas Iherusalem descendit ad terram, per ipsos cotidie ascendit ad caelos.'

The Heyday of Paradise on Maps

For the whole sensible world is like a kind of book written by the finger of God.
Hugh of Saint Victor, *Didascalicon*, VII, *De tribus diebus*, 4 (late 1130s)

'Where is paradise?' the disciple asks his master in a twelfth-century dialogue by the theologian Hugh of St Victor. To answer the question, the teacher refers to a map: 'Why do you ask what you can see? You begin in the east; what you see here is the Tree of Life.'[1] The dialogue seems to take place in front of a map of the world that showed the earthly paradise – 'the beginning' – in the east.[2] The eager disciple was being urged here to look to the eastern edge of the map to see the place of mankind's origin. At the dawn of time, a specific event had taken place in a specific place, מִקֶּדֶם (*miqedem*), that is, *in oriente*, in the east, and *a principio*, at the beginning of history. As an event/place, paradise was a place simultaneously in the past and in the present, believed to exist still, somewhere on earth in a virtual present. Only a map including both the temporal and the spatial dimensions could accommodate this inaccessible paradise, impossible to pinpoint in any other way.

Any world map could be used to combine time and space, and from about the seventh century, if not before, the names of the three sons of Noah were sometimes added to the classical T-O diagram to indicate the historical notion of the post-diluvian division of the earth.[3] It was, however, the detailed *mappa mundi* that provided room for the depiction, in text or as an image, of the sites of past events. The format of the *mappa mundi* allowed the combination of the geographical features of the physical landscape with representations of the intangible features of history.

The Ordering of Space and Time

The reading of *mappae mundi* as pictorial representations combining time and space is not a modern construct. The point that space and time were interdependent, and that geographical space was the stage of salvation history, had been made particularly clearly and explicitly by Hugh of St Victor in the twelfth century.[4] Both during his lifetime and afterwards, Hugh gained widespread admiration not only for his learning and piety and the profundity of his theology, but also for his didactic commitment and outstanding ability to communicate his thinking through tables, diagrams, maps, and any mnemonic he could devise to clarify meaning, aid explanation and promote learning. Not surprisingly, he rose to become master of the abbey's school, which his reputation packed with students.[5] The study of history was for Hugh more than an exercise in the training of memory;[6] it was of paramount importance in understanding God's world. History narrated the works performed by God in time and space to redeem mankind.[7] God was both creator and redeemer of the world. God's *opus conditionis* (work of creation) had sprung from his love for humanity. When Adam and Eve's sin ruined the perfection of God's creation, God initiated his *opus restaurationis* (work of restoration), a process intended to restore humans to their divine likeness that culminated in the Incarnation of Christ. Each historical event was part of God's restorative plan as it unfolded through time and in space.[8] Hugh's conviction that secular knowledge and understanding of the world had to inform knowledge of the sacred history orchestrated by God and recorded in the Bible led to his interest in geographical learning.[9]

Hugh's portrait of the earth is found in his *Descriptio mappe mundi*, written about 1130–5. Hugh described in detail, possibly for students, a map of the world, listing the seas, islands, mountains, rivers, regions, cities, and peoples of Asia, Europe and Africa.[10] There is no mention in this text of the earthly paradise, and the Tigris, Euphrates, Nile and Ganges are not distinguished from the other rivers of the inhabited earth. Paradise is alluded to in Hugh's theological work dealing with the construction of the Ark, the *Libellus de formatione arche*, also known as *De archa Noe mystica* (1128–9). This treatise contains instructions for creating a drawing or diagram to show Noah's Ark superimposed on a map of the world and to present God's acts of redemption in their phases – past, present and future – and as they occur in space.[11] At the centre of the model was a map of the earth (*mappa mundi*) on which the beginning, Hugh said, was to be marked by the Garden of Eden and placed in the east, and the end, to be marked by the Last Judgement, located in the west.[12] The whole structure was to be embraced by the figure of Christ, from whose mouth descended a string of six circles representing the six days of creation.[13] The diagram was to show how the Church (signified by Noah's Ark, that those wanting to be saved were urged to board) was a manifestation of the divine plan unfolding in human history from the beginning to the end of time. In another text, *De arca Noe* (1124–7), Hugh synthesized the entire history of salvation, urging the believer to meditate upon God's work of restoration throughout human history, a work that was even more wonderful than his work of creation.[14]

The thrust of Hugh's argument was that God's *opus restaurationis* should also be considered in its spatial aspects.[15] Hugh explained in *De arca Noe* that God's work of restoration took place as a spatially ordered sequence of historical events. Much earlier, Augustine had already drawn attention to the way the end of the Assyrian Empire in the east had been followed by the beginning of the Roman Empire in the west, suggesting that the providentially determined succession of earthly kingdoms followed a specific temporal and spatial order.[16] Orosius's review of the whole of human history, *Historia adversus paganos* (c.418), had persuaded him too of a preordained transfer of imperial power (*translatio imperii*) and culture (*translatio studii*) from east to west, through Assyria, Macedonia, and Carthage to Rome.[17] The succession of the four world powers was suggested by the Christian reading of Daniel's interpretation of King Nebuchadnezzar's dream of a large statue made out of four different materials and smashed into pieces by a divine rock (Daniel 2.26–45).[18] In turn, Hugh developed the idea of the four empires more explicitly as a theology of geo-history. He explained that history had begun in the east, where paradise was, and that the weight of events was proceeding from the Orient to the Occident:

> *In the succession of historical events the order of space and the order of time seem to be in almost complete correspondence. Therefore, divine providence's arrangement seems to have been that what was brought about at the beginning of time would also have been brought about in the east – at the beginning, so to speak, of the world as space* [mundus] *– and then, as time proceeded towards its end, the centre of events would have shifted to the west, so that we may recognize out of this that the world* [saeculum] *nears its end in time as the course of events has already reached the extremity of the world in space* [mundus]. *Indeed, the first man was placed after his creation in the east, in the Garden of Eden, so that his progeny should spread throughout the orb from that origin. Likewise, after the Flood, the earliest kingdoms and the centre of the world* [mundus] *were in the*

> *eastern regions, amongst the Assyrians, the Chaldaeans and the Medes. Afterwards, dominion passed to the Greeks; then, as the end of the world* [saeculum] *approached, supreme power descended in the Occident to the Romans – who inhabited at the extremity, so to speak, of the earth* [mundus].[19]

It is significant that Hugh used two quite different words to refer to the world: *mundus* to indicate the world in its spatial extension, and *saeculum* to indicate the world in its temporal development. He made it clear that the two worlds were coincidental; as the beginning of time was in the east and the end of time would be in the west, so when the course of major events of human history reaches the end of the world as space (*mundus*), it also reaches the end of the world as time (*saeculum*). All historical events were distributed in the world according to this linking of temporal sequence and spatial order. What happened in the east, happened first. Adam's Fall in the earthly paradise had taken place in the extreme east of the earth (as Hugh had placed it on his mental map). As it moved westwards from the Orient in an echo of the sun's daily course, the Ark (the Church, existing since the beginning of the world) carried with it the centre of gravity of human history, the most important events defined from a global point of view. Not only political history, but also the arts and sciences followed the same east–west path. Cultural excellence had passed in turn from Chaldaea to Egypt, to Greece, and thence to Rome on the way towards the culmination of human history in the Incarnation of Christ and the ultimate revelation of the truth of the Kingdom of Heaven.

The idea of the progression of history from east to west lies at the heart of the *mappa mundi*. The importance of Hugh's mental map lies in its focus on the containment of the time–space continuum between an original paradise, only vaguely located in space, and the paradise to come, only vaguely located in time, and the way it provides us with a unique key for understanding the medieval *mappa mundi* as a map of historical process and not a static facsimile of the earth. *Mappae mundi* combined time and space to present an overview of human history dynamically open to the future. In portraying paradise, on the eastern edge of Asia (to mark the beginning of time) and simultaneously showing the main sites of God's intervention in the course of human history, the maps were anticipating the end of that history.

The fundamental east–west progression underlying the *mappae mundi* gives them a more or less standard internal structure and a common basic content. The order of the event/places marked on the maps records the order of human history as this proceeds from east to west. The starting point for appreciating the conceptual layout of a *mappa mundi* thus has to be the top of the map, where paradise is shown in the extreme east of the world, marking the start of time. From here to the Mediterranean Sea, a selection of geographical features of historical significance forms an axis of Christian belief and traces a path, as it were, down the centre of the map, linking the remote past and a distant eastern paradise with the contemporary and familiar western regions of the Mediterranean and northern Europe, as well as the six ages (from Adam to Christ) and the four world empires (from the early kingdoms in the eastern regions to the Roman Empire). Jerusalem, navel of the world, marks the end of the fifth age and the start of the sixth age, the present, and final, phase and the time of waiting, which thus coincides with the Mediterranean basin. At the bottom of the map, the Straits of Gibraltar – or the Pillars of Hercules, which are depicted on some maps – mark a geographical boundary and the western edge of the physical world. The central axis is strengthened by other details of the map's content. From paradise westwards down the map, places of biblical significance unfold in historical sequence. Below the Garden of Eden, some maps show the first city, Enoch, founded by Cain before the Flood.[20] Farther on is Noah's Ark, usually in Armenia, then the Tower of Babel in Mesopotamia and Abraham's city of Ur, in Chaldaea.[21] Farther west still are, usually, Joseph's barns in Egypt, the Exodus of the Israelites from Egypt to the Promised Land, Mount Sinai, the division of the land of Canaan among the Twelve Tribes of Israel, Jerusalem and features associated with the New Testament, such as the Nativity, the Crucifixion, and Resurrection of Christ.[22]

The historical succession of the four world empires further emphasizes the east–west ordering of human history. There was no general consensus about the identity of the four empires, but the various models all included the Assyrian/Babylonian, the Median/Persian, the Greek/Macedonian, and the Roman empires.[23] The importance of these kingdoms in human history was indicated on the maps by visual or textual references to their rulers, people or monuments. The foundation of the city of Babylon, for example, would be ascribed to Nimrod and its restoration to Ninus and Semiramis, or the great empire of Babylonia described as the land in which the people of Israel were held captive. The Persian king, Cyrus, might be recorded as the liberator of the exiled Jews, or a map sign might mark the ancient Persian city of Persepolis. Significantly, no attempt was made to represent Asia in contemporary terms. There was no need to do so; by the time of the *mappae mundi*, the weight of history had already long shifted west to Europe. The ancient empire of Rome had superseded that of Greece and Macedonia, the tangible vestiges of which were to be seen all around the shores of the eastern Mediterranean. Features such as the Delphic oracle of Apollo (often located, confusedly, on the island of Delos), Minos's labyrinth on Crete, and the camps of Alexander the Great were also indicated, as on the Ebstorf and Hereford *mappae mundi* (see Figures 6.15, 6.17, 6.20 and Plates 7 and 9). For many medieval chroniclers, the contemporary Holy Roman Empire was the current manifestation of the fourth kingdom, centred on Rome and ordained to last until the end of the world. Accordingly, the cartographical representation of Western Europe and the Mediterranean basin referred to the epoch in which the map maker himself was living. The physical relicts of Rome were part of the medieval landscape, often personally familiar to the compilers of the maps. Proximity in space went hand in hand with proximity in time, and Roman provincial boundaries could sit happily on the map alongside the sites of recent events and places of contemporary political, commercial or religious significance.

Medieval *mappae mundi* reflect a Eurocentric viewpoint. The historical perspective of the cartographical image was so ordered that it changed with increasing distance from Europe, and placed an Asia (and Africa) of the past – for which ancient sources were quarried – adjacent to a Europe of the present. The point was that in the total human history displayed on the map in the context of the whole world, each region had its own place on the stage and its own moment of importance on the continuum of world history.[24]

The Dominance of History over Geography

It was not by accident that, in putting his name to the Hereford *mappa mundi*, the presumed author, Richard of Haldingham, referred to his work as a 'history'.[25] History was the crucial factor in the structure and content of medieval *mappae mundi*. This was because map makers in the Middle Ages drew their maps following a temporal geography (different from the modern notion of historical geography, which has tended to be concerned with the influence of human agents in shaping the physical and cultural landscape).[26] Temporal geography implied that the more elaborate temporal pattern structured the less articulated spatial dimension. The medieval notion of time was more sharply defined and clearly articulated than the medieval notion of space.[27] Historical time had a beginning, went through a number of ages, and was heading for a determined end. Geographical space had not been surveyed with equal completeness. Europeans knew their neighbourhoods, read about the features of distant lands in classical sources, speculated on the zones of the globe, but had not established definitive boundaries around the space they inhabited. Failing a rigorously structured system of spatial relationships, the more developed system of temporal relationships inevitably prevailed. The map thus became a chronicle, and the spatial extension represented on it was a projection of human history.[28]

It needs to be borne in mind, moreover, that the medieval conception of time and its interplay with space was different from that of the post-Renaissance period, when time and space were defined according to principles established towards the end of the seventeenth century by Isaac Newton.[29] For Newton, time and space were permanent and universal properties to be measured in and quantified as equal units. In the Middle Ages, in contrast, time and space were not empty and infinite containers waiting to be filled with a series of events and a range of locations. Medieval space was what it contained, and time was what had happened. A place was important in itself, as the stage of an event, and not because of its situation within an abstract spatial system or because of its spatial relationship with other places. The notion of a paradise contiguous with yet inaccessible from the inhabited earth becomes contradictory only when space is conceived of as an abstract and infinite container, but not when it is conceived of as part of a well-defined temporal structure in which the places are linked to each other by historical, rather than spatial, relationships and do not, therefore, have to be accommodated within a grid constructed from measurements.

Map signs on *mappae mundi* represented not only localities or natural features but also historical events taking place there. They marked event/places. A great number of event/places were displayed on *mappae mundi* to form the spatio-temporal aggregate that made up mankind's universal history in its relation to geography. There were, however, sets of event/places that belonged to the same stratum of time or the same cultural tradition and that were spatially contiguous. When mapped, each set formed a spatio-temporal unit that we may call an 'epochal zone'. The ancient Persian or Roman empires, for example, were made manifest through the depiction on the map of their monuments, cities, and respective event/places; citations from Greek myths placed in Greece and the eastern Mediterranean evoked the Hellenistic Empire; event/places from the Old and New Testament, marked in Palestine, western Asia and northern Africa, stood for the story of salvation recorded in the Bible, just as the portrayal of Saint Anthony in the Egyptian desert represented the age of monasticism. An epochal zone, which embraced a whole age, was a wider spatio-temporal unit than the event/place, which was a single deed that had taken place in a particular location. On the map, however, a single map sign was usually all that was necessary to call to mind a wider epochal zone, irrespective of its duration or complexity. The city sign for Troy, for example, referred to the epochal zone of the ten-year war between Greeks and Trojans in Asia Minor. The Red Sea crossing could stand alone for all the events of the Israelites's forty-year Exodus from Egypt. Jerusalem, which had many different layers of association, was represented above all as the city of the time of Christ, and the fifth-century north African town of Hippo signified all that Saint Augustine represented. Likewise, the mountains behind which Gog and Magog were imprisoned by Alexander the Great could suggest the whole series of events that would take place at the end of the world. When each epochal zone was placed on the map in relation to its position on the space–time continuum, the result was a mosaic of event/places and epochal zones ordered historically from east to west.

Different places constituted medieval space, giving it heterogeneity and making it possible to talk about an edge, a centre, and qualitatively different parts. Similarly, medieval time was a sequence of different ages that were not structured according to a regular distribution of years, but were defined by key events. It was this particular conception of crucial happenings that bestowed importance on the areas where the events had occurred, allowed the interlinking of the temporal and the spatial dimensions, and enabled Hugh of St Victor, for example, to envision the world simultaneously as time (*saeculum*) and as space (*mundus*). As is often pointed out, much of our vocabulary for time is spatial and that for space is temporal: we talk casually about, for instance, a 'distant' event and a 'succession' of objects.[30] In the Middle Ages, however, the correspondence between time and space was far more than a metaphor.

The overriding importance of the historical dimension in the medieval concept of space accounts for the manner of presenting each event/place on the map as well as allowing a whole range of places of time-specific importance to be shown on the same map. Whereas the grid of astronomical coordinates of a Renaissance Ptolemaic map specified exactly where each place was to be plotted, there were no systematic or all-embracing spatial rules to impose order on the *mappa mundi*, only history. Historically important event/places were positioned on the map irrespective of the size of the features themselves or the distance on the ground between them. In the absence of an abstract spatial system that measured with universal standards the regions portrayed on the map, the space within each local area remained, so to speak, private and independent, and within each 'disconnected' area the significance of a place in world history could be emphasized by exaggerating the size of the sign. What mattered to the compiler of a *mappa mundi* was the intrinsic importance of each event/place. *Where* it was also mattered greatly, but only in terms of the identity of neighbouring places, not mathematically measured distance or astronomically defined direction.[31] On medieval world maps, epochal zones were represented according to their spatial *and* temporal contiguity. Their relative importance in history determined both their inclusion on the map and the size, and perhaps the nature, of the sign used to represent them. Geographical features were wholly subordinate to history, for which they were normally merely part of the backdrop.

The medieval vision of history and historical order that governed every *mappa mundi* set its internal structure, which contained several time layers. On reaching the Mediterranean Sea, the east–west axis of human history that ran from paradise to Jerusalem expanded to include the parts of the inhabited northern hemisphere – Europe and North Africa – that were closest in both time and space to the compilers of the *mappae mundi*. This was the final epochal zone of world history, that of the sixth age, the age that was inaugurated at the event/place of the Passion of Christ in Jerusalem and that concludes the march of history begun in the earthly paradise. The much wider geographical spread of this last epochal zone gave the map maker more room on the map for features of physical and human geography, current politics, or purely local detail, but now human history had reached its peak and the western progression of history had ceased, and the epochal zone of the sixth age filled its space to the limit.[32] Time could still pass, but it was different; nothing crucial was due to happen until the Second Coming. Meanwhile, a reflux movement was taking place. The process of Christianization was being radiated back to the areas covered by the epochal zones of the five preceding ages.

This reverse movement broke the map's east–west temporal structure, and added yet another layer of meaning to the cartographical representation. From the epochal zone of the sixth age, the Church was expanding to claim the whole earth, just as the rock of King Nebuchadnezzar's dream had broken the statue of many metals to become a huge mountain that filled the earth. As a sign that the sixth and final age had already arrived, the burial places of the apostles – who had been sent out to convert the peoples of the world to Christianity – are distributed over all three parts of the world.[33] Other signs on the maps indicated the sixth age and the waiting for the final revelation of the Apocalypse and the Second Coming. On the Hereford map, for example, Augustine is seen in Hippo, north Africa; in Egypt, a horned satyr points out to the temptations suffered by Saint Anthony; and a legend on a peninsula in northern Asia describes how the tribes of Gog and Magog were imprisoned behind the mountains by an earthquake and by Alexander the Great's indestructible wall (see Figure 6.15 and Plate 7). The note also tells of the horrific customs of this fearsome people, who fed on human flesh and blood, and were destined to break out of their enclosure at the time of the Antichrist.[34] An underlying contra-flow, on at least the largest *mappae mundi*, related to man's exploration of the wonders of God's creation. This additional layer is illustrated by a vast and weird array of animals, plants and natural wonders drawn from the accounts of the animal and the human kingdoms in late classical sources such as Pliny the Elder's *Naturalis historia* and Solinus's *Collectanea rerum memorabilium*, in which, as we have seen in Chapter 3, the marvels of nature were described in reverse order compared to the east–west descriptions of Christian historiography, that is, from Iberia in the west to India in the east. On the Hereford *mappa mundi*, for instance, texts in Asia report on Chinese silk traders and an abundance of snow in the mountains of the Far East, and on Indian elephants used in warfare.[35] In India, though, the author of the map seems to have been overwhelmed by all the monsters and marvels of the subcontinent's geography that he simply noted 'about all of which more may be read than depicted'.[36]

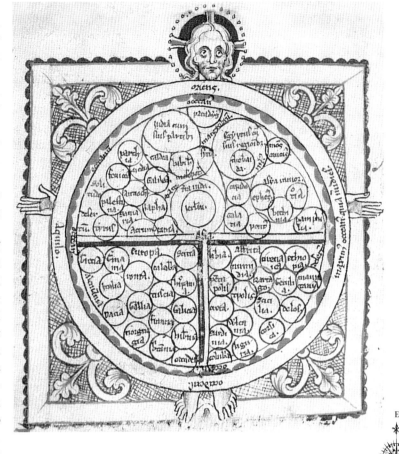

Fig. 6.1. The Lambeth Palace, or Reading, world map. Reading Abbey, c.1300. 7.7 × 7.5 cm (with Christ's head: 8.7 × 7.5 cm). London, Lambeth Palace Library, MS 371, fol. 9v. The text adjoining the map describes how after the Flood the earth was divided amongst the sons of Noah: Asia was given to Sem, Africa to Cham, and Europe to Japhet. The text also lists the provinces of the three parts of the earth (15 in Asia, 12 in Africa and 14 in Europe) and the peoples that descended from Japhet and inhabited Europe. The map, which includes more place names than those mentioned in the text, follows the tripartite division of the earth described in it. The earth is shown as embraced by Christ. A similar superimposition of the earth on the body of Christ is found on the Ebstorf map (Fig. 6.17 and Plate 9) and on the verso of the Psalter map (Fig. 6.16a shows the recto).

The Prime Epochal Zone: Paradise

The epochal zone of the earthly paradise was different from all others. It comprised not several, but a single event/place. It pointed to a place that was inaccessible from (although connected to) the inhabited earth. It was represented on the very edge of the world map expressly to underline the point that there was nothing beyond it. Paradise was the first event of world history and the easternmost place on earth. The exact position of paradise within an abstract spatial system was not an issue for the map maker, who was unconcerned with the precise measurement of distance and for whom the propinquity of the Garden with the boundary of the inhabited earth was a sufficient condition. The status of paradise governed the way the world's topography was presented. Other epochal zones could be placed 'next' to it, as to each other, but only in the topological sense. We see this, for example, on the tiny Lambeth Palace (or Reading) map of *c*.1300, where the world, embraced by a figure of Christ, is presented in a T-O structure, as an agglomerate of provinces, cities and islands, instead of through pictorial signs (Figure 6.1). Each place name has a circle around it. A semicircle in the Far East, at a tangent to the circle of India and containing the word *paradisus*, indicates the Garden of Eden. Jerusalem is near the centre of the map, within the land of Judah.

Places are distributed not according to scale or geographical coordinates, but according to the tripartite division of the world amongst the sons of Noah and the diffusion of their descendants over Europe, Asia, and Africa, which is described in detail in the adjoining text.[37] Status also accounted for the visual pre-eminence of paradise on most *mappae mundi*, on which it was privileged in terms of the size and the composition of the icon representing it in a variety of ways. On the early thirteenth-century *mappa mundi* now in Vercelli, Italy, for example, the Garden of Eden is represented at the top of the map, just below the word *Asia*, by a simple square (Figure 6.2). The square is outlined by double lines, indicating its walls, and contains nothing apart from a cross at the centre, probably signifying the single source and the four rivers that issued from it, but also a reminder of Christian redemption. The square is completely surrounded by block of text, and it is its emptiness that gives it its prominence.[38]

A legend on the Vercelli map explains that paradise is the most eastern region of Asia, that the Tree of Life and the source of the four rivers are found within it, and that the biblical Pishon is the Ganges and the Gihon is the Nile. Next to paradise are indicated the countries that border it, amongst which are scattered various exotic animals. Paradise appears to be located near India, not far from the tomb of Saint Thomas. Adjacent to paradise itself is another prominent sign reinforcing the point that

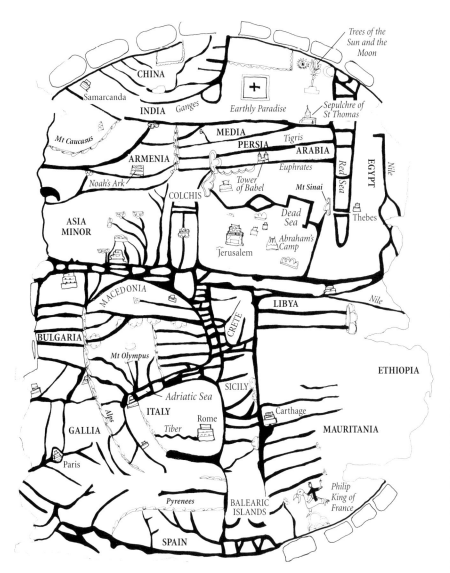

Fig. 6.2a (opposite). The Vercelli world map. Italy, France or England, c.1191–1218. Reproduced from Youssuf Kamal, *Monumenta cartographica Africae et Aegypti* (Cairo: 1926–51), III/5 (1935), p. 997. Original in Vercelli, Archivio e Biblioteca Capitolare, 80 × 70–2 cm. The Vercelli map was discovered in 1908 and is thought to have been brought to Italy by Cardinal Guala Bicchieri, papal legate to England from 1216 to 1218. Paradise is shown in Asia, next to India. The legend referring to the Garden of Eden describes several Indian phenomena, such as polygamy, suttee, precious stones (sapphires, emeralds, beryls, carbuncles), and mountains of gold (*Montes Aurei*) that were guarded by dragons and serpents.

Fig. 6.2b. Diagram of Fig. 6.2a.

The Heyday of Paradise on Maps   133

Fig. 6.3a. Map of the world from Ranulf Higden's *Polychronicon*. Ramsey, England, *c.*1350. 46 × 34 cm. London, British Library, Royal MS 14.C.IX, fols 1v–2r. In Book One of the chronicle, Higden described the various provinces of the earth in terms of the historical events that took place in them. The first province, he wrote, is paradise. In the second version of the text (*c.*1340), Higden declared his intention to include a map (see also Plate 4). The large rectangle at the top of the map reproduced here has been left blank, presumably in anticipation of a specialist illuminator who would have drawn Adam and Eve in paradise. The maps that were created for Higden's text appear to belong to the same tradition as the Sawley and Hereford maps (see Figs 6.10 and 6.15).

the Garden of Eden is on the eastern edge of the earth, that of the Trees of the Sun and of the Moon. According to medieval legend, these two oracular trees, which warned Alexander the Great of his imminent death, grew where the sun and moon rise and thus marked the easternmost margin of Asia.[39] Despite the proximity of neighbouring India, the isolation of paradise from the earth is reiterated in the surrounding text. The medieval reader is warned that the wall of fire surrounding paradise reached up to heaven and that, since Adam's Fall, no one was permitted to pass through to enter the Garden of Eden. A wealth of other event/places are also shown on the Vercelli map.[40] Thebes (seat of the Pharaohs) and Colchis (from where the Argonauts seized the Golden Fleece) are named. A tent in Syria indicates one of Abraham's camps on his journey to the Promised Land, and a rubric near the Dead Sea relates the destruction of Sodom and Gomorrah because of the sins of their inhabitants. Places such as the Tower of Babel,

134 Mapping Paradise: A History of Heaven on Earth

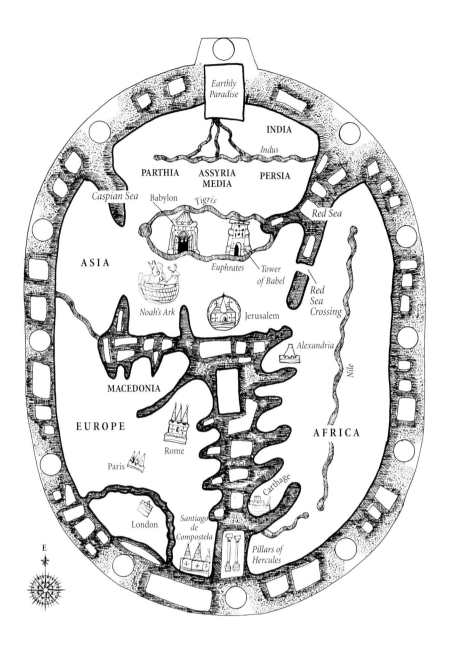

Fig. 6.3b. Diagram of Fig. 6.3a.

Noah's Ark, Bethlehem, Jericho, and the burial places of some of the apostles are also indicated. As intimated by a drawing of a multi-storied building labelled *Sepulcrum*, Jerusalem was to be thought of as the place of Christ's Resurrection. Contemporary references include the city of Alessandria, Italy (founded 1167) and, in Africa, a figure identified as Philip 'king of France', presumably Philip II (1165–1223) who had participated in the Third Crusade (1190–1).[41] In accordance with the underlying topological principle, no feature or region is shown on the map in proportion to its size or distance in measurable terms from its neighbours; Sicily is as big as India, and the Balkans appear to the north of Italy.

The two key criteria in the mapping of paradise, that it must be shown as adjacent to, but not part of, the inhabited earth and that it should be prominent, is seen no less clearly on other medieval world maps, such as those that illustrate copies of Ranulf Higden's *Polychronicon* (written between the 1320s and 1363) and the separate Evesham map (drawn between 1390 and 1415). Two maps included in the mid fourteenth-century copy of Higden's universal history produced at Ramsey Abbey, England, are tellingly incomplete.[42] On the larger and more detailed map, the large

The Heyday of Paradise on Maps 135

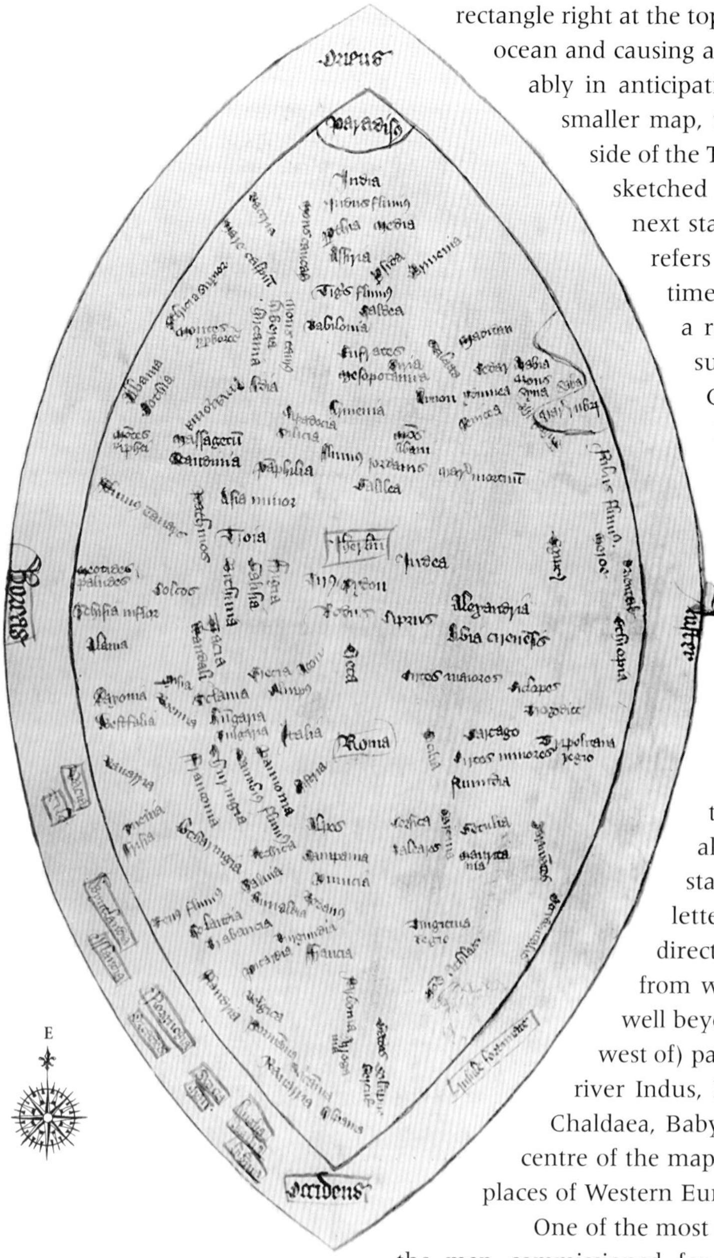

rectangle right at the top of the map, protruding through the encircling outer ocean and causing a projection of the map itself, was left blank, presumably in anticipation of a specialist illuminator (Figure 6.3). On the smaller map, the vignette of paradise, with Adam and Eve either side of the Tree of Knowledge of Good and Evil, has already been sketched out in pencil, although it too was left awaiting the next stage of illumination (Plate 4). In Higden's text, which refers to the standard arguments and authorities of his time, paradise is described as the first locality of the earth, a region as large as India situated in the farthest east, surrounded by a wall of fire, and out of which flow the Ganges, the Nile, the Tigris and the Euphrates. In Higden's view, the fact that the Garden of Eden had maintained its celebrity for more than six thousand years was itself testimony to its existence.[43] On both of Higden's maps, paradise is placed beyond India as the easternmost region of Asia. Amongst the most important event/places depicted on this map are Noah's Ark, the Tower of Babel, the passage of the Red Sea, the four world empires, the journeys of Alexander the Great, and Jerusalem (depicted here as a church). A later (after 1342), more stylized, variant of Higden's world map took the form of a mandorla (Figure 6.4).[44] Even on this almost wholly verbal map, paradise has been made to stand out clearly. The name *paradisus*, written in larger letters than anything on the map other than the cardinal directions, occupies the eastern extremity of the earth, from which it is closed off by a curved line, and is placed well beyond the closest place name, that of India. Below (i.e. west of) paradise and India, a succession of names follows – the river Indus, Parthia and Media, Assyria, Persia, the river Tigris, Chaldaea, Babylonia – and so on to Jerusalem (which is near the centre of the map and individualized by a large place sign) and to the places of Western Europe.

Fig. 6.4. Mandorla-shaped variant of the Higden world map. Hospital of St Thomas of Acon (Acre), Cheapside, London, mid-fourteenth century. 36 × 21 cm. London, British Library, Royal MS 14.C.XII, fol. 9v. The map features more than 150 toponyms, mostly classical and biblical. As on the Hereford map, paradise, Jerusalem and Rome are aligned on an east–west axis. The Garden of Eden is mapped as an inaccessible region both in the east and also at the beginning of human history.

One of the most eye-catching representations of paradise is found on the map commissioned for Evesham Abbey, Gloucestershire, in about 1390 (Figure 6.5 and Plate 5a). Paradise is represented by a disproportionately large rectangular inset superimposed on eastern Asia, and protruding beyond the edge of the map, containing a boldly executed scene of the Fall in the Garden of Eden.[45] Below (i.e. to the west), the Tower of Babel is depicted, no less bold but somewhat smaller than the figure of Adam in the inset. The four rivers are also shown in the inset. As on the other maps, the size, style and above all positioning of the vignette of paradise allows no doubt that paradise was the first earthly region in the east. The impression of primacy is also heightened here by the way the biblical scene is elaborately framed by what appears to be a throne – similar to the Abbot of Evesham's throne – possibly as a general symbol of divine authority and/or a specific reference to the Last Judgement.[46] Admittedly, though, it may have been that the artist was simply depicting, with Gothic exuberance, the entrance to a walled paradise through which the observer could see the scene of the Fall and glimpse part of the Garden.[47] Whatever the artist's intention, the map as a whole offers a dramatic depiction of the earthly paradise as of paramount importance, far away in the east, beyond India.

136   Mapping Paradise: A History of Heaven on Earth

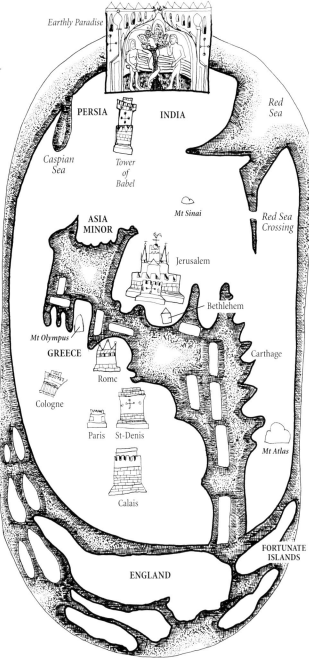

Fig. 6.5a. The Evesham world map. Evesham Abbey, Gloucestershire, c.1390–1415. 94 × 46 cm. London, College of Arms, Muniment Room 18/19. Paradise is represented by a large inset in the farthest east (see also Plate 5). In its rendering of the Mediterranean, the Red Sea and the British Isles, as well as in the general arrangement of the parts of the world, the Evesham map shows some similarity with the eighth-century Pseudo-Isidorean map in the Vatican Library (see Fig. 5.8). The map includes Mount Olympus, the Tower of Babel, the passage of the Israelites across the Red Sea (marked as *Transitus Ebreorum*), the cities of Jerusalem, Bethlehem, Rome, Paris and Cologne. The lower, western part includes several references to the Hundred Years' War. Calais, conquered by the English troops in 1347, and St-Denis, the burial place of the French kings, are given greater pictorial emphasis than either Paris or Rome.

Fig. 6.5b. Diagram of Fig. 6.5a.

The Heyday of Paradise on Maps   137

The Invisible Paradise on Maps

Not every map of the world drawn in the Middle Ages shows paradise. A medieval map of the world without direct reference to paradise by means of a special icon or place-name did not necessarily mean ignorance or denial of an earthly Garden of Eden, but simply that some map makers preferred to represent only the inhabited portion of the globe. Even these map makers, however, would refer indirectly to paradise by indicating the four rivers on the map. Sometimes, an associated text describing the topography of the world would include the earthly paradise. The four rivers were the crucial features on the map that linked the known regions to the remote Garden of Eden.

The earliest known map that includes only indirect references to paradise is the eighth-century map now in Albi, in southern France (Figure 6.6).[48] The Albi map, which is in a codex containing a miscellany of texts, is followed by a list of the twenty-five seas and the twelve winds of the world and then by the geographical chapter from Orosius's *Historia adversus paganos* (I.2).[49] The codex also contains a version of Iulius Honorius's fourth-century *Cosmographia*, which reported the traditional identification of the Nile with the Gihon but also identified the Pishon with the Danube.[50] The map itself seems to reflect in part Orosius's world view, and thus could be associated with the extract from the *Historia* included here, but it also contains place names taken from Book XIV of Isidore's *Etymologiae* and from Polemius Sylvius's *Nomina omnium provinciarum*, a brief geographical text that is also found in the codex, so it is difficult to be sure for which text, if any, the map was intended. The map shows the borders of Roman provinces, together with the major cities of classical Antiquity, such as Athens and Carthage, and the four empires of Babylon, Persia, Macedonia and Rome are named.

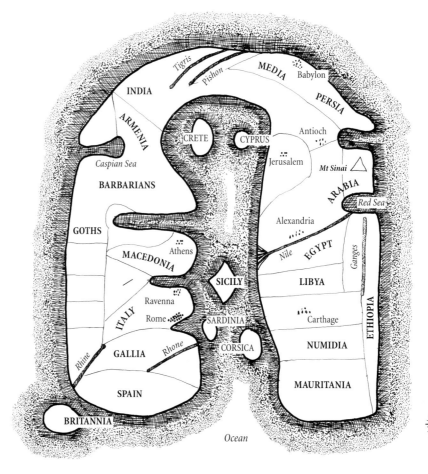

Fig. 6.6. Diagram of the Albi world map. Southern France or Spain, late eighth century. Size of the original 29 × 23 cm. Albi, Bibliothèque Municipale, MS 29, fol. 57v. The map is thought to have been based on a map dating back to classical times. The city of Ravenna in Italy features prominently as the capital of the Western Roman Empire. Armenia is marked, probably as a reference to Noah's Ark. Paradise is referred to only indirectly through the name of the Pishon river, shown here flowing between India and Media (at the top of the map). If the Pishon is to be identified with the Ganges and not, as Iulius Honorius thought, with the Danube, the Tigris and the Pishon are marked in reverse order. The Ganges, however, is shown as flowing south of the Nile, perhaps confused with the Gihon.

138 Mapping Paradise: A History of Heaven on Earth

Prominent in the Arabian peninsula is Mount Sinai, represented by a stemmed triangle. There is no explicit reference to the Garden of Eden. Instead, two rivers between India and Media are named as the Pishon and the Tigris. The Euphrates is only implied by the presence of Babylon, marked by a large sign, but the Nile is shown between the Red and the Mediterranean seas.

Another map lacking a sign for the earthly paradise is the Anglo-Saxon (Cotton) map of the world of about 1025–50 (Figure 6.7 and Plate 6).[51] This map is included in a copy of Priscian's fifth-century Latin translation of the descriptive poem *Periegesis* (*Journey around the World*) written by Dionysius of Alexandria in the first half of the second century. It is a hybrid of secular and Christian geography, and is generally accepted as containing elements deriving from Roman mapping traditions.[52] Besides features of human and physical geography, the map depicts places associated with biblical events such as Noah's Ark in Armenia, the territories of nine of the Twelve Tribes of Israel, the passage of the

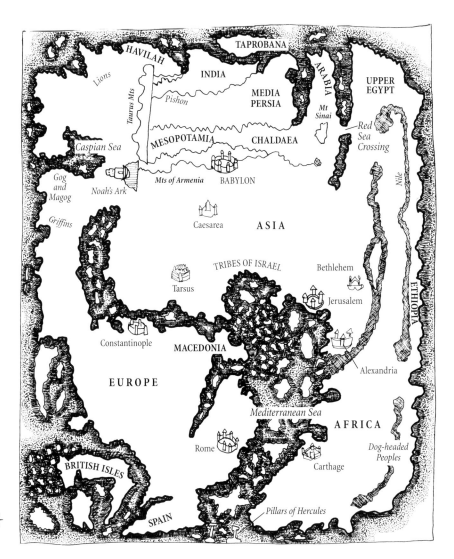

Fig. 6.7a. The Anglo-Saxon, or Cotton, world map. Canterbury, England, *c.*1025–50. 21 × 17 cm. London, British Library, Cotton MS Tiberius B. V., fol. 56v (see also Plate 6). The Anglo-Saxon map may have originated in a map in the Roman tradition, but it also can be seen to have been the object of careful Christianization. Mountains, rivers and cities are marked and the last are named. The British Isles are portrayed in the north-west (i.e. bottom left). Beside geographical features, historical time is represented in the portrayal of biblical events and characters from classical myths.

Fig. 6.7b. Diagram of Fig. 6.7a.

The Heyday of Paradise on Maps 139

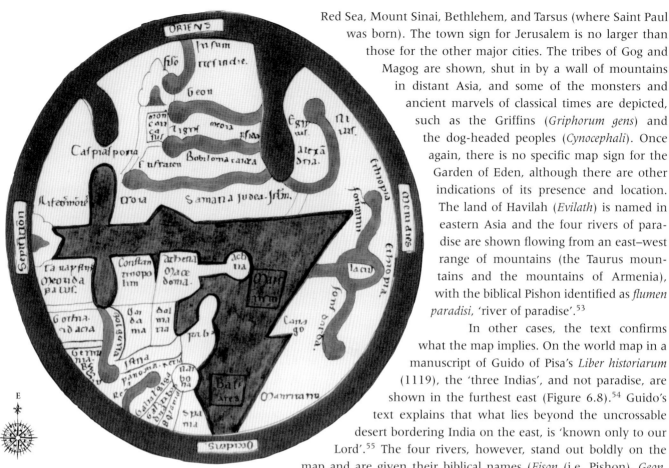

Red Sea, Mount Sinai, Bethlehem, and Tarsus (where Saint Paul was born). The town sign for Jerusalem is no larger than those for the other major cities. The tribes of Gog and Magog are shown, shut in by a wall of mountains in distant Asia, and some of the monsters and ancient marvels of classical times are depicted, such as the Griffins (*Griphorum gens*) and the dog-headed peoples (*Cynocephali*). Once again, there is no specific map sign for the Garden of Eden, although there are other indications of its presence and location. The land of Havilah (*Evilath*) is named in eastern Asia and the four rivers of paradise are shown flowing from an east–west range of mountains (the Taurus mountains and the mountains of Armenia), with the biblical Pishon identified as *flumen paradisi*, 'river of paradise'.[53]

In other cases, the text confirms what the map implies. On the world map in a manuscript of Guido of Pisa's *Liber historiarum* (1119), the 'three Indias', and not paradise, are shown in the furthest east (Figure 6.8).[54] Guido's text explains that what lies beyond the uncrossable desert bordering India on the east, is 'known only to our Lord'.[55] The four rivers, however, stand out boldly on the map and are given their biblical names (*Fison* (i.e. Pishon), *Geon*, *Tigris*, *Eufraten*). Another passage in Guido's manuscript, a quotation from Isidore's *Etymologiae*, includes a detailed description of the Garden of Eden as the first region of Asia.[56] Particularly interesting, perhaps, for their intimation of early antecedents, are the so-called 'Jerome maps' of Asia and Palestine (Figure 6.9). The two maps are bound in with a twelfth-century French manuscript containing three texts by Jerome: his *Liber locorum* (389–91), the translation into Latin of Eusebius's *Onomasticon* (c.324) and two other reference works providing lists of names of persons and places mentioned in the Bible.[57] The *Liber locorum* includes an entry for paradise, which Jerome placed vaguely in the east, describing it as the source of the four rivers.[58] Neither map contains a sign for paradise, but the Trees of the Sun and Moon, marking the eastern edge of the earth, are shown on the map of Palestine. Only three of the rivers appear on the Asia map, where the Ganges is identified as *Ganges vel Fison flumen*, but all four are shown and named on the map of Palestine, the Nile – like the Ganges – with its biblical as well as its geographical name.

Maps without a vignette of or direct reference to paradise may thus have lacked the explicit reference to the prime epochal zone, but they nonetheless reflected its presence. Their scope was different, inasmuch as they portrayed only the inhabited earth. The map maker was not particularly concerned with charting human history from the very beginning of the *saeculum*, the world as time, nor was he trying to map the *mundus*, the world as space, to its uttermost eastern limit and to show what lay beyond the boundary of the inhabited earth, namely the earthly paradise. Yet the presence of the four rivers, especially when the biblical names Pishon and Gihon were included, was sufficient to betray the existence, somewhere on the earth, of the inaccessible Garden of Eden. Given the topological structure of the medieval map, the four rivers served perfectly adequately as a signifier of the proximity of the earthly paradise.

Fig. 6.8. The world map in Guido of Pisa's *Liber historiarum*. ?Northern Italy, 1119. Diameter 13 cm. Brussels, Bibliothèque Royale de Belgique/Koninklijke Bibliotheek van België, MS 3897–3919, fol. 53v. Guido explains in the prologue that he wanted to obtain eternal fame through his work, in which he gave a number of maritime itineraries and descriptions of Italy and the world as well as historical material including a history of Rome, a chronicle from the time of Augustine to the year 1108, an account of the deeds of Alexander the Great and of the Trojan war in relation to Italy, and extracts from the history of Paul the Deacon (eighth century). The world map describes the three parts of the world arranged along a triangular Mediterranean Sea and watered by heavily coloured rivers: the four rivers of paradise – with the Gihon flowing as the Nile in Egypt and Ethiopia – the Tanais, or Don, and the Rhine. The manuscript now in Brussels once belonged to the Renaissance cardinal Nicholas of Cusa (1401–64).

Mapping God's Creation and Redemption

By the thirteenth century, the cartographical genre of *mappae mundi* was reaching its peak as an illustrated compendium of the history of mankind from the beginning in Eden to the sixth and final age that was to close with the end of the world. More than a century earlier, Hugh of St Victor had shown how knowledge of all earthly phenomena was essential for gaining a transcendental vision that would embrace the totality of history. The most detailed *mappae mundi* offered a graphic synthesis of human knowledge on the geographical stage, although it was simply impossible to accommodate on a single map all phenomena that were addressed in geographical treatises or historical works. Such a map did make it possible, however, for the human eye to embrace, as if from a higher viewpoint, what otherwise it could not see, namely the physical world created by God as the arena of the human history orchestrated by God.[59]

One feature on a *mappa mundi* that could be read as an explicit intimation of the end of time was the representation in north-east Asia of Gog and Magog, the dreaded apocalyptic tribes associated with Alexander the Great and his Asian journey. Gog and Magog were approached in various ways on different maps, and were identified with various ethnic groups in different texts.[60] Belief in a remote and isolated dwelling place of Gog and Magog coexisted with the widespread notion derived from the mainstream allegorical reading of the Book of Revelation that Gog and Magog could represent also threats from within the Church and within Christian Europe.[61] The inscriptions on some maps, however, warned that Gog and Magog would storm the world immediately prior to the Day of Judgement (Revelation 20.7–8).

Thus, the inclusion of Gog and Magog on the late twelfth-century Sawley map (as it is called, from the abbey in Yorkshire to which it was presented), inserted in a copy of the twelfth-century world chronicle by Honorius Augustodunensis, can be seen as an explicit allusion to the ultimate ending of the whole course of human history (Figure 6.10).[62] The compiler of the Sawley map placed Gog and Magog on a remote peninsula, ringed by mountains on three sides and locked in by a wall on the fourth side. Another apocalyptic reference on the Sawley map is found in the four angels that flank it. Various passages in the New Testament could have been the inspiration for these figures. They could be interpreted, for instance, as the four angels that according to the

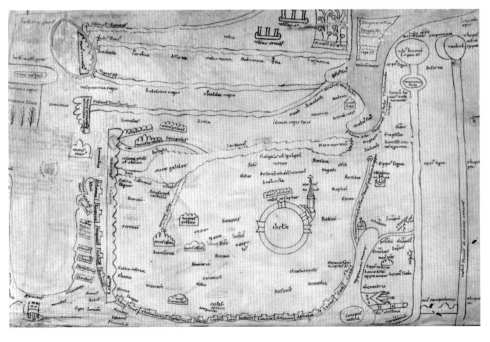

Fig. 6.9. The so-called map of Palestine, from a codex containing Jerome's *Liber locorum* (written *c.* 389–91). Tournai, France, late twelfth century. 35.5 × 23 cm. London, British Library, Add. MS 10049, fol. 64v. The map is centred on the Holy Land, but it also includes the rivers Ganges (Pishon) and the Indus – with the altars of Hercules and Alexander the Great between them – the Tigris and Euphrates in the east (top); the Nile, here given its biblical name Gihon as well, in the south (right); and Asia Minor and the Caspian Sea in the north (left). Only a thin strip of the Mediterranean Sea is shown in the west (bottom). The map features 69 places in Palestine. The sign for Jerusalem is a gated circle with the Tower of David, the Valley of Josaphat and Mount Sion.

The Heyday of Paradise on Maps 141

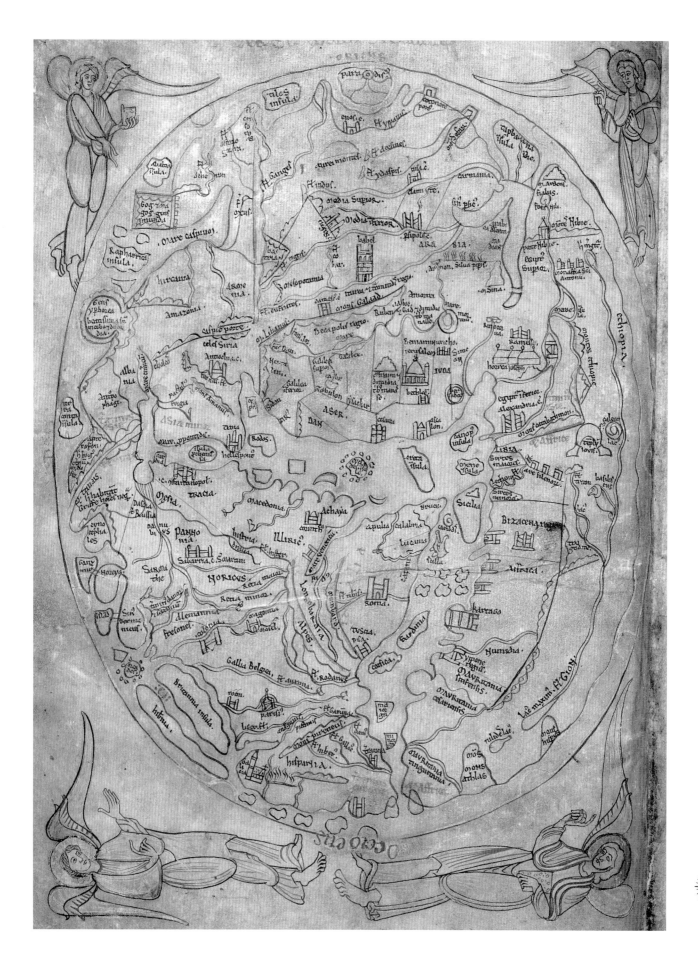

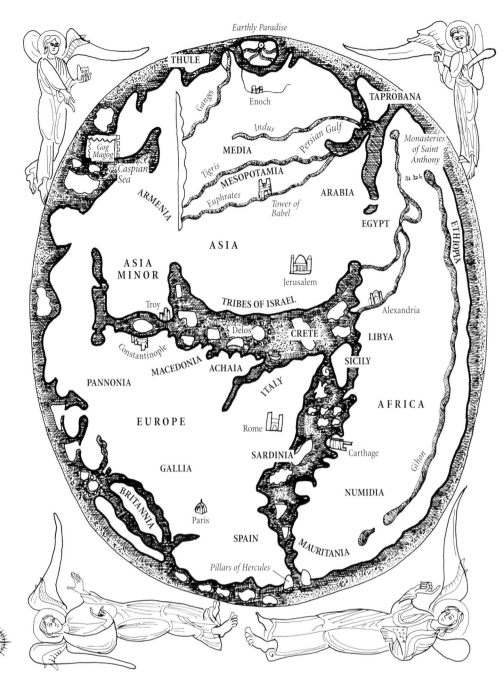

Fig. 6.10a (opposite). The Sawley world map. Durham, England, late twelfth century. 29.5 × 20.5 cm. The Master and Fellows of Corpus Christi College, Cambridge, Parker Library, Corpus Christi College, MS 66, p. 2. The eastern island of the earthly paradise has been placed at the top of the map. The geometrical centre is marked not by Jerusalem but by the island of Delos, giving rise to speculation among some scholars that a Graeco-Roman source underlies this map and others in the same tradition.

Fig. 6.10b. Diagram of Fig. 6.10a.

gospel were destined to gather up the elect from the four winds (Matthew 24.31).[63] Or, as is more common, they could be seen as the four angels who restrain the winds after the opening of the sixth seal (Revelation 7.1).[64] The usual iconography of the relevant verse, however, where each angel has its hand over the mouth of a wind head and a fifth angel bears a cross – the sign of the living God mentioned in the Book of Revelation – either in its hand or marked on its mantle (or, in an alternative portrayal, blows into a trumpet to signify that it stands for one of the four winds), does not fit the figures on the map and a more appropriate reading may be suggested.[65] The angel in the upper left-hand corner would seem to relate to the one described in yet another passage of Revelation (14.6–8) in which an angel, bearing a scroll that represents the eternal gospel and saying 'Fear the Lord', announces that the hour of judgement has come. The angels in the other three corners could be conveying other messages from the same

The Heyday of Paradise on Maps   143

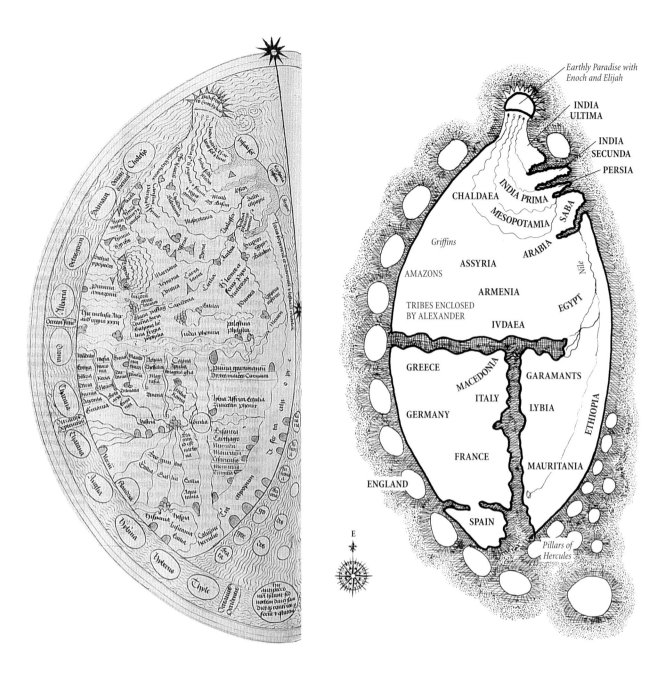

Fig. 6.11a. World map from a manuscript of Lambert of St-Omer's *Liber floridus*. Flanders, second half of the fifteenth century. 40 × 30 cm. Genoa, Biblioteca Durazzo Giustiniani, MS A IX 9, fols 67v–68r (see also Plate 5b). Lambert of St-Omer was writing about 1120. The world maps found in the various manuscripts of Lambert's work show how geography corresponds to the history of mankind from its dawn to its imminent end.

Fig. 6.11b. Diagram of Fig. 6.11a.

passage of Revelation, one announcing the reward of the righteous, and the other two conveying warnings of the fall of Babylon and the judgement of those who worship the beast.[66] The weight of the argument seems to rest on the index finger of the angel in the upper left corner, pointing to the inscription *Gog and Magog, gens immunda*, which identifies on the map the warring tribe in its peninsular prison, a concrete reference to the impending replacement of time by eternity. The general message the four angels of the Sawley map appear to be conveying is that the history of mankind soon would reach its fulfilment. That history, however, had begun in paradise, which is marked at the top of the map as an island and described in Honorius's text as the first and easternmost region of Asia, beautiful and yet inaccessible because of a wall of fire.[67]

Event/places such as those referring to the Apocalypse were reminders of the historical ordering of *mappae mundi*. The end of the axis along which human history unfolded, from the beginning of time in paradise to the fullness of time in Judaea, was reached with the Incarnation and Passion of Christ. The sixth and final age, the ultimate

144 Mapping Paradise: A History of Heaven on Earth

Fig. 6.12. The Hereford world map (see Fig. 6.15). Detail of the earthly paradise. Paradise appears as a walled garden located on an island in the farthest east, where Adam and Eve eat the forbidden fruit. The insularity of paradise was suggested by Isidore's statement in the seventh century that the Garden of Eden was not to be confused with the classical Fortunate Islands, and confirmed in the following century by Bede. The island of paradise marks the event that caused the place to be forbidden and inaccessible, and that initiated the process of human history that is to be fulfilled with the Second Coming of Christ, which is represented at the top of the map.

Fig. 6.13. The Hereford world map (see Fig. 6.15). Detail of Christ in Majesty. Christ is coming to judge the world. On the left are the saved. On the right (at Christ's left hand) are the damned. Below the figure of Christ, the Virgin Mary is represented with uncovered breasts saying to her son: 'See dear son, my bosom within which you became flesh, and the breasts from which you sought the Virgin's milk. Have mercy, as you yourself have promised, on all those who have served me, who have made me their way to salvation' (see Westrem, *The Hereford Map* (2001), p. 7 for the Anglo-Norman text).

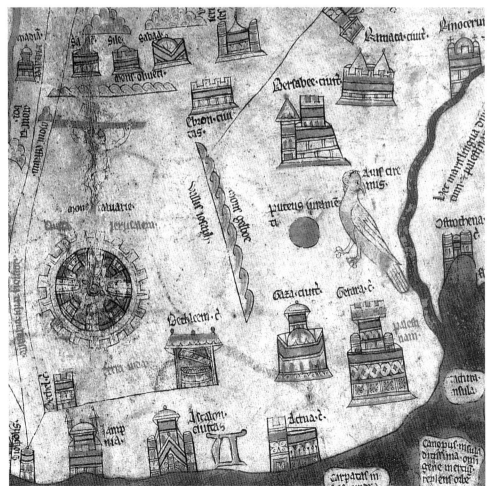

Fig. 6.14. The Hereford world map (see Fig. 6.15). Detail of Jerusalem and part of the Holy Land. The holy city of Jerusalem, which Jerome called 'the navel of the earth' in his translation of the Bible (Ezekiel 5.5), is placed at the centre of the inhabited earth. Above Jerusalem is a representation of the Crucifixion. The Hereford *mappa mundi* depicts the totality of the inhabited earth, including a wide range of geographical information relating to mountain ranges, islands, rivers and lakes, but it is above all an historical account depicted spatially.

The Heyday of Paradise on Maps 145

Fig. 6.15a. The Hereford world map. Lincoln (?and Hereford). *c.*1300. 158 × 133 cm. Hereford Cathedral (see also Plate 7). By permision of the Dean and Chapter of Hereford and the Hereford Mappa Mundi Trust. East is at the top, where the earthly paradise is also found. The map portrays the three parts of the world as it was then known (Asia, Europe, Africa) surrounded by an outer ocean, and surrounding a Mediterranean Sea packed with islands. One of these is the island of Cadis, on which are the Pillars of Hercules, another is a triangular Sicily, and further east is the island of Crete. In the north-west (i.e. bottom left) are the British Isles. The colours of the map have darkened over the centuries. Images of castles, churches and fortified cities indicate towns and built-up areas. The map, which is well known for its representation of monstrous races, is also rich in material drawn from natural science.

revelation of God (the 'Apocalypse'), had arrived, but was not yet complete; as the world awaited Christ's Second Coming, the Church was reclaiming the world. The axis between paradise and Jerusalem worked in both directions, for Jerusalem signified both the arrival of the Church on earth and the Heavenly Jerusalem due at the end of time. The relationship between the earthly paradise and the new paradise in heaven is particularly evident on certain *mappae mundi*, notably the maps in the various codices of Lambert of St-Omer's *Liber floridus* (*c.*1120), some of which include an allusion to the placing of Enoch and Elijah in the earthly paradise (an island in the ocean, protected by a ring of flames or mountains and connected to eastern Asia by the four rivers), as well as references to Gog and Magog (Figure 6.11 and Plate 5b).[68]

The relationship between the earthly paradise, Jerusalem and the new paradise in heaven was similarly expressed on the *mappa mundi* that is now in Hereford Cathedral (Figure 6.15). The Hereford map is thought to have been made about 1300.[69] The three elements of the Christian vision of world's history – Adam's sin in the Garden of Eden, the sacrifice of Christ in Jerusalem, and Christ's Second Coming – are clearly discernible, locked into an elaborate network of space–time relationships (see, for

146   Mapping Paradise: A History of Heaven on Earth

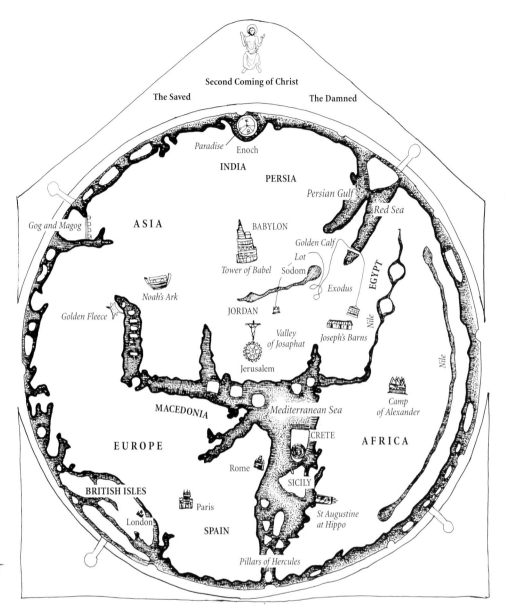

Fig. 6.15b. Diagram of Fig. 6.15a.

example, Plate 7). The earthly paradise stands for the first element, portrayed as a walled garden on an island at the easternmost extremity – and thus the apex – of the map. Within the Garden, Adam and Eve partake of the forbidden fruit of the Tree of the Knowledge of Good and Evil, which overlooks the spring from which emerge the four rivers (Figure 6.12). To the right, outside the Garden, a scene represents the Expulsion. Nearby, marking the limit of the inhabited earth, is a tree labelled *Arbor Balsami, id est, Arbor sicca*, a conflation of the balsam tree mentioned in the Alexander story of the Trees of the Sun and Moon with the Dry Tree that came from the branch of the Tree of Knowledge brought by Seth from paradise.[70] The second element, the sacrifice of Christ, focuses attention on Jerusalem, which together with the earthly paradise defined with a single event/place a crucial shift in the relationship between God and mankind, and is therefore positioned at the exact centre of the map (Figure 6.14).[71] Immediately above the city is the Crucifixion. These signs point to the scene of the Son of God's sacrifice and his Resurrection at the centre of the earth, singling out Jerusalem as the inauguration of the Church's eschatological time (the wait for Christ's final advent). Jerusalem was the meeting point of heaven and earth, the historical place of

The Heyday of Paradise on Maps 147

Fig. 6.16a. The Psalter map. London, *c.*1265. Diameter 9 cm. London, British Library, Add. MS 28681, fol. 9r (see also Plate 8). This tiny, but highly detailed map is thought to have been copied from a larger original, and fills the page in an English Psalter. Jerusalem (a darkened circle within two concentric circles) is at the centre of the map, paradise at the top. The Dead Sea is prominent in Asia. The south of Africa is peopled by monstrous races. Interspersed among the islands of the Ocean that encircle the earth are twelve heads, which represent the principal and secondary winds.

the Crucifixion and Resurrection of Christ as well as the eschatological place of his Second Coming: according to the prophet Joel, the Lord will gather together all his dispersed people and lead them in the Valley of Josaphat (3.2, 12), which is marked on the map and which Jerome located between Jerusalem and the Mount of Olives.[72]

The third element, the final coming of Christ in glory, dominates the map from the apex of the entire composition. The majestic figure of Christ appears through an aperture in heaven as the consummation of world history articulated in geographical space (Figure 6.13). Christ displays the marks of his Passion and, as the scrolls in his hands tell us, a divine voice is saying: 'Behold my witness.'[73] The moment portrayed anticipates the moment of the end of the world, when the Son of Man 'shall come in his glory, and all the holy angels with him' (Matthew 25.31–46). To the right of Christ are the saved, rising from their graves and guided upwards to heaven by an angel while a second angel displays a scroll reading: 'Arise! You shall come to joy everlasting.'[74] To

148   Mapping Paradise: A History of Heaven on Earth

the left of Christ are the damned, rejected by another angel and led to the mouth of hell by demons, as forewarned by the words on the scroll that tumbles from the trumpet of yet another angel: 'Arise! You are going to the fire established in hell.'[75] Other angels surround the figure of Christ, some holding the instruments of the Passion as a reminder of the source and the nature of his triumph. Below, the Virgin Mary reveals her breasts to plead the cause of those who have prayed to her.[76]

The progression from the first to the second pivotal event/place, from the earthly paradise to the earthly Jerusalem, was given expression on *mappae mundi* through the polarization of the relevant epochal zones along the east–west axis. The relationship between the second event/place, Christ's Passion in Jerusalem, and the third and final event, his Second Coming, presented a greater challenge that was resolved in various ways on different maps. On the Hereford map, the glorious Second Coming of Christ that crowns the world is placed immediately above the earthly paradise as – it may be suggested – an all-embracing reiteration of the essence of Christian belief: paradise lost because of Adam's sin becomes paradise regained through the sacrifice of Christ, and paradise granted to the righteous at the end of time.

The idea that it would be possible to regain paradise was emphasized in geographical terms by the notion that the earthly paradise was simultaneously inaccessible and contiguous to the inhabited earth. This sort of connectivity, implying both a barrier and a way through, is repeated in the relationship between the centre and the periphery of the *mappa mundi*. Inaccessible at the extremity of the earth, at the limits of the *mundus* and the *saeculum*, the Garden of Eden stands for the act that had brought about man's fallen state. At the same time, it points to Jerusalem, the centre of the earth and the place of Christ's Crucifixion and Resurrection. Centre and periphery complement each other as parts of the same structure.[77] On the so-called London or Psalter map, for example, Christ dominates the world below in much the same way as on the Hereford map, reclaiming it through his embrace (Figure 6.16).[78] More clearly than most *mappae mundi*, this tiny (diameter 9 cm) map demonstrates how the route of salvation history moved in time and over space from periphery to centre, and from the primordial Garden of Eden to Jerusalem, the capital city of Judaea at the time of Christ's death and Resurrection. The vignette of the earthly paradise presents the faces of Adam and Eve either side of the Tree of the Knowledge of Good and Evil, the whole surrounded by a double ring marked to indicate, perhaps, high mountains or a wall of fire (Plate 8).

One of the largest – but no longer extant – and visually richest and most detailed of all known *mappae mundi*, the Ebstorf map, developed the interrelationship between the earthly paradise, the Jerusalem of the Holy Land, and the Heavenly Jerusalem in yet another way. This huge map, originally nearly 13 square metres and possibly made in Ebstorf, in northern Germany, in 1239 or thereabout, contains some 1200 entries (Figure 6.17 and Plate 9).[79] The Garden of Eden is represented by a rectangle within which are Adam and Eve, the four rivers (shown disappearing into the ground),

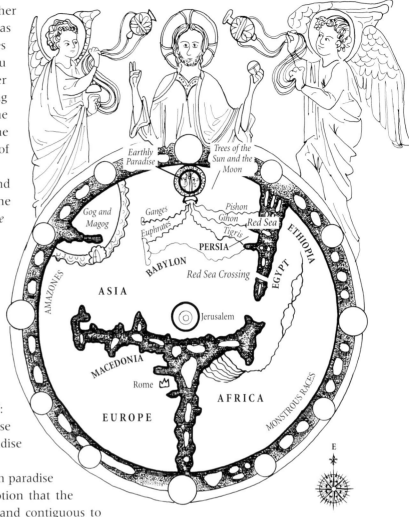

Fig. 6.16b. Diagram of Fig. 6.16a. The figure of Christ, flanked by two incense-swinging angels above the map proper, raises his right hand in a gesture of blessing, while holding in his left hand a small tripartite globe.

The Heyday of Paradise on Maps 149

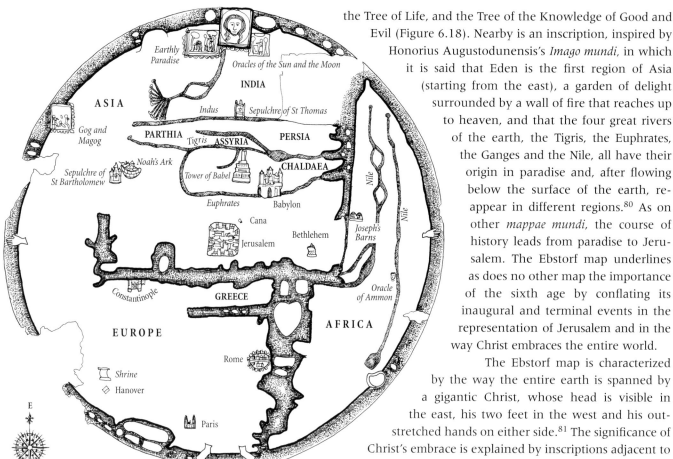

the Tree of Life, and the Tree of the Knowledge of Good and Evil (Figure 6.18). Nearby is an inscription, inspired by Honorius Augustodunensis's *Imago mundi*, in which it is said that Eden is the first region of Asia (starting from the east), a garden of delight surrounded by a wall of fire that reaches up to heaven, and that the four great rivers of the earth, the Tigris, the Euphrates, the Ganges and the Nile, all have their origin in paradise and, after flowing below the surface of the earth, reappear in different regions.[80] As on other *mappae mundi*, the course of history leads from paradise to Jerusalem. The Ebstorf map underlines as does no other map the importance of the sixth age by conflating its inaugural and terminal events in the representation of Jerusalem and in the way Christ embraces the entire world.

The Ebstorf map is characterized by the way the entire earth is spanned by a gigantic Christ, whose head is visible in the east, his two feet in the west and his outstretched hands on either side.[81] The significance of Christ's embrace is explained by inscriptions adjacent to his head, feet, and hands. They point to the double structure of Christian time (as discussed in Chapter 4) and to the east–west course of salvation history from paradise to Jerusalem (as described above). Christ is depicted as the Alpha and the Omega, letters that appear on either side of his head, close to the prime epochal zone of Eden, together with the words *primus et novissimus* ('the first and the last'; Revelation 1.17).[82] By Christ's right hand, which is marked by the stigmata, is a quotation from Psalm 118.16: 'The right hand of the Lord doeth valiantly.'[83] By his left hand we read: 'He holds the earth in his hand.'[84] This could be an allusion to Isaiah 40.12,[85] but the exact wording is found in a liturgical text, the Antiphonary of the Office.[86] Another quotation, by his feet, in the extreme west, confirms that the map maker's intention was to demonstrate God's arrangement of human history, from its beginning in the east to its end in the west, and that in fulfilling the prophecies of the Old Testament the Incarnation of Christ had inaugurated the wait for the end. The quotation reads 'Mightily to the end, sweetly ordering all things', and is inspired by the Book of Wisdom 8.1: 'Wisdom reacheth from one end to another mightily: and sweetly doth she order all things.'[87] The Wisdom of God referred to, and illustrated by the compiler of the Ebstorf map through his particular portrayal of the figure of Christ superimposed on the whole world, is Christ himself, as Saint Paul made clear (1 Corinthians 1.24).

The standard reading of the verse at Christ's feet, expounded by the compilers of the *Glossa ordinaria*, was that the Wisdom of God extended from the beginning of the world to the advent of Christ (the period accounted for in the Old Testament) and from the Incarnation of Christ to the end of the world (as related in the gospels).[88] The positioning of the inscription at Christ's feet indicates both the western limits of the earth and the impending end of time. Christ's global embrace points up the way God has structured salvation history between Adam's Fall and the advent of his son, and directs

Fig. 6.17. Diagram of the Ebstorf world map. ?Ebstorf, Lower Saxony, between 1235 and 1240. 3.56 × 3.58 m. Destroyed in an Allied bombing raid over Hanover on the night of 8/9 October 1943, and now only available as a prewar facsimile (see Plate 9). The adventures of Alexander the Great in Asia (for example, his encounter with the oracle of the sun and the moon and the oracle of Ammon) bear witness to classical history. Biblical history is represented by the Fall in the Garden of Eden, Noah's Ark in Armenia, the Tower of Babel in Mesopotamia, Joseph's barns in Egypt and the sacred places in Palestine, while the burial places of the apostles are an indication of the diffusion of the gospel over the whole world. The tribes of Gog and Magog are shown in northern Asia. The depiction of a shrine in northern Germany (not found on any other extant map) suggests a connection with Ebstorf.

150 Mapping Paradise: A History of Heaven on Earth

Fig. 6.18. The Ebstorf world map. Detail of the earthly paradise (see Fig. 6.17.). The vignette of the Garden of Eden shows Adam and Eve, the four rivers plunging underground (to resurface in the inhabited world), and the Trees of Life and of the Knowledge of Good and Evil. Acting as a counterpoint to the sin committed by Adam and Eve in paradise, the most important event in salvation history is the advent of Christ.

attention towards the end, and his son's final coming. The nature of the sign at the centre of the map representing Jerusalem conflates even more strikingly the first advent of Christ with his Second Coming (Figure 6.19). The city is depicted, in plan, as a square. It has gilded walls and twelve gates, in accordance with the apocalyptic description of the Heavenly Jerusalem (Revelation 21.11–12, 16, 18). The way in which the Ebstorf map's author thought to present in an earthly context the establishment of the Heavenly Jerusalem confirms yet again that the Book of Revelation was read as referring to the sixth age as a whole, that is both to the spiritual reality of the Church on earth and to the end of the world.[89] The historical and earthly Jerusalem was assimilated with the celestial Jerusalem precisely because the former marked the beginning of the sixth and final age, of which the latter was the end.[90]

The uniqueness of Jerusalem, as both the historical place of the Crucifixion and the Resurrection of Christ and the meeting point of heaven and earth, is well described on the Ebstorf map. All around the Holy Land are allusions to the earthly life of Christ: the ox, the ass, and the star in Bethlehem allude to the Nativity; six jars in Galilee attest Christ's first miracle, the turning of water into wine at a wedding feast at the village of Cana (John 2.1–12). The depiction of the Garden of Gethsemane, to which Christ retired to pray just before his arrest (Matthew 26.36–46; Mark 14.32–42; Luke 22.39–46), evokes the Passion. The figure of Christ rising from the grave that dominates Jerusalem, however, points to a different dimension and to the fundamental quality of the city as the stage of the Resurrection. This event was the fulfilment of God's plan of salvation, as is confirmed by the inscription to the right of the city, which reads: 'Jerusalem, the glorious sepulchre of the Lord, according to Isaiah's testimony' [Psalm 73(74).12].[91] The unique character of the city of Jerusalem is explained in a longer text:

> *This most famous city is the capital of all the cities of the whole world, for it was there that mankind's redemption was accomplished through the death and Resurrection of the Lord, as the psalmist says* [Psalm 73(74).12]: *'For God is my King of old, etc* [i.e., working salvation in the midst of the earth]*'. That great city contains the sepulchre of the Lord, which the whole world desires to visit with pious eagerness, because Christ, the conqueror of death, has ennobled it by rising from the dead. Hence Sedulius says: 'There is found the holy place that received the great treasure of the body of the Lord, a place ennobled by the Deposition, but even more ennobled by the Resurrection.'*[92]

Fig. 6.19. The Ebstorf world map. (see Fig. 6.17 and Plate 9). Detail of Jerusalem. The representation of Jerusalem borrows from the iconography of the Heavenly Jerusalem as described in the Book of Revelation (21.11–12, 16, 18). The figure of the resurrected Christ appears to be facing north, at a right angle to the viewer. It has been suggested that this is a function of the biblical notion of a threat that would come out of the north, the home of Israel's foes and the quarter whence evil arises, an idea recurrent in the Old Testament prophets Isaiah, Jeremiah and Zechariah. Other scholars have concluded that the rotation is due to the belief that Christ rose towards the east and that his Second Coming would take place again from the east. The Saviour holds the banner of the Resurrection with a cross and is stepping out of the open sarcophagus. His head is surrounded by a crossed nimbus. The scene includes the soldiers guarding the tomb in the attitude of sleep.

Jerusalem, the city in Judaea ennobled by Christ's Resurrection, had come to stand for the Heavenly Jerusalem on earth. The two-humped camel below the text, indicating the presence of camels in Palestine (see Plate 9) probably also refers to the humility of Christ.[93] Medieval bestiaries made the point that Christ lowered himself, as does the camel, to take on the burden of human weakness. Like the camel referred to in the gospel (Matthew 19.24), Christ too passed through the eye of a needle – his Passion – to gain the Kingdom of Heaven for mankind.[94] The camel on the Ebstorf map may also point to the glory of a Jerusalem open to all nations, as celebrated by the prophet Isaiah (60.6) in anticipation of Christ's advent: 'The multitude of camels shall cover thee, the dromedaries of Madian and Ephah; all they from Sheba shall come: … they shall shew forth the praises of the Lord.' Jerusalem at the centre of the map, together with the body of Christ that spans the entire world, points to the eventual reclamation of the whole world by the end of the sixth age for the eternal reign of Christ. In the period of waiting, what is already on earth, the city in Judaea, anticipates and makes almost present what has not yet occurred.

The Utopian Search for a Place

All the maps discussed so far contain some sort of reference to the impending end of the world. Mankind had reached the western limits of the earth; the tribes of Gog and Magog, enclosed by Alexander the Great, were on the point of invading the inhabited world; Enoch and Elijah were waiting patiently in paradise, ready to face death and the Last Things. Angels warned that the Last Judgement was near and, on the Hereford map, we see Christ coming to judge the world. Medieval maps disclose in their portrait of the world a clear emphasis on its anticipated ending. On the Hereford map, the letters 'M', 'O', 'R' and 'S' are spaced out beyond the boundary of the world to spell *MORS* (death), reminding the observer of human mortality and of the world's eventual passing away. As Saint Paul noted, the only permanent dwelling is heaven and Christians are encouraged to set their minds on things above, not on earthly things (Colossians 3.2). A parallel reminder of the transitory nature of the world, and of the Church's role in dispatching souls to heaven, is found along the bottom of the surviving fragment of the Duchy of Cornwall map. Here, circles indicate the later ages of man, prior to purgatory and the ultimate release of the soul in angelic form.[95]

The picture conveyed by *mappae mundi* is not quite so simple as the summary above might imply. Medieval maps also reveal the complexity of the Christian notion of space and time that we referred to in the early chapters of this book. Not only are both the Garden of Eden and the Heavenly Jerusalem conceived of as being on the spatio-temporal borders of the human realm, but there is also the idea that the kingdom of God has already been established on earth. For Christians, the Incarnation of Christ and his redeeming sacrifice on Golgotha – portrayed with emphasis on the Hereford and the Ebstorf maps to highlight the event/place of Jerusalem – have sanctified earthly space and historical time once and for all. Christian eschatological hopes are founded on an essentially historical faith, and the space–time structure of medieval *mappae mundi* underlines the importance of the various geographical places in relation to the events

152 Mapping Paradise: A History of Heaven on Earth

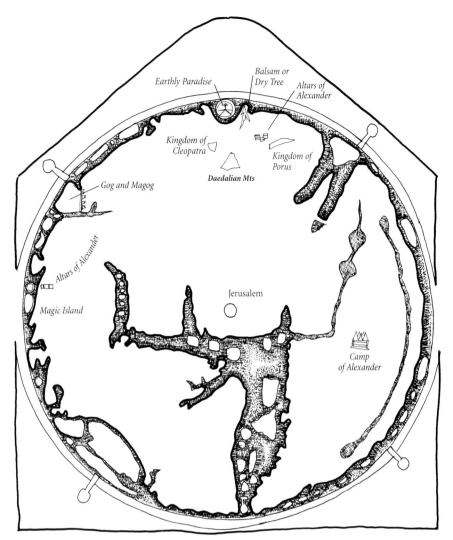

Fig. 6.20. Diagram of the Hereford world map (see Fig. 6.15), showing features relating to Alexander the Great.

that occurred there. More generally, the unified picture of reality offered by a medieval *mappa mundi* showed how the representation of the earth, and of man's place within it, could be seen as a sign pointing towards God. On the maps, the Christian challenge to transcend the world coexisted with interest in its beauty and variety.

Mapping paradise on earth may be considered as one of the most powerful expressions of the fundamental tension between the locative and utopian tendencies in Christianity.[96] The depiction of the Garden of Eden on maps marked the duality of the Christian attitude towards life on earth, as it pointed to both the reality and the loss of a perfect human nature in paradise. It is significant that, according to a popular medieval legend, it was in front of the earthly paradise that Alexander the Great, hero *par excellence* of utopian anxiety, whose exploits were abundantly celebrated on medieval *mappae mundi* (Figure 6.20), came to understand the vanity of his ambition and thus ended his otherwise never-ending journey.[97] The point about the frailty of human condition, and the call to eternal life, was confirmed on maps by the inclusion of the Trees of the Sun and the Moon, often, as already noted, portrayed close to paradise at the easternmost border of the earth. By depicting the earthly paradise among other geographical features and together with the journeys of the indefatigable Alexander the Great, the medieval makers of maps seem already to have articulated the urge expressed in the twentieth century by the Jewish-German poet Paul Celan: 'Research for a place, a *topos*? Sure! But in the light of what must be searched for: in the light of U-topia.'[98]

The Heyday of Paradise on Maps   153

1. Hugh of St Victor, *De sacramentis legis naturalis et scriptae dialogus*, PL CLXXVI, col. 24: 'D. Ubi est paradisus? M. Cur quaeris quod vides? Ad orientem ingressus tuus, hoc est lignum vitae quod conspicis.'
2. Patrick Gautier Dalché, *La 'Descriptio Mappe Mundi' de Hugues de Saint-Victor* (Paris: Études Augustiniennes, 1988), p. 101; Joachim Ehlers, *Hugo von St Viktor: Studien zum Geschichtsdenken und zur Geschichtsschreibung des 12. Jahrhunderts* (Wiesbaden: Steiner Verlag, 1973), p. 128; according to Patrice Sicard, *Diagrammes médiévaux et exégèse visuelle: Le Libellus de formatione arche de Hugues de Saint-Victor* (Paris–Turnhout: Brepols, 1993), pp. 267–8, the dialogue does not refer to a *mappa mundi sensu stricto*, but to a visual representation used for teaching purposes in the abbey school of Paris.
3. See above, Chapter 5, p. 89.
4. Hugh, a theologian of, probably, Saxon origin, had embarked on his teaching and writing career at the Abbey of St Victor, Paris, some time before 1125. It is likely that the so-called 'Isidore map' now in Munich (Bayerische Staatsbibliothek, MS Clm. 10058, fol. 154v) was closely related to Hugh's writings: Patrick Gautier Dalché, *La 'Descriptio Mappe Mundi'* (1988), pp. 81–6.
5. On Hugh of St Victor's life see Rebecca Moore, *Jews and Christians in the Life and Thought of Hugh of St. Victor* (Atlanta, GA: Scholars Press, 1998), pp. 15–21; Roger Baron, *Études sur Hugues de Saint-Victor* (Paris: Desclée, 1963); Jerome Taylor, *The Origin and Early Life of Hugh of St. Victor: An Evaluation of the Tradition* (Notre Dame, IN: University of Notre Dame Press, 1957); F. E. Croydon, 'Notes on the Life of Hugh of St. Victor', *Journal of Theological Studies*, 40 (1939), pp. 232–53.
6. Hugo of St. Victor, *De tribus maximis circumstantiis gestorum*, ed. by William M. Green, *Speculum*, 18 (1943), pp. 488–92; English transl. in Mary J. Carruthers, *The Book of Memory: A Study of Memory in Medieval Culture* (Cambridge–New York: Cambridge University Press, 1990), pp. 261–6. On Hugh and the art of memory applied to history see Grover A. Zinn, 'Hugh of Saint Victor and the Art of Memory', *Viator*, 5 (1974), pp. 211–34.
7. For Hugh's combination of the stress on human history and contemplation, see Grover A. Zinn, '*Historia fundamentum est*: The Role of History in the Contemplative Life according to Hugh of St. Victor', in George H. Shriver, ed., *Contemporary Reflections on the Medieval Christian Tradition: Essays in Honor of Ray C. Petry* (Durham, NC: Duke University Press, 1974), pp. 135–58. See also Joachim Ehlers, '*Historia, allegoria, tropologia*: Exegetische Voraussetzungen der Geschichtskonzeption Hugos von St. Viktor', *Mittellateinisches Jahrbuch*, 7 (1972), pp. 153–60. Hugh's devotion to the historical sense of Scripture marked a new attitude towards Scripture, see Beryl Smalley, *The Study of the Bible in the Middle Ages*, 2nd edn (Oxford: Blackwell, 1952), pp. 83–106.
8. Hugh of St Victor, *In scripturam sacram*, II, PL CLXXV, col. 11; *Commentariorii in hierarchiam coelestem S. Dionysii Areopagitae*, I.1–2, PL CLXXV, cols 926–30; *De sacramentis fidei christianae*, 'Prologus', II–III; I.10.5; II.1.1, PL CLXXVI, cols 183–4; 333–4; 371; *De archa Noe*, IV.3, ed. by Patrice Sicard, CCCM CLXXVI (Turnhout: Brepols, 2001), pp. 92–4; *De vanitate mundi*, III, PL CLXXVI, cols 721–3. See Ehlers, *Hugo von St Viktor* (1973), pp. 51–71.
9. See Patrick Gautier Dalché, 'L'Espace de l'histoire: Le Rôle de la géographie dans les chroniques universelles', in Jean-Philippe Genet, ed., *L'Historiographie médiévale en Europe* (Paris: Centre National de la Recherche Scientifique, 1991), pp. 287–300.
10. Patrick Gautier Dalché, 'Nouvelles Lumières sur la *Descriptio mappe mundi* de Hugues de Saint-Victor', in *Géographie et culture: La Représentation de l'espace du VIe au XIIe siècle* (Aldershot, Hants: Ashgate, 1997), XII, pp. 1–27, edited version of 'La "Descriptio Mappe Mundi" de Hugues de Saint-Victor: retractatio et additamenta', in Jean Longère, ed., *L'Abbaye parisienne de Saint-Victor au Moyen Âge* (Paris–Turnhout: Brepols, 1991), pp. 143–79; and Gautier Dalché, *La 'Descriptio Mappe Mundi'* (1988), with Hugh's text at pp. 133–60; the text is dated at pp. 55–8.
11. Hugh of St Victor, *Libellus de formatione arche*, ed. by Patrice Sicard, CCCM CLXXVI (Turnhout: Brepols, 2001), pp. 119–62. See also Sicard, *Diagrammes médiévaux* (1993); Danielle Lecoq, 'La Mappemonde du *De Arca Noe Mystica* de Hugues de Saint-Victor (1128–1129)', in Monique Pelletier, ed., *Géographie du monde au Moyen Âge et à la Renaissance* (Paris: Comité des Travaux Historiques et Scientifiques, 1989), pp. 9–31; Joachim Ehlers, '*Arca significat ecclesiam*: Ein theologisches Weltmodell aus der ersten Hälfte des 12. Jahrhunderts', *Frühmittelalterliche Studien*, 6 (1972), pp. 171–87; Marie-Therèse d'Alverny, 'Le Cosmos symbolique du XIIIe siècle', *Archives d'Histoire Doctrinale et Littéraire du Moyen Âge*, 28 (1953), pp. 31–81. In the above studies Patrice Sicard, Danielle Lecoq and Joachim Ehlers have attempted to reconstruct the diagram of the Ark.
12. Hugh of St Victor, *Libellus de formatione arche*, XI (2001), p. 157.
13. Hugh, ibid., also observes that paradise appears as a sort of bosom of Abraham in its position both in the eastern part of the earth and in the centre of the bosom of Christ. See Sicard, *Diagrammes médiévaux* (1993), pp. 265–6; Gautier Dalché, *La 'Descriptio Mappe Mundi'* (1988), pp. 19–20.
14. For an introduction to Hugh of St Victor see Dominique Poirel, *Hugues de Saint-Victor* (Paris: Cerf, 1998); Vincenzo Liccaro, *Studi sulla visione del mondo di Ugo di S. Vittore* (Trieste: Del Bianco, 1969); on the importance of the ecclesiological theme in Hugh see Roger Baron, *Science et sagesse chez Hugues de Saint-Victor* (Paris: Lethielleux, 1957), pp. 128–32; for a detailed discussion of the motif of the ark as an image of the Church in Hugh see Ehlers, *Hugo von St Viktor* (1973), pp. 120–55. Hugh's world diagram is discussed by Evelyn Edson, *Mapping Time and Space: How Medieval Mapmakers Viewed their World* (London: British Library, 1997), pp. 159–63. As highlighted by Moore, *Jews and Christians in the Life and Thought of Hugh of St. Victor* (1998), pp. 116–28, for Hugh the Church began to exist at the beginning of the world and will last until the end. See his *De arca Noe*, I.3 (2001), p. 16.
15. Hugh of St Victor, *De arca Noe*, IV.9 (2001), p. 111.
16. Augustine, *De civitate Dei*, XVIII.2, ed. by Bernard Dombart and Alfons Kalb, CCSL XLVIII (Turnhout: Brepols, 1955), p. 593.
17. Orosius, *Historiae adversum paganos*, II.1.4–6; VII.2.1–16, ed. by Marie Pierre Arnaud-Lindet, 3 vols (Paris: Belles Lettres, 1990), I, pp. 84–5; III, pp. 17–20. See Joseph Ward Swain, 'The Theory of the Four Monarchies Opposition History under the Roman Empire', *Classical Philology*, 35/1 (1940), pp. 20–1.
18. Daniel describes the statue as having a head made of gold, chest and arms of silver, belly and thighs of bronze, legs of iron, and feet of iron and clay; the rock that smashed it is 'cut out of the mountain without hands', which became a huge mountain filling the whole earth. See also Chapter 4 above, p. 81 n. 13.
19. Hugh of St Victor, *De archa Noe*, IV.9 (2001), pp. 111–12: 'Ordo autem loci et ordo temporis fere per omnia secundum rerum gestarum seriem concurrere videntur. Et ita per divinam providentiam videtur esse dispositum, ut que in principio temporum gerebantur, in oriente – quasi in principio mundi – gererentur, ac deinde ad finem profluente tempore usque ad occiden-

tem rerum summa descenderet, ut ex hoc ipso agnoscamus appropinquare finem seculi, quia rerum cursus iam attigit finem mundi. Ideo primus homo in oriente in hortis Eden conditus collocatur ut ab illo principio propago posteritatis in orbem terrarum proflueret. Item post diluvium principium regnorum et caput mundi in Assyriis et Chaldeis et Medis in partibus orientis fuit, deinde ad Grecos venit; postremo circa finem seculi ad Romanos in occidente – quasi in fine mundi habitantes – potestas summa descendit.' See also *Libellus de formatione arche*, XI (2001), p. 157; *De vanitate mundi*, II, *PL* CLXXVI, col. 720, and *Praesens saeculum*, in Dominique Poirel, *Livre de la nature et débat trinitaire au XIIe siècle: Le De tribus diebus de Hugues de Saint-Victor* (Turnhout: Brepols, 2002), pp. 446–51. See Gautier Dalché, *La 'Descriptio Mappe Mundi'* (1988), pp. 109–11. In his *Libellus de formatione arche*, II (2001), p. 128, Hugh also points out that mankind had spread out from paradise in all four cardinal directions and this is why the initials of the Greek names for the directions spelt the name ADAM. The same idea, as we have already seen, was expressed by Honorius Augustodunensis and on the thirteenth-century T-O map from Gautier de Metz's *Imago mundi*: see above, Chapter 5, pp. 91 and 93, Figure 5.7.
20   Genesis 4.17.
21   Genesis 8.1–19; 11.1–9; 11.31.
22   Genesis 41.53–7; Exodus 14.15–30, 20.1–18; Joshua 13–19; Matthew 2.1–12; Luke 2.1–20; Matthew 27.33–56; 28.1–10; Mark 15.22–41; 16.1–8; Luke 23.33–49; 24; John 19.17–37; 20.
23   In fact, the geographical sequence of the empires from east to west mirrorred their historical succession. The exception was Persia, which was known to be east of Babylon and which was described as the second kingdom, after the Babylonian. For the many identifications of the four empires see Harold H. Rowley, *Darius the Mede and the Four World Empires in the Book of Daniel: A Historical Study of Contemporary Theories* (Cardiff: University of Wales Press, 1935) and Swain, 'The Theory of the Four Monarchies' (1940), pp. 1–21.
24   I have discussed some of the points made in this chapter in 'À la recherche du paradis perdu: Les Mappemondes du XIIIe siècle', in Jean Galard, ed., *Ruptures: De la discontinuité dans la vie artistique* (Paris: Musée du Louvre-École Nationale Supérieure des Beaux-Arts, 2002), pp. 17–57; and 'Les Colonnes d'Hercule dans la cartographie médiévale: Limite de la Méditerranée et porte du paradis', in Bertrand Westphal, ed., *Le Rivage des mythes: Une géocritique meditérranéenne*, I, *Le Lieu et son mythe* (Limoges: Presses Universitaires de Limoges, 2001), pp. 339–65.
25   On the question of authorship see below, n. 69. The inscription is on the bottom left of the map, in the space between the frame and the round description of the world: Scott D. Westrem, *The Hereford Map: A Transcription and Translation of the Legends with Commentary* (Turnhout: Brepols, 2001), p. 11: 'Tuz ki cest estorie ont / Ou oyront ou lirront ou veront, / Prient a Jhesu en deyte / De Richard de Haldingham o de Lafford yet pite, / Ki lat fet e compasse, / Ki joiie en cel li seit done.' 'Let all who have this history – or who shall hear, or read, or see it – pray to Jesus in his divinity to have pity on Richard of Holdingham, or of Sleaford, who made it and laid it out, that joy in heaven may be granted to him.'
26   For a discussion on modern historical geography see Leonard Guelke, 'The Relations between Geography and History Reconsidered', *History and Theory: Studies in the Philosophy of History*, 36/2 (1997), pp. 216–34.
27   Bernard Guenée, *Histoire et culture historique dans l'Occident médiéval* (Paris: Aubier Montaigne, 1980; repr. 1991), pp. 22, 166–72. See also Hans-Werner Goetz, 'On the Universality of Universal History', in *L'Historiographie médiévale en Europe* (Paris: Centre National de la Recherche Scientifique, 1991), pp. 247–61.
28   Anna-Dorothee von den Brincken sees *mappae mundi* as visual chronicles paralleling the textual chronicles. Peter Barber, Paul D. A. Harvey, David Woodward, Evelyn Edson, Catherine Delano-Smith and Roger Kain have pointed out that geographical exactitude was not the object of these maps: Catherine Delano-Smith and Roger J. P. Kain, *English Maps: A History* (London: British Library, 1999), pp. 5, 37–40; Edson, *Mapping Time and Space* (1997), pp. 13–16; Peter Barber, 'The Evesham World Map: A Late Medieval English View of God and the World', *Imago Mundi*, 47 (1995), pp. 13–33; Paul D. A. Harvey, *Medieval Maps* (London: British Library, 1991), p. 19; David Woodward, 'Medieval *Mappaemundi*', in Brian Harley and Woodward, eds, *The History of Cartography*, I, *Cartography in Prehistoric, Ancient and Medieval Europe and the Mediterranean* (Chicago: University of Chicago Press, 1987), pp. 286–92; Woodward, 'Reality, Symbolism, Time, and Space in Medieval World Maps', *Annals of the Association of American Geographers*, 75/4 (1985), pp. 510–21; Anna-Dorothee von den Brincken, 'Mappa mundi und Chronographia: Studien zur *imago mundi* des abendländischen Mittelalters', *Deutsches Archiv für Erforschung des Mittelalters*, 24 (1968), pp. 118–86.
29   'Absolute, true, and mathematical time', to quote Isaac Newton's own words, 'from its own nature, flows equably without relation to anything external', and absolute space 'in its own nature, without relation to anything external, remains always similar and immovable'. Isaac Newton, *Mathematical Principles of Natural Philosophy and his System of the World*, transl. by Andrew Motte and Florian Cajori (Berkeley: University of California Press, 1934), p. 6. See Alfred W. Crosby, *The Measure of Reality: Quantification and Western Society, 1200–1600* (Cambridge: Cambridge University Press, 1997), pp. 93, 108. Isaac Newton's *Philosophiae naturalis principia mathematica* was published in 1687.
30   See G. Malcolm Lewis, 'The Origins of Cartography', in Harley and Woodward, eds, *The History of Cartography*, I (1987), pp. 50–3, esp. p. 51 n. 6, quoting Alan Roberts Lacy, *A Dictionary of Philosophy* (London: Routledge & Kegan Paul, 1976), p. 204.
31   On topological arrangement of medieval maps see above, Chapter 5, p. 117 n. 21.
32   On the inclusion of local information on *mappa mundi*, see, for example, Barber, 'The Evesham World Map' (1995), pp. 13–33, esp. pp. 24–7, and Armin Wolf, 'Neues zur Ebstorfer Weltkarte: Entstehungszeit – Ursprungsort – Autorschaft', in Klaus Jaitner and Ingo Schwab, eds, *Das Benediktinerinnenkloster Ebstorf im Mittelalter* (Hildesheim: August Lax, 1988), pp. 75–109.
33   The Christianizing mission of the apostles had already been recorded on the map in Beatus of Liébana's commentary: see above, Chapter 5, pp. 108–13.
34   Westrem, *The Hereford Map* (2001), pp. 68–71.
35   Ibid., p. 47: 'Seres primi homines post deserta occurrunt, a quibus serica vestimenta mittuntur'; p. 23: 'Hic inicium orientis estivi, ubi immensas esse nives, Marcianus et Solinus dicunt'; p. 43: 'Yndia mittit eciam elephantes maximos quorum dentes ebur esse creditur, quibus Yndei turribus inpositis in bellis utuntur.'
36   Ibid., p. 29: 'Decies sepcies centena et quinquaginta milia passuum longitudo Indie tenet, teste Solino. Item: quinque milia civitatum et diversissimo gentes monstruoso vultu, ritu, et habitu vario, plus quam credi possit. Gemmarum et metallorum affluencia cum periculo tocius generis bestiarum et serpencium, que omnia plus legenda sunt quam pingenda.'
37   I am grateful to Scott D. Westrem for drawing

The Heyday of Paradise on Maps   155

attention to this map in the course of a lecture on the 'Calculation, Delineation, Depiction, Inscription: Practicalities of Medieval Mapmaking', presented at The Warburg Institute in the *Maps and Society* series on 27 May 2004, and especially for a transcription of the adjacent text. On the Lambeth Palace, or Reading, map in general see his 'Geography and Travel', in Peter Brown, ed., *A Companion to Chaucer* (Oxford: Blackwell, 2000), pp. 195–217.

38 On the Vercelli map see Anna-Dorothee von den Brincken, 'Monumental Legends on Medieval Manuscript Maps: Notes on Designed Capital Letters on Maps of Large Size Demonstrated from the Problem of Dating the Vercelli Map (Thirteenth Century)', *Imago Mundi*, 42 (1990), pp. 9–23; Woodward, 'Medieval *Mappaemundi*' (1987), pp. 306–9; Carlo F. Capello, *Il mappamondo medievale di Vercelli (1191–1218?)* (Turin: C. Fanton, 1976); Marcel Destombes, *Mappemondes AD 1200–1500: Catalogue preparé par la Commission des Cartes Anciennes de l'Union Géographique Internationale* (Amsterdam: N. Israel, 1964), pp. 193–4.

39 Pietro Gribaudi, 'Il mito degli alberi del Sole e della Luna e dell'Albero secco nella cartografia medievale', in *Atti del V. Congresso Geografico Italiano* (Napoli: Tocco e Salvietti, 1905), pp. 828–42.

40 Capello, *Il mappamondo medievale di Vercelli* (1976), pp. 83–4.

41 Ibid., pp. 119–22.

42 London, British Library, Royal MS 14. C. IX, fols 1v–2r and 2v. On the Higden maps see Edson, *Mapping Time and Space* (1997), pp. 126–31; Barber, 'The Evesham World Map' (1995), pp. 13–17. On Higden's historical chronicle, John Taylor, 'Ranulf Higden', in John B. Friedman, Kristen Mossler Figg, and others, eds, *Trade, Travel, and Exploration in the Middle Ages: An Encyclopedia* (New York–London: Garland, 2000), pp. 252–4; Taylor, *The Universal Chronicle of Ranulf Higden* (Oxford: Oxford University Press, 1966), with a discussion of the map at pp. 63–8. For a printed edition, with an early translation of Higden's *Polychronicon*, see *Polychronicon Ranulphi Higden, monachi cestrensis, together with the English Translations of John Trevisa and of an Unknown Writer of the Fifteenth Century*, ed. by Churchill Babington (vols I, II) and Joseph R. Lumby (vols III–IX), RBMAS XLIV, 9 vols (London: Longmans, Green, and Co., 1865–86). See also Vivian H. Galbraith, 'An Autograph Manuscript of Ranulf Higden's "Polychronicon"', *Huntington Library Quarterly*, 34 (1959), pp. 1–18.

43 Higden, *Polychronicon*, I.10 (1865), I, pp. 66–78. The world map is referred to in the text: I.3, p. 26.

44 London, British Library, Royal MS 14. C. XII, fol. 9v.

45 On the Evesham map see Barber, 'The Evesham World Map' (1995), pp. 13–33; Barber, 'Die Evesham-Weltkarte von 1392: Eine mittelalterliche Weltkarte im College of Arms in London: Von der Universalität zum Anglozentrismus', *Cartographica Helvetica*, 9 (January 1994), pp. 17–22.

46 Ibid. A throne is appropriate in the context of the Fall. The verse in Isaiah – 'Thus saith the Lord, The heaven is my throne, and the earth is my footstool' (66.1) – was read throughout the Middle Ages as a celebration of God's majesty on the one hand and as a call to humility on the other. In this light, the earthly paradise would have been seen as intermediate between heaven (God's throne) and earth (God's footstool), the field of his work of salvation, as Isaiah prophesies later in the same chapter, where the extension of God's glory to all nations and to the most distant islands is celebrated (66.18–19).

47 Note that the view of the Garden is blocked by the external wall block, so that part of the Tree of Knowledge of Good and Evil and (to the right of the scene) the single source of the rivers are invisible.

48 Albi, Bibliothèque municipale, MS 29, fol. 57v (the map has almost disintegrated and is no longer exhibited); see Peter Barber, 'Mito, religione e conoscenza: La mappa del mondo medievale', in *Segni e sogni della terra: Il disegno del mondo dal mito di Atlante alla geografia delle reti* (Novara: De Agostini, 2001), p. 65; Edson, *Mapping Time and Space* (1997), pp. 32–3; John Williams, 'Isidore, Orosius and the Beatus Map', *Imago Mundi*, 49 (1997), pp. 7–32; Konrad Miller, *Mappaemundi: Die ältesten Weltkarten*, 6 vols (Stuttgart: J. Roth, 1895–8), III, *Die kleineren Weltkarten* (1895), pp. 57–9.

49 For a transcription of both the legends on the map and Orosius's text in the codex see François Glorie, 'Mappa mundi; Indeculum quod maria vel venti sunt et (Pauli Orosii) Discriptio Terrarum e codice Albigensi 29', in *Itineraria et alia geographica*, CCSL CLXXV (Turnhout: Brepols, 1965), pp. 467–94.

50 Iulius Honorius, *Cosmographia*, 24, 45, in *Geographi latini minores*, ed. by Alexander Riese (Heilbronn: Henninger, 1878), pp. 38, 48.

51 London, British Library, Cotton MS Tiberius B.V, fol. 56v. See Delano-Smith and Kain, *English Maps* (1999), pp. 34–6; Edson, *Mapping Time and Space* (1997), pp. 74–80.

52 The Anglo-Saxon map of the world, like some other, later, English maps of the British Isles, is widely accepted as containing within it elements of the tradition of Roman geographical maps of the world, which presumably did not disappear with the collapse of the Empire, but lingered on in Western Europe well into the Middle Ages. See Delano-Smith and Kain, *English Maps* (1999), pp. 34–6; Harvey, *Medieval Maps* (1991), p. 21.

53 An explicit sign for paradise is also missing from the world map in the twelfth-century manuscript in Munich, Bayerische Staatsbibliothek, MS Clm. 10058, fol. 154v, but the four rivers of paradise are clearly marked. See above, n. 4.

54 Manuscripts of Guido's *Liber historiarum* are in Brussels, Bibliothèque Royale de Belgique/ Koninklijke Bibliotheek van België, MS 3897–3919 (twelfth century), and Florence, Biblioteca Riccardiana, MS 881 (thirteenth century). For a description of the content of the manuscript see Patrick Gautier Dalché, *Carte marine et Portulan au XIIe siècle: Le 'Liber de existencia riveriarum et forma maris nostris Mediterranei' (Pise, circa 1200)* (Rome: École Française de Rome, 1995), pp. 255–61; see also Edson, *Mapping Time and Space* (1997), pp. 117–18; Anna-Dorothee von den Brincken, *Fines terrae: Die Enden der Erde und der vierte Kontinent auf mittelalterlichen Weltkarten* (Hanover: Hahnsche Buchandlung, 1992), pp. 62–3; Woodward, 'Medieval *Mappaemundi*' (1987), pp. 348, 350; Destombes, *Mappemondes* (1964), pp. 48, 167; Miller, *Mappaemundi*, III, *Die kleineren Weltkarten* (1895), pp. 54–7.

55 Guido of Pisa, *Liber historiarum*, Brussels, Bibliothèque Royale de Belgique/ Koninklijke Bibliotheek van België, MS 3897–3919, fol. 54 r: 'Asia habens fines ab oriente eremum Indiae Dimericae intransmeabilem ultraque ea domino nostro tantummodo cognitum.' Cfr. the correspondent passages in *Ravennatis Anonymi Cosmographia et Guidonis Geographica*, ed. by Moritz E. Pinder and Gustav Parthey (Berlin: Fridericus Nicolaus, 1860; repr. Aalen: Zeller, 1962), p. 547. See also Youssuf Kamal, *Monumenta cartographica Africae et Aegypti*, 5 vols (Cairo: 1926–51), III/3 (1933), p. 773v.

56 Guido of Pisa, *Liber historiarum*, fol. 46v. See Isidore, *Etymologiae*, XIV.3, ed. by Wallace Martin Lindsay, 2 vols (Oxford: Clarendon Press, 1911).

57 London, British Library, Add. MS 10049, contains *De hebraicis quaestionibus* (fols 2–21); *De interpretationibus nominum Veteris et Novi Testamenti* [*Liber nominum*] (fols 22–48); and *De nominibus*

locorum [or *De locis hebraicis*] (fols 44–63). The maps are on fol. 64, recto (map of Asia) and verso (map of Palestine). On the 'Jerome' maps, see Edson, *Mapping Time and Space* (1997), pp. 26–30; Woodward, 'Medieval *Mappaemundi*' (1987), pp. 288–92, 322–5; Miller, *Mappaemundi*, III, *Die kleineren Weltkarten* (1895), pp. 1–21. On Jerome's *Liber locorum*, published in *Onomastica sacra*, ed. by Paul de Lagarde (Göttingen: Horstmann, 1887; repr. Hildesheim: Olms, 1966), pp. 117–90, see also above, Chapter 5, p. 118 n. 30. I am grateful to Paul D. A. Harvey for having provided me with the text of his paper 'The Twelfth-Century Jerome Maps of Asia and Palestine', delivered at the XVII International Conference on the History of Cartography, Lisbon, 10 July 1997, where he refers to the content of the manuscript as a palimsest and investigates the changes and the erasures on the maps. On the regional maps of the Holy Land see above, Chapter 5, p. 120 n. 72, and Patrick Gautier Dalché, 'Cartes de Terre Sainte, cartes de pèlerins' in Massimo Oldoni, ed., *Fra Roma e Gerusalemme nel Medioevo: Paesaggi umani ed ambientali del pellegrinaggio meridionale*, 3 vols (Salerno: Laveglia, 2005), I, pp. 573–612.

58 Jerome, *Liber locorum* (1887; 1966), p. 150.

59 See Patrick Gautier Dalché, 'De la glose à la contemplation: Place et fonction de la carte dans les manuscrits du haut Moyen Âge', in *Testo e immagine nell'alto Medioevo*, 2 vols (Spoleto: Centro Italiano di Studi sull'Alto Medioevo, 1994), II, pp. 751–7.

60 Scott D. Westrem, 'Against Gog and Magog', in Sylvia Tomasch and Sealy Gilles, eds, *Text and Territory* (Philadelphia: University of Pennsylvania Press, 1998), pp. 54–75. See also above, Chapter 5, p. 119 n. 52.

61 As the discussion on Beatus's map should have made clear, the Apocalypse was believed to have already begun with the first advent of Christ, which inaugurated the sixth and final age of history. Beatus was among those who interpreted Gog and Magog allegorically. See above, Chapter 5, p. 113.

62 The Sawley map is in a copy of the world chronicle of Honorius Augustodunensis, twelfth century, now in Cambridge, Corpus Christi College, MS 66, p. 2. See Edson, *Mapping Time and Space* (1997), pp. 111–17; Paul D. A. Harvey, 'The Sawley Map (Henry of Mainz) and Other World Maps in Twelfth-Century England', *Imago Mundi*, 49 (1997), pp. 33–42; Danielle Lecoq, 'La Mappemonde d'Henri de Mayence ou l'image du monde au XIIIe siècle', in Gaston Duchet-Suchaux, ed., *Iconographie médiévale: Image, texte, contexte* (Paris: Centre Nationale de la Recherche Scientifique, 1990), pp. 155–207; Miller, *Mappaemundi*, III, *Die kleineren Weltkarten* (1895), pp. 21–9.

63 'And he shall send his angels with a great sound of a trumpet, and they shall gather together his elect from the four winds, from one end of heaven to the other.'

64 This is the interpretation of Edson, *Mapping Time and Space* (1997), pp. 113–15 and Lecoq, 'La Mappemonde d'Henri de Mayence' (1990), pp. 170–1.

65 For examples of the iconography of Revelation 7.1 in medieval manuscripts see John Williams, *The Illustrated Beatus: A Corpus of Illustrations of the Commentary on the Apocalypse*, 5 vols (London: Harvey Miller, 1994–2003), II (1994), Figures 49, 173; III, Figure 406; Gilberte Vezin, *L'Apocalypse et la fin des Temps: Étude des influences égyptiennes et asiatiques sur les religions et les arts* (Paris: Éditions de la Revue Moderne, 1973), Figure 112; Nigel J. Morgan, *The Lambeth Apocalypse: Manuscript 209 in Lambeth Palace Library* (London: Harvey Miller, 1990), pp. 150–1 (text) and 288 (Figures), Figure 25; Anne de Egry, *Um estudo de o Apocalipse de Lorvão e a sua relação com as ilustrações medievais do Apocalipse* (Lisbon: Fundação Calouste Gulbenkian, 1972), Plate IIIa; Clement Gardet, *L'Apocalypse figurée des ducs de Savoie* (Annecy: Gardet, 1969), reproduction of Madrid, El Escorial, Real Biblioteca di San Lorenzo, MS E. Vitr. 5, fol. 9v; Wilhelm Neuss, *Die Apokalypse des Hl. Johannes in der altspanischen und altchristlichen Bibel-Illustration (das Problem der Beatus-Handschriften)*, (Münster in Westphalia: Aschendorff, 1931), Figure 110.

66 The angels on the Sawley map may be compared with those in other medieval illustrations of the Apocalypse text: see, for example, Williams, *The Illustrated Beatus*, I (1994), Figures 15, 29; II (1994), Figures 130, 134; III (1998), Figures 125, 131; Morgan, *The Lambeth Apocalypse* (1990), pp. 303–4 (Figures), Figures 13–17. Another world map featuring angels is the Psalter map. See the discussion below.

67 Honorius Augustodunensis, *Imago mundi*, I.8, ed. by Valerie I. J. Flint, *Archives d'Histoire Doctrinale et Littéraire du Moyen Âge*, 49 (1982), p. 52. See Kamal, *Monumenta cartographica*, III/3 (1933), p. 786.

68 Paradise is described in Lambert of St-Omer's text as an island in the eastern ocean: see *Lamberti S. Audomari canonici liber floridus codex autographus Bibliothecae Universitatis Gandavensis*, ed. by Albert Derolez (Ghent: Story-Scientia, 1968), p. 104 (facsimile edition and transcription of Ghent, Rijksuniversiteit Bibliotheek, MS 92, fol. 51v); see also Kamal, *Monumenta cartographica*, III/3 (1933), p. 776. On the content of the manuscript and the relationship between the text and the maps, see Gautier Dalché, 'De la glose à la contemplation' (1994), pp. 740–9. Lambert of St Omer gave a lucid account of the division of world history into the six ages, combined with the pattern of the four world empires suggested by the Book of Daniel (see above, Chapter 4, p. 65): *Lamberti S. Audomari canonici liber floridus* (1968), pp. 40, 66–95, 464 (facsimile edition and transcription of Ghent, Rijksuniversiteit Bibliotheek, MS 92, fols 19v, 32v–47r, 232v). See Edson, *Mapping Time and Space* (1997), pp. 100, 105–11; Danielle Lecoq, 'La Mappemonde du *Liber Floridus* ou la vision de Lambert de Saint-Omer', *Imago Mundi*, 39 (1987), pp. 25–8, 34–7; Albert Derolez, 'Lambert van Sint-Omaars als kartograaf', *De Franse Nederlanden/Les Pays-Bas français, Jaarboek Annales* (1976), pp. 15–30.

69 A legend on the map mentions the name of a 'Richard de Haldingham o de Lafford' who 'made it and laid it out', but it remains unclear whether Richard was the map maker, the scribe, the patron or the mind behind the project; see Westrem, *The Hereford Map* (2001), pp. XXIV, 11. On the Hereford map see also Paul D. A. Harvey, ed., *The Hereford World Map: Medieval World Maps and their Context* (London: British Library, 2006); Naomi R. Kline, *Maps of Medieval Thought* (Woodbridge: Boydell Press, 2001); Paul D. A. Harvey, *Mappa Mundi: The Hereford World Map* (London: Hereford Cathedral–British Library; Toronto: University of Toronto Press, 1996). As Harvey notes, p. 7, the inscriptions on the map itself are in Latin, but those in the margin are in Anglo-Norman, the language of the English aristocracy of the time, which indicates that the author of the map intended to reach a wider audience than the learned milieu of Latin-speaking scholars.

70 Westrem, *The Hereford Map* (2001), pp. 38–9. See above, p. 134 and n. 39, and Chapter 3, p. 55.

71 The event/place of Jerusalem was made up of a single event – Christ's Passion – in one place, Golgotha, a hill in Jerusalem. The Passion could be signified on different maps by features of the Crucifixion or the Resurrection.

72 Jerome, *Liber locorum*, III.13–14 (1887; 1966), p. 145; see Westrem, *The Hereford Map* (2001), pp. 166–7.

73 Westrem, *The Hereford Map* (2001), p. 5: 'Ecce Testimonium [me]um.'
74 Ibid.: 'Levez! Si vendrez a joie pardurable.'
75 Ibid., p. 7: 'Levez! Si alez au fu de enfer estable.'
76 Ibid.: 'Veici, beu fiz, mon piz, de deinz la quele chare preistes, / E les mamelectes, dont leit de Virgin queistes. / Eyez merci de touz si com vos memes deistes, / Ke moy ont servi, kant Sauveresse me feistes'; 'See dear son, my bosom, from which you took on flesh, / And the breasts at which you sought the Virgin's milk; / Have mercy – as you yourself have pledged – on all those / Who have served me, since you made me the way of salvation.' See also Harvey, *Mappa Mundi: The Hereford World Map* (1996), p. 54.
77 On the notion of boundary (Latin *fines*) with its double role of defining the edges of a place and simultaneously creating the space within, as concerns paradise, see my 'Mapping Eden: Cartographies of the Earthly Paradise', in Denis Cosgrove, ed., *Mappings* (London: Reaktion Books, 1999), pp. 56–8.
78 London, British Library, Add. MS 28681, fol. 9r. The map is in a psalter and is thought to have been reduced, probably about the 1260s, from a now lost much larger original: see Barber, 'Mito, religione e conoscenza' (2001), p. 68; Edson, *Mapping Time and Space* (1997), pp. 135–7; Harvey, *Mappa Mundi: The Hereford World Map* (1996), p. 29; Peter Barber and Michelle P. Brown, 'The Aslake World Map', *Imago Mundi*, 44 (1992), pp. 31–3.
79 On the Ebstorf map, see Jürgen Wilke, ed., *Kloster und Bildung* (Göttingen: Vandenhöck und Ruprecht, forthcoming); Jürgen Wilke, *Die Ebstorfer Weltkarte*, 2 vols (Bielefeld: Verlag für Regionalgeschichte, 2001), bibliography, II, pp. 307–31; Hartmut Kugler and Eckhard Michael, eds, *Ein Weltbild vor Columbus: Die Ebstorfer Weltkarte: Interdisziplinäres Colloquium 1988* (Weinheim: VCH, Acta humaniora, 1991); Rolf Lindemann, 'A New Dating of the Ebstorf Mappamundi', in Pelletier, ed., *Géographie du monde* (1989), pp. 45–50; Wolf, 'Neues zur Ebstorfer Weltkarte. Entstehungszeit – Ursprungsort – Autorschaft' (1988); Birgit Hahn-Woernle, *Die Ebstorfer Weltkarte* (Ebstorf: Kloster Ebstorf, 1987), bibliography pp. 100–1; Hartmut Kugler, 'Die Ebstorfer Weltkarte: Ein europäisches Weltbild im deutschen Mittelalter', *Zeitschrift für Deutsches Altertum und Deutsche Literatur*, 116/1 (1987), pp. 1–29; Wolf, 'Die Ebstorfer Weltkarte als Denkmal eines Mittelalterlichen Welt- und Geschichtsbildes', *Geschichte in Wissenschaft und Unterricht*, 8 (1957), pp. 204–15; Walter Rosien, *Die Ebstorfer Weltkarte* (Hanover: Niedersächsisches Amt für Landesplanung und Statistik, 1952); Richard Uhden, 'Das Weltbild von Ebstorf', *Niedersachsen*, 33 (1928), pp. 179–83; Konrad Miller, *Die Ebstorfkarte: Eine Weltkarte aus dem 13. Jahrhundert* (Stuttgart: J. Roth, 1900). Richard Uhden, 'Gervasius von Tilbury und die Ebstorfer Weltkarte', *Jahrbuch der Geographischen Gesellschaft zu Hannover* (1930), pp. 185–200, was the first to associate the Ebstorf map with the *Otia imperialia* by Gervasius of Tilbury, a link also suggested by later scholars, but without conclusive evidence. See Armin Wolf, 'Gervasius von Tilbury und die Welfen: Zugleich Bemerkungen zur Ebstorfer Weltkarte', in Bernd Schneidmüller, ed., *Die Welfen und ihr Braunschweiger Hof im hohen Mittelalter* (Wiesbaden: Harrassowitz, 1995), pp. 407–38.
80 Miller, *Die Ebstorfkarte* (1900), p. 48. See Honorius Augustodunensis, *Imago mundi*, VIII, IX (1982), p. 52. Another inscription explains why the Tree of the Knowledge of Good and Evil is so called with an explanation drawn from Augustine's reading of Genesis: the tree is a material tree and by picking its fruit man learned the difference between good and evil: ibid., pp. 7–8. See Augustine, *De Genesi ad litteram libri duodecim*, VIII.6, ed. by Joseph Zycha, *CSEL* XXVIII/1 (Prague–Vienna–Leipzig: F. Tempsky–G. Freytag, 1894), pp. 239–40.
81 The intriguing construction has been variously interpreted as Christ supporting the earth or even as a gigantic eucharistic Host formed by the body of Christ: Wolf, 'Die Ebstorfer Weltkarte' (1957), p. 213. See also Jörg-Geerd Arentzen, *Imago mundi cartographica: Studien zur Bildlichkeit mittelalterlicher Welt- und Ökumenekarten unter besonderer Berücksichtigung des Zusammenwirkens von Text und Bild* (Munich: Fink, 1984), pp. 267–74. The idea that Christ supports the earth is put forward, for instance, by Uhden, 'Das Weltbild von Ebstorf' (1928), p. 179, and Rosien, *Die Ebstorfer Weltkarte* (1952), p. 39. Armin Wolf has compared the Ebstorf map with medieval representations of microcosm and macrocosm in order to explain the image of the earth as the body of Christ, and his interpretation has been widely accepted, for example by Uwe Ruberg, '*Mappae mundi* des Mittelalters in Zusammenwirken von Text und Bild', in Christel Meier and Uwe Ruberg, eds, *Text und Bild: Aspekte des Zusammenwirkens zweier Künste in Mittelalter und früher Neuzeit* (Wiesbaden: Reichert, 1980), p. 571 and von den Brincken, 'Mappa mundi und Chronographia' (1968), p. 146. Wolf's thesis has been rejected by Barbara Bronder, 'Das Bild der Schöpfung und Neuschöpfung der Welt als "orbis quadratus"', *Frühmittelalterliche Studien*, 6 (1972), p. 209 n. 88.
82 Miller, *Die Ebstorfkarte* (1900), p. 10.
83 Ibid.: 'Dextera Domini feci(t virtutem).' Psalm 118.16: 'Dextera Domini fecit virtutem: dextera Domini exaltavit me, dextera Domini fecit virtutem.' The compilers of the Gloss, *Biblia latina cum Glossa ordinaria* facsimile of the *editio princeps* … (Strasbourg: Adolph Rusch, ed. by Margaret T. Gibson and Karlfried Fröhlich, 2 vols 1480/1), (Turnhout: Brepols, 1992), I, p. 605, explain that the right hand of Yahvé cited in Psalm 118 (117) is Christ.
84 Miller, *Die Ebstorfkarte* (1900), p. 10: 'Terram palmo concludit.'
85 'Quis mensus est pugillo aquas, et coelos palmo ponderavit? Quis appendit tribus digitis molem terrae, et libravit in pondere montes, et colles in statera?' 'Who hath measured the waters in the hollow of his hand, and meted out heaven with the span, and comprehended the dust of the earth in a measure, and weighed the mountains in scales, and the hills in a balance?'
86 *Corpus antiphonalium officii*, ed. by Réné Jean Hesbert, *RED SM, Fontes* IX (Rome: Herder, 1968), III, p. 424: 'Qui coelorum contines thronos et abyssos intueris, Domine Rex regum, montes ponderas, terram palmo concludis; exaudi nos, Domine, in gemitibus nostris.' 'O Lord, king of kings, you who contain the throne of the heavens and see the abyss, who ponder the mountains and hold the earth in your hand, Lord, listen to our lamentations.' In Christian liturgy, the Antiphonary of the Office is an inventory of all prayers and psalms sung during the Office, the Church's public daily prayer. The text quoted above is found in all the medieval Antiphonaries surveyed by Hesbert, which date from the ninth to the twelfth centuries in both parishes and monasteries.
87 Respectively, 'Usque ad finem fortiter; Suaviter disponensque omnia' (Miller, *Die Ebstorfkarte* (1900), p. 10) – and 'Attingit ergo a fine usque ad finem fortiter, et disponit omnia suaviter.' See also *Corpus antiphonalium officii* (1968), p. 376: 'O Sapientia, quae ex ore Altissimi prodisti, attingens a fine usque ad finem, fortiter suaviter disponensque omnia; veni ad docendum nos viam prudentiae.' 'O Wisdom, coming from the mouth of the Most High, you who reach from one end of the world to the other and govern the whole world with benevolence, come to teach us the way of prudence.'

88 *Biblia latina cum Glossa ordinaria* (1480/1; 1992), II, p. 730: 'Attingit a fine ..., id est a principio mundi usque ad adventum christi mirifica opera et sincera testimonia per vetus testamentum fortiter asserit; et ab incarnatione verbi usque ad finem mundi suavitatem evangelii exponit.' 'She [Wisdom] reaches from one end ...: that is, from the beginning of the world to the advent of Christ, through the Old Testament she declares strongly prodigious deeds and true testimonies, and, from the Incarnation of the Word to the end of the world, she expounds the sweetness of the gospel.'

89 On Jerusalem on the Ebstorf map see Ingrid Baumgärtner, 'Die Wahrnehmung Jerusalems auf mittelalterlichen Weltkarten', in Dieter Bauer, Klaus Herbers, and Nikolas Jaspert, eds, *Jerusalem im Hoch- und Spätmittelalter: Konflikte und Konfliktbewältigung-Vorstellungen und Vergegenwärtigungen* (Frankfurt am Main: Campus, 2001), pp. 299–301; Kerstin Hengevoss-Dürkop, 'Jerusalem – Das Zentrum der Ebstorf-Karte', in Kugler and Michael, eds, *Ein Weltbild vor Columbus* (1991), pp. 205–22; Arentzen, *Imago mundi cartographica* (1984), p. 269. According to Étienne Gilson, *Les Métamorphoses de la Cité de Dieu* (Louvain: Publications Universitaires de Louvain, 1952), the tension between heaven and earth during the Middle Ages led the notion of the City of God – the mystical community of mankind saved in Christ – to a series of crucial 'metamorphoses' and to the dream of an earthly commonwealth of mankind enlightened by the Church's teaching.

90 The iconographical assimilation of the historical and earthly Jerusalem with the Heavenly Jerusalem was also a persistent theme in medieval art: see above, Chapter 4, p. 82 n.61. Not far from Jerusalem on the Ebstorf map is the island of Patmos, where, according to Christian tradition, John had his vision of the celestial city.

91 Miller, *Die Ebstorfkarte* (1900), p. 41: 'Iherusalem, Sepulchrum Domini gloriosum teste Ysaia'. See Isaiah 11.10: 'And in that day there shall be a root of Jesse, which shall stand for an ensign of the people; to it shall the Gentiles seek: and his rest shall be glorious.'

92 Miller, *Die Ebstorfkarte* (1900), p. 41: 'Hec civitas celeberrima capud omnium civitatum toti mundo extat, quia in ea salus humani generis morte et resurrectioni Domini consumata est dicente psalmista: Rex noster ante saecula et cetera. Ipsa etiam civitas magna continet sepulchrum Dominicum, quod pia aviditate querere desiderat totus orbis, quia nobilitavit illud resurgens a mortuis victor mortis, unde Sedulius: Ubi, inquid, depositi thesaurum corporis amplum nobilis accepit Domino locus ille iacente, nobilior ipso resurgente sanctus est.' Sedulius, a fifth-century Christian poet, describes the life of Christ from his virginal conception to his Resurrection and Ascension: see Sedulius, *Paschale carmen*, V.295–7, in *Opera omnia*, ed. by Johannes Huemer, CSEL X (Vienna: C. Geroldi filius, 1885; repr. New York: Johnson Reprint, 1967), p. 136; see also the version in prose, *Paschalis operis libri quinque*, V.25, ibid., p. 294.

93 A camel in Palestine is shown also on the mid-thirteenth-century map of Matthew Paris (on this map see above, Chapter 5, pp. 85 and 116–7 nn. 8–9). The interpretation of the camel as a sign of humility may come from Gregory the Great: see Uwe Ruberg, 'Die Tierwelt auf der Ebstorfer Weltkarte im Kontext mittelalterlicher Enzyklopädik', in Kugler and Michael, eds, *Ein Weltbild vor Columbus* (1984), p. 344 n. 53. Ruberg also hypothesized that the camel could refer to the humility of the authors of the map who did not sign their work.

94 See, for example, Richard Barber, ed., *Bestiary: Being an English Version of the Bodleian Library, Oxford M. S. Bodley 764 with All the Original Miniatures reproduced in Facsimile* (Woodbridge: Boydell Press, 1993), pp. 95–6; Terence H. White, *The Book of Beasts: Being a Translation from a Latin Bestiary of the Twelfth Century* (London: Cape, 1954), pp. 79–80.

95 See Edson, *Mapping Time and Space* (1997), pp. 137–8; Barber and Brown, 'The Aslake World Map' (1992), pp. 31–2, 36, 38; Graham Haslam, 'The Duchy of Cornwall Map Fragment', in Pelletier, ed., *Géographie du monde* (1989), pp. 33–44. For an overview of the merging of time and eternity on twelfth- and thirteenth-century maps see Danielle Lecoq, 'Le Temps et l'intemporel sur quelques représentations médiévales du monde au XIIe et au XIIIe siècles', in Bernard Ribémont, ed., *Le Temps: Sa mesure et sa perception au Moyen Âge* (Caen: Paradigme, 1992), pp. 113–49.

96 It has been suggested that in order to comprehend the opposition between an eschatological understanding of the biblical narrative unconnected with any specific locations and a topographical approach that developed the idea of holy places on earth, the terminological distinction between 'locative' and 'utopian' tendencies in religion should be adopted. According to this distinction, 'locative' refers to a vision that emphasizes place and installation within the world, while 'utopian' refers to a more open view that seeks absolute freedom and transcendence. The value of such terminology is that it enables the acknowledgement that in early and medieval Christianity these two world views could, and did, coexist as part of a fundamental dichotomy. See Robert A. Markus, 'How on Earth Could Places Become Holy? Origins of the Christian Idea of Holy Places', *Journal of Early Christian Studies*, 2/3 (1994), pp. 257–71, esp. 264–6; Jonathan Z. Smith, *Maps Is Not Territory* (Leiden: Brill, 1978), pp. 88–104; 104–28; 172–189 and other works quoted in Markus, p. 264 n. 33.

97 See above, Chapter 3, p. 51.

98 Paul Celan, 'Rede anlässlich der Verleihung des Georg-Büchner-Preises, am 22. Oktober 1960', in *Gesammelte Werke in sieben Bänden*, III (Frankfurt am Main: Suhrkamp, 2000), p. 199 : 'Toposforschung? Gewiß! Aber im Lichte des zu Erforschenden: im Lichte der U-topie.'

7

Where Is Nowhere?

If some people do not hesitate to go as far as the borders of the earth, looking for silk, with the purpose of a miserable trade, how anyone could hesitate to set out on the journey to contemplate paradise?

Cosmas Indicopleustes ('Mr World who travels to the Indian sea'), *Christian Topography* (sixth century)

The paradox of the Christian notion of an earthly paradise, a heavenly locality on earth, was that it was a 'nowhere' that was 'somewhere'. However different from the rest of this world, that no-place (*u-topia*) was part of real geography, and mappable. Christianity acknowledged in many different ways the intangible relationship between the post-lapsarian human realm and the perfect communion between man and God that was implied in the idea of an earthly paradise; but the mapping of the Garden of Eden in the Christian Middle Ages points with the utmost clarity to the medieval tendency, generally overlooked in modern scholarship, to conceive of a state of human perfection on earth. Paradise was not only a vague condition in a future heaven, or an original harmony forever lost in the past, but also an earthly place, different from this world and yet part of it, situated and indicated on maps.

The idea of an earthly paradise distinguishable from the heavenly paradise had taken several centuries to evolve. Following Augustine's influential treatment of the subject, the belief that Adam and Eve had lived in a specific region before the Fall, and that this region still existed on earth, was accepted as beyond question. Then, once it was acknowledged that the Garden of Eden remained physically present on earth, and once the difference between the terrestrial and the celestial paradises had been established, the difficulty had to be faced as to how to explain paradise's intermediate character. Thus, the focus of the debate switched to the problem of distinguishing Eden from the earth on which it was literally located. The notion of an earthly paradise involved imagining a place lying somewhere between heaven and earth and that managed to escape the natural conditions of earth. Mapping the Garden of Eden presented the ultimate cartographical paradox: how to map a place that was *on* earth but not *of* earth.

Paradise between East and West

Augustine had relatively little impact in the Christian East, where the allegorical reading of the paradise narrative of earlier times continued to predominate and to encourage a symbolic approach. The world map drawn in the sixth century by Cosmas Indicopleustes, the Greek-speaking Alexandrian merchant who seems to have become a monk, shows a conception of the world much indebted to biblical symbolism unknown in the Latin West (Figure 7.1).[1] In his *Christian Topography*, Cosmas intended to prove that any inheritance from the pagan ancients was inappropriate for a Christian audience and that true Christian teaching about the shape of the universe should come only from Scripture. He thus explained that the earth could not be spherical, as those

Fig. 7.1a (opposite). World map from a manuscript of Cosmas Indicopleustes's *Christian Topography*. Constantinople, ninth century. 23.3 × 31.5 cm. Vatican City, Biblioteca Apostolica Vaticana, MS Vat. Gr. 699, fol. 40v. Cosmas intended to prove that Moses's wilderness tabernacle (Exodus 25–7) was a model for the structure of the entire universe. His map shows a rectangular flat earth with four gulfs and surrounded by the ocean, in its turn encircled by rectangular land. The four rivers of paradise flow out of the 'land of the beyond' to cross the ocean and gush out again in the inhabited earth. The four heads blowing trumpets indicate the four main winds. Cosmas relates that he was once sailing by India, not far from the entrance of the ocean, but the crew of his ship was terrified by the unhealthy air, the strong currents and the huge backwash. Wisely, the pilot veered back towards the gulf to avoid being carried away into the ocean and finding death there.

Fig. 7.1b (opposite). Diagram of Fig. 7.1a.

160

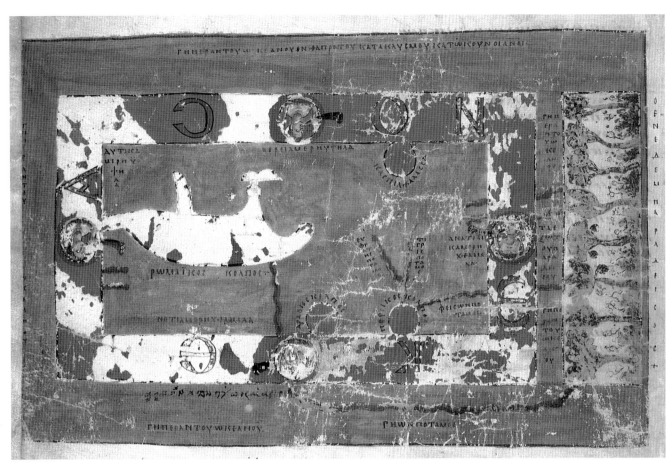

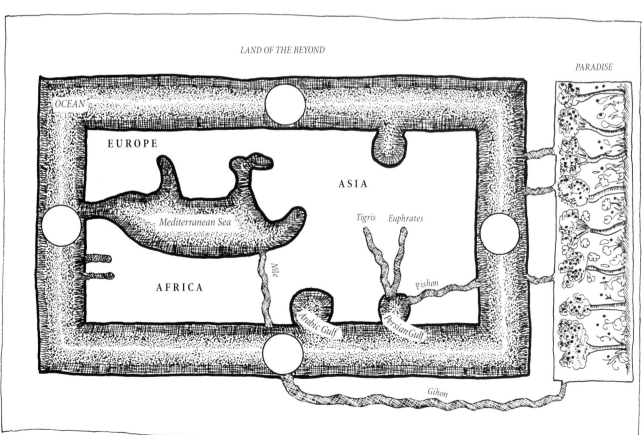

Where Is Nowhere? 161

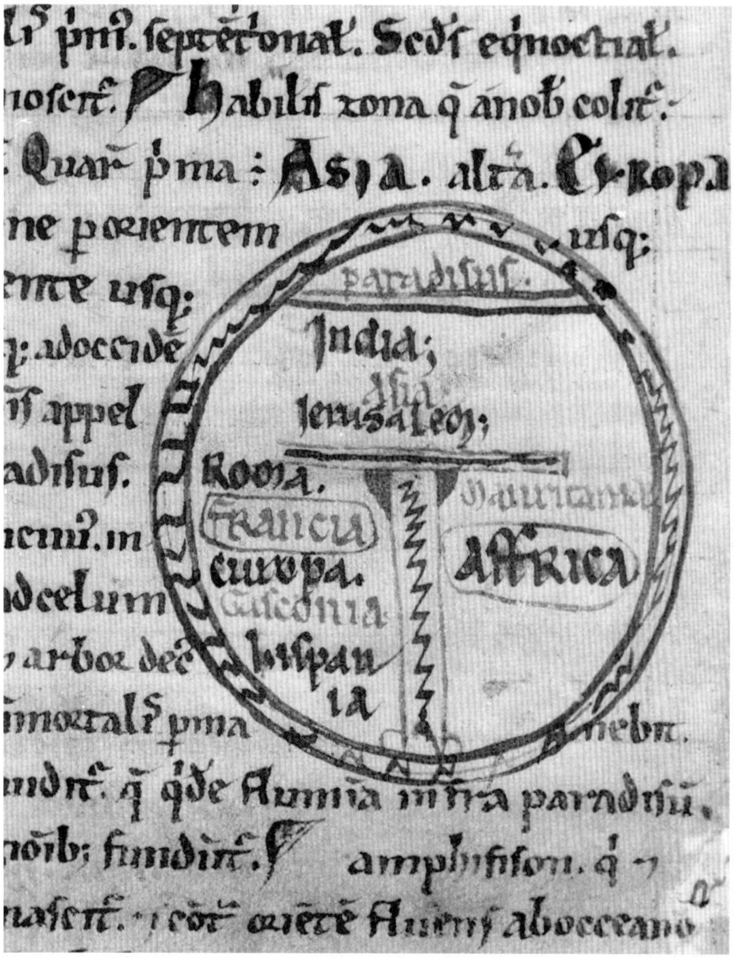

Fig. 7.2. World map illustrating an excerpt from Honorius Augustodunensis's *Imago mundi*. ?England, thirteenth century. 19 × 13 cm. London, British Library, Harley MS 218, fol. 104v. The map shows the three parts of the inhabited earth arranged in traditional T-O structure. India is marked in Asia, France, Gascony and Spain in Europe, Mauritania in Africa. Jerusalem features at the centre of the map. Paradise, described in the adjoining text as the first region of Asia, is indicated by a vast area in the farthest east and next to India. The text also explains that within that delightful paradise is the Tree of Life, which is able to grant eternal life to anyone eating its fruit.

who only pretended to be Christians claimed, but was structured like a tabernacle. The rectangular inhabited land was surrounded by the ocean, and the terrestrial paradise lay to the east, beyond the ocean, on a 'land of the beyond'. The 'land of the beyond' was another earth surrounding the post-diluvian world, and separated from it by the impassable ocean. Its extremities were attached to the extremities of heaven, and prior to the Flood, mankind had been living here after the expulsion from paradise. According to Cosmas, navigation was possible only on the four gulfs he had marked on the map – the Mediterranean Sea, the Arab or Red Sea, the Persian Sea, and the Caspian Sea. The ocean could not be navigated because of powerful currents, the vapours arising from it that totally obscured the light of the sun, and the excessive distances involved. During the Flood, however, Noah had miraculously crossed the ocean in the ark, arriving on Mount Ararat and populating our present land with his offspring.[2]

A textual example of the Christian East's different approach to paradise is found in the treatment of the paradise narrative advanced in the fourth century by the theologian and religious poet Ephrem the Syrian. Ephrem approached the mystery of paradise in symbolical mode, poetically and by way of a series of paradoxical assertions. He made no clear distinction between the Garden of Eden, the paradise of the Church, and the Kingdom of Heaven.[3] His idea was that paradise was a reality that had existed in the beginning, still existed in the present, and would exist until the end of time. In his fifteen *Hymns on Paradise*, Ephrem describes paradise as a mountain that rises to the divine sphere and consists of separate levels. On the summit dwells the divine presence. In the higher circle of the mountain is the Tree of Life, like the Holy of Holies of the Temple, where Adam and Eve were forbidden to go. This is the region destined for the glorious. Midway up to the highest level is the Tree of the Knowledge of Good and Evil, a sort of veil hiding the Holy of Holies, for it leads to the Tree of Life. Between the Tree of Knowledge and the boundary of paradise is the region inhabited by Adam and Eve and destined for the just. On the foothills of the mountain are placed the Cherubim with the flaming sword, but the fiery barrier has now been removed by Christ's sacrifice.[4] This division corresponds, in Ephrem's view, to the structure of the human being – consisting of intellectual spirit, soul and body; to the various levels in the Ark – Noah, the birds, the animals; and, following Exodus 19, to the different stages of the approach to Mount Sinai – that of Moses, Aaron, and the Jewish people.[5] Ephrem thought of Eden as a structure that had to match all levels of reality since it was the prototype of everything in existence, not merely a region of the earth, but the pinnacle of all creation.[6] Although the Garden of Eden was described as an earthly mountain, it seemed to transcend and surround the physical world according to a mythical geography. The mountain was excessively high and surrounded the ocean or 'Great Sea', which itself surrounded the earth. In this way, paradise itself encircled both land and sea.[7] Ephrem's profound theological synthesis seems to contain a fun-

162 Mapping Paradise: A History of Heaven on Earth

damental contradiction between the idea of paradise as an earthly garden and its role as a symbolic region belonging to a different mode of existence. He claimed to have seen the lofty mountain of paradise 'with the eye of the mind'.[8] He also compared paradise to the halo of light around the moon and to the golden wreath of Moses around the altar; in like manner, he believed, paradise surrounded the whole of creation.[9] At the same time, Ephrem said, to enter paradise was to enter the inner sanctuary of the Temple.[10] According to Ephrem, paradise was simultaneously at the summit, the periphery and the centre of the universe: an earthly region and yet not part of the normal world, formed by a type of spiritual matter constituting a divine and everlasting 'Garden of splendours' blessing and purifying the earth.[11]

Western Christian exegetes, in contrast, had developed the geographical side of Augustine's vision of history, and by the thirteenth century were less inclined to see Eden 'with the eye of the mind'. They acknowledged the symbolic role of paradise, but they were more interested in the question of its location on earth. True, their map makers found that, like Cosmas Indicopleustes, who had located paradise on a 'different' earth beyond the ocean, they had to emphasize the *passage* to a dimension different from the time and space of ordinary human experience if they wanted to represent the geography of paradise on a map. So Eden was often portrayed on the edge of the habitable world as an island in the outer ocean, as on the Sawley and the Hereford *mappae mundi*, for example (Figures 6.10, 6.12 and 6.15). When paradise was represented within the habitable world, it was always shown to be insulated from the world by some sort of daunting barrier, such as a high mountain chain or a wall of fire, as on the Psalter and Ebstorf maps (Figures 6.16, 6.17, 6.18 and Plates 8 and 9). On a rarely noticed map illustrating a passage from Honorius Augustodunensis's *Imago mundi* dating from the thirteenth century, paradise is shown as a vast region in the east separated from Asia by a double black and red line (Figure 7.2); an adjacent text makes it clear that a wall of fire reaching up to heaven renders paradise absolutely inaccessible.[12] A similar, but smaller, enclosure is found on the fourteenth-century world map in *Les Grandes Chroniques de Saint Denis du temps de Charles V* (Figure 7.3).[13] It was not only map makers who recognized the separateness of the earthly paradise; exegetes had always made it clear that their descriptions of the unlocatable and unknown Garden of Eden were inaccurate and incomplete. Once, however, thirteenth-century Western Christian scholars embarked on an elaboration of Augustine's literal reading for their treatises on Genesis, they found they had set themselves an uphill task to unravel the geographical contradictions of a paradise on earth.

A Garden between Heaven and Earth

The biblical description of the earthly paradise is of a garden, and that is almost always (but not inevitably) how paradise was represented on the *mappae mundi*. What was open for discussion, however, was what the garden was actually like. What was its climate, what plants grew there, could the trees named in Genesis be identified with anything in the contemporary world? On such matters, the opinions of medieval exegetes varied as they struggled to match their concept of the perfection represented by the earthly paradise with what they knew of the physical world around them.

Many medieval maps, such as the one in *Les Grandes Chroniques de Saint Denis du temps de Charles V*, included in their depiction of the world a reference to at least the most important winds that blew over the earth and that were held to give the various earthly regions their physical characteristics (Figure 7.3). The idea of the winds and their climatic effects went back to classical Antiquity, but it was Isidore's *De natura rerum* (written between 612 and 615) that gave the Middle Ages a standard classification and the names by which the winds came to be generally known.[14] On medieval maps the winds are either pictured as figures or heads, blowing onto the earth from the appropriate direction, or simply named. Very often the eastern wind – the *Subsolanus* – appears close to the eastern paradise at the top of an east-orientated map. On the Hereford map, for example, a demonic-looking figure hangs upside down over the island of paradise (Figure 6.12). On the Psalter map, a rotund human face blows wind onto the circle enclosing Adam and Eve (Figure 6.16 and Plate 8). On the early fourteenth-century map in a manuscript of William of Tripoli's *De statu Sarracenorum*, the eastern wind is marked just above the map sign for paradise (Figure 7.4).[15] Each wind was thought to have a different effect on the world. As explained on the Hereford map, the *Subsolanus* produced seasonal rain. The east-north-east *Vulturnus* desiccated everything. The western *Favonius* tempered the harshness of winter and brought forth flowers. The north-north-east *Boreas* restrained clouds, whereas the southern *Auster* produced them. The north-north-west *Circius* brought clouds and hail.[16] By definition, though, paradise was subject neither to any of the winds nor to their effects. Notwithstanding its presence on the map of the world, the Garden of Eden was held to be beyond the reach of earthly climatic forces. Unlike normal regions, the climate within paradise was permanently, and enviably, temperate.

For centuries, Christian theologians and poets had sought to make sense of Eden's atmospheric conditions. The usual way of describing the Edenic climate emulated the classical descriptions of primordial perfection and blessed lands.[17] Plato had celebrated the Age of Kronos as a period when 'people spent most of their time roaming around in the open air without clothes or bedding, since the climate was temperate and caused them no distress', and Ovid had confirmed that the age 'was a season of everlasting spring, when peaceful zephyrs, with their warm breath, caressed the flowers that sprang up without having been planted'.[18] In telling how Kronos had transferred his kingdom to the Elysian Fields, Hesiod expressed the idea of a primordial perfection in spatial terms.[19] When late antique and medieval Christian thinkers came to describe their earthly paradise, they also negated all the unpleasant features of worldly climate. For them, there was no bitter cold in paradise, no torrid heat, no winds, no storms, no rain, no hail, no snow, and no clouds. On the contrary, the Garden of Eden was always described as having luxurious vegetation, abundant fruit, fertile soil, richly perfumed flowers, and pure, wholesome air. John of Damascus, for instance, spoke of 'a region with a temperate climate, everywhere illuminated by a very subtle and most pure air, blooming with evergreen plants, filled with the sweetest odour and permeated by light, beyond any idea of physical grace and beauty'.[20] Isidore of Seville's celebration of a delightful garden that knew neither extreme cold nor extreme heat, only permanently temperate weather, was repeated for centuries.[21]

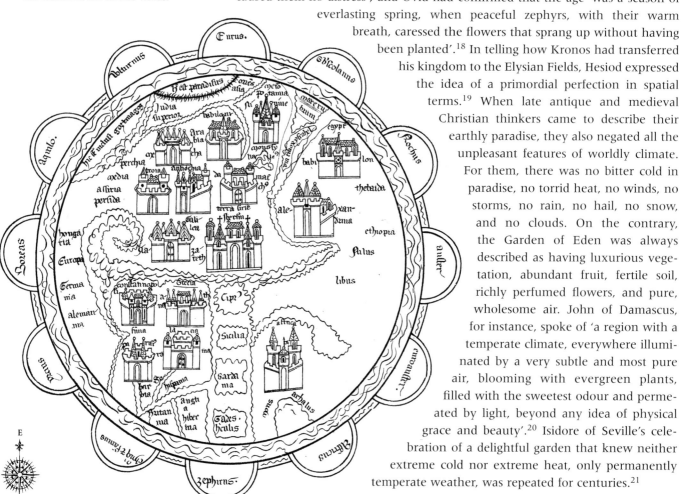

Fig. 7.3. World map in *Les Grandes Chroniques du temps de Charles V*. France, c.1370. Diameter 16 cm. Paris, Bibliothèque Ste-Geneviève, MS 782, fol. 374v. Redrawn from Manuel Francisco de Barros e Sousa, Viscount of Santarém, *Atlas composé de mappemondes, de portulans et de cartes hydrographiques et historiques depuis le VIe jusqu'au XVIIe siècle* (Paris: Maulde et Renou, 1849), Plate 21. The map has a T-O structure. Larger signs point to the most important regions and towns, such as Rome, Constantinople, Alexandria and Jerusalem, situated in the centre. Gog and Magog, surrounded by mountains, and paradise, encircled by a wall of fire, painted in red, are in the east. Around the encircling ocean are twelve semicircles with the names of the twelve winds.

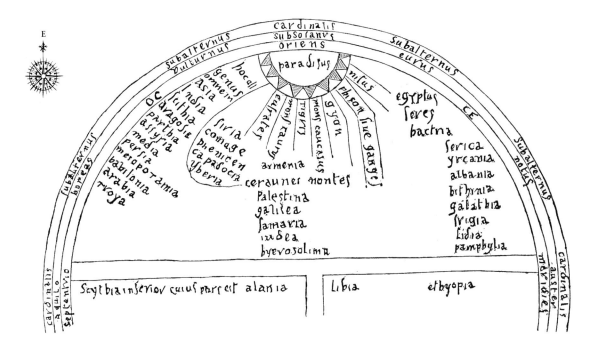

Paradise, then, was a physical garden. It contained living plants, but there were no seasons; it was eternally spring, fruit and flowers never faded or failed, and the harvest was plentiful all the year around. The crucial question for late medieval scholars was where on earth was such a delightful and perfectly temperate place to be found. To answer the question, the paradoxical and mythical condition of paradise had to be framed within the geographical learning of the time.

Paradise and the Climatic Zones

East-orientated and topological *mappae mundi* were not the only kind of maps circulating in the West during the Latin Middle Ages to show the world. Zonal maps – diagrammatic maps showing the sequence of five climatic belts encircling the earth as defined in classical times (two frigid, two temperate, and a torrid zone along the equator) – were drawn to illustrate medieval copies of the ancient geographical texts. Unlike *mappae mundi*, zonal maps were usually south- or north-orientated (see Figure 7.5).[22] Also unlike *mappae mundi*, which portrayed the inhabited northern hemisphere as neighbouring paradise and on which history was the overriding factor, the zonal map depicted the entire terrestrial globe in a non-topological and entirely ahistorical manner, and showed the earth divided into zones determined through astronomical observation. Both *mappae mundi* and zonal maps were part of Christian medieval culture, but the vast majority of zonal maps lack a sign for paradise. There is nothing surprising in this. A zonal map of the earth was by definition focused on the physical structure of the globe – its division into climatic zones – and not on the development of mankind's history in earthly space. Moreover, it was difficult to associate paradise with any of the world's climatic zones. There were zones known as 'temperate' in both the northern and the southern hemispheres but they were supposed to be subject to changing atmospheric conditions. Not only was the location of paradise unknown, but, with its perpetually temperate weather, it fitted none of the climatic belts, unless it was some quite special area independent from 'normal' weather conditions. Whereas the essentially topological T-O map, similarly a descendant of the diagrams in classical texts, but which served as a summary of the human geography of the world, could readily be absorbed

Fig.7.4. 'Il mappamondo del *De statu Sarracenorum*', detail from William of Tripoli's world map. France, early fourteenth century. Diameter 9.5 cm. Paris, Bibliothèque Nationale de France, MS Lat. 5510, fol. 118r. Redrawn from Edoardo Coli, *Il Paradiso terrestre dantesco* (Florence: Carnesecchi, 1897), p. 109. William of Tripoli was a thirteenth-century Dominican living in Acre, Palestine. He wrote to promote the conversion of the Muslims to Christianity. The map is placed in the codex just after William's text and is possibly linked to a passage where William accounts for the passage of the Saracens from Africa to Europe to conquer Spain. It refers to the tripartite earth. The map contains some 85 toponyms, 45 of which are found in Asia. Paradise is indicated by the word *paradisus* within a semicircle surrounded by a ring of mountains and the four rivers flowing from it. The Taurus and Caucasus mountains are shown next to paradise. Above paradise the eastern wind (*Subsolanus*) is indicated. Imagining the landscape and environment of Eden was a challenge. Was the earthly paradise subject to the passing of the seasons? Did it rain there? Did frost ever whiten its lawns? By definition, simply because it was paradise, the weather in the Garden of Eden had to be assumed to be perfect.

Where Is Nowhere? 165

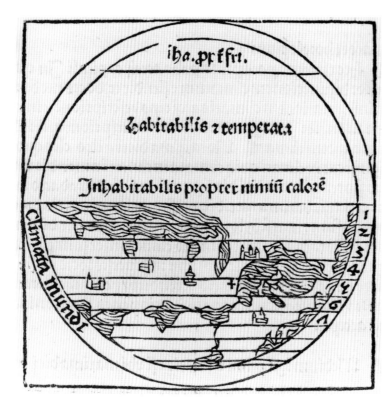

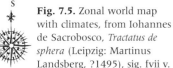
Fig. 7.5. Zonal world map with climates, from Iohannes de Sacrobosco, *Tractatus de sphera* (Leipzig: Martinus Landsberg, ?1495), sig. fvii v. Diameter 14 cm. London, British Library, IA 11977. Sacrobosco's thirteenth-century treatise on the *Sphera* was the most influential work on astronomical geography in the Middle Ages. Sacrobosco maintained that the region at the equator could not be temperate because of the excessive heat produced by the sun, as explained by the inscription on the map (*inhabitabilis propter nimium calorem*). In contrast, the two zones lying between the tropics and the poles were tempered by the heat emanating from the torrid zone and the cold emanating from the polar circles, and were thus habitable. The south-orientated map shows the two habitable zones, the temperate southern zone (*habitabilis temperata*) and the temperate northern zone, together with the seven climates of classical Antiquity.

into a Christian cartographical discourse, there was no obvious place for paradise on the astronomically defined zonal map. All the same, it would be wrong to conclude from the absence of paradise on a zonal map a lack of belief in its existence. The zonal map served a purpose – the definition of climatic zones – which was inconsistent with the inclusion of the Garden of Eden. Nevertheless, a number of medieval maps survive on which the two cartographical genres of the topological *mappa mundi* and the zonal map have been merged in such a way as to demonstrate quite positively belief in an earthly paradise. Consider, for example, the group of maps penned in the same hand on two facing folios at the end of a twelfth-century Bible from Arnstein, Germany.[23] On one page, a topological map of the northern hemisphere has been superimposed onto a zonal map to produce a T-O/ zonal hybrid on which paradise and other event/places are marked (Figure 7.6a). On the facing page a second zonal map lacks any mention of paradise, but gives a broader perspective by showing the heavenly spheres and the signs of the zodiac as well as the earth (Figure 7.6b). Both maps depict the earth (*terra*), of which only the northern hemisphere was held to be inhabited, but the second map includes it in a wider representation of the universe (*mundus*).[24] In the centre of each map, the earth is shown divided into five zones: the arctic and antarctic zones, uninhabitable because of the extreme cold; the equatorial zone, occupied by the sea; and the two inhabitable temperate zones, one in the northern hemisphere, known and inhabited, and the unknown, and presumed uninhabited, temperate zone of the southern hemisphere. Unusually for zonal maps, the two Arnstein Bible maps are east-orientated, which means the zones run vertically down the page. On the second map, the zones are simply outlined and identified, and only the inhabited zone has been subdivided to show the three parts of the world. On the first map, though, an expanded inhabited zone takes up most of the map, making room for a more detailed presentation of the world in T-O fashion and a list of the most important classical and biblical sites. Listed first, in east Asia, is *Paradisus*, followed by India, Parthia, Assyria, Persia, Media, Mesopotamia, and other places associated with the east–west axis of human history. That this map was intended to allude to God's work of restoration as well as creation is suggested by the tree diagram drawn on the same page that summarizes Hugh of St Victor's classification of the different branches of science, all of which he saw as remedies of one sort or another for humanity's post-lapsarian failings.[25] The second map, in contrast, which also includes the heavens, switched the emphasis from spatialized human history to the entire cosmos as the work of God's creation.

The detailed rendering of the northern, inhabited, temperate zone on these relatively uncommon hybrid zonal-topological maps introduced a historical dimension to a spatial representation ruled otherwise only by the astronomical zones of classical science. It was this fusion of two cartographical genres that made it possible to refer to the geographically unlocatable paradise on a zonal map. For another example, we can turn to the east-orientated zonal map bound in a fifteenth-century manuscript that illustrates an excerpt from the universal chronicle by Girard of Antwerp (compiled *c*.1272). The map bears a significant title, *Tabulata Biblia*, which may be rendered

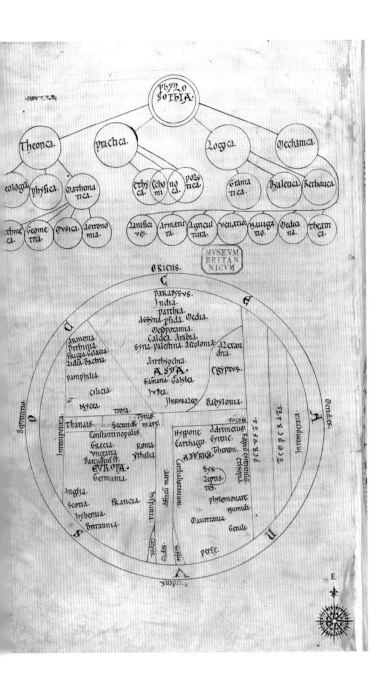

Fig. 7.6a. World map from the Arnstein Bible. Arnstein, Germany, 1172. Diameter 28 cm. London, British Library, Harley MS 2799, fol. 241v. The terrestrial globe is divided vertically into five zones. The northern inhabited zone takes up most of the map so as to accommodate a detailed presentation that includes paradise in the farthest east. The tree diagram above the map summarizes Hugh of St Victor's classification of the different branches of science. In Hugh's view, all four branches of philosophy (theoretical, practical, mechanical, and logical) helped to restore in man the lost image and likeness of God. Logic – divided into grammar, dialectic, rhetoric – aided the understanding of the other three branches. Human infirmity and weakness were alleviated by the seven mechanical arts, which were devoted to the occupations of this life: cloth making, armaments, agriculture, hunting, transport, medicine, and the performing arts. Concupiscence was countered by practical philosophy, which was subdivided into ethics, economics and politics, and ordered by morality. Finally, ignorance was illuminated by theoretical science, which was divided into theology, physics and mathematics, and directed at the contemplation of truth.

Fig. 7.6b. Zonal map from the Arnstein Bible. Arnstein, Germany, 1172. Diameter 32 cm. London, British Library, Harley MS 2799, fol. 242r. The simple north-orientated map shows the five zones of the globe with the inhabited, northern zone subdivided into the three parts of Asia, Africa and Europe. The diagram lacks any indication of paradise, but it gives a comprehensive view of the seven heavenly spheres that surround the earth, with their planets, and the outermost sphere of the heaven of the fixed stars, complete with the signs of the zodiac. The two smaller diagrams on the same folio illustrate the tropics (left) and the relationship between micro- and macro-cosmos (right). The final diagram, usually found as an illustration of Isidore's *De natura rerum*, describes the interrelationship of the four elements paired with their opposing qualities (hot and cold, moist and dry). To show the link between the world (*mundus*), man (*homo*) and the cyclical succession of time (*annus*), the view of cosmic harmonies is completed by reference to the four ages of man, the four seasons of the year, and the four humours of the human body.

Where Is Nowhere?   167

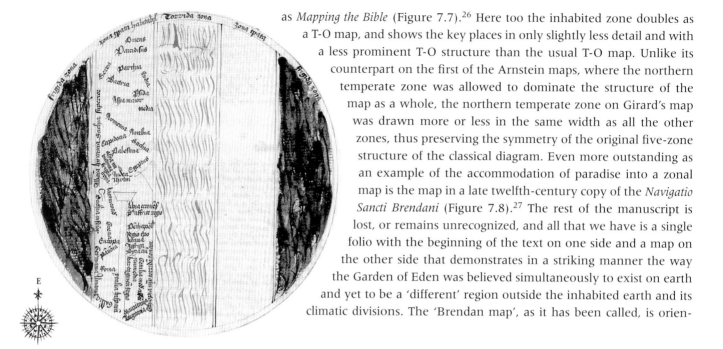

as *Mapping the Bible* (Figure 7.7).[26] Here too the inhabited zone doubles as a T-O map, and shows the key places in only slightly less detail and with a less prominent T-O structure than the usual T-O map. Unlike its counterpart on the first of the Arnstein maps, where the northern temperate zone was allowed to dominate the structure of the map as a whole, the northern temperate zone on Girard's map was drawn more or less in the same width as all the other zones, thus preserving the symmetry of the original five-zone structure of the classical diagram. Even more outstanding as an example of the accommodation of paradise into a zonal map is the map in a late twelfth-century copy of the *Navigatio Sancti Brendani* (Figure 7.8).[27] The rest of the manuscript is lost, or remains unrecognized, and all that we have is a single folio with the beginning of the text on one side and a map on the other side that demonstrates in a striking manner the way the Garden of Eden was believed simultaneously to exist on earth and yet to be a 'different' region outside the inhabited earth and its climatic divisions. The 'Brendan map', as it has been called, is orien-

Fig. 7.7 (above). *Tabulata Biblia, Mapping the Bible*. World map from a manuscript containing an excerpt from Girard of Antwerp's *Historia figuralis ab origine mundi usque ad 1272*. Flanders, fifteenth century. Diameter 18.3 cm. Utrecht, Rijksuniversiteit Bibliotheek, MS 737, fol. 49v. Paradise is placed in the habitable temperate zone (*zona temperata habitabilis*) north of the equatorial ocean. The presence of paradise on the zonal map bears witness to the European search for an earthly Garden of Eden that could be given an exact location, and to the increasing importance of geographical space in cartographical representations.

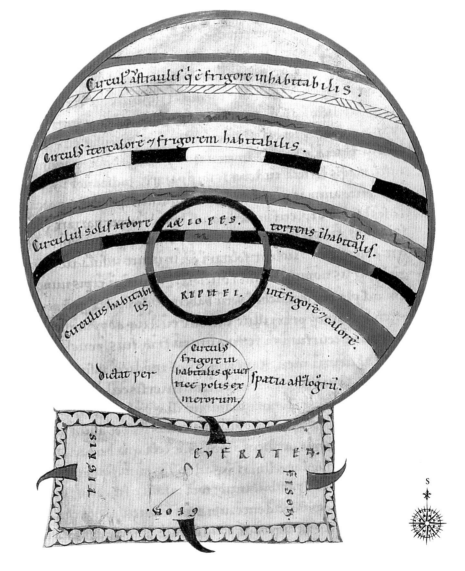

168   Mapping Paradise: A History of Heaven on Earth

tated to the south, and is a representation of the terrestrial sphere, which shows the five climatic zones. A circle represents the *oikumene* (the inhabited earth) with its southern and northern boundaries, named *Aetiopes* and *Riphei* respectively. Paradise is represented by a large rectangle drawn below the terrestrial globe. The rectangle appears to touch the spherical globe, but it is shown on a different plane to emphasize the separateness of paradise from the earth and its climatic divisions. Only the four rivers of paradise, indicated and named on the rectangle, connect the two otherwise unconnectable planes.

The ingenuity of the maker of the Brendan map highlights the reason why medieval zonal maps did not normally include paradise: the two cartographical discourses, that of the topological *mappa mundi* and the astronomical zonal map, were on different planes, as it were, and incompatible. It was theoretically impossible to show the earthly paradise on a map structured on astronomical measurements. As these examples show, however, paradise was occasionally included on a zonal map, either by pasting a simplified *mappa mundi* onto the zonal configuration (as on the Arnstein map and Girard's map) or by finding a way of overcoming the contradictions inherent in conflating two cartographical genres based on quite different principles (as on the Brendan map). The hybrid T-O/zonal maps reflect how geography was beginning to

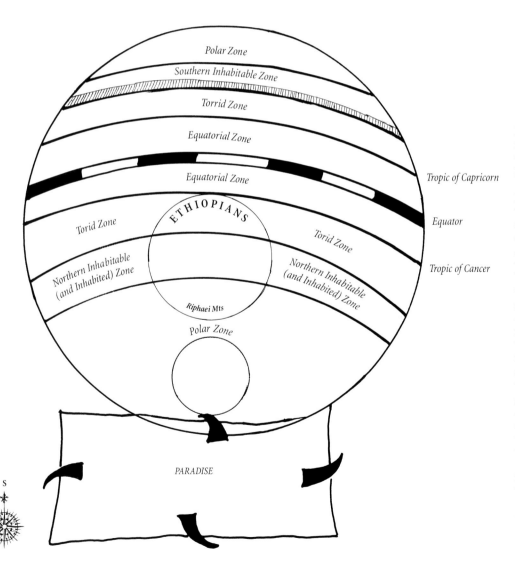

Fig. 7.8a (opposite). The 'Brendan map'. Southern Germany, late twelfth century. 21 × 16.5 cm. Bischofszell (Canton of Thurgovia, Switzerland), Ortsmuseum, Dr-Albert-Knoepfli-Stiftung. The south-orientated map represents the five zones of the earth: two polar zones, uninhabitable because of excessive cold; the equatorial belt, uninhabitable because of excessive heat; and the two temperate zones, south and north of the equator. The larger of the inner circles represents the inhabited part of the northern temperate zone. The smaller one represents the northern polar zone as a circle. Paradise, which was regarded as an uncontroversial geographical feature, is shown as a rectangle lying outside the climates that, according to medieval tradition, encircled the whole terrestrial globe.

Fig. 7.8b. Diagram of Fig. 7.8a. The Ethiopians, at the southern extremity of the inhabitable (and inhabited) zone, and the Riphaei mountains at the northern extremity, are shown as extending into the torrid and polar zones.

Where Is Nowhere? 169

catch up with history in the later Middle Ages. Paradise on a zonal map was a geographical statement, suggesting a specific location associated with global space measured through astronomical observation and divided into climatic zones, quite unlike the topological intimation of paradise on a *mappa mundi*. In fact, the whole of the thirteenth century was spent in lengthy scholarly debate on how to reconcile the location of the earthly paradise with the astronomical division of the globe.

Grasping the Mystery: Paradise and the Geographical Renaissance

Augustine's vision of history allowed the geographical location of paradise to be named as *somewhere* in eastern Asia. No scholar had ventured to suggest exactly where, and Isidore of Seville, Bede and Peter Lombard had discussed the issue only in the most general of terms. Now, in the thirteenth century, such vagueness was no longer tenable and the problem of the location of the earthly paradise turned to a specific geographical conundrum. Any attempt to express the paradise question in purely rational and physical terms, however, or to tie paradise down to a recognizable geographical location was doomed to failure. Eventually, deprived of its aura of mystery by being dissected as a topographical feature, the Garden of Eden was to prove to be in an inaccessible *nowhere*.

The recourse to scientific speculation to confirm the truthfulness of Holy Scripture was a natural development from Augustine's vision of history and his literal reading of Genesis. Once the earthly dimension of Eden that he had postulated had been accepted, it was inevitable that belief in it would be followed by confronting the geographical and astronomical implications. Instead of dwelling on the idea of Eden as a region belonging to a different mode of existence, thirteenth-century exegetes tried to handle the problem of its location by taking into account the physical structure of the globe. One idea that emerged from their interest in the climatic zones, for example, was that the fiery sword of the Cherubim guarding the entrance to the Garden of Eden (Genesis 3.24) could be a reference to the torrid zone, traditionally regarded as impassable because of excessive heat and beyond which lay the earthly paradise.

Between the middle of the twelfth and the end of the thirteenth century, a complex system of thought was transmitted by Arabic and Jewish commentators to the West. The new learning was based on Greek science and Aristotelian philosophy.[28] It encouraged Christian authors to face up to the challenge of a purely rationalistic approach to their religious beliefs and to the Church's teaching, asserted by Scripture and tradition, that there was a Garden of Eden on earth. Western exegetes and theologians were obliged to reconcile the biblical account of Eden with philosophical, physical and geographical learning derived from the works of Aristotle and his commentators. What was the thirteenth-century scholar to make of, for example, the statement in the *Glossa*, repeated by Peter Lombard, that paradise reached all the way to the sphere of the moon, an idea which was visualized on some maps by the contiguity of paradise to the lunar sphere (Figure 7.9), but difficult to defend in the light of the new science?[29] The credibility of the Garden of Eden and its earthly location was being put to the test.

The revival of astronomical geography lent added poignancy to the problem of the location of the terrestrial paradise. The thirteenth century in Western Europe was witnessing what has been described as a 'geographical renaissance' within the general revival of learning.[30] In Paris and Oxford, where the new teaching orders of the Franciscans and Dominicans contributed importantly to the intellectual outlook of the universities, curiosity about geography and cosmography was further stimulated by the availability of Arabic mathematical and astronomical works, as well as Aristotle's geographical material, in Latin translation. The Aristotelian corpus included treatises such as the *Physica*, *De caelo*, *Metereologica* and *De generatione et corruptione*, and the pseudo-Aristotelian *De causis proprietatum et elementorum*.[31] The Arab corpus included Arab commentaries on Aristotle and a number of astronomical tables, the most important of which were the so-called Khorazmian Tables of Al-Khwarizmi, translated into Latin by Adelard of Bath in 1126. The origins of these Tables lay in Hindu sources, whence the idea that the centre of the world was in the Indian city of Arin. Knowledge of Ptolemy's *Geography* had also provided Arab geographers with the coordinates of a great number of localities.[32] No less widely diffused were the astronomical tables compiled in the Spanish city of Toledo by Al-Zarqali between 1061 and 1080. As well as the raw data of the tables, there were astronomical texts. In 1135, John of Seville translated Al-Farghani's *On the Elements of Astronomy*, which dealt with the climatic zones; about 1140 Plato of Tivoli translated Al-Battani's *Astronomy*, which offered a new description of the inhabited regions; and in 1175 Gerard of Cremona translated an Arabic version of Ptolemy's *Almagest* into Latin. Translations were also made of astrological texts, which included descriptions of the climatic zones and their cities.[33]

The torrent of new scientific material reaching the Latin West could be channelled to the benefit of Christian doctrine. One theologian who accepted the challenge some of it posed was William of Auvergne, bishop of Paris from 1228 to 1249 and an open-minded eclectic who made use of the new learning to stress the rationality of the notion of an earthly paradise.[34] For William, belief in the existence of 'a region of such pleasantness and vernal beauty' in the farthest east did not run counter to reason.[35] Noting that everybody gave credence to reports of historians and geographers describing the pleasantness and fruitfulness of certain regions in India, he pointed out that this

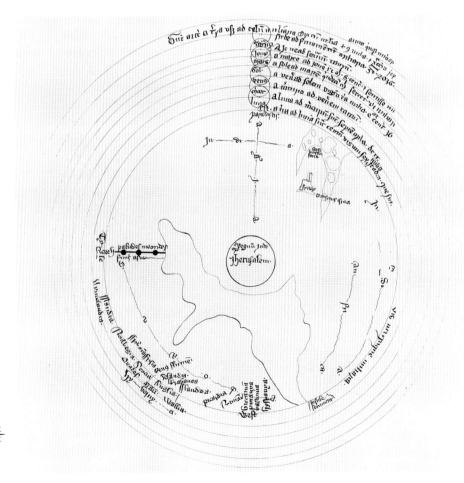

Fig. 7.9. Map of the world from Ranulf Higden's *Polychronicon*. York, fourteenth century. Diameter 17 cm (including the celestial spheres, 27 cm). Paris, Bibliothèque Nationale de France, MS Lat. 4126, fol. 1v. Redrawn from Manuel Francisco de Barros e Sousa, Viscount of Santarém, *Atlas composé de mappemondes, de portulans et de cartes hydrographiques et historiques depuis le VI jusqu'au XVII siècle* (Paris: Maulde et Renou, 1849), plate 19. On this version of the world map that illustrates a number of copies of the universal chronicle compiled by Ranulf Higden (see Figs 6.3, 6.4 and Plate 4), the earth is surrounded by seven concentric rings representing the planetary spheres. On each sphere the distance of the planet from the planet below is indicated, giving first the distance of the moon from the earth, then of Mercury from the moon, followed by Venus from Mercury, and so on. Jerusalem is located in the middle of the map. The inscription *paradisus*, to the east of India, has been placed close to the lunar sphere.

meant there was all the more ground for believing Moses, a prophet inspired by God, when he spoke about paradise. In the universities, professors 'of natural science and astronomy' confirmed that paradise was in the eastern regions of the world, where the sun rose and nurtured plant life. Because paradise was to be the dwelling place of all mankind, it had to be a region at least as big as India or Egypt. False human opinions, William concluded, usually go out of fashion. This had not happened in the case of the earthly paradise, and anyway the word of God was not to be doubted.[36]

William's aim was to prove that the Christian belief in the earthly paradise did not contradict either human reasoning or nature. If it appeared to do so, it was because people were mistaken in their understanding; 'neither nature nor reason' could allow, for example, that paradise touched the sphere of the moon.[37] Referring to the Aristotelian model of an earth surrounded by water, air, fire, and the celestial spheres (the first of which was the lunar), William reasoned that if paradise did reach the sphere of the moon, it would be within the region of fire and there would be no vegetation. Moreover, were paradise to touch the sphere of the moon, the shape of the earth would not be the perfect sphere that eclipses revealed it to be and it would be hard to imagine how the four rivers of paradise could reach the earth from such an extremely high altitude.[38]

William's objections opened up the question of the altitude of the mountain of paradise. Aristotle had taught that the region of the air was divided into three parts, the first of which was too hot because of its proximity to the sun, the second too cold because of its distance from the sun, but that the third, the closest to the earth and warmed by the reflection of the solar rays, was temperate and habitable.[39] William explained that paradise could not be in the middle region of the air, for its great weight would have caused it to fall to the ground and its height would have caused both lunar and solar eclipses, for which there was no evidence. Paradise, he concluded, was a vast part of the entire body of the earth in the inaccessible farthest east.[40] William's line of argument was typical of the way the challenge posed by the geographical renaissance of the thirteenth century was met by theologians. The new thinking was an opportunity for Christian scholars to demonstrate that their faith provided credible insights into the physical structure of the world. Henceforth, the debate on the location of the Garden of Eden became increasingly dependent on the recently acquired understanding of the size and the shape of the globe and on new knowledge of its relationship to the sun and the other celestial bodies.

Paradise and the Terrestrial Globe

Medieval scholars accepted that paradise was somewhere in the unknown parts of the world, but exactly where was another matter. The problem of its location was linked to the debate over the habitability of the equatorial regions and the southern hemisphere, which provoked further reflection and speculation in the thirteenth century. The idea of parallel zones, beginning from the equator and going towards the poles, had come to the Middle Ages from Greek mathematical and astronomical geography.[41] The region near the equator, called the torrid zone, and the polar circles were considered equally uninhabitable because of either excessive heat or cold. The region known from experience to be inhabited was the temperate region in the northern hemisphere. Controversy, however, surrounded the equatorial region and the whole of the southern hemisphere. Aristotle had held that the equatorial region was so hot a part of the world as to be uninhabitable because the sun passed directly overhead during the spring and autumn equinoxes, subjecting the region to the burning heat of its perpendicular rays – the strongest of all – twice a year.[42] Eratosthenes and Polybius, however, considered the regions at the equator to be more temperate than those above the tropic of Cancer and below the tropic of Capricorn because the sun, which remained almost stationary over the tropics for over two months, passed over the equator much more quickly at the equinoxes.[43] Ptolemy too had thought that the equatorial region had a milder climate than the tropical latitudes for similar reasons, although he admitted that the lack of exploration and the absence of sufficiently reliable evidence from the south beyond the established *oikumene* allowed no conclusions, only guesses.[44]

The influx of Arabic astronomy into Western Europe lent urgency to the question of the habitability of the equatorial regions. Aristotle's idea that the equatorial zone was too hot for human habitation was disproved by the astronomical tables, which indicated that in fact a number of cities were to be found there; the Toledan Tables gave the coordinates, for example, of the Indian city of Arin as latitude 0° and longitude 90°.[45] According to ninth- and tenth-century leading Muslim scholars such as Al-Battani and Avicenna the sun did not stand still over the equator nor did even the most vertical of the sun's rays linger overhead to overheat the ground. Avicenna's view was that, since it was the duration of the period when the sun was directly overhead that resulted in greatest heat, and not its proximity, the equatorial regions enjoyed the most temperate climate. Other arguments for the habitability of the equatorial region included the idea that its heat was tempered by the extreme cold of the polar regions, that the equal length of day and night throughout the year meant that the night was sufficiently chilled to moderate the heat of day, and that local topography could diffuse the rays of the sun.[46]

Those who insisted that the equatorial zone was habitable were also inclined to argue that life was impossible south of the equator. In the ninth century, Al-Farghani, for example, suggested that the southern hemisphere was covered with water.[47] Another explanation for an uninhabitable southern hemisphere was the eccentricity of the sun's orbit. The fact that the circle described by the sun was not perfectly concentric in relation to the earth, with the consequence that the revolving sun came closer to the southern hemisphere than to the northern, was thought to result in seasonal extremes. The heat was most intense when the sun was closest to the southern hemisphere, and the cold at its bitterest bitter when the sun was in the northern hemisphere. On these grounds, the learned Spanish astronomer and geographer Petrus Alphonsus explained in the early twelfth century that the southern hemisphere was uninhabitable.[48] Astronomy confirmed the rejection of the idea of the Antipodes, already denied on theological grounds.

Aristotle's persuasion that the equator was uninhabitable was adopted by his most loyal medieval followers, however. There were those who favoured the idea of zonal symmetry, which meant an uninhabitable middle (equatorial) zone and which accommodated the possibility of human life in an antipodean temperate zone. These scholars also favoured Aristotle's idea of an innately superior southern hemisphere, where the stars moved (when viewed from the South Pole) from right to left, thus bestowing a more noble character on that hemisphere.[49] Moreover, the argument ran, for the entire southern hemisphere not to have been inhabited would mean that the hemisphere existed in vain when nothing in nature is in vain. In the Middle Ages, support for the idea of the Antipodes was drawn mainly from the late antique works of Macrobius and Martianus Capella, both of whom had written about a temperate and inhabited southern hemisphere.[50] Macrobius's commentary on the *Dream of Scipio* was richly illustrated with zonal maps. There were also those who took a middle position, mediation being a good scholarly convention. One of these was William of Conches, who had suggested in the twelfth century that the southern hemisphere was habitable, but that it was not actually inhabited.[51]

Paradise On or Beyond the Equator

The medieval hypothesis that paradise might be located at the equator was not new. Already in the fifth century, the Arian ecclesiastical historian Philostorgius had suggested that paradise lay on the equator.[52] The idea was embraced in the thirteenth century as a way of explaining the terrestrial location of paradise in a manner that did not conflict with the new scientific learning. When Alexander of Hales, the first Franciscan professor at the University of Paris in the 1220s and one of the most celebrated teachers of his day, said that paradise was in the eastern corner of the inhabited world, he had in mind Macrobius's description of a globe divided in the middle by a equatorial sea that opened into the ocean.[53] Alexander placed paradise in the east exactly at the point where the equatorial sea merged with the outer ocean, saying that such an equatorial location did not deny the pleasantness of the place or its suitability for human habitation for, although the sun's rays fell directly on the region twice a year, other factors mitigated the burning heat of the zone. By definition, day and night were of equal length at the equator, and the spring that watered the garden had a also cooling effect; the air was pure, and the rounded, smooth surface of the mountain top on which paradise was sited reflected the sun's rays and dispersed the heat that, in the dissected terrain along the rest of the equator, was trapped in the valleys.[54]

Alexander also articulated and expanded the belief in the great altitude of paradise, already asserted by Bede and the compilers of the *Glossa*.[55] For him, it was above all the extreme altitude that countered the heat of the equatorial regions. Whereas Aristotle had stated that only the lowest part of the air could be inhabited, the middle being too cold and the highest too hot, Alexander envisaged a tepid layer within the middle part, where the air was temperate and calm, and neither excessively hot nor excessively cold, and that this was where paradise was to be found (Figure 7.10). Aristotelian philosophy, in other words, allowed the climatic negatives of paradise – no cold, no heat, no winds, no storm, no rain, no clouds in paradise – to be translated by Alexander into scientific terms. The air in the delightful region of Eden was tranquil and perfect because the mountain of paradise penetrated beyond the turbulent atmosphere of the perpetually cold and dark lower part of the middle region where winds and rain originated, the wind formed under the influence of the sun from the dry vapour given off by the earth and the rain from the humid vapour given off by rivers, lakes and seas. At the same time, the heat coming from the highest region of the air prevented any of these vapours from rising further so that, at the level of paradise, the air was calm and pleasant.[56]

After introducing his scientific model, Alexander turned his attention to the statement in the *Glossa* that paradise reached all the way to the sphere of the moon. He suggested that the statement was either a metaphor for paradise's eternity and incorruptibility or an indication that it was the highest place on earth. Then, more specifically, he referred to his model to show how the circle of the moon (the heavenly sphere that lay, according to Aristotle, beyond the sphere of fire) could also be considered to include the part of the air that came under the moon's influence as it lay beyond the reach of the ascending vapours. To say that paradise reached the sphere of the moon, Alexander pointed out, was not to say that it actually extended into the heavenly sphere of the moon, above the sphere of fire, but that it reached the tranquillity of the highest part of the middle region of the air.[57] It was well known, however, that it was impossible to live on the Greek mountain of Olympus, which was much lower than paradise, because of the thinness of the air – Augustine had reported that to climb Olympus it was necessary to apply a moist sponge to the nose – and to counter that argument Alexander noted that before the Fall man had been able to breathe in a much thinner air than he could thereafter.[58] Another objection dealt with by Alexander was that the shadow cast by such an excessively high mountain would delay the rising of the sun. His answer was that the mountain of paradise may indeed have been large, but it took up only a relatively small area of the globe. He had an answer, too, for those who objected that the excessive height of such a mountain would have

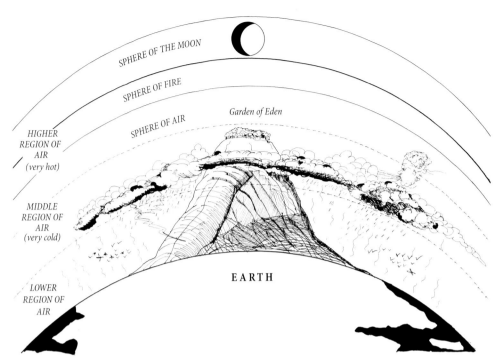

Fig. 7.10. Diagram to represent Alexander of Hales's location of paradise in the middle region of the air. According to Aristotle, the sphere of fire, which is both hot and dry, comes immediately below the lunar sphere. Then comes the sphere of air, which is hot and moist and subdivided into three regions. The highest of these regions is immediately adjacent to the sphere of fire and is accordingly hot and dry. Winds, rain and storms originate in the middle region of the air, which is, in contrast, excessively cold because of its distance from the sphere of fire and from the ground. The lowest region of the air is heated by the reflection of the sun's rays. After the sphere of air comes the element of water, which is cold and humid, and finally the cold and dry earth.

altered the shape of the earth or displaced its centre, which was also the centre of the heavens. Paradise, Alexander asserted, was not as high as the sphere of the moon, and its altitude was of no consequence in comparison to the vastness of the heavens.[59]

Alexander of Hales turned the old commonplace, that paradise was a region nobler than any other on earth although inferior to heaven, into a question of altitude. The mystery of the biblical paradise became a theme for scientific discourse on its location on a mountain peak. One of Alexander's pupils in Paris was Bonaventure, who adopted his master's views on paradise.[60] However, when Bonaventure described paradise as 'situated in a very prominent and high place, near the equinoctial circle in the east', he added a potentially crucial qualification. The location of paradise, he said, was 'situated in a certain way towards the south'.[61] This allusion to the south implied that paradise did not lie precisely on the equator but was located on the southern side of the equatorial belt, an important point in the context of the thirteenth-century debate about the location of paradise and the habitability of the southern hemisphere.

A close connection between the location of paradise and scientific theories about the different zones of the earth is seen in the writings of leading figures of the medieval 'geographical renaissance', many of whom were Englishmen.[62] An outstandingly influential work of astronomical geography in the Middle Ages was the *Tractatus de sphera* of Iohannes de Sacrobosco (or of Holywood (*sacro bosco*), in Yorkshire). Sacrobosco drew on classical authors such as Lucan and Virgil, and on Al-Farghani to conclude that the heat of the sun rendered the equator uninhabitable, but that the two zones between the tropics and the poles, tempered by the heat of the torrid zone and the cold of the polar circle respectively, were habitable.[63] Sacrobosco's enormously popular *Sphera,* written before 1235, gave rise to a number of commentaries. Unlike Sacrobosco, though, most of his commentators thought that the regions at the equator were temperate and habitable and some of them placed paradise there. Robert Grosseteste, for example, who had also been influenced by Muslim geography and astronomy, leant on the authority of Ptolemy and Avicenna to explain that, while the subtropical regions were indeed very hot, and the southern hemisphere uninhabitable because of the eccentricity of the sun's orbit, the subequatorial zone was quite temperate and that this was why theologians placed paradise on the equator.[64] Another thirteenth-century commentator who knew the work of Avicenna was Robertus Anglicus.[65] In presenting, in 1271, all the usual arguments in support of the temperate climate of the equatorial regions, Robertus also referred to Isidore, adding that heat at the equator was tempered by vapours rising from the extensive water bodies of the southern hemisphere, and explaining that the reflection of solar rays on the convex surface of the land at the equator did not generate fire in the same way as it did when the rays fell on the concave surface of a highly polished mirror. Heat was, after all, he remarked, a source of life. The proximity of the sun and the other planets to earth generated life, and it had to be acknowledged that the equator was the best place for life. The further the lands were from the equator, the poorer was their vegetation. Robertus's final point was that not only the mountain of paradise but also the Indian city of Arin were said to be at the equator, which thus could not be dismissed as uninhabitable.[66] The reason there was no communication with the inhabitants of the equatorial lands was because of certain dangerous mountains that exerted a powerful attraction on human flesh in the same way as a magnet attracts iron. Robertus Anglicus's opinions on the question of the habitability of the southern hemisphere differed slightly from Grosseteste's in that he was willing to countenance the idea of habitable land south of the equator. He felt, though, that the extent of land left uncovered by water, if there were any at all, would be minimal. The problem was lack of evidence.[67]

Not all thirteenth-century scholars thought that paradise was at the equator. There was Cecco d'Ascoli, for example, who suggested that the uninhabitable equator lay between the two temperate hemispheres and who decided that the eastern paradise was in the northern hemisphere.[68] Another author, possibly Michael Scot, had a similar view of the globe and of the excessive heat at the equator, but placed paradise in the southern rather than the northern hemisphere.[69] The symmetry of the zones of the globe implied that the southern hemisphere had to be habitable, although it could not be reached across the intervening torrid zone. Excluding, for theological reasons, the possibility of inhabitants in the Antipodes, the author of the text ascribed to Scot argued that since the southern hemisphere would not have been created for no purpose, it must accommodate the earthly paradise and that the torrid zone was the flaming sword of the Cherubim guarding the entrance to the Garden of Eden.[70]

The Mathematics of Paradise
In the context of strong interest in the thirteenth century in geographical issues, the terrestrial paradise needed to be located where the combination of climatic elements and geographical factors was *optimal*, according to available scientific data, in the same way as, for example, the preponderance of monsters in and the general fecundity of east Asia and Africa could be explained by the dominance of the hot and moist qualities of the air there. The confidence of natural philosophers that it was worth speculating about the actual location of a physical paradise on earth even if it were not possible to discover its location is amply demonstrated in the works of the thirteenth-century scholar Roger Bacon.[71] The issue of paradise entered Bacon's scientific writings through his discussion of the utility of mathematical and astronomical knowledge for theology and the understanding of Holy Scripture. Bacon's scientific arguments were not aimed at indicating the exact situation of paradise. He avoided endorsing any of the opinions he discussed, explaining them instead with caution and objectivity to underline his conviction that theologians interpreting the geographical references in the Bible needed a sound knowledge of the principles of astronomy.[72] Bacon systematically analysed the

different theories. He asked himself whether paradise was on the equator, concluding that the equatorial regions could be temperate, but not necessarily 'very temperate', and that it was by no means certain that paradise was there.[73] He also considered the possibility that paradise was in the southern hemisphere, acknowledging that a number of factors, such as the eccentricity of the solar orbit, could affect the habitability of the different zones of the earth. Referring to the common belief that the equatorial regions were too hot for human habitation, Bacon repeated the argument that at the equator the sun's rays fell perpendicularly onto the ground twice a year (at the spring and autumn equinoxes) whereas at the tropics this happened only once a year, during the summer solstice on the tropic of Cancer and the winter solstice on the tropic of Capricorn. He announced that the idea was confirmed by his own study of the refraction of light: solar rays reaching the ground at right angles are not scattered, but bounce back at the same angle, thus doubling their intensity. He then discussed the opposite opinion, held by Ptolemy and Avicenna, that the equatorial regions were exceptionally temperate, an idea that led many of the theologians of his day to reckon that the earthly paradise was situated at the equator.[74]

In explaining the details of Avicenna's scientific arguments for a temperate equator, Bacon quoted Ptolemy's table of inclination in the *Almagest*.[75] Bacon found that the declination of the sun – that is, its distance from the celestial equator – reached its maximum of nearly 24 degrees at the two solstices (Figure 7.11). Between the solstices, the angle of the sun in relation to the equator varied. On its way through the equinoctial

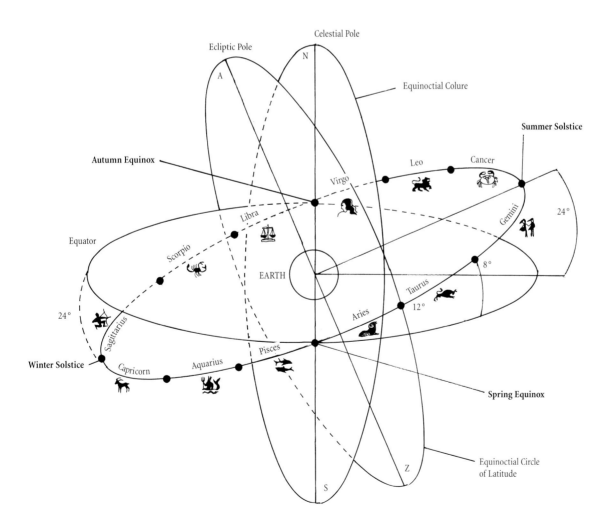

Fig. 7.11. Diagram of the celestial circles and the orbit of the sun, to illustrate the medieval argument for a temperate equator.

Where Is Nowhere? 177

signs (Pisces, Aries, Virgo and Libra), the sun's declination increased by 12 degrees; on its way through Taurus, Leo, Scorpio and Aquarius, it was increased by just under 8 degrees; and on its way through Gemini, Cancer, Sagittarius and Capricorn by a mere 4 degrees. The progression showed that the sun passed over the equatorial regions relatively quickly, and that during the summer solstice in the tropic of Cancer and the winter solstice in the tropic of Capricorn its declination remained practically unchanged for forty days, as the sun ceased further movement north or south before starting its return towards the equator. Another factor considered by Bacon was the equal length of day and night at the equator, which also helped temper the air.[76] He concluded his discussion on the equatorial regions by admitting that 'there is no doubt that the place is temperate; but whether it is extremely temperate, I do not yet know. And therefore it is not proved with certainty that paradise must be located there.'[77]

Bacon was not in a hurry to draw conclusions. Faithful to his habit of scrutinizing received opinion before accepting anything, he considered another aspect of the equatorial question, that of the eccentricity of the sun's orbit. He explained that two points of the solar orbit mark the sun's greatest and least distance from the centre of the heavens, the apogee at Gemini, and the perigee at Sagittarius respectively. At its perigee, the sun burns violently because of its proximity, but when it has moved on, through Libra and Aries, it is still close enough to the earth to create excessive heat. Bacon concluded that were all this to be true, it was clearly impossible for the equatorial regions to be sufficiently temperate for the earthly paradise. Paradise – Bacon reminded his readers – 'has a perfect climate'.[78] Interestingly, he also referred to the theory that immediately before Adam's sin and expulsion from the Garden of Eden the sun had been in Libra and that, after the Fall, Adam was prevented from returning to paradise by the heat generated by the sun's increasing proximity to the earth.[79]

Turning next to the question of the habitability of the southern hemisphere, Bacon considered the possibility that paradise could be there. As he had done for the equator, Bacon systematically explained the astronomical arguments. He pointed out that the eccentricity of the sun's orbit should render the southern hemisphere uninhabitable. At the same time, it could be that certain particular characteristics made these southern regions habitable, such as the presence of mountains that acted in the same way as concave mirrors to deflect the heat of the sun or that were high enough to shade the ground.[80] He was reassured by accounts in authoritative sources that were based on experience and that confirmed the existence of habitable land south of the equator; Ptolemy, for instance, had claimed that there were two races of Ethiopians, one beyond each of the tropics.[81] Other reports, for example in the writings of Pliny the Elder and Ambrose, also told of people living south of the equator.[82] As for himself, Bacon said, since the effect of the sun and the planets on the two hemispheres was supposed to be identical, climatic symmetry must imply the existence of an extensive habitable land south of the equator. Furthermore, the eccentricity of the sun's orbit would have meant that there was a higher proportion of land in the southern hemisphere than in the northern, because in the south the proximity of the sun to the earth would have considerably reduced the area under water through evaporation.[83] Bacon also introduced the idea that in many southern regions winter was better than summer for plant life, and that the land produces crops in spring as well as autumn because of the temperate influence of the equinoxes.[84] No author had described these regions, however, since nobody had ever visited them. Nevertheless, confident that the regions south of the tropic of Capricorn were suitable for human habitation, Bacon cited Aristotle's and Averroes's arguments that the southern half of the globe was the best part of the earth, observing that 'for these reasons some feel that paradise is situated there'.[85]

Bacon himself gave no hint of a definitive conclusion as to the precise location of paradise. He seems to have thought of paradise as somewhere at the longitude of the furthest part of India. East and west, he said, were points on the equator corresponding on one hand to the easternmost limit of habitable land in India, and on the other hand to the westernmost limit in Spain. Paradise in the farthest east must thus be at the longitude that corresponded to the easternmost extent of the habitable earth.[86] As for the latitude of paradise, it may be assumed that Bacon held the southern hemisphere as a likely location, putting paradise possibly at a southern latitude equivalent to that of Babylon in the north, the place where Noah and his children had lived after the Flood, and in the fourth climatic zone, which Bacon defined as the mildest.[87] Irrespective of Bacon's real thoughts – whether the Garden of Eden was at the equator or in the southern hemisphere – his treatment of the issue testified to the contemporary view that a physical paradise, distant from any known region, existed somewhere on the globe. Bacon's primary concern was to emphasize the need to use astronomy and geographical knowledge in attempting to determine its location. In claiming the existence of secluded zones on the earth favoured by temperate climate, the astronomical theories of the thirteenth century, instead of contradicting Scripture, served only to fuel belief in the existence of an earthly paradise.

An Unresolved Debate

Scientific speculation might be thought to confirm the truthfulness of the Bible and to reinforce religious belief, but it could not provide the final answer to the problem of the exact location of the earthly paradise. For the whole of the thirteenth century the issue of whether paradise was at the equator or south of it remained an unresolved and much debated topic. Not every scholar could convince even himself. The theologian Albert the Great was a pioneer in a number of scientific disciplines who paid particular attention to the Aristotelian heritage.[88] Albert's inquisitive mind allowed him to see the problems he was examining with a clarity of perception that in the end prevented him from reaching any firm conclusion about the earthly paradise and that left him wavering between an Eden in the southern hemisphere and an Eden at the equator.

Albert's views on a southern paradise were first expressed in *De homine*, part of his *Summa de creaturis* (1246). He argued against the interpretation that before the Fall the earthly paradise had comprised the entire world by accepting that it was a specific part of the globe destined by God for man while the remaining part was for the animals.[89] As to its precise location, Albert wrote that he was 'speaking without preconception' when he asserted that Eden was 'in the south-east, beyond the equinoctial line'.[90] He produced a number of arguments in support of his view, starting with Bede's statement that the earthly paradise was separated from the inhabited world by a broad stretch of sea (or land), which he thought was confirmed by the belief that an unnavigable salty sea lay along the equator.[91] Only a miracle had allowed Enoch and Elijah to cross it. He also agreed with the view that living conditions in the southern part of the globe were excellent, an idea that was 'very much according to reason'.[92] Against those who asserted that the southern hemisphere was uninhabited, Albert quoted Ptolemy's statement that there was no evidence to confirm or deny the issue.[93] His final line of reasoning was rather original and referred to the flooding of Egypt by the Nile in the dry season as evidence that paradise was in the southern hemisphere. The Nile's floods, he observed, occur when the sun is in Cancer, that is, during summer in the northern hemisphere when it is the rainy season in the southern hemisphere, which was why the Nile has an overabundance of water and floods Egypt: 'No other logical reason for the flooding of the Nile when the sun is in Cancer, a period when its waters should instead decrease, can be found. Since the Nile is a river of paradise, paradise must be in the same place.'[94] Anticipating the objection that the Tigris and the Euphrates, which also rose in paradise, did not flood at the same time as the Nile, Albert explained that their courses ran underground, while the Nile remained above ground all the way from paradise.[95]

Three years later, Albert seems to have changed his mind. In his commentary on Peter Lombard's *Sentences*, completed in 1249, he reverted to the idea of an equatorial paradise, again saying that he was 'speaking without preconception' when he said that 'it seems to me that paradise is at the equinoctial line and towards the east in relation to our inhabited region.'[96] In support of the possibility of life in the equatorial zone, Albert cited the arguments of Ptolemy and Avicenna, and concluded that 'that place is the most favourable and suitable'.[97] What seems to have induced Albert's change of mind was learning from 'those who came from that land' that the Nile flooded not once but twice a year, when the sun was in Cancer and when it was in Capricorn or, in other words, during both summer and winter.[98] Instead of coming from a paradise in the southern hemisphere, Albert concluded, the Nile must come from a paradise on the equator, for only at the equator did the rainy season occur twice a year. Notwithstanding all these scientific arguments, Albert felt he should warn his readers that 'what has been said about the location of paradise is by way of conjecture not assertion'.[99]

Albert was not alone in linking the rise in the level of the Nile with a rainy season somewhere in Africa. He could have read in a brief treatise attributed to Aristotle about the regular inundations as the result of heavy summer rains in Ethiopia.[100] In the middle of the twelfth century, Benjamin of Tudela had noted that rising levels in the Nile coincided with heavy rain in the land of Al-Habash (Abyssinia), which he identified with the biblical Havilah. William of Tyre (1130–90) had written about the annual rising levels of the Nile in his *Historia rerum in partibus transmarinis gestarum* (1170–84).[101] Given the closeness of commercial contacts between Europe and Egypt in the thirteenth century and the large presence of Frankish merchants in the Nile area, Albert could easily have heard new information from a wide range of witnesses about the flooding of the Nile that persuaded him to change his mind. He could have been told about the biannual floods by pilgrims coming from the Holy Land, or by veterans of the Crusade launched in the area by Louis IX of France in 1248.[102]

In a broader sense, Albert's vacillation over the location of the Garden of Eden can be explained by reference to the different intellectual contexts in which he wrote *De homine* and his commentary on the *Sentences*. Despite difficulties in dating many of Albert's works, it has been established that he finished the *De homine* in 1246 and that in the same year he was already involved in his commentary on the second book of Peter Lombard's *Sentences*, which he had completed by 1249.[103] The difference between the two overlapping works can be accounted for by a personal event of crucial importance, that of Albert's move to Paris. From 1 September 1245 to 29 June 1248, when he returned to Cologne, Albert was teaching in Paris. Prior to his arrival in France, he already knew a good deal about Aristotelian science, but it was in Paris, one of the foremost intellectual centres in Christendom, that he became familiar with the 'new learning'.[104] He seems to have found it difficult to decide between Aristotle's theory about the 'nobility' of the southern hemisphere and the uninhabitability of the equatorial region – which led him to a southern paradise – and Ptolemy's belief that the climate was 'extremely temperate' at the equator, and thus eminently suitable for paradise. Torn between two apparently equally convincing theories, and challenged by the new geographical material that recorded the presence of cities in equatorial zones, Albert was reluctant to commit himself to either one or the other.[105]

After completing the *De homine* and the commentary on the *Sentences*, Albert continued to seek to deepen his understanding of the equatorial and southern regions. He did not, though, return to the specific problem of the location of paradise. Another work, the *De natura locorum*, was compiled in Cologne between 1248 and 1257. In this treatise, Albert described how the globe was divided into five zones and cited the opinion that the equatorial zone was uninhabitable because of the presence of an exceptionally salty sea (the theory of an equatorial sea had been developed in ancient times by Crates of Mellos).[106] Although Albert had adopted the idea of the equatorial sea in the *De homine*, he now cast doubts on it, maintaining instead that the equatorial zone could be habitable and agreeable. He quoted Ptolemy and Avicenna to draw attention to the fact that India and Ethiopia were said to lie at the equator, and that the weather in these regions was known to be pleasant and delightful, and to arrive at the conclusion that 'it is reasonable to assume that [the equator] is inhabited'.[107] While Albert appears here to be inclined to think that the area around the equator was habitable, he remained equivocal as to whether the equatorial regions were the most temperate.[108] Nor could he bring himself to abandon the view that the southern hemisphere might be habitable, saying that 'failing a better judgement, we say that the quarter of the globe which is to the south of the equator is habitable, in accordance with nature, and, as we think, inhabited.'[109] Here again, Albert was following Aristotle in his praise of the southern hemisphere as pleasantly habitable, and more 'noble' than the northern hemisphere.[110] At the same time, he reiterated that the source of the Nile (and thus, by implication, paradise) was at the equator.[111] He remained silent, though, throughout the *De natura locorum* about the location of the earthly paradise although it is likely that the issue stimulated his interest in the habitability of the different parts of the globe. In his final work, the *Summa theologiae*, written after 1270 and towards the end of his life, Albert gave – as Gilson has put it – 'his last word on all the questions with which he deals'.[112] He returned to the question of paradise as a specific region on earth, extremely temperate and separate from the inhabited regions. He stated again that the idea that paradise reached as far as the sphere of the moon was to be taken figuratively, but he still held back from volunteering a definitive statement as to whether the Garden of Eden was at the equator or in the southern hemisphere.[113]

Thomas Aquinas, another open-minded scholastic philosopher, studied under Albert in Paris and followed him to Cologne in 1248. Like his master, Thomas combined old theology with the new scientific learning, but his main concern was to prove that the Bible was consistent with knowledge gained through the natural sciences.[114] In his opinion, if there were different views of past authors on certain issues, no one theory could be accepted as true or dismissed as false unless it contradicted reason or the Holy Bible.[115] In view of Aristotle's comment that the east was on the right-hand side of the heavens, a 'nobler' side than the left, Thomas thought that the Garden of Adam and Eve was suitably somewhere in the east. He accepted that no traveller had ever reported a sight of paradise because it was closed off from the inhabited world by impassable barriers – mountain ranges, seas, or torrid regions – and that the four rivers of paradise travelled underground, as Augustine had said, before emerging at their individual sources as known to Aristotle.[116]

When discussing the penalties of original sin, Thomas interpreted the flaming sword of the Cherubim in geographical terms as the violence of the equatorial heat, quoting Augustine's point that through the ministry of the angels some sort of protection by fire was provided for paradise.[117] Thomas was uncertain whether the Garden of Eden was actually on or to the south of the equator. He seems to have bowed to Aristotle, his respected scientific authority, who had stated clearly that the equatorial regions were too hot to be inhabited. He quoted those authors who placed paradise at the equator, but left the question open, saying that Aristotle's view was the 'more likely' and concluding simply that paradise was in a 'most temperate' place, either at the equator or somewhere else.[118] What was certain from Thomas's point of view was that paradise belonged to the sublunary world, a region in the cosmos that was subject to 'generation, change and corruption', and not to the world of the incorruptible stars.

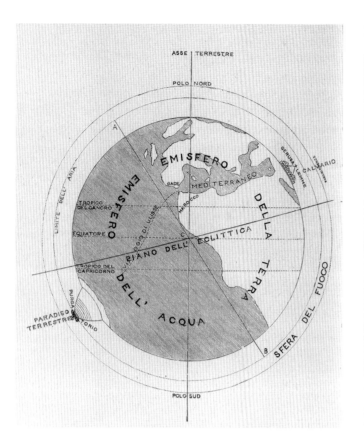
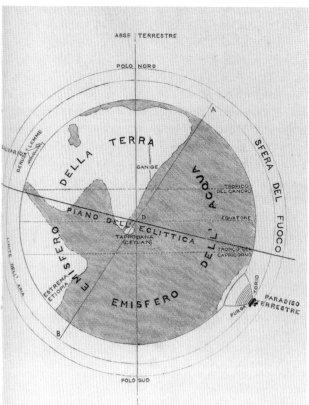

Fig. 7.12. *Il globo terracqueo secondo l'Alighieri*, from Edoardo Coli, *Il paradiso terrestre dantesco* (Florence: Carnesecchi, 1897), Fig. 22. Dante thought of Jerusalem and the mountain-island of Purgatory, on whose peak was the earthly paradise, as exactly antipodal to each other. Coli's illustration shows Dante's Garden of Eden in the temperate zone south of the equator and the tropic of Capricorn.

Paradise was an earthly garden, composed of the four elements (earth, water, air and fire), and not a heavenly abode made up of ether, the indestructible and eternal fifth element.[119] Thomas rejected also the idea that paradise reached the sphere of the moon. Instead, he subscribed to the idea that the moon, the celestial body closest to the earth, was mentioned only to indicate that the perpetually temperate climate of paradise was similar to the stable atmosphere of the stars and planets. He found untenable the proposition put forward by commentators such as Alexander of Hales that paradise was in the layer of the middle region of the air influenced by the moon. Ignoring Alexander's expedient of a tepid zone, Thomas objected that a paradise in the middle region of the air could not have a temperate climate.[120] He made no mention at all of the Flood, though (Aristotle, of course, had made no reference to it), which was the main reason for believing in the exceptional altitude of a mountain that kept paradise high and dry above water level during the global inundation.

A Poetic Flight to Paradise

The theologians' and natural philosophers' interest in geography was shared by the Italian poet Dante Alighieri. Dante had wide-ranging interests, and his *Divina Commedia* (c.1305–20), a literary work of the imagination, was an account of his journey through the three realms of the other world, Hell, Purgatory and Heaven.[121] The poem voiced the geographical and cosmographical knowledge of his age, even though Dante elaborated it in a strikingly original manner. For Dante, the earthly paradise was an intermediate region that lay between the corporeal world and immortal eternity. The poem describes the situation of paradise on the peak of an exceptionally high mountain (the mountain of Purgatory) that bordered on heaven and was kept separate from the inhabited regions of the earth by an impassable ocean. Dante's visionary mountain was diametrically opposite the Holy Land, which meant it was in the southern hemisphere. Virgil, Dante's guide through Hell and Purgatory, explains to him that since Jerusalem and the mountain of Purgatory are exactly opposite each other on the globe, the observer in Purgatory sees the sun moving in a direction that appears to be contrary to that seen by the observer in Jerusalem.[122] Adam and Eve, in their austral paradise, would have seen stars that nobody else had ever seen.[123] When Dante meets Ulysses in Hell, where the latter endures eternal punishment for fraudulence, Ulysses tells Dante that when he left Gibraltar for the Atlantic, on a south-westerly course, he was sailing towards paradise (Figure 7.12).[124] After sailing for five months, the Greek hero caught sight of the high mountain of paradise in the distance, but his ship was immediately overwhelmed by a storm, for no human was allowed to complete the crossing of that sea.[125]

Other passages in the *Commedia* hint at the longitude of Eden. Dante observes that the difference in time between Italy and Purgatory was 135°, or nine hours.[126] From allusions such as this, modern scholars have attempted to identify the precise coordinates of Dante's earthly paradise with, however, little agreement.[127] Attempts have also been made to calculate the dimension and altitude of the mountain of Purgatory from hints such as the observation that neither vapours nor cold exhalations were able to rise higher than the gate of Purgatory.[128] In conceiving a paradise above the region of the air where rain and winds originate, Dante was agreeing with Alexander of Hales. He did not specify the mountain's altitude, stating only that he climbed up, with Beatrice, his guide to Heaven, from the region of pure air through the sphere of fire.[129] By putting the inaccessible Eden at the antipodes of Mount Calvary in Jerusalem, it should be noted, Dante was emphasizing the link between Adam's Fall and Christ's redemption. Such a link had already been highlighted when Dante started his imaginary otherworldly adventure, in the holy year 1300, on the 25 March, the day of the creation and of the Fall of Adam, of the Incarnation and Passion of Christ.[130] The idea of an island lost in the ocean of the southern hemisphere also accorded with the view that the southern hemisphere was covered by water.[131] Dante's vision was also in harmony with Aristotle's notion that the southern hemisphere was the noblest part of the whole earth. From wherever Dante derived his learning – Orosius, Isidore, Alexander of Hales, Albert the Great, Brunetto Latini and, possibly, Roger Bacon – it cannot be denied that his knowledge of geographical lore was deep, rich and varied, and that his blend of all the traditional features of paradise was unique.[132] In the final analysis, though, his imaginary guided tour of the three realms of Hell, Purgatory, and Heaven, treated in the early fourteenth century as domains of cosmographical and geographical science, was a literary expression of a spiritual itinerary. It was not the journey of a foot-weary traveller, mile by mile, across the physical world. For Dante geography was always subservient to poetry. In Canto XXVI of the *Inferno*, he refers to Ulysses's 'folle volo', his 'mad flight'.[133] The moral of the story of the Greek sailor who went beyond the Pillars of Hercules and glimpsed the mountain of paradise from a distance before being prevented from approaching any nearer by being wrecked in a storm, was that man on his own, and lacking the support of divine revelation, is not allowed to penetrate the mystery of the earthly paradise.

1   Some internal references to sixth-century events help us to date Cosmas's work; Cosmas himself refers to his maps and drawings in the text, so that it is assumed that the illustrations in the extant manuscripts (dating from the ninth and eleventh centuries) are reasonably faithful to the original: see *Cosmas Indicopleustès: Topographie chrétienne*, ed. by Wanda Wolska-Conus, 3 vols, *SC* CXLI, CLIX, CXCVII, (Paris: Cerf, 1968–73), I, *SC* CXLI (1968), pp. 16, 124–39. The passage with the world map is in IV.7–10, I, *SC* CXLI (1968), pp. 544–9. The three extant manuscripts of Cosmas's *Christian Topography* where the map is found are Vatican City, Biblioteca Apostolica Vaticana, MS Vat. Gr. 699, fol. 40v (ninth century); St Catherine's at Mt Sinai, MS Gr. 1186, fol. 66v (eleventh century); Florence, Biblioteca Medicea Laurenziana, Plut. IX.28, fol. 92v (eleventh century). On Cosmas's map see also Evelyn Edson, *Mapping Time and Space: How Medieval Mapmakers Viewed their World* (London: British Library, 1997), pp. 145–9; John O. Madathil, *Kosmas der Indienfahrer: Kaufmann, Kosmologe und Exeget zwischen alexandrinischer und antiochenischer Theologie* (Thaur–Vienna: Kulturverlag, 1996); Kunio Kitamura, 'Cosmas Indicopleustès et la figure de la terre', in Alain Desrumeaux and Francis Schmidt, eds, *Moïse géographe: Recherches sur les représentations juives et chrétiennes de l'espace* (Paris: Vrin, 1988), pp. 79–98; Oswald A. W. Dilke, 'Cartography in the Byzantine Empire', in Brian Harley and David Woodward, eds, *The History of Cartography*, I, *Cartography in Prehistoric, Ancient and Medieval Europe and the Mediterranean* (Chicago: University of Chicago Press, 1987), pp. 261–3. For an English translation, *The Christian Topography of Cosmas, an Egyptian Monk*, ed. by John M McCrindle (London: Hakluyt Society, 1987). See also *The Christian Topography of Cosmas Indicopleustes*, ed. by Eric Otto Winstedt (Cambridge: Cambridge University Press, 1909); and Cosimo Stornajolo, *Le miniature della Topografia cristiana di Cosma Indicopleuste: Codice vaticano greco 699* (Milan: Hoepli, 1908). Little is known of Cosmas Indicopleustes. His Greek name may be translated in English as 'Mr World who travels to the Indian sea'. From a scatter of allusions in his book, we gather that Cosmas was a native of Egypt, that early in his life he was a merchant travelling far and wide, and that he eventually settled in Alexandria, probably as a monk.

2   Cosmas's map has been often cited by nineteenth- and twentieth-century historians as the quintessential example of medieval geographical ignorance, despite the fact that Cosmas's writing were completely unknown in the Latin West in the Middle Ages: see Edson, *Mapping Time and Space* (1997), p. 149 n. 5; Jeffrey B. Russell, *Inventing the Flat Earth: Columbus and Modern Historians* (New York: Praeger, 1991), pp. 31–5; Patrick Gautier Dalché, *La 'Descriptio Mappe Mundi' de Hugues de Saint-Victor* (Paris: Études Augustiniennes, 1988), p. 119 n. 8. Examples of the inappropriate emphasis given to Cosmas's conceptions in a Latin medieval context include Corin Braga, *Le Paradis interdit au Moyen Âge: La Quête manquée de l'Eden oriental* (Paris: L'Harmattan, 2004), pp. 130, 378; Claudine Poulouin, *Le Temps des origines: L'Eden, le Déluge et 'les temps reculés': De Pascal à l'Encyclopédie* (Paris: Honoré Champion, 1998), p. 229; Jean Delumeau, *Une histoire du paradis*, I, *Le Jardin des délices* (Paris: Fayard, 1992), pp. 62–5, 198; *History of Paradise: The Garden of Eden in Myth and Tradition*, transl. by Matthew O'Connell (New York: Continuum, 1995), pp. 41–3, 151; Fred Plaut, 'General Gordon's Map of Paradise', *Encounter* (June/July 1982), pp. 26–7.

3   Ephrem the Syrian, *Hymns on Paradise*, transl. by Sebastian Brock (Crestwood, NY: St Vladimir's Seminary Press, 1990): see the Introduction, pp. 7–75, and the bibliography, pp. 228–33. See also Robert Murray, *Symbols of Church and Kingdom: A Study in Early Syriac Tradition* (London–New York–Cambridge: Cambridge University Press, 1975), pp. 29–32, 337–40; Murray, 'The Theory of Symbolism in St. Ephrem's Theology', *Parole de l'Orient*, 6–7 (1975–6), pp. 1–20; Louis Leloir, *Doctrines et méthodes de S. Éphrem après son commentaire de l'Évangile concordant*, CSCO CCXX (Louvain: CSCO, 1961); Edmund Beck, 'Symbolum-Mysterium bei Aphrahat und Ephräm', *Oriens Christianus*, 42 (1958), pp. 19–40.

4   Ephrem the Syrian, *Hymns on Paradise*, I.4, I.16, II.7, II.11, III, IV.1, X.14 (1990), pp. 78–9, 83–4, 87–9, 90–97, 152–3. It seems that the penitent are outside the barrier formed by the Cherubim, but this is not stated explicitly; see p. 57. The idea of paradise as a mountain – a mountain-island, according to Ephrem the Syrian, *The Commentary on Genesis*, II.6, in *Hymns on Paradise* (1990), p. 201 – is of great interest and was to be developed by many later thinkers. Robert Murray speculates about Ephrem's sources for this notion, suggesting perhaps Ezekiel's description (28.12–18) or the intertestamental literature; see Murray, *Symbols of Church and Kingdom* (1975), pp. 306–10.

5   Ephrem the Syrian, *Hymns on Paradise*, II.10–13, IX.20 (1990), pp. 143, 88–9.

6   See Riccardo Terzoli, *Il tema della beatitudine nei padri siri: Presente e futuro della salvezza* (Brescia: Morcelliana, 1972), pp. 85–7; Jean Daniélou, 'Terre et paradis chez les Pères de l'Église', *Eranos-Jahrbuch* 22 (1953), pp. 450–1.

7   Ephrem the Syrian, *Hymns on Paradise*, I.8, II.6 (1990), pp. 80, 87.

8   Ibid., I.4, p. 78.

9   Ibid., I.8–9, pp. 80–1.

10 Ibid., III.14–17, pp. 95–6.

11 Ibid., V.14, XI, pp. 107, 154–9. See also Kronholm Tryggve, *Motifs from Genesis 1–11 in the Genuine Hymns of Ephrem the Syrian with Particular Reference to the Influence of Jewish Exegetical Tradition* (Lund: Gleerup, 1978), p. 69, and Ephrem the Syrian, *Hymnen über das Paradies*, ed. and transl. by Edmund Beck (Rome: Herder, 1951), p. 2.

12 London, British Library, Harley MS 218, fol. 104v (I am grateful to Scott D. Westrem for drawing my attention to this diagrammatic map).

13 Paris, Bibliothèque Ste-Geneviève, MS 782, fol. 374v. See Marcel Destombes, *Mappemondes AD 1200–1500: Catalogue preparé par la Commission des Cartes Anciennes de l'Union Géographique Internationale* (Amsterdam: N. Israel, 1964), pp. 177–8.

14 Isidore, *De natura rerum*, XXXVI–XXXVII, ed. by Antonio Laborda (Madrid: Instituto Nacional de Estadística, 1996), pp. 149–51; Isidore, *Etymologiae*, XIII.2.2–14, ed. by Wallace Martin Lindsay, 2 vols (Oxford: Clarendon Press, 1911). Aristotle, *Metereologica*, II.6, had already accounted for the winds and their effects. On the windrose and the most commonly listed winds in the Middle Ages, see Helen Wallis, ed., *Cartographical Innovations: An International Handbook of Mapping Terms to 1900* (Tring: Map Collector–International Cartographic Association, 1987), pp. 245–6.

15 Paris, Bibliothèque Nationale de France, MS Lat. 5510, fol. 118r. See William of Tripoli, *De statu Sarracenorum*, in William of Tripoli, *Notitia de Machometo/De statu Sarracenorum*, ed. by Peter Engels, CISC SL IV (Würzburg–Altenberge: Echter–Oros, 1992), pp. 263–71, on the map pp. 131, 421; Anna-Dorothee von den Brincken, '"Ut describeretur orbis universus": Zur Universalkartographie des Mittelalters', in Albert Zimmermann, ed., *Methoden in Wissenschaft und Kunst des Mittelalters* (Berlin: De Gruyter, 1970),

16 pp. 249–78, esp. 268–9; Destombes, *Mappemondes* (1964), p. 176. The map, which comes immediately after William's work, is likely to be linked to a passage referring to the tripartite earth, XV.3–4, p. 305.

16 Scott D. Westrem, *The Hereford Map: A Transcription and Translation of the Legends with Commentary* (Turnhout: Brepols, 2001), pp. 12–19.

17 For early writers such as the Pseudo-Tertullian, Cyprian, Prudentius, Falconia Proba, Dracontius, Avitus and Claudius Marius Victor, see Howard Rollin Patch, *The Other World, according to Descriptions in Medieval Literature* (Cambridge, MA: Harvard University Press, 1950), pp. 134–74.

18 Plato, *Statesman*, 272a–b, ed. and transl. by Julia Annas and Robin Waterfield (Cambridge–New York: Cambridge University Press, 1995), p. 25; ed. by Christopher J. Rowe (Warminster: Aris and Phillips, 1995), p. 68: γυμνοὶ δὲ καί ἄστρωτοι θυραυλοῦντες τὰ πολλὰ ἐνέμοντο· τὸ γὰρ τῶν ὡρῶν αὐτοῖς ἄλυπον ἐκέκρατο. Ovid, *Metamorphoses*, I.107–8, transl. by Mary M. Innes (Harmondsworth: Penguin, 1955), p. 32. Publius Ovidius Naso, *Metamorphoses*, ed. by William S. Anderson (Leipzig: Teubner, 1977), pp. 4–5: 'Ver erat aeternum, placidique tepentibus auris mulcebant Zephyri natos sine semine flores.'

19 Hesiod, *Works and Days*, 170–5. Homer, *Odyssey*, IV.563–8, had already sung of the Elysian Fields, located beyond the ocean.

20 John of Damascus, *De fide orthodoxa*, II.11, *PG* XCIV, col. 464: 'In Oriente omni terra sublimior positus fuit, probeque temperatus, ac subtilissimo purissimoque aere undique collustratus, plantis nunquam non floridis vernans, suavissimo odore et lumine plenus, elegantiae omnis, quae quidem in sensum cadat, et pulchritudinis cogitatum superans, divina plane regio.' See the edition by Eligius M. Buytaert (St Bonaventure, NY: Franciscan Institute, 1955), pp. 106–7.

21 Isidore, *Etymologiae*, II, XIV.3.2–3 (1911): 'non ibi frigus, non aestus; sed perpetua aeris temperies'. Isidore's dictum was repeated, for instance, by a number of encyclopedists: Vincent of Beauvais, *Speculum maius*, XXXII.11, 4 vols (Douai: Baltazar Bellerus, 1624; repr. Graz: Akademische Drück und Verlagsanstalt, 1965), IV, p. 24; *Polychronicon Ranulphi Higden, Monachi Cestrensis, together with the English Translations of John Trevisa and of an Unknown Writer of the Fifteenth Century*, ed. by Churchill Babington (vols I, II) and Joseph R. Lumby (vols III–IX), *RBMAS* XLIV, 9 vols (London: Longmans, Green, and Co., 1865–6), I (1865), pp. 74–6; Brunetto Latini, *Li Livres dou tresor*, ed. by Francis J. Carmody (Berkeley: University of California Press, 1948), p. 114; Bartholomeus Anglicus, *De proprietatibus rerum*, XV.112 (Frankfurt am Main: Wolfang Richter, 1601; repr. Frankfurt am Main: Minerva, 1964), pp. 680–3.

22 See the discussion of zonal maps in David Woodward, 'Medieval *Mappaemundi*', in Harley and Woodward, eds, *The History of Cartography*, I (1987), pp. 296–7, 353–5. Reproductions of several medieval zonal maps may be found, for example, in Patrick Gautier Dalché, '*Mappae mundi* antérieures au XIIIe siècle dans les manuscrits latins de la Bibliothèque Nationale de France', *Scriptorium*, 52/1 (1998), pp. 102–61, Plates 25–7, 30–2, and John E. Murdoch, *Album of Science*, I, *Antiquity and The Middle Ages* (New York: Scribner, 1984), pp. 281–3.

23 London, British Library, Harley MS 2799, fols 241v–242r. Note that the second page has been bound upside down. On the Arnstein Bible maps, see Edson, *Mapping Time and Space* (1997), pp. 92–4; Jörg-Geerd Arentzen, *Imago mundi cartographica: Studien zur Bildlichkeit mittelalterlicher Welt- und Ökumenekarten unter besonderer Berücksichtigung des Zusammenwirkens von Text und Bild* (Munich: Fink, 1984), pp. 72, 102–6.

24 The distinction between *mundus*, the Latin translation of the Greek κόσμος (*kosmos*), referring to the universe, and *terra* as referring to the earth, appears in Isidore *Etymologiae*, XIV.I (1911): 'Terra est in media mundi regione posita.' Isidore also explains that the term *terra*, when singular, refers to the terrestrial globe, when plural (*terrae*) indicates particular earthly regions. See also Anna-Dorothee von den Brincken, 'Mappe del Medio Evo: Mappe del cielo e della terra', in *Cieli e terre nei secoli XI–XII: Orizzonti, percezioni, rapporti* (Milan: Vita e Pensiero, 1998), pp. 31–2.

25 The *Didascalicon* contained Hugh's ideas for the best programme of study. See Rebecca Moore, *Jews and Christians in the Life and Thought of Hugh of St Victor* (Atlanta, GA: Scholars Press, 1998), pp. 44–50.

26 Utrecht, Rijksuniversiteitsbibliotheek, MS 737, fol. 49v. Gerard of Antwerp's *Historia figuralis ab origine mundi usque ad ... 1272* is found between fols 47 and 68. See Anna-Dorothee von den Brincken, *Fines terrae: Die Enden der Erde und der vierte Kontinent auf mittelalterlichen Weltkarten* (Hannover: Hahn, 1992), pp. 84–5; von den Brincken, '"Ut describeretur orbis"' (1970), p. 275; von den Brincken, 'Mappa mundi und Chronographia: Studien zur *imago mundi* des abendländischen Mittelalters', *Deutsches Archiv für Erforschung des Mittelalters*, 24 (1968), pp. 150–1; Destombes, *Mappemondes* (1964), p. 186; Arno Borst, *Der Turmbau von Babel: Geschichte der Meinungen über Ursprung und Vielfalt der Sprachen und Völker*, 4 vols (Stuttgart: Hiersemann, 1957–63), II/2 (1959), p. 795.

27 Bischofszell (Canton Thurgovia, Switzerland), Ortsmuseum, Dr Albert-Knoepfli-Stiftung. See Anna-Dorothee von den Brincken, 'Das Weltbild des irischen Seefahrer-Heiligen Brendan in der Sicht des 12. Jahrhunderts', *Cartographia Helvetica*, 21 (January 2000), pp. 17–21; von den Brincken, 'Mappe del cielo e della terra nel Medioevo' (1998), pp. 46–8. On Saint Brendan and paradise see above, Chapter 3, pp. 52–3.

28 There is an abundant literature on the thirteenth-century philosophical and scientific revolution. See, for example, the bibliography cited in Michael Haren, *Medieval Thought: The Western Intellectual Tradition from Antiquity to the Thirteenth Century*, 2nd edn (Toronto–Buffalo: University of Toronto Press, 1992), pp. 267–8, 276–80, 288–90, 301–5. The assimilation of Aristotelianism, considered by many to be a potential threat to the Christian faith, was by no means an untroubled process, as can be seen from the series of ecclesiastical condemnations, culminating in the 219 doctrines condemned by the bishop of Paris in 1277.

29 See above, Chapter 3, pp. 50–1 and pp. 59, 60 nn. 41, 50.

30 David Woodward, with Herbert M. Howe, 'Roger Bacon on Geography and Cartography', in Jeremiah Hackett, ed., *Roger Bacon and the Sciences: Commemorative Essays* (Leiden–New York–Cologne: Brill, 1997), p. 213; see also George H. T. Kimble, *Geography in the Middle Ages* (London: Methuen, 1938), pp. 69–99.

31 John K. Wright, *The Geographical Lore of the Time of the Crusades: A Study in the History of Medieval Science and Tradition in Western Europe* (New York: American Geographical Society, 1925; repr. New York: Dover, 1965), pp. 95–9. On the *De causis proprietatum et elementorum*, falsely attributed to Aristotle in the Middle Ages, see the edition of Stanley L. Vodraska, PhD Dissertation, The Warburg Institute, University of London (1969).

32 See *The Astronomical Tables of Al-Khwarizmi: Translation with Commentaries of the Latin Version*, ed. by Heinrich Suter and Otto Neugebauer (Copenhagen: Kommission hos Munksgaard,

1962); Raymond Mercier, 'Astronomical Tables in the Twelfth Century', in Charles Burnett, ed., *Adelard of Bath: An English Scientist and Arabist of the Early Twelfth Century* (London: Warburg Institute, 1987), pp. 87–118. See also Abu Ma'shar, *The Abbreviation of the Introduction to Astrology together with the Medieval Latin Translation of Adelard of Bath*, ed. and transl. by Charles Burnett, Keiji Yamamoto and Michio Yano (Leiden–New York: Brill, 1994). Ptolemy's *Geography* was translated into Arabic in the ninth century by Al-Khwarizmi. On its influence on Arab geographical writing, see Evelyn Edson and Emilie Savage-Smith, *Medieval Views of the Cosmos* (Oxford: Bodleian Library, 2004), esp. pp. 30–43, 61–6; Jeremy Johns and Emilie Savage-Smith, 'The Book of Curiosities: A Newly Discovered Series of Islamic Maps', *Imago Mundi*, 55 (2003), pp. 7–24; Edward S. Kennedy and Mary Helen Kennedy, *Geographical Coordinates of Localities from Islamic Sources* (Frankfurt am Main: Institut für Geschichte der Arabisch-Islamischen Wissenschaften, 1987).

33   See, e.g., Abu Ma'shar, *On Historical Astrology: The Book of Religions and Dynasties (On the Great Conjunctions)*, ed. and transl. by Charles Burnett and Keiji Yamamoto, 2 vols (Leiden: Brill: 2000). The 'geographical renaissance' was not, however, a radical break with the earlier tradition, since the process of assimilation of Aristotelian cosmology lasted for several decades, and geographical studies still depended on the same classical and patristic sources that were already known. An important innovation, however, was that contemporary travellers' accounts began to be regarded as providing valuable information; see Clarence J. Glacken, *Traces on the Rhodian Shore: Nature and Culture in Western Thought from Ancient Times to the End of the Eighteenth Century* (Berkeley: University of California Press, 1967), p. 262, cited by Woodward, 'Roger Bacon' (1997), p. 202. Moreover, the precise time at which Aristotelianism began to influence Christian cosmology is not certain; see Reuven S. Avi-Yonah, *The Aristotelian Revolution: A Study of the Transformation of Medieval Cosmology, 1150–1250*, PhD Dissertation, Harvard University (1986).

34   On William of Auvergne see Roland J. Teske, 'Introduction' to William of Auvergne, *The Trinity or the First Principle*, ed. and transl. by Teske and Francis C. Wade (Milwaukee, WI: Marquette University Press, 1989), pp. 1–5; Steven P. Marrone, *William of Auvergne and Robert Grosseteste: New Ideas of Truth in the Early Thirteenth Century* (Princeton, NJ: Princeton University Press, 1983), pp. 27–134; Jan Rohls, *Wilhelm von Auvergne und der mittelalterliche Aristotelismus* (Munich: Kaiser, 1980); Noel Valois, *Guillaume d'Auvergne, évêque de Paris (1228–1249): Sa vie et ses ouvrages* (Paris: Irvington, 1980); Aimé Forest, 'Guillaume d'Auvergne, critique d'Aristote', in *Études médiévales offertes à Augustin Fliche* (Paris: [n. pub.], 1952), pp. 67–79. For an introduction to William's view of nature and creation see Albrecht Quentin, *Naturkenntnisse und Naturanschauungen bei Wilhelm von Auvergne* (Hildesheim: Gerstenberg, 1976). On the *De universo* see Amato Masnovo, *Da Guglielmo d'Auvergne a S. Tommaso d'Aquino*, 3 vols (Milan: Vita e Pensiero, 1945–6), I (1945), pp. 131–43.

35   William of Auvergne, *De universo*, I, 1a, 56, in *Opera omnia* (Paris: E. Couterot, 1674), I, p. 674: 'non mirabitur in extremis partibus orientis, regionem esse tantae amaenitatis atque vernantiae, ut hortus voluptatis, et paradisus terrestris a tanto propheta, et legislatore merito sit vocata'.

36   Ibid.

37   Ibid., I, 1a, 57, p. 674: 'hoc autem nec natura patitur, nec ratio'.

38   Ibid., I, 1a, 57, pp. 674–5. According to Aristotle, beneath the celestial sphere is the sublunary sphere, within which the four terrestrial elements, earth, water, air and fire, are arranged in four concentric spheres. The sphere of fire surrounds the air and is contiguous with the celestial regions. See Aristotle, *Meteorology*, I.3, 340b24–341a13.

39   See above, n. 38.

40   William of Auvergne, *De universo*, I, 1a, 56 (1674), pp. 674–5.

41   Ptolemy's division of the inhabited part of the globe into seven *climata*, from the equator to the polar circle, was known during the time of the Roman Empire and reappeared in the Middle Ages. On Ptolemy and the contribution of ancient geography to medieval geographical knowledge see Ernst Honigmann, *Die sieben Klimata und die ΠΟΛΕΙΣ ΕΠΙΣΗΜΟΙ: Eine Untersuchung zur Geschichte der Geographie und Astrologie im Altertum und Mittelalter* (Heidelberg: Winter, 1929), pp. 58–60; Oswald A. W. Dilke, 'The Culmination of Greek Cartography' and 'Cartography in the Ancient World: A Conclusion', in Harley and Woodward, eds, *The History of Cartography*, I (1987), pp. 177–200 and 276–9; Wright, *Geographical Lore* (1925), pp. 9–42. On Ptolemy's *Almagest* see Olaf Pedersen, *A Survey of the Almagest* (Odense: Odense Universitetsforlag, 1974); on his *Geography* see Otto Neugebauer, *A History of Ancient Mathematical Astronomy*, 3 vols (Berlin–Heidelberg–New York: Springer, 1975), II, pp. 934–41.

42   Aristotle, *Meteorology*, II.5, 362b6–9.

43   Polybius's views on the inhabited world below the equator are known through Geminus's *Elementa astronomiae*; see Geminus, *Introduction aux phénomènes*, XVI.32–8, ed. and transl. by Germaine Aujac (Paris: Belles Lettres, 1975), pp. 82–3. Eratosthenes's views are reported by Strabo, *Geography*, II.3.2; see Eratosthenes, *Die geographischen Fragmente*, ed. by Hugo Berger (Leipzig: Teubner, 1898). See also Geminus, *Introduction aux phénomènes*, pp. 152–3 n. 1; and Brian Harley and David Woodward, 'Greek Cartography in the Early Roman World', in Harley and Woodward, eds, *The History of Cartography*, I (1987), p. 162 n. 1.

44   Ptolemy, *Almagest*, II.6, ed. and transl. by Gerald J. Toomer (London: Duckworth, 1984), p. 83. See also Dilke, 'Culmination of Greek Cartography' (1987), p. 182.

45   Wright, *Geographical Lore* (1925), pp. 162–3.

46   Al-Battani, *Opus astronomicum*, ed. by Carlo A. Nallino, 3 vols (Milan: Hoepli, 1899–1907), I (1899), p. 14; Avicenna, *The Canon of Medicine*, I.2.8, ed. and transl. by Oskar Cameron Gruner (London: Luzac, 1930), p. 197. See also Mazhar H. Shah, *The General Principles of Avicenna's Canon of Medicine* (Karachi: Naveed Clinic, 1966), pp. 169–70.

47   See Francis J. Carmody, *Arabic Astronomical and Astrological Sciences in Latin Translation* (Berkeley: University of California Press, 1956), p. 14.

48   Petrus Alphonsus, *Dialogus contra Iudaeos*, I, in *Diálogo contra los Judíos*, introd. by John Tolan, Latin text ed. by Klaus-Peter Mieth, Spanish transl. by Esperanza Ducay, gen. ed. by Jesús Lacarra (Huesca: Instituto de Estudios Altoaragoneses, 1996), pp. 22–3. On the Antipodes in the Middle Ages see above, Chapter 5, p. 119 n. 50.

49   Aristotle, *De caelo*, II.2, 285b15–28. For a discussion of this problem see John Pecham, *Tractatus de sphera*, ed. and transl. by Bruce R. MacLaren, PhD Dissertation, University of Wisconsin, Madison, WI (1978), p. 249 n. 33. Pecham thought that the equator was suitable for human habitation and the southern hemisphere uninhabitable because of the eccentricity of the solar orbit.

50   Macrobius, *Commentarii in Somnium Scipionis*, II.5.22–36, *Commentary on the Dream of Scipio*, ed. and transl. by William H. Stahl (New York:

Columbia University Press, 1952), pp. 48–53; Martianus Capella, *De nuptiis philologiae et Mercurii*, VI.602–8; VIII.874, ed. by Adolfus Dick (Stuttgart: Teubner, 1978), pp. 298–300, 461. See Wright, *Geographical Lore* (1925), pp. 160, 258, 386.

51 William of Conches, *Dragmaticon philosophiae*, VI.3.6, ed. by Italo Ronca, in *Opera omnia*, ed. by Édouard Jeauneau, CCCM CLII (Turnhout: Brepols, 1997), p. 189; see Wright, *Geographical Lore* (1925), p. 160.

52 Philostorgius, *Historia ecclesiastica*, III.10, *PG* LXV, cols 491–5. Philostorgius had explained that the river Pishon was the Hyphase, a tributary of the Indus, that the Hyphase was the river where Alexander the Great terminated his journey; and that, as it flowed out of paradise and crossed the deserts, the river carried a flower thought to come from Eden. Its waters had miraculous powers.

53 In fact, Alexander quoted Isidore, *Etymologiae*, XIII.15–16. On Alexander's life see Venicio Marcolino, *Das alte Testament in der Heilsgeschichte: Untersuchungen zum dogmatischen Verständnis des alten Testaments als Heilsgeschichtliche Periode nach Alexander von Hales* (Münster in Westphalia: Aschendorff, 1970), pp. 9–20. On the *Summa* see Elisabeth Gössmann, *Metaphysik und Heilsgeschichte: Eine theologische Untersuchung der Summa Halensis (Alexander von Hales)* (Munich: Hueber, 1964); Efrem Bettoni, *Il problema della conoscibilità di Dio nella scuola francescana* (Padua: CEDAM, 1950), pp.1–92; Meldon C. Wass, *The Infinite God and the Summa Fratris Alexandri* (Chicago: Franciscan Herald Press, 1964).

54 Alexander of Hales, *Summa theologica*, II, inq.IV, tract.2, sect.2, q.1, 2.462, 4 vols (Ad Claras Aquas, Quaracchi, Florence: Apud Collegium S. Bonaventurae, 1924–48; repr. 1979), II (1928), pp. 607–9.

55 See above, Chapter 3, p. 49 and p. 59 nn. 33, 40.

56 Alexander of Hales, *Summa theologica*, II, inq.IV, tract.2, sect.2, q.1, 2.462, II (1928), p. 606.

57 Ibid., pp. 605–7.

58 Ibid., p. 606. Compare Augustine, *De Genesi contra Manichaeos*, I.15.24, ed. by Dorothea Weber, CSEL XCI (Vienna: Österreichische Akademie der Wissenschaften, 1998), p. 90; see Charles Burnett, 'High Altitude Mountaineering 1600 Years Ago', *The Alpine Journal*, 88/332 (1983), p. 127. See also Solinus, *Collectanea rerum memorabilium*, VIII.5–6 ed. by Theodor Mommsen, 2nd edn (Berlin: Weidmann, 1958), p.62; Lucan, *Pharsalia*, II.271–3; Peter Comestor, among others, reported these observations, in his *Historia scholastica*, *Genesis*: XXXIV, PL CXCVIII, col. 1084.

59 Alexander of Hales, *Summa theologica*, II, inq.IV, tract.2, sect.2, q.1, 2.462, II (1928), pp. 606–7.

60 Étienne Gilson, e.g., in his *La Philosophie de saint Bonaventure* (Paris: Vrin, 1924; repr. 1943), insists on Bonaventure's anti-Aristotelianism, whereas Fernand Van Steenberghen, *Aristotle in the West* (Louvain: Nauwelaerts, 1955), pp. 147–62, maintains that Bonaventure greatly respected Aristotle's philosophy. For an analysis of the scholarly debate, which has mainly followed the lines of these two influential views, see John F. Quinn, *The Historical Constitution of St Bonaventure's Philosophy* (Toronto: Pontifical Institute of Medieval Studies, 1973), pp. 17–99, 841–96.

61 Bonaventure, *Commentaria in quatuor libros Sententiarum*, II, d.17, dub.3, in *Opera omnia*, 10 vols (Ad Claras Aquas, Quaracchi, Florence: Apud Collegium S. Bonaventurae, 1882–1902), II (1885), p. 427: 'sicut doctores dicunt, paradisi corporalis situs est valde eminens et altus et iuxta aequinoctialem in oriente, quodam modo vergens ad meridiem.'

62 Woodward, 'Roger Bacon' (1997), p. 213; Wright, *Geographical Lore* (1925), pp. 408–9 n. 97.

63 Iohannes de Sacrobosco, *Tractatus de spera*, II, in Lynn Thorndike, *The Sphere of Sacrobosco and its Commentators* (Chicago: University of Chicago Press, 1949), pp. 94 (Latin text), 129 (English translation). On Sacrobosco see Thorndike, *Sphere of Sacrobosco* (1949), pp. 1–21. Thorndike argues that Sacrobosco's treatise on the sphere was written before 1235: ibid., pp. 10–14.

64 Robert Grosseteste, *De sphera*, in *Die Philosophischen Werke des Robert Grosseteste, Bischofs von Lincoln*, ed. by Ludwig Baur (Münster in Westphalia: Aschendorff, 1912), pp. 23–4; Robert Grosseteste, *De natura locorum*, in ibid., pp. 66–7. In *De sphera* Grosseteste deals with problems of astronomical geography, while in *De natura locorum* he discusses the influences of celestial rays upon the earth; many of his views were later elaborated by Roger Bacon. See Wright, *Geographical Lore* (1925), pp. 163–4.

65 Anglicus's comments on Sacrobosco were probably first delivered as lectures: Thorndike, *Sphere of Sacrobosco* (1949), p. 28.

66 Robertus Anglicus, *Commentary to the Sphere of Sacrobosco*, XIII, in Thorndike, *Sphere of Sacrobosco* (1949), pp. 188–91 (Latin text), 237–40 (English translation). Robertus Anglicus also reported a story (which he heard from 'a wise man in England who lived a good and holy life') of a demon supplying ripe figs all the year around from the equator.

67 Ibid., pp. 193 (Latin text), 241–2 (English translation).

68 Cecco d'Ascoli, *Commentary to the Sphere of Sacrobosco*, II, in Thorndike, *Sphere of Sacrobosco* (1949), pp. 400–1.

69 Michael Scot, *Commentary to the Sphere of Sacrobosco*, XI, in Thorndike, *Sphere of Sacrobosco* (1949), pp. 317–19. The attribution of this commentary to Michael Scot is far from certain. In fact, in his *Liber introductorius*, Michael Scot located paradise in the equatorial zone and not in the southern hemisphere: see Patrick Gautier Dalché, 'Le Paradis aux antipodes? Une *Distinctio divisionis terre et paradisi delitiarum (XIVe siècle)*', in Dominique Barthélemy and Jean-Marie Martin, eds, *Liber largitorius: Étude d'histoire médiévale offertes à Pierre Toubert par ses élèves* (Paris: Droz, 2003), p. 633 n. 45. Gautier Dalché, ibid., has discussed a fourteenth-century treatise on the location of paradise in the southern hemisphere.

70 Michael Scot, *Commentary to the Sphere of Sacrobosco*, XI, in Thorndike, *Sphere of Sacrobosco* (1949), pp. 321–2.

71 See Jeremiah Hackett, 'Bacon: His Life, Career and Works', and Woodward, 'Roger Bacon' (1997), pp. 9–23 and 199–222. On the complex interplay between scientific and religious beliefs in Roger Bacon's approach to natural science see David C. Lindberg, 'The Medieval Church Encounters the Classical Tradition', in David C. Lindberg, Ronald L. Numbers, eds, *When Science and Christianity Meet* (Chicago–London: University of Chicago Press, 2003), pp. 7–32.

72 Roger Bacon, *Opus maius*, ed. by John H. Bridges, 2 vols (Oxford: Clarendon Press, 1897–1900), I (1897), pp. 175–404, esp. 175, 180–93. On the issue of paradise, Bacon writes that 'there are many astronomical considerations here. And therefore it is necessary that theologians dealing with this area have a good knowledge of the foundations of astronomy'. Ibid., p. 193: 'Et multae considerationes astronomiae sunt hic. Et ideo oportet theologum in hac parte scire bene radices astronomiae.'

73 Ibid., p. 137.

74 Ibid., pp. 135–7, 305–7.

75 Ptolemy, *Almagest*, I.15 (1984), p. 72. Bacon frequently quotes Ptolemy and states that he follows the teachings of the *Almagest*: see, e.g., *Opus maius*, I (1897), p. 298.

76 Roger Bacon, *Opus maius*, I (1897), pp. 135–7.

77 Ibid., p. 137: 'Nec est dubium quin locus sit temperatus, sed an sit temperatissimus, non

78   percipio adhuc. Et ideo non est certificatum an paradisus debeat ibi esse.'
78   Ibid., p. 137: 'paradisus habet plenum temperamentum.'
79   Ibid., p. 193.
80   Ibid., pp. 305–7. Students of nature, says Bacon, speculate about the orbit of the sun, whether it is perfectly centred round the earth but then moves in epicycles, or whether it is simply eccentric. Ptolemy in the *Almagest* acknowledged both possibilities but opted for the hypothesis of eccentricity, a much simpler model, as Bacon notes in *Communia naturalium, liber secundus: De celestibus*, ed. by Robert Steele, *Opera hactenus inedita*, fasc. IV (Oxford: Clarendon Press, 1913), pp. 420–1. There he points out that Ptolemy calculated that the sun takes 94½ days to cover the quarter from the spring equinox until the summer solstice; from the summer solstice until the autumn equinox it takes 92½ days; from the autumn equinox until the winter solstice, 88 days, 7 minutes and 30 seconds; and from the winter solstice to the spring equinox, 90 days, 7 minutes and 3 seconds. See Ptolemy, *Almagest*, III.3 (1984), pp. 141–53. Roger Bacon, *Opus maius*, I (1897), p. 294: 'Et propter hoc ex ordinatione naturae erit quod impedimenta habitationis magis excludantur.' 'There will be something ordained by nature so that forces hampering habitation are less likely to be present.'
81   Roger Bacon, *Opus maius*, I (1897), pp. 293–4. Bacon's quotes from Ptolemy's work *De dispositione spherae*, a passage which also appears in *De natura locorum* of Albert the Great. As Woodward points out, 'Roger Bacon' (1997), p. 207 n. 25, Ptolemy's *De dispositione spherae* is an introduction to the *Almagest*, based on Geminus's *Introduction to the Phenomena*, and the passage on the two races of Ethiopians is found in Paris, Bibliothèque Nationale de France, MS Lat. 16198, fol. 174v.
82   Roger Bacon, *Opus maius*, I (1897), p. 305–8; see Pliny, *Naturalis historia*, II.75.183–5; Ambrose, *Hexameron*, IV.5.23, ed. by Karl Schenkl, *CSEL* XXXII/1 (Prague–Vienna–Leipzig: Tempsky–Freytag, 1897; repr. New York–London: Fathers of the Church, 1962), pp. 130–1.
83   Roger Bacon, *Opus maius*, I (1897), p. 293.
84   Ibid., p. 192.
85   Ibid., p. 307: 'Et ideo opinio aliquorum est quod ibi sit paradisus, cum sit locus nobilissimus in hoc mundo secundum Aristotelem et Averroem secundo Coeli et Mundi.' See also ibid., p. 294. See Aristotle, *De caelo*, II.2, 285b15–28 and Averroes's commentary on this passage in Aristotle, *Opera cum Averrois commentariis*, 11 vols (Venice: apud Iuntas, 1562–74; repr. Frankfurt am Main: 1962), V, fols 104v–105v.
86   Roger Bacon, *Opus maius*, I (1897), pp. 298–300.
87   Ibid., p. 333.
88   Hieronymus Wilms, *Sant'Alberto Magno: Scienziato, filosofo e santo* (Bologna: Edizioni Studio Domenicano, 1992), pp. 29–70; Edward A. Synan, 'Albertus Magnus and the Sciences', and James A. Weisheipl, 'The Life and Works of St Albert the Great', in Weisheipl, ed., *Albertus Magnus and the Sciences* (Toronto: Pontifical Institute of Medieval Studies, 1980), pp. 1–12, 13–51; Heribert Ch. Scheeben, *Albertus Magnus* (Bonn: Verlag der Buchgemeinde, 1932).
89   Albert the Great, *De homine*, II.79, in *Opera omnia*, ed. Auguste C. Borgnet, 38 vols (Paris: Vivès, 1890–9), XXXV (1896), p. 639.
90   Ibid., p. 640: 'Sine praeiudicio loquendo, credo, quod ultra aequinoctialem in Oriente versus Meridiem.'
91   On Bede see above, Chapter 3, pp. 48–9.
92   Albert the Great, *De homine*, II.79 (1896), p. 640: 'valde secundum rationem'. This was also Aristotle's view; see n. 49 above.
93   See above, n. 44.
94   Albert the Great, *De homine*, II.79 (1896), p. 640: 'Alia autem causa inundationis Nili cum sol est in Cancro, cum potius deberet diminui, non potest rationabiliter assignari. Cum autem Nilus sit fluvius paradisi, oportet paradisum esse in eodem loco.'
95   Ibid., pp. 640–1.
96   Albert the Great, *Commentarii in II Sententiarum*, d.17, q.5, in *Opera omnia*, XXVII (1894), p. 304: 'ille locus ab habitabili nostra secretus est: non quod non sit in habitabili nostra, sed quod Dominus clausit eum a nobis. Et sine praeiudicio loquendo videtur mihi, quod sit in linea aequinoctiali versus Orientem in comparationem ad nostram habitationem.'
97   Ibid.: 'locus ille optimus et paratissimus sit'. On Ptolemy and Avicenna see above, p. 173.
98   Albert the Great, *Commentarii in II Sententiarum*, d.17, q.5 (1894), p. 305: 'illi qui venerunt de terra illa'.
99   Ibid.: 'Hoc igitur opinando, non asserendo, dictum sit de paradisi loco.'
100   Pseudo-Aristotle, *De inundatione Nili*, transl. and ed. by Danielle Bonneau, *Études de Papyrologie*, 9 (1971), pp. 1–33; see Janine Balty-Fontaine, 'Pour une Édition nouvelle du *Liber Aristotelis de inundatione Nili*', *Chronique d'Égypte*, 34 (1959), pp. 95–102. On the problem of the flooding of the Nile and the various explanations offered from antiquity to the Middle Ages see Bonneau, *La Crue du Nil, divinité égyptienne* (Paris: Klincksieck, 1964); Kimble, *Geography in the Middle Ages* (1938), p. 175; Wright, *Geographical Lore* (1925), pp. 30–1, 60, 206–7, 300.
101   Benjamin of Tudela, *The Itinerary*, ed. and transl. by Ezra H. Haddad (Baghdad: Eastern Press, 1945), pp. 71–3; William of Tyre, *Historia rerum in partibus transmarinis gestarum*, XIX.26/7, *PL* CCI, cols 774–5. See also Wright, *Geographical Lore* (1925), p. 300.
102   Wright, *Geographical Lore* (1925), pp. 299–300. The city of Damietta was held by Louis IX of France from 1248 to 1249. However, when Albert completed his commentary on the second book of Peter Lombard's *Sentences* in 1249, Louis's ill-fated campaign had not yet come to its sorry conclusion at al-Mansurah (1250).
103   Weisheipl, 'Life and Works of St Albert' (1980), p. 22; Franz Pelster, *Kritische Studien zum Leben und zu den Schriften Alberts des Grossen* (Freiburg im Breisgau: Herder, 1920), pp. 94–130.
104   Weisheipl, 'Life and Works of St Albert' (1980), pp. 21–3; Scheeben, *Albertus Magnus* (1932), pp. 45–56. Albert arrived in Paris around 1244 and became a master in the spring of 1245. The chronology of Albert's so-called Aristotelian paraphrases remains, however, controversial: Weisheipl, op. cit., pp. 27–8. While in Paris, Albert was probably seeking a compromise between the Aristotelian and Ptolemaic cosmologies. Pierre Duhem, *Le Système du monde: Histoire des doctrines cosmologiques de Platon à Copernic*, 10 vols (Paris: Hermann, 1913–59), III (1915), pp. 327–45, has described the ambivalence of Albert's astronomical views; see also William A. Wallace, 'The Scientific Methodology of St Albert the Great', in Gerbert Meyer and Albert Zimmermann, eds, *Albertus Magnus Doctor Universalis 1280/1980* (Mainz: Matthias Grünewald, 1980), p. 399.
105   As we shall see below, Albert's student Thomas Aquinas also admitted that both of the opposing opinions were possible, although he described Aristotle's view as 'more likely'. See above, p. 182.
106   See Brian Harley and David Woodward, with Germaine Aujac, 'Greek Cartography in the Early Roman World', in Harley and Woodward, eds, *The History of Cartography*, I (1987), pp. 162–4, 243–4.
107   Albert the Great, *De natura locorum*, I.6, in *Opera omnia*, IX (1891), p. 540: 'est enim rationabile ibi esse habitationem'. 'India' and 'Ethiopia' were considered to be huge land areas, not cotermi-

108 nous with what we call today 'India' and 'Ethiopia'. See below, Chapter 8, pp. 218–19.
108 Albert the Great, *De natura locorum*, I.6 (1891) p. 541.
109 Ibid., I.7, pp. 543–4: 'Nos autem salvo meliori iudicio, dicimus aliquam quartam partem quae est ultra aequinoctialem ad Meridiem, esse habitabilem secundum naturam et habitatam, ut putamus.' See also Karl Klauck, 'Albertus Magnus und die Erdkunde', in Heinrich Ostlender, ed., *Studia Albertina* (Münster in Westphalia: Aschendorff, 1952), pp. 234–48, and Wright, *Geographical Lore* (1925), pp. 406–7.
110 Albert the Great, *De natura locorum*, I.7 (1891), p. 544: 'spatium quod est post tropicum Capricorni usque ad latitudinem septim[i] climatis in Meridie mensurando, hoc est, usque ad latitudinem quadraginta et octo vel quinquaginta graduum habitabile est secundum delectationem et continue, sicut et nostrum, et forte plus quam nostrum'. Albert also discusses some of the theories which attempt to explain the lack of communication between the two hemispheres: ibid., pp. 544–5.
111 Ibid., III.4, p. 575.
112 Étienne Gilson, *History of Christian Philosophy in the Middle Ages* (London: Sheed and Ward, 1985), p. 290. See also the chronological list of Albert's works in Alain de Libera, *Albert le Grand et la philosophie* (Paris: Vrin, 1990), p. 21, and Pelster, *Kritische Studien* (1920), pp. 169–72.
113 Albert the Great, *Summa theologiae, pars secunda*, q.79, in *Opera omnia*, XXXIII (1895), pp. 110–13. Here Albert states that the theory of paradise reaching up to the sphere of the moon went back to the Apostle Thomas.
114 Hugh Pope, *St. Thomas Aquinas as an Interpreter of Holy Scripture* (Oxford: Blackwell, 1924). For bibliography on Thomas see Norman Kretzman and Eleonore Stump, eds, *The Cambridge Companion to Aquinas* (Cambridge: Cambridge University Press, 1993), pp. 269–80. On Aristotle and Thomas see, among many other studies, Jan A. Aertsen, 'Aquinas's Philosophy in its Historical Setting' and Joseph Owens, 'Aristotle and Aquinas', in Kretzman and Stump, eds, *The Cambridge Companion to Aquinas* (1993), pp. 12–37 and pp. 38–59; Frederick C. Copleston, *Aquinas* (Harmondsworth: Penguin, 1955), pp. 63–5. On Thomas and the paradise question, Joseph E. Duncan, *Milton's Earthly Paradise* (Minneapolis: University of Minnesota Press, 1972), pp. 69–75.
115 See, for example, Thomas Aquinas, *Quaestiones quodlibetales*, III, q.4, a.10, ed. by Pierre Felix Mandonnet (Paris: Lethielleux, 1926), p. 86. See also Thomas Aquinas, *Summa theologiae*, Blackfriars edn, 61 vols (London–New York: Eyre and Spottiswoode–McGraw-Hill, 1964–81), X (1967), Appendix 3, p. 182, and Appendix 9, pp. 221–4, and Pope, *Thomas as an Interpreter of Holy Scripture* (1924) p. 20.
116 Thomas Aquinas, *Summa theologiae*, 1a, q.102, a.1, ed. by Pietro Caramello, Leonine edn, 3 vols (Turin–Rome: Marietti, 1948–50), I (1950), pp. 483–4. See Aristotle, *De caelo*, II.2, 285b16–23; Aristotle, *Meteorology*, I.13, 350b30–351a18. For Augustine, see above, Chapter 3, p. 46.
117 Thomas Aquinas, *Summa theologiae*, 2a 2ae, q.164, a.2, Leonine edn, II (1948), pp. 771–3. See Augustine, *De Trinitate*, III.4, ed. by William J. Mountain and François Glorie, *CCSL* L (Turnhout: Brepols, 1968), pp. 135–6. Corporeal motions in creation are governed by the angels, as Augustine explained.
118 Thomas Aquinas, *Summa theologiae*, 1a, q.102, a.2, Leonine edn, I (1950), pp. 484–5.
119 According to Aristotle's view of the cosmos, the four elements account for everything in the sublunar realm, whereas the ether constitutes the entire supralunar sphere of the cosmos, including the stars. Unlike the naturally centrifugal movement of fire and air and the naturally centripetal movement of water and earth, the movement of ether is circular. The final cause of its movement is the Unmoved Mover. Aristotle, *De caelo*, II.3, 289b35. See Thomas Aquinas, *Summa theologiae*, Blackfriars edn, X (1967), Appendix 3, pp. 182–4; Thomas S. Hall, *Ideas of Life and Matter: Studies in the History of General Physiology, 600 BC–1900 AD*, 2 vols (Chicago–London: University of Chicago Press, 1969), I, pp. 104–19.
120 Thomas Aquinas, *Summa theologiae*, 1a, q.102, a.1, Leonine edn, I (1950), p. 483.
121 Dante deals with geographical and astronomical problems in the *Convivio*; he discusses linguistic geography in *De vulgari eloquentia*; and a treatise on the distribution of seas and lands, the *Quaestio de aqua et terra*, has been attributed to him (although the attribution is by no means secure): see Kimble, *Geography in the Middle Ages* (1938), pp. 240–4, and Wright, *Geographical Lore* (1925), pp. 106–7, 410. The literature on Dante's geography and astronomy is vast. See, e.g., Brenda D. Schildgen, *Dante and the Orient* (Urbana: University of Illinois Press, 2002); Peter S. Hawkins, 'Out upon Circumference: Discovery in Dante', in Scott Westrem, ed., *Discovering New Worlds: Essays on Medieval Exploration and Imagination* (New York–London: Garland, 1991), pp. 193–220, republished in Peter S. Hawkins, ed., *Dante's Testaments: Essays in Scriptural Imagination* (Stanford, CA: Stanford University Press, 1999), pp. 265–83; James Dauphiné, *Le Cosmos de Dante* (Paris: Belles Lettres, 1984); Beniamino Andriani, *Aspetti della scienza in Dante* (Florence: Le Monnier, 1981); Patrick Boyde, *Dante Philomythes and Philosopher: Man in the Cosmos* (Cambridge: Cambridge University Press, 1981), pp. 96–111; Ideale Capasso, *L'astronomia nella Divina Commedia* (Pisa: Domus Galilaeana, 1967); Bruno Nardi, 'Il mito dell'Eden', in his *Saggi di filosofia dantesca*, 2nd edn (Florence: La Nuova Italia, 1967), pp. 311–40; Giovanni Buti and Renzo Bertagni, *Commento astronomico della Divina Commedia* (Florence: Sandron, 1966); Charles S. Singleton, *Dante's 'Commedia': Elements of Structure* (Baltimore: Johns Hopkins University Press, 1954) ('The Return to Eden'); Edward Moore, *Studies in Dante, Third Series: Miscellaneous Essays* (Oxford: Oxford University Press, 1903; repr. 1968), pp. 109–43, esp. 134–9. On Dante's cantos describing the earthly paradise see also Peter Armour, *Dante's Griffin and the History of the World: A Study of the Earthly Paradise (Purgatorio, Cantos XXIX–XXXIII)* (Oxford–New York: Clarendon Press, 1989).
122 Dante, *Purgatorio* IV.67–71; see also XXVII.1–5. In *De vulgari eloquentia* I.VIII.6–10, Dante says that the root of mankind was planted in the east, but Dante's comment in fact refers to Adam's home after his expulsion from paradise: see Mary A. Orr, *Dante and the Early Astronomers* (London–Edinburgh: Gall and Inglis, 1913), p. 345 n. 4.
123 Dante, *Purgatorio* I.22–7.
124 Dante, *Inferno* XXVI.124, 126. Ulysses's fraudulence was held by Dante to be not only his lies about the Trojan horse but also the way he spoke to his men, speech luring them to go west through the Straits of Gibraltar.
125 Dante, *Inferno* XXVI.130–42; see also XXXIV.112–26. Adam in the heavenly paradise says that he lived on the mount that rises highest from the sea, 'sul monte che si leva più dall'onda': *Paradiso* XXVI.139.
126 Dante, *Purgatorio* III.25; XV.1–6; see also Orr, *Dante* (1913), p. 345.
127 Orr has put its latitude at 32° S, at the antipodes of Jerusalem, and its longitude at 180° W: Orr, *Dante* (1913), p. 350. Edoardo Coli, too, locates Dante's Eden in the temperate zone south of the

equator and of the tropic of Capricorn: *Il paradiso terrestre dantesco* (Florence: Carnesecchi, 1897), pp. 185–207, esp. 195–7. Paolo Pecoraro, however, *Le stelle di Dante* (Rome: Bulzoni, 1987), pp. 63–5, 86, 93–125, disagrees, claiming that, in Dante's view, the mountain of Purgatory, and therefore also the earthly paradise, is located in the tropic of Capricorn, at a latitude of 23°30' S, at the antipodes, not of Jerusalem, but of Mount Sinai.

128   Dante, *Purgatorio* III.15; XXI.43–57; and XXVIII. 97–102.

129   Dante, *Paradiso* I.76–81.

130   Dante, *Inferno* I.1, XXI.112–13. The date, still discussed among Dante's scholars, is inferred from a series of passages in the poem: see Dante Alighieri, *La Divina Commedia*, ed. by Natalino Sapegno, 3 vols (Florence: La Nuova Italia, 1979), I, p. 4 n. 1. See also above, Chapter 4, p. 67; Moore, *Studies in Dante* (1903; 1968), pp. 144–77.

131   The problem of the distribution of seas and lands puzzled scientists of the time, a matter which was discussed in the *Quaestio de aqua et terra*, attributed to Dante. See Wright, *Geographical Lore* (1925), pp. 186–7.

132   Wright, *Geographical Lore* (1925), pp. 106–7; Moore, *Studies in Dante* (1903; 1968), pp. 110–11; Coli, *Il paradiso terrestre dantesco* (1897), p. 195. For Albert the Great and Dante see Cesare Vasoli, 'Fonti albertine nel *Convivio* di Dante', in Maarten J. F. M. Hoenen and Alain de Libera, eds, *Albertus Magnus und der Albertismus: Deutsche philosopische Kultur des Mittelalters* (Leiden–New York–Cologne: Brill, 1995), pp. 33–49. Brunetto Latini, Dante's teacher, located paradise in India: see his *Li Livres dou tresor* (1948), p. 114.

133   Dante, *Inferno* XXVI.125.

8

The Twilight of Paradise on Maps

Too much curiosity lost paradise.
Aphra Behn, *The Lucky Chance* (1686)

It was not only Ulysses's vessel that was wrecked within sight of the mountain-island that was crowned by paradise. Thirteenth-century scholars had enthusiastically espoused the new Aristotelian sciences as a means of unlocking the mystery of the location of the earthly paradise, but these efforts too were doomed to eventual failure. Theologians had been trying to harmonize Christian faith with the contemporary scientific world view without betraying the Church's traditions, but their synthesis of natural philosophy and theology in the end provided neither a stable nor a long-lasting solution. No consensus had been reached on the whereabouts of the earthly paradise. Over the next century, scholars would find themselves obliged to abandon a rational approach to the paradise question and give up trying to define their faith in terms of Aristotelian science and mathematical astronomy. As Étienne Gilson puts it, in the course of the fourteenth century 'the honeymoon of theology and philosophy hadcome to an end'.[1]

At the same time, however, a new conceptual framework was developing in the field of map making, heralded by the occasional merging of the astronomical zonal maps and the topological *mappae mundi*. While some fourteenth-century map makers continued to replicate the earlier *mappae mundi* model, others were more open to new influences and absorbed these into their maps of the world. It was above all the appearance in the world of navigation of the nautical chart, possibly around 1200, that introduced a completely different cartographical mindset and by the end of the fifteenth century radically changed European mapping.[2] The challenge of the nautical chart, however, and all that it implied in terms of measurable distance and direction, did not necessarily spell the end of paradise on maps of the world. Nor did the generation of new geographical information from travel and exploration in Asia and along the African coasts immediately pose undue problems for continuing to show paradise on world maps. It is fair to say, however, that the cartographical context of the fourteenth and fifteenth centuries was fundamentally different from the one that had produced the *mappae mundi* of the earlier Middle Ages.

The End of the Honeymoon

The search on a purely geographical level for a rational explanation of paradise and the application of the new tools of science to examine sacred geography – the great undertaking of the thirteenth century – proved unsatisfactory. A fundamental problem was that characteristics had been ascribed to the earthly paradise that did not seem to correspond to any place on earth. It had been said, for example, that to be untouched by the Flood (and thus to remain pristine), paradise must be the summit of a high mountain. It was now difficult to accept the idea of excessive altitude in the light of contemporary knowledge. Earlier, opinions had been divided as to whether this mountain was on the equator or in the southern hemisphere, but by the end of the thirteenth century the tendency was to accept that it was on the equator. A theological handbook thought to be closely connected to the school of Albert the Great, the *Compendium theologicae veritatis* (c.1268), describes the equatorial earthly paradise as a place without storms, with two winters and two summers, where trees produced fruit twice a year.[3] The entry on *paradisus* in the *Catholicon*, a widely diffused encyclopedic dictionary compiled in 1286 by the Genoese Dominican Iohannes Balbus, states that paradise was 'thought to be located at the equator in the east because certain philosophers say that that location is the most temperate'.[4] In fact, though, the problem of the location of paradise was far from being solved and whichever way the problem was tackled in the fourteenth century, as the arguments of the Dominican theologian Durandus of Saint-Pourçain reveal, the various components always seemed to be irreconcilable. The answer, as Nicholas of Lyre's exegesis showed, was to return to the text of the Bible.

The impossibility of identifying precisely the location of paradise, and the difficulty of coming up with a plausible altitude for the mountain on which paradise was thought to be situated, were faced by Durandus of Saint-Pourçain. Durandus was teaching theology in Paris in 1312, and is best known for his commentary on Peter Lombard's *Sentences*, in which he concluded that, despite the excessive heat of the equatorial zone, this was where the Garden of Eden was to be found, its climate tempered by altitude.[5] He explained that air cooled with increasing altitude, and that the rarefied air at the level of paradise absorbed less heat. The overheated air at the bottom of the mountain, suggested Durandus, could account for the biblical description of the Cherubim's 'flaming sword, which turned every way' (Genesis 3.24).[6] Moreover, the height of the mountain would have ensured that paradise had remained above the level of the Flood. At the same time, the notion of an excessively high mountain had to be compatible with the Aristotelian model of the regions of the air. This Durandus achieved by reinforcing Alexander of Hales's solution of placing paradise in a tepid zone between the excessively cold middle region and the excessively hot highest region with the comment that in nature there has always to be an intermediate state between any pair of extremes. Were paradise not at such a high altitude, Durandus felt compelled to admit, it would be necessary to reconsider *de novo* the entire problem of its location and even to acknowledge that it could be in a completely different part of the world.[7]

The conception of a high-altitude Garden of Eden climatically independent of its zone illustrates the way fourteenth-century scholars tended to shy away from the exigencies of geographical debate by distancing paradise as far as possible from the earth. As attempts to explain the location of paradise through Aristotelian science crumbled, exegetes turned back to the Bible. By the beginning of the 1320s, the French Franciscan Nicholas of Lyre was at work on what would be a lastingly popular biblical commentary, the multi-volumed *Postillae* (1323–33).[8] Not for nothing was Nicholas known, only half a century after his death, as *Doctor planus et utilis* ('the plain-speaking and useful doctor').[9] Deeply committed to extracting every ounce of the literal sense of Holy Scripture, Nicholas rigorously avoided embracing any speculation not based on the words on the written page. Whereas others, Bacon for example, could believe that God's revelation was to be found in Greek philosophy as well as in the Bible and that mathematical astronomy was essential for the understanding of the Scriptures, Nicholas maintained that theology surpassed all other sciences and that the Bible, which surpassed all other writings, was the only real text for theologians. The Bible proceeded from God and contained nothing false.[10] Complementing his desire to understand clearly, Nicholas aimed at an equal clarity of exposition. To help himself reach the meaning of the biblical text, he went back to the original Hebrew.[11] To communicate his findings

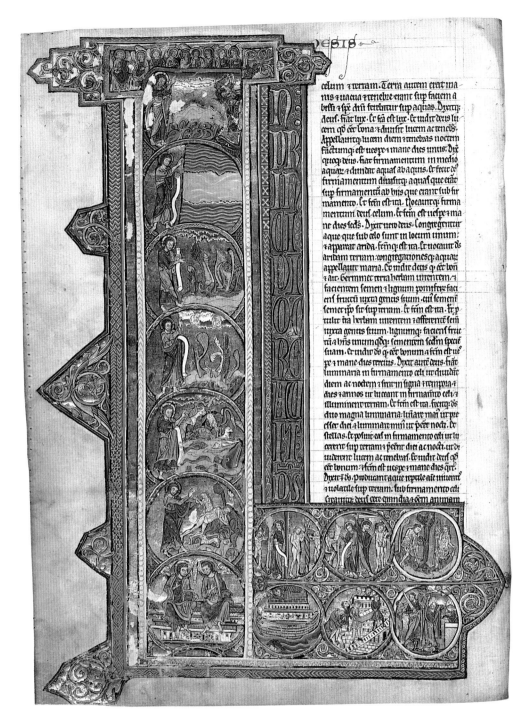

Plate 1. The opening of the Book of Genesis in a thirteenth-century Bible made for Robertus de Bello, Abbot of St Augustine, Canterbury. 27.5 × 20 cm. London, British Library, Burney MS 3, fol. 5v. Contrary to appearances, the incipit is not an 'L', but a giant 'I'. In the vertical panel immediately left of the text are the first four words of Genesis 1.1: 'In principio creavit Deus', which, together with the words along the top, 'celum et terram', mean 'In the beginning God created the heaven and the earth.' The first verses of Genesis are illustrated on the left with seven medallions, corresponding to the week of creation: the division of light from darkness; the division of the waters above and below the firmament; the creation of dry land and vegetation; the creation of the sun, moon and stars; the creation of animals; the creation of man and woman; the Trinitarian God resting on the seventh day. The illuminations on the bottom right account for the earliest events in human history, the paradise story in the upper register and in the lower register Noah's Ark, the construction of the Tower of Babel, and the sacrifice of Abraham.

Plate 2a. Giovanni Leardo, *Mapa Mondi. Figura Mondi*, 1442. Verona, Biblioteca Civica, MS 3119, unfoliated. Detail of the earthly paradise next to India (see Fig 1.1). Instead of representing the earthly paradise as a garden, as described in Genesis, Leardo used the same fortified city sign that he employed elsewhere on the map to represent regions, towns and cities. In fact, there was nothing unusual in Leardo's visualization of the earthly paradise as a city, for this is how it was sometimes described in medieval manuscripts.

Plate 2b. Map of the world from a manuscript of Beatus of Liébana's *In Apocalypsin*. Catalonia, early twelfth-century. 37 × 54 cm. Ministero per i Beni e le Attività Culturali. Turin, Biblioteca Nazionale Universitaria, MS I.II.1 (old D.V.39), fols 45v–46r. At the time of Beatus's map, the authority of Revelation extended over historical and geographical data.

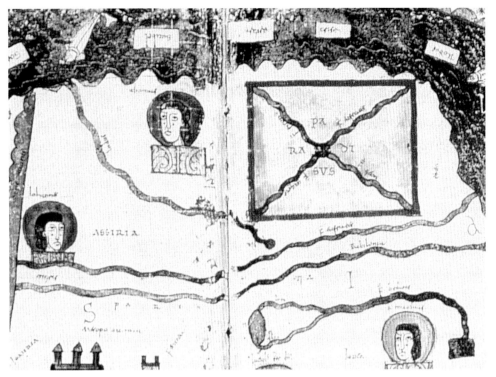

Plate 3a. Map of the world from a manuscript of Beatus of Liébana's *In Apocalypsin*. Shahagún, Navarre, 1086. Burgo de Osma, Archivo de la Catedral, MS 1, fols 34v–35r. Detail of the earthly paradise (see Fig. 5.11). Paradise is shown as a rectangle with the four rivers flowing from the centre of the rectangle to its four corners. The rectangle recalls the spread of Christianity to the four corners of the earth. The rivers Indus, Tigris, Euphrates and Jordan are seen flowing in Asia. Not far from paradise are shown the heads of three apostles: they mark the burial places of Thomas in India, John in Assyria and Matthias in Judaea.

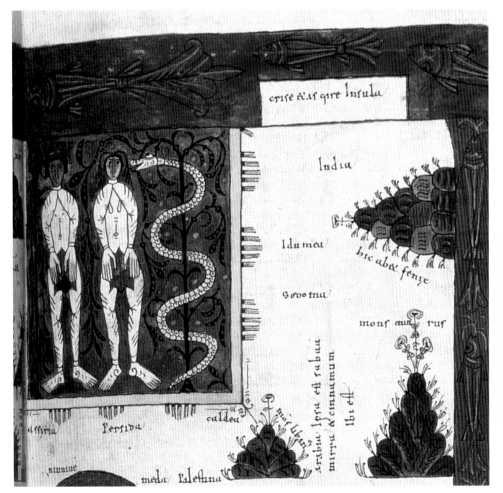

Plate 3b. Beatus of Liébana's map of the world from a manuscript of Beatus of Liébana's *In Apocalypsin*. Monastery of Santo Domingo de Silos, 1109. London, British Library, Add. MS 11695, fols 39v–40r. Detail of the earthly paradise (see Fig. 5.14). Paradise is represented by a vignette showing Adam and Eve in a rectangular Garden of Eden immediately after their act of disobedience. The representation of Adam and Eve on medieval maps highlights the moment of their Fall, regarded as the first 'historical' event and the temporal and moral barrier that, for medieval Christians, barred mankind from re-entering the earthly paradise.

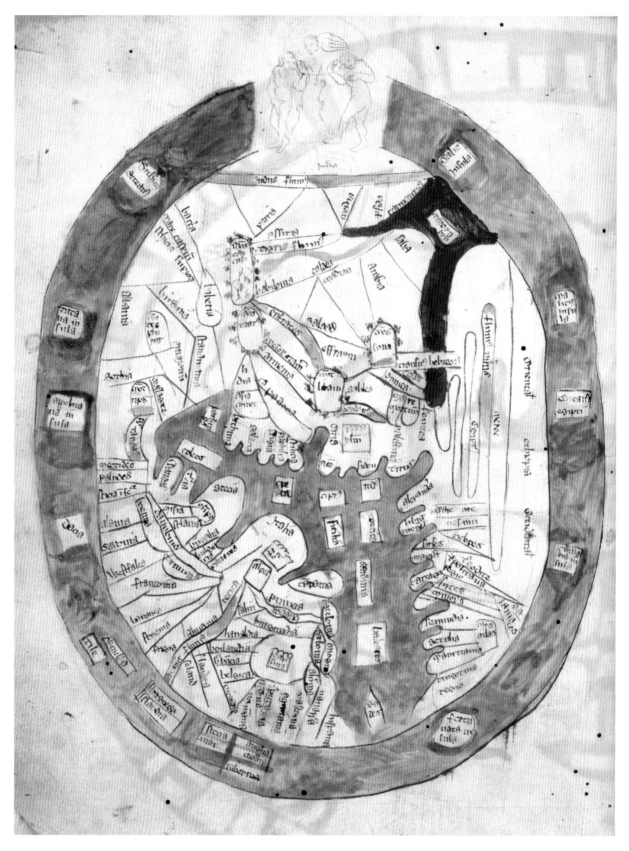

Plate 4. The smaller map of the world from Ranulf Higden's *Polychronicon*. Ramsey, England, c.1350. 37 × 25 cm. London, British Library, Royal MS 14.C.IX, fol 2v. In the second, less detailed version of Higden's world map (see also Fig. 6.3), the vignette of paradise, which shows Adam and Eve flanking the Tree of Knowledge, has been pencilled in next to India, whose border is marked by the river Indus.

Plate 5a. The Evesham world map. Evesham Abbey, Gloucestershire, c.1390–1415. London, College of Arms, Muniment Room 18/19. Detail of the earthly paradise with Adam and Eve eating the forbidden fruit (see Fig. 6.5). Paradise is represented by a large inset in the farthest east. The garden was believed to occupy a different dimension from that of the 'ordinary' world with its known regions.

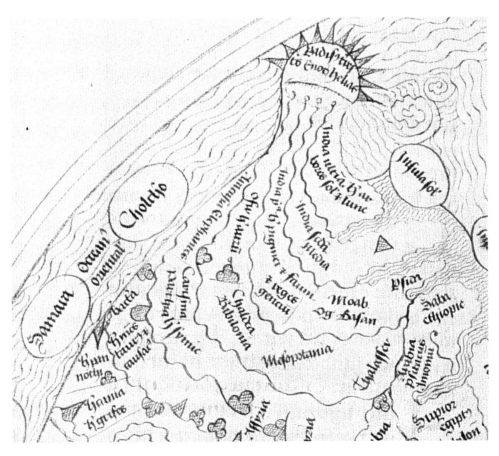

Plate 5b. World map from a manuscript of Lambert of St-Omer's *Liber floridus*. Flanders, second half of the fifteenth century. Genoa, Biblioteca Durazzo Giustiniani, MS A IX 9, fols 67v–68r. Detail of eastern Asia (see Fig. 6.11). The inscription on the island of paradise tells us that the Garden of Eden is the resting place of Enoch and Elijah, who were waiting in the place where human history had begun for the coming of the Antichrist at the very end of time. The map also shows Gog and Magog, confined here to an island, in a corner of northeast Asia, as on certain other maps.

Plate 6. The Anglo-Saxon, or Cotton, world map. Canterbury, England, c.1025–50. London, British Library, Cotton MS Tiberius B.V., fol. 56v. Detail of Asia, part of Europe and the Mediterranean basin (see Fig. 6.7). The map depicts the whole of the inhabitable world, from the Atlantic to the Far East, with east at the top. Included (from left to right) are the cities of Constantinople, Tarsus, Caesarea, Babylon, Jerusalem, Bethlehem and Alexandria. Noah's Ark is shown in Armenia. The Red Sea is shown, prominently coloured in red, as is the Nile. The map does not present an explicit depiction of the earthly paradise, but includes references to Havilah (*Evilath* on the map) and the four rivers.

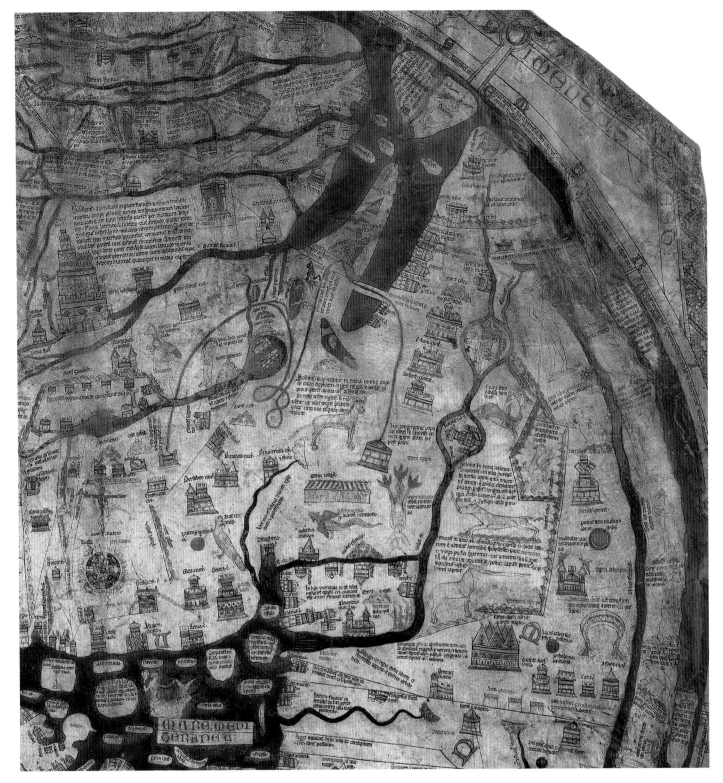

Plate 7. The Hereford world map. Lincoln (?and Hereford), *c*.1300. Hereford Cathedral. By permission of the Dean and Chapter of Hereford and the Hereford Mappa Mundi Trust. Detail of parts of Asia and Africa (see Fig. 6.15). Included (from left to right, starting at the bottom) are Athens, the island of Delos, Jerusalem, the Tower of Babel, the city of Enoch (at the top, in eastern Asia), Jericho, Sodom and Gomorrah in the Dead Sea (Lot's wife is turned into a pillar of salt near Sodom), the path of the wandering Jews, the Red Sea, the 'granaries of Joseph', and, on the border between Asia and Africa, the camp of Alexander the Great (see also Fig. 6.20). Moses is shown on Mount Sinai receiving the tablets of the Law from God and the Israelites are depicted worshipping the golden calf. Other Old Testament references include Noah's Ark in Armenia, the Golden Fleece of the Argonauts in Asia and Minos's labyrinth in Crete.

Plate 8. The Psalter map. London. c.1265. London, British Library, Add. MS 28681, fol. 9r. Detail of Asia and the eastern Mediterranean (see Fig. 6.16). The double ring around paradise emphasizes its inaccessibility. The four rivers are shown both flowing out of Eden and originating from their known sources. The course of the Ganges/Pishon gives the impression that there is a fifth river. A large part of northeast Asia is isolated from the rest by a wall of mountains interrupted only by a closed gate. This is the land of Gog and Magog. To the right of paradise are the Trees of the Sun and the Moon, which whispered to Alexander the Great the prophecy of his forthcoming death.

Plate 9. The Ebstorf world map. ?Ebstorf, Lower Saxony, between 1235 and 1240. 3.56 × 3.58 m. Destroyed in an Allied bombing raid over Hanover on the night of 8/9 October 1943, and now available as a pre-war facsimile (see also Figs 6.17, 6.18). The Ebstorf map is one of the largest and most richly detailed maps known to have been produced in Europe during the Middle Ages. The author of the Ebstorf map has represented the earth within the embrace of a gigantic Christ, whose head is in the east, near paradise, whose feet are in the west, and whose outstretched arms encompass the middle of the earth along a north-south axis. This intriguing construction has been variously interpreted. For instance, is Christ supporting the earth or is the map adapted to the body of Christ, forming a gigantic eucharistic Host?

Plate 10. Illuminated page from a manuscript of Augustine's *De civitate Dei*. Île de France or Normandy, *c.*1473–80. 46.5 × 31 cm. Macon, Bibliothèque Municipale, MS Franç. 2, fol. 19r. The Fall is taking place in the Garden of Eden (in the farthest east of the east-orientated world map in the centre). Outside paradise, Eve is portrayed looking after her two sons, and Adam is shown digging. Cain is seen on the left killing Abel; on the right he is building the first city in the world (Enoch); in the part of the map that would correspond to Europe, he leads a fugitive and nomadic life among wild beasts and demons until he is killed, in the part of the map that would correspond to Africa, by Lamech. Above the map is Christ's Second Coming. Flanking the map on the right is a scene from Eden, with the rivers that are the source of life, animals peacefully grazing, and the creation of Eve. Below, Augustine comments on the Bible before a group of scholars who are quoting from the Holy Scriptures. On the left is a fool with the inscription from Psalm 80.16 [81.15]: 'Inimici Domini mentiti sunt ei' ('The haters of the Lord should have submitted themselves unto him'). At the bottom left of the page, Augustine is responding to a group of philosophers by reasserting the uniqueness of the creation of the world by God, a point confirmed by another figure of Augustine in the historiated initial on the same page.

Plate 11a. World map. From the *Rudimentum noviciorum* (Lübeck: Lucas Brandis, 1475). London, British Library, C3.d.11. Detail of the earthly paradise (see Fig. 8.3). One of the earliest printed maps made for a wide audience, this map was also one of the last examples of the cartographical genre of medieval *mappae mundi*. The earthly paradise is represented in the farthest east, at the top of the map, next to Havilah, India and Taprobana. The Garden is rich in vegetation, enclosed by a wall and inhabited by Enoch and Elijah. The four rivers pour through four arcaded openings in the bastions of the walled enclosure to flow out over all the world.

Plate 11b. World map by Hanns Rüst. Separately printed broadside. Augsburg, *c.*1480. Reproduced as a separate sheet by Ludwig Rosenthal's Antiquariat (Munich, 1924). London, British Library, Maps 856(5). Detail of the earthly paradise (see Fig. 8.4). By the end of the fifteenth century the cartographical genre of medieval *mappae mundi* had already begun to destabilize, but a number of world maps continued to show the earthly paradise in its traditional location in the farthest east. On Rüst's map, paradise is represented as a hexagonal walled garden within which Adam and Eve stand by the Tree of Knowledge. The round-faced eastern windhead overlooks the garden. The land of Gog and Magog is shown not far from paradise. The four rivers flow through arches in the enclosing, towered wall to give life to the whole earth, as well as shape and structure to the map.

Plate 12a. Vesconte Maggiolo, map of Europe, Africa and part of Asia, first chart of his *Atlante nautico*. Naples, 1512. Size of the original 48 × 34 cm. Parma, Biblioteca palatina, MS 1614. Detail of Africa. Paradise is located on the Mountains of the Moon, from which flows the river Nile, believed to be the biblical Gihon. It is represented as a fortified castle and is recognisible not through an inscription, but by means of the image of the Tree of the Knowledge of Good and Evil represented above it.

Plate 12b. World map by Fra Mauro. Venice, *c*.1450. Venice, Biblioteca Nazionale Marciana. Vignette of the earthly paradise (see Fig. 8.25). In representing the earthly paradise in one of the eastern corners as a circular garden lying outside the inhabited earth, Fra Mauro was adhering to Christian tradition. Paradise is not placed anywhere on the map of the inhabited earth, because it is located in an inaccessible beyond, but the ocean surrounding both paradise and the whole earth and the four rivers connecting the otherwise inaccessible paradise with the known world confirm its earthly existence.

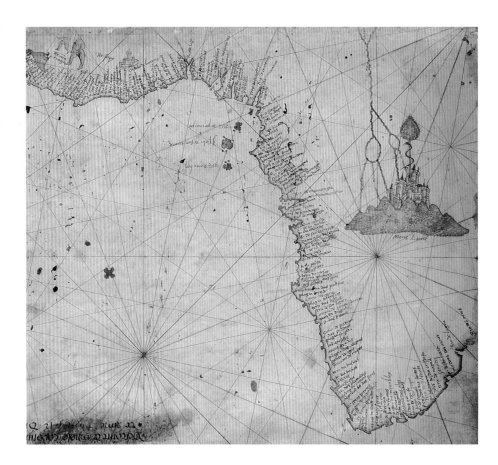

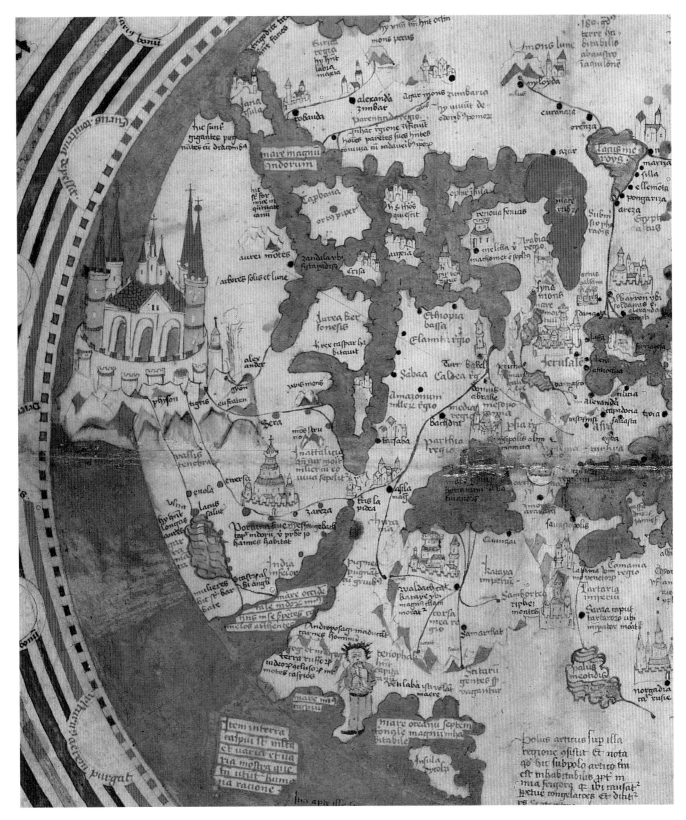

Plate 13. World map by Andreas Walsperger. Constance, southern Germany, 1448. Vatican City, Biblioteca Apostolica Vaticana, MS Pal. Lat. 1362b, unfoliated. Detail of Asia and the earthly paradise (see Fig. 8.24). Paradise is depicted as a fortified castle in the farthest east, out of which flow the four rivers. The castle appears to be not far from the Mongol Empire, but the Trees of the Sun and the Moon represented nearby indicate that paradise is east of the easternmost margin of Asia, and thus well beyond the inhabited and known part of the earth. According to medieval legend, the Trees of the Sun and the Moon that informed Alexander the Great of his imminent death grew where the sun and moon rose, thus marking the limits of the known world. Paradise is separate from the ordinary world, hidden beyond the impenetrable barriers of a mysterious castle, which is guarded by towered walls. At the same time, the connection with the inhabited earth is assured by the four rivers.

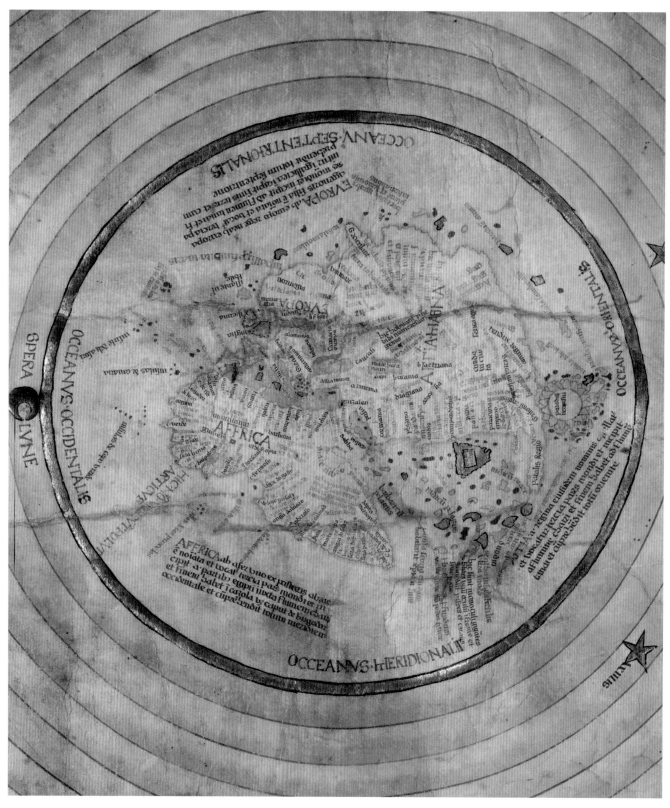

Plate 14. World map accompanying the nautical chart formerly attribted to Christopher Columbus. ?Genoa, c.1492. Diameter 20cm. Paris, Bibliothèque Nationale de France, Cartes et Planes, Rés. Ge AA 562 (see Fig. 8.12). The map is orientated to the north and features paradise as an island opposite Cathay, in the farthest east. Iceland in the north and the African regions south of the equator are described as inhabitable. Jerusalem is at the centre of the map. The Indian Ocean is portrayed, against Ptolemaic tradition, as open to the south.

Plate 15. The Garden of Eden, in *Biblia, das ist, die gantze Heilige Schrifft Deudsch* (Wittenberg: Hans Lufft, 1536), I, adjacent to Genesis 1–2. 22 × 14.5 cm. London, British Library, I.b.9. This woodcut, devised for the first complete edition of the Lutheran Bible, features the perfect and beautiful world that was later corrupted by the Fall and transformed by the Flood. Paradise, which once was but which no longer exists, is a large part of this pristine and perfect world. God the Father gives his blessing to his own creation. The earth, on which we see the first human couple, is surrounded by the zone of air, with its birds, clouds, moon, sun, and stars, and then by that of the waters above the heaven, as described in Genesis 1.6–8.

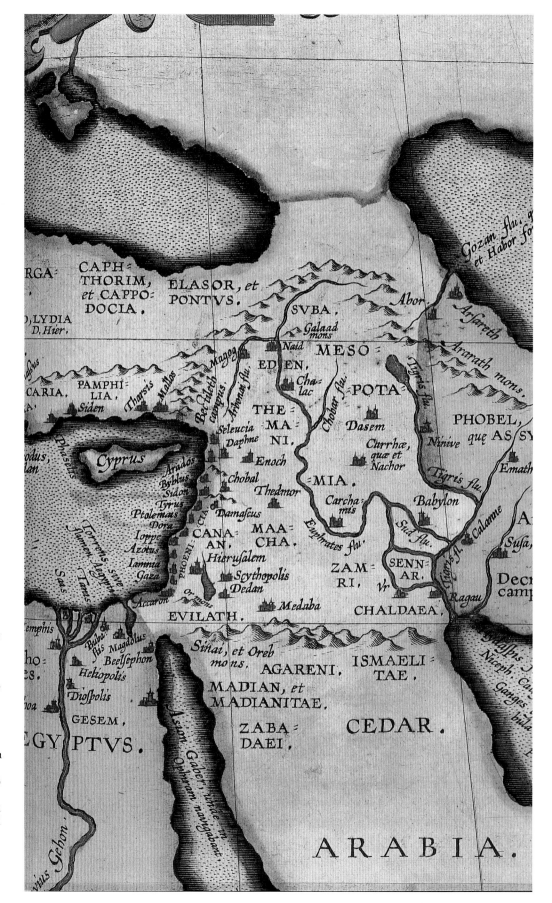

Plate 16. Abraham Ortelius's *Geographia sacra*, a map of Eden in Armenia, was published in the *Theatrum orbis terrarum* (Antwerp: ex officina Plantiniana, Ioannes Moretus, 1601), after sig. 1. Size of the original 35 × 48 cm. Detail of the Middle East. Eden is rather vaguely marked in Armenia, where the Tigris and Euphrates have their sources. Although Ortelius shows the Gihon as the Nile, he identifies the Pishon both in Europe (as the Danube) and in Mesopotamia (as the Hydaspis) even while referring as well, on the map, to the traditional notion that the Pishon is the Ganges and showing the land of Havilah (*Evilath* on the map), which the Bible describes as encompassed by the Pishon, in the Middle East near Mount Sinai.

to a wide readership, he illustrated his exegesis liberally with drawings, plans and maps (but not a map of paradise), most of which gave the Jewish as well as the Christian interpretation. His lucid prose contains neither irrelevant theological or philosophical digression nor any hasty judgement.[12]

Nicholas's exegetical principles are seen at work in his comments on paradise. Noting that paradise was said in the Vulgate to have been planted *a principio* (Genesis 2.8), he explained that *a principio* (from the beginning) actually meant on the third day, when God created the plants. He rejected Augustine's allegorical view of the week of creation as too remote from the text. Referring to the original Hebrew term, מִקֶּדֶם, *miqedem* (that could also mean 'in the east' as well as 'from the beginning'), he said that the eastern location of paradise was confirmed by Aristotle's notion of the orient as on the right-hand side of the heavens and was unanimously supported by Christian doctors, Greek as well as Latin. As regarded the actual location of paradise, though, he had to admit that this remained uncertain. He discussed the two principal theories: that paradise was at the equator and that paradise was in the temperate zone in either the northern or the southern hemisphere. He started by citing Aristotle's insistence on the uninhabitability of the equatorial zone and the nobility of the southern hemisphere, then turned to John Pecham as an example of the opposite view, that the equatorial region was the most temperate region and the southern hemisphere had to be uninhabitable because of the eccentricity of the sun's orbit.[13] Nicholas's characteristically even-handed approach was deployed in his explanation of the flaming sword of the Cherubim. If, he said, paradise was in the equatorial region, the flaming sword should be interpreted as the torrid zone that separated paradise from the inhabited world. In this case, the explanation of *versatilis* (Genesis 3.24: 'which turned every way') would be that the sun, source of the impenetrable heat, was nearer to the earth in summer and further away in winter. If, however, paradise was not at the equator, the flaming sword should be interpreted as referring to flames of real fire, shaped like swords, that rose from the ground just as they do on Mount Etna, in Sicily.[14] There are hints in the *Postillae* that the eastern paradise was probably in the northern hemisphere (possibly not far from India). Like references to an Indian location for gold-rich Havilah, such hints are likely to have come from Bede and the *Glossa*.[15]

Nicholas of Lyre's interest was focused exclusively on the text of the Bible. He did not intend to put forward new doctrines, even though some of his analysis was novel as a consequence of his distillation of the work of other scholars. What he wanted to offer was a straightforward and consistent exposition of the meaning of that text, with all attention given to the literal sense of Scripture. So, for Nicholas, paradise was a real region on earth, and most probably in the east. From start to finish, the *Postillae* testify to Nicholas's determination to protect both the literal and the historical meanings of Holy Scripture from the manipulation of human reason. His renewed appeal to the word on the page paved the way for a fideistic acceptance – by faith and not by reason – of Christian teaching. Revelation would come from Scripture alone.

Paradise East of Anywhere

An early fourteenth-century thinker who left a particularly profound mark on the later Middle Ages and the early Renaissance was John Duns Scotus.[16] Duns Scotus placed great emphasis on the impossibility of discovering the location of the Garden of Eden and on the miraculous aspect of its geography. Paradise was indeed part of the earth, he believed, but mathematical astronomy had failed to determine its precise location. Duns Scotus's treatment of the subject put paradise firmly back into the court of faith and biblical exegesis. His commentary on Peter Lombard's *Sentences* included a sharply worded critique of the thirteenth-century debate on the Garden of Eden.[17] Highlighting the weakness of opinions based on Aristotelian science through reference to both theological argument and geographical evidence, he revealed the unscientific character of some of the most commonly held assumptions about the earthly paradise.[18] Unlike Thomas Aquinas, who thought that natural philosophers and theologians would eventually reach a common conclusion even when they came from different directions, Duns Scotus insisted that philosophers inevitably arrived at unacceptable results because they shut themselves up in a purely material world and claimed that nature was perfect, whereas 'theologians, by contrast, are aware of the deficiency of nature and of the necessity of grace and of supernatural perfections.'[19] The path to certainty about the location of the Garden of Eden could not possibly lie in natural philosophy, for without God's aid human intelligence was incapable of grasping the mystery of paradise. The mystery could be approached only through faith and by bowing to the authority of Holy Scripture.

Whenever geographical arguments served his purpose, however, Duns Scotus did not hesitate to deploy them to challenge the competence of contemporary geographical science to deal with the question of paradise, and he was quick to put his finger on the weak spots of established theories about paradise. A fundamental flaw, he pointed out, was that 'east' was a relative notion depending on the observer's position:

> *As for those who say that paradise is in the east, I say that wherever it is, it is possible to understand it as in the east. With the exception of the two poles, which are fixed, any point on earth may be understood to be in the east in relation to the heavens, or with regard to different places on earth.*[20]

The eastern location of paradise had been seen for centuries as having been sanctioned by the word of God, and as one of the few absolutes in the Genesis story. Yet Duns Scotus realized that there is nothing ultimately 'absolute' about 'east'. He stated that paradise could be situated anywhere: 'Therefore, by saying that paradise is in the east, one does not have a certain and precise determination of the region of the earth in which paradise is located.'[21] The equatorial paradise received similar treatment. Duns Scotus noted that the argument that the climate at the equator was temperate because of the speed of the sun in its course lacked solidity, which was why supporters of the equatorial paradise hid behind the authority of Ptolemy and Avicenna. In fact, he countered, since movement caused heat, it was precisely the speed of the sun revolving around the centre of the earth that produced the greatest heat along the equator, from which the sun never migrated far either north or south. He was no more inclined to accept theories about the high altitude of Eden. He dismissed the belief that paradise could reach the moon (the sphere of fire was too hot) or that it could be in the middle region of the air (the cold was excessive). He was prepared to concede that there could be particular local conditions that favoured a temperate climate, but insisted that an excessively high altitude would be unsuitable for habitation because of the thinness of the air.[22]

The real target of Duns Scotus's objections, however, was the heart of all the usual arguments for a high paradise, namely the idea that the Flood had failed to reach the top of the mountain.[23] For Duns Scotus, the Flood was a miraculous act of God, and should not be explained by an appeal to natural laws. God's all-transcending power allowed him to disregard all natural laws. The miraculous inundation of the earth was intended as a punishment for sinners. Where there were no sinners, no miracle was necessary. Hence paradise had remained untouched by the disaster and it was irrelevant to speculate about its altitude. Those who went against Scripture and the Church's teaching to argue that the Flood was a natural phenomenon brought about by some astral conjunction were mistaken.[24] Scientific speculation about the natural world should not be used either to question or to support the veracity of the Bible. There was no need to see, for example, the torrid zone in the flaming sword of the Cherubim.[25]

With arguments such as these, Duns Scotus was at pains to distance himself from the thirteenth-century enthusiasm for the new learning and that century's optimistic belief in the capacity of natural philosophy to confirm theological teaching. In his view, the whole edifice of theology had to rest squarely on the Bible. The theologian's calling was to understand and explain the literal meaning of the Bible. To these ends, it was legitimate, for Duns Scotus, to combine scriptural exegesis with other disciplines. Revealed knowledge would always remain superior to human reasoning, and Christian belief did not need the validation of empirical evidence. For Duns Scotus it was not at all surprising that the earthly paradise eluded all rational attempt to locate it; faith alone would bridge the gulf between human knowledge and God's plan as revealed in Holy Scripture. The only traditional notion associated with the geography of paradise that Duns Scotus was prepared to accept was that the climate in Eden was exceptionally temperate.[26] He was unwilling either to confirm or deny that paradise was on a high mountain or that it was on the equator. Paradise could be east of anywhere.

A Humanist Search for Paradise

The fifteenth century thus inherited a bundle of unresolved problems from the medieval debate on the geography of paradise. The writings of the French cardinal Pierre d'Ailly serve to illustrate how traditional geographical lore about paradise was passed on in the new century. D'Ailly's account of the world, *Ymago mundi,* contained no original insights and his arguments on paradise mostly rehearse ideas already at least a century old. He marshalled authorities such as Isidore of Seville, John of Damascus, Bede, Strabo and Peter Comestor in support of the idea that paradise was 'a very pleasant place, somewhere in the east, separated from our inhabited regions by a long tract of land and sea'.[27] He took from Alexander of Hales the view that the mountain on which paradise stood reached as high as the upper part of the middle region of the air.[28] He reported an old story about the four rivers of paradise that, falling from the high mountain into a large lake with a great roar, resulted in the local inhabitants being born deaf.[29] He quoted Roger Bacon almost verbatim (without acknowledgement) regarding the inhabitability of the equatorial regions and the southern hemisphere.[30] In repeating the standard identification of the four rivers and the notion of their underground courses, d'Ailly added that, given that paradise was in the east, it was likely that the origins of the Nile lay in the east, not the west.[31] The possibility that the Nile rose from Mount Atlas, in west Africa had been contemplated by Orosius.[32] Finally, in the *Compendium cosmographiae,* as well as in *Ymago mundi,* he repeated Bacon's argument that the equatorial regions were likely to be temperate, although this did not mean that they were necessarily the *most* temperate place on earth and that it was far from certain that paradise was located there.[33] The maps inserted into the printed edition of his work, however, did not feature the Garden of Eden at all.

Whereas Pierre d'Ailly repeated ideas already in circulation, Aeneas Sylvius Piccolomini (who ruled as Pope Pius II, 1458–64) questioned the established view on the earthly paradise. Piccolomini's ideas are fully developed only in the unfinished *Dialogus de somnio quodam* (*Dialogue about a Dream*), a work he wrote immediately after the fall of Constantinople in 1453 as 'a reprimand for the lack of action on the part of the Christian world and an exhortation to take revenge for the outrage'.[34] The *Dialogus* was cast along the lines of Dante's *Divina Commedia* and as a dream fantasy. Piccolomini imagined that he, as secretary to Holy Roman Emperor Frederick III, together with his friend Pietro da Noceto, as secretary to Pope Nicholas V, were being conducted by Saint Bernardino to a meeting in the earthly paradise that was to be chaired by Emperor Constantine's soul. The main item on the agenda was the fall of Constantinople, but the imaginary trip provided Piccolomini with the opportunity to discuss all sorts of matters along the way, theology, philosophy, history, political science and, in particular, where was paradise. Instead of employing the scholastic method of systematically refuting each objection to belief in an earthly paradise, Piccolomini dramatized his discussion in the classical format of a dialogue. On their way to Eden, the character Aeneas Sylvius confessed his doubts to Saint Bernardino, who dispelled them by explaining, amongst other things, that the assertion that the earthly paradise reached the moon was intended as a metaphor. On arrival in paradise, Piccolomini described what he saw: a wonderful garden with luxurious vegetation. From the way the shadows fell on the ground, the character Aeneas Sylvius and Pietro realized they were still in the northern hemisphere.[35]

Responding to their surprise at this discovery, Saint Bernardino went through the various theories about the location of paradise. Finally he confirmed that paradise was neither at the equator nor in the southern hemisphere: 'Who, indeed', he remarked, 'would believe that the most temperate place is found under the hottest sun?'[36] The climates in the southern hemisphere were presumed to be identical to those of the northern hemisphere, whereas paradise was thought to enjoy a climate unlike that of any other on earth. It was also far more appropriate for paradise to be on the right side of east in the northern part of the earth, than on the left side of east.[37] The little group, Saint Bernardino informed his companions, was now in paradise, somewhere in an unknown eastern region that Ptolemy had situated far beyond China. He also reminded them that Adam had been expelled from paradise to the Damascene Fields, in the northern hemisphere (where he had been created), and that the whole point of placing a guard at the entrance of paradise – the Cherubim – was that paradise was adjacent to the habitable portion of the earth, which meant that it was in the northern hemisphere. Moreover, the inhabited earth was a gift from God to mankind. It was improbable that the Creator would have given another, southern, habitable region to a quite different human race.[38] In Piccolomini's dialogue, the character Aeneas Sylvius then takes up the discourse to explain why Albert the Great's views about the southern Eden and the flooding of the Nile were 'ridiculous'.[39] He said that the Nile's summer floods were the result of melting snow in the mountains of Lower Mauritania, which was where Solinus said the river began.[40] Even were Solinus to be disregarded, the character Aeneas Sylvius continued, the fact remained that the three other rivers of paradise flowed in the northern hemisphere and from north to south. Saint Bernardino closed the conversation by remarking that Albert the Great had been too preoccupied with his other studies to go more deeply into the question; after all, even the incomparable Homer had sometimes nodded off.[41]

Piccolomini reflected on the paradise question as a humanist. He was anxious to revive an idea originally promoted by Justin, Tertullian and Clement of Alexandria that had become entrenched in the patristic literature before being taken up by Isidore of Seville. The idea was that when the classical poets and philosophers described the Elysian Fields they had in fact been registering an imperfect knowledge of biblical Eden.[42] In Piccolomini's *Dialogus*, Saint Bernardino, Aeneas Sylvius and Pietro da Noceto encounter Enoch, Elijah and the Apostle John in Eden together with Cheremon and his wife, who have been brought to the Garden by an angel of God to protect them from the persecutions of the Roman emperor Decius.[43] The character Aeneas Sylvius then asks to meet both Romulus, one of the founders of Rome, and his namesake, progenitor of the Roman nation, because it was said that their bodies had mysteriously disappeared and that they had been worshipped as gods. John the Apostle informs him, however, that Romulus has been killed in a plot by the Senate, the Trojan Aeneas has drowned in a river, and both souls have gone to the underworld.[44]

Like his contemporaries, Piccolomini was conscious of the radical difference between the earthly paradise and the rest of the world. The separateness of paradise from the inhabited earth emerges strongly from Piccolomini's account of the journey to paradise. He made Saint Bernardino tell the character Aeneas Sylvius that nothing on earth could equal the marvels of the Garden of Eden and that these were beyond description.[45] He turned Bede's assertion that Eden was separated from the inhabited regions by a long tract of sea and land, and that it was on an exceedingly high and inaccessible peak, into a narrative. When the character Aeneas Sylvius and his friends found their way blocked by a 'very steep mountain of incredible altitude, whose peak extends beyond the clouds', Saint Bernardino worked a miracle with his girdle and they were all instantly whisked up onto the summit of the mountain.[46] Once up there, they still had to cross great expanses of open plain, desert and wood before they saw 'a very large and deep mass of water, which was either a river or a sea, whose other shore was invisible'.[47] Once again, the pilgrims needed a miracle if they were to progress any farther. On this occasion Saint Bernardino had only to place his mantle on the water for it to be used as a boat. In a trice they were on the other shore, where they found themselves in a land 'far more beautiful than any before'.[48] Still they had to walk on and on, until eventually they saw the strongly fortified wall of paradise, entirely surrounded by a moat. A fearsome angel with flaming eyes stood on the drawbridge leading to the single entrance, brandishing a sword. Only by divine permission and 'the decree of the supreme ruler' were Piccolomini's pilgrims allowed to enter Eden.[49]

Piccolomini searched for the truth restlessly, and through the characters of the *Dialogus* he was expressing his own critical opinions. When the character Aeneas Sylvius admitted his confusion in face of contradictions in the ancient authorities as to the habitability of the equator, Saint Bernardino urged in reply that 'one should seek the truth and not agreement' and scrutinize every written statement, even those of highly respected authors such as Ptolemy and Albert the Great.[50] Only the Bible had its own authority.[51] And in fact, through his focus on the problem of paradise, the ever-questioning Piccolomini became the first Western scholar since the patristic age to doubt the identification of the Gihon with the Nile, an identification that went back to Flavius Josephus in the first century AD.[52] Writing to Gregory Heimburg from Vienna on 31 January 1449, Piccolomini, who was then bishop of Trieste, reported that he had been discussing the Nile and the rivers of paradise and failed to see 'how it is possible for this river to issue from the mountain of paradise'.[53] The fact that the Nile was supposed to rise in Mauritania, in the west, or that it had an unknown source elsewhere in Africa, led him to doubt that the Nile was the river Gihon from Genesis 2.13. Piccolomini's letter reveals how he took for granted only what was absolutely unambiguous in the biblical text, in this particular case the statements that the Tigris and Euphrates flowed out of paradise and that the latter was an earthly region. In contrast, the identification of the Gihon with the Nile was open to investigation.

A fruitful comparison can be made between the literary *Dialogus de somnio quodam* and another of Piccolomini's works, the unfinished *Cosmographia* (never released as a complete work in Piccolomini's lifetime, and first printed in 1477). The *Cosmographia* is a geographical work in two parts, one on Europe (which Piccolomini compiled in 1458 but began to rewrite after 1461), and the other on Asia (compiled in 1461).[54] Both parts reflect Piccolomini's intention to provide a geographical and historical framework for the extraordinary happenings of his own time – the fall of Constantinople, above all – that were tainting relationships between Europe and Asia. He amalgamated geographical material, which he took mostly from ancient authors such as Solinus, Pliny the Elder, Ptolemy, Strabo, and the classical poets as well as the Church Fathers, with contemporary information and with personal observation.[55] In the preface, he noted that his aim was to offer a complete, systematic and critical description of the earth and its peoples, from east to west, avoiding affirming false things as true, for he knew 'that nothing is so contrary to history as a lie'.[56] For this reason, he dismissed the account of the Hyperboreans as a fairy tale, and that of the Golden Fleece as a 'made up story'.[57]

Piccolomini could not, however, dismiss the Genesis account of paradise quite so easily. After all, he was now pope. In the section of the *Cosmographia* devoted to Asia, he discussed the habitability of the equatorial regions, concluding that they might indeed be habitable, but that it was also known that there were extensive deserts along the equator. Paradise could not be an equatorial region, anyway, because Holy Scripture described the Tigris and the Euphrates as rivers of paradise, and it was well known that these rivers flowed from north to south through the northern hemisphere, far away from the equator (and, as we have seen, Piccolomini did not think that the Nile flowed out of paradise).[58] He said nothing specific in this work about a southern paradise, only that the same arguments he had used against an equatorial paradise applied equally to a southern paradise. He offered no suggestions for another location of the Garden of Eden, and did not even mention it when describing the easternmost limits of Asia.[59] Nor did he refer to the usual arguments for the underground courses of the four rivers, or specifically associate the Ganges with the biblical Pishon.[60] He could hardly discuss the Nile at any length since the *Cosmographia*, as he left it, did not cover Africa. He did volunteer some remarks on the geographical origin of mankind, but without relating them directly to a particular, literal paradise.[61]

Like the *Dialogus de somnio quodam*, Piccolomini's geographical work was left unfinished. The description of Asia is incomplete and the description of Africa was never begun. We may doubt that it was simply Piccolomini's death in Ancona on 14 August 1464 that prevented him from completing the work. There is reason to believe that he gave up his ambition to produce an exhaustive description of the entire earth because he found the task impossible. He could not decide between the disagreeing opinions of the ancients or resolve their conflicts with the Christian tradition.[62] It may be significant that Piccolomini chose to discuss the problem of the location of the earthly paradise in his fictional *Dialogus de somnio quodam* while avoiding the matter in the *Cosmographia*. It may even be speculated that the geographical conundrum of the Garden of Eden was one of the intractable difficulties that prevented him from completing the *Cosmographia*. Whatever the reason, Piccolomini's writings testify to the tension that was emerging by the second half of the fifteenth century between different ways of combining the tenets of Christian faith with the data of geographical science. Discussing paradise at length in the *Dialogus de somnio quodam*, Piccolomini maintained that 'mortals do not know all things that happen on earth' and that 'the inscrutable depth of divine providence produces wonderful effects when it wants and in the way it wants'.[63] His aim in the *Cosmographia* was a purely geographical description, and paradise was accorded only a few lines. Just as Piccolomini left both texts unfinished, so the geographical problem of paradise was also left open.

Paradise and Cartography in Flux

By the beginning of the fifteenth century, the exegetical stage was set for crucial changes in the thinking about paradise. There were also major changes in cartography.[64] Various factors were involved. One was the appearance in the Mediterranean world, by the beginning of the thirteenth century at the latest, of an entirely new kind of map, the nautical chart, which was for use in navigation but which was also often copied by those compiling world maps. Another was the arrival in 1406 in Venice, from

Constantinople, of a Byzantine manuscript of Ptolemy's *Geography* (often known until the mid-1500s as *Cosmographia*), which Jacopo Angelo had turned from Greek into Latin by the next year. By about 1415 there was not only a Latin translation of the text, but also copies of the 27 maps that went with it.[65] The maps, drawn on geometrical projections, were unlike anything seen before in the medieval West. A third factor was the flood of new information about distant parts of the world, most especially about Asia and the Far East, reaching Western Europe and altering the way those regions were perceived and portrayed on maps. A final factor was the shift, which had been taking place since the thirteenth century, of the centre of gravity of book and map production away from the predominantly rural monastic *scriptoria* to the university towns and mercantile cities. A much more broadly based system of book production, involving guilds of lay and ecclesiastical scribes and artists had, by the end of the fourteenth century, challenged the traditional monopoly of the monasteries.[66] The single most far-reaching cartographical result of the interplay of all these factors was that the traditional intimate connection between the historical dimension and geographical description was gradually lost from fourteenth- and fifteenth-century world maps. Now the *mappae mundi* (and not only the zonal maps) were reorientated, and east was no longer inevitably at the top of the map. Their historico-geographical comprehensiveness was slowly being reduced to the purely geographical dimension, with an emphasis on contemporary topography. Yet, paradise continued to be shown on the new maps, although in a now different cartographical context.

One of the most important factors behind the transformations in fourteenth- and fifteenth-century mapping of the world was the influence of the nautical chart. Nautical charts, maps of a quite different genre, were made for practical use, to guide navigation within the enclosed or nearly enclosed waters of the Mediterranean and Black seas (or perhaps to show the navigational skill of their owners), and, as Mediterranean sailors ventured with increasing regularity into the Atlantic, along the oceanic shores of Europe and Africa. From a purely geographical point of view, the charts portrayed these coastlines with remarkable accuracy, together with the names of major ports and prominent coastal features. Ruled lines (rhumbs) radiated from wind roses to provide the means of working out the shortest compass course between different places. Prominent in each set of rhumbs were those indicating the north–south and east–west axes. Few of the charts that were used on board ship have survived; the majority of extant nautical charts represent those that, richly embellished, were made to enhance the collections of kings and wealthy patrons.[67]

The earliest surviving example of a nautical chart is the so-called Carte Pisane, thought to have been created between 1275 and 1291, but the majority of surviving charts – mostly ornate copies made for presentation – date from the fifteenth century.[68] It seems, though, that charts had sometimes been used in the fourteenth century in the compilation of *mappae mundi*, as indicated by the delineation of the Mediterranean coastline. The earliest known world maps to show the influence of a nautical chart in this way are thought to be those constructed about 1320 by Pietro Vesconte.[69] Vesconte's maps have survived in a copy of the *Liber secretorum fidelium crucis*, the treatise written by Marino Sanudo calling for a new crusade to recover the Holy Land, and in a world history written by the Minorite friar, Paolino of Venice.[70] A map in the British Library's copy of Sanudo's *Liber secretorum fidelium crucis* shows the characteristic criss-crossing lines of a nautical chart (Figure 8.1). The map also shows Jerusalem in the centre, Gog and Magog in eastern Asia, the kingdom of Prester John – the powerful priest thought to rule a vast Christian empire beyond Islam – not far from China, and the seat of the Great Khan.[71] There is no explicit indication on this particular map, however, of the Garden of Eden, even though the biblical Gihon is marked as flowing in India. Vesconte's rhumb lines cover the entire map to lock, as it were, every part of the

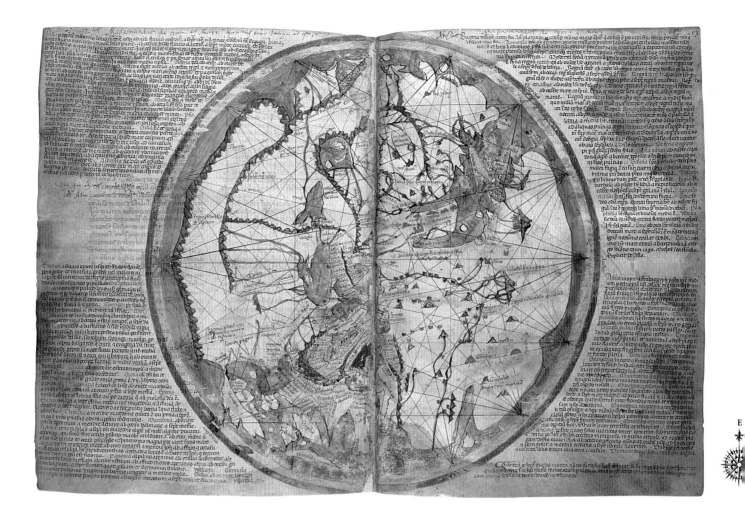

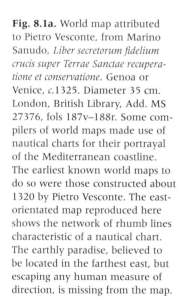

Fig. 8.1a. World map attributed to Pietro Vesconte, from Marino Sanudo, *Liber secretorum fidelium crucis super Terrae Sanctae recuperatione et conservatione*. Genoa or Venice, c.1325. Diameter 35 cm. London, British Library, Add. MS 27376, fols 187v–188r. Some compilers of world maps made use of nautical charts for their portrayal of the Mediterranean coastline. The earliest known world maps to do so were those constructed about 1320 by Pietro Vesconte. The east-orientated map reproduced here shows the network of rhumb lines characteristic of a nautical chart. The earthly paradise, believed to be located in the farthest east, but escaping any human measure of direction, is missing from the map.

known earth into a network of measurable directions. It would have seemed to him more difficult or inappropriate to indicate on this sort of map the earthly paradise, by definition beyond any human measure of direction. The transfer of the Mediterranean coastline from nautical chart to world map, it becomes clear, implied far more than cartographical refinement. It intimated a different approach to the representation of space. Measurable distance and direction were essential for a sailor who needed to work out his route from the map. The insertion of coastal outlines prepared for navigation or to celebrate the skill of sailors into a topologically structured *mappa mundi* announced a critical break. Something was changing in the perception of space. Henceforth, the mapping of the world was becoming a matter of increasingly accurate measurement of angles and directions and, eventually, distance.[72] The plotting of places according to contiguity, that is without taking into account their measured interrelationship, was losing ground as a way of representing the world cartographically. For example, as nautical charts themselves came to include scale bars, so too did maps of the world, even if sometimes the scale bar might have had only a symbolic and not practical meaning.

New geographical information prompted other changes to *mappae mundi*. Before the fourteenth century, world maps had been essentially historical representations, referring to the six ages and only partially mirroring contemporary features inasmuch as these related to the present state of the Christian world. Their portrayal of lands in southern Africa and eastern Asia had been taken mainly from ancient sources. European travel in, and trade contacts with, the East, were by no means completely blocked during the period of Islamic dominance in central Asia, and it would not have

Fig. 8.1b. Diagram of Fig. 8.1a.

been impossible to include much more about the remote lands than was actually shown on European maps.[73] After the fourteenth century, however, maritime cartography offered the compilers of *mappae mundi* an alternative perspective. Given the function of their maps, it is unsurprising that the chart makers represented the Asian and African parts of the known world as contemporary and not historical. Some aspects of their maps were still based on biblical and classical sources, but much contemporary information was now being taken from other sources, such as Arabic maps and works of art, Latin translations of astronomical/geographical tables and, above all, travellers' accounts. Just as the view of an eastern Asia of the past fitted the medieval vision of history and accommodated the geographical inventories of classical Antiquity, so a sharper interest in contemporary non-Christian lands was appropriate in the context of the fourteenth-century renewal of the Christian Church's missionary efforts in the face of the expansion of Islam, a renewal amply documented in the works of Marino Sanudo, Pietro Vesconte and Fra Paolino, for example.[74]

After the Mongolian invasion of Central Asia and the establishment of Mongolian rule, sufficient security returned to the region to allow the full renewal of contacts with Europe. For a century or so, from about 1250 to 1350, certain European travellers, mostly diplomats, traders and missionaries, were allowed to journey through Persia to India and China. Some of these travellers recorded their experiences. Sometimes their descriptions were used to correct long-established errors. The report from the Franciscan William of Rubruck that the Caspian Sea was not open to the ocean, for example, was based on his personal experiences between 1253 and 1255, as

The Twilight of Paradise on Maps 201

related in his *Itinerarium*.[75] By far the greater part of the general information about Mongolian Asia that would appear on maps a century and more later came from Marco Polo (*Il Milione* or *Le Divisament du monde*, c.1298–9), John of Monte Corvino (*Epistolae*, 1305–6) and Odoric of Pordenone (*Relatio*, 1330), who had visited Sumatra, Java and Borneo in addition to China and India between 1318 and 1330.[76] Material from the writings of John of Plano Carpini, the Franciscan provincial minister of Saxony, who compiled his *Historia Mongalorum* in 1247, and the Dominican Simon of St-Quentin (*Historia Tartarorum*, c.1248) was included in the third part of Vincent of Beauvais's *Speculum maius* (final version c.1260), the *Speculum historiale*. Through Vincent's popularity, Carpini's text enjoyed a far wider circulation than it might otherwise have done.[77] All this new material filtered through to European map makers only slowly. After the collapse of the Mongolian Empire and the rise of the Ottoman Turks, late fourteenth-century map compilers began to borrow from sources made available in the previous century, and in relation to the thirteenth-century *mappae mundi*, their creations represented a relative advance in the state of knowledge about contemporary central Asia. Fifteenth-century map makers were able to draw on sufficient new information to be encouraged to portray non-European lands as regions of the present, not of the past, and no longer just as the stage for a display of Christian universal history.[78]

Of all the characteristics of fifteenth-century world maps, by far the most significant was their reorientation. The dissemination of the new north-orientated Ptolemaic maps weakened the privileging of east found on earlier *mappae mundi*.[79] Once east was no longer at the top of the world map, the visual impact of the east–west progression of history was lost. The constituent elements of that history – Rome, Greece/Macedonia, Mesopotamia, Persia – were all still to be found on the maps, but represented as contemporary regions, not as the four great empires of world history. Likewise, the general historical content of the *mappae mundi* lost much of its relevance on maps of the modern world. In particular, the cartographical framework that had enabled the mapping of paradise was declining. The growing concern for direction, distance and contemporary geography, which had always coexisted in European cartography with topological principles, began to displace, on world maps, the former close interrelationship of history and geography. Above all, though, it was the reorientation of the map that deprived the eastern paradise of its visual pre-eminence.

Conventional Paradise in the Far East

The process of change from medieval *mappa mundi* to Renaissance map of the world was gradual. Maps of the world were created in the fifteenth century in a variety of formats, on many of which the earthly paradise continued to feature. Some of the world maps of this period, printed as well as manuscript, continued earlier models in every respect. They still had east at the top, and they still showed the earthly paradise in the farthest east and at the top, as does, for instance, the map reproduced in Plate 10, which comes from a fifteenth-century manuscript of Augustine's *De civitate Dei*.[80] Christ is shown seated in majesty in God's jewel-encrusted city and above a tripartite earth with, in the distance at the extreme edge of Asia, the earthly paradise. The main episodes of the Genesis story after the Fall are depicted on all three parts of the world. In the world history by Jean Mansel, *La Fleur des histoires* (*The Flower of Histories*, late fifteenth century) is another map of the world, also preserving the established arrangement (Figure 8.2). The map is placed at the start of the prologue to Book IV, in which a description of the regions of the world is given in alphabetical order.[81] Mansel's description includes ancient places such as Carthage and Delos, biblical places such as Babylon and Judaea, and lands of contemporary importance such as Westphalia and Gascony.[82] His aim was to show how Rome developed from a small town to a global empire, almost every corner of which had been penetrated by Christianity before the empire passed to the

Fig. 8.2. World map, from Jean Mansel, *La Fleur des histoires*. France, c.1460–70. 38 × 28 cm. Brussels, Bibliothèque Royale de Belgique/Koninklijke Bibliotheek van België, MS 9260, fol. 11r. The east-orientated map features the inhabited earth crowned by the earthly paradise. It immediately precedes the prologue to Book IV, which includes a regional description of the world in alphabetical order. Within each letter, though, places are not described in strict alphabetical order. Under the letter 'A', for example, we read first about Asia and Assyria, then Arabia, Armenia and Albania. Paradise is listed before Parthia, Pamphilia and Pannonia. The text also retains Isidore's distinction between the Fortunate Islands and the earthly paradise and repeats the belief that Enoch and Elijah dwelt in the Garden and that the noise of the waters of paradise falling to earth from a such a great height caused the local population to be born deaf.

Franks.[83] The map is circular. Heads symbolizing the four principal winds surround the habitable world, on which a number of historical or encyclopedic features are portrayed, such as the Tower of Babel, the Trees of the Sun and the Moon, the river Jordan, Jerusalem with Mount Calvary, the Red Sea, Rome, and some of the monstrous races. At the apex of the circular terrestrial landmass, in the distant east, is paradise. Adam and Eve, the tree and the serpent are shown within an ornate architectural frame, as if to emphasize the uniqueness and splendour of the Garden. From paradise the four rivers flow out to give life to the earth and to establish a mysterious yet material connection between paradise and the human realm. In his book, Mansel described paradise as a wonderful region surpassing all other earthly lands, fit for man's initial perfection, surrounded by a wall of fire and situated on an exceptionally high mountain that reached the sphere of the moon.

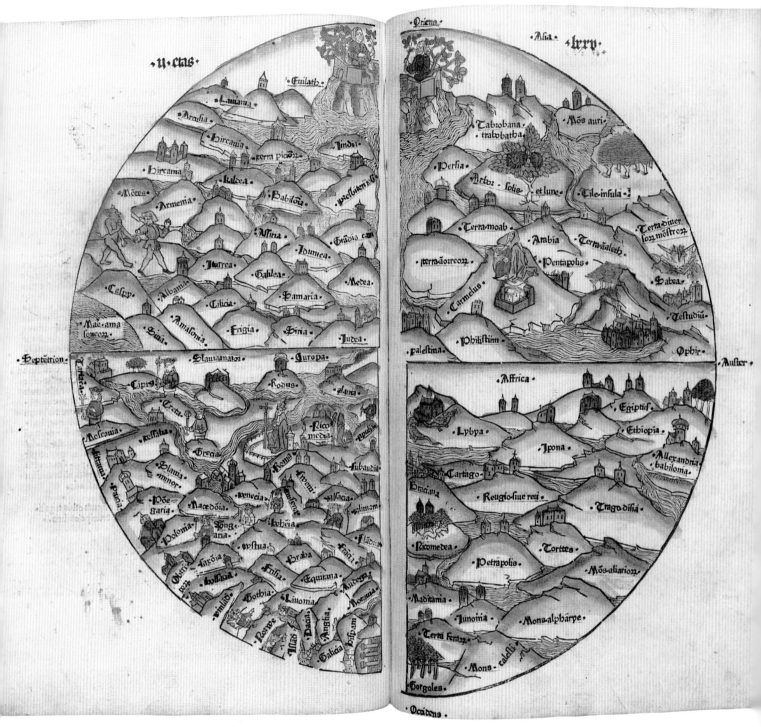

Fig. 8.3. World map, from the *Rudimentum noviciorum* (Lübeck: Lucas Brandis, 1475), fols 85v–86r. Diameter 38 cm (see also Plate 11a). London, British Library, C3.d.11. This *Elementary Book for Beginners* was intended as a handy summary of historical learning for the education and edification of a wide readership. The chronicle follows the sixfold division of human history from Adam to Christ, and carries this forward to the fifteenth century. The map is in the geographical section that accompanies the report of the second age of human history and opens with Noah and the division of the earth between his three sons. The map is circular and, albeit divided into four quarters, follows the T-O pattern, with the world divided into its three parts and with east at the top. Whereas Asia and Africa show place names belonging largely to a biblical or classical past, the representation of Europe is mostly based on contemporary geographical knowledge. The map, however, does not register the rise of Islam.

204 Mapping Paradise: A History of Heaven on Earth

What can be considered as the swansong of the medieval genre of *mappa mundi* was also the earliest printed detailed map of the world, the woodcut map in the *Rudimentum noviciorum* (*Elementary Book for Beginners*, 1475) (Figure 8.3).[84] The *Rudimentum noviciorum* was printed in Lübeck, Germany, by Lucas Brandis as a summary designed to synthesize all historical learning within, as noted in the colophon, a single and affordable book destined for a broad readership.[85] Whoever compiled the work drew on a wide range of patristic and medieval sources that included Jerome, Augustine, Orosius, Bede, Isidore, Peter Comestor, Jacques de Vitry, Hugh of St Victor, Nicholas of Lyre and Vincent of Beauvais. Peter of Poitiers's genealogy of Christ is included, updated with the names of recent emperors, popes, writers and scholars. The map illustrates the description of the world given in the account of the second age of human history, which opens with Noah and the division of the earth amongst his three sons. The account, which appears to have been modelled on Bartholomaeus Anglicus's *De proprietatibus rerum*, takes the form of an alphabetical listing of countries, mountains, islands, and rivers.[86] Neither the author of the text nor the map maker is known. Discrepancies between the place names in the text and on the map suggest that different people were responsible for text and map.[87]

It may not be easy immediately to see that the map of the world in the *Rudimentum noviciorum* is structured on the basis of the T-O model. The map is circular, and it does have east at the top, but it gives the appearance of being divided into quarters rather than the usual three parts of the world. The three parts are named, but neither the Mediterranean Sea nor the two rivers (Nile and Don) that separate them are shown, although it is not difficult to imagine, despite the map's topographical density, where they should be, that is, at the divisions of the quarters. Also unusual is the style of the map, for hill signs, each crowned with a settlement or a figure standing for the ruler, have been used to represent towns and regions, over a hundred of which are named without any clear distinction between the categories. Typically, though, Judaea and Palestine are close to the centre of the map, and classical, biblical and contemporary places are intermixed. Nearly all the 46 places in Europe were of contemporary importance, and eight countries are identifiable from their rulers, the pope for Rome, for example. In contrast, most of the 57 places shown in Asia and Africa come from biblical or classical history (Sodom and Gomorrah; the Roman provinces; the Amazones) or from thirteenth-century travel narratives (the Asian kingdom of the Great Khan). Other features include Prester John, the Tree of the Sun and the Moon, the Phoenix amid flames, the Valley of Devils, and a 'Land of Diverse Monsters' (*Terra diversorum monstrorum*) in Asia. In Africa, a 'Land of Wild Beasts' (*Terra ferarum*) is shown and, in the extreme west, the Pillars of Hercules are represented by three columns.

There is nothing in the *Rudimentum noviciorum* map to suggest any influence of contemporary nautical charts, still less of Ptolemy's *Geography*. The underlying spatial organization is topological, just as it was on the earlier *mappae mundi*.[88] The emphasis is on the juxtaposition of well-delineated regions, rather than on the distribution of settlements over a measurable surface, in an Aristotelian vision of space as a conglomeration of places. It is not difficult to see how each place or region is correctly shown next to its neighbour – Germany is next to Flanders, which is next to France – although direction is ignored, so that France appears to be north of Flanders. Libya is properly next to Egypt, India next to Persia, and the kingdom of Prester John next to the Great Khan and to India (but Ethiopia appears to the west of Egypt and Chaldaea to the north of India). The map maker's aim was to inform the reader of the general arrangement of the most important historical place names, not to specify their position in a mathematically defined space, and it was this objective that allowed him to incorporate the earthly paradise into the map, at the top and next to Havilah, India and Taprobana (see also Plate 11a). Paradise is represented as a walled garden on a craggy island, within which

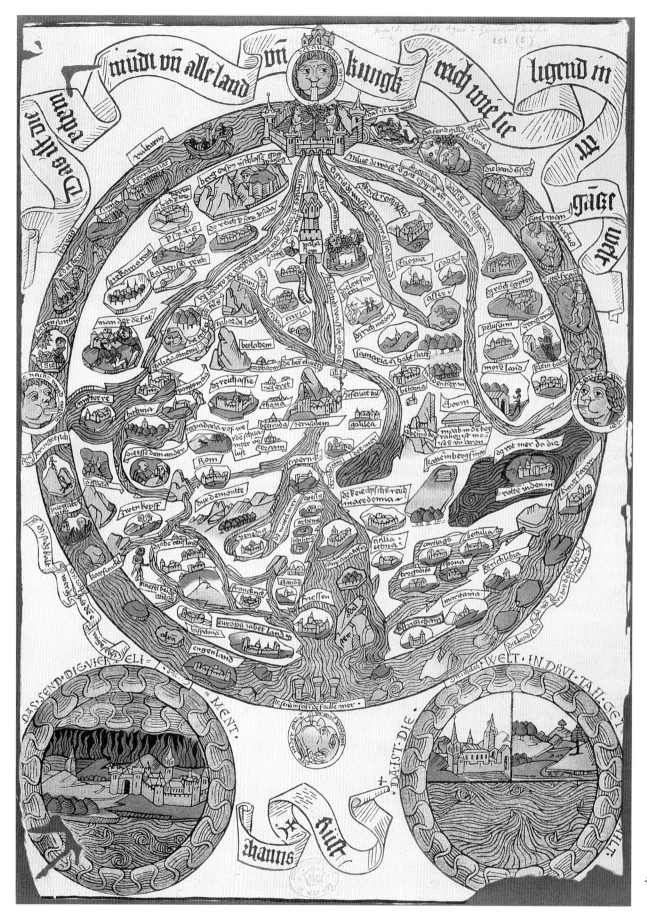

are Enoch and Elijah and from which flow the four rivers to pour their waters over the whole earth.[89]

Hanns Rüst's woodcut map of the world was produced in Augsburg about 1480 (Figure 8.4).[90] The immediate impression is of a quite different map from the one in the *Rudimentum novicorum*, but a second look shows it to be fundamentally similar. Like the earlier map, Rüst's map was intended for the use of educated readers, although in this case it was produced as a separate map, printed on a single sheet. Here again, as the announcement in the scroll above the map makes clear, what mattered was to show the countries and kingdoms 'with their positions in the whole world', meaning, we may infer, in relation to each other and not to some external or mathematical framework.[91] Like many larger *mappae mundi*, Rüst's map portrays the inhabited earth surrounded by an ocean full of islands and linked to paradise by the meanderings of the four rivers.[92] The identity of each river across the inhabited earth is given in the text that runs beside it: the Gihon is the Nile, flowing through Ethiopia and Egypt; the Pishon is the Ganges; the Tigris flows towards Assyria and the Euphrates across Chaldaea and Mesopotamia. The earthly paradise is shown at the top of the map, in the farthest east, as a hexagonal walled enclosure, inside which Adam and Eve stand by the Tree of Knowledge (Plate 11b). Close by, next to Persia, is the land of Gog and Magog, represented as a mountain chain, as the German label confirms: 'the Caspian Mountains, Gog and Magog confined'.[93] Not far from paradise are the Trees of the Sun and the Moon, and next to them is the Tower of Babel with a river flowing through it labelled as the Pishon (wrongly; it should be the Euphrates). Four wind heads mark the cardinal points. The three parts of the world are named and associated with Noah's sons: Asia with Sem (*asia sem*); Europe with Japhet (*Europa iabet land*); Africa with Cham (*affrica cham*). At the centre of the inhabited earth, and the map, Jerusalem is marked by a place sign, a mountain, and the figure of Christ on the cross surmounted by the letters *IHS* (the abbreviation of the name Jesus in Greek, Ἰησοῦς, which also acquired the meaning, in Latin, *Iesus hominum salvator*, 'Jesus, the Saviour of men'.[94] Other significant biblical places are shown in the vicinity; Bethlehem, Nazareth, the Mount of Olives, and Mount Tabor, for instance. In the extreme west, balancing paradise, are three Pillars of Hercules.

Other maps continuing the tradition of the medieval *mappae mundi* did reflect something of the changes taking place in contemporary cartography. There was a tendency to keep to tradition as regards the most remote and least-known parts of the old world, by putting vignettes of Gog and Magog and Prester John in north-east Asia, for example, but the Mediterranean outlines had usually been copied from a nautical chart. Andrea Bianco's world map of 1436 was a map in this mould (Figure 8.5).[95] Bianco was a Venetian sailor, a *comito di galia* (officer on a galley) and chart maker, and it is hardly surprising that when he came to compile a map of the world he was able to provide it with the outlines of the Mediterranean basin and the Atlantic coasts of Europe and north Africa typical of the charts of his day.[96] At the same time, he saw no contradiction in finding room on his map for paradise, placing it on a peninsula in the far east of Asia. From paradise, the four rivers flow out across the rest of the world. Bianco's paradise is not enclosed within walls, but, as if to remind those looking at the map that the Garden of Eden was unapproachable to all living persons, he placed beside it the cave of Saint Macarius, who had been warned by an angel of the inaccessibility of paradise.[97] On a neighbouring peninsula are Gog and Magog, cut off from the rest of the world by Alexander's wall.[98] Prester John is placed in eastern Africa instead of Asia – a move facilitated by the way Bianco stretched eastern Africa into a peninsula that almost reached eastern Asia – and Noah's Ark in Armenia. The Nativity and the Baptism of Christ are portrayed in Judaea. Another traditional aspect of Bianco's map was the labelling of Europe as *Imperium romanorum* (the Roman Empire), thus linking contemporary Europe with the last of the world's great empires.

Fig. 8.4 (opposite). World map by Hanns Rüst. Separately-printed broadside. Augsburg, *c*.1480. Reproduced as a separate sheet by Ludwig Rosenthal's Antiquariat (Munich, 1924). 40 × 28 cm (see also Plate 11b). London, British Library, Maps 856(5). On this east-orientated map neither distance nor direction has been taken into account: Venice, for example, appears to the west of Rome, Athens in Italy, and Macedonia in Africa. Toponyms are given in Middle High German. A scroll above the map states that 'This is the *mappa mundi* and [a map] of all countries and kingdoms with their positions in the whole world'. A scroll between the two small circles below the circular world map contains Rüst's signature. As noted in the inscriptions, the circle on the left represents the four elements – earth, water, air, and fire – and the circle on the right the threefold division of the earth. The division probably refers to the three parts of the world, but it is possible that the right-hand circle in fact indicates the inhabited and uninhabited zones of the known hemisphere (at the top) and the sea that occupies the unknown hemisphere (at the bottom), thus showing, in other words, the geographical types of town, country, and sea.

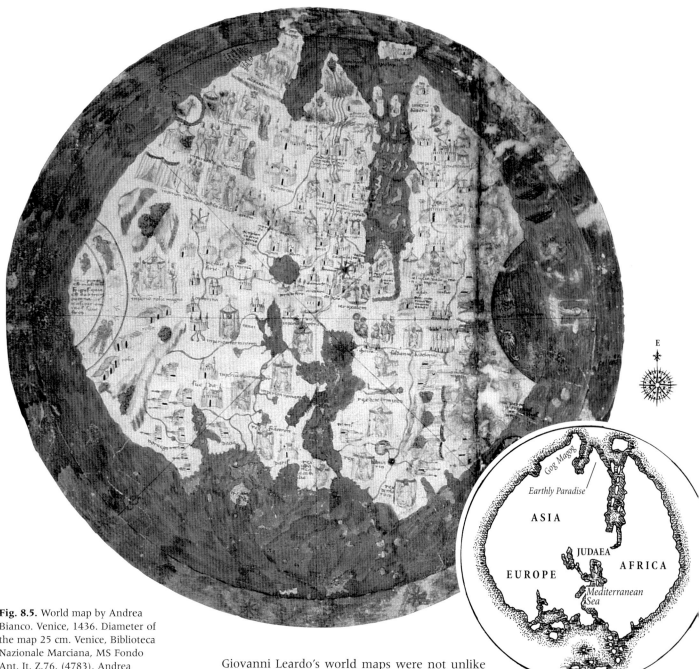

Fig. 8.5. World map by Andrea Bianco. Venice, 1436. Diameter of the map 25 cm. Venice, Biblioteca Nazionale Marciana, MS Fondo Ant. It. Z.76. (4783). Andrea Bianco compiled a number of nautical charts in the tradition of Pietro Vesconte; he would not have had any difficulty in adopting for his east-orientated map of the world the outline of the Mediterranean he had drawn for his nautical charts. As on many other fifteenth-century cartographical documents, the topological approach that allowed the depiction of the earthly paradise next to the inhabited earth on *mappae mundi* coexists with the geometrical accuracy of nautical charts, in which distance and direction were important elements in the representation of space.

Giovanni Leardo's world maps were not unlike Bianco's. Both – the first, completed in 1442, described in Chapter 1 (Figure 1.1 and Plate 2a), and the second he finished in 1448 (Figure 8.6) – bear the hallmarks of fifteenth-century changing cartographical context.[99] In their essentials, the two maps are similar, although the later map is more detailed. Both are circular; both are surrounded by a calendar; both have east, with paradise, at the top and Jerusalem at the centre; and both show the *oikumene* surrounded by the outer ocean. Other features on the 1448 map include Gog and Magog in eastern Asia, the site of Saint Thomas's preaching in India, the Red Sea crossing, Prester John in Ethiopia, and the monstrous races in Africa. Characteristically, the map also includes new material, taken mostly from thirteenth-century travel accounts. It was Marco Polo, for instance, who had reported on the Great Khan's tomb in Northern Asia and commented on the number of people who had to be killed to provide an escort for the emperor's soul, and who described the nuts found in India.[100] Other aspects, notably the writing of place names at right angles to the coastline, especially in the

Mediterranean, reflect the use of nautical charts in compiling the map. Leardo's second world map, like his first, reflected the efforts he made to accommodate geographical material that, increasingly, had nothing to do with the east–west chronological progression of human history. It was appropriate that he presented paradise on both maps as concealed behind the impenetrable fortifications of a medieval castle, or city, instead of simply enclosed by a wall of stone or fire. It is true that the map sign of a castle or a fortified city is used elsewhere on the map – to indicate India and the kingdom of Prester John, for example – but as in the case of other fifteenth-century maps, the representation of paradise as a highly fortified settlement without the individualizing depiction of Adam and Eve could have been a new way of emphasizing the isolation and separateness of the post-lapsarian Garden of Eden and highlighting a new cartographical disjunction between the earthly paradise and the earth. There was still room on fifteenth-century maps for the epochal zone of the Garden of Eden, but the primary concern had switched from showing event/places to the mapping of geographical space, and paradise had to disappear behind its enclosure once temporal distance did not distinguish it too clearly from the contiguous inhabited earth.

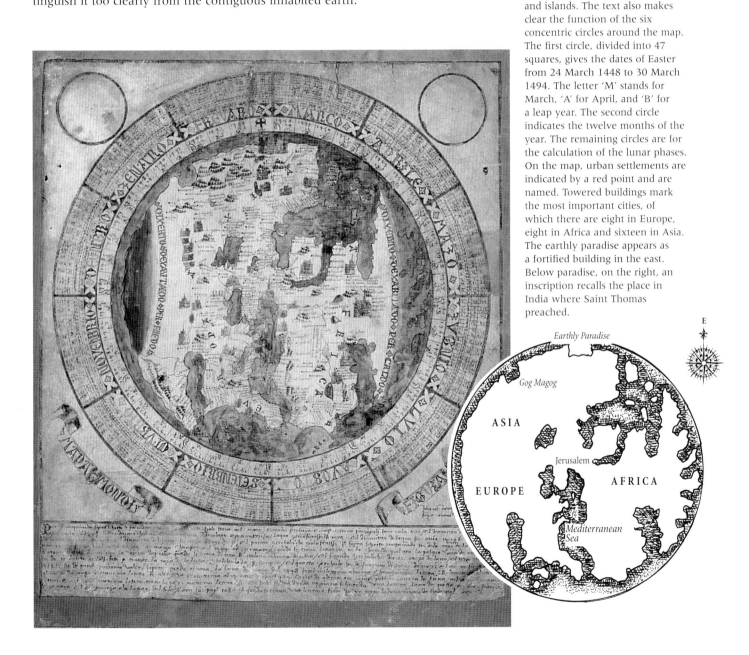

Fig. 8.6. Giovanni Leardo, *Mapa mondi. Figura mundi*. Venice, 1448. 35 × 31 cm. Vicenza, Biblioteca Civica Bertoliana, MS 598a. The east-orientated map is signed and dated below the scroll on the bottom right. The text below the map gives the measure of the diameter of the earth according to Macrobius, and explains that the aim of the map is the description of the seas and the earth with its main regions, rivers, mountains and islands. The text also makes clear the function of the six concentric circles around the map. The first circle, divided into 47 squares, gives the dates of Easter from 24 March 1448 to 30 March 1494. The letter 'M' stands for March, 'A' for April, and 'B' for a leap year. The second circle indicates the twelve months of the year. The remaining circles are for the calculation of the lunar phases. On the map, urban settlements are indicated by a red point and are named. Towered buildings mark the most important cities, of which there are eight in Europe, eight in Africa and sixteen in Asia. The earthly paradise appears as a fortified building in the east. Below paradise, on the right, an inscription recalls the place in India where Saint Thomas preached.

Fig. 8.7a. The Borgia world map. Austria, southern Germany or Bohemia, *c*.1430. Diameter *c*.63 cm. Vatican City, Biblioteca Apostolica Vaticana, Borgiano XVI. Reproduced from Manuel Francisco de Barros e Sousa, Viscount of Santarém, *Atlas composé de mappemondes, de portulans et de cartes hydrographiques et historiques depuis le VIe jusqu'au XVIIe siècle* (Paris: Maulde et Renou, 1849), Plate 23. The world map, once owned by Cardinal Stefano Borgia, has south at the top. It shows a number of influences from contemporary Catalan-style maritime charts, most notably in the way early fifteenth-century political events, relatively recent current affairs, and features of contemporary life are all shown on the map. The Great Khan and the fortified camp of the Mongols, for example, are shown in Asia. The legends are in Latin. Vignettes and scenes, such as Gog and Magog, the earthly paradise and the pillars of Alexander the Great in the Far East, are presented according to the medieval encyclopedic tradition.

210   Mapping Paradise: A History of Heaven on Earth

Conventional Paradise on Disorientated Maps

A number of world maps (not zonal maps) produced in the course of the fifteenth century had south, north, or west at their top, rather than east. Irrespective of their orientation, however, many Latin world maps continued to show the Garden of Eden in the farthest east even though this meant a displacement of the primary event/place from the top of the map to left, right, or the bottom of the map. The visual disjuncture, the shift of paradise from a dominant to a visually much weaker position, marked the break with the tradition of creating maps to communicate the east–west progression of history. The downplaying of the historical dimension meant that the inclusion of paradise on fifteenth-century maps of the world now had to be justified in spatial terms alone. Paradise began to lose its role as a point of reference, still less as the easternmost boundary at the top of the earth.[101]

The common factor in all the maps described below is the combination of historical features of medieval cartography with a new, underlying concern with geographical accuracy and a new orientation. The resulting mixture of characteristics is the hallmark of several fifteenth-century *mappae mundi*. Thus, for example, the eastern paradise appears on south-orientated maps, as on the map engraved on copper, once owned by Cardinal Stefano Borgia, which presented a circular structure in its portrayal of the earth surrounded by an outer ocean at the same time as absorbing novelties taken from contemporary Catalan maritime charts (Figure 8.7). Borgia's densely packed and largely pictorial map is thought to have been compiled about 1430 somewhere in Austria, Bohemia, or southern Germany.[102] Curiously, while the alignment of the Atlas mountains across North Africa, the well-marked gap in the range representing the way

Fig. 8.7b. Diagram of Fig. 8.7a.

Fig. 8.8. The Borgia world map (see Fig. 8.7a). Detail of the earthly paradise. On the Borgia map, the Garden of Eden is portrayed not far from the mouth of the Ganges, near *India superior*, a land of marvels and precious stones, and near China, peopled on the map by tiny figures collecting silk from the trees.

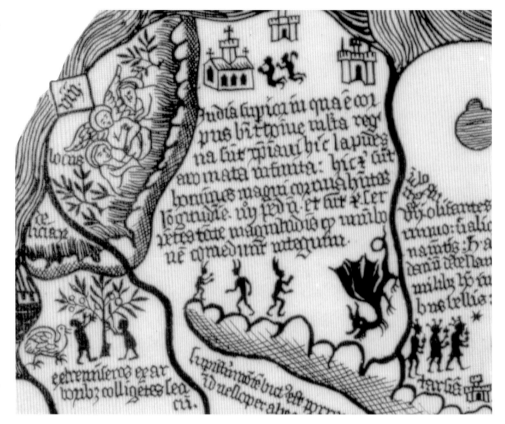

Fig. 8.9 (opposite). World map by John of Udine. Monastery of Comburg, Schwäbisch Hall, southern Germany, fifteenth century. 19 × 19.5 cm. Stuttgart, Württembergische Landesbibliothek, YMS Theol. Fol. 100, fol. 3v. John of Udine's mid-fourteenth-century chronicle was an expanded version of the genealogical tree from Adam to Christ compiled by Peter of Poitiers (discussed in Chapter 4). Udine's world map was intended to provide the geographical background for the historical account. The map is structured on the T-O division of the known world. South is at the top, so that Africa is in the upper right quarter, Europe in the lower right, and Asia on the left. Jerusalem is in the centre, and Gog and Magog are in northern Asia. Among some fifty locations marked on the map, the Red, Dead, and Mediterranean seas are demarcated. The legends at the extreme north and south refer to the torrid zone to the south of the inhabited earth, and the cold zone to the north, both uninhabitable. The eastern paradise is also located beyond the boundaries of the inhabited earth, its presence suggested by the four rivers that are shown rising in the Far East.

through to the *terra nigrorum* ('the land of the blacks') beyond, and the wavy fashion of depicting the sea were all features associated with the Catalan school of chart making, the outline of the Mediterranean Sea appears to owe nothing to a nautical chart. Many vignettes and scenes labelled in Latin, such as Gog and Magog and the pillars of Alexander in Asia and Prester John in Africa, point to the continuing relevance in late medieval mapping of historical and encyclopedic material. Among the relatively modern features identified on the Borgia map are the fortified camp of the Mongols, the land of the Great Khan, and Cathay (China). The 37 holes pierced through the copper and coinciding with the landmasses are thought to have been made by the pins of medallions on which were portrayed local rulers and cities.[103] It is less clear what was the function of the 24 small squares containing Roman numerals that divide the circumference into 24 segments of equal size, but the regularity of their spacing hints at the new emphasis on geometrical precision and at some kind of coordinate system.

The Borgia map contains other significant features. Europe occupies a large proportion of the mapped space and the map is not centred on Jerusalem, but on a point much further north, close to the mouth of the Danube. Paradise is not aligned with the map centre, and it is found instead somewhat to the south of it, not far from China, where two people are shown collecting silk from a tree (as the text below explains), and India, where the burial place of the Apostle Thomas is shown, as well as dragons and horned men (Figure 8.8). Described as a *locus deliciarum* ('place of delights'), paradise is surrounded by both mountains and a fiery wall. Within it are the waiting Enoch and Elijah, accompanied by a guardian angel.[104] There is no sign of the single source or of the four rivers springing from it. On the contrary, the impression is gained, from the way one of the two branches of the Ganges has been allowed to flow through the Garden into the ocean beyond, that the map maker placed paradise on the map without attempting to adapt the hydrography of Asia to its presence. Other south-orientated world maps continued to indicate an eastern paradise. A fifteenth-century

212 Mapping Paradise: A History of Heaven on Earth

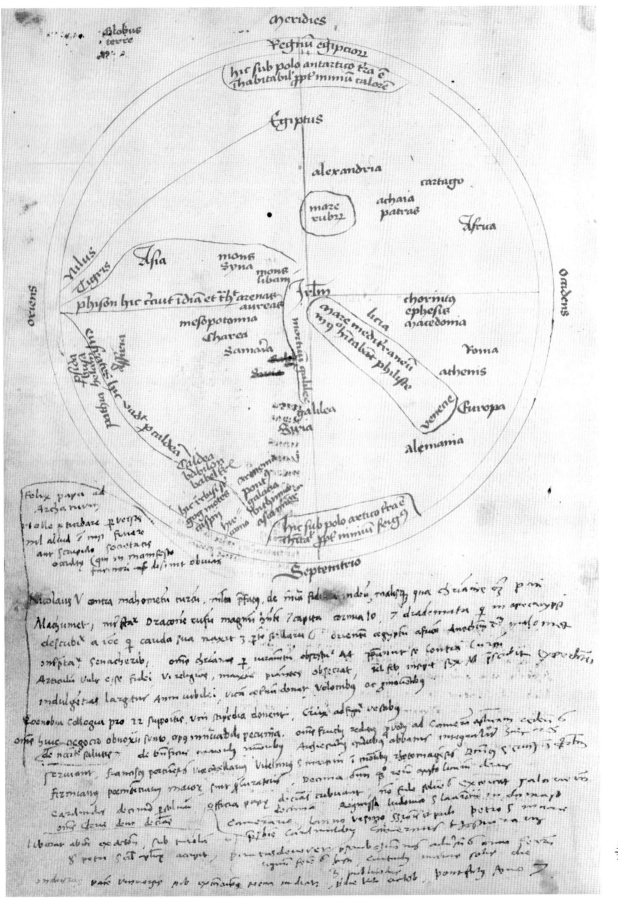

Fig. 8.10. World map. Central Europe, *c*.1450. Formerly in Olomouc (now in the Czech Republic) Studienbibliothek, MS g/9/155. Reproduced from Anton Mayer, *Mittelalterliche Weltkarten aus Olmütz* (Prague: Geographisches Institut der Deutschen Universität in Prag, 1932), Plate 1 (size of the reproduction 21.5 × 16 cm). This west-orientated anonymous *mappa mundi* is thought to have been lost after the Second World War. The outline of the Mediterranean Sea clearly has not been taken from a nautical chart. The Caspian Sea is open to the outer ocean. The Holy Land, at the centre of the map, is heavily emphasized. The four rivers flow out over the world from a peripheral paradise situated in the farthest east.

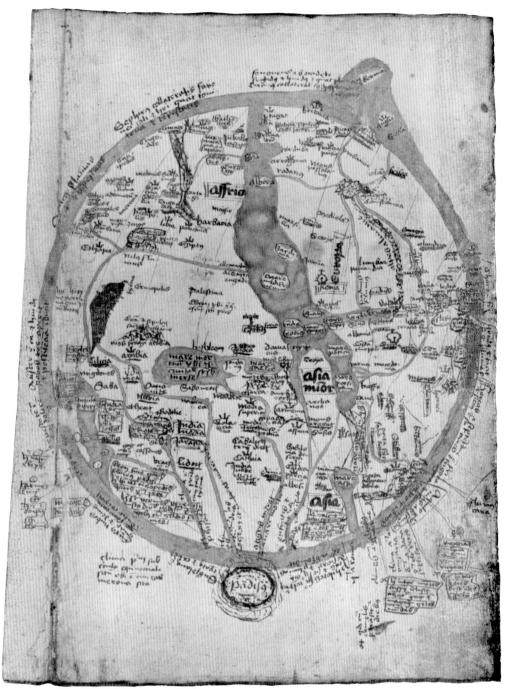

copy of John of Udine's expanded version of Peter of Poitiers's genealogical tree, the *Compilatio librorum historialium totius Bibliae ab Adam ad Christum*, contains a map of the world on which paradise is not actually marked, but its presence is hinted at by the four rivers that emanate from the eastern cardinal point to spread out over the world, and confirmed in the adjacent text: 'the Nile, the Tigris, the Pishon, which encircles India and carries golden sand, the Euphrates, which flows through Chaldaea' (Figure 8.9).[105] The author of the map was focusing on the inhabited part of the earth, the parts lying between the torrid zone in the south and the frigid zone in the north, which are labelled on the map.

An eastern paradise could still be shown on maps orientated in other directions. For example, an anonymous map of the world dating, probably, from about 1450 that

214   Mapping Paradise: A History of Heaven on Earth

is now lost, but that up to the Second World War survived in Olomouc (now in the Czech Republic), has west at the top. The compiler of this map placed paradise in the east even though this meant it was now at the bottom of the map, beyond the outer ocean (Figure 8.10).[106] Another anonymous map, the so-called Wieder-Woldan map, has north at the top (Figure 8.11). This is an engraved map produced in Venice between 1485 and 1490.[107] The map shows signs of the influence of medieval *mappae mundi*, contemporary nautical charts, and Ptolemy's *Geography*. The medieval paradise, labelled *Paradisus*, occupies a curving peninsula in the east. The outline of the Mediterranean Sea and the coasts of the North Sea and the Caspian Sea are typical of nautical cartography. The depiction of western Africa reflects what had been learnt from recent voyages of exploration in the years immediately preceding Bartolomeu Dias's rounding of the Cape of Good Hope in 1488. Most striking, however, is a distinctly Ptolemaic feature, a landlocked Indian Ocean. Three of the rivers of paradise are shown flowing from the Garden of Eden across central Asia, but for the fourth the map maker took advantage of the continuous land on the southern side of the Indian Ocean, between eastern Asia and southern Africa, to show the whole course of the Gihon or Nile, from the spring within paradise to Egypt and the Mediterranean Sea. Before reaching Africa, the river passes through two mountain chains as if in allusion to the underground

Fig. 8.11. The Wieder-Woldan map of the world. Venice, *c.*1485–90. Diameter 175 cm. Vienna, Österreichische Akademie der Wissenschaften, Sammlung Woldan. K-V(Bl): WE 3. The map is known as the Wieder-Woldan map from the previous owners of the only two extant copies. It was probably engraved in Venice, given the marking of the province of *Venetie*. It has north at the top and the earthly paradise in the farthest east, with a lake, from which flow the four rivers. Names are in Latin. The Ptolemaic notion of a landlocked Indian Ocean has been adopted, but there is no geographical grid or division into climatic zones. The cardinal points are indicated in the outer ocean. The map does not include any geographical information about Asia brought by late medieval Italian explorers, but it does take into account Portuguese discoveries in western Africa. It shows Greenland as an Eurasian peninsula bearing the name of *Engrovelant*. The Caspian Sea is correctly represented as a lake.

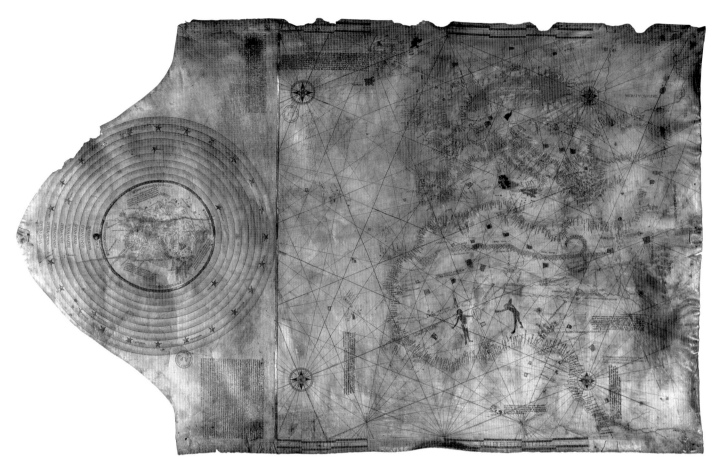

Fig. 8.12a The nautical chart formerly attributed to Christopher Columbus. ?Genoa. *c.*1492. 111 × 70 cm (world map, diameter 20 cm). Paris, Bibliothèque Nationale de France, Cartes et Planes, Rés. Ge AA 562. Instead of the more common arrangement, whereby an outline of the Mediterranean area was inserted into fifteenth-century maps of the world, on the so-called Columbus chart a small circular world map drawn on the neck of the parchment accompanies the chart. The world map complements the nautical chart, which represents familiar and well-navigated seas, with a view of the wider world lying beyond the economic circuit of European trade. Paradise is shown in the farthest east as an island opposite Cathay. The Mediterranean area, the Atlantic coast, southern and eastern Africa are all described with a reasonable knowledge of recent Portuguese voyages of discovery.

Fig. 8.12b Diagram of the world map in Fig. 8.12a (see Plate 14).

216 Mapping Paradise: A History of Heaven on Earth

courses of the paradisiacal rivers, but in Africa, the biblical Nile is joined by the Ptolemaic Nile, flowing north from its origins in the Mountains of the Moon.

A different sort of relationship between the nautical chart, commonly used in the fifteenth century for the outlines of the Mediterranean Sea and Atlantic coasts, and the world map is found on a tiny north-orientated circular map of the world of about 1492 (Figure 8.12 and Plate 14).[108] The map has a diameter of 20 centimetres and fills the narrow neck end of a parchment nautical chart formerly associated with Christopher Columbus, and complements the detailed portrayal of well-navigated seas at the heart of the old world with a view of the whole cosmos that occupies the rest of the skin. It shows the earth, with Jerusalem at the centre, surrounded by the outer ocean and nine concentric circles representing the nine heavenly spheres. Paradise, absent from the adjacent nautical chart, is strikingly displayed as a large circular island surrounded by high mountains off the coast of Asia. The outlines of the Mediterranean, Atlantic Europe, and southern and eastern Africa are in nautical style. Their place names reflect local knowledge gained from recent Portuguese voyages of discovery. At the same time, the islands visited by Saint Brendan (in the surrounding Ocean) and Gog and Magog (in eastern Asia) are also shown on the map.

A relatively late example of a north-orientated world map on which an eastern paradise is marked is found in the sixteenth-century encyclopedia compiled by Antoine de la Sale, entitled *La Salade* (Paris, 1521) (Figure 8.13). De la Sale was at pains to emphasize the altitude of paradise, describing it as 'the head of the earth', and as

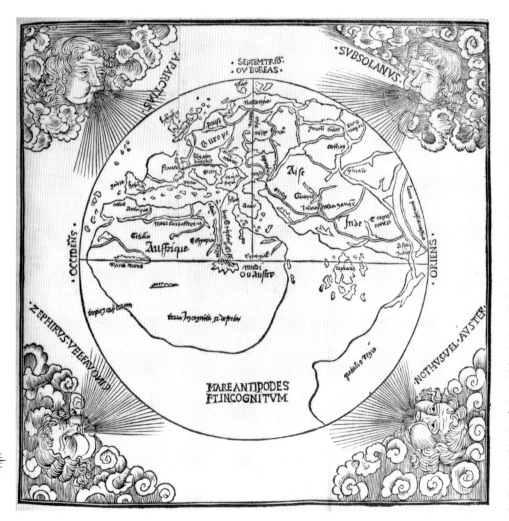

Fig. 8.13. World map, from Antoine de la Sale, *La Salade* (Paris: 1521). Inserted after fol. XXVII. 25 × 25.5 cm. London, British Library, 216.a.8. Antoine de la Sale was tutor to John of Calabria, son of René, King of Sicily and Count of Anjou. A Provençal by birth, he had taken up writing late in his life after having been a soldier and a courtier in the Angevin household. He was in his fifties when he composed *La Salade* (1437–44) for the education of his ten-year-old princely pupil. *La Salade* is a moral and historical miscellany in thirty books. The title is a pun on de la Sale's name, adopted 'because in a salad one puts many good herbs'. The world map illustrating the sixteenth-century printed edition is north-orientated and surrounded by four wind heads. The eastern paradise is located north of the equator, identified by an inscription on the extreme right of the map.

The Twilight of Paradise on Maps 217

surrounded by high mountains peopled by dragons and wild beasts.[109] No mountains are portrayed on his map, but a large area in the northern hemisphere is marked off as paradise and made to look completely separate from the rest of the world. No other historical features are shown on the map. De la Sale's labelling of *locus paradisi terrestris* ('the place of the earthly paradise') confirms that by his time the showing of paradise on a map of the world was a cartographical residual and a relict from the heyday of the east-orientated encyclopedic *mappae mundi* of the Middle Ages.

Mapping Paradise in Africa

While some fifteenth-century map makers abandoned the eastern orientation but continued to show an Asian paradise on their world maps, others preferred to change the location of the Garden of Eden entirely, placing it instead in Africa, with a paradise at the tip of southern Africa (south of the equator) or in the Horn of Africa (at the equator). An African location for the Garden of Eden suited various arguments. It had the convenience that, while eastern Asia was being increasingly explored by Europeans, sub-Saharan Africa remained largely unknown. There were various factors to take into account in connection with an African paradise.

First of all, the Vulgate's translation of the Hebrew מִקֶּדֶם (*miqedem*) as 'from the beginning' rather than as 'eastward' had left the way open for locating paradise in a different direction. Although the translators of the Greek Septuagint and the *Vetus Latina* had opted for the 'eastward' version, Duns Scotus's argument in the fourteenth century that 'east' was a relative notion depending on the position of the observer had weakened the long-standing tradition of an eastern, i.e. Asian, location. After Duns Scotus, paradise not only could be situated elsewhere, but also often was, on maps as well as in the texts. A southern paradise had not been an entirely new idea in the fourteenth century. In the previous century Bonaventure, Roger Bacon and Albert the Great all seem in their different ways to have favoured it, even as they pondered over the possibility of an equatorial paradise. Thomas Aquinas, too, had admitted the possibility that paradise could lie on, or beyond, the torrid zone. Moreover, the Nile, long identified as one of the rivers of paradise (the Gihon) was known to flow through Africa, and the land of Cush – that Genesis described as encircled by the Gihon – had been identified with Ethiopia.[110] In the fourteenth century reference was made to a paradise 'in the high sierras of the Antarctic Pole' in *El Libro del conosçimiento de todos los reinos* (*c*.1350–60), an anonymous travelogue compiled by an unknown Spanish author.[111] The fifteenth-century theologian Jacob Perez de Valencia tried to be more precise. Blending Ptolemaic geography with the Bible he suggested that the Garden of Eden was at the inaccessible top of the Mountains of the Moon, where, according to Ptolemy, the Nile had its source.[112]

Secondly, the shift from an Asian to an African paradise was less crucial than it might at face value seem because of the old association between India and Ethiopia. Depending on the authority, the large territory generally alluded to in the Middle Ages vaguely as 'India' was sometimes subdivided into regions, one of which was Ethiopia. Most commonly there were three subregions, a division found already in the early Christian apocryphal literature. The fifth- (or sixth-) century *Passio Bartholomaei*, for example, described the three regions of India.[113] The threefold division is also found on a number of maps, including the Leardo world maps discussed in Chapter 1 and above in this chapter (Figures 1.1 and 8.6; see also 6.11). In the later Middle Ages, India was distinguished as *India inferior* (*prima* or *minor*, east of Persia), *India superior* (*secunda* or *maior*, the subcontinent and southeast Asia) and *India tertia* (or *ultima*, usually covering Ethiopia and eastern Africa). For this reason, the Ethiopians were often called Indians.[114]

Another factor leading to the association of the remotest eastern part of Asia with the equally remote parts of Africa was the idea that Africa extended towards the east, as shown on Andrea Bianco's map (see Figure 8.5). The association between eastern Africa and Asia was also supported by references, which went back to classical learning, to the Nile as marking the border between Africa and Asia, as the T-O maps demonstrated. Since Ethiopia was east of the Nile, it is easy to understand why Ethiopia was often thought to be part of Asia. The classical theory that the Nile ran in an east–west direction across sub-Saharan Africa also contributed to the linking of Asia with all sub-Saharan territories.[115]

The medieval conflation of eastern Africa and eastern Asia is reflected in the removal, in the fourteenth and fifteenth centuries, of Prester John from Asia to Africa.[116] From the twelfth century onwards, the link between the legendary Christian sovereign and the earthly paradise had been a close one. The late twelfth-century text of the *Epistola*, the letter believed to have been sent by Prester John to the Byzantine emperor Manuel I, reports a spring in the land of Prester John distant only three days' journey from paradise.[117] At first, Prester John's kingdom was thought to be, like the Garden of Eden, in eastern Asia, but in the fifteenth century his kingdom, together with paradise, was beginning to be placed on maps in eastern Africa. To be precise, Prester John made his move first, before paradise. About 1330 a Dominican missionary, Jordan of Sévérac, claimed that paradise lay between *India tertia* and Ethiopia, land of Prester John.[118] This intermediate situation is seen on Bianco's map of 1436 and Leardo's map of 1442, which both show east Africa and east Asia as two big peninsulas separated by a narrow arm of the sea and Prester John in Africa and paradise in Asia (Figures 8.5 and 1.1). Other fifteenth-century maps, however, show the move completed, with both Prester John and the Garden of Eden placed in Africa.

Paradise in the South

Perhaps the earliest map of the world to show both the Garden of Eden and Prester John in southern Africa is the one made by the Venetian Albertin de Virga in 1411 or 1415 (Figure 8.14).[119] Virga's map has been described as 'a compromise between the classical medieval *mappa mundi* and the nautical chart'.[120] His circular map is contained within a rectangular frame occupying three-quarters of the folio. In the remaining space is a circular diagram showing the relationship between the signs of the zodiac and the various parts of the human body. The diagram is flanked by two square tables. One table, composed of a hundred black or red letters of the alphabet, was for the calculation of the date of Easter (from 2 April 1301 to 18 April 1400). The other table is a *tola di Salamun* (a Table of Solomon), for calculating the position of the moon.[121] Both tables and diagram seem to have been taken from early thirteenth-century astronomical tables. The map features a vermilion-coloured Red Sea, Prester John in Ethiopia (indicated by an inscription and a crown), and Gog and Magog in northern Asia. The influence of nautical charts is seen in the eight rays of a wind rose that radiate from the middle of the map, which lies west of the Caspian Sea and not in the Holy Land. The outline of the Mediterranean Sea, the Atlantic coast of Europe and north-west Africa, together with the recently discovered Atlantic islands (e.g. the Canaries and Azores) and the Atlas Mountains, are also in nautical chart style. It is difficult to be certain of the intended orientation of Virga's map, since place names and texts are written in different ways up, so that whereas the signature indicates an eastern orientation, and the tables have to be read with west at the top, some of the inscriptions on the map can be read only with north at the top while others have to have south at the top to be readable. Pictorial signs, of crowns, palaces or castles, indicate the major countries on all three parts of the world. A peninsula to the north-west of Europe is identified as Norway, and a large island to the south-east of Asia is labelled both *caparu sive iava*

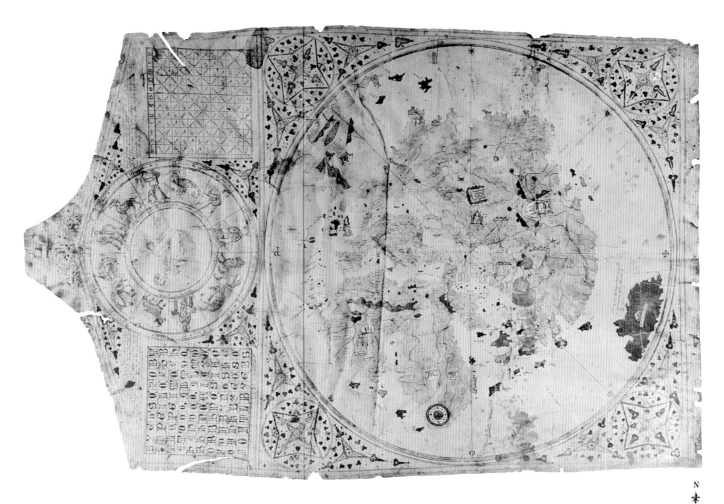

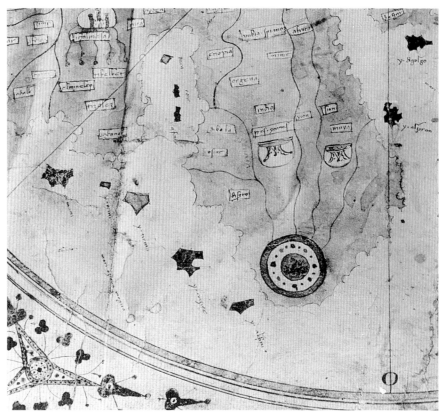

Fig. 8.14 (opposite). World map by Albertin de Virga. Venice, 1411 or 1415. Dimensions of the whole parchment 69.6 × 44 cm; diameter of the map 41 cm. Location unknown. Albertin de Virga was one of a number of fifteenth-century map makers who located the Garden of Eden in the south. His circular map is flanked by a round diagram, showing the relationship between the signs of the zodiac and the various parts of the human body, and two square tables for the calculation of the dates of Easter and the positions of the moon. The map depicts Europe, Africa and Asia surrounded by the ocean. A star indicates the north (the polar star), and a cross points to the east. The initials of the names of the winds – in Italian – are given around the map: P (*Ponente*), O (*Ostro*), G (*Greco*), S (*Scirocco*), A (*Africus*), M (*Maestro*). Paradise, identified by a large circle (shown in the enlarged detail), appears in southern Africa. The map has been missing since it was sold by auction in Lucerne, Switzerland, in 1932.

Fig. 8.15. Giovanni di Paolo, *The Creation of the World and the Expulsion of Adam and Eve from Paradise*. Siena, 1445. 46.5 × 52 cm. New York, Metropolitan Museum, Robert Lehman Collection. On the left God the Father, surrounded by his angels, presides over a schematic rendering of the Ptolemaic geocentric universe. After the earth, shown in the centre, come the zones of water, air, and fire; and then the seven circles representing the planets. The next circle contains the zodiac, while the outermost ring is the *primum mobile*, which divides the material world from the empyrean, the measureless region of God. The right half of the painting illustrates the expulsion of Adam and Eve from Eden. Below the garden are four streams representing the four rivers of paradise. The same rivers are depicted flowing down a mountain located at the top of the map of the earth that occupies the centre of the sphere to the left in the panel. The map is south-oriented and the mountain of paradise appears to be located in southern Africa.

The Twilight of Paradise on Maps   221

magna and *Zipangu* in a conflation of Marco Polo's islands of Japan and Java.[122] The majority of Asian place names seem to have been derived from thirteenth-century travel accounts. *Catajo* and *Borgar tartarorum*, for example, referred to the Mongolian Empire. Paradise on Virga's map is found in the extreme south of Africa, marked prominently by a circular sign from which four rivers flow, one of which is identified as the biblical Gihon.[123]

Another southern paradise is found on the map in Giovanni di Paolo's panel painting of *The Creation of the World and the Expulsion of Adam and Eve from Paradise* (1445) (Figure 8.15).[124] On the left of the painting, above the universe, is God the Father, surrounded by twelve angels. At the centre of the universe is the earth with its lands and seas. Surrounding the earth are the concentric spheres of, first, the elements of air and fire, then those of the seven planetary circles, the zodiac and, finally, that of the *primum mobile*, the sphere that divides the physical world from the empyrean, God's

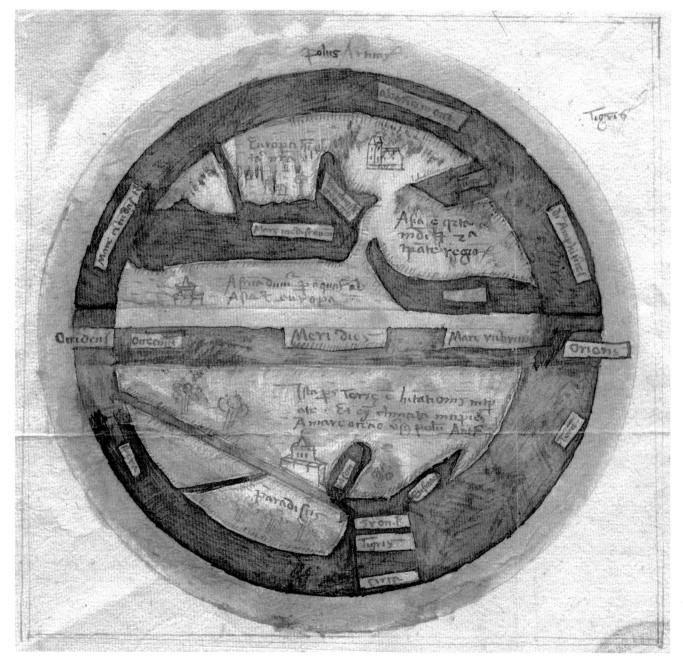

realm. The image depicts the creation of the world in the form of a *thema mundi*, a sort of natal chart representing the moment the world was born.[125] The map of the earth has south at the top, so that the mountain of paradise in southern Africa appears to overlook the rest of the earth. The four rivers, source of all water on earth, flow down the sides of the mountain. They are seen again in the bottom right-hand corner, flowing out of paradise in the scene of the Expulsion. Giovanni's preference for a southern paradise was likely to have been inspired by his familiarity with Dante's *Commedia*, part of which he had been illuminating between 1438 and 1444, not long before he completed his panel painting.[126]

Giovanni di Paolo adopted the symbolism of altitude to emphasize paradise's intermediate situation, halfway to heaven from earth, and gave paradise its traditional pre-eminence at the top of – in his case – a south-orientated map. Four decades or so later, the idea that the Garden of Eden was located on the top of the Mountains of the Moon, out of which the Nile flowed, was also illustrated by the Genoese Vesconte Maggiolo on the first of the four maps forming his atlas of 1512 (Plate 12a).[127] Maggiolo was active in Naples. In the same years, north of the Alps, in Germany, an anonymous map maker placed paradise not in the south of Africa, but more generally in the southern hemisphere (Figure 8.16).[128] The map came to light in a library in Constance (Konstanz) in 1991 in a copy of the 1515 edition of Gregor Reisch's *Margarita philosophica*.[129] The connection between the map and Reisch's text is unclear. The map is small and on a separate piece of paper. It is thought to have come from a monastic school in southern Germany and to date from the last years of the fifteenth century.[130] It resembles the kind of rough sketch often made as a learning aid. It presents an imperfect

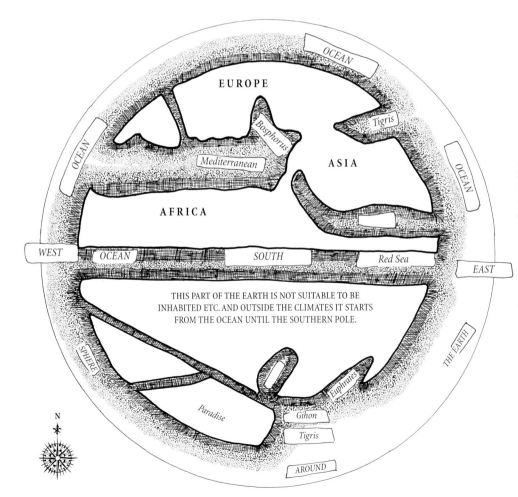

Fig. 8.16a (opposite). Manuscript map of the world. From a loose sheet of paper found in a copy of Gregor Reisch's *Margarita philosophica* (Strasbourg: Grüningerus, 1515). Southern Germany, *c*.1500. External diameter 13 cm, internal diameter 11.5 cm. Constance, Heinrich-Suso-Gymnasium, Bibliothek, Da 165. The map does not seem to have any connection with Reisch's *Margarita philosophica*. Its unknown author was probably looking at a Macrobian zonal map (see Fig. 8.17) as he compiled his own sketch map, from which he excluded the polar extremities, uninhabitable because of the cold, while simplifying, and distorting, the standard zonal division. The map has north at the top. A reference to Saint George's arm indicates the Bosphorus.

Fig. 8.16b. Diagram of Fig. 8.16a.

rendering of what could have already been considered an old-fashioned image of the world. It is as if the anonymous map maker had used as his base map one of the zonal maps in medieval and Renaissance versions of Macrobius's commentary on Cicero's *Somnium Scipionis* (see Figures 8.17a and 17b). His sketch map has north at the top, legends and place names are in Latin, and there is a modest scattering of tree signs and place signs, the latter made up of various combinations of towers and buildings. The uninhabitable polar zones are not shown. The terrestrial sphere is divided across the middle into northern and southern hemispheres, separated by the equatorial ocean. In the north is the inhabited part of the earth with Europe, Asia, and Africa identified by name. As in the case of all medieval *mappae mundi*, no part of the known world falls south of the equator. A vacant band between the habitable world and the equatorial ocean may indicate the torrid and uninhabitable zone of the northern hemisphere. The general oversimplification and distortion of the standard zonal division is particularly marked south of the equator. The torrid zone of the southern hemisphere seems to have been given the inlets usually found in the northern torrid zone, where they represent the Red and Indian seas. A three-line Latin inscription written just below the equatorial ocean announces that: 'This part of the earth is not suitable to be inhabited etc. And outside the climates it starts from the

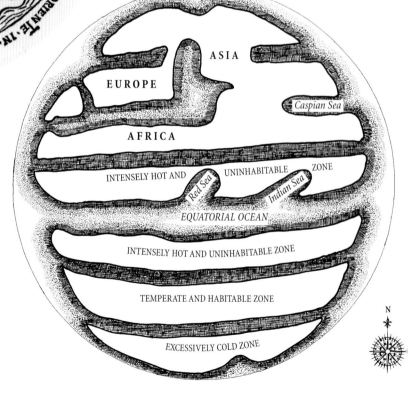

Fig. 8.17a. Zonal world map, from Macrobius, *In Somnium Scipionis expositio* (Brescia: Boninus de Boninis, 1483). Diameter 14 cm. London, British Library, 1B 31072. Zonal maps were included in medieval manuscripts and Renaissance printed editions of Macrobius's commentary to illustrate the division of the terrestrial sphere into five climatic zones. In the course of the fifteenth century, exploration of the Atlantic coast of Africa ensured once and for all that the notion of an equatorial ocean was untenable. However, the idea that paradise could lie beyond the torrid zone, as claimed by some eminent thirteenth-century scholars, such as Albert the Great and Roger Bacon, was accepted by a number of later thinkers and it was placed on maps on the equator, in the Horn of Africa, or to the south of the equator, in the extreme south of Africa.

Fig. 8.17b. Conjectural diagram of the kind of Macrobian zonal map that could have inspired the author of the map preserved in a copy of Gregor Reisch's *Margarita philosophica* (see Fig. 8.16).

224   Mapping Paradise: A History of Heaven on Earth

ocean until the southern pole.'[131] For this map, the climatic zones have probably been interpreted according to the geographical material gathered by members of the Vienna–Klosterneuburg circle in the first half of the fifteenth century, in which all the world's place names and topographical features were located within specific *climata*, but which also left empty, as uninhabitable, terrestrial space beyond them.[132] The statements in the inscription, however, contradict the Macrobian model, and it is difficult to know how to interpret the rest of the map in the light of them.

If the map found in the *Margarita philosophica* were really intended to reflect Macrobius's divisions, the southern landmass, which stretches from the equatorial ocean to the polar zone, should contain two zones, an intensely hot and uninhabitable zone close to the equator and a temperate, habitable zone south of it. The first part of the Latin inscription would then refer to the uninhabitability of only the torrid zone by the equator. Paradise, which is indicated by the word *paradisus*, would thus have been placed in the temperate zone of the southern hemisphere, assuming, that is, that the straight line running obliquely from north-west to south-east through the southern landmass was meant to separate the torrid from the temperate southern zones. It may well be that the different colour washes differentiate uninhabitable and habitable land, the buff indicating the uninhabitable regions and the green (where the tree signs are found) the habitable regions. If the second part of the inscription is taken as following on from the first, the assumption has to be that whoever was drawing the map believed that the entire southern hemisphere was uninhabitable, in which case paradise has to be seen as situated in some exceptionally temperate part of an otherwise uninhabitable hemisphere. Another interpretation of the map would be to see the oblique line as marking the limit not just of the southern landmass but the entire southern hemisphere. In this case, the two land areas south of that line would represent the northern and southern parts of the *western* hemisphere (land on the other side of the globe, in other words). The quadripartite division of the globe was another idea that went back to classical

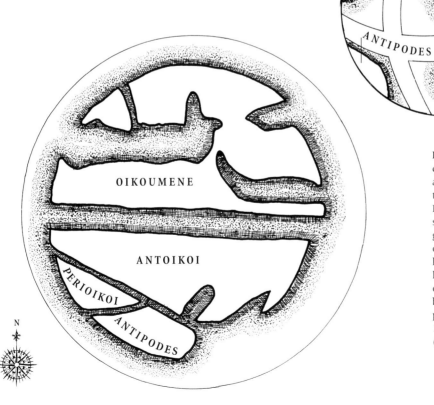

Fig. 8.18. Diagram comparing the quadripartite division of the globe according to Crates of Mallos with the map preserved in a copy of Reisch's *Margarita philosophica* showing the four parts of the globe: οἰκουμένη (inhabited part of the world); ἄντοικοι (those who live between the same meridian lines, but on the opposite sides of the equator); περίοικοι ('neighbours', those who live on the same parallel of latitude as ourselves but 180°E or W of us); ἀντίποδεως (the Antipodes).

times (Figure 8.18).[133] The map maker would then have been envisaging paradise as in the southern part of the western hemisphere or, as Dante imagined, at the antipodes of the inhabited earth. There is a place sign on the map just above the name *paradisus*, and it could be thought that the buildings indicated paradise were it not for the fact that a similar sign appears on each of the three parts of the world in the northern hemisphere. Stretching the argument, it is not inconceivable that this fourth place sign indicates the habitability of this particular part of the southern landmass.

Whichever way the zones of the map found in the *Margarita philosophica* are interpreted, the fact remains that the earthly paradise was placed in the southern hemisphere. The names of the four rivers of paradise are given nearby, each separately boxed so that it stands out from the coloured background of the sea. The name of the river Tigris is also given in a major indentation in Asia, which could be seen as the Persian Gulf, and the word *Tigris* has been written for a third time close by, but in the margin as if to draw attention to it. The anonymous compiler of the map was fully aware that the four rivers of paradise were thought to re-emerge in the northern and inhabited part of the world, but perhaps uncertain as to how best to envisage the situation of the Garden of Eden on the physical globe.

Paradise in Equatorial Africa

Sometimes, in the fifteenth century, paradise was moved from its pre-eminence at the top of the map (as when the map was turned to any orientation except east) or from its traditional eastern location (as in Giovanni di Paolo's depiction). In other cases, paradise was also removed from its rather symbolic position at the edge of the map and put somewhere else within the African landmass, so that it was no longer bordering the outer ocean or nudging a cardinal point. Fifteenth-century map makers who did not want to put the Garden of Eden on the periphery faced even more of a problem than their predecessors as paradise now had to be fitted into a plurality of spatial relationships with neighbouring territories. The Estense, or Catalan, world map of the 1450s is one of these cases, featuring paradise in equatorial Africa.

The Estense map illustrates quite clearly the contradiction inherent in showing the earthly paradise – an event/place contiguous to the known part of the earth yet unlocatable – together with a dense network of lines taken from nautical charts – structured according to distance and direction (Figure 8.19).[134] The Estense map depicts the three parts of the old world with paradise in the heart of equatorial Africa, and contains a mixture of navigational data, classical mythology, biblical history and medieval legend. Gog and Magog and the exploits of Alexander the Great, for instance, are recorded in the Far East. The Fortunate Islands are shown off the west African coast, accompanied by a long legend summarizing Isidore's discussion of the difference between the Fortunate Islands and paradise, and referring to the journeys of Saint Brendan. Another text alludes to the Platonic myth of Atlantis.[135] A bicoloured island near Cape Verde represents, according to the adjacent text, the Pillars of Hercules, relocated from Gibraltar. Jerusalem is represented by a pictorial sign, showing a tall church-like building enclosed within high walls, labelled *Santo Sepolcro*, representing Christ's tomb. Other biblical features include Noah's Ark on Mount Ararat, the ruins of Nineveh, the Queen of Sheba, and Tarsia, in Persia, the home of the Magi. Classical and mythological elements include the phoenix in Arabia and the dog-headed monsters (*Cynocephali*) in western Africa. Two scale bars are reminders that nautical charts provided the model for the outlines of the Mediterranean Sea and the names of ports and coastal towns, and account has also been taken of the Portuguese explorations along the Atlantic coast of Africa before 1446.

The Estense map is particularly interesting because of the way in which it highlighted the current commercial significance of biblical places, and for the way that

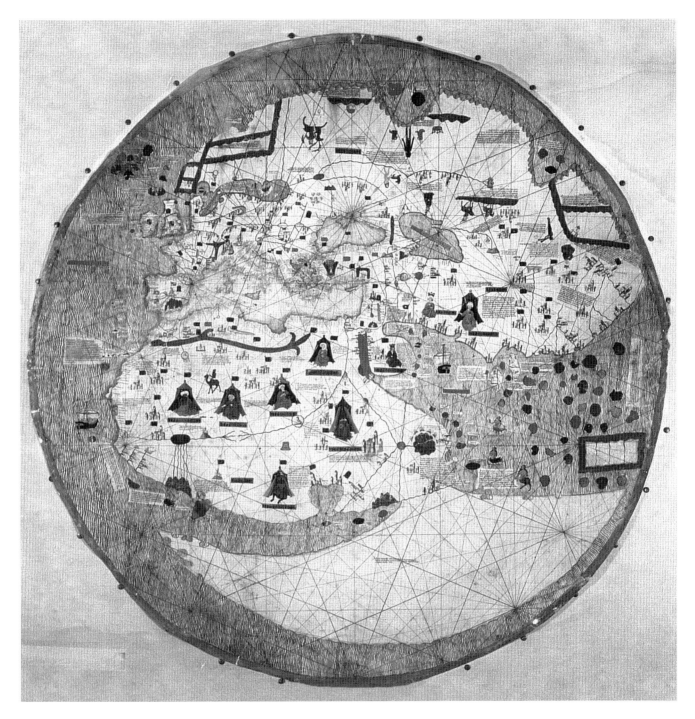

Fig. 8.19. The Catalan Estense world map. ?Majorca, c.1450–60. 112 × 100 cm. Modena, Biblioteca Estense, C.G.A. 1. The Estense map is also known as the 'Catalan map' from its affinity with Catalan charts, evidenced in the compass lines, the delineation of the Mediterranean with navigational accuracy, the use of flags and shields to identify cities and kingdoms, and the portraits of exotic rulers in their tents. All legends are in Catalan, apart from the Latin caption of the Fortunate Islands and a few Portuguese names in west Africa. The southernmost part of Africa forms a huge crescent, partly isolated from the rest by an arm of the Atlantic. A canal or river appears to run across the neck of land connecting the Atlantic and Indian oceans. The centre of the rhumb system coincides with the vertical axis of the map, but lies south of the horizontal axis to coincide instead with an east–west line defined by four inscriptions as the equator to mark a point close to the mythical Indian city of Arin. Paradise is located on the equator in east Africa.

Fig. 8.20. The Catalan Estense world map (see Fig. 8.19). Detail of the earthly paradise. Paradise is near the territory of Prester John, between Nubia and the city of Arin (*Civitasarim*), the latter prominently marked and centrally placed in the Horn of Africa, not far from the Indian Ocean, in which six islands of various sizes and colours are depicted. Paradise is guarded by five high 'Diamond Mountains' surmounted by flames. Within paradise Adam and Eve are shown standing on either side of the Tree of Life. The single river originates in the middle of the Garden before flowing out of it into a lake, thereafter to separate into four streams. One legend, near Cape Verde, explains the equal duration of night and day at the equator and another, close to paradise, emphasizes that the delights of the Garden of Eden are incomparable with the features of any other earthly region.

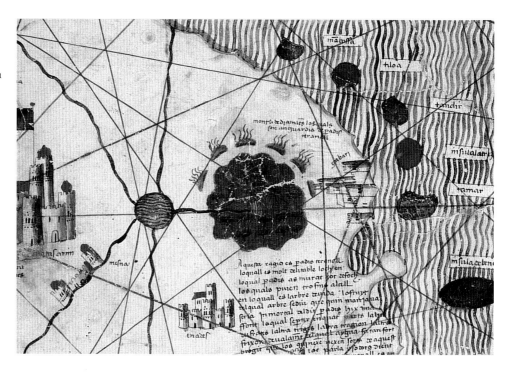

the requirements of contemporary trade were gradually eclipsing the motif of the four world empires. For example, Babylon is indicated on the Estense map as Nebuchadnezzar's great city and as an important trading centre for Indian spices. The Red Sea, painted red as on the medieval *mappae mundi*, is associated in one legend with the Exodus, while in another legend it is noted that the colouring of the waters depends on the nature of the sea floor and that most of the Indian spice trade passed through that sea. An inscription in the Gulf of Aden records that merchants coming from India were obliged to hand over a tenth of their spices as a toll. Chinese junks are depicted in the Indian Ocean, with a legend explaining their use in commerce. Other contemporary features shown on the map came from accounts of recent travel and exploration. As on Leardo's map of 1448, the Estense map pictures the Great Khan's tomb and gives the story of the sacrifice of his subjects. The coronation of the first Tartar king in 1187 is shown, and the contemporary importance of Islam is acknowledged, not least by the reference to the Muslim pilgrim route across the mountains of the interior of Libya towards lower Egypt and thence to Mecca.

Another significant characteristic of the Estense map is the reversal reflected in it of the relationship between paradise and the cartographical portrayal of the earth. On medieval *mappae mundi*, paradise was depicted as contiguous to the inhabited earth within a historical and topological definition of space. On the Estense map, in contrast, paradise was employed to define a key astronomical measure of an earthly space no longer structured by history, namely the equator. Paradise is situated right on the equator, not far from the legendary city of Arin (*Civitasarim*) that Indian and, later, Arabic geographical tradition also placed on the equator to mark the point exactly midway between the westernmost and easternmost meridians. For Arab map makers and scholars, Arin was the centre of the earth. Also shown, but farther along the equator, is the well of Meroe, in which the sun was said to be reflected when at its zenith: its perfectly perpendicular position confirmed that the well marked the equinoctial line. On the Estense map, paradise is thus found in the Horn of Africa, close the Indian Ocean to mark the line of the equator. Within paradise, Adam and Eve are shown standing by the Tree of Life (Figure 8.20).[136] The inaccessibility of paradise is emphasized by the five

Fig. 8.21a. Genoese world map. Genoa, 1457. 82×41 cm. Florence, Biblioteca Nazionale Centrale, MS Portolano 1. The ellipse-shaped map shows the three parts of the known world. North is at the top. The map retains some of the characteristics of the medieval *mappa mundi*, with its encyclopedic features. At the same time, nautical charts have obviously provided the model for most of the Mediterranean basin, western Africa and Europe, and for an incomplete system of rhumbs. The map also contains material on Asia and Africa drawn from recent and contemporary travellers' accounts. Not all the colourful vignettes and elegant legends that decorate the map are legible.

Fig. 8.21b. Diagram of Fig. 8.21a.

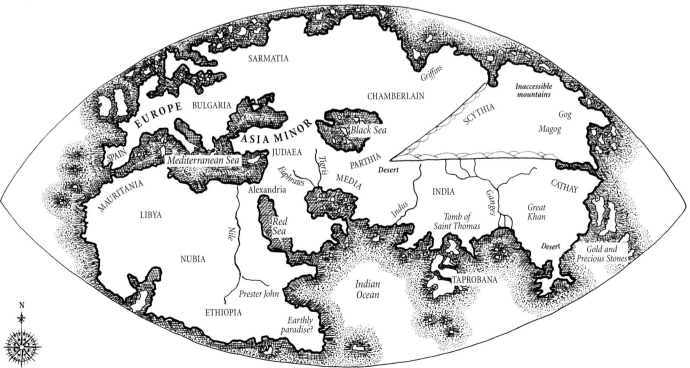

The Twilight of Paradise on Maps

high *Monts de diamants* ('Diamond Mountains'), surmounted by flames, that guard it. A text refers to its equatorial situation, its isolation by a wall of fire that reaches up to heaven, its extreme altitude, and its incomparability with any other region on earth. It also refers to the legend of the noise made by the four rivers of paradise that caused the inhabitants of the region to be born deaf.[137] Also marked on the map is the lake that was fed by the single river that flowed from paradise. To the west of paradise is portrayed the Christian ruler of the Indies, as Prester John is described in the relevant text in confirmation of the traditional association between India and Ethiopia. Prester John is shown as bearded, sporting a tall hat, and sheltering under a tent.

A map made at much the same time as the Estense map also shows Prester John, in Ethiopia, together with other features such as the tomb of Saint Thomas in India and the apocalyptic peoples of Gog and Magog within a territory occupying the whole of the north-eastern quarter of Asia (Figure 8.21).[138] The intention of the anonymous author of the so-called Genoese map, dating from 1457, was to display the 'true description of the cosmographers' as used on sea charts and to reject all 'frivolous tales'.[139] His elliptical-shaped map has north at the top. It shows the known world. It is largely a visual encyclopedia, with a full complement of precious stones, strange races, exotic animals, pigmies and mermaids. At the same time, though, and true to the map maker's promise, the influence of navigational cartography is clearly visible. The outlines of the Black and the Mediterranean seas and the Atlantic coast of Europe have been taken from charts, as have the rhumb lines. Four heads outside the map indicate the winds and the cardinal directions. Two scale bars highlight the decline of topological representation and the rise of mathematically based cartography that is represented by the Genoese map.

As is usual with several fifteenth-century world maps, the Genoese map is packed with detail derived from a range of sources, including thirteenth-century travel narratives. Marco Polo's account of his journey to China has provided much of the information about Asia; China on the Genoese map is almost wholly filled with a portrait of the Great Khan. The prominence of Asia also reflects the incorporation in the fifteenth century of new information about south-east Asia, most of which would have come from the travels of Niccolò de Conti. Thus the configuration of the Ganges delta and the outlines of the islands of Borneo, Java and the Moluccas have all been updated. Reference is made in west Africa to a trade route between Spain and India, and the text next to a ship portrayed in the Indian Ocean points out that sailors navigate in that southern sea without sight of the northern stars.[140] Paradise is not shown on the Genoese map, and a text in eastern Africa explains that the map maker was unwilling to commit himself to plotting a place whose location was uncertain. In placing his inscription where he did, however, the map maker recognized that there were those who 'depicted the paradise of delights in this region', although others 'say that it is in the east beyond the Indies', but since the authorities he consulted made no mention at all of paradise, he himself preferred to abstain from accounting for it.[141] Here the map maker has been consciously avoiding placing his 'paradise of delights' with any exactitude either in the heart of the African landmass or anywhere else.

The Coordinates of the Earthly Paradise

Paradise continued to be a feature on maps throughout the fifteenth century, placed variously in east Asia, equatorial or southern Africa, or in the remotest part of the southern hemisphere. Wherever paradise was put, the medieval cartographical genre of *mappa mundi* was obviously destabilizing. A new method, taken from nautical charts, of locating places in mathematical relationship to each other was coming to be used in the compilation of world maps. The Garden of Eden, now altered from 'event/place' to 'place', had to be described much more precisely than in traditional topological terms. Saying simply that it was 'next' to the inhabited earth was no longer enough. Early in the century, a circle of mathematicians and astronomers at the Benedictine monastery of Klosterneuburg, near Vienna, were busy compiling a list of coordinates of latitude and longitude for as many places in the world as possible. They appear to have had no direct knowledge of Ptolemy's *Geography*, which had only just arrived in Venice, but they were familiar with a method of using latitude and longitude to describe places on earth with reference to tables of coordinates. These tables had been introduced into Western Europe in the late twelfth century through translations into Latin of Greek and Arabic works such as Ptolemy's *Almagest*, the *Toledo Tables* and Al-Farghani's *Elements of Astronomy*.[142] As new centres of astronomical study arose throughout Western Europe, attempts were made to increase the number of coordinates available for places in Europe and supplement the data in the Arabic sources. Astronomical research in Germany in the following century paved the way for the achievements in and around Vienna and at Klosterneuburg during the fifteenth century.[143]

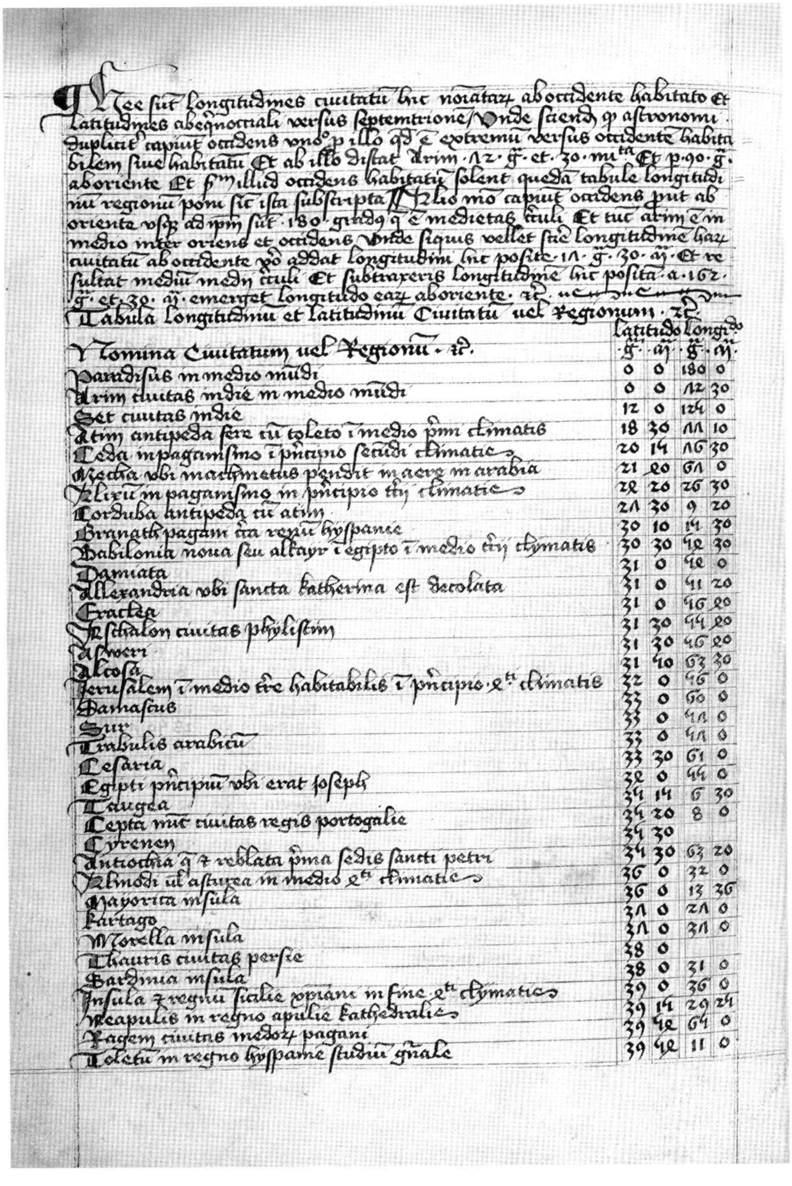

Fig. 8.22. Iohannes Gmunden's tables of latitude and longitude. Vienna, c.1436–9. Full page 28 × 21 cm. London, British Library, Add. MS 24070, fol. 74v. The list is one of the so-called 'University Tables', a type frequently copied by Gmunden and other members of the Vienna school (which extended well beyond Vienna) and so called from the annotation of major university towns in the list with the words *studium generale* (as is Toledo here). The table includes the place where the calculations were gathered, Vienna (47°46' N, 31°30' E), and the university centres of Toledo (39°54' N, 11°0' E), Bologna (44°0' N, 32°35' E), and Oxford (51°50' N, 19°14' E). Important historical sites include Jerusalem (32°0' N, 56°0' E), Carthago (37°0' N, 27°0' E), Armenia (47°0' N, 77°0' E), and Babylon (45°0' N, 78°0' E). Longitude and latitude are assigned also to important political centres such as Paris (48°0' N, 23°0' E), Rome (41°50' N, 35°25' E), Naples (39°15' N, 29°25' E), Constantinople (43°40' N, 56°40' E). Paradise on earth, the first place listed, was given the coordinates 0° N, 180° E.

The Twilight of Paradise on Maps   231

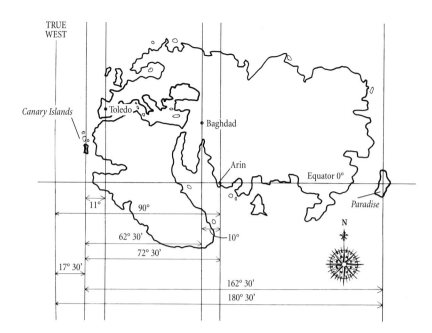

Fig. 8.23. Diagram of the latitude and longitude of paradise in relation to other earthly localities according to the list of places in Iohannes Gmunden's tables of latitude and longitude.

The first place listed in the Vienna–Klosterneuburg coordinate tables is paradise. Iohannes de Gmunden, a leading scholar in the Klosterneuburg group, devoted himself to the calculation of latitude and longitude and he gave paradise its coordinates.[144] His continuously expanding and Europeanized Alphonsine Tables were widely disseminated amongst the libraries of Europe. The opening page from one of his *Tabulae astronomiae*, that of 1436, is reproduced in Figure 8.22.[145] Here we can see how he set out the latitude, in degrees and minutes, for each place in the first pair of columns (reading left to right) and longitude in the second pair of columns. He has ranked his list by latitude, starting from the prime parallel, latitude 0° or the equator. The first two places on the list, at latitude 0°, are paradise and the mythical city of Arin, the central places of two merging, and conflicting, traditions; one, the prime historical event at the geographical periphery, and the other marking the geographical centre of the world. Both paradise and Arin are described as being *in medio mundi*, 'in the middle of the world', in other words at the equator. Significantly, however, paradise is placed first in the list.

Gmunden's head note explains the problem of selecting a prime meridian. Longitude could be calculated with reference either to an imaginary line through the Canary Islands, at the westernmost point of the known world, or to an abstract 'true west' (see Figure 8.23). The Canary prime meridian, 11° west of Toledo, had been adopted by Ptolemy and Al-Khwarizmi. The 'true west', the supposed occidental limit of the habitable earth, well to the west of the Canaries, had been put forward by Al-Zarqali in his comments in the Toledo Tables. The 'true west' was believed to be 90° west of Arin, which lay equidistant between the extreme east and the extreme west. Arin was also understood to be 10° east of Baghdad, which was at 62°30′ east of the Canaries. Subtracting from 90° (the central longitude of Arin) the longitudinal distance between Arin and the Canary Islands (72°30′) left a discrepancy of 17°30′, which meant that the 'true west' had to be 17°30′ to the west of the Canaries.

Both Al-Khwarizmi's and Al-Zarqali's prime meridians had been transmitted to Western Europe in the Latin translations of the so-called Khorazmian Tables and the Toledo Tables. Although the difference between the two systems was clearly stated in the introductory canons, confusion was introduced in later versions of the astronomical tables when longitude would sometimes be calculated from one prime meridian, sometimes from the other. In his 1436 *Tabulae astronomiae*, Gmunden indicated that his longitudes were calculated from the Canary meridian and that anyone who wanted to know the longitude from true west of any location should add 17°30′ to the longitudes presented in the table. Consistently, Gmunden gave Arin the longitude from the Canaries, that is 72°30′, instead of 90°. In the case of paradise, however, Gmunden was reluctant to abandon the highly symbolical figure of 180° that expressed its position in the extreme east in terms of mathematical astronomy. Given that the distance between east and west was half the total terrestrial circle (360°) and that Arin was at 90° from true west, true east should be 180° east of true west. Gmunden, accordingly, instructed those wanting to calculate any longitude from true east to subtract the longitude of that place from the Canaries from the figure of 162°30′, the distance of the true east from the Canaries (see Figure 8.23). The fact that Gmunden provided a way to calculate the longitude from true east seems again to underline the importance of paradise. It is clear, however, that in Gmunden's 1436 tables the longitude of paradise – alone out of all the places listed there – was reckoned not from the Canaries but from true west, in contradiction to what was said in the head note. For Gmunden and his contemporaries, however, paradise was a real place on earth, unlike all astronomical abstractions, such as the calculation of true west and true east. By adhering to the round figure of 180° longitude (instead of 162°30′) for paradise, Gmunden was underlining in astronomical terms the traditional pre-eminence of the Christian prime epochal zone. In later versions of his tables, paradise was even more explicitly privileged over the Indo-Arabic Arin by being made the prime meridian. Longitude was then calculated from the east, Jerusalem (instead of Arin) was accorded the central longitude of 90°, and the longitude of all other places was recalculated in relation to Jerusalem and thus paradise.[146] The earthly paradise, which on medieval *mappae mundi* had marked the beginning of history, now became at Klosterneuburg a quantified and spatially locatable place that functioned as an astronomical point of departure.

The geographical and astronomical data tabulated by the Vienna scholars was used to produce maps. In 1448, in the southern German town of Constance, a Benedictine monk from Salzburg compiled a south-orientated map of the world, thought to be based on the data from Klosterneuburg tables (Figure 8.24 and Plate 13).[147] The monk, Andreas Walsperger, depicted paradise as a fortified castle in the farthest east, and showed the four rivers flowing from it. On the opposite side of the world, in the extreme west, he placed the Pillars of Hercules in the Canary islands. Walsperger's map incorporated medieval encyclopedic and historical features while also showing a concern for measured distance and direction. Beneath the map, a text presents the work as having taken into account latitude and longitude and the climates found in Ptolemy's *Geography*, as well as new data from nautical charts.[148] Instructions are also given here for calculating the position of places and the distances between them by comparing the interval between any two places on the map with those marked off on the scale bar of German miles immediately below the text.[149]

The map is surrounded by nine concentric circles representing the two spheres of air (with the twelve winds) and fire and the seven planetary spheres (with the twelve zodiac signs). Beyond these is the outermost circle, that of the empyrean, the seat of God and the angelic hierarchies, which in effect encloses the entire universe. Regular marks around the innermost circle, subdividing it into 360 degrees, suggest that Walsperger's original intention had been to construct a system of coordinates over the map.[150]

Despite its air of mathematical precision, Walsperger's map retains the encyclopedic nature of a *mappa mundi*. Legends refer to Amazons and Pigmies, the Trees of the Sun and the Moon (not far from paradise), and the monstrous races of the extreme south.[151] Places mentioned in the Bible are also mapped. Jerusalem is marked in the middle of the inhabited landmasses, but because the earth is slightly displaced towards the south in the surrounding ocean, it does not indicate the geometric centre of the map. Prester John is shown north of paradise, in Asia, with Gog and Magog farther north still and enclosed by the Caspian mountains. Not far from these apocalyptic destroyers, a cannibal is shown eating human flesh, a reminder of the terrible times they were destined to bring to the rest of humanity (see Plate 13).[152] What has been completely lost in Walsperger's map is the east–west chronological structure of the archetypal *mappa mundi*. Not only have the outlines of the Mediterranean Sea come from a nautical chart, and not only does the map portray contemporary Asia with, for example, the residence of the Great Khan in Cathay, but the non-European lands are no longer the site of the succession of past empires. Indeed, these lands are openly acknowledged as non-Christian. As explained in the text, Walsperger colour-coded his place signs, using red for Christian cities and black for non-Christian cities.

Fig. 8.24a. World map by Andreas Walsperger. Constance, southern Germany, 1448. Diameter 42.5 cm (57.5 cm with surrounding circles). Vatican City, Biblioteca Apostolica Vaticana, MS Pal. Lat. 1362b, unfoliated (see also Plate 13). On the world map drawn by the Benedictine Andreas Walsperger, the earth is embraced by nine concentric circles, starting with the spheres of air and fire and going to the empyrean, the seat of God. The final circle contains the words alfa et omega – a reference to Christ, who is the beginning and end of the universe – and the names of the various angelic powers: *Cherubim, Seraphin, Dominationes, Virtutes, Potestates, Principati, Throni, Archangeli*. The map has south at the top and shows Africa, Asia and Europe. Paradise is a castle in the farthest east. The text below the map acknowledges that account has been taken of latitude and longitude, the climatic divisions from Ptolemy's *Geography*, and new data provided by navigational cartography. Clear instructions are given for the calculation of distances and the exact positions of the various places shown on the map with the help of a scale bar expressed in German miles.

The fact that Walsperger bothered to make the distinction between Christian and non-Christian parts of the world underlines the continuing importance of Christianity for European map makers. In the new cartographical context, however, contemporary geography was given more prominence than universal history in the construction of the map. Cartographically speaking, paradise, which Gmunden confirmed to be the first region of the earth and placed accordingly at the top of his tables, seems from Walsperger's map to be no more than just another eastern place, however beautiful, and close, moreover, to the Mongolian Empire. It is true that paradise has been given an unparalleled size and sophisticated design; but, instead of being the first link in the mapped chain of event/places that made up salvation history, on Walsperger's map of the world the Garden of Eden appears surrounded by the non-Christian places of contemporary Asia. If one relates its situation to the seven climates marked on the right-hand side of the map, it would seem that Walsperger believed the Garden of Eden was in the northern hemisphere. However, irrespective of the scientific precision with which the scholars of the Vienna–Klosterneuburg School defined the location of paradise, on the equator and in the extreme east, the question remained: what astronomical instrument, itinerary or empirical data could really confirm its coordinates? The move towards quantified and mathematical mapping was forcing European map makers to confront a number of new issues. Foremost among these was the exact location of the earthly paradise.

Paradise Cornered

At first sight, the coordinates of paradise in Gmunden's astronomical tables look convincing enough. In theory, the numbers pinned paradise down to a specific spot, latitude 0° (i.e. on the equator) and longitude 180° E. In practice, they were no more than abstract values, indicating an absolute east that could not be expressed in any other way. The giving of coordinates did not alter the fact that paradise was by definition elusive: it could not be found, even intellectually. It was indefinable. For all the modernity of their cartographical approach, the fifteenth-century map makers still faced the difficulty of having to find room for an unlocatable event/place on maps which had to show plottable and contemporary places. Individual historical sites could also be included, but no longer as part of the structural east–west progression of universal history.

Fra Mauro found a solution to the problem of mapping the earthly paradise. He was a Venetian monk in the Camaldolese monastery of San Michele on the island of Murano.[153] His depiction of the earthly paradise fills one of the four corner spaces left between the circular map of the inhabited earth and its square frame (Figure 8.25 and Plate 12b).[154] Although Fra Mauro has been hailed as the herald of modern cartography, and his cornered paradise as an anticipation of the imminent triumph of Renaissance science and empirical geography over medieval credulity, Fra Mauro's intention was certainly not to downgrade paradise into a decorative detail, as is often assumed.[155] He has been rightly praised by modern scholars for recording the contemporary impressive advance in geographical knowledge and for devising a new image of the world. At the same time, however, in his treatment of paradise, Fra Mauro was both true to medieval Christian tradition and an acute and ingenious cartographer, with the intellectual flexibility necessary to handle the different mindsets of his day.[156]

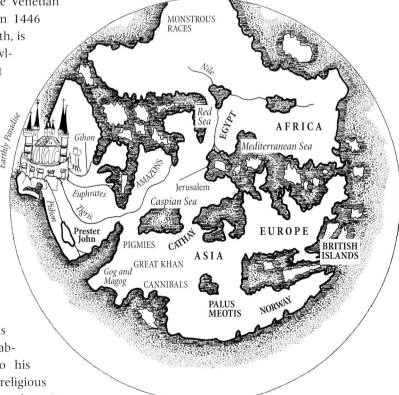

Fig. 8.24b. Diagram of Fig. 8.24a.

Fra Mauro's map was made for the Venetian *Signoria*, and was probably drawn between 1446 and 1453.[157] The map, orientated to the south, is an impressive synthesis of European knowledge about the world on the eve of the great discoveries. Fra Mauro made use of all that he could lay hands on that would serve his purposes, for example Ptolemy's *Geography* and nautical charts. His sources also included the descriptions of Roman and late antique geographical writers, such as Pliny the Elder, Solinus and Pomponius Mela, the writings of the Fathers of the Church and medieval theologians, the works of natural philosophers, in particular Aristotle and his commentators, travellers' accounts, such as those of Marco Polo and Niccolò de' Conti, and the oral reports of the sailors and merchants of his time.[158] Popular belief and regard for established authority were equally subject to his perceptive and critical approach and strong religious faith. Fra Mauro's cartographical work is the only major map to entertain – within its very legends – an engaging

Fig. 8.25a. World map by Fra Mauro. Venice, *c*.1450. Diameter 193 × 196 cm (outer frame, approx. 223 cm). Venice, Biblioteca Nazionale Marciana (see also Plate 12b). The circular map is surrounded by a square outer frame. Each of the four corner spaces contains a geographical or cosmographical diagram or image showing (clockwise from top left) the number of heavens, the flood tides and the relationship between ocean and land, the relationship between the elements and the different parts of the globe, and the location of the earthly paradise. The map is an amalgamation of different conceptual frameworks and ideas drawn from a wide variety of sources including Ptolemy's *Geography*, nautical charts and travel narratives. Little is known about the author. Documents dated 1409 and 1433 record that Fra Mauro was a Venetian monk in the Camaldolese monastery of San Michele on the island of Murano. It is known that he made a map of a monastic estate in Istria and that in 1444 he was a member of a commission charged with diverting the Brenta river. In 1448 he was apparently at work on world maps, as can be inferred from the monastery's account books, where two collaborators are also mentioned, Francesco da Cherso and Andrea Bianco. It is thought he died in 1459.

236   Mapping Paradise: A History of Heaven on Earth

dialogue with its reader, for example in the discussion of the location of Gog and Magog with reference to the works of Augustine and Nicholas of Lyre.[159] Another peculiarity is that, unlike other contemporary world maps, Fra Mauro's map, packed as it may be with textual and graphic detail (there are scores of miniature landscapes, towered cities, rivers, trees and ships), has no human figures, nor is any single feature allowed to dominate the image. His long explanatory texts, written in Venetian vernacular, are not the impersonal descriptions found on earlier *mappae mundi*, but are full of personal comments culled from every source conceivable, including people – priests from Ethiopia, for example – whom he himself had met.[160]

The individuality of Fra Mauro's approach is seen in a number of aspects of his map, but most powerfully in the way he dealt with Jerusalem and paradise. By the middle of the fifteenth century, it was becoming increasingly difficult to keep Jerusalem at the centre of a map of the world. Travel and exploration had extended European knowledge of eastern Asia, which was occupying an increasingly greater proportion of the map to the extent that the rest of the old world had to be shifted westwards, displacing Jerusalem from the centre of the map. Fra Mauro pointed out, however, that were the density of the population of the whole world to be taken into account as well

Fig. 8.25b. Diagram of Fig. 8.25a.

The Twilight of Paradise on Maps 237

as the extent of geographical space, Jerusalem would still be in the centre of the habitable world.[161] His attitude towards Ptolemy was no less revealing. Apologizing for not marking on his map the parallels, meridians, and degrees of the *Geography*, Fra Mauro commented that he found them much too much of a constraint. They made it difficult to show lands that the ancient Alexandrian had not known about. To avoid being blamed for not following Ptolemy to the letter, Fra Mauro quoted Ptolemy's own comment that he expected the *Geography* to be improved with time as new information became available.[162] He thus proved to be truly Ptolemaic precisely in his critical reading of Ptolemy's work. In the same spirit, he opted for modern rather than Ptolemaic place names. Like other map makers, he was also ready to dismiss Ptolemy's landlocked Indian Ocean in the light of what he had heard from Portuguese contacts with Arab traders.

At the same time, Fra Mauro classed all that might be seen by his readers as fanciful as part of the secrets and mysteries of God's creation. Faith in God had to come before intellectual curiosity: 'he who wants to understand has to first believe.'[163] In this context, he felt free to show the Fortunate Islands, Saint Patrick's Purgatory, the marvels of Taprobana, Java, the blessed kingdoms of Prester John and the Queen of Sheba, and let his map carry references to creatures such as dog-headed and cannibal peoples, the Phoenix and the Old Man of the Mountain.[164] He did occasionally have his doubts. He questioned the existence of monsters in Asia and Africa, and he pointed out that the Pillars of Hercules were just rocks.[165] Most interestingly, Fra Mauro did not believe that Gog and Magog were the wild hordes said to be locked up by Alexander the Great behind the Caspian mountains awaiting the Antichrist, which were by now being identified with the peoples of Ung and Mongul in distant eastern Asia. The Caspian area, he said, had been visited by Venetian merchants who had seen no apocalyptic tribes. Above all, following Augustine, Fra Mauro denied that Gog and Magog could ever be identified with a specific ethnic group, let alone a Mongolian tribe. They were to be interpreted allegorically.[166]

When, some six hundred years earlier, Beatus of Liébana omitted Gog and Magog from his world map, he did so because of the function of the map, which was to illustrate an allegorical reading of the Book of Revelation and to portray the universal Church in the sixth age.[167] Fra Mauro also had in mind a specific purpose for his map. His intention was to map the regions of his own day, not to chart the process of mankind's history, and to this end he filtered the established features, selecting only those that he considered acceptable, once updated, as part of contemporary human geography. He excluded Gog and Magog but included Persia, for example, not however as one of the four world empires, but as a contemporary territory. In a note on the map he described Persia as a country of eight kingdoms, that bred good horses and donkeys for trade with India, and that cultivated cotton. Although the majority of the population had converted to Islam, he noted, various forms of idolatry and a variety of faiths were to be found there.[168]

Given the contemporary emphasis of Fra Mauro's map of the world, the question was how to represent paradise. Fra Mauro's solution was straightforward and wholly logical. In the main circle of his map he showed the inhabited earth in considerable detail. Around it, in the four corners of the square frame, he placed four supplementary diagrams and seven texts to provide the wider context. The two upper corners demonstrated the earth's relationship with the heavenly spheres and astral bodies, and the two lower corners indicate what was thought to lie on the terrestrial globe beyond the inhabited regions of the earth. In the top left-hand corner is a diagram of the universe. Thirteen concentric circles surround the earth: first the three elements of water, air, and fire; then the seven planetary spheres; the heaven of the fixed stars; the crystalline heaven; and the empyrean. The adjacent text gives figures for the distance between the earth and each sphere as far as Saturn. The diagram in the top right-hand corner is also composed of concentric circles. It shows the influence of the moon on flood tides. In the text next to it, Fra Mauro reminded the observer that divine providence had made it possible for the heavy element of earth to emerge above the water in order to provide mankind with dry land. The circles and texts in the two bottom corners refer to the terrestrial globe. In the bottom right-hand corner, the globe is shown divided by the equator and the two tropics, and crossed by the ecliptic.[169] The text next to it contains Fra Mauro's argument for the habitability of both the equator and the southern hemisphere. In the opposite corner, bottom left, a complex vignette and a text relate to the earthly paradise, both making the point that paradise was situated far from the inhabited earth in some unknown region of the globe (Plate 12b).

The text next to the vignette of paradise is a summary of the theological and geographical debate on its location. Fra Mauro refers to Augustine, Bede, Peter Lombard and Albert the Great. Augustine is quoted to make the fundamental point that the Garden of Eden was a real place somewhere on earth, although 'very far from our habitable world and remote from human knowledge'. Fra Mauro then mentions Bede's and Peter Lombard's idea that paradise lies in the east, as well as giving Albert the Great's alternative argument that paradise is located in the south-east.[170] Fra Mauro avoided taking sides.[171] To be consistent with the explanation that paradise existed *somewhere* in the east, either on the equator or south of it, Fra Mauro placed the vignette of paradise (possibly having a professional artist to execute the painting) in one of the eastern corners of the map.[172] The vignette shows paradise as a circular walled garden surrounded by sea and containing the Tree of the Knowledge of Good and Evil, God the Father and Adam and Eve.[173] The water from the single spring flows from the middle of the Garden to separate into the four rivers immediately outside the walls. Fra Mauro showed that the otherwise inaccessible Garden, quite separate from the world of the geographers and with the entrance to it guarded by an angel, was nonetheless connected to the inhabited earth by the four rivers. In the text he described how the four rivers originated in paradise and flowed underground before re-emerging on the earth's surface, the Tigris and the Euphrates in Armenia, the Ganges in India, and the Nile in Ethiopia. In a legend on the map proper, Fra Mauro discussed the Nile's source on the inhabited earth without any direct reference to paradise, although he gave it its biblical name (Gihon) in Egypt, and called the Ganges, in India, the Pishon.[174] Fra Mauro did not place paradise on the map of the inhabited earth simply because he knew it to be outside the inhabited earth, and that it belonged to an inaccessible eastern *nowhere*, not to somewhere in the known world. Through his naming of the biblical rivers on the map, however, as well as his textual description of paradise, he made it abundantly clear that he believed in the literal existence of a terrestrial Garden of Eden. In the text by the paradise vignette he also reported the belief that Enoch and Elijah had dwelt in the Garden of Eden until Christ set them free from original sin with his Passion and, at the moment of his Ascension, took them up to the heavenly paradise.[175]

Significantly, each of the four corner diagrams on Fra Mauro's map of the world itself contains, in one way or another, the inhabited earth that is represented in full cartographical detail in the main map. It would be a mistake to regard the visual eccentricity of the vignette of paradise as an indication merely of the separateness of paradise from the world. What Fra Mauro was illustrating in the vignette, topologically, was the proximity of the inaccessible paradise to the rest of the earth, which is represented by the mountainous landscape next to the circular Garden. Both paradise and the rest of the earth are embraced by the same outer ocean. It should also be noted that the vignette on Fra Mauro's map was not a miniature *mappa mundi*, replete with history. The world portrayed in the bottom left corner of his map was the world before the Fall.

Inside paradise, God is seen commanding Adam and Eve, who are still naked and without a sense of shame, not to touch the Tree of the Knowledge of Good and Evil. The angel standing by the entrance hints at the Fall that is still to come, and the landscape outside the Garden of Eden is an empty one, not yet inhabited by Adam and Eve's descendants. Nor are Enoch and Elijah to be seen within Eden, as they were not yet there at the time of the Fall and by Fra Mauro's time they had been already taken up to heaven. In any case, there was no need to link the original paradise with the end of the world. On his map of the inhabited earth, Fra Mauro privileged geographical information over universal history, and instead of portraying a single and comprehensive space–time continuum from beginning to end, he displayed the original paradise and the contemporary human realm – geographically connected by the four rivers but historically separated by original sin and a succession of ages – on two quite different planes. Fra Mauro did not give an exact site for paradise either in the vignette or in the text because he knew that paradise belonged to the wider cosmos and to the past. His consummate skill lay in the way he managed to portray the ambiguous condition of paradise, relating its relevance in human history to contemporary geographical reality in a new fashion. His cartographical solution, that of 'cornering' paradise, has to be seen neither 'in contrast to earlier medieval maps' nor as showing 'how paradise could be banished from the world'.[176] On the contrary, the Camaldolese monk was observing medieval Christian tradition when he paired the contemporary inhabited earth, displayed in detail on the central map, with the vignette's topological portrayal of the primordial earth to be inhabited after sin, watered by the four rivers, surrounded by the ocean, and contiguous to the earthly paradise. In short, Fra Mauro had not only found a way to map the unlocatable Garden of Eden as an earthly region, but he had also shown an acute awareness of the temporal chasm decisively separating the lost paradise of delights from the human realm.

Paradise and the Geographical Discoveries

As Fra Mauro's map shows, at the end of the Middle Ages and at the beginning of the Renaissance the earthly paradise was still believed to exist somewhere on earth and yet in an inaccessible beyond. A commonplace of modern scholarship is that in the fifteenth century, and at the opening of the Age of Discovery, Western travellers embarked on voyages in the express hope of seeing for themselves the Garden of Eden. A few researchers would even like to see the search for paradise as a major motive behind the Portuguese voyages of discovery.[177] Others draw attention to the Edenic aspects of the geographical discoveries, arguing that paradise shifted from being an unattainable aspiration to becoming a geographical reality subject to empirical proof.[178] It is true that the Renaissance travel literature is peppered with references to the earthly paradise, but all such references merit a good deal more scrutiny than they usually receive. Some Portuguese explorers did see themselves as following in the footsteps of Saint Brendan, the monk who was believed to have sailed to the island of paradise, but that is a different matter.[179] Pêro Vaz de Caminha, for example, wrote to King Manuel in 1500 describing the land and people of Brazil in terms of the earthly paradise, and the sixteenth-century Portuguese humanist, Pedro Margalho, author of the *Phisices compendium* (Salamanca, 1520), posed the question whether the Portuguese, who had sailed all over the world, had discovered the earthly paradise.[180] Most famous of all, and quoted by almost every historian who has written about paradise, is the passage in Christopher Columbus's journal from his third voyage to the New World in 1498, when, on reaching the coast of South America, he found himself in a huge estuary, thought now to be the Orinoco:

> *I believe that this water may originate from [paradise], though it be far away and may come to collect there where I came and may form this lake. These are great indications of the earthly paradise, for the situation agrees with the opinion of those holy and wise theologians, and also the signs are very much in accord with this idea, for I have never read or heard of so great a quantity of fresh water coming into and near the salt. And the very mild climate also supports this view.*[181]

It has to be asked whether Christopher Columbus or any other Renaissance explorer really intended to, or thought they could, discover the location of Eden and recapture for mankind the lost paradise of delights. Once full account has been taken of the context in which those travellers lived and wrote, the answer, which is 'no', is to be found in the texts themselves. None of the earlier, thirteenth-century travellers such as Marco Polo, William of Rubruck, John of Plano Carpini and Odoric of Pordenone ever suggested that they had actually seen (or had wanted to see) the Garden of Eden itself.[182] What they referred to were other marvels. Marco Polo described not the Garden of Eden, but the mysterious Old Man of the Mountain who 'dwelled in a most noble valley shut in between two very high mountains where he had made them make the largest garden and the most beautiful that ever was seen in this world'.[183] The fabulous Christian kingdom of Prester John, a place often associated with paradise, was mentioned in almost every report from Asia (other than Marco Polo's), but not one of the authors intimated that they had set their own eyes on paradise. One of the most widely circulated narratives was *The Book of John Mandeville*, a fourteenth-century encyclopedia of geographical lore presented as the account of a real voyage to India and China but almost certainly a fiction, based on earlier, true accounts.[184] In the course of this instructive tour of the medieval East, Mandeville mentioned the earlier trip made by Adam's son, Seth, to the earthly paradise, but made it quite clear that he personally had not been there.[185] He had only heard about Eden, and indeed was unworthy even to talk about it. Admittedly, he thought of paradise as geographically in the proximity of India.[186] However, he said that paradise was in the east, that it was the highest land on earth, that it reached the sphere of the moon, and that it lay beyond dark regions, vast deserts and impassable mountains and was cut off by a wall of fire so that nobody could enter it. He warned his readers that those who had in the past tried to enter the earthly paradise had died in the attempt, and that the special grace of God was needed if anyone were to be permitted to penetrate into the Garden. John of Marignolli, another fourteenth-century writer, also reported nothing more definite than that he had been informed by the inhabitants of Ceylon that Adam's Peak, a mountain on the island, was 40 miles away from paradise and that on certain days it was possible to hear the crashing of water from the rivers that flowed out of the Garden.[187] Marignolli believed that the earthly paradise existed. He had seen for himself, he said, the rivers flowing from the Garden of Eden and had discovered in the island of Ceylon visible traces of Adam's sojourn there after his expulsion from Eden.[188] Not once, though, did he say anything that could be construed as suggesting that he had seen paradise itself. On the contrary, when he wrote of his intention to plant a flag bearing the papal coat of arms 'on the highest peak of the earth facing paradise', he did not say *in* paradise; and his expression was intended metaphorically, to indicate 'the back of beyond'.[189] Other travellers or travel writers, happy to confirm the existence of a paradise on earth, reported hearsay, and rumours of 'glimpses' of paradise, as proof of its existence. Iohannnes Witte de Hese, for example, another late fourteenth-century author who claimed to have been in Jerusalem on a pilgrimage, but, again, who was compiling an imaginary journey from other sources, claimed to have seen from a distance the walls of paradise and the sunset reflecting on it.[190] To all these writers, the earthly paradise was known to be inaccessible and beyond the horizon for ordinary humans; not one of them would have seriously thought of it as a place to travel to in this life. As the Ravenna cosmographer put it, some 800 years before Columbus:

> *We regard it as a wrong idea to think that a man, made of corruptible matter, sent on a journey, or marching by himself, could or would be able to see with his own corporeal eyes the very noble paradise or that he could or would walk on its sacred soil with his dirty feet.*[191]

The Ravenna cosmographer did believe in the earthly existence of the Garden of Eden, which he located east of India, but insisted that nobody could ever enter paradise.[192] Paradise could be approached in the visionary or mystical literature, but throughout the Middle Ages up to the Renaissance there was never a real claim made to have seen the Garden of Eden, or hope to see it, at first hand.

Like his predecessors, then, Christopher Columbus both believed in the reality of the earthly paradise and knew it to be inaccessible. On his third voyage across the Atlantic he reflected that 'Holy Scripture testifies that Our Lord made the earthly paradise and in it placed the Tree of Life, and from it issues a fountain from which flow four of the chief rivers of this world ...'[193] Then he added the critical qualification that so many commentators seem to have missed:

> *Not that I believe that to the summit of the extreme point is navigable, ... or that it is possible to ascend there, for I believe that the earthly paradise is there and to it, save by the will of God, no man can come.*[194]

Understandably overwhelmed by the sight of the great watery expanse at the mouth of the Orinoco, Columbus might well have wondered if he were close to paradise. He would never, however, have entertained any hope that he himself could enter the Garden of Eden. Amerigo Vespucci also had moments of euphoria when he too might have felt *close* to the earthly paradise, but neither he, any more than Columbus, nor anybody else was saying or intending to say that he thought he was on the very threshold of paradise.[195] For them, the mystery that paradise was simultaneously here on earth and in the beyond was sufficient in itself. The situation is not much different today, when most of us have little difficulty in accepting the idea of a remote and unreachable galaxy in deepest space.[196] Throughout the Middle Ages and into the Renaissance, Christian belief in the earthly paradise rested on the duality of the place, its separateness and (through the four rivers) connectedness with the inhabited earth. This is the paradox that provides the key to the presence of the earthly paradise on medieval maps of the world. We must recognize as misguided the eagerness of modern scholars to call attention to the 'naïveté' of the medieval and Renaissance hunt for paradise and its eventual failure. The medieval notion of an earthly paradise that was in a sense *nowhere* was an integral element of a theological and geographical debate of great complexity and subtlety. Late medieval exegesis and fifteenth-century exploration and discovery made that *nowhere* even more elusive. Above all, the Garden of Eden was located at the head of the time flow as well as in an inaccessible eastern location, and no physical journey could ever allow the return to the original innocence. Christians knew that the only possible journey to paradise was along the temporal flow that began in Eden and would end in heaven, through the death and Resurrection of Christ in Jerusalem. By the sixteenth century, the teacher in Reisch's *Margarita philosophica* was instructing his student not to worry about the precise location of the earthly paradise, but to do all that was necessary to be sure of entering the heavenly one.[197] Christians did continue, though, to concern themselves with the location of paradise. They did so, however, in terms completely different from those of their medieval predecessors.

1   Étienne Gilson, *History of Christian Philosophy in the Middle Ages* (London: Sheed and Ward, 1985), p. 465.
2   On the dating of portolan charts see above, Chapter 5, p. 116 n. 6.
3   [Pseudo-]Albert the Great, *Compendium theologicae veritatis*, II.64, in Albert the Great, *Opera omnia*, ed. by Auguste C. Borgnet, 38 vols (Paris: Vivès, 1890–9), XXXIV (1895), p. 86. The handbook was included by Borgnet in Albert the Great's *Opera omnia*, but was later attributed to Hugh Ripelin of Strasbourg (d. 1268): see Georg Steer, 'Das *Compendium theologicae veritatis* des Hugo Ripelin von Strassburg', in Maarten J. F. M. Hoenen and Alain de Libera, eds, *Albertus Magnus und der Albertismus: Deutsche philosophische Kultur des Mittelalters* (Leiden–New York–Cologne: Brill, 1995), pp. 133–54, esp. 135.
4   Ioannes Balbus, *Catholicon* (Mainz: [n. publ.] 1460; repr. Westmead, England: Gregg, 1971), entry *paradisus*: 'paradisus creditur esse quidam locus corporalis in determinata parte terre situs temperatissimus et amenus ut homo nullis perturbacionibus impeditus, spiritualibus deliciis quiete fruetur, hunc autem locum extimant sub equinoctiali esse versus partem orientalem: eo quod locum illum quidam philosophi temperatissimum asserunt.' The *Catholicon* was widely consulted by fourteenth-century scholars, including Petrarch and Boccaccio, and was still popular in the late fifteenth century; twelve editions were printed before 1500: *Corpus glossariorum latinorum*, ed. by Georg Goetz, 7 vols (Leipzig: Teubner, 1888–1923; repr. Amsterdam: Hakkert, 1965), I, pp. 215–17.
5   On the *Sentences* commentary of Durandus of St-Pourçain see Josef K. Koch, *Durandus de S. Porciano, O.P.: Forschungen zum Streit um Thomas von Aquin zu Beginn des 14. Jahrhunderts* (Münster in Westphalia: Aschendorff, 1927), pp. 5–85; on his views on the relationship between theology and science see Maria Teresa Beonio-Brocchieri Fumagalli, *Durando di S. Porziano: Elementi filosofici della terza redazione del 'Commento alle Sentenze'* (Florence: La Nuova Italia, 1969).
6   Durandus of Saint-Pourçain, *In Petri Lombardi Sententias theologicas commentariorum libri IIII*, d.17, q.3 (Venice: ex typographis Guerraea, 1571; repr. Ridgewood, NJ: Gregg, 1964), I, fol. 160v.
7   Ibid., d.17, q.3, I, fols 160v–161r.
8   Ceslaus Spicq, *Esquisse d'une histoire de l'exégèse latine au Moyen Âge* (Paris: Vrin, 1944), pp. 7, 143. On Nicholas of Lyre's life see Henri Labrosse, 'Sources de la biographie de Nicolas de Lyre', *Études Franciscaines*, 16 (1906), pp. 383–404; Labrosse, 'Biographie de Nicolas de Lyre', *Études Franciscaines*, 17 (1907), pp. 489–505, 593–608. For his popularity see also Henri de Lubac, *Exégèse médiévale: Les Quatre Sens de l'Écriture*, 4 vols (Paris: Aubier, 1959–64), IV (1964), pp. 354–5. The *Postillae perpetuae*, composed from 1323 to 1333, were later complemented by the *Postillae morales*, 35 books of brief 'moral' notes on passages of Scripture, written in 1339. The edition consulted here is the Antwerp edition of 1617, where the *Postilla litteralis* is printed together with the *Postilla moralis* and the *Glossa ordinaria*, with the controversial additions of Paul of Burgos, and the *Replicae defensivae* of Mathias Doering.
9   See, e.g., a letter written in 1406 by the Carthusian Henri Egher from Kalkar, quoted in Heinrich Rüthing, 'Kritische Bemerkungen zu einer mittelalterlichen Biographie des Nikolaus von Lyra', *Archivum Franciscanum Historicum*, 60 (1967), pp. 52–4.
10 Nicholas of Lyre, *Postilla super totam Bibliam*, 'Prologus primus', in *Biblia sacra cum Glossa ordinaria*, ed. by Leander de Sancto Martino, Ioannes Gallemart, and others, 6 vols (Douai: Baltazar Bellerus; Antwerp: Ioannes Keerbergius, 1617), I, sig. +3v E, +4r A.
11 On the extent of Nicholas's knowledge of Hebrew see Herman Hailperin, *Rashi and the Christian Scholars* (Pittsburgh: University of Pittsburgh Press, 1963), pp. 144, 290 nn. 71–2.
12 Spicq, *Histoire de l'exégèse latine* (1944), p. 337.
13 Nicholas of Lyre, *Postilla super totam Bibliam* (1617), cols 69–70. For Pecham's views on the subject see his *Tractatus de sphera*, ed. and transl. by Bruce R. MacLaren, PhD Dissertation, University of Wisconsin, Madison, WI (1978), pp. 120–8 (Latin text); pp. 194–203 (English translation). See also above, Chapter 7, pp. 172–3.
14 Nicholas of Lyre, *Postilla super totam Bibliam* (1617), cols 109–10. The Cherubim signify in both cases the angels who govern corporeal nature, and in particular the movement of the heavenly bodies.
15 Ibid., col. 73. See above, Chapter 3, pp. 49–50.
16 On the life of Duns Scotus see William A. Frank and Allan B. Wolter, *Duns Scotus, Metaphysician* (West Lafayette, IN: Purdue University Press, 1995), pp. 1–16; Carolus Balic, 'The Life and Works of John Duns Scotus', in Bernardine M. Bonansea and John K. Ryan, eds, *John Duns Scotus: 1265–1965* (Washington, DC: Catholic University of America Press, 1965), pp. 1–27.
17 Duns Scotus's commentary on the *Sentences* survives in three forms: his own original notes for the lectures, the *Lectura*; copies of his students's notes, the *Reportatae*; and his own final revision, the *Ordinatio*. The Vatican edition of the *Ordinatio* was begun in 1950 and will eventually run to fifteen volumes; see Carolus Balic, 'The Nature and Value of a Critical Edition of the Complete Works of John Duns Scotus', in Bonansea and Ryan, eds, *John Duns Scotus* (1965), pp. 368–79. The Latin text is available in the Vivès reprint of Wadding's 1639 edition of Scotus's *Opera omnia* and in Felix Alluntis's 1963 edition in the series *Biblioteca de autores cristianos*. References here are to the original Wadding edition.
18 Duns Scotus's attitude towards Aristotelianism and his view of the relationship between philosophy and theology have been widely debated. See, e.g., Richard Cross, *The Physics of Duns Scotus: The Scientific Context of a Theological Vision* (Oxford–New York: Clarendon Press, 1998); Stephen Brown, 'Scotus's Method in Theology', Daniele Crivelli, 'La teologia come scienza pratica nel prologo dell'"Ordinatio" di Duns Scoto', and Roberto Zavalloni, 'Ragione e fede in Duns Scoto nel contesto del pensiero medievale', in Leonardo Sileo, ed., *Via Scoti: Methodologica ad mentem Ioannis Duns Scoti*, 2 vols (Rome: Antonianum, 1995), I, pp. 229–43, II, pp. 593–609, and 611–32; Orlando Todisco, 'Il metodo della concordanza e l'unione attitudinale dei fenomeni naturali: Progetto scotista di una scienza cristiana', in Camille Bérubé, ed., *Homo et mundus* (Rome: Societas Internationalis Scotistica,1984), pp. 301–20; Pietro Scapin, 'Capisaldi di un'antropologia scotista', in *Deus et homo ad mentem Ioannis Duns Scoti* (Rome: Societas Internationalis Scotistica, 1972), pp. 269–91; Efrem Bettoni, 'Duns Scoto nella scolastica del secolo XIII', Edward D. O'Connor, 'The Scientific Character of Theology according to Scotus', and Agostino Trapè, 'La nozione della teologia presso Scoto e la scuola agostiniana', in *De doctrina Ioannis Duns Scoti*, 4 vols (Rome: Cura Commissionis Scotisticae, 1968), I, pp. 101–11, III, pp. 3–50, IV, pp. 73–81; Efrem Bettoni, 'The Originality of the Scotistic Synthesis', in Bonansea and Ryan, eds, *John Duns Scotus* (1965), p. 36.
19 John Duns Scotus, *Quaestiones in librum I Sententiarum*, 'Prologus', q.1, in *Opera omnia*, ed. by Lucas Wadding, 12 vols (Leiden: [n.p.], 1639), V/1, p. 4: 'Tenent enim philosophi perfectionem naturae, et negant perfectionem supernaturalem. Theologi vero cognoscunt defectum naturae, et necessitatem gratiae, et perfectionum supernaturalium.' This passage is quoted, e.g., by Étienne

Gilson, *Jean Duns Scot: Introduction à ses positions fondamentales* (Paris: Vrin, 1952), p. 14, and Efrem Bettoni, 'Duns Scoto denuncia l'insufficienza dell'antropologia filosofica', in *Deus et homo ad mentem Ioannis Duns Scoti* (1972), p. 245.

20 John Duns Scotus, *Quaestiones in librum II Sententiarum*, d.17, q.2, in *Opera omnia* (1639), VI/1, p. 789: 'Quantum ad hoc, quod dicunt, quod est ad Orientem, dico, quod in quocunque situ sit, potest intelligi esse ad Orientem, quilibet enim punctus in terra, potest intelligi esse ad Orientem in comparatione ad coelum, vel respectu diversorum situum terrae, praeterquam respectu duorum polorum, qui sunt immobiles.'

21 Ibid.: 'Unde per esse ad Orientem, non habetur determinatio terrae certa, et precisa, in qua scilicet parte terrae sit paradisus.'

22 Ibid., d.17, q.2, pp. 788–90. Scotus refers to the story on Mount Olympus: see above, Chapter 7, pp. 175 and 187 n. 58. He does not, however, mention Alexander of Hales's theory that there was a 'tepid' area between the middle and the higher zone of the air.

23 Peter Lombard, for instance, following Bede and the Gloss, had said that paradise was higher than any other mountain, and this was why the Flood had not reached it. See above, Chapter 3, pp. 49–50 and 59 nn. 33, 40, 41.

24 John Duns Scotus, *Quaestiones in librum II Sententiarum*, d.17, q.2, in *Opera omnia* (1639), VI/1, pp. 789–90. See Pietro Scapin, 'Il significato fondamentale della libertà divina secondo Giovanni Duns Scoto', in *De doctrina Ioannis Duns Scoti* (1968), II, pp. 519–66.

25 John Duns Scotus, *Quaestiones in librum II Sententiarum*, d.29, q. unica, in *Opera omnia* (1639), VI/1, p. 918. Thomas Aquinas, for instance, had discussed the geographical interpretation of the flaming sword of the Cherubim as referring to the torrid zone: see above, Chapter 7, pp. 181–2 and 189 n. 117.

26 John Duns Scotus, *Quaestiones in librum I Sententiarum*, 'Prologus', q.1, 3, in *Opera omnia* (1639), V/1, pp. 25–6, 37, 102. See O'Connor, 'Scientific Character of Theology' (1968), p. 34; Bettoni, 'Originality of Scotistic Synthesis' (1965).

27 Pierre d'Ailly, *Ymago mundi*, ed. by Edmond J. P. Buron, 3 vols (Paris: Maisonneuve Frères, 1930), II, p. 458: 'locus amenissimus in partibus orientis longo terre et maris tractu a nostro habitabili segregatus'. For the relevant passage in Isidore see above, Chapter 3, pp. 47–8. D'Ailly, however, might also have referred to Pseudo-Isidore, *Liber de ordine creaturarum*, X, ed. by Manuel C. Díaz y Díaz (Santiago de Compostela: Universidad de Santiago de Compostela, 1972), pp. 156–67. For John of Damascus see *De fide orthodoxa: Versions of Burgundio and Cerbanus*, 25.1, ed. by Eligius M. Buytaert (St Bonaventure, NY: Franciscan Institute, 1955), pp. 106–7; *PG* XCIV, col. 915. For Bede see above, Chapter 3, pp. 48–9. Strabo is the supposed author of the *Glossa* (see above, Chapter 3, p. 50 and 59 n. 41, also for the relevant passage). For Peter Comestor see *Historia scholastica, Liber Genesis*, XIII, *PL* CXCVIII, col. 1067. D'Ailly, *Ymago mundi* (1930), I, p. 240, also describes the climate in paradise as very mild and temperate.

28 Pierre d'Ailly, *Ymago mundi* (1930), II, p. 458; see above, Chapter 7, pp. 174–5. D'Ailly refers to Alexander of Hales and not to Alexander of Aphrodisias, as Buron, pp. 458–9 note n. and Jean Delumeau, *Une histoire du paradis*, I, *Le Jardin des délices* (Paris: Fayard, 1992), p. 78 (p. 54 in the English edition, 1995), have wrongly assumed.

29 Pierre d'Ailly, *Ymago mundi* (1930), II, p. 460. D'Ailly took this story, like various other material on paradise, from Bartholomaeus Anglicus, *De proprietatibus rerum*, XV.112 (Frankfurt am Main: Wolfgang Richter, 1601; repr. Frankfurt am Main: Minerva, 1964), p. 683. Following Bartholomaeus Anglicus, he quotes Basil and Ambrose; but no reference to this story is found in Basil's *Hexaemeron* or in Ambrose's *Hexaemeron*, *De paradiso* or *Epistula* to Sabinus. A reference to the extraordinary noise of the waters falling from the mountain of paradise, however, is found in *The Book of Sir John Mandeville*, XXXIII, ed. by Michael C. Seymour (Oxford: Clarendon Press, 1967), p. 222.

30 Pierre d'Ailly, *Ymago mundi* (1930), I, pp. 230–43. For the relevant passages from Roger Bacon see above, Chapter 7, pp. 177–9.

31 Pierre d'Ailly, *Ymago mundi*, II (1930), pp. 460–3.

32 See above, Chapter 3, pp. 49 and 59 n. 35 and Chapter 5, pp. 102 and 121 n. 81.

33 Pierre d'Ailly, *Ymago mundi* (1930), III, p. 648.

34 Aeneas Sylvius Piccolomini, *Opera inedita*, ed. by Giuseppe Cugnoni (Rome: Salviucci, 1883; repr. Farnnborough: Gregg, 1968), p. 234: 'reprehensio ignavie christianorum, et exhortatio ad ulciscendam iniuriam'. Piccolomini's work was first published as a *Dialogus* in Rome in 1475; Giuseppe Cugnoni included it, under the title *Tractatus*, in his edition of Aeneas Sylvius Piccolomini, *Opera inedita*, pp. 234–99. See Enea Silvio Piccolomini, *Dialogo su un sogno: Dialogus de somnio quodam*, ed. by Alessandro Scafi (Turin: Aragno, 2004).

35 Aeneas Sylvius Piccolomini, *Opera inedita* (1883; 1968), pp. 273–4.

36 Ibid., p. 276: 'Quis enim temperatissimum locum inveniri sub ferventissimo sole crediderit?'

37 Ibid., pp. 279–80. In Scripture, says Saint Bernardino, the right side is preferred to the left.

38 Ibid., pp. 280–2.

39 Ibid., p. 281: 'ridiculum illud dictum est.'

40 Solinus, *Collectanea rerum memorabilium*, XXXII.1–15 ed. by Theodor Mommsen, 2nd edn (Berlin: Weidmann, 1958), p.138. Solinus, as well as King Juba II of Mauritania, as reported by Pliny, *Naturalis historia*, V.9.10, and other ancient writers, held that the Nile rose in Lower Mauritania, flowed in an easterly direction, was engulfed by the sands of the Sahara, reappeared as the Niger, again sunk into the earth and finally came to light once more near the Great Lake of Debaya as the Nile itself. See also above, Chapter 3, pp. 49 and 59 n. 35), and Chapter 5, pp. 102 and 121 n. 81.

41 Aeneas Sylvius Piccolomini, *Opera inedita* (1883; 1968), p. 282. See Horace, *De arte poetica*, 359: 'quandoque bonus dormitat Homerus'.

42 Aeneas Sylvius Piccolomini, *Opera inedita* (1883; 1968), pp. 250–1, 274. On the amalgamation of the classical and biblical paradise see Delumeau, *Une Histoire du paradis* (1992), pp. 21–7 (pp. 10–15 in the English edition, 1995). On Isidore see above, Chapter 3, pp. 47–8.

43 Aeneas Sylvius Piccolomini, *Opera inedita* (1883; 1968), p. 273. The story of Cheremon, bishop of Nilopolis, who at the time of the persecutions of Decius in Egypt escaped with his wife to a mountain in Arabia and disappeared – their relatives never found their bodies – is narrated by Dionysius of Alexandria, *Epistola III ad Fabium episcopum Antiochiae*, IX, in *Epistolae*, *PG* X, col. 1306; and by Eusebius, *Historia ecclesiastica*, VI.42, *PG* XX, col. 614. See also Hippolyte Delehaye, 'Les Martyrs d'Egypte', *Analecta Bollandiana*, 40 (1922), p. 16.

44 Aeneas Sylvius Piccolomini, *Opera inedita* (1883; 1968), p. 283.

45 Ibid., p. 274.

46 Ibid., p. 271: 'mons preruptus incredibilis altitudinis, cuius cacumen nubile trascendit'.

47 Ibid., p. 272: 'sive illud flumen sive mare fuit, aquam latissimam atque profundissimam … cuius non erat ulteriorem ripam videre'.

48 Ibid.: 'que longe venustior prioribus fuit'. The character Aeneas Sylvius remembers at this point that also Giovanni da Capestrano, a disciple of

49 Ibid.: 'summi regis decretum'.
50 Ibid., p. 277: 'Non hic concordia, sed veritas querenda est.'
51 Ibid., p. 255. Saint Bernardino makes these remarks when commenting on writings about the life of Constantine and his supposed donation of temporal power to the Church.
52 See above, Chapter 2, pp. 35 and 41 n. 9.
53 Aeneas Sylvius Piccolomini, *Selected Letters*, transl. and ed. by Albert R. Baca (Northridge, CA: San Fernando Valley State College, 1969), p. 43 (English translation); for the original text see Aeneas Sylvius Piccolomini, *Der Briefwechsel*, ed. by Rudolf Wolkan, II, *Briefe als Priester und als Bischof von Triest (1447–1450)* (Vienna: Hölder, 1912), p. 80: 'ceterumque de Nilo fuit hodie nobis sermo. non video, quo pacto ex paradisi monte progredi possit.' On the problem of the source of the Nile in classical literature see Brigitte Postl, *Die Bedeutung des Nil in der römischen Literatur* (Vienna: Notring, 1970).
54 The full title of the work is *Cosmographia seu rerum ubique gestarum historia locorumque descriptio*. See *Opera geographica et historica (Cosmographia seu rerum ubique gestarum historia locorumque descriptio)* (Helmstadt: Sustermann, 1699). A recent edition of *Europa* has appeared, ed. by Adrianus van Heck (Vatican City: Biblioteca Apostolica Vaticana, 2001). For the structure of the work, the date of composition and the relationship between the two parts, *Asia* and *Europa*, see Luigi Guerrini, 'Geografia e politica in Pio II', in Claudio Crescentini and Margherita Palumbo, eds, *Nymphilexis: Enea Silvio Piccolomini, l'umanesimo e la geografia* (Rome: Shakespeare and Company2, 2005), pp. 27–52, esp. 30–3, and Nicola Casella, 'Pio II tra geografia e storia: La "Cosmographia"', *Archivio della Società Romana di Storia Patria*, 3rd series, 26 (1972), pp. 35–112, esp. 35–56.
55 In his *Commentarii rerum memorabilium quae temporibus suis contigerunt*, V, ed. by Adrianus van Heck (Vatican City: Biblioteca Apostolica Vaticana, 1984), p. 248, Pius II lists the sources of his description of Asia. See also Guerrini, 'Geografia e politica in Pio II' (2005); Casella, 'Pio II tra geografia e storia' (1972), pp. 66–80; Remo Ceserani, 'Note sull'attività di scrittore di Pio II', in Domenico Maffei, ed., *Enea Silvio Piccolomini Papa Pio II* (Siena: Accademia Senese degli Intronati, 1968), pp. 99–115.
56 Aeneas Sylvius Piccolomini, *Cosmographia* (1699), p. 4: 'nec nos falsa pro veris astruemus, scientes nil tam contrarium esse historiae quam mendacium'. See also p. 5.
57 Ibid., XIV–XV, XXIV, pp. 24–6, 46: 'aurati velleris ficta fabula'.
58 Ibid., V, pp. 11–12. Delumeau's analysis of Piccolomini's views on the habitability of the equator and the location of paradise is not entirely accurate: *Une histoire du paradis* (1992), pp. 187, 205 (pp. 143, 156 in the English edition, 1995).
59 Moreover, when considering the size of Asia, calculated as 45,000 stadia wide (a stade was for Piccolomini 177.6 metres), from Rhodes, in the west, to India and Scythia, in the east, Piccolomini does not mention the earthly paradise, which was traditionally thought to be the first and easternmost region of Asia: VII, p. 14.
60 Aeneas Sylvius Piccolomini, *Cosmographia* (1699), VII, pp. 15–16.
61 Ibid., XCVIII, p. 208. First he referred to the theory that the earliest inhabitants of Asia Minor must have come from Syria since they were descendants of Adam, who had been expelled from the earthly paradise to the Damascene Fields, and then he considered the possibility that the human race had expanded into the rest of the habitable earth from Armenia, where Noah's Ark had come to rest.
62 Casella, 'Pio II tra geografia e storia' (1972), pp. 80–5. In a pamphlet written in 1461, when he was pope, to Mohammed II, *Epistola ad Mahometem II (Epistle to Mohammed II)*, XVII.194, ed. by Albert R. Baca (New York: P. Lang, 1990), pp. 94–5 (English translation); p. 203 (Latin text), Piccolomini also complained of the difficulty of arriving at a figure for the circuit of the earth, citing the different figures given by Eratosthenes, Hipparchus, Dionysius of Cnido, and Pliny the Elder. Giovanni Antonio Campano, 'Vita Pii secundi', in Piccolomini, *Opera omnia*, ed. by Marcus Hopperus (Basle: Henricus Petri, 1551), sig. c4r, wrote that Piccolomini's work on the *Cosmographia* was interrupted by his death. According to Casella, 'Pio II tra geografia e storia' (1972), p. 102, Piccolomini anticipated the era of the geographical discoveries with his unquenchable desire for the truth.
63 Aeneas Sylvius Piccolomini, *Opera inedita* (1883; 1968), pp. 251–2: 'Non omnium, que sub celo fiunt, mortales notitiam habent … inscrutabilis altitudo divini consilii cum vult et quo modo vult mirabiles operatur effectus.'
64 On the so-called 'transitional period' in the history of cartography (1300–1460) see David Woodward, 'Medieval *Mappaemundi*', in Brian Harley and David Woodward, eds, *The History of Cartography*, I, *Cartography in Prehistoric, Ancient and Medieval Europe and the Mediterranean* (Chicago: University of Chicago Press, 1987), pp. 314–18.
65 The first Latin manuscript with maps is Biblioteca Vaticana, MS Vat. Lat. 5698. On the fifteenth-century 'rediscovery' of Ptolemy's *Geography* see below, Chapter 9, pp. 254–8.
66 See Patrick Gautier Dalché, 'D'une technique à une culture: Carte nautique et portulan au XIIe et au XIIIe siècle', in *L'uomo e il mare nella civiltà occidentale: Da Ulisse a Cristoforo Colombo, Atti della Società ligure di storia patria*, 32 (1992), pp. 284–312; Gautier Dalché, *Carte marine et Portulan au XIIe siècle: Le 'Liber de existencia riveriarum et forma maris nostris Mediterranei' (Pise, circa 1200)* (Rome: École Française de Rome, 1995), has discussed the portolano he edited (c.1200) as the result of a compromise between technical and clerical culture. See also Catherine Delano-Smith, 'Maps and Religion in Medieval and Early Modern Europe', in David Woodward, Catherine Delano-Smith, and Cordell D. K. Yee, *Plantejaments i objectius d'una història universal de la cartografia/Approaches and Challenges in a Worldwide History of Cartography* (Barcelona: Institut Cartogràfic de Catalunya, 2001), p. 246, where she quotes Jonathan J. G. Alexander, *Medieval Illuminators and their Methods of Work* (New Haven: Yale University Press, 1992), pp. 22–3.
67 The origin of the navigational charts is uncertain, and much discussed. The genre is far less homogeneous than the impression given in the brief description here, and a number of changes took place in the course of the fourteenth and fifteenth centuries. See Tony Campbell, 'Census of Pre-Sixteenth-Century Portolan Charts', *Imago Mundi*, 38 (1986), pp. 67–94; Campbell, 'Portolan Charts from the Late Thirteenth Century to 1500', in Harley and Woodward, eds, *History of Cartography*, I (1987), pp. 371–463; as noted above, Patrick Gautier Dalché, *Carte marine et Portulan au XIIe siècle* (1995), suggests that the charts emerged a century earlier than is generally thought.
68 Paris, Bibliothèque Nationale de France, Rés. Ge. B 1118. About 180 sailing charts and bound atlases have survived, of which a few dozen were drawn before 1400, but many more must have existed.
69 For a list of Vesconte's maps see Woodward, 'Medieval *Mappaemundi*' (1987), pp. 363–4. On the inclusion in world maps of the shape of the

Mediterranean basin as featured in portolan charts see above, Chapter 1, p. 29 n. 4.

70 See Marino Sanudo, *Liber secretorum fidelium crucis super Terrae Sanctae recuperatione et conservatione ...* (*The Book of the Secrets for Crusaders* [lit. *for the Faithful of the Cross*] *on the Recovery and Keeping of the Holy Land*), published in Jacques Bongars, *Gesta Dei per Francos*, II, *Liber secretorum fidelium crucis super Terrae Sanctae recuperatione etc.* (Hanover: Typis Wechelianis apud heredes Ioannis Aubrii, 1611; repr. Jerusalem: Masada Press, 1972). Paolino of Venice's *Chronologia magna* covered the period between Adam and Eve and his own day (*c.*1320). See Evelyn Edson, 'Reviving the Crusade in the Fourteenth Century: Sanudo's Schemes and Vesconte's Maps', in Rosamund S. Allen, ed., *Eastward Bound: Medieval Travel and Travellers 1050–1500* (Manchester: Manchester University Press, 2004), pp. 131–4; Oswald A. W. and Margaret Dilke, 'Mapping a Crusade', *History Today*, 39 (1989), pp. 31–5; Woodward, 'Medieval *Mappaemundi*' (1987), p. 314; Jürgen Schulz, 'Jacopo de' Barbari's View of Venice: Map Making, City Views, and Moralized Geography before the Year 1500', *Art Bulletin*, 60 (1978), pp. 425–74, esp. 445, 452; Bernard Degenhart and Annnegrit Schmitt, 'Marino Sanudo und Paolino Veneto', *Römisches Jahrbuch für Kunstgeschichte*, 14 (1973), pp. 1–137, esp. 64–71; Konrad Miller, *Mappaemundi: Die ältesten Weltkarten*, 6 vols (Stuttgart: J. Roth: 1895–8), III, *Die kleineren Weltkarten* (1895), p. 132. It is noteworthy that for his map of the Holy Land, Sanudo located the most significant places within a reference grid system, a device to allow the reader to find on the map the place listed in the text.

71 There is a vast literature on Prester John: see, e.g., István P. Bejczy, *La Lettre du prêtre Jean: Une Utopie médiévale* (Paris: Imago, 2001); Gioia Zaganelli, ed., *La lettera del Prete Gianni* (Milan: Luni, 2000); Robert Silverberg, *The Realm of Prester John* (Athens, OH: Ohio University Press, 1996); Jacqueline Pirenne, *La Légende du 'Prêtre Jean'* (Strasbourg: Presses Universitaires de Strasbourg, 1992); Charles F. Beckingham and Bernard Hamilton, eds, *Prester John, the Mongols, and the Ten Lost Tribes* (Brookfield, VT: Variorum, 1996); Vsevolod Slessarev, *Prester John: The Letter and the Legend* (Minneapolis: University of Minnesota Press, 1959).

72 Alfred W. Crosby, *The Measure of Reality: Quantification and Western Society, 1250–1600* (Cambridge: Cambridge University Press, 1997), describes the gradual emergence of a quantitative model of reality. His comparison between the Middle Ages and the Renaissance, however, is informed by a firm belief in historical progress.

73 For example, in the twelfth century, Benjamin of Tudela had travelled to Baghdad. See *The World of Benjamin of Tudela: A Medieval Mediterranean Travelogue*, ed. by Sandra Benjamin (Madison, WI–London: Fairleigh Dickinson University Press–Associated University Presses, 1995); *The Itinerary*, ed. and transl. by Ezra H. Haddad (Baghdad: Eastern Press, 1945). Before the thirteenth century, various Italian cities and the Catalans had trading posts in various parts of the Islamic world. On the European view of the Muslim world see Michael Frassetto, ed., *Western Views of Islam in Medieval and Early Modern Europe: Perception of Other* (New York: St. Martin's Press, 1999); Norman Daniel, *The Arabs and Mediaeval Europe* (London: Longman; Beirut: Librairie du Liban, 1975), pp. 218–20; Daniel, *Islam and the West: The Making of an Image* (Edinburgh: Edinburgh University Press, 1966) and Richard W. Southern, *Western Views of Islam in the Middle Ages* (Cambridge, MA: Harvard University Press, 1962).

74 On the influence of Arabic mapping on Western cartography, see, for example, Carsten Drecoll, *Idrīsī aus Sizilien: Der Einfluss eines arabischen Wissenschaftlers auf die Entwicklung der europäischen Geographie* (Egelsbach–Frankfurt am Main–Munich–New York: Hänsel-Hohenhausen, 2000); I am grateful to Sonja Brentjes for her comments on the Arabic, Persian and Ottoman material incoporated into the Western maps and portolan charts and for providing me with the text of her paper 'Revisiting Catalan Portolan Charts: Do They Contain Elements of Asian Provenance?' presented at the conference *Maps and Images: How They Have Transmitted Visual Knowledge along the Silk Road*, Zurich 14–15 May 2004. On the missionary and crusading late medieval approach to the East see Kenneth M. Setton, gen. ed., *A History of the Crusades*, 6 vols (Madison: University of Wisconsin Press, 1969–89), III, *The Fourteenth and Fifteenth Centuries*, ed. by Harry W. Hazard (1975); and V, *The Impact of the Crusades on the Near East*, ed. by Harry W. Hazard and Norman P. Zacour (1985), esp. Marshall W. Baldwin, 'Missions to the East in the Thirteenth and Fourteenth Centuries', pp. 452–518.

75 William of Rubruck, *Itinerarium*, XVIII.5, in *Sinica Franciscana*, 3 vols (Ad Claras Aquas, Quaracchi, Florence: Apud Collegium S. Bonaventurae, 1929–36), I (1929), *Itinera et relationes fratrum minorum saeculi XIII et XIV*, ed. by Anastasius van den Wyngaert, p. 211; see also *The Mission of Friar William of Rubruck: His Journey to the Court of the Great Khan Möngke, 1253–1255*, transl. by Peter Jackson, ed. by Peter Jackson and David Morgan (London: The Hakluyt Society, 1990), p. 129. Isidore, *Etymologiae*, XIII.15.2 and 17.1, ed. by Wallace Martin Lindsay, 2 vols (Oxford: Clarendon Press, 1911), had maintained the received opinion that the Caspian sea was a gulf of the outer ocean. See also Pliny, *Naturalis historia*, VI.15.36–7 and Strabo, *Geography*, II.5.14.

76 Marco Polo, *Il Milione*, ed. by Luigi Foscolo Benedetto (Florence: Olschki, 1928); Marco Polo, *The Description of the World*, ed. by A. C. Moule and Paul Pelliot (London: Routledge, 1938); John of Monte Corvino, *Epistolae*, in *Sinica Francescana*, I, *Itinera et relationes* (1929), pp. 333–55; John of Monte Corvino, *Letters*, in Christopher Dawson, ed., *The Mongol Mission: Narrative and Letters of the Franciscan Missionaries in Mongolia and China in the Thirteenth and Fourteenth Centuries* (London–New York: Sheed and Ward, 1955; repr. New York: AMS Press, 1980), pp. 224–31; Odoric of Pordenone, *Relatio*, in *Sinica Francescana*, I, *Itinera et relationes* (1929), pp. 379–495; Odoric of Pordenone, *The Eastern Parts of the World Described*, in Henry Yule, ed. and trans., *Cathay and the Way Thither, being a Collection of Medieval Notices of China*, new ed. by Henri Cordier, 4 vols (London: Cambridge University Press, 1913–16), II (1913).

77 John of Plano Carpini, *Ystoria Mongalorum*, in *Sinica Francescana*, I, *Itinera et relationes* (1929), pp. 1–130; John of Plano Carpini, *The Story of the Mongols*, transl. by Erik Hildinger (Boston: Branden, 1996); Simon of St-Quentin, *Histoire des Tartars*, ed. by Jean Richard (Paris: Geuthner, 1965). On Vincent of Beauvais and the travel literature on the Mongols see Anna-Dorothee von den Brincken, 'Die Mongolen im Weltbild der Lateiner um die Mitte des 13. Jahrhunderts unter besonderer Berücksichtigung des "Speculum Historicale" des Vincenz von Beauvais OP', *Archiv für Kulturgeschichte*, 57 (1975), pp. 117–40; Gregory G. Guzman, 'The Encyclopedist Vincent of Beauvais and his Mongol Extracts from John of Plano Carpini and Simon of St-Quentin', *Speculum*, 49 (1974), pp. 287–307. See also Michèle Guéret-Laferté, *Sur les routes de l'empire mongol: Ordre et rhétorique des relations de voyage aux XIIIe et XIVe siècles* (Paris: Honoré Champion, 1994); Scott D. Westrem, ed., *Discovering New Worlds: Essays on Medieval Exploration and Imagination* (New York–London: Garland, 1991); Igor de Rachewiltz, *Papal Envoys*

78 *to the Great Khans* (London: Faber and Faber, 1971); Giotto Dainelli, *Missionari e mercadanti rivelatori dell'Asia nel Medio Evo* (Turin: Unione Tipografico-Editrice Torinese, 1960).

78 A fresh influx of information on central Asia and the Far East was eventually provided by the renewed contact and revitalized trade that followed the Portuguese discovery in 1497 of a sea route to the east around the Cape of Good Hope.

79 South-orientated Arabic maps could also have exercized an influence on the southern orientation of fifteenth-century European maps but there is no conclusive evidence to support the suggestion.

80 Macon, Bibliothèque Municipale, MS Franc. 2, fol. 19r. The manuscript is thought to have been copied by an artist from the Île de France or Normandy between 1473 and 1480 (information supplied by the Bibliothèque Municipale, Macon). See Lars Ivar Ringbom, *Paradisus terrestris: Myt, bild och verklighet* (Copenhagen–Helsinki: Ejnar Munksgaards–Akademiska Bokhandeln–Nordiska Antikvariska Bokhandeln, 1958), pp. 23–4; Alexandre de Laborde, *Les Manuscrits à peintures de la Cité de Dieu de Saint Augustin*, 2 vols (Paris: Societé de Bibliophiles François-Rahir, 1909), II, pp. 448–64, esp. 461–2.

81 See Camille Gaspar and Frédéric Lyna, *Les Principaux Manuscrits à peintures de la Bibliothèque Royale de Belgique*, 3 vols (Brussels: Bibliothèque Royale Albert 1er, 1937–45; repr. 1984–9), III (1989), pp. 95–7, 215–29, 238–40, 386–90; Guy de Pörck, *Introduction à La Fleur des histoires de Jean Mansel (XVe siècle)* (Ghent: E. Claeys-Verheughe, 1936). *La Fleur des histoires* opens with the Creation, and includes the Old Testament narrative, a history of various ancient peoples, a history of the Roman Empire, a history of the Church – up to Pope Clemente V (1305–14) – and a history of France up to King Charles VI (1368–1422).

82 Brussels, Bibliothèque Royale de Belgique/Koninklijke Bibliotheek van België, MS 9260, fol. 11r. In the versions of the so-called first family of manuscripts, the list of Roman provinces is at the end of Book II: de Pörck, *Introduction à La Fleur des histoires* (1936), pp. 21–42.

83 The text of the Prologue has been transcribed by de Pörck, *Introduction à La Fleur des histoires* (1936), pp. 66–7.

84 *Rudimentum noviciorum* (Lübeck: Lucas Brandis, 1475). The world map is at fols 85v–86r. See Wesley A. Brown, *The World Image Expressed in the Rudimentum novitiorum* (Washington, D.C.: Geography and Map Division, Library of Congress, 2000); Tony Campbell, *The Earliest Printed Maps: 1472–1500* (London: British Library; Berkeley: University of California Press, 1987), pp. 144–51; Anna-Dorothee von den Brincken, 'Universalkartographie und geographische Schulkenntnisse im Inkunabelzeitalter (Unter besonderer Berücksichtigung des "Rudimentum Noviciorum" und Hartmann Schedels)', in Bernd Moeller, Hans Patze, and Karl Stackmann, eds, *Studien zum städtischen Bildungswesen des späten Mittelalters und der frühen Neuzeit* (Göttingen: Vandenhoeck-Ruprecht, 1983), pp. 398–411.

85 *Rudimentum noviciorum* (1475), fol. 473. See von den Brincken, 'Universalkartographie' (1983), pp. 401–2. For the authorship of the text, a link with the Franciscan monastery at Lübeck has been proposed, ibid., pp. 403–5. French editions were printed in 1488 and 1491 under the title *La Mer des hystoires* ('The Sea of Histories').

86 The geographical section immediately follows the world map at fols 87r–118r. The account of the third age begins with Abraham and contains a description of the Holy Land, at fols 176r–199v, based on the *Descriptio Terrae Sanctae* of Burchard of Monte Sion (c.1283). This section is accompanied by a map of Palestine, at fols. 174v–175r. See von den Brincken, 'Universalkartographie' (1983), pp. 406–11.

87 Brown, *The World Image Expressed in the Rudimentum novitiorum* (2000), pp. 21–2; Campbell, *Earliest Printed Maps* (1987), pp. 144–5. For the attribution to Brandis, see also Leo Bagrow, 'Rust's and Sporer's World Map', and 'Essay of a Catalogue of Map-Incunabula', *Imago Mundi*, 7 (1950), pp. 35, 107. For the attribution to Johannes Columpna, see Heinrich Winter, 'Notes on the World Map in the "Rudimentum novitiorum"', *Imago Mundi*, 9 (1952), p. 102.

88 The map in the *Rudimentum noviciorum* presents a spatial structure similar to that of the Lambeth Palace or Reading map (see Figure 6.1).

89 Winter, 'Notes on the World Map in the "Rudimentum novitiorum"' (1952), thought that the two men in paradise were a Jew and a Christian living in mutual tolerance, an image that would be inspired by the prologue of Lull's *De adventu Messie*.

90 Hanns Rüst was a woodcutter active in Augsburg between 1477 and 1484. His map was copied by Hanns Sporer. For dating and existing exemplars, see Rodney W. Shirley, *The Mapping of the World: Early Printed World Maps, 1472–1700* (London: Early World Press, 2001), pp. 5–8; Campbell, *Earliest Printed Maps* (1987), pp. 79–84, with ample bibliography; Destombes, *Mappemondes* (1964), p. 253; Klaus Stopp, 'Relation between the Circular Maps of the World of Hanns Rüst and Hans Sporer', *Imago Mundi*, 18 (1964), p. 81; Erwin Rosenthal, 'Concerning the Dating of Rüst's and Sporer's World Maps', *Papers of the Bibliographical Society of America*, 47 (1953), pp. 156–8; Bagrow, 'Rust's and Sporer's World Map' (1950), pp. 32–6; Michael C. Andrews, 'An Early Printed Map in the Pierpont Morgan Library', *Geography Journal*, 65 (1925), pp. 469–70. Rüst's map survives in a unique example formerly owned by the geography professor Franz von Wieser, who sold it to the Pierpont Morgan Library in New York in 1911 through the intermediary of the antiquarian Ludwig von Rosenthal of Munich. It was reproduced in colour as a separate sheet by Ludwig Rosenthal's Antiquariat, Munich 1924, with a text by von Wieser. For a description of the map, Hugo Hassinger, 'Deutsche Weltkarten-Inkunabeln', *Zeitschrift der Gesellschaft für Erdkunde zu Berlin* (1927), pp. 455–82.

91 'Das ist die mapa mundi und alle land und kungkreich wie sie ligend in der gantze welt.' See Campbell, *Earliest Printed Maps* (1987), p. 81. The nomenclature on the map is in German.

92 Gerald A. Danzer, *Images of the Earth on Three Early Italian Woodcuts: Candidates for the Earliest Printed Maps in the West* (Chicago: Newberry Library, 1991), p. 29, saw the four rivers of paradise giving shape and structure to the map.

93 'Berg Caspij verschlossen Gog Magog'. See Andrew Gow, 'Gog and Magog on *Mappaemundi* and Early Printed World Maps: Orientalizing Ethnography in the Apocalyptic Tradition', *Journal of Early Modern History*, 2/1 (1998), pp. 61–88, for the Rüst map, p. 83, and Anna-Dorothee von den Brincken, 'Gog und Magog', in Walter Heissig and Claudius C. Müller, eds, *Die Mongolen* (Innsbruck: Pinguin; Frankfurt am Main: Umschau, 1989), pp. 27–9.

94 Another explanation of the letters IHS is that they stand for *In hoc signo* [vinces], 'By this sign shall thou conquer', the words that the emperor Constantine heard as he dreamt of a cross in the sky on the eve of the battle of the Milvian bridge, over the Tiber, against Maxentius in 312: see Eusebius, ΕΙΣ ΤΟΝ ΒΙΟΝ ΤΟΥ ΜΑΚΑΡΙΟΥ ΚΩΝΣΤΑΝΤΙΝΟΥ ΒΑΣΙΛΕΩΣ, I.27–32, ed. by Ivar A. Heikel, in *Eusebius Werke*, I (Leipzig: Hinrichs, 1902), pp. 21–3; *Life of Constantine*, transl. by Averil Cameron and Stuart G. Hall (Oxford: Clarendon Press, 1999), pp. 79–82.

95 Venice, Biblioteca Nazionale Marciana, MS Fondo Ant. Ital. Z 76 (4783), fol. 10. Andrea Bianco worked with Fra Mauro, whose world map is discussed below.

96 In 1436 Andrea Bianco produced an atlas containing various nautical drawings, seven maritime charts and two world maps, now in the Marciana, Venice. See Woodward, 'Medieval *Mappaemundi*' (1987), pp. 317, 432–3; Destombes, *Mappemondes* (1964), p. 246. In 1448 he compiled a portolan chart: Milan, Ambrosiana, MS F. 260 inf. (1), fol. 10.

97 On Saint Macarius and paradise, see above, Chapter 3, p. 52.

98 Gog and Magog are shown on a peninsula in north Asia: 'Gog Magog chest Alexander gie ne roccon ecarleire de tribus iudeoron'. 'Gog and Magog of the Jewish tribes whom Alexander enclosed in the rocks (mountains) ages ago'. See Gow, 'Gog and Magog' (1998), p. 78, and Valerie I. J. Flint, *The Imaginative Landscape of Christopher Columbus* (Princeton, NJ: Princeton University Press, 1992), pp. 19–20.

99 For bibliographical references to Leardo maps, see above, Chapter 1, p. 29 n. 1. On the 1448 map, now in Vicenza, see Giovanni Dal Lago, 'Giovanni Leardo, *Mapa Mondi*', in Guglielmo Cavallo, ed., *Due mondi a confronto, 1492–1728: Cristoforo Colombo e l'apertura degli spazi*, 2 vols (Rome: Istituto Poligrafico e Zecca dello Stato-Libreria dello Stato, 1992), I, pp. 159–62, with bibliography.

100 Marco Polo, *The Description of the World* (1938), pp. 167–8, 374. See also Dal Lago, 'Giovanni Leardo' (1992), I, p. 162; Pompeo Durazzo, *Il planisfero di Giovanni Leardo* (Mantua: Eredi Segna, 1885), pp. 16–17. That the Apostle Thomas preached in India had been claimed by Jerome, *Epistula* LIX.4, ed. by Isidorus Hilberg, *CSEL* LIV (Vienna– Leipzig: Tempsky, 1910; repr. Vienna: Österreichische Akademie der Wissenschaften, 1996), p. 546; it was also restated by Beatus in his description of the preaching of the apostles throughout the earth (see above, Chapter 5, p. 109), and mentioned by Marco Polo, *The Description of the World* (1938), p. 388, when talking about his tomb in the Indian province of Maabar.

101 On the eastern orientation of medieval *mappae mundi* see above, Chapter 5, p. 117 n. 22.

102 The Borgia map was so called because it was acquired in Portugal by Cardinal Stefano Borgia in 1794. Destombes, *Mappemondes* (1964), pp. 239–40; Roberto Almagià, *Monumenta cartographica vaticana*, I, *Planisferi carte nautiche e affini dal secolo XIV al XVII* (Vatican City: Biblioteca Apostolica Vaticana, 1944), pp. 27–9.

103 Almagià, *Monumenta cartographica vaticana*, I, (1944), p. 27.

104 Almagià suggested that the two human figures are those of Adam and Eve. This would seem unlikely for, although only one figure is bearded, both have halos.

105 Renate Frohne, ed., *Die Historienbible des Johannes von Udine (Ms 1000 Vad)* (Bern: P. Lang, 1992), p. 54: 'Nilus, Tigris, Phison hic circumit Indiam et trahit arenas aureas, Eufrates hic vadit per Caldeam.' The map is in Stuttgart, Württembergische Landesbibliothek, YMS Theol. Fol. 100, fol. 3v. Similar world maps are found in Munich, Bayerische Staatsbibliothek, MS Clm. 721, fol. 3v, and Wolfenbüttel, Herzog August Bibliothek, YMS Helmst. 442. See Destombes, *Mappemondes* (1964), pp. 182, 186, 189; Jörg-Geerd Arentzen, *Imago mundi cartographica: Studien zur Bildlichkeit mittelalterlicher Welt- und Ökumenekarten unter besonderer Berücksichtigung des Zusammenwirkens von Text und Bild* (Munich: Fink, 1984), pp. 127–9. John of Udine's historical work was completed in 1349.

106 The Olomouc (now in the Czech Republic) *mappa mundi* was formerly in the Studienbibliothek, MS g/9/155. See Anna-Dorothee von den Brincken, 'Mappamundi', in John B. Friedman and Kristen Mossler Figg, eds, *Trade, Travel, and Exploration in the Middle Ages: An Encyclopedia* (New York–London: Garland, 2000), p. 365; Scott D. Westrem, 'Against Gog and Magog', in Sylvia Tomasch and Sealy Gilles, eds, *Text and Territory* (Philadelphia: University of Pennsylvania Press, 1998), pp. 62–5; Anton Mayer, *Mittelalterliche Weltkarten aus Olmütz* (Prague: Geographisches Institut der Deutschen Universität in Prag, 1932), Plate 1.

107 The Wieder–Woldan map takes its name from the former owners of the two known extant copies. The Venetian origin has been argued because of the marking of the province of *Venetia*: Arthur M. Hind, *Early Italian Engraving: A Critical Catalogue with Complete Reproduction of All the Prints Described*, 7 vols (London: Knoedler; New York: Quaritch, 1938–48), I (1938), pp. 291, 296–7. See also Shirley, *Mapping of the World* (2001), pp. 13–4; Campbell, *Earliest Printed Maps* (1987), pp. 23–6; Arentzen, *Imago mundi cartographica* (1984), p. 232; Destombes, *Mappemondes* (1964), pp. 90–1; Erich Woldan, 'A Circular, Copper-Engraved, Medieval World Map', *Imago Mundi*, 11 (1954), pp. 13–6; Youssuf Kamal, *Monumenta cartographica Africae et Aegypti*, 5 vols (Cairo: 1926–51), I (1926), p. 1507.

108 Paris, Bibliothèque Nationale de France, Cartes et Planes, Rés. Ge AA 562. The map was first attributed to Columbus by Charles de la Roncière, *La Carte de Christophe Colomb* (Paris: Champion, 1924); 'Une Carte de Christophe Colomb', *Revue des Questions Historiques*, 7 (1925), pp. 27–41. He suggested that the work had been inspired by Christopher Columbus for presentation to the Spanish monarchs. He compared the legends on the map with Columbus's glosses to d'Ailly's *Ymago mundi* and pointed to the emphasis given on the chart to Genoa and the Genoese explorers. In Roncière's view, knowledge of Portuguese discoveries in Africa could be explained by Columbus's sojourn in Lisbon before 1484 or 1485, and the Spanish flag over Granada could suggest that the map had been completed after the Spanish reconquest of Granada from the Moors in 1492. The attribution to Columbus, however, has been questioned: see Monique Pelletier, 'Peut-on encore affirmer que la BN possède la carte de Christophe Colomb?' *Revue de la Bibliothèque Nationale*, 45 (1992), pp. 22–5; Gaetano Ferro, entry on the map in Cavallo, ed., *Due mondi a confronto* (1992), I, pp. 506–11.

109 Antoine de la Sale, *La Salade*, in *Oeuvres complètes*, ed. by Fernand Desonay, I (Paris: Droz, 1935), p. 139: 'Auquel paradis n'est personne qui y peust entrer ne monter, tant pour les roydes montaignes dont il est tout enclos, fors de l'entree. Et la sont tant de diverses conditions de dragons, de serpens, de cucques et d'aultres tresfieres bestes venimeuses qui la sont et par celles treshaultes montaignes, que les bestes d'icelles se approchent a le element du feu; pour ce sont tant fieres et ardans, comme le plaisir de Dieu est qui l'a ainsi ordonné. Et pour ce disent les maistres que, ainsi que ledit paradis terrestre est le chief de la terre pour sa treshaulte haulteur, sont les enfers en la plus basse parfondeur du corps de la terre.'

110 The identification of the Nile with the biblical Gihon flowing out of paradise was confirmed on maps coming from the portolan chart tradition, such as the world map made in 1385 by the Majorcane William Soler and the Pizzigani world map (1367). See Theobald Friedrich Fischer, *Sammlung mittelalterlicher Welt- und Seekarten italienischer Ursprungs und aus italienischen Bibliotheken und Archiven* (Venice: Ferdinand Omgania, 1886), p. 171.

111 *El libro del conoscimiento de todos los reinos (The Book of Knowledge of all Kingdoms)*, ed. and transl. by

Nancy F. Marino (Tempe, AZ: Arizona Center for Medieval and Renaissance Studies, 1999), p. 57: '… Nilo, el qual nasçe de las altas sierras del polo Antartico do dizen que es el Paraisso Terrenal'. Although most commonly described in the literature as a Franciscan or a mendicant friar, the most recent edition of the manuscript suggests that the author was in fact a herald: ibid., pp. XXXVIII–XLIV.

112 Jacob Perez de Valencia, *Commentaria in Psalmos* (Valencia: [n.p.], 1484), Psalm 103.5–9. See also William G. L. Randles, *De la terre plate au globe terrestre: Une Mutation épistémologique rapide, 1480–1520* (Paris: Armand Colin, 1980), pp. 22–6, and Laurinda S. Dixon, 'Giovanni di Paolo's Cosmology', *Art Bulletin*, 67/4 (1985), p. 610. A similar location of paradise in the Mountains of the Moon is advocated by Arnold von Harff in his *Pilgerfahrt*, see *The Pilgrimage of Arnold von Harff Knight from Cologne, through Italy, Syria, Egypt, Arabia, Ethiopia, Nubia, Palestine, Turkey, France and Spain, Which he Accomplished in the Years 1496 to 1499*, ed. by Malcolm Henri Letts (London: Hakluyt Society, 1946), pp. 173–4. See also Francesc Relaño, *The Shaping of Africa: Cosmographic Discourse and Cartographic Science in Late Medieval and Early Modern Europe* (Aldershot, Hants: Ashgate, 2002), pp. 197–202 and 202–4 nn. 1–35.

113 *Passio Sancti Bartholomaei Apostoli*, 1, in *Acta Apostolorum Apocrypha*, ed. by Constantinus Tischendorf, Ricardus Adelbertus Lipsius, and Maximilianus Bonnet, 2 vols (Hildesheim: Georg Olms, 1959), II/1, ed. by Maximilianus Bonnet, p. 129. For the Leardo maps see Figures 1.1, 8.6 and Plate 2a.

114 For the tripartite division of India see Relaño, *The Shaping of Africa* (2002), pp. 53–5, who quotes Armando Cortesão, *History of Portuguese Cartography*, 2 vols (Coimbra: Junta de Investigações do Ultramar, 1969–71), I (1969), p. 256, for the first appearance of the term 'three Indies' in Guido of Pisa (*c*.1118); Scott D. Westrem, *Broader Horizons: A Study of Johannes Witte de Hese's Itinerarius and Medieval Travel Narratives* (Cambridge, MA: The Medieval Academy of America, 2001), p. 246; John K. Wright, *The Geographical Lore of the Time of the Crusades: A Study in the History of Medieval Science and Tradition in Western Europe* (New York: American Geographical Society, 1925; repr. New York: Dover, 1965), pp. 272–303.

115 On the Nile as border between Asia and Africa see also above, Chapter 5, p. 89; for the theory on the western course of the Nile, see Chapter 3, pp. 49 and 59 n. 35 and Chapter 5, pp. 102 and 121 n. 81. On India embracing sub-Saharan Africa seee Relaño, *The Shaping of Africa* (2002), pp. 53–5, and William G. L. Randles, 'Notes on the Genesis of the Discoveries', *Studia*, 5 (1960), pp. 20–46. How an Asian Ethiopia could be seen distinct from Africa is particularly clear, for example, on a map in Florence, Biblioteca Medicea Laurenziana, MS Med. Pal. 89, fol. 7v. See the reproduction in Cavallo, ed., *Due mondi a confronto* (1992), I, p. 69 and Mario Tesi, ed., *Monumenti di cartografia a Firenze (Secc. X–XVII)* (Florence: Biblioteca Medicea Laurenziana, 1981), Plate IX.

116 Relaño, *The Shaping of Africa* (2002), pp. 51–72.

117 Friedrich Zarncke, ed., 'Der Priester Johannes', *Abhandlungen der Philologisch-Historischen Classe der Königlich Sächsischen Gesellschaft der Wissenschaften*, 8 (1876), pp. 912–13. On Prester John see above, n. 71.

118 Jordan of Sévérac, *Mirabilia descripta*, in Wilhelm Baum and Raimund Senoner, eds, *Indien und Europa im Mittelalter* (Klagenfurt: Kitab, 2000), p. 136; see also Jordan of Sévérac, *Les Merveilles de l'Asie*, ed. by Henri Cordier (Paris: Paul Geuthner, 1925), p. 119. For the move of Prester John from Asia to Africa see Relaño, *The Shaping of Africa* (2002), pp. 51–74.

119 The map was once in the collection of Albert Figdor in Vienna but its present location is unkown. See Arthur Dürst, 'Die Weltkarte von Albertin de Virga von 1411 oder 1415', *Cartographica Helvetica*, 13 (1996), pp. 18–21; Destombes, *Mappemondes* (1964), pp. 205–7; Kamal, *Monumenta cartographica Africae et Aegypti*, IV/3 (1938), pp. 1376–7; Roberto Almagià, 'Il mappamondo di Albertin de Virga (1415)', *Rivista Geografica Italiana*, 21 (1914), pp. 92–6; Franz R. von Wieser, *Die Weltkarte des Albertin de Virga aus dem Anfange des XV. Jahrhunderts in der Sammlung Figdor in Wien* (Innsbruck: Schwick, 1912). The map has been missing since an auction in 1932: H. Gilhofer and H. Ranschburg, 'Albertin de Virga: Weltkarte auf Pergament, mit Feder und in Farben gezeichnet, Venedig 141(5)', in *Versteigerungs-Katalog No. VIII zur Auktion am 14. und 15. Juni 1932 in Luzern* (Lucerne: H. Gilhofer & H. Ranschburg, 1932), pp. 17–18 n. 56. The map is signed and dated (the last figure of the year is partially illegible): von Wieser, *Die Weltkarte des Albertin de Virga* (1912), p. 5: 'A. 141?: Albirtin di virga me fezit. in vinexia.' Albertin de Virga is known to have made a portolan chart in 1409: Paris, BN, Rés. Ge. D 7900. See Giuseppe Caraci, 'Un'altra carta di Albertin da Virga', *Bollettino della Reale Società Geografica Italiana*, 63 (1926), pp. 781–6.

120 Von Wieser, *Die Weltkarte des Albertin de Virga* (1912), p. 9: 'Die Karte des Albertin de Virga ist ein Mittelding zwischen mittelalterlicher Weltkarte und Portulan-Karte.' Von Wieser associated the map with the Medici Atlas, Florence, Biblioteca Laurenziana. See also George H. T. Kimble, 'The Laurentian World Map with Special Reference to its Portrayal of Africa', *Imago Mundi*, 1 (1935), pp. 29–33.

121 On the Table of Solomon see von Wieser, *Die Weltkarte des Albertin de Virga* (1912), pp. 6–7.

122 According to Gunnar Thompson, *America's Oldest Map – 1414 AD: First Edition Draft of Technical Report* (Seattle: Misty Isles Press–The Argonauts, 1995), p. 4, the map displays the first representation of America. In Thompson's view, the large peninsula north-west of Europe would represent North America, the large island south-east of Asia, South America.

123 According to Destombes, *Mappemondes* (1964), p. 206, followed by Dürst, 'Die Weltkarte von Albertin de Virga von 1411 oder 1415' (1996), p. 20, and others, the sign indicates the Mountains of the Moon, but the map sign for mountain all over the map is completely different.

124 See Kristen Lippincott, 'Giovanni di Paolo's "Creation of the World" and the Tradition of the "Thema Mundi" in Late Medieval and Renaissance Art', *Burlington Magazine*, 132/1048 (July 1990), pp. 460–8.

125 László Baránszky-Jób, 'The Problems and Meaning of Giovanni di Paolo's Expulsion from Paradise', *Marsyas*, 8 (1957–9), pp. 1–6, argued that God the Father is shown here not in the act of creating the universe, but pointing to the terrestrial orb to which Adam and Eve are being banished, an interpretation which neglects the basic fact that the Garden of Eden was thought to be located on earth. Laurinda Dixon, 'Giovanni di Paolo's Cosmology' (1985), subsequently proposed that the subject of this picture is not the Creation, but God setting the date of the Incarnation at the very moment of the Expulsion. She has observed that God is indicating the point on the zodiac, between Aries and Taurus, that marks the period of the feast of the Annunciation, which is on 25 March. Kristen Lippincott, however, 'Giovanni di Paolo's "Creation of the World"' (1990), has more convincingly argued that Giovanni di Paolo was depicting the creation of the world in the form of a *thema mundi*. Confirmation that the panel would have been interpreted as the Creation

126. Public readings from Dante were also a regular part of Siena's cultural life: see, for example, John Pope-Hennessy, *Paradiso: The Illuminations to Dante's Divine Comedy by Giovanni di Paolo* (London: Thames and Hudson, 1993), p. 13. Thus, Dantesque allusions in this panel would hardly have been lost on a Sienese public well versed in the *Commedia*.

comes also from a copy of Giovanni's painting, a choir-book illumination by Pellegrino di Mariano that illustrates Genesis 1.1.

127. Parma, Biblioteca Palatina, MS 1614. See Relaño, *The Shaping of Africa* (2002), p. 199; Alberto Capacci, 'Vesconte Maggiolo, *Atlante nautico*', in Cavallo, ed., *Due mondi a confronto* (1992), I, pp. 335–8 (where the map sign for paradise has not been recognized); Georges Grosjean, ed., *Vesconte Maggiolo, "Atlante nautico del 1512": Seeatlas vom Jahre 1512* (Dietikon-Zurich: Urs Graf, 1979); Giuseppe Caraci, 'La produzione cartografica di Vesconte Maggiolo (1511–49) ed il Nuovo Mondo', in *Memorie Geografiche*, 4 (1958), pp. 223–90, esp. 236–40. Vesconte's chart shows Europe and the Mediterranean, Africa, part of Asia, various islands in the Atlantic Ocean (Azores, Madeira, Cape Verde, Ascension and St Helena) and the easternmost part of southern America (labelled as *tera de brazile/jamase santa croxe*). There are 25 views of cities and 18 portraits of rulers, three of which are naked black figures shown in western Africa, the chiefs of local tribes mentioned in fifteenth-century sources. Vesconte drew mostly on Portuguese information for his nomenclature in Africa.

128. Zonal maps are described in Chapter 7 above, pp. 165–6.

129. The first edition of the *Margarita philosophica* was published in Freiburg im Breisgau in 1503 (see below n. 197). I am grateful to Jürgen Baumgart, who shared with me the material he gathered on the map that he discovered in the library of the Heinrich-Suso-Gymnasium, Constance, Germany (a collection derived from the old library of the local Jesuitic college), which he presented at the *8. Kartographiehistorische Colloquium*, Bern, 1996, and which is the subject of an article due to be published in *Cartographica Helvetica*.

130. Jürgen Baumgart, personal comunication.

131. 'Ista pars terre est habitationis intemperate. Et extra climata incipiens A mare oceano usque polum Ant[arcticum].'

132. Dana Bennett Durand, *The Vienna–Klosterneuburg Map Corpus of the Fifteenth Century: A Study in the Transition from Medieval to Modern Science* (Leiden: Brill, 1952), pp. 180–9.

133. The Greek philosopher Crates of Mallos is supposed to have constructed a globe in the second century BC showing a quadripartite division. See Brian Harley and David Woodward, with Germaine Aujac, 'Greek Cartography in the Early Roman World', in Harley and Woodward, eds, *History of Cartography*, I (1987) pp. 162–3.

134. Modena, Biblioteca Estense, C.G.A. 1. See Mauro Bini, 'Dalla cosmografia classica alla cartografia del Quattrocento', in *Alla scoperta del mondo: L'arte della cartografia da Tolomeo a Mercatore* (Modena: Il Bulino, 2001), pp. 25–32; Ernesto Milano, *Il mappamondo Catalano Estense*, facsimile edition with commentary and transcription of the toponyms, ed. by Annalisa Battini (Dietikon-Zurich: Urs Graf, 1995); Ilaria Luzzana Caraci, 'Mappamondo detto Estense o Catalano', entry in Cavallo, ed., *Due mondi a confronto* (1992), I, pp. 494–5; Milena Ricci, 'Mappamondo anonimo catalano', entry in Ernesto Milano, *La carta del Cantino e la rappresentazione della terra nei codici e nei libri a stampa della Biblioteca Estense e Universitaria* (Modena: Il Bulino, 1991), pp. 58–62; George H. T. Kimble, *The Catalan World Map of the R. Biblioteca Estense at Modena* (London: Royal Geographical Society, 1934).

135. See Honorius Augustodunensis, *Imago mundi*, I,35, ed. by Valerie I. J. Flint, *Archives d'Histoire Doctrinale et Littéraire du Moyen Âge*, 49 (1982), p. 66.

136. It would seem that the map maker interpreted the biblical account as meaning that the division of the single stream into the four rivers of paradise took place outside the Garden of Eden. As we shall see, the question of the division of the single river (and of the distinction between the garden of paradise and the wider region of Eden) will become crucial in post-Renaissance mapping of paradise.

137. Milano, *Il mappamondo Catalano Estense* (1995), p. 185. Quoting Isidore, the legend rehearses the traditional remarks compiled by Honorius Augustodunensis and Gautier de Metz.

138. See Gow, 'Gog and Magog' (1998), pp. 78–9, and Heinrich Wuttke, *Über Erdkunde und Karten des Mittelalters* (Leipzig: Mahler, 1853), p. 46. On Gog and Magog see above, Chapter 5, pp. 95 and 119 n. 52 and Chapter 6, pp. 141 and 157 n. 60.

139. Florence, Biblioteca Nazionale Centrale, Port. 1. The inscription is not easy to read and has been variously interpreted. Fischer, *Sammlung mittelalterlicher Welt* (1886), p. 156: 'Hec est vera cosmographorum cum marino accordata terra, quorundam frivolis narrationibus reiectis 1447.' Sebastiano Crinò, 'La scoperta della carta originale di Paolo dal Pozzo Toscanelli che servì di guida a Cristoforo Colombo per il viaggio verso il Nuovo Mondo', *L'Universo*, 22 (1941), pp. 379–405: 'hec est vera cosmographorum cum marino accordata descricio quorundam frivolis narracionibus reiectis 1457'. *Marino* has also been read as both a reference to Marino di Tiro or Marino Sanudo and as a generic sailor (*marinus*) to mean cartography for the use of navigation. See, for example, Giovanni Battista Baldelli Boni, *Il Milione di Marco Polo*, 2 vols (Florence: Giuseppe Pagani, 1827), I, p. XXXI n. 1: 'Haec est vera Cosmographorum cum Marino accordata descriptio, quottidie frivolis narrationibus iniectis 1417.' See Alberto Capacci, 'Anonimo, *Planisfero* detto *Genovese*', in Cavallo, ed., *Due mondi a confronto* (1992), I, pp. 491–4. Crinò's claim that the Genoese world map was the map by Paolo dal Pozzo Toscanelli that was used by Christopher Columbus has been confuted.

140. Fischer, *Sammlung mittelalterlicher Welt* (1886), p. 185: 'In hoc mari australis poli aspectu navigant septentrionali absconso.'

141. Ibid., p. 171: 'In hac regione depinxerunt quidam paradisum deliciarum. Alii vero ultra Indias ad orientem eum esse dixerunt. Sed quoniam haec est cosmographorum descriptio qui nullam de eo fecerunt mentionem, ideo obmittitur hic de eo narratio.'

142. See, for example, José Chabás and Bernard R. Goldstein, *The Alfonsine Tables of Toledo* (Dordrecht–Boston–London: Kluwer, 2003) and bibliography therein.

143. Bennett Durand, *The Vienna–Klosterneburg Map Corpus* (1952), pp. 93–113. In England in the thirteenth century, Roger Bacon had tried to use coordinates for a map that is now lost: see below, Chapter 9, pp. 255 and 278 n. 3.

144. On Gmunden see Maria G. Firneis, Paul Uiblein, and Hans Kaiser, 'Johannes von Gmunden um 1384–1442', in Günther Hamann and Helmut Grössing, eds, *Der Weg der Naturwissenschaft von Johannes von Gmunden zu Johannes Kepler* (Vienna: Verlag der Österreichischen Akademie der Wissenschaften, 1988), pp. 9–100; Rudolph Klug, *Der Astronom Johannes von Gmunden und sein Kalender* (Linz: Pirngruber, 1912). I am grateful to Darin Hayton for his help and advice on Gmunden's tables.

145. London, British Library, Add. MS 24070, fol. 74v.

146. Bennett Durand, *The Vienna–Klosterneuburg Map Corpus* (1952), p. 112. On an astronomical table, dated 1438 and included among other tables attributed to William Worcester, now in Oxford, Bodleian Library, MS Laud. Misc. 674, fols

73r–74r, paradise is given latitude 0° and longitude 0°, indicating that it was situated on both the equator and the prime meridian (i.e. where the prime meridian crossed the equator). Longitude in this table, however, is calculated from the west. See John D. North, *Horoscopes and History* (London: Warburg Institute, 1986), pp. 186–95.

147 Rome, Biblioteca Apostolica Vaticana, MS Palat. Lat. 1362b, unfoliated. See Almagià, *Monumenta cartographica vaticana*, I (1944), pp. 30–1. The map is signed. Bennett Durand, *The Vienna–Klosterneuburg Map Corpus* (1952), pp. 209–13, has noticed the analogies between the work of the Vienna School and the Walsperger map. See also Franz Wawrik, 'Österreichische kartographische Leistungen im 15. und 16. Jahrhundert', in Hamann and Grössing, eds, *Der Weg der Naturwissenchaft* (1988), p. 111; Woodward, 'Medieval *Mappaemundi*' (1987), pp. 316–17, 325–7, 358; Karl-Heinz Meine, 'Zur Weltkarte des Andreas Walsperger, Konstanz 1448', in Wolfgang Scharfe, Hans Vollet and Erwin Herrmann, eds, *Kartenhistorisches Colloquium Bayreuth 1982* (Berlin: Dietrich Reimer, 1983), pp. 17–30; Paul Gallez, 'Walsperger and his Knowledge of the Patagonian Giants', *Imago Mundi*, 33 (1981), pp. 91–3. Closely related to Walsperger are the Zeitz map and the James Ford Bell map but neither show paradise: see Scott D. Westrem, *Learning from Legends on the Bell Library Mappamundi* (Minneapolis: Associates of the James Ford Bell Library of the University of Minnesota, 2000), and Bennett Durand, *The Vienna–Klosterneuburg Map Corpus* (1952), pp. 213–15.

148 Bennett Durand, *The Vienna–Klosterneuburg Map Corpus* (1952), p. 210: 'mappa mundi sive descripcio orbis geometrica facta ex cosmographya ptolomey proporcionabiliter secundum longitudines et latitudines et divisiones climatum. Et cum vera et integra cartha navigationis marium'.

149 The scale bar on Walsperger's world map may not have been all that effective on this sort of small-scale map but the notion of showing it to estimate distance, and Walsperger's consciousness of his readership are noteworthy. For the earliest printed maps with scale bars (the first probably was Erhardt Etzlaub's map of Nürenberg, 1492) see Campbell, *Earliest Printed Maps* (1987), pp. 56–8, and Figure. 5, and Catherine Delano-Smith, 'Cartographic Signs and their Explanation', *Imago Mundi*, 37 (1985), pp. 9–27.

150 Almagià, *Monumenta cartographica vaticana*, I (1944), p. 30.

151 A legend states: 'Et circa hunc polum sunt mirabilissima monstra non solum in feris etiam in hominibus'; 'And in the area of this [Antarctic] pole are most amazing monsters not only of the animal variety but even among humans'.

152 See Gow, 'Gog and Magog' (1998), p. 78.

153 Little is known about Fra Mauro. Various fifteenth-century documents recently surveyed by Angelo Cattaneo, 'Fra Mauro *Cosmographus Incomparabilis* and his *Mappamundi*: Documents, Sources, and Protocols for Mapping', in Cattaneo, Diego Ramado Curto, and André Ferrand Almeida, eds, *La cartografia europea tra primo Rinascimento e fine dell'Illuminismo* (Florence: Olschki, 2003), pp. 19–48, esp. 21–32, record that Fra Mauro was in the Camaldolese monastery of San Michele, made a map of a monastic estate in Istria, and in 1444 was a member of a commission charged with deviating the Brenta river. In the monastery's account books two collaborators are mentioned: Francesco da Cherso and Andrea Bianco. It is thought he died in 1459.

154 On Fra Mauro's map see Piero Falchetta, *The Fra Mauro World Map* (Turnhout: Brepols, 2006); Cattaneo, 'Fra Mauro *Cosmographus Incomparabilis* (2003); Cattaneo, *La Mappamundi di Fra Mauro Camaldolese: Venezia 1450*, PhD Dissertation, Istituto Universitario Europeo, Florence (2005), to be published as *Fra Mauro's Mappamundi and Fifteenth-Century Venetian Culture* (Turnhout: Brepols, 2007); Wojciech Iwánczak, 'Entre l'espace ptolémaïque et l'empirie: Les Cartes de Fra Mauro', *Médiévales* 18 (1990), pp. 53–68; Woodward, 'Medieval *Mappaemundi*' (1987), pp. 286–370; Destombes, *Mappemondes* (1964), pp. 223–6; Tullia Gasparrini Leporace, *Il mappamondo di Fra Mauro* (Rome: Istituto Poligrafico dello Stato, 1956); Giuseppe Caraci, 'The Italian Cartographers of the Benincasa and Freducci Families and the So-Called Borgiana Map of the Vatican Library', *Imago Mundi*, 10 (1953), pp. 23–49; Almagià, *Monumenta cartographica vaticana*, I (1944), pp. 32–40; Enrico Cerulli, 'Fonti arabe del mappamondo di Fra Mauro', *Orientalia Commentarii Periodici Pontifici Instituti Biblici*, 4 (1935), pp. 335–8; Placido Zurla, *Il mappamondo di Fra Mauro Camaldolese* (Venice: [n. pub.], 1806). I am grateful to Angelo Cattaneo for having discussed with me Fra Mauro's *mappa mundi*.

155 See, for example, Peter Whitfield, *The Image of the World: 20 Centuries of World Maps* (London: British Library, 1994), p. 13; Delumeau, *Une Histoire du paradis* (1992), p. 93 (pp. 65–7 in the English edition, 1995); Fred Plaut, 'Where is Paradise? The Mapping of a Myth', *The Map Collector*, 29 (1984), pp. 2–7; Plaut, 'General Gordon's Map of Paradise', *Encounter* (June/July 1982), pp. 28–9.

156 Piero Falchetta, personal communication, sees Fra Mauro's intellectual flexibility as a mark of Venetian open-mindedness and dynamism, in a city always ready to include diverse and even contrasting cultural contributions. See also Cattaneo, 'Fra Mauro *Cosmographus Incomparabilis*' (2003), pp. 37–48, and Iwánczak, 'Entre l'Espace ptolémaïque et l'Empirie' (1990).

157 For the dating of Fra Mauro's map see, Cattaneo, 'Fra Mauro *Cosmographus Incomparabilis*' (2003), pp. 29–32. A now lost copy of Fra Mauro's map is thought to have been commissioned by Alfonso V, king of Portugal, who would have supplied Fra Mauro with the most recent Portuguese sea charts. Mauro's map appears to be closely related to a portolan chart now in the Vatican, considered by Almagià to be a copy of the chart Fra Mauro used. While Almagià dated (on grounds of content) the original *mappa mundi* to before 1450, other scholars disagreed, preferring a later date (1459): see Woodward, 'Medieval *Mappaemundi*' (1987), p. 315; Caraci, 'The Italian Cartographers' (1953), pp. 33–4. A date on the reverse of the map (26 August 1460) follows Mauro's death and probably refers to the map's final framing.

158 See Cattaneo, 'Fra Mauro *Cosmographus Incomparabilis* (2003), pp. 37–41.

159 Gasparrini Leporace, *Il mappamondo di Fra Mauro* (1956), pp. 56, 61.

160 Ibid., p. 26.

161 Ibid., pp. 38, 46.

162 Ibid., p. 62.

163 Ibid. p. 53: 'e perhò queli che vol intender prima creda azò le intenda'.

164 On Prester John and the Old Man of the Mountain, see ibid., pp. 26, 44.

165 Ibid., pp. 32, 34.

166 Ibid., pp. 56, 61; see Westrem, 'Against Gog and Magog' (1998), p. 68.

167 See above, Chapter 5, p. 113.

168 Gasparrini Leporace, *Il mappamondo di Fra Mauro* (1956), pp. 36, 44. See also my 'The Image of Persia in Western Medieval Cartography', in *Proceedings of the V European Conference of Iranian Studies* (Bologna: Societas Iranologica Europaea, Università di Bologna, 2006) pp. 223–34.

169 What appear to be design or verbal vestiges in

the right half of the diagram are in fact patches of lost paint: it is likely that Fra Mauro had initially placed the circle showing the influence of the moon on the tides here, before moving it to the upper right-hand corner.

170 Gasparrini Leporace, *Il mappamondo di Fra Mauro* (1956), p. 22: 'El paradiso de le delicie non solamente ha sentimento spirituale ma etiam quello essere uno luogo ne la terra situado mette sancto Augustino sopra el Genesis et ancora nel libro De Civitate Dei, el qual luogo è molto remoto da la habitation e cognition humana, posto ne le parte oriental, segondo la doctrina del sacro dottor Beda per la cui autorità el maistro de le sentencie tal oppinion aferma, avegna ch'el comentator Alberto Magno nel libro de la natura di luogi metta quello oltra el circulo equinotial, pur ne la region oriental'; 'The paradise of delight not only has a spiritual meaning but is also a place situated on earth as Saint Augustine put it in his work on Genesis as well as in his *The City of God*. This place is very far from our habitable world and remote from human knowledge. According to the doctrine of the holy doctor Bede it is located in the east. Following his authority, the Master of the Sentences [Peter Lombard] confirms this opinion, even though the commentator Albert the Great in his book *On the Nature of Places* puts it beyond the equinoctial circle, yet still in the east.' Fra Mauro mentions Albert's *De natura locorum* for the idea of a paradise in the southern hemisphere, an idea that Fra Mauro only cites and does not embrace. Albert, as we have seen in Chapter 7, pp. 179–81, argued for the southern location of paradise in his *De homine* (and not in his *De natura locorum*, as Fra Mauro recalls). See also my 'Il paradiso terrestre di Fra Mauro', *Storia dell'Arte*, 93/94 (January–June 1999), pp. 219–27 and 'Fra Mauro's World Map', in Friedman, Mossler Figg, and others, eds, *Trade, Travel, and Exploration in the Middle Ages* (2000), pp. 383–6.

171 Cattaneo, 'Fra Mauro *Cosmographus Incomparabilis*' (2003), pp. 26, 32–7, rightly observes that Fra Mauro, who dealt with both small-scale cosmography and big-scale topography, presented in the inner circle of his map a chorographical representation of the *oikumene* and in the four corners outside it a cosmographical space that also includes paradise.

172 Gasparrini Leporace, *Il mappamondo di Fra Mauro* (1956), p. 22. Angelo Cattaneo, 'God in This World: the Earthly Paradise in Fra Mauro's Mappamundi Illuminated by Leonardo Bellini', *Imago Mundi*, 55 (2003), pp. 121–6, and Plates 9–11; and Susy Marcon, 'Il mappamondo di Fra Mauro e Leonardo Bellini', in Mario Piantoni and Laura de Rossi, eds, *Per l'arte da Venezia all'Europa: Studi in onore di Giuseppe Maria Pilo* (Venice: Edizioni della Laguna, 2001), pp. 103–8, independently suggested that Fra Mauro did not paint the vignette himself but asked the Venetian illuminator Leonardo Bellini to do it.

173 Gérard de Champeaux and Sébastien Sterckx, *Introduction au monde des symboles* (Yonne: Atelier Monastique de l'Abbaye Ste-Marie de la Pierre-qui-vire, 1966), p. 77, have compared the circular image of paradise in Fra Mauro's map with another circular Garden of Eden illustrating *Les Très Riches Heures du Duc de Berry* in the Musée Condée at Chantilly.

174 Gasparrini Leporace, *Il mappamondo di Fra Mauro* (1956), pp. 30, 34, 37.

175 Ibid., p. 22. On the Garden of Eden as a staging place for the souls of the righteous on their way to heaven, see Delumeau, *Une Histoire du paradis* (1992), pp. 37–57 (pp. 23–38 in the English edition, 1995).

176 Whitfield, *The Image of the World* (1994), p. 13: 'Paradise is located outside the real world, in contrast to earlier medieval maps'; Plaut, 'General Gordon' (1982), p. 29: 'Fra Mauro of Venice showed as early as 1459 how paradise could be banished from the world, and yet be honourably placed.' This was the general opinion found also in the other studies quoted above, n. 155.

177 John M. Prest, *The Garden of Eden: The Botanic Garden and the Re-creation of Paradise* (New Haven: Yale University Press, 1981), pp. 27–37.

178 Sérgio Buarque de Holanda, *Visão do paraíso: Os motives edênicos no descobrimento e colonização do Brasil*, 2nd edn (São Paulo: Brasiliense, 1969), pp. 153, 208; Ernest H. P. Baudet, *Paradise on Earth: Some Thoughts on European Images of Non-European Man* (New Haven: Yale University Press, 1965), p. 15; Charles L. Sanford, *The Quest for Paradise: Europe and the American Moral Imagination* (Urbana: University of Illinois Press, 1961), pp. 17, 39–40. See also Danielle Lecoq, 'Saint Brandan, Christophe Colomb, et le paradis terrestre', *Revue de la Bibliothèque Nationale*, 45 (1992), pp. 14–21; Plaut, 'Where is Paradise?' (1984), pp. 2–7.

179 See Gomes E. de Zurara, *Chronique de Guinée (1453)*, transl. by Léon Bourdon, and others (Paris: Chandeigne, 1994), p. 52. Zurara suggests that paradise might be located in inner Guinea. On the voyage of Saint Brendan see above, Chapter 3, pp. 52–3. I have discussed this point in 'Paradise Found and Lost around 1500', in *Acts of the International Conference 'Vasco Da Gama: Men Voyages and Cultures'*, ed. Joaquim R. Magalhães, 2 vols (Lisbon: Comissão Nacional para as Comemorações dos Descobrimentos Portugueses, 2002), II, pp. 435–52.

180 Pêro Vaz de Caminha, 'Lettre au Roi Dom Manuel', in Pêro de Magalhães de Gândavo, *Histoire de la province de Santa Cruz que nous nommons le Brésil*, transl. by Henri Ternaux, ed. by Philippe Billé (Nantes: Le Passeur-Cecofop, 1995), pp. 125–49. See Luís de Matos, *L'Expansion portugaise dans la littérature latine de la Renaissance* (Lisbon: Fundação Calouste Gulbenkian, Serviço de Educação, 1991), p. 427.

181 *Select Documents Illustrating the Four Voyages of Columbus*, transl and ed. by L. Cecil Jane, 2 vols (London: The Hakluyt Society, 1930–3), II (1933), p. 38.

182 On European travel literature see, e.g., Peter Wunderli, ed., *Reisen in reale und mythische Ferne: Reiseliteratur in Mittelalter und Renaissance* (Düsseldorf: Droste, 1993); Mary B. Campbell, *The Witness and the Other World: Exotic European Travel Writing, 400–1600* (Ithaca, NY: Cornell University Press, 1988); Arthur Percival Newton, ed., *Travel and Travellers of the Middle Ages* (London–New York: Routledge & Kegan Paul, 1926).

183 Marco Polo, *The Description of the World* (1938), p. 129.

184 See Iain Macleod Higgins, *Writing East: The 'Travels' of Sir John Mandeville* (Philadelphia: University of Pennsylvania Press, 1997).

185 See above, Figures 3.1 and 3.2.

186 *The Book of Sir John Mandeville* (1967), pp. 195–217 (XXX–XXXII). Mandeville describes the Christian Kingdom of Prester John, ibid., pp. 218–22 (XXXIII).

187 'John of Marignolli's Recollections of Eastern Travels', in Yule, ed. and trans., *Cathay and the Way Thither*, new ed. Cordier, III (1914), p. 220. See also Ananda Abeydeera, 'In Search of the Garden of Eden: Florentine Friar Giovanni dei Marignolli's Travels in Ceylon', *Terrae Incognitae*, 25 (1993), pp. 1–24.

188 'John of Marignolli's Recollections of Eastern Travels' (1914), pp. 220–44.

189 'In cono mundi contra paradisum', quoted in Anna-Dorothee von den Brincken, 'Mappa mundi und Chronographia: Studien zur *imago mundi* des abendländischen Mittelalters', *Deutsches Archiv für Erforschung des Mittelalters*, 24 (1968), p. 118

190 See Westrem, *Broader Horizons* (2001), pp. 150, 199–200, 223; Zarncke, ed., 'Der Priester Johannes', 8 (1876), p. 170.

191 *Ravennatis Anonymi Cosmographia et Guidonis Geographica*, I.7, ed. by Moritz E. Pinder and Gustav Parthey (Berlin: Fridericus Nicolaus, 1860; repr. Aalen: Zeller, 1962), p. 17: 'velut unam de iniquis cogitationibus esse decernimus, quod corruptibilis missus fuerat homo aut sponte perambulans suis corporalibus oculis potuisset vel modo possit nobilissimum videre paradisum aut pollutis terram sanctam perambulare pedibus.'

192 On the Ravenna cosmographer, see above, Chapter 5, pp. 89 and 118 n. 33.

193 *Select Documents Illustrating the Four Voyages of Columbus*, II (1933), p. 34.

194 Ibid., p. 36.

195 Vespucci's letter to Lorenzo de' Medici is quoted in Henry Vignaud, *Americ Vespuce, 1451–1512: Sa biographie, sa vie, ses voyages, ses découvertes, l'attribution de son nom à l'Amérique, ses relations authentiques et contestées* (Paris: Leroux, 1917), p. 410.

196 It may be helpful in this context to refer to the modern concepts of *microspace* and *macrospace*: *microspace* is the empirical known world around us; while *macrospace* is a cosmological category. These concepts are developed by Dick Harrison, *Medieval Space: The Extent of Microspatial Knowledge in Western Europe during the Middle Ages* (Lund: Lund University Press, 1996), pp. 2–7. See also my 'Mapping Eden' (1999), p. 55.

197 Gregorius Reisch, *Margarita Philosophica*, ed. by Adam Wernherus (Freiburg im Breisgau: Ioannes Schottus Argen., 1503), VII.I.45: 'Ceteris hac supputatione dimissa maior nobis cura sit ita vivere ut illuc angelico conductu tandem perduci mereamur: vel potius illo transcurso ad celestem perveniamus.'

9

Paradise Lost and Found

The devil has not been able, by his astuteness, to prevent the light of God from shining still in the midst of the shadows ...
John Calvin, *Sermon on John*, 1.1–5 (1537)

From the time of the Creation, it was believed, the Garden of Eden continued to exist somewhere on the globe, but in an inaccessible and unlocatable nowhere. The Garden of Eden had linked heaven and earth and had survived the Flood. As we have seen, the paradoxical notion about the location of the earthly paradise was an important component of the medieval world view. In their commentaries on the Book of Genesis, medieval and early Renaissance biblical exegetes had been pointing to the hidden yet still existing paradise as both a mark of divine goodness and as a reminder of the tragic consequences of sin. In the course of the sixteenth century, however, belief in an existing Garden of Eden declined, and the general tendency in exegesis was to admit that Eden was no longer part of the present world and that paradise had now disappeared; its ambiguous condition had always involved a kind of virtual presence projected backwards in time, but now the inhabited earth was consciously deprived of an adjacent and contemporary garden of delight. The change from paradise present to paradise past is mirrored in maps. From about 1500 onwards no map of the world showed the earthly paradise. Paradise was shown only on historical and regional maps drawn specifically as aids to biblical exegesis in printed Bibles and biblical literature.

The conclusion to Chapter 8 noted that modern researchers have been too quick to credit the decline of belief in a contemporary paradise to the impact of the geographical discoveries of the Renaissance. In fact, as noted in that chapter, early European explorers would never have dreamt of trying to get past the flaming sword of the Cherubim to penetrate the mystery of the Garden of Eden. The decline of belief in an extant earthly paradise represented a major shift in religious thought that paralleled changes in mapping practice and accompanied the opening of new geographical horizons. It was a culmination of a long process, not the result of the disappointment of naïve travellers. The changed attitude towards the mapping of paradise in the Renaissance was a reflection of profound changes in theological and philosophical thinking that had been taking place since the thirteenth century. Once the format of world maps changed radically with the introduction of the Ptolemaic grid, paradise could be, and was, included only on those maps that detailed the relevant part of the world and that referred to a geography of the past.

Changing Cartography

The Alexandrian astronomer Claudius Ptolemaeus (Ptolemy) compiled his work devoted to the description of the earth in the second century AD. Ptolemy's *Geography* included detailed instructions for different ways of portraying the spherical earth on a flat surface by means of mathematical and astronomical principles. His book also included lists of coordinates of latitude and longitude for thousands of cities, towns, and major natural features.[1] Ptolemy's *Geography* is generally assumed to have been lost to Western Europe until the fifteenth century, but hints have been accumulating to suggest that some knowledge of the text may have survived in closed scholarly circles from Late Antiquity up to the twelfth century.[2] It is not difficult to demonstrate the keen interest shown by medieval scholars in the mathematical and astronomical sciences. Geographical coordinates had for some time already been used by astrologers in connection with their study of the influence of the heavens. In the thirteenth century the English scholar Roger Bacon discussed the use of latitudes and longitudes in mapping procedure, and may even have constructed a map on the basis of some sort of grid.[3] The system of coordinates, however, was not rigorously applied in cartography, as medieval map makers had been drawing their *mappae mundi* according to a qualitative conception of space and its subordination to historical time, and it was only when new horizons emerged in the fifteenth century that a renewed alliance was formed between mathematics and map making. The first Latin translation of the Greek text of Ptolemy's *Geography*, without maps, was made in Italy in 1407 by the Florentine Jacopo Angelo, and maps (one world map and 26 regional maps) were soon being drawn to illustrate subsequent manuscripts.[4] In 1475 Ptolemy's *Geography* was printed in Vicenza but without maps. In 1477 the first printed edition with maps was published in Bologna, followed by an edition printed in Rome in 1478, and the edition with some modern maps compiled by Nicholaus Germanus and printed in Ulm by Lienhart Holle in 1482 and 1486.[5]

The recovery in the early fifteenth century by the Latin West of the *Geography* and its maps was in the longer term of unparalleled significance, for the Ptolemaic model became the basis of modern cartography. By the beginning of the fifteenth century, some of the most fundamental characteristics of *mappae mundi* had already become diluted, starting with the gradual loss of their eastward orientation. On the old *mappae mundi*, where the placing of east at the top of the map was linked to the space–time configuration of the world, the presence of paradise crowned the unfolding of world history from east to west. Ptolemy's instructions were for the construction of maps with north at the top, and the growing influence of the *Geography* effectively sanctioned the loss of that eastern privilege on world maps. The final change of orientation marked the definitive divorce between history and geography. Consequently, the representation of historical time on maps shifted from its former all-embracing nature to an episodical significance.

Ptolemaic maps, that is, maps based on projections modelled on Ptolemy's or constructed in such a way as to give at least the illusion of a projection or a graphic perspective, came to express the temporal dimension in a markedly different way from medieval *mappae mundi*.[6] On a Ptolemaic map, geography was cleared from the domination of historical time. Time did not disappear from the map, but it played a different role. The relationship between space and time remained intimate, but it was now mathematically defined. The astronomical measurement of time determined the parallels and meridians that structured the map. Degrees of longitude and latitude were defined temporally; in the case of longitude, according to the hour at which the sun reached the same position on the horizon in different places; in the case of latitude, they were defined by the duration of daylight, so that space itself was now configured by a grid of numbered parallels and meridians measured by time. The biblical past was still relevant, but the Ptolemaic grid cast over the earth reduced the possibility of interpreting that past cartographically. Whereas a fully fledged medieval *mappa mundi* depicted past events and places – including those mentioned in the Bible – alongside contemporary features diachronically, the development of a methodical system of astronomical coordinates favoured the inclusion on maps of synchronous features only. At the same

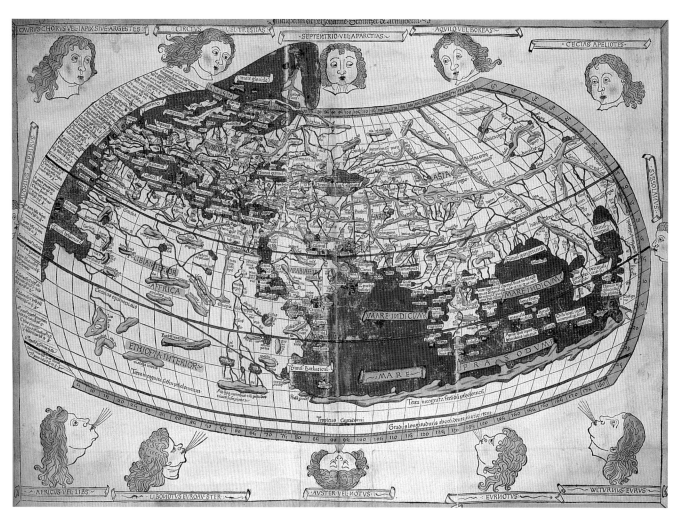

Fig. 9.1. World map on Claudius Ptolemy's second projection, from Claudius Ptolemy, *Cosmographia* (Ulm: Lienhart Holle, 1486). 40 × 55 cm. London, British Library, IC 9303. In his *Geography* (entitled *Cosmographia* in some Latin printed editions), Ptolemy had described two possible projections that would allow the known part of the terrestrial globe to be configured on a plane surface. In the first projection, the meridians are drawn as straight lines, in the second as curved lines. In both, the parallels form concentric circular arcs. The adoption of Ptolemaic principles in the course of the fifteenth century produced a radical change in European cartography as the space on the map became ordered by a geometrical grid of coordinates.

time, maps from the second half of the fifteenth century onwards had to take into account the increased rate at which geographical information about the world was reaching the West through the search for new commercial routes and other voyages (the process that culminated in the discovery of a new world). Maps no longer needed to present a stable image of a world into which the history of salvation could be incorporated, as in the medieval tradition. They became provisional documents, showing, for a specific moment in time, a world that needed to be constantly redrawn in the light of further discoveries.[7] Any transhistorical quality was lost. Instead of a comprehensive space–time image, in which geography was combined with historical time, maps of the world from the late fifteenth century onwards mirrored the surface of an earth measured in units of astronomical time alone.

On medieval *mappae mundi*, history had privileged individual places according to the events that had occurred there. Core and periphery were distinguished. The central place was not necessarily geometrically centred, but was made identifiable through the map sign. Some towns, such as those in the vicinity of the map's origin, and Jerusalem, were more important than others. On a Ptolemaic map, in contrast, space defined by mathematical astronomy was homogeneous and indifferent to human history. No one point on the map was any more important than any other. The absence of a centripetal focus reduced the possibility of portraying any earthly navel. Jerusalem, for example, included on Ptolemaic maps as an administrative municipality in the Roman Empire, was assigned its position on the map according to its latitude and longitude in

256 Mapping Paradise: A History of Heaven on Earth

exactly the same way as were all other places. Nonetheless, there was still a centring on Ptolemaic world maps inasmuch as a prime meridian had to be selected and parallels were measured from the equator. The choice of a projection and other cartographical techniques could also determine the focus of the map.[8]

The change in cartography that followed the gradual adoption of Ptolemaic principles in the course of the fifteenth century becomes obvious when we look at the woodcut world map in one of the earliest printed Latin editions of the *Geography* to contain maps, the *Cosmographia* published by Lienhart Holle in Ulm, Germany, in 1486 (Figure 9.1). The map fills a double-page spread of a large folio volume. It is constructed on Ptolemy's second projection, which gives it a frame of curved meridians and parallels. At first glance, the general content of the map is much the same as that of the old *mappae mundi*, depicting the three parts of the old world (Asia, Europe, Africa), and with twelve wind heads indicating the directions (four principal and eight secondary winds), but the modified spherical projection, as well as the map's landscape format, stretches the image of the world along a horizontal (east–west) axis.[9] The edges of the map (and the part of the world it shows) are no longer defined by an ocean ring, for the map portrays only the inhabited or known parts of the globe as defined by Ptolemy. The latitudinal dimension of the *oikumene* is bounded by the parallel of Thule in the north (63° N), where the longest day has twenty hours, and in the south by the parallel opposite Meroë (16°25′ S), where the longest day has thirteen hours. Longitudinally, the map reaches from the principal meridian, which passes through the Canaries in the extreme west (0°), to the meridian that passes through Sera and Cattigara in the extreme east (180° E), a distance incorporating about 12 hours' difference in time.[10] Thus defined by mathematical astronomy, the edges of the map cut arbitrarily across lands and seas and include unknown territories in the south, labelled *Terra incognita secundum ptholomeum* ('Unknown Land according to Ptolemy'), which entirely close off the Indian Ocean in that direction.

The world map in the Ulm *Geography*, like its counterparts in other editions and in the earlier Greek manuscripts, does not show the Garden of Eden. Nor is the earthly paradise mentioned in the gazetteer. This is hardly surprising. First of all, the earliest Renaissance Ptolemaic maps were just illustrations of the work of an Alexandrian geographer who did not read the Bible, and their compilers must have hesitated before altering an ancient and authoritative text. Moreover, on a map of the world ruled by measurable coordinates, defined by astronomical time, and whose function was to give accurately measured and correctly proportioned distances between places, there would be no room for an inaccessible and unlocatable earthly paradise. The compilers of the *mappae mundi* had accepted the paradox of mapping the unlocatable location of the Garden of Eden within the topological structure of the *mappae mundi* because of its importance as the scene of the Fall. Renaissance map makers, inheriting the same Christian tradition, knew very well that the crucial event had taken place many thousands of years earlier. Once they began to draw their own 'modern' world maps using Ptolemaic principles, however, they would probably have found it difficult to include that primordial feature on a map that transfixed the world displayed on an east–west span of twelve hours (the interval between the two extreme meridians) and a north–south span based on the variation in length of day and the climatic zones, criteria which in both cases were natural and not historical. Most definitively, the earthly paradise, which had been shown on medieval maps as beyond the boundaries of the inhabited earth, could not be included on a map of the *oikumene*.

It is an unproductive exercise, then, to seek for the earthly paradise on a world map made after around 1500.[11] That the absence of Eden was not the result of a sudden indifference to Christian theology or of any loss of belief can be seen in the Ptolemaic world map that was included in the immensely popular world chronicle

Fig. 9.2. World map in Hartman Schedel's *Liber chronicarum* (Nuremberg: Anton Koberger, 1493). 31 × 43.5 cm. London, British Library, 1C 7452. The world history compiled by the German humanist Hartmann Schedel was richly illustrated. Among the many woodcuts illustrating the book is a world map, which has been compiled from Ptolemy's instructions for the construction of a map from coordinates. In addition to the twelve wind heads surrounding the inhabited earth, Schedel's map maker has put the figures of Noah's three sons into three of the corners of the map, a reminder that, as the Bible relates, Japhet, Shem and Cham inherited Europe, Asia and Africa, respectively, after the Flood. In the fourth corner, a text explains the effects of the winds on the lands.

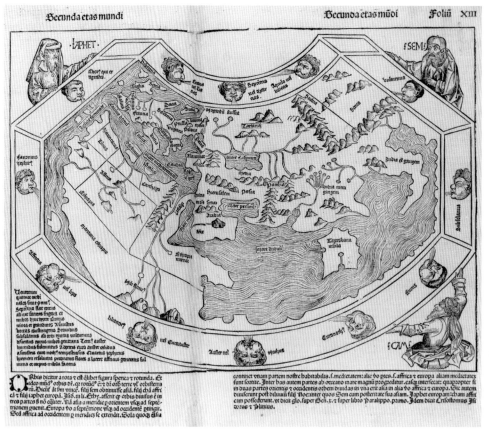

compiled by Hartman Schedel and printed in Nuremberg in 1493. Schedel's map shows three prominent biblical figures surrounding the inhabited world. The figures are those of Shem, Japheth and Ham, the three sons of Noah who, according to the Bible, inherited the three parts of the known world (Figure 9.2).[12] Even so, the key to understanding the absence of Eden on Renaissance world maps lies not in cartographical developments alone, but above all in contemporary changes in theological thinking.

Changing Theology

The disappearance of the earthly paradise from world maps at the end of the fifteenth century was certainly part of a changing cartographical context, but it was also a function of changes in theological thinking and, above all, of the slow decline of the idea that the Garden of Eden was still surviving somewhere on earth. Renaissance explorers were disclosing new lands where there was no trace of the biblical paradise – either in India, Ethiopia, China, Japan or the New World – but it was not this discovery alone that eventually expunged the presence of Eden from the cartographers' maps of the world. The intellectual search for the site of the Garden of Eden had already presented biblical exegetes with huge difficulties well before fifteenth- and sixteenth-century merchants and travellers started to bring news of their experiences back to Europe. The problem for the theologians was the wide range of contradictions and paradoxes that were contained in the very idea of an earthly Eden. Despite the thirteenth-century scholars' scientific approach to the Bible in their quest for the exact site of paradise, their later medieval successors failed to reach a consensus, and in the fourteenth century discussion about Eden retreated into the realm of faith and religion. As it developed in the fifteenth century, the theological debate simply confirmed the awkwardness of trying to establish a consistent geographical account of the location of a terrestrial paradise. Medieval debate had left unresolved all sorts of problems relating to the geography of paradise, most notably the sheer impossibility of identifying its exact site and of finding a credible idea about its altitude. Instead of an agreed location, there was only a host of contradictory suppositions embedded in theories about the extreme altitude of the mountain of Eden and about how paradise had escaped the Flood.

The Flemish theologian Dionysius the Carthusian, a prolific writer and serious student of the early Christian authors, played an important role in fifteenth-century religious and intellectual life.[13] More than anybody in his day, perhaps, Dionysius attempted to extract something coherent from the records of centuries of debate about the location of Eden, but even he found himself mired in seemingly insoluble problems. Reporting Alexander of Hales's idea that paradise was on the equator at an exceptionally high altitude – as high as the middle region of the air – Dionysius had to admit that the possibility of such an altitude had been questioned by many scholars and conclude that the problem of the altitude of paradise could not be easily solved through human speculation: 'I think that … one cannot obtain certainty about this matter without revelation or can only do so with great difficulty.'[14] Turning to differences in opinion about the location of paradise – at the equator or to the south of it – he also found himself obliged to admit that 'we can only know about it through revelation, which is why philosophers and historians have not spoken about it'.[15] For Dionysius, unaided human reason could not provide a final answer to the paradise question despite all the best efforts of earlier thinkers. Others fared no better. In his own time the German scholastic Gabriel Biel was considered to be 'the king' of theologians, but modern research has classed his philosophy as the 'harvest and swan-song of medieval theology'.[16] As he accepted Duns Scotus's ideas on the location and altitude of paradise, Biel nonetheless had to agree with Duns Scotus that 'the exact location of the earthly paradise is hidden and unknown to us'.[17] The geographical debate about Eden that had intrigued Latin scholars since Augustine's authoritative reading and that had been pursued with such energy between the thirteenth and the end of the fifteenth century had failed to unravel the mystery of the location of the earthly paradise, as it was destined to, for theological reasoning had to be based solely on faith. Theologians such as Dionysius and Gabriel Biel had to admit defeat; it was impossible to discover where paradise might be, and its existence just had to be accepted on the authority of divine revelation.

The Waning of the Idea of Corporeal Perfection

The geographical conundrum of paradise was only part of the challenge of having to explain the intermediate character of an earthly heaven. At the same time as they developed the geographical side of Augustine's reading of Genesis, trying to make the inaccessible place locatable, medieval theologians had also been struggling to maintain the idea of Adam and Eve as creatures physically embodied in the corporeal paradise. For Augustine, the Garden of Eden was a geographical location chiefly because Adam in his state of perfection was endowed with a body which meant he interacted with his spatial environment, which was also perfect.[18] After Augustine, though, his notion of a perfect Adam in a perfect Eden had undergone important changes. The twelfth-century compilers of the *Glossa*, for example, accepted that Adam had a natural and immortal body in an earthly garden, but their long marginal glosses on the second account of his creation (Genesis 2.7) bear witness to the difficulty of making sense of the paradise narrative without raising new problems.[19] Their contemporary Peter Lombard's treatment of these issues also indicates the difficulties that arose when speculating about Eden.[20] A century later, Thomas Aquinas turned to Aristotelian learning in an attempt to make sense of Adam and Eve's perfect yet physical life in the Garden of Eden.[21] Their immortality, he explained, depended on a supernatural element in their souls, and not on their bodies, which were composed of the four elements and therefore perishable. In paradise man had to eat to prevent the loss of body fluids through the effects of heat, but he was able to avoid the decay of his body thanks to the perfect submission of body to soul, and thanks also to the fruit of the Tree of Life. In this way, according to Thomas, Adam and Eve were able to eat, sleep, work, and enjoy sexual pleasure with their bodies while remaining immortal and full of God's grace.[22]

As already noted, by the fourteenth century the 'honeymoon' of theology and philosophy had come to an end.[23] John Duns Scotus, the theologian who denounced the uselessness of geographical theories about paradise, stated firmly that perfection in Eden was a supernatural gift from God that supplemented the naturally imperfect human condition. Duns Scotus dismissed Thomas's arguments about the potency of the soul and its effect on the body and denied wholly the possibility of an immortal physical life, which Thomas, referring to Aristotle, had striven so hard to explain. In appealing to Aristotelian principles, Duns Scotus was emphasizing that, even in paradise, Adam and Eve could not avoid death, which came from the reciprocal action of the counteracting elements of which their bodies were composed and from the decay of the flesh induced by natural heat. Had they not sinned, God would have transferred them to heaven before their physical bodies had decayed.[24] It is true that earlier theologians, including Thomas, had already pointed out that Adam's immortality was simply a divine gift, not something that belonged to him through nature – which was why he was created outside Eden – but the notion that the human body in paradise was not immortal in itself marked a major shift in theological attitudes. Physical death had been thought of as a consequence of sin and not as something inherent to the human condition.[25] Duns Scotus also stressed that concupiscence was an unavoidable part of human nature, even in paradise.[26] Augustine claimed that human nature came perfect from the hands of God, but Scotus transformed the notion that human perfection was naturally possible in Eden into the idea that man was intrinsically imperfect, naturally mortal, subject to error and inclined to pursue disorderly pleasures. Duns Scotus's shift of perspective now seemed to revive the earlier tendencies of some Christian theologians to conceive of human perfection in merely celestial terms, outside earthly time and space. Paradoxically, Duns Scotus had come to the same conclusion as had the allegorizing theologians of the early centuries of Christianity, although by a different route. He believed that the Fall did not involve any real alteration of the human condition (now regarded as imperfect). In his view, the Fall only deprived man of God's supernatural gifts. Like Origen, he portrayed man's corporeal condition as in itself fallen.

The difficulty of conceptualizing a Garden of Eden situated in some exact spot on earth paralleled the difficulty of imagining a state of natural human perfection in that Garden. In the fifteenth century, Dionysius the Carthusian looked for the answer in accumulated tradition, but even by this means he could not avoid the recurring paradoxes and problems.[27] In his turn, Gabriel Biel followed Duns Scotus in playing down the difference between the original integrity of man and his fallen human condition. Biel explained the earthly paradise as a place prepared by God for *viatores* (travellers), a term that in his vocabulary described human beings as they journeyed towards their eternal destiny.[28] Like Duns Scotus's, Biel's vision implied that the only perfect state available to man was in heaven. After mentioning the traditional idea of the earthly paradise as an intermediate region between heaven and earth, he added, significantly, that in a way Eden was closer to earth than it was to heaven.[29]

The concept of an earthly paradise was also challenged in the fifteenth century by the growing importance of Neo-Platonism, which openly promoted a revival of the purely allegorical tradition of biblical exegesis. The Italian Giovanni Pico della Mirandola was an aristocrat famous for his learning. In a detailed commentary on a friend's Platonic poem, Pico described paradise as the angelic mind, a garden whose fruitful and evergreen trees are the ideas planted by God and where men are elevated to the condition of angels.[30] The term 'paradise', he says, was used by Christian theologians to mean 'heaven', and stood for the intellectual goodness to be regained by abandoning the corporeal imperfection of human nature.[31] In another work, the *Oratio de dignitate hominis* (*Speech on the Dignity of Man*), Pico also referred to paradise as a mystical condition and emphatically *not* as a geographical place. For Pico, the Garden of Eden of Genesis was not a physical garden somewhere in the east (or south) but a spiritual ethos to which Apollo, Pythagoras, Plato, Zoroaster and Saint Paul all referred, comprising the purification of the human soul and its perfection through divine enlightenment and love.[32] In his nine hundred theses, which he intended to debate publicly in Rome in 1486 (an event that never took place) and in which he claimed that the Jewish Cabala, together with other ancient writings by Zoroaster, Hermes Trismegistus, Plato and Plotinus, contained many of the same mysteries that are revealed in Christianity, Pico described paradise as a stage in the process of divine emanation through which the unknowable Creator makes himself known. In Pico's view, to enter paradise is to contemplate the progressive movement of the hidden life of God, symbolized by the four rivers flowing out of Eden.[33]

For Pico, nothing connected to matter and belonging to the sublunary world could possibly be perfect. His exclusion of the notion of a physical paradise on earth represented an abandonment of Augustine's vision and a return to the thinking of Philo and Origen. Once again, amongst Neo-Platonists, human nature was considered to be intrinsically unstable and evil, while perfect human nature was thought of as belonging in every aspect to the divine. As we have seen, not all theologians who saw a gulf between divine perfection and human nature necessarily described it in Neo-Platonic terms. The *possibility* of human perfection, however, was an essential part of the concept of a localized and physical paradise, and the denial of that possibility was in effect a denial of the existence of an earthly paradise. Medieval and early Renaissance maps reflected the concept of an earthly and physical Garden of Eden, and the belief that the Bible had to be taken above all literally and that the original state of human perfection was entirely compatible with a natural body. Pico's Eden was not a geographical region, for his only concern was man's spiritual journey towards heaven and divinity. Adam in paradise was a being of godlike beauty, not yet embodied in the flesh. Had Adam started with an angelic nature, being given a body only as a result of sin, there was no need to imagine, still less to map, a physical paradise for him to inhabit.

Changing Exegesis: Paradise Lost in Mesopotamia

By 1500 the geographical debate about the location of Eden had reached stalemate: no credible earthly location had been found for the Garden of Eden and a radical change of approach was needed if the authority of Scripture was to be preserved. In much the same way as the geographical discoveries were altering the map of the world, so too was the literal and historical sense of Scripture being challenged by the re-emergence of spiritualizing and allegorical tendencies. To deal with the unresolved conundrums, biblical commentators turned away from the geographical horizons of the present and looked instead for a geography of the past. The idea of a contemporary and inaccessible paradise – distant in the remote east or south – shifted to the notion of a paradise distant in the past, now lost but formerly in a region that was still within reach. Paradise had always existed in the past, but its temporal remoteness was henceforth taken in a more explicit and exclusive way: Eden had once been on earth but now had disappeared. Explaining the problem of its location in this way provided the only means of preserving the integrity of a literal and historical reading of Genesis and thus of leaving intact the authority of Scripture.

The view that the Garden of Eden had been totally destroyed by the Flood was advanced by both Catholics and Protestants. The fiercest defenders of the traditional idea of a paradise still on earth were, on the whole, Catholics; Protestants tended to denounce it as a popish fancy. Paradoxically, though, one of the first steps towards the decline of the idea of a paradise still on earth had been taken in 1449 when the humanist Aeneas Sylvius Piccolomini, the future Pope Pius II, expressed a doubt about the traditional identification of the biblical Gihon with the Nile.[34] The first exegete to claim explicitly that the earthly paradise had actually disappeared was a Catholic, Augustinus Steuchus. Steuchus, who was also known as Eugubinus (from his native town of Gubbio), was an influential Roman churchman and head of the Vatican Library.[35] As the Protestant Marmaduke Carver would point out, over a century later, Steuchus took advantage of his position in one of the best-endowed libraries in Christendom to produce what he hoped was a reasonable view on the location of paradise.[36] Steuchus's writings proved to be a key factor in the mapping of paradise from the mid sixteenth century onwards.

Steuchus was well versed in Greek and Hebrew and wrote a number of works on sacred antiquities and Biblical exegesis. Commenting on the biblical description of paradise, he declared that he preferred not to derive his arguments from the vast amount of earlier literature in Hebrew, Arabic, Chaldaean, Greek and Latin. In his view, the numberless works already written on the subject all showed sheer ignorance and a confusing variety of opinions. For his part, he promised to attempt to shed light on a problem that was, he admitted, 'obscure and remote from human understanding', but that was susceptible of explanation if the meaning of Scripture were carefully enough weighed.[37] For, as he reassured his readers, 'God was not so inhuman as to want mankind to be tortured down the generations by its ignorance, for there was no passage in Scripture that could not be understood if pondered accurately.'[38]

Steuchus argued that paradise did not exist in some mystical way beyond the terrestrial orb, but that Adam had been expelled from Eden to another earthly region, which was why the Cherubim had to block the way back to the Tree of Life. Scholars who situated paradise at the equator, high in the air and reaching the stars, or in some otherworldly location, had not paid due attention to the biblical text.[39] Eden was a region in the inhabited part of the earth. The idea that the flaming sword of the Cherubim stood for the torrid zone, for example, was vain and ungrounded.[40] As a skilled Hebraicist, Steuchus commented on the various aspects of the Latin translation of the Hebrew original. Eden was the proper name of a region in which the earthly paradise was situated, and this Edenic region was to be found in Mesopotamia. To support his argument, Steuchus remarked that Adam had been created in the Damascene Fields and had lived in that part of the earth after the Expulsion; that Cain had settled in the land of Nod to the east of Eden (Genesis 4.16); that the prophet Ezekiel had associated Eden with Charan (or Haran), a real city in Mesopotamia (Ezekiel 27.23); and that, finally, the word 'paradise' was of Persian origin. All Steuchus's arguments pointed to the location of the earthly paradise in what is called today the Middle East. Before the Flood, Noah had also lived in the area. After the Flood, his Ark had come to rest on the mountains of Armenia, from which, in neighbouring Chaldaea, mankind spread out to populate the entire earth.

Steuchus's Garden of Eden was on some delightful hill in Mesopotamia, celebrated as the most fertile land on earth.[41] Paradise had lost its original beauty because of human sin, and this was why it was impossible to recognize it anywhere in Mesopotamia. After the Fall, argued Steuchus, the garden remained uncultivated and had become as wild and overgrown as would any abandoned plot. Up to the Flood, the location of Eden was known to man and the way to it was guarded by the Cherubim. The waters of the Flood, however, drowned the Garden, destroyed everything, and left not a single trace of it. To counter objections that, were this to have been the case, the Old Testament patriarchs Enoch and Elijah and the Apostle John would have had nowhere in which to wait for the end of the world, Steuchus said that they had been translated into a divine condition and not carried to the Garden of Eden. Had Adam not sinned, mankind too would have inherited the whole earth in a state of peaceful perfection before being directly transferred to heaven. Thanks to the sacrifice of Christ, however, those who led a holy life were still given the opportunity to reach the same condition.

The inclusion of paradise in a region that was part of the inhabited earth raised a question that had not been considered important so long as paradise was believed to have been outside the known world: the distinction between Eden and paradise, and the problem of the point of origin of the single river and the point at which it separated into the four streams. The Genesis text, 'And a river went out of Eden to water the garden; and from thence it was parted, and became into four heads' (Genesis 2.10), was ambiguous. Steuchus elucidated the text, explaining that the great river of paradise was formed by the confluence of the Tigris and the Euphrates, and that the Gihon and the Pishon branched off from the main river outside the Garden of Eden. He pointed to the example of the Po in northern Italy, which he knew from his frequent sojourns in the Veneto and Emilia regions and which branched into two rivers first above Ferrara and then again below the city to reach the Adriatic Sea in two streams. He insisted that the interpretation of the biblical verse indicated that the river left the region of Eden, not paradise itself, and that there was, accordingly, no need to imagine a single source within paradise with four rivers flowing from it. This meant that the sources of the rivers could have been at some distance from paradise, elsewhere in the vast region of Eden. In fact, Steuchus noted, the sources of the Tigris and Euphrates were far apart from each other in the Taurus mountains. The Gihon and the Pishon had nothing to do with the Nile and the Ganges, which flowed northwards and had no single source (as Piccolomini had already noted), and the land of Havilah and Cush were not in India and Ethiopia, but in Mesopotamia and Arabia.

An early echo of Steuchus's arguments is found in the Italian translation of Piccolomini's unfinished geographical work by the humanist Fausto da Longiano, which was printed in Venice in 1544. Fausto added a section on Africa and expanded Piccolomini's section on Asia to include a description of the Holy Land. He also included a detailed account of the location of Eden, saying that 'it seemed too inappropriate to leave out a discussion of the earthly paradise'.[42] Fausto argued that paradise had been destroyed either by God's curse on the earth, after Adam's sin, or later by the Flood, and that it had consequently totally disappeared. It had probably been somewhere in the east, as the names of the rivers indicated, and must have been very large for, in any case, the entire earth had been destined by God to be man's dominion. Beyond that, though, it was difficult to point to the exact spot where Adam and Eve had lived so blissfully. Noting that the Pishon and the Gihon were not to be identified with the Ganges and the Nile, and that there were two Armenian candidates (the Araxes and the Cyrus), Fausto speculated that a single source of the four rivers might have been the ocean, as John of Damascus had thought.[43]

Steuchus's arguments opened the door for a new round of discussion and controversy over the lost paradise. He overtly encouraged scholars to work further on the problem of the earthly paradise:

> *These things are so obscure, and so remote from our understanding, that they do not allow us to assert anything certain. Therefore, it will be the task of every keen student of Sacred Scripture to inquire diligently after these things. For now, we wish that there may be many of them. We have offered on the subject what we could.*[44]

Scholars did indeed take up Steuchus's challenge. Steuchus may have circumvented the problem of identifying an extant Eden in the contemporary world, but other questions remained, not least whether it was possible to agree on a more exact location. One could insist on placing paradise in Mesopotamia or one could site it further south, along the confluence of the Tigris and the Euphrates, privileging the idea that it was watered by a single stream. One might also wonder whether paradise was a specific place, or whether, prior to the Fall, the entire earth had enjoyed paradisiacal qualities. In the Middle Ages, the belief that Eden had survived the Flood was accompanied by a strong persuasion that the Garden was spatially remote, separate from the known or knowable terrestrial landmass, and inaccessible. Steuchus's thesis postulated a wholly vanished paradise. Thought once to have been enclosed by a wall of fire, or isolated by an impassable ocean, in the sixteenth century the Garden of Eden was believed to be divided from the inhabited earth only by time. Paradise was engulfed in the mists of its origins, and needed no boundary to indicate its separation from the rest of the earth.

Paradise as the Whole Earth

In an attempt to provide a literal and rational interpretation of Genesis that matched the geographical lore of their age, some Renaissance scholars argued that paradise on earth referred to nothing more concrete than the blessed state experienced by Adam and Eve before their sin, and that Eden, instead of being a specific location, had once encompassed the entire earth before God cursed it. For this reason no precise site for Eden had ever been or could ever be found. Whereas in the Middle Ages it was thought that the divine curse had spared a still existing Eden, it was now suggested that the curse had ruined paradise itself, meaning the whole earth. The idea was not a complete novelty, but it reached its most consistent formulation in the middle of the sixteenth century in the heady intellectual atmosphere of Renaissance humanism.[45] The list of those who believed in such a paradise-like primitive earth is long. It includes Vadianus (Joachim Van Watt), the Swiss humanist and a follower of Zwingli; Goropius Becanus (Jan van Gorp), the Flemish philosopher, physician, antiquarian and linguist; Ludovicus Fidelis, a Flemish professor of theology; Ludovico Nogarola, a humanist and scholar from Verona who attended the Council of Trent; and Juan de Pineda, a Spanish Jesuit and a biblical exegete and historian.[46]

Common to all these thinkers was the belief that the Garden of Eden did not comprise a limited and particular area, and that the beauty, plenty and delights associated with it in Scripture were common to the whole earth before sin had occurred. According to Scripture and tradition, the four rivers of paradise were the Tigris, the Euphrates, the Ganges and the Nile. The exegetes holding the belief that paradise was the whole world, however, did not need to worry about finding a single source for the four rivers of paradise. They understood the ocean to be the source of all water. The second chapter of Genesis, where paradise on earth is described as a garden, was interpreted in the light of the first chapter, where God is said to have given the whole world to Adam and his descendants. The fact that the first man had been told by his Creator to people the whole earth and to rule over all living things (Genesis 1.28–30) was seen as reinforcing the idea that the garden intended for Adam and his children was the whole earth and not just a secluded portion of it. Supporters of the theory of a whole-earth paradise acknowledged that Adam and Eve had themselves occupied a precise site, which was particularly pleasant and delightful, but they insisted that the rest of the earth was equally beautiful and fruitful and destined for their progeny.

All the details of the paradise narrative were reinterpreted in the light of this new framework. According to Goropius Becanus, the biblical reference to God's planting of paradise 'in the east' indicated that Adam had been created and placed in the east and that the creation of plants had begun from the east.[47] Juan de Pineda speculated that, had Adam not sinned and had his progeny persevered in the state of innocence, other Trees of Life would have been provided for those descendants who lived far away from Adam's original dwelling place or that, alternatively, the fruit of the Tree of Life would have been in some way transported to them.[48] Certainly, the idea of paradise as the whole world solved all the difficulty inherent in the theory of a secluded paradise. Had the Garden of Eden been only a limited place, and had Adam and Eve not sinned, their children would have been confined within a specific area and the rest of the earth would have been created in vain. Future generations would have found paradise far too small a place in which to live and the entire earth would eventually have been needed anyway in which to accommodate an increasing population. Moreover, since Adam and Eve did sin, the idea that a localized Garden of Eden had been left empty and unused after the Expulsion seemed untenable.

The doctrine of a whole-earth paradise was rationally appealing to some Renaissance exegetes, but went against the letter of Scripture. According to this new reading of Genesis, Adam and Eve's expulsion from paradise was not banishment from a localized spot. The whole earth, once a garden of delights, became a place of thorns, labour and sweat, where mankind was exposed to sorrow, infirmity and death. In this case, how was sense to be made of the explicit reference in Genesis to the expulsion of Adam and Eve from paradise by the Cherubim? This crucial problem was addressed in various ways. The Expulsion could be interpreted literally as part of a historical narrative or allegorically as a figurative reference to the idea of a change in the human condition. Ludovicus Fidelis and Goropius Becanus were the most radical in considering the Expulsion solely as a change of state, associated with the curse on the earth. In their view, Adam and Eve remained after the Fall where they already were. They were ejected from paradise in the sense that they lost their immortality, their dominion over nature and the bliss of their former innocence. Eden stood for the lost generosity of nature and the fiery sword guarding it for God's justice and anger.[49] Other writers, not wanting to depart too far from the letter of the biblical text, interpreted the Expulsion and the posting of the Cherubim as a change of place as well as an alteration in state. They were not always consistent. In one of his works, Vadianus considered the Expulsion as a change in condition, but in another he admitted a change of location, although he always stressed the altered state and advocated a figurative interpretation of the Cherubim – in his view an image used by Moses to accommodate his Jewish audience.[50] Exegetes such as Ludovico Nogarola and Juan de Pineda viewed the Expulsion literally, as a change of place intended to keep Adam and Eve away from the Tree of Life and its beneficial properties, but the Expulsion by the Cherubim figuratively, as a man's fallen condition.[51]

Many sixteenth- and seventeenth-century writers, both Catholic and Protestant, rejected the idea of a whole-earth paradise as absurd and dangerous on the grounds that a doctrine that conceived the expulsion of Adam and Eve from Eden as a change of state departed from the biblical text. Catholic commentators, such as Benedictus Pererius and Cornelius a Lapide, insisted that Adam had been placed within a specific paradise, in the region of Eden, and that he was later ejected from this garden and not from the whole world.[52] Walter Raleigh and his contemporary Samuel Purchas criticized the idea of a whole-earth paradise and, significantly, blamed those who denied a specific paradise in the same way as they blamed the early Christian dualistic heresy of the Manichaeans, who denied the goodness of material creation.[53] Those who adhered to the notion of a whole-earth paradise were looking for a literal reading of the biblical text that would remain consistent and convincing under the spotlight of contemporary geographical knowledge. The vision of a vanished blessed world of unlimited beauty and plenty that was no longer to be found indeed presented, as Raleigh and Purchas hinted, a striking similarity to the allegorical and philosophically orientated exegesis long ago disposed of by Augustine. Philo, who had interpreted the Bible in terms of Greek philosophy, had thought of paradise as the whole universe.[54] In their different ways, by claiming that the whole earth was paradise and dramatizing the gravity of the Fall, the Renaissance writers were also idealizing a more beautiful and purer lost earth. They read in the classical descriptions of blessed lands and a vanished Golden Age references to the lost paradise-like earth.[55] For Renaissance proponents of the idea that paradise had once comprised the entire earth, the definitive and impermeable barrier preventing access to paradise was time. Spatial contiguity between a specific paradise and the rest of the earth was not lost: it had never existed. Had the whole-earth paradise theory prevailed, paradise would have become unmappable. In the course of the sixteenth century, however, exegetes gradually returned to the notion of paradise as a particular piece of land, while map makers found new and different ways to map paradise on earth.

Paradise Flooded

It was Martin Luther who initiated a new approach to the problem of pinpointing Eden on a map. For Luther, the past, with a primordial Eden, was radically different from the present, with a fallen earth. Paradise, he said, had vanished because of the tragedy of human sin. He shared with those who believed in a whole-earth paradise the idea that sin had brought about the ruin of nature and the vision of a radical difference between the primitive, Edenic, world and the present, corrupted earth. He acknowledged that it was the event of human sin rather than a spatial barrier that deprived fallen mankind of the delights of paradise. Luther, however, was a passionate defender of the historical meaning of Scripture and found it hard to reject the letter of the text. When he read in Genesis that Adam was driven out of a particular place and that an angel had been put on guard at the entrance to paradise, he was prompted to suggest that paradise had once occupied a specific part of the earth, but was later wiped out by the Flood, and that it was therefore pointless to speculate about its exact location.

Commenting on the biblical account of God planting the Garden of Eden 'towards the east' (Genesis 2.8), Luther remarked that from that passage 'a sea of questions concerning paradise arises'.[56] There was no need in his view, however, to be drowned in this sea of geographical conundrums, as the earth itself had been radically changed by the Flood:

> *At this point people discuss where paradise is located. The interpreters torture themselves in amazing ways. Some favour the idea that it is between the two tropics under the equinoctial point. Others think that a more temperate climate was necessary, since the place was so fertile. Why waste words? The opinions are numberless. My answer is briefly this: it is an idle question about something no longer in existence. Moses is writing the history of the time before sin and the Deluge, but we are compelled to speak of conditions as they are after sin and after the Deluge. And so I believe that this place was called Eden either by Adam or at the time of Adam because of its fertility and the great charm which Adam beheld in it. And the name of the lost place persisted among*

his descendants, just as the names of Rome, Athens and Carthage are still in existence today, although hardly any traces of those great states are apparent. For time and the curse which sins deserve destroy everything. Thus when the world was obliterated by the Deluge, together with its people and cattle, this famous garden was also obliterated and became lost.[57]

Like sixteenth-century rational thinkers who claimed that paradise was the whole earth, Luther was going against traditional ideas by insisting that paradise had completely vanished because of sin instead of agreeing that paradise had survived the Flood and still existed somewhere.[58] We may suppose that he would have rather liked to embrace the notion of a lost whole-earth paradise himself: in the record of his private and domestic conversations, *Table Talk*, he shows that once he did in fact support that idea.[59] His loyalty to the biblical text, however, must have prevented him from embracing wholeheartedly the idea of a whole-earth paradise. In his public *Lectures on Genesis*, he was careful to pay greater attention to the importance of preserving the letter of the biblical account of the Expulsion, which implied the existence of some sort of boundary between paradise and the rest of the earth. He hastened to say that, if the impossibility of finding a common source for the four rivers of paradise was leading many exegetes to imagine Eden as the entire earth, this was 'obviously wrong'.[60] *Eden* was a proper noun, the name of a place, and *miqedem* qualified the eastern location of paradise.[61] This delightful garden, situated in a particular place, had an entrance (road or gate) facing the east, and after the sin angels were placed to guard it.[62] Paradise on earth was intended by God to be 'a palace' for man, 'the temple of the entire world'.[63] It had its own magnificence, being a most excellent part of the earth, with the best crops supplying the most delightful foods. The biblical text, Luther insisted, explicitly distinguished the Edenic Garden from the rest of the earth.[64]

Once a boundary was established separating Eden from the surrounding regions to conform with the Genesis narrative, Luther could insist on the pristine beauty of the whole earth: 'Eden was a choice garden in comparison with the magnificence of the whole earth, which itself also was a paradise compared with its present wretched state.'[65] He suggested too that paradise must have been of considerable size, as it was destined to become the dwelling place not only for Adam but for all his descendants also, and that had Adam and Eve remained in the state of innocence, God would have extended the garden to accommodate their progeny. Luther was trying to overcome the difficulty of conceiving of a garden too small to contain the whole of mankind, even though the text urged him to accept the notion of a secluded area.[66]

After the Fall, the Garden of Eden – for Luther a vast but specific piece of land – remained inaccessible to mankind, guarded by the Cherubim and the flaming sword. The divine curse on the whole earth after the Fall applied also to the Garden of Eden, which brought forth thorns and thistles and lost its fertility, although its boundary and the angelic guard remained.[67] Eventually, at the time of the Flood, even the remains of paradise were completely destroyed. By way of a summary, Luther added:

All this, I say, is historical. Therefore we ask in vain today where or what that garden was. ... My opinion ... is, first, that paradise was closed to man by sin, and, secondly, that it was utterly destroyed and annihilated by the Flood, so that no trace of it is visible any longer. ... I am fully of the opinion that after Adam's fall paradise remained in existence and was known to his descendants, but was inaccessible because of the angel who kept watch over the garden with his flaming sword, as the text states. But the Flood laid everything waste, just as it is written that all the fountains and abysses were torn open (Genesis 7.11). Who, then, would doubt that these sources, too, were rent and thrown into confusion?[68]

The problem of the identification of the four rivers of paradise was solved by the argument – already put forward by Steuchus – that the rivers were affected by the Flood's destruction. Thus Luther was able to discard the various allegorical interpretations of paradise which, in his view, strayed too far from the Genesis text.[69]

It was Luther's insistence on both keeping to the letter of the Bible and describing Eden as a vast and yet specific region no longer in existence that prepared the ground for a new cartography of paradise. The Garden of Eden, for which there was no room on the sixteenth-century maps of the *oikumene*, was shown in a woodcut in the first complete edition of Luther's translation of the Bible, published by Hans Lufft in Wittenberg in 1534 (Plate 15). The image, which faces the first chapter of Genesis, corresponds to Luther's reading of the biblical text.[70] The illustration is a portrait of the perfect and uncorrupted universe at end of the week of creation. God the Father, radiating light and clothed in a royal mantle, with long hair and flowing beard, blesses the universe below him. He has already divided light from darkness, the waters above and below the firmament, created dry land and vegetation, the sun, moon, stars, birds and fishes, man and woman. Adam and Eve are in paradise, themselves pure, perfect and innocent like the world in which they dwell. Eden is a pleasant landscape, rich in flora and fauna and watered by the four streams.

The illustration derives from the tradition of depicting the creation of Eve in Eden at the centre of the cosmos, an iconography common in late fifteenth-century Bibles and also taken up by Hans Holbein in the early sixteenth century.[71] Whereas earlier images focused on the action of God within paradise creating the first woman, the image in the Wittenberg Bible presents a description of the Edenic landscape with the upright figures of Adam and Eve and God blessing his creation from outside. The Garden of Eden is represented as a huge region, part of a paradise-like earth with clearly depicted coastal outlines, islands and mountains. The lushly vegetated landmass on which Adam and Eve are standing seems to be part of a vast land, surrounded by sea and watered by four rivers. Luther's image gives emphasis to the figure of the Creator and might appear to be more of a pictorial and fictional representation of a lost world than a map locating Eden. Nonetheless, the illustration can be considered a map of the world at the time of creation, featuring the Garden of Eden as part of the paradise-like world before the Fall and the Flood. Luther's Eden was, as he said, 'certainly not a narrow garden of a few miles in circumference'.[72] It occupied an area corresponding, as his lectures made clear, to present-day Syria, Mesopotamia and Arabia. The illustration in the Wittenberg Bible conveys something of this idea, but it would be vain to attempt to recognize any familiar geographical features; Luther's point was that the face of the earth had dramatically changed since the Flood. So, in the Wittenberg illustration, the four rivers are portrayed winding their ways towards the sea from a single source in the middle of the garden, without concern for post-diluvian geography. For Luther, the hydrography described in Genesis was not to be found anywhere because the rains of the Flood had thrown the four rivers into disorder and had changed their courses.[73] The Mediterranean Sea and other present-day geographical features such as the Persian and the Arabian gulfs were produced by the Flood. In the image in the Wittenberg Bible, paradise cannot be situated in relation to ordinary geography, because the world depicted therein no longer exists.

In the Wittenberg woodcut, the whole earth is pictured at a particular moment in time, before the dramatic changes brought about by sin, and possibly on the seventh day of creation. Although Luther had made it clear that nothing certain could be said about the timing of original sin, he allowed himself to speculate that Eve was created toward the end of the sixth day and that original sin had taken place in the afternoon of the seventh day.[74] In the woodcut Adam and Eve are represented naked, in peaceful harmony with their world, surrounded by all kinds of obedient animals. They had not yet sinned. A snake is visible, standing upright to the left of the couple, not yet crawling on the ground as it was condemned to do after the Fall (Genesis 3.14). The scene of the Fall, here missing, was often depicted on medieval *mappae mundi* in order to connect the space–time reality of Eden with human history, while the four rivers of paradise were represented as flowing through the enclosure protecting Eden to water the known regions of the earth. Here, in contrast, there is no link between paradise, its perfect and beautiful world, and post-diluvian geography. For Luther, no map could feature both the original Edenic perfection and the earth cursed by God on account of sin. Paradise was lost because of human wickedness, and the Flood had destroyed all remains of the uncorrupted creation of God. As it happens, the Wittenberg Bible contains another portrayal of the world that does feature the post-diluvian earth. This map was originally produced by Philipp Melanchthon for his commentary on the Book of Daniel and printed in 1529 (Figure 9.3).[75] Melanchthon's image illustrates Daniel's prophetic dream of the four beasts, symbols of the succession of the four heathen kingdoms before God's approaching victory. Although loaded with religious significance, the image does not feature any paradise on earth.

The T-O map in a late fifteenth-century copy of Bartholomaeus Anglicus' encyclopedic work, the *De proprietatibus rerum*, shows God blessing the world. Here, a surviving paradise is depicted in the Far East as a golden and splendid palace presiding over

Fig. 9.3. Daniel's prophetic dream of the four beasts, in *Biblia, das ist, die gantze Heilige Schrifft Deudsch* (Wittenberg: Hans Lufft: 1536), II, sig. C2r, Daniel, fol. 14r. 15 × 12 cm. London, British Library, 1.b.10. On this north-orientated map, illustrating a passage of the Book of Daniel in later edition of Luther's Bible the earth is portrayed in its post-diluvian condition. The beasts symbolized the succession of four heathen kingdoms before the establishment of the kingdom of God. The woodcut also includes the detail of the four winds of heaven stirring up the sea (Daniel 7.2), and, in northern Asia, what are, presumably, the enclosed and warlike tribes of Gog and Magog as they await the appearance of Satan (Revelation 20.7–8).

the earth.[76] The divine blessing shines down over both paradise and the post-lapsarian earth. The image presents the medieval world view of a surviving Garden of Eden, but places it firmly in a fifteenth-century world of geographical exploration (as indicated by four large sailing vessels). Also in the traditional manner, paradise is shown connected to the rest of the earth by the four rivers yet separated from it by its impenetrable architecture. In contrast, in the Wittenberg Bible woodcut – created only a few dacades later – God blesses only paradise, the world of the beginning, before the corruption of human sin, and before the destruction of the Flood.

Luther's belief that paradise had disappeared and that the surface of the earth had radically changed as a consequence of sin was consistent with his more general theological views.[77] God had created a righteous and perfect Adam, but 'the image of God in man disappeared after sin in the same way the original world and paradise disappeared.'[78] The garden of delight vanished, the four rivers of paradise became corrupted, and the earth lost its fertility.[79] Luther went so far in his description of God's curse on the ground (Genesis 3.17–19) as to say that before sin even the sun's light had been clearer, the air purer and the water more bountiful.[80] Likewise, man had lost all his innocence and immortality, while his body became wretchedly corrupted. God's punishment was made even more severe through the Flood, which completely ruined and destroyed the pristine beauty of the world. In Luther's view, God flooded the first and original world when sin prevailed. The second, post-diluvian world, however, proved to be even more iniquitous and devoted to the worship of idols, and this is why God overthrew one monarchy after the other until the advent of Christ. The third world was the world of God's grace, but was still full of sins, blasphemies and abominations and would soon be consumed by a flood of fire, the eternal punishment awaiting anything sinful and ungodly.[81] The entire creation had been corrupted by sin and was now awaiting final restoration. Indeed, the gospel had brought about the restoration of the divine image in man, but Luther believed that it was important to acknowledge the tragedy of the Fall: 'There is a great danger that if we forget our former sins, we shall be overcome by them again.'[82] Depicting the perfection of the beginning brought home the weight of the curse that was inflicted on both nature and mankind. It would be impossible for a Lutheran to expect to find paradise on maps of the earth after the Fall. The Flood was a restoration of sorts and purged the earth, Christ redeemed mankind with his sacrifice on the Cross, but the world was still too sinful to accommodate the garden of delights.[83]

Paradise Mapped in the Middle East

The idea that paradise had once been on earth, but had disappeared after the Flood allowed Luther to defend the literal truth of the biblical narrative against any criticism based on modern geographical knowledge. By saying that the garden had disappeared, Luther was simply dismissing enquiries about Eden's geographical position. It was the theological thinking of another sixteenth-century Reformer, John Calvin, that allowed paradise to be mapped in a different form from anything seen previously: that is, on a regional and historical map. Calvin believed, as did Luther, that the beauty and perfection of the earthly paradise had vanished because of sin, but his comprehensive knowledge of classical authors – and his familiarity with Steuchus's arguments – led him to conclude that Mesopotamia was in fact fertile and beautiful, and a fitting place for Eden. After the Fall, when the effect of the curse rendered paradise barren and unattractive, the Cherubim still demarcated the boundary of the Garden of Eden but, according to Calvin, only in order to remind Adam and Eve and their descendants of the gravity of sin and of God's anger.[84] To prevent mankind from being completely overwhelmed by despair, the Creator had scattered signs of his goodness over the world. One of these signs was a remnant of the earthly paradise, and it was this remnant that could be shown on a map. The composite nature of Calvin's theological thought and his rhetorical use of Renaissance mapping brought about a new and long-lasting type of paradise map.

Calvin saw post-lapsarian life on earth as governed by the depravity of man and the hostility of nature. The world for him was a chaotic and uncertain place because of sin:

> *After man's rebellion our eyes, wherever they turn, encounter God's curse, [which] must overwhelm our souls with despair. For even if God wills to manifest his fatherly favour to us in many ways, yet we cannot, by contemplating the universe, infer that he is Father. Rather, conscience presses us within and shows in our sin just cause for his disowning us and not regarding or recognizing us as his sons. Dullness and ingratitude follow, for our minds, as they have been blinded, do not perceive what is true. And as all our senses have become perverted, we wickedly defraud God of his glory.* [85]

Calvin was horrified by God's punishment of a sinful humanity and tormented by a profound sense of guilt.[86] Anxiety itself was a source of sin, betraying a lack of faith and trust in God's promises. The Fall had subverted man's moral life as well as the whole order of reality. Nature taught that man was powerless. Calvin used the images of the abyss and the labyrinth to convey his sense of void and oppression. He also saw in the most violent atmospheric phenomena – winds, clouds, storms, lightning and rainfall – a symbol of the supreme power of God and of the impotence of man. He wrote, for example, that although the air might now be tranquil at any moment it could change suddenly, and that a storm always arises when the sky appears calm and serene.[87] Air, for the poetically inclined eastern theologian Ephrem the Syrian, had a pleasant symbolic value in paradise, being 'the glorious air whose heavenly breath restores humanity to life'; air was referred to by medieval and early Renaissance exegetes to reinforce religious belief in a perfectly temperate Eden; for Calvin, air became, because of its frightful turbulence, a potent symbol of paradise lost.[88]

Calvin tried to relieve his consternation over the tumultuous abyss into which mankind was thrown after sin. A contradictory impulse in his theological thought urged him to allay his anxiety by adopting an intense moralism that would make sense of the universe. He insisted that, despite human sin, divine providence controlled everything in a fair and rational way, even though its powerful governance was often beyond human understanding. God cared about man and the world. If God's power was behind all events in history, preserving order in the universe, nature was there to reassure man and to teach him about divine grace. The universe was like a mirror in which to contemplate God, who was otherwise invisible: 'We have been placed here, as in a spacious theatre, to behold the works of God; and there is no work of God so small that we ought to pass it lightly – they should all be observed carefully and diligently.'[89] Mankind was thus called to contemplate the beauty of the created world in order to acknowledge its author and rejoice at the manifestation of God's glory in even the tiniest blade of grass.[90] One of the comforting traces of divine care for mankind was what was left of the earthly paradise.

For Luther, the entire surface of the earth had been destroyed by the Flood and nothing remained of the ante-diluvian life, apart from a few fossils of living organisms destroyed by the Flood.[91] While influenced by Luther's exegesis, Calvin insisted that the original geography of Eden had not been dislocated by the Flood, but that only its delight had been lost, in the same way as human nature, however deformed and damaged as a result of the Fall, nevertheless preserved some aspects of God's image. Some sparks of the glory and divinity of God can be discerned in every part of the human being; in for instance man's intelligence and reason, in his moral behaviour or even in his body. Satan was not able to extinguish God's light completely even in the most wicked and reprobate men. Preaching on the opening verse of John's Gospel, Calvin reminded his listeners that the evangelist's proposition that the light shines in the darkness was true.[92] The persistence of God's light in man corresponded to the way in which the earth itself retained sparks of the grace that God had put into Adam's garden:

> *For although I acknowledge that the earth from the moment it was cursed was reduced to a miserable and hideous state, and has been dressed as if in a mourning veil, and has since been corrupted in many places, nevertheless I say that it is the same earth that was created at the beginning.*[93]

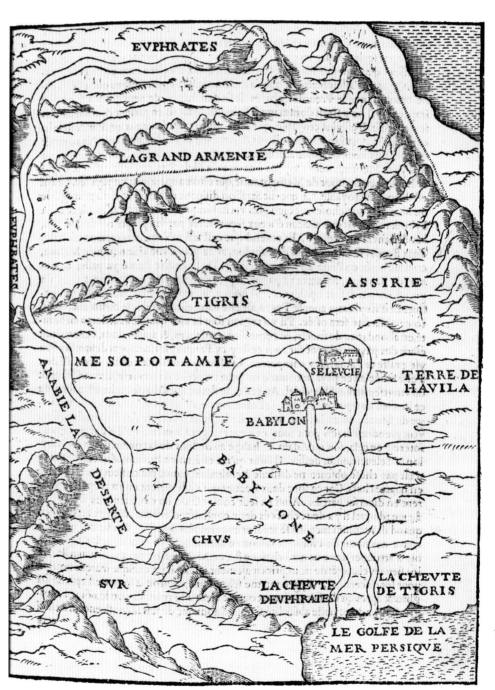

Fig. 9.4a. John Calvin's map of the location of Eden, from his *Commentaire … sur le premier livre de Moyse, dit Genèse* (Geneva: Ian Gerard, 1553, 2nd edn, 1554), p. 33. 15.7 × 12 cm. London, British Library, 1016.1.3. By refining Augustinus Steuchus's inventive reading of Genesis by means of a map, Calvin initiated the transfer of the earthly paradise from maps of the world to regional and historical maps. In Calvin's view, the Garden of Eden had once existed, in Mesopotamia, but the question of its precise location was a point that paled in comparison with the events with which it was associated. Despite its description as a map of the location of the Garden of Eden, in fact the map does not show the exact location of the Mesopotamian earthly paradise but only the course of the Tigris and the Euphrates rivers, and, among the surrounding territories, the lands of Havilah and of Cush (*Chus* on the map), described in the Book of Genesis as in the vicinity of the Garden of Eden (Genesis 2.11–14). The crucial message of his map was that the four rivers of paradise still flowed in Mesopotamia as a reassuring sign of God's continuing benevolence.

272   Mapping Paradise: A History of Heaven on Earth

Calvin's geography of paradise followed from his conviction that God's light still shone in the midst of darkness. The earth, once a delightful mirror of God's paternal indulgence towards man, had been cursed on account of the Fall, when God withdrew his blessing and his favour. Signs of divine anger were everywhere. But from a geographical point of view, the earth had not changed and still carried significant marks of divine benevolence to urge mankind to seek a remedy in Christ.[94] The rivers of paradise were still there and a map would help readers to reconcile the biblical account with the geographical science of the day.

The map of the location of the Garden of Eden that Calvin produced to illustrate his *Commentary on Genesis* was printed in Geneva in 1553. Despite the fact that paradise was not explicitly indicated, the map proved to be the first in a long series of regional maps dealing with the location of Eden (Figure 9.4).[95] In compiling it, Calvin made use of Ptolemaic cartographical principles, classical learning, and the arguments already put forward by Augustinus Steuchus. In fact, Steuchus's *Recognitio Veteris Testamenti* was one of the main, if unacknowledged, sources of Calvin's *Commentary*.[96] Calvin may have criticized the Vatican librarian for his views on issues such as the eternity of the world and the Donation of Constantine, but this did not prevent him from basing his own discussion of the geography of Eden on Steuchus's ground-breaking idea.[97] Following Steuchus, Calvin thought that Eden had been located in Mesopotamia, a well-known and already well-mapped region of the earth, praised by the ancient authorities as exceptionally fertile: 'If ever a region existed beneath the skies which topographers celebrated above all other places as excelling in beauty, in abundance of fruits, in fertility, in delights and other gifts, this is the one.'[98] Calvin's map was not merely an adjunct but an integral part of his theological discourse, as he said in the *Commentary*: 'Let me put here before your eyes a figure by means of which you may understand where I think Moses situates paradise.'[99] The map was introduced, in other words, to display the location of paradise according to Moses (the traditional author of the first five books of the Bible, including Genesis) and to show the continuing presence of the rivers of paradise, as a reassurance to mankind of God's benevolence and of the importance of earthly life. In all his commentaries, Calvin drew on philological, historical and geographical information to urge his readers to apply the teaching of the Bible to their own lives, and not only to elucidate the meaning of the text. Mapping paradise was for Calvin part of a rhetorical effort.[100] Moreover, a device such as a map must have suited a theologian whose aim was always to organize human phenomena into a coherent structure of thought and to make visible the principles of order hidden in a chaotic world.

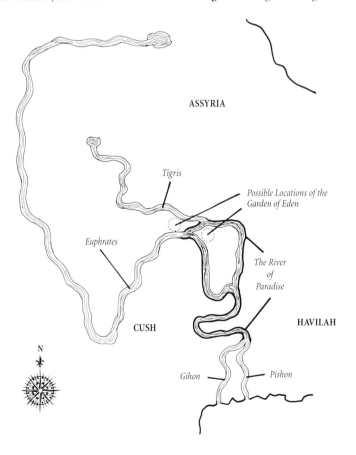

Fig. 9.4b. Diagram of Fig. 9.4a.

Calvin argued against those authors who claimed that paradise had once included the whole earth.[101] The earthly paradise was a specific region intended by God for Adam. The entire earth enjoyed the bliss of a perfect condition of fruitfulness, but Eden was its richest and most fertile region. He took issue, too, with allegorical interpretations of the paradise narrative such as Origen's, which, in his view, departed too far from the letter of the biblical text. He also attacked medieval theories about the altitude of paradise, ridiculing the debate over its location in the higher or middle regions of the air: how could Adam cultivate the fields or eat certain types of fruit if he was close to the circle of the moon? It was essential, in Calvin's view, to acknowledge that Eden existed on earth, so that after the Fall the benevolence of God could be seen as still anchoring mankind to the earth for the duration of his pilgrimage to heaven:

> *What is to be gained from soaring into the air, and leaving behind the very earth where God has given testimony of his benevolence to mankind? But now someone will say that the more subtle interpretation of this account is that it refers to heavenly bliss. To this I answer that it is indeed appropriate to long for heaven, since heaven is the eternal inheritance of humanity. However, in so far as God has willed that mankind is to use the earth for a while as its dwelling place, we must settle our feet on that earth for a while. For we read in this account that Adam was ordered by God to be an inhabitant of the earth, so that by spending his temporal life there, he could nevertheless contemplate the glory of heaven; and we read that the Lord generously bestowed upon him numberless gifts, so that by enjoying them he might infer his fatherly benevolence.*[102]

The rigour of the punishment inflicted upon Adam was mitigated by the goodness of God, who still provided him with a home on earth. Mankind was urged by divine care and paternal love to live on earth in acceptance of the anomalies and contradictions of life and in anticipation of heavenly reward.[103] All this was to be corroborated by the map.

Calvin argued that earlier thinkers had turned to odd and ungrounded theories because they could not identify on earth the place described in Genesis. Such difficulties would be eliminated, he said, if the biblical text were to be sufficiently scrutinized. Genesis specifies that the earthly paradise was in the east. He pointed out that Jerome's translation of the Hebrew term מִקֶּדֶם (*miqedem*) as 'from the beginning' was wrong. Moreover, Moses was near the Holy Land when he was writing the Pentateuch, and, obviously, Eden had to be 'to the east' of the Holy Land.[104] In discussing Moses's description of the rivers – which Calvin considered had been simplified for the sake of the Jewish readership – Calvin exhorted his own readers that they should:

> *Note, in the first place, that there is nothing mentioned here about either a source or a fountain, but only of one river, and I understand the four heads to be both the origins, or beginnings, where those rivers have their source, and the mouths where they flow into the sea. In ancient times the Tigris flowed into the Euphrates and the two rivers formed one course, such that one could justly regard it as one river branching into four heads.*[105]

In this passage, Calvin was underlining, as had Steuchus, the fact that Genesis makes no explicit mention of a source within paradise itself. The text says only that 'a river went out of Eden to water the garden'. That river, according to Calvin and Steuchus, flowed through paradise as a single stream, dividing outside paradise – both upstream and downstream – into two branches, making four rivers in all. To support this interpretation, Calvin went back to the term used in the Vulgate to indicate the four rivers, namely 'heads' (*capita*). Heads of the rivers, for Calvin, meant the two channels that brought the water to paradise *and* the two channels which discharged it into the sea. This notion of a single stream branching into four courses was a reversal of the medieval concept of a single source from which four rivers flowed out in four directions. The model envisaged by Calvin (and by Steuchus, before him) described four 'heads' (*capita*) branching from the single river within paradise, with the two upper streams flowing from their sources into the confluence and the two lower ones flowing from it towards the Persian Gulf. Calvin refined Steuchus's ingenious interpretation by means of a map that clarified his idea (Figure 9.4).

Steuchus's and Calvin's model had the advantage of making the question of the source of the four rivers irrelevant. Only two rivers, whose courses first came together and then separated, needed to be found, and these were at hand in the two rivers uncontroversially named in Genesis as the Tigris and the Euphrates and readily recognized from any map of Mesopotamia. Calvin quoted a range of classical authorities – Strabo, Pliny, Arrian and Pomponius Mela – who confirmed that the Tigris and the Euphrates united for some distance near Babylon to form one river and then separated into two branches before reaching the sea. Calvin identified these two branches as the third and fourth 'heads' mentioned in Genesis. It was not unusual, Calvin pointed out, for rivers to change their names, which was why the western lower channel, corresponding to the Euphrates, was called Gihon, whereas the eastern lower channel, corresponding to the Tigris, took the name of Pishon. According to Pliny, the Tigris was called *Diglito* near its source and after forming many channels, took the name of *Pasitigris*, known locally as *Pasin* (according to a reference in Quintus Curtius). The affinity between Pasin and Pishon made it likely that the name Pasitigris was a vestige of the ancient name Pishon. A similar change of name is likely to have applied to the Euphrates.[106] Classical geography was called into service to explain sacred geography.

Calvin's map shows the confluence of the Tigris and the Euphrates north of Seleucia and their joining again south of the city of Babylon, creating an island. Downstream the two rivers again divided to reach the Persian Gulf in two outfalls marked on the map as *la cheute d'Euphrates* and *la cheute de Tigris*. The lands of Havilah and Cush, which had been placed in far eastern Asia and equatorial Africa by earlier Christian scholars, were situated by Calvin in the Middle East. The huge bend made by the Tigris south of Babylon effectively encompasses the land of Havilah in the same way as the Pishon was said to do in the Genesis text. On Calvin's map, Havilah is a region to the west of Persia, reputedly rich in gold and precious stones, as he noted in the *Commentary*. The middle eastern location of Havilah is confirmed on the map by the presence of Shur (*Sur*), the land described in Genesis 25.18 as situated between Havilah and Egypt. Next to Shur is the land of Cush (*Chus*) that is described in Genesis as watered by the Gihon and that is shown on the map as a region neighbouring *Arabia deserta*. Assyria, towards which the Tigris bends (Genesis 2.14), is also marked.

What is missing from Calvin's map of the location of the Garden of Eden, however, is the Garden of Eden itself. Calvin avoided pinpointing the exact site of paradise, saying that the precise location of Adam's dwelling was of no great interest.[107] For him, it was sufficient to show the general area of the remnant of Eden and how the single river divided into four heads, and to suggest that Adam was put into a well-watered region. This region, Calvin explained, was either on the island formed by the convergence of the Tigris and the Euphrates or in the area immediately north of the branch that connected the two rivers. The main purpose of the map was not so much to pinpoint paradise, but to show that its rivers had remained unchanged despite the curse on the earth and the destruction brought about by the Fall and the Flood. The map identified the region where paradise had existed and showed that this region was geographically real and that it continued to exist, as indicated by the historical cities of Babylon and Seleucia. In his effort to trace the courses of the four rivers, however, Calvin incidentally found himself giving at least an approximate indication where the garden was most likely to have been situated. Whereas Steuchus had indicated just the region of Mesopotamia, Calvin had gone further in his pointing to the area of the Garden of Eden.

While Calvin's exegetical source was Steuchus, his cartographical source was Ptolemy. The Fourth Map of Asia from printed editions of Ptolemy's *Geography*, showing

Fig. 9.5. Sebastian Münster's version of Ptolemy's *Tabula Quarta Asiae*, from Claudius Ptolemaeus, *Geographia universalis*, ed. by Sebastian Münster and Bilibaldus Pirckheimer (Basle: Henricus Petrus, 1540). London, British Library, Maps C.1.C2(1). Calvin may have used either this edition or the one published in Basle in 1545, under the title of *Cosmography*, a few years before Calvin produced his *Commentary* on Genesis (1553). Münster was a well-known Hebrew scholar and geographer whom Calvin would have respected as an authority. The summary of the different theories about the location of paradise held in the Middle Ages that Calvin presented in his *Commentary* was probably inspired by the section on the earthly paradise in Münster's 1545 edition of the *Cosmography*.

the Eastern Mediterranean and the Middle East from Cyprus to Babylon, provided Calvin with the two key points of his Edenic geography: the four river heads, and the great bend of the river Pishon. Calvin took other features, such as the north-eastern mountain range near the Caspian Sea, from the Third and the Fifth Maps of Asia. Two sets of Ptolemaic maps would have become recently available to Calvin. Sebastian Münster's edition of Ptolemy's *Geography*, complete with maps, had been published in Basle first in 1540, and then in 1545. Münster was an outstanding Hebraicist and geographer for whom Calvin would have had great respect. It is known that Calvin used the second edition of Münster's bilingual (Hebrew–Latin) Bible, printed in Basle in 1546, and that the library of Calvin's Academy in Geneva possessed Münster's Hebrew and Chaldaean dictionaries.[108] Moreover, Münster had a section in his *Cosmography* on the earthly paradise in which he summarized the various medieval opinions about the location of paradise, and which Calvin may have had in mind when he stated in his own *Commentary* that 'the garden was situated on earth, and not, as some have dreamed, in the air', and when he preached in Geneva, in September 1559, against phantasizing that the Garden of Eden floated in the air, reached the sphere of the moon or was the same thing as eternal life.[109] Like Luther, Münster held that no trace of paradise had been left on earth after the Flood and accordingly he had nothing to say about Eden or its location in the section of the *Commentary* devoted to Asia. Paradise is dealt with only in the general introduction where the Creation of the world and various aspects of the structure of the globe are discussed. For Münster, paradise was a feature of the past. How else, he asked, can it be explained that its site cannot be found? Reading Münster's words, Calvin insisted on mapping the remnants of paradise – the first Ptolemaic map of paradise – as an act of his own faith in God's concern for human life on earth.

Calvin's configuration of the river courses with the confluence of the Tigris and the Euphrates is identical to that on Ptolemy's *Tabula Quarta Asiae* inserted in Münster's version (Figure 9.5). For Calvin, it was critical that the single river of paradise should flow through Mesopotamia. By definition, however, the very name Mesopotamia designates territory 'between [two] rivers'. Had Calvin adhered closely to the map in Ptolemy's *Geography*, the single stream made by the confluence of the Tigris and the Euphrates, and therefore paradise, would not, of course, have been actually in Mesopotamia. To show a Mesopotamian paradise, it was necessary to have a Mesopotamian single river, a contradiction in terms. By linking the two rivers just north of Babylon, Calvin at a stroke created a cartographical feint to depict a single river flowing through Mesopotamia, which now included an island surrounded by the two rivers. Inspiration for this crucial addition could have come to Calvin from Michael Servetus's edition of Ptolemy's *Geography* (1541), which is recorded as having been in the Academy's library in 1572.[110] The Fourth Map of Asia in Servetus's edition shows a complex river network formed by the maze of braided channels and islands in the Euphrates just above its confluence with the Tigris (Figure 9.6). Seeing these streams, and needing to make Mesopotamia extend south of the confluence, Calvin would have realized that all he had to do was to show a link between the two rivers to make his point. Calvin's geography of paradise, inspired by Steuchus's exegesis and Ptolemy's cartography, left the issue of the exact location of Eden open, for he indicated only the area near Babylon – either to the south or to the north of the stream linking the Euphrates and the Tigris – as the likely abode of Adam. Later exegetes and map makers would propose various locations and compile new maps of paradise, trying to say what Calvin had left unsaid.

The change already noted from the *mappae mundi*, with their depiction of a whole earth that included Eden, to the regional maps of the sixteenth century and later reflects above all else a theological shift. The medieval idea that the Garden of Eden belonged to the present world had been largely abandoned. Preoccupation with the original site of paradise would be left to 'biblical archaeologists' who, ignoring the efforts of their medieval predecessors, would attempt to excavate that site from a few lines of biblical text. The rupture between medieval and modern mapping of paradise would soon be completed. Paradise was both lost from the face of the earth, and found in Mesopotamia. Instead of a paradise vaguely indicated in a remote location on a medieval *mappa mundi*, Renaissance historical and regional maps defined – with Ptolemaic precision, climatic zones, latitude and longitude – a region where in the remote past the earthly paradise had once existed.

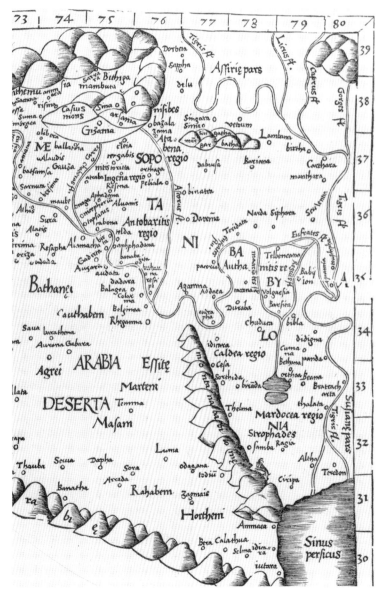

Fig. 9.6. Michael Servetus's version of Ptolemy's *Tabula Quarta Asiae*, from Claudius Ptolemaeus, *Geographia*, ed. by Michael Servetus and Bilibaldus Pirckheimer (Lyons: Hugues de La Porte, 1541). Maps engraved by Gaspar Treschel. London, British Library, 215.f.9. Detail showing Mesopotamia. The key aspect of Calvin's Edenic geography – the Mesopotamian pattern of the four river heads – clearly derives from Ptolemy's Fourth Map of Asia. The cartographical inspiration for the idea of a single river flowing in Mesopotamia could have been Servetus's rendering of the Fourth Map of Asia, which shows two islands formed by the Euphrates before its junction with the Tigris, shown on the map at the extreme right, adjacent to latitudinal markers 34/35.

Paradise Lost and Found 277

1. In the *Geography*, Ptolemy describes in detail how to construct three different projections to enable the spherical (i.e. three-dimensional) globe to be represented to scale on a flat (i.e. two-dimensional) surface. Tables of the necessary astronomical coordinates of latitude and longitude, already given in shorter form in the *Almagest* and the *Handy Tables*, fill the rest of the *Geography*. On Ptolemy's *Geography* see Oswald A. W. Dilke, 'The Culmination of Greek Cartography in Ptolemy', in Brian Harley and David Woodward, eds, *The History of Cartography*, I, *Cartography in Prehistoric, Ancient and Medieval Europe and the Mediterranean* (Chicago: University of Chicago Press, 1987), pp. 177–200. For a critical introduction to the three books giving instructions for the construction of maps on a projection, and an English translation, see J. Lennart Berggren and Alexander Jones, *Ptolemy's Geography: An Annotated Translation of the Theoretical Chapters* (Princeton, NJ–Oxford: Princeton University Press, 2000).

2. For details, see Berggren and Jones, *Ptolemy's Geography* (2000), pp. 50–5, and Patrick Gautier Dalché, 'Le Souvenir de la *Géographie* de Ptolémée dans le monde latin médiéval (VIe–XIVe siècles)', *Euphrosyne*, 27 (1999), pp. 79–106. In recent years, Gautier Dalché has made a particular study of the transmission of Ptolemy from Antiquity to the Renaissance. See, for example, his 'Connaissance et usages géographiques des cordonnées dans le Moyen Âge latin (du Vénérable Bède à Roger Bacon)', in Louis Callebat and Olivier Desbordes, eds, *Science antique, science médiévale (autour d'Avranches 235)* (Hildesheim–Zurich–New York: Olms–Weidmann, 2000), pp. 401–36; and, most importantly, his forthcoming 'The Reception of Ptolemy's Geography', in David Woodward, ed., *The History of Cartography*, III, *Cartography in the European Renaissance* (Chicago: University of Chicago Press, in press). In 827 Ptolemy's *Almagest* was translated from Greek into Arabic (neither language was well known at the time in the West) and in 1175 into Latin.

3. David Woodward, with Herbert M. Howe, 'Roger Bacon on Geography and Cartography', in Jeremiah Hackett, ed., *Roger Bacon and the Sciences: Commemorative Essays* (Leiden: Brill, 1997), pp. 199–222; Woodward, 'Roger Bacon's Terrestrial Coordinate System', *Annals of the Association of American Geographers*, 80 (1990), pp. 109–22. There is growing evidence that the use of a mathematical grid in the construction of maps may never have been entirely lost since Ptolemy's day: see Jeremy Johns and Emilie Savage-Smith, '*The Book of Curiosities*: A Newly Discovered Series of Islamic Maps', *Imago Mundi*, 55 (2003), pp. 7–24, esp. 12–13, and Plates 1–7, and Evelyn Edson and Emilie Savage-Smith, *Medieval Views of the Cosmos* (Oxford: Bodleian Library, 2004), p. 62, where the authors argue that 'the carefully executed graphic scale on the rectangular map in the eleventh-century *Book of Curiosities* … suggests that such maps [with a horizonal and vertical scale constructed according to instructions given in a book of geographical coordinates compiled by the Arab writer Suhrab in the late tenth century] may have been in circulation.'

4. Jacopo Angelo's translation, entitled *Cosmographia*, was followed in 1415 by a manuscript with maps with Latin legends. See Germaine Aujac, 'Ptolemy', in John B. Friedman, Kristen Mossler Figg, and others, eds, *Trade, Travel, and Exploration in the Middle Ages: An Encyclopedia* (New York–London: Garland, 2000), pp. 507–11 and bibliography; Aujac, *Claude Ptolémée, astronome, astrologue, géographe* (Paris: Comité des Travaux Historiques et Scientifiques, 1993), pp. 165–78; Aujac, 'Les Images du monde entre rêve et réalité', *Revue de la Bibliothèque Nationale*, 45 (1992), pp. 2–13, esp. 9. See also Sebastiano Gentile, 'Umanesimo e cartografia: Tolomeo nel secolo XV', in Angelo Cattaneo, Diego Ramado Curto, and André Ferrand Almeida, eds, *La cartografia europea tra primo Rinascimento e fine dell'Illuminismo* (Florence: Olschki, 2003), pp. 3–18. On the vexed question of whether Ptolemy himself compiled the maps, see the remarks by Dilke, 'The Culmination of Greek Cartography in Ptolemy', (1987), pp. 177–200, esp. 177–8, and Berggren and Jones, *Ptolemy's Geography* (2000), pp. 45–50. See also above, Chapter 8, pp. 198–9. French and German scholars were the first to show a keen interest in the scientific aspects of Ptolemy's *Geography*: see Patrick Gautier-Dalché, 'Un astronome, auteur d'un globe terrestre: Jean Fusoris à la découverte de la *Géographie* de Ptolémée', and 'L'Oeuvre géographique du cardinal Fillastre († 1428): Représentation du monde et perception de la carte à l'aube des découvertes', in Didier Marcotte, ed., *Humanisme et culture géographique à l'époque du Concile de Constance, autour de Guillaume Fillastre* (Turnhout: Brepols, 2002), pp. 161–75, esp. 173, and 293–355.

5. The Bologna edition included 26 copperplates and was produced by Dominicus de Lapis. On Ptolemy's printed editions see Tony Campbell, *The Earliest Printed Maps: 1472–1500* (London: British Library; Berkeley: University of California Press, 1987), pp. 122–38; Carlos Sanz, *La Geographia de Ptolomeo* (Madrid: Victoriano Suarez, 1959).

6. On the new image of the map as abstract geometric transformation see David Woodward, 'The Image of the Map in the Renaissance', in David Woodward, Catherine Delano-Smith, and Cordell D. K. Yee, *Plantejaments i objectius d'una història universal de la cartografia/Approaches and Challenges in a Worldwide History of Cartography* (Barcelona: Institut Cartogràfic de Catalunya, 2001), pp. 133–52, esp. 140, where he stresses the importance of numbered (i.e. measured) grid lines in distinguishing projections and perspective drawings. See also his 'The Image of the Spherical Earth', *Yale Architectural Journal*, 25 (1991), pp. 4–15, and Jean-Marc Besse, *Les Grandeurs de la terre: Aspects du savoir géographique à la Renaissance* (Lyons: ENS, 2003), pp. 111–49.

7. Christian Jacob, 'Il faut qu'une carte soit ouverte ou fermée: Le Tracé conjectural', *Revue de la Bibliothèque Nationale*, 45 (1992), pp. 35–40.

8. On the impact of a map's projection and technical presentation see, for example: Denis Cosgrove, *Apollo's Eye: A Cartographic Genealogy of the Earth in the Western Imagination* (Baltimore–London: Johns Hopkins University Press, 2001), esp. pp. 110–14; Mark Monmonier, *How to Lie with Maps* (Chicago–London: University of Chicago Press, 1991), pp. 5–18; Brian Harley, 'Maps, Knowledge and Power', in Denis Cosgrove and Stephen Daniels, eds, *The Iconography of Landscape: Essays on the Symbolic Representation, Design and Use of Past Environments* (Cambridge: Cambridge University Press, 1988), pp. 277–312, esp. 287–90; Arthur H. Robinson, *The Look of Maps* (Madison: University of Wisconsin Press, 1952), esp. pp. 58–9.

9. Bergrenn and Jones, *Ptolemy's Geography* (2000), p. 15.

10. Ibid., p. 59. *Oikumene* in Greek means 'the inhabited part of the world', but it sometimes also indicates the known part of the world, a similar but not identical concept, ibid., p. 58 n. 3.

11. For a useful compendium of world maps see Rodney W. Shirley, *The Mapping of the World: Early Printed World Maps, 1472–1700* (London: Early World Press, 2001). There were, of course, exceptions. See, for instance, the world map by Antoine de la Sale, already discussed in Chapter 8, Figure 8.13 (Shirley, Entry 50, p. 56), and the map reproduced in the frontispice of Walter Raleigh's *History of the World*, reproduced below in Figure 10.12.

12 See above Chapter 5, p. 89.
13 For his life and works see Dirk Wassermann, *Dionysius der Kartäuser: Einführung in Werk und Gedankenwelt* (Salzburg: Institut für Anglistik und Amerikanistik, Universität Salzburg, 1996), pp. 7–15. Denys Turner, *The Darkness of God: Negativity in Christian Mysticism* (Cambridge–New York: Cambridge University Press, 1995), p. 216, describes him as 'perhaps the last medieval theologian to try to do *everything*'. See also Kent Emery, 'Preface', to Dionysius the Carthusian, *Opera selecta*, ed. by Emery, CCCM CXXI–CXXIA (Turnhout: Brepols, 1991), CXXI, p. 5; Henri de Lubac, *Exégèse médiévale: Les Quatre Sens de l'Écriture*, 4 vols (Paris: Aubier, 1959–64), II (1959), pp. 363–7.
14 Dionysius the Carthusian, *In librum II Sententiarum*, d.17, q.5, in *Opera omnia*, 42 vols (Montreuil-sur-mer–Tournai–Parkminster: typis Carthusiae Sanctae Mariae de Pratis, 1896–1935), XXII (1903), p. 154: 'Puto ... in isto certitudinem sine revelatione haberi aut non vel vix posse.'
15 Dionysius the Carthusian, *Enarratio in Genesim*, II, a.19, in *Opera omnia*, I (1896), p. 74: 'Est autem paradisus a terris habitabilibus remotus atque secretus, montibus, aquis, aut etiam intemperie aeris interpositis in mediis locis, ita ut adiri non possit; nec notitia eius poterit haberi, nisi ex revelatione. Idcirco philosophi et historiographi non sunt de eo locuti.'
16 Maria Luisa Picascia describes Biel's theology as a 'swansong': *Un occamista quattrocentesco: Gabriel Biel* (Florence: La Nuova Italia, 1979), p. 9. Heiko Obermann's monograph on Biel is entitled *The Harvest of Medieval Theology: Gabriel Biel and Late Medieval Nominalism*, 2nd edn (Grand Rapids, MI: Eerdmans, 1967). Henry Bebel defined Biel as 'the king of contemporary theologians': see Charles Ruch, 'Biel Gabriel', in *Dictionnaire de théologie catholique*, ed. by Alfred Vacant, Eugène Mangenot, Émile Amann and others, 17 vols (Paris: Letouzey et Ané: 1923–72), II/1 (1923), col. 814.
17 Gabriel Biel, *Collectorium circa quattuor libros Sententiarum*, II, d.17, q.2, ed. by Wilfridus Werbeck and Udo Hofmann, 2 vols (Tübingen: Mohr, 1973–92), II (1992), p. 404: 'certus paradisi situs nobis occultus est et ignotus.'
18 See above Chapter 3, p. 46. See also Augustine, *De civitate Dei*, XIV.11, ed. by Bernard Dombart and Alfons Kalb, CCSL XLVIII (Turnhout: Brepols, 1955), pp. 328–9: 'vivebat itaque homo secundum Deum in paradiso et corporali et spiritali.' I have elaborated on the issue of the condition of Adam and Eve in relation to paradise as expounded in medieval exegesis in my *The Notion of the Earthly Paradise from the Patristic Era to the Fifteenth Century*, PhD Dissertation, The Warburg Institute, University of London (1999).
19 *Biblia latina cum Glossa ordinaria*, facsimile of the *editio princeps* ... (Strasbourg: Adolph Rusch 1480/1), ed. by Margaret T. Gibson and Karlfried Fröhlich, 2 vols (Turnhout: Brepols, 1992), I, pp. 20–1.
20 Peter Lombard, *Sententiae in IV libris distinctae* II, d.16–29, ed. by Ignatius C. Brady, 3rd edn, 2 vols (Grottaferrata, Rome: Editiones Collegii S. Bonaventura Ad Claras Aquas, 1971), I, pp. 406–95.
21 Joseph E. Duncan emphasizes the importance of Thomas's approach to paradise as an influential compromise between Christian theology, on the one hand, and rational methodology and scientific knowledge, on the other: Duncan, *Milton's Earthly Paradise* (Minneapolis: University of Minnesota Press, 1972), pp. 69–75. Frederick C. Copleston, however, *Aquinas* (Harmondsworth: Penguin, 1955), pp. 63–5, points out that Thomas was not primarily concerned with putting together Aristotle and Christian theology, but simply adopted Aristotelian theories because he believed them to be true. Aristotle described the process of physical decay in his *De generatione et corruptione*, I.5, 322a28.
22 Thomas Aquinas, *Summa theologiae*, 1a, q.91–102, ed. by Pietro Caramello, Leonine edn, 3 vols (Turin–Rome: Marietti, 1948–50), I (1950), pp. 446–84. Aristotle had explained how heat is responsible for generation and growth, see Thomas Steele Hall, *Ideas of Life and Matter: Studies in the History of General Physiology, 600 BC–1900 AD*, 2 vols (Chicago–London: University of Chicago Press, 1969) I, pp. 110–19; Everett Mendelsohn, *Heat and Life: The Development of the Theory of Animal Heat* (Cambridge, MA: Harvard University Press, 1964), pp. 1–26. See also Robert Edward Brennan, *Thomistic Psychology: A Philosophic Analysis of the Nature of Man* (New York: Macmillan, 1941), pp. 85–110, and George Sarton, *Introduction to the History of Science*, 3 vols (Baltimore, MD: Williams & Wilkins, 1927), I, pp. 373–4.
23 See above, Chapter 8, p. 191.
24 John Duns Scotus, *Quaestiones in librum II Sententiarum*, d.17–19, q.2, in *Opera omnia*, ed. by Lucas Wadding, 12 vols (Leiden: [n.p.], 1639), VI/1, pp. 789–813.
25 See Augustine, *De civitate Dei*, XIII.15 (1955), pp. 184–5.
26 John Duns Scotus, *Quaestiones in librum II Sententiarum*, d.29–32, in *Opera omnia* (1639), VI/1, pp. 921–47, and *Reportata Parisiensa*, II d. 29–33, in *Opera omnia* (1639), XI/1, pp. 380–5.
27 Dionysius the Carthusian, *In librum II Sententiarum*, d.17, q.5, in *Opera omnia*, XXII (1903), pp. 151–7. Dionysius tried to present a summary of the medieval theological debate, but the time for producing a comprehensive synthesis was over: see Turner, *Darkness of God* (1995), pp. 216–17.
28 Gabriel Biel, *Collectorium*, III, d.16, q. unica, II (1992), p. 406: 'propter peccatum donum illud perdidit, et ideo relicta est humana natura ... in statu, qui debetur ei ex natura suorum principiorum'; 'because of sin man lost that gift, and thus human nature is left ... in the condition which is proper to it in accordance with the nature of its own principles,' English transl. by John L. Farthing, *Thomas Aquinas and Gabriel Biel: Interpretations of St Thomas Aquinas in German Nominalism on the Eve of the Reformation* (Durham, NC–London: Duke University Press, 1988), p. 52. See also Biel, *Collectorium*, II, d.16–17, 20–4, 28, II (1992), pp. 403–6, 425–34, 466–79, 527–45, and Oberman, *Harvest of Medieval Theology* (1967), pp. 57–68, 128–31. For Biel, the *viator* is neither an angel nor a *beatus*, since he has not yet achieved his eternal destiny and may always fall *en route* to it, a condition shared by both the prelapsarian Adam and the postlapsarian *viator*: Gabriel Biel, *Collectorium*, 'Prologus', q.1, I, (1973), pp. 8–9. See Oberman, *Harvest of Medieval Theology* (1967), pp. 39, 227–8, 329.
29 Gabriel Biel, *Collectorium*, II, d.23, q. unica, II (1992), p. 467: 'locus paradisi medius est inter hanc miseriae vallem et patriam caelestem. Et quemadmodum paradisus terrestris plus se tenet cum terra quam cum caelo'.
30 Giovanni Pico della Mirandola, *Commento sopra una canzone d'amore*, ed. by Paolo De Angelis (Palermo: Novecento, 1994), pp. 25, 48, 62–74, 102–3. The *canzone d'amore*, by his friend Girolamo Benivieni, was a poem based on Marsilio Ficino's 1469 commentary on Plato's *Symposium* and intended to celebrate the Platonic doctrine of the divine love which is aroused in the rational soul.
31 Ibid., p. 48.
32 *Discorso sulla dignità dell'uomo*, ed. by Giuseppe Tognon and Eugenio Garin (Brescia: La Scuola, 1987), pp. 26–8.
33 *Conclusiones nongentae: Le novecento tesi dell'anno 1486*, ed. by Albano Biondi (Florence: Olschki, 1995), p. 56. Pico's aim in the *Heptaplus*, his commentary on the week of creation as narrated

in Genesis, was to discover the true philosophy hidden behind the letter of the Bible, which could also be found in Cabalistic and pagan writings: *Heptaplus*, transl. by Douglas Carmichael (Indianapolis, IN: Bobbs; New York: Merrill, 1965), pp. 167–383.
34   See above, Chapter 8, p. 197.
35   On Steuchus see Mariano Crociata, *Umanesimo e teologia in Agostino Steuco* (Rome: Città Nuova, 1987); Theobald Freudenberger, *Augustinus Steuchus aus Gubbio: Augustinerchorherr und päpstlicher Bibliothekar (1497–1548) und sein literarisches Lebenswerk* (Münster in Westphalia: Aschendorff, 1935).
36   Marmaduke Carver, *A Discourse of the Terrestrial Paradise Aiming at a More Probable Discovery of the True Situation of that Happy Place of our First Parents Habitation* (London: James Flesher, 1666), sigs A6r–v.
37   Augustinus Steuchus, *Recognitio Veteris Testamenti ad hebraicam veritatem* (Venice: Aldus, 1529), fol. 22v: 'Quanquam multa ac varia ab Hebraeis, Arabibus, Chaldaeis, Graecis, ac Latinis de paradiso referantur, nihil tamen eorum attingere decrevimus, cum nostro id refragetur instituto, qui circa simplices voces tantum, nos versaturos sumus polliciti. Quod si quis ea de re accuratius cognoscere voluerit, ab his petat necesse est, quorum est munus cupiosius haec explicare. Apud quos quanta est rei ignoratio, atque obscuritas, tanta est varietas sententiarum. Neque tamen haec tam sicco pede praeterire voluimus, quin tam obscurae, et ab humana intelligentia submotae rei aliquid luminis afferremus.' The discussion on the geography of paradise is found on fols 22v–29r.
38   Ibid., fol. 23v: 'Neque adeo inhumanus fuit deus, ut voluerit huius rei ignoratione per omnes aetates homines torqueri, cum neque ullum in sacris scripturis esse passus sit locum, quem si accurate pensitemus, interpretari non possimus.'
39   Ibid., fol. 24r.
40   Ibid., fols 35r–v.
41   Ibid., fol. 23v.
42   Fausto da Longiano, 'Del paradiso terrestre', in Aeneas Sylvius Piccolomini, *Discrittione de l'Asia et Europa di papa Pio II e l'historia de le cose memorabili fatte in quelle, con l'aggiunta de l'Africa, secondo diversi scrittori, con incredibile brevità e diligenza*, transl. by Fausto da Longiano (Venice: Vincenzo Vaugris, 1544), fol. 375v: 'Pareva troppo sconvenevole lasciar da parte il ragionare de'l paradiso terrestre.' On the various editions of the *Cosmographia* and Fausto da Longiano's translation, see Nicola Casella, 'Pio II tra geografia e storia: La "Cosmographia"', *Archivio della Società Romana di Storia Patria*, 3rd series, 26 (1972), pp. 48–9. In the dedicatory letter addressed to Giulia Trivulzio (fols 2r–v), Fausto states that Piccolomini's work is incomplete. The title of the new section (fol. 307r) is: 'Discrittione dell'Africa terza parte del mondo, e d'altre parti dell'Asia tralasciate da Papa Pio, e di tutta Terra Santa, raccolta da diversi scrittori, con la citatione de luochi de la scrittura, e dichiaratione d'essi, e del Paradiso Terrestre'. Fausto da Longiano translated another work by Piccolomini, the *Historia bohemica*, into Italian. On Fausto da Longiano's life see *Dizionario biografico degli Italiani*, XLV (Rome: Istituto dell'Enciclopedia Italiana, 1995), pp. 394–8.
43   Fausto da Longiano, 'Del paradiso terrestre', fol. 379v. See John of Damascus, *De fide orthodoxa: Versions of Burgundio and Cerbanus*, 25.1, ed. by Eligius M. Buytaert (St Bonaventure, NY: Franciscan Institute, 1955), pp. 106–7; *PG* XCIV, col. 915. Dionysius the Carthusian, *Enarratio in Genesim*, II, a.19, and *In librum II Sententiarum*, d.17, q.5, in *Opera omnia*, I (1896), pp. 76–7, and XXII (1903), pp. 152–3, 158, mentioned that John of Damascus's idea that the source of paradise arises from the ocean was confirmed by Thomas Aquinas: cfr. Thomas Aquinas, *Summa theologiae*, 1a, q.102, a.1, Leonine edn, I (1950), pp. 483–4.
44   Augustinus Steuchus, *Recognitio Veteris Testamenti ad hebraicam veritatem* (1529), fol. 29r: 'Haec igitur sunt adeo obscura, adeoque a nostra intelligentia remota, ut non permittant nos quicquam certi asserere. Erit igitur sacrarum literarum studiosissimi cuiusque ea perquirere, quos plurimos iam fore auguramur, nos quod potuimus, in hac re praestitimus.'
45   As early as the twelfth century, Hugh of St Victor observed that 'some affirm that the whole earth was to be paradise, had Adam not sinned but the whole of it became a place of exile for him because of sin. However, we affirm nothing else than what the holy [Fathers] unanimously affirm, namely that paradise is a certain specific place in a part of the earth'; Hugh of St Victor, *Adnotationes elucidatoriae in Pentateuchon*, *PL* CLXXV, col. 39: 'quidam affirmant totam terram futuram paradisum, si homo non peccasset, totam autem factam exsilium per peccatum. Nos vero … non asserimus nisi quod sancti communiter asserunt, scilicet paradisum esse quemdam locum determinatum in parte terrae.' Albert the Great mentioned the theory only to criticize it: see above Chapter 7, p. 179. For the Renaissance debate on this topic see Joseph Duncan, 'Paradise as the Whole Earth', *Journal of the History of Ideas*, 30 (1969), pp. 171–86; Duncan, *Milton's Earthly Paradise* (1972), pp. 199–202.
46   See Duncan, 'Paradise as the Whole Earth' (1969), pp. 173–5. Amongst others who could be cited were Noviomagus (Johann Bronchorst), professor of philosophy at various Protestant universities, and Wolfgang Wissenburg, the Swiss Protestant professor of mathematics and theology who compiled a large regional map of the Holy Land showing the Exodus. The ideas of all these writers were taken up in the late seventeenth century by Thomas Burnet, who, in his *Telluris theoria sacra* (1681), described the antediluvian earth in detail.
47   Jan van Gorp (Goropius Becanus), *Origines antwerpianae, sive Cimmeriorum becceselana novem libros complexa* (Antwerp: Christophorus Plantinus, 1569), pp. 481–2, 495–8.
48   Juan de Pineda, *Los treynta libros de la monarchia ecclesiastica o historia universal del mundo*, I.1, 5 vols (Barcelona: Hieronymo Margarit, 1620), I, pp. 16–17.
49   Ludovicus Fidelis, *De mundi structura* (Paris: [n.p.], 1556), p. 294; Jan van Gorp, *Origines antwerpianae* (1569), p. 482.
50   Vadianus (Joachim Van Watt), *Pomponii Melae de orbis situ libri tres* (Paris: Christianus Wechelus, 1540), sig. U2; Vadianus, *Epitome trium terrae partium, Asiae, Africae et Europae compendiarum locorum descriptionem continens* (Zurich: Christophorus Frosch, 1534), p. 192.
51   Ludovico Nogarola, *Dialogus qui inscribitur Timotheus, sive de Nilo* (Venice: V. Valqrysius 1552), pp. 20–1; Juan de Pineda, *La monarchia ecclesiastica* (1620), pp. 18, 28–9. For different views on the Expulsion, see Duncan, 'Paradise as the Whole Earth' (1969), pp. 179–81.
52   Benedictus Pererius, *Commentariorum et disputationum in Genesim tomi quattuor*, 4 vols (Ingolstadt: David Sartorius, 1590), I, pp. 367–8; Cornelius a Lapide, *Commentaria in Pentateuchum Mosis* (Paris: E. Martin, 1630), p. 56.
53   Walter Raleigh, *The History of the World* (London: Walter Burre, 1614), pp. 29, 36; Samuel Purchas, *Purchas his Pilgrimage; or, Relations of the World and the Religions Observed in all Ages …* (London: W. Stansby, for H. Fetherstone, 1613), p. 13. See Duncan, 'Paradise as the Whole Earth' (1969), pp. 172–3.
54   Philo, *De plantatione*, II.111, [Works], Loeb Classical Library, transl. by Francis H. Colson and George H. Whitaker, 10 vols and 2 supplementary vols (London: Heinemann; New York: Putnam's Sons, 1929–62), III (1930), pp. 235–7.

Also Clement of Alexandria, *Stromata*, V.11.72.2, in *Les Stromates: Stromate V*, ed. by Alain Le Boullec, transl. by Pierre Voulet, 2 vols, SC CCLXXVII, CCLXXIXX (Paris: Cerf, 1981), I, SC CCLXXVIII, p. 145, thought that paradise could have been the whole world.

55 See, for example, Duncan, *Milton's Earthly Paradise* (1972), pp. 200–1 and n. 21.

56 Martin Luther, *Lectures on Genesis Chapters 1–5*, in *Luther's Works*, ed. by Jaroslav Pelikan (St Louis, MO: Concordia, 1958), I, p. 87; *Genesisvorlesung*, ed. by Gustav Koffmane and Otto Reichert, in *D. Martin Luthers Werke*, XLII (Weimar: Hermann Böhlaus Nachfolger, 1911), p. 66: 'Hic mare questionum nascitur de Paradiso.' Luther lectured on the first chapters of Genesis between 1535 and 1536: see Martin Luther, *Lectures on Genesis Chapters 1–5* (1958), pp. IX–X.

57 Martin Luther, *Lectures on Genesis* (1958), p. 88; *Genesisvorlesung* (1911), p. 67: 'Hic disputatur: Ubi sit Paradisus? Et miris modis torquent se interpretes. Quibusdam placet esse intra duos tropicos sub aequinoctiali. Alii temperatiorem aerem requirunt ad tantam loci foecunditatem. Quid multis? Opinionum non est numerus. Ego breviter sic respondeo: Ociosam esse quaestionem de re, quae amplius non est. Nam Moses scribit res gestas ante peccatum et diluvium. Nos autem cogimur de rebus loqui, sicut sunt post peccatum et post diluvium. Credo igitur vel ab Adam, vel tempore Adae hunc locum fuisse appellatum Eden, a foeconditate et ingenti amoenitate, quam Adam in ea vidit. Ac mansit nomen amissae rei apud posteros, sicut hodie Romae, Athenarum, Carthaginis nomina extant, sed tantarum rerumpublicarum vix vestigia quaedam conspiciuntur. Nam tempus et maledictio, quam peccata merentur, consumunt omnia. Cum igitur mundus per diluvium una cum hominibus et pecudibus aboleretur, hic tam nobilis hortus etiam est abolitus et periit.'

58 Duncan, *Milton's Earthly Paradise* (1972), pp. 199–202, and 'Paradise as the Whole Earth' (1969), pp. 171–86, has put the two theories of the destruction of paradise by the Flood and paradise as the whole earth in sharper contrast.

59 Martin Luther, *D. Martin Luthers Werke: Tischreden*, ed. by Ernst Kroker, I (Weimar: Hermann Böhlaus Nachfolger, 1912), I.1093, p. 549: 'Summa summarum, constans mea opinio est paradisum fuisse totum mundum, et movet me illud, quod quatuor ista flumina, de quibus textus dicit, hunc totum orbem circumfluunt. Eici autem ex paradiso est destitui voluptate istorum bonorum propter maledictionem: Terra germinet tibi spinas et tribulos etc.' See also Martin Luther, *Table Talk*, transl. by William Hazlitt (London: Harper Collins, 1995), 122, p. 60. The first German edition of Luther's *Tischreden* (*Table Talk*) was published in Eisleben in 1566 (the editor was Iohannes Aurifaber).

60 Luther, *Lectures on Genesis* (1958), p. 98; *Genesisvorlesung* (1911), p. 74: 'Sed hic cum palam falsum sit.' Interestingly, when discussing the creation of the world, Luther did not distinguish between the earth and paradise: he pointed out that, had man not sinned, God would have transferred him to heaven 'from paradise, or from the earth': *Lectures on Genesis* (1958), p. 78; *Genesisvorlesung* (1911), p. 59: 'donec Deus hominem transtulisset de Paradiso seu terra'.

61 Luther, *Lectures on Genesis* (1958), pp. 87–8; *Genesisvorlesung* (1911), pp. 66–7.

62 Luther, *Lectures on Genesis* (1958), p. 230; *Genesisvorlesung* (1911), pp. 171–2.

63 Luther, *Lectures on Genesis* (1958), p. 231; *Genesisvorlesung* (1911), p. 172: 'Templum totius mundi'.

64 Luther, *Lectures on Genesis* (1958), pp. 89–92, 98; *Genesisvorlesung* (1911), pp. 68–70, 74.

65 Luther, *Lectures on Genesis* (1958), p. 89; *Genesisvorlesung* (1911), p. 68: 'hic locus habuit suum meliorem cultum, ut Eden esset delectus hortus prae cultu totius terrae, quae ipsa quoque, si ad hodiernam miseriam conferas, Paradisus fuit'.

66 Luther, *Lectures on Genesis* (1958), p. 97; *Genesisvorlesung* (1911), p. 74.

67 Luther, *Lectures on Genesis* (1958), pp. 77–8; *Genesisvorlesung* (1911), pp. 58–9.

68 Luther, *Lectures on Genesis* (1958), pp. 89–90, 98; *Genesisvorlesung* (1911), p. 68, 74–5: 'Omnia haec, inquam, sunt historica. Frustra itaque hodie querimus, ubi aut quid fuerit hortus ille? … Mea igitur haec sententia …: Quod primum Paradisus per peccatum homini clausa, Deinde per diluvium tota vastata sit et disiecta, ut iam nullum eius appareat vestigium. … omnino existimo, Paradisum post Adae lapsum extitisse, et notam fuisse posteritati eius, sed tamen inaccessibilem propter custodiam Angeli, qui cum gladio flammeo hortum custodivit, ut textus dicit. Sed diluvium omnia vastavit, sicut scriptum est: ruptos esse omnes fontes et abyssos. Quis igitur dubitet, etiam hos fontes ruptos et confusos esse?'

69 In *Lectures on Genesis* (1958), Luther describes Origen's allegorical speculations as 'senseless discussions' (p. 88), a 'twaddle unworthy of theologians' (p. 90), and 'most silly allegories' (p. 99); while praising Nicholas of Lyra as 'among the best, because throughout he carefully adheres to, and concerns himself with, the historical account' (p. 93); *Genesisvorlesung* (1911), p. 67: 'absurda'; p. 68: 'istae nugae Theologo indignae sunt'; p. 74: 'unde Origeni et aliis praebita occasio est mirabilia fabulandi'; p. 71: 'Ego Lyram ideo amo et inter optimos pono, quod ubique diligenter retinet et peraequitur historiam.'

70 *Biblia, das ist, die gantze Heilige Schrifft Deudsch* (Wittenberg: Hans Lufft, 1534), I, sig. 1v. See Catherine Delano-Smith and Elizabeth M. Ingram, *Maps in Bibles, 1500–1600: An Illustrated Catalogue* (Geneva: Droz, 1991), p. 2, and Albert Schramm, '*Luther und die Bibel*', I, 2 vols (Leipzig: Hiersemann, 1923), I, *Die Illustration der Lutherbibel*, p. 23.

71 Delano-Smith and Ingram, *Maps in Bibles* (1991), p. 2, give the example of the Heinrich Quentell's Cologne Bible of 1478–90. See also Walter Eichenberger and Henning Wendland, *Deutsche Bibeln vor Luther* (Hamburg: Wittig, 1977), Figure 107, p. 73. For Holbein's woodcuts see *Hans Holbein d.J.: Die Druckgraphik im Kupferstichkabinett Basel* (Basle: Schwabe, 1997), Figures 95–6, pp. 131–2.

72 Luther, *Lectures on Genesis* (1958), p. 89; *Genesisvorlesung* (1911), p. 68: 'frustra est, ut imaginemur angustum hortum, aliquot milliarium spatio'.

73 Luther, *Lectures on Genesis* (1958), pp. 99–100; *Genesisvorlesung* (1911), pp. 74–5. Luther did not identify the land of Havilah with a place in India, but with the region known by the geographers of his day as 'Arabia Felix'.

74 Luther, *Lectures on Genesis* (1958), pp. 81–2; *Genesisvorlesung* (1911), pp. 61–2.

75 *Biblia, das ist, die gantze Heilige Schrifft Deudsch* (1534), II, sig. C.2r, Daniel, fol. 14r. See Shirley, *Mapping of the World* (2001), pp. 71–2; Uta Lindgren, 'Die Bedeutung Philipp Melanchthons (1497–1560) für die Entwicklung einer naturwissenschaftlichen Geographie', in *Gerard Mercator und seine Zeit: 7. Kartographisches Colloquium Duisburg 1994: Vorträge und Berichte*, ed. by Wolfgang Scharfe (Duisburg: Walter Braun, 1996), pp. 1–2; Schramm, *Luther und die Bibel* (1923), p. 26.

76 Bartholomaeus Anglicus, *Livre des propriétés des choses* (French translation by Jean Corbechon), Paris, Bibliothèque Nationale de France, MS Franç. 9140, fol. 226v.

77 See, e.g., Elizabeth M. Ingram, 'Maps as Readers' Aids: Maps and Plans in Geneva Bibles', *Imago Mundi*, 45 (1993), pp. 34–5.

78 Luther, *Lectures on Genesis* (1958), p. 90; *Genesisvorlesung* (1911), p. 68: 'Imaginem Dei in homine ita post peccatum periisse, sicut originalis mundus et Paradisus perierunt.'
79 Luther, *Lectures on Genesis* (1958), pp. 100–1; *Genesisvorlesung* (1911), pp. 75–6.
80 Luther, *Lectures on Genesis* (1958), pp. 77–8, 204; *Genesisvorlesung* (1911), pp. 58–9, 152–6.
81 Luther, *Lectures on Genesis* (1958), pp. 355, 358–9; *Genesisvorlesung* (1911), pp. 260–1, 263.
82 Luther, *Lectures on Genesis* (1958), p. 225; *Genesisvorlesung* (1911), p. 168: 'Magnum enim periculum in eo est, si obliti priorum, iterum in ea demergamur.' For the restoration of the divine image, see *Lectures on Genesis* (1958), p. 64; *Genesisvorlesung* (1911), p. 168: 'Magnum enim periculum in eo est, si obliti priorum, iterum in ea demergamur.'
83 Note that Melanchthon, who played down Luther's dark vision of the corruption of human nature, was interested in geography and cartography and favoured the inclusion of maps in Lutheran Bibles: Lindgren, 'Die Bedeutung Philipp Melanchthons (1497–1560) für die Entwicklung einer naturwissenschaftlichen Geographie' (1996), pp. 1–12.
84 John Calvin, *Commentaire ... sur le premier livre de Moyse, dit Genese* (Geneva: Jean Gerard, 1554), pp. 64–5. The first English translation, *A Commentarie ... upon the first booke of Moses called Genesis*, was produced by Thomas Tymme (London: John Harison and George Bishop, 1578) The originals have been abridged in the recent ed. by Alister McGrath and J. I. Packer (Wheaton, IL–Nottingham: Crossway Books, 2001). Calvin also expressed his views on the location of the Garden of Eden in his sermons on Genesis: *Sermons sur la Genèse: Chapitres 1,1–11,4*, ed. by Max Engammare, *Supplementa Calviniana* XI/1 (Neukirchen-Vluyn: Neukirchener Verlag, 2000), Sermon 9, Genesis 2.7–15, pp. 100–5.
85 Quoted in William J. Bouwsma, *John Calvin: A Sixteenth-Century Portrait* (New York–Oxford: Oxford University Press, 1988), pp. 139–40; John Calvin, *Institution de la religion chrestienne*, II.6.1, (Geneva: Jean Martin, 1565), p. 257: 'Mais depuis la cheute et revolte d'Adam, quelque part que nous tournions les yeux, il ne nous apparoist haut ne bas que malediction: laquelle estant espandue sur toutes creatures, et tenant le ciel et la terre comme enveloupez, doit bien accabler nos ames d'horrible desespoir. Car combien que Dieu desploye encores en plusieurs sortes sa faveur paternelle, toutesfois par le regard du monde nous ne pouvons pas nous asseurer qu'il nous soit Pere: pource que la conscience nous tient convaincus au dedans, et nous fait sentir qu'à cause du peché nous meritons d'estre reiettez de luy, et n'estre point tenus pour ses enfans. Il y a aussi la brutalité et ingratitude: pource que nos esprits, selon qu'ils sont aveuglez, ne regardent point à ce qui est vray: et selon que nous avons tous les sens pervertis, nous fraudons iniustement Dieu de sa gloire.'
86 My understanding of Calvin here largely follows the interpretation of William J. Bouwsma in his *John Calvin: A Sixteenth-Century Portrait* (1988). See also Max Engammare, 'D'une Forme l'autre: Commentaires et sermons de Calvin sur la Genèse', in *Calvinus Praeceptor Ecclesiae* (Geneva: Droz, 2004), pp. 107–37, and Donald K. McKim, ed., *The Cambridge Companion to John Calvin* (Cambridge: Cambridge University Press, 2004).
87 John Calvin, *A Commentary on Daniel*, 4.34 (Edinburgh: Calvin Translation Society, 1852–3, 2 vols; repr. Edinburgh, Banner of Truth Trust, 1966, 1986), I, p. 296; quoted in Bouwsma, *John Calvin* (1988), p. 82.
88 Ephrem the Syrian, *Hymns on Paradise*, X.4, transl. by Sebastian Brock (St Vladimir's Seminary Press: Crestwood, NY, 1990), p. 149.

On Alexander of Hales's theory of the location of paradise in the middle region of the air see above, Chapter 7, pp. 174–5.
89 John Calvin, *Isaiah*, 57.1, ed. by Alister McGrath and J. I. Packer (Wheaton, IL–Nottingham: Crossway Books, 2000), p. 341; John Calvin, *Commentarii in Isaiam prophetam* (Geneva: I. Crispinus, 1559), 57.1, p. 485: 'Hic enim, velut amplo in theatro, ad contemplanda Dei opera constituti sumus: nec ullum iam minutum est opus Dei, quod leviter a nobis praetereundum sit: sed omnia accurate et diligenter observari debent.' The passage is referred to in Bouwsma, *John Calvin* (1988), p. 103. On the notion of nature as a theatre for contemplating the works of God see Susan E. Schreiner, *The Theater of His Glory: Nature and the Natural Order in the Thought of John Calvin* (Durham, NC: Labyrinth Press., 1991).
90 See Bouwsma, *John Calvin* (1988), pp. 103, 134–5.
91 Luther, *Lectures on Genesis* (1958), p. 98: 'And so, just as there are mountains after the Flood where previously there were fields in a lovely plain, so undoubtedly there are now springs where there were none before, and vice versa. For the entire surface of the earth was changed. I have no doubt that there are remains of the Flood, because where there are now mines, there are commonly found pieces of petrified wood. In the stones themselves there appear various forms of fish and other animals'; *Genesisvorlesung* (1911), p. 75: 'Sicut itaque post diluvium montes sunt, ubi antea agri in amoena planicie fuerunt, ita quoque non est dubium nunc esse fontes, ubi antea nulli fuerunt, et e contra. Tota enim terrae facies mutata est. Nec dubito ego reliquias diluvii esse, quod, ubi fodinae metallicae nunc sunt, ibi non raro reperiuntur ligna durata in saxum. In ipsis saxis conspiciuntur formae variae piscium et aliarum bestiarum.'
92 See Bouwsma, *John Calvin* (1988), pp. 142–3.
93 John Calvin, *Commentaire* (1554), p. 32: 'Car combien que ie confesse que la terre, depuis qu'elle a esté maudite, a esté reduite en un miserable et hideux estat, et a vestu comme un habit de deuil, et que depuis elle a esté gastée en plusieurs lieux: ie dy toutesfois que c'est la mesme terre qui avoit esté creée au commencement.'
94 John Calvin, *Commentaire* (1554), p. 61.
95 For sixteenth-century examples see Delano-Smith and Ingram, *Maps in Bibles* (1991), pp. 2–24. For later examples, see below, Chapter 10.
96 Anthony N. S. Lane, *John Calvin: Student of the Church Fathers* (Edinburgh: Clark, 1999), p. 233.
97 In discussing the issue of the eternity of the world, Calvin referred to Steuchus's commentary on Genesis 1.1. Calvin attacked Steuchus's defence of the Donation of Constantine in *Institution de la religion chrestienne*, IV.11.12 (1565), p. 1012: 'Et d'autant plus grande a esté la vilainie du bibliothecaire du Pape, Augustin Steuche, lequel a esté si effronté de se faire advocat d'une cause si desesperee, pour gratifier à son maistre.'
98 John Calvin, *Commentaire* (1554), p. 34: 'S'il y a region sous le ciel qui soit excellente en beauté, en abondance de fruicts, en fecondité, en delices, et autres dons, ceux qui ont escrit des pays celebrent sur tous cestuy-cy.'
99 John Calvin, *Commentaire* (1554), p. 33: 'Ie mettray icy une figure devant les yeux par laquelle on pourra entendre où i'estime que Moyse met Paradis.'
100 See Max Engammare, 'Portrait de l'exégète en géographe: La Carte du paradis comme instrument herméneutique chez Calvin et ses contemporains', *Annali di Storia dell'Esegesi*, 13/2 (1996), pp. 565–81.
101 John Calvin, *Commentaire* (1554), p. 30.
102 John Calvin, *Commentaire* (1554), p. 30: 'Que

profite-t-il de voler en l'air, et laisser la terre, où Dieu a donné tesmoignage de sa bienvueillance au genre humain? Mais quelcun dira, que l'exposition est bien plus subtile, d'entendre cecy de la beatitude celeste. Ie respon, Puis que l'heritage eternel de l'homme est au ciel, que c'est chose convenable que nous tendions là. Toutesfois il nous faut arrester le pied en terre, tant que nous considerions le logis, duquel Dieu a voulu que l'homme usast pour un temps. Car nous sommes maintenant en ceste histoire, laquelle enseigne qu'Adam a esté ordonné de Dieu pour estre habitant de la terre: afin que passant icy la vie temporelle, il meditast cependant la gloire du ciel. Que le Seigneur l'a enrichy largement de biens sans nombre, afin que par le goust d'iceux il recueillist sa bienvueillance paternelle.'

103   John Calvin, *Commentaire* (1554), pp. 64–5.
104   John Calvin, *Commentaire* (1554), pp. 29–31.
105   John Calvin, *Commentaire* (1554), pp. 32–3: 'Premierement il faut noter, qu'il n'est icy fait nulle mention de source ne de fontaine: mais qu'il est seulement dit Un fleuve. Et enten par les quatre chefs, tant les entrées, ou commencemens où lesdits fleuves naissent, que les issues, par lesquelles ils se deschargent en la mer. Ia autresfois a esté Euphrates conioint avec le Tygre qui tomboit dedans luy: tellement qu'on pouvoit dir à bon droit que c'estoit un fleuve separé en quatre chefs.'
106   John Calvin, *Commentaire* (1554), pp. 34–5. See Strabo, *Geography*, XV.3.4; Pliny, *Naturalis historia*, VI.9; Arrian, *Historia Indica*, XL.8; Pomponius Mela, *De Chorographia*, III.8.77. On Calvin's use of ancient sources see Engammare, 'Portrait de l'exégète en géographe' (1996), pp. 570–4.
107   John Calvin, *Commentaire* (1554), p. 34: 'il n'y ait pas grand interest en cela'.
108   Jean-François Gilmont, *Jean Calvin et le livre imprimé* (Geneva: Droz, 1997), p. 195. Alexandre Ganoczy, *La Bibliothèque de l'Académie de Calvin* (Geneva: Droz, 1969), p. 164. Admittedly, it is unclear which of Calvin's books now in the library of Calvin's Academy in Geneva came to the library on Calvin's death in 1564.
109   John Calvin, *Commentaire* (1554), p. 30: 'le iardin a esté situé en la terre, et non pas en l'air, comme aucuns songent'; *Sermons sur la Genèse: Chapitres 1,1–11,4* (2000), Sermon 9, Genesis 2.7–15, p. 101. Sebastian Münster, *Cosmographia* (Basle: Henricus Petri, 1545), p. 28. See also below, Epilogue, p. 365. On Münster's *Cosmographia* see Besse, *Les Grandeurs de la Terre* (2003), pp. 151–259; on Münster and paradise see Arianne Faber Kolb, *Jan Brueghel the Elder: The Entry of the Animals into Noah's Ark* (Los Angeles: J. Paul Getty Museum, 2005), pp. 47–59.
110   Ganoczy, *La Bibliothèque de l'Académie de Calvin* (1969), p. 255. On Servetus see below, Chapter 10, p. 323.

10

The Afterlife of Paradise on Maps

Atheists and Scoffers, whom the Psalmist calls Pests, usually demand, What's become of paradise? Shew us the place in the Maps? *And if this be not done for them (they are generally lazy) with all exactness, butted and bounded by Longitude and Latitude; hedged in with Degrees, and Minute Measures; attested also by Strabo and Ptolemy; they will slide into a disbelief first of Genesis, then of the whole Bible, and lastly of all revealed Religion.*

Thomas Gale, 'To the Reader', in Pierre-Daniel Huet, *A Treatise of the Situation of the Terrestrial Paradise* (1694)

The map that Calvin produced for his *Commentary on Genesis* in 1553–4 provided the basis for one of the commonest map subjects in the second half of the sixteenth century and throughout the seventeenth century. Indeed, one of the five maps regularly included in various editions of the Geneva Bible from 1560 on was modelled on his.[1] Whereas Calvin had managed to combine advanced cartography with advanced exegesis, however, the challenge of placing the Garden of Eden on a map that also satisfied the increasingly exacting requirements of contemporary map making was to vex a long succession of commentators. In a way, Calvin's role in the early modern mapping of the Garden of Eden can be likened to Augustine's in the Middle Ages. For, just as medieval exegetes had failed to embrace fully the implications of Augustine's vision of history in their attempts to locate the earthly paradise, so, from the Reformation onwards, those who created maps showing the location of the Garden of Eden in the wake of Calvin's exegesis did so with little of Calvin's theological sophistication.

Calvin had avoided specifying the exact site of paradise on his map. His primary aim was to persuade humans on their earthly pilgrimage of God's benevolence and concern for them. For this, it was sufficient to make the point that traces of the Garden were recognizable in the appropriate region and that this region was within reach, well to the west of its far-eastern location in many medieval texts and maps. Unmindful of the full thrust of Calvin's argument, later scholars and exegetes did not resist the temptation of indicating the site of paradise within the general region of Eden. In this process the wider region of Eden became separated from paradise itself and the inclusion of a vignette portraying Adam and Eve transformed the map into a more explicit record of an historical event. Attempting to identify both Eden and the Garden of Eden within it developed into an exacting preoccupation, especially as the ability to point to the historical site of the biblical narrative came to be seen as a crucial defence against either allegorical interpretation or sceptical disbelief in the later sixteenth and the seventeenth century. Long-'lost' sites of biblical and classical history were being discovered or more accurately interpreted. If it were possible to locate Jericho's walls or to identify the Egyptian pyramids as tombs (rather than Joseph's granaries), why then should one not attempt to establish the location of a place as fundamental to the faith as the past and lost Garden of Eden? Mapping in support of the literal meaning of the Bible blossomed. New cartographical techniques were introduced, but there was no slackening in the zeal with which the authority of Scripture was defended with maps.[2]

Everything Changes, Nothing Changes

There is a quality to the changes in European cartography in the sixteenth century that reminds one of a sentence in Giuseppe Tomasi di Lampedusa's *The Leopard*: 'Everything has to change so that nothing changes.' The adoption of Ptolemy's cartographical techniques in the Renaissance allowed map makers to represent the temporal dimension on world maps in a completely different way from that deployed on the old *mappae mundi*. Astronomical measurement of time and not the narrative of historical progression now governed the representation of space, and the mapping of biblical history appeared to be threatened by the mathematical structure of the Ptolemaic map. It was not long, however, before Ptolemaic geography was Christianized.

The commonest way in which Ptolemaic geography was made to accommodate Christian belief was by the inclusion of biblical subjects in an otherwise wholly secular map, even those of Ptolemy's *Geography* itself. Among the 'modern' maps (*tabulae novae*) that started to be added to the older ones in manuscript copies of the *Geography* in the second half of the fifteenth century, was a map closely related to the Sanudo-Vesconte map of *c*.1320 (see Figure 8.1a). This showed an explicitly historical *Terra Sancta* packed with places and features from both Old and New Testaments.[3] Even the 'old' maps, though, could be made to relate to Christian belief by including details of biblical history. The Third Map of Asia in Sebastian Münster's 1540 edition of the *Geography* shows Noah's Ark floating on the Caspian Sea, and the legend describes its resting place on the mountains of Armenia (*Arca Noe quae quievit in montibus Armeniae*); the Second Map of Africa portrays the drama of Saint Paul's shipwreck off Malta, and the Third Map of Africa shows a whale swallowing the prophet Jonah. Münster was a noted scholar and Hebraicist. Giacomo Gastaldi was primarily a map maker and engraver, but he too added the Ark and the shipwreck on the appropriate 'old' maps in his elegant pocket edition of the *Geography* (1548).[4] A second way of reconciling Christian belief with Ptolemaic mathematics was by the deployment of the principles of astronomically defined location – measured degrees and lines of latitude and longitude – in order to make a theological point. Of all the early modern authors who strove to re-arrange Christian beliefs in the context of newly opened intellectual and geographical horizons, the sixteenth-century Jesuit philosopher and historian Guillaume Postel remains outstanding for the originality of his thought. Postel's vision of history was based largely on his reading of ancient cosmography and Christian Cabalism. Recognizing that the Ptolemaic grid made all points on earth geometrically equivalent and historically neutral, and that the lines of latitude and longitude created circles around the earth that were measured by the daily revolution of the sun, Postel ingeniously manipulated the Ptolemaic notion of the longitudinal plane to explain his idea of the imminent return of the universe to God.[5]

Postel is usually remembered for the theory of a North Pole paradise he expounded in his unpublished *De paradisi terrestri loco* (1561).[6] Among his varied intellectual activities, Postel was a compiler of maps, and a printed world map on a polar projection produced in 1581 may indeed have been connected with his argument that paradise was at the Arctic Pole.[7] Paradise is not explicitly marked on the printed map, but in the manuscript Postel described how it had once been at the very point in the centre of the northern hemisphere through which, at the time of the world's creation and at the moment of separation of the earth from the waters, God had drawn out the earthly element from the waters below.[8] Postel saw paradise as situated on the top of an extremely high mountain at the northern Pole that reached up to the middle region of the air, and as a place entirely different from any other place on earth. Its favourable climatic conditions assured eternal life to both humans and plants. Postel claimed that this Arctic paradise was in an 'absolute and eternal Orient', where the sun would never rise more than 23 degrees above the horizon. Postel adopted his theory of an Arctic paradise to promote a vision of the history and politics of his own day, pointing to the importance and responsibility of north European peoples such as the Germans. A corollary of his view of a polar Eden was that the language of the Goths had been the first language of mankind and that the Scythians, who were able to endure cold, labour and privations, had preserved more than other races the physical features of the inhabitants of the Arctic paradise.[9]

Postel's Arctic paradise was an afterthought, however, as he himself admitted in his manuscript treatise of 1561.[10] A decade earlier, in 1553, he had situated paradise on the meridian that passes through the Moluccan Islands. This he baptized 'the meridian of paradise' in the belief that physical evidence of paradise was found along it. After all, for centuries travellers' reports identified east Asia as a region of marvels. One of these marvels, the bird of paradise (the *manucodiata*), corroborated for Postel the veracity of Scripture claiming an eastern location of paradise, for the bird was said to be able to fly only in the pure air of paradise, which was preserved in the highest part of the sky along the Moluccan meridian. Furthermore, it was said, the bird lacked claws, which forced it to remain perpetually suspended in the air since there was no place on earth where it could perch. It had never been seen alive because if it left the radiant and limpid air of paradise, it would fall to the ground and die. The existence of a bird unable to come to terms with the earth was for Postel simultaneously a proof of the lost perfection of the Garden of Eden and a prophecy of the return of mankind to God.[11]

It is not for his belief in the bird of paradise, however, that Postel deserves to be remembered, but rather for the remarkable way he manipulated the quantitative Ptolemaic description of the earth into a qualitative differentiation that suggested a location for the earthly paradise. Instead of highlighting a single historical site or geographical region, as everybody else did, Postel started by dividing the earth into quarters by means of four meridians (Figure 10.1). One meridian, in the extreme east, passed through the Moluccas (and paradise). Another, in the east, passed through the Holy Land. The third, in the west, passed through France. And the fourth, in the extreme west, passed through Mexico. Allowing for a margin of error of about 20 degrees along each meridian, Postel then likened the difference in time produced by the movement of the sun from east to west to the process of mankind's spiritual enlightenment throughout history. When it was noon in the Moluccas, it was dawn in the Holy Land; when it was noon in the Holy Land, it was dawn in Europe and noon in Europe meant sunrise in the New World. These four meridian lines, which marked the sun's daily passage, Postel associated with the four ages of human life and the world: childhood, youth, adulthood and old age. By ascribing a historical and symbolic value to the dimension of longitude, Postel was able to demonstrate the progress of the divine work of restoration in geographical terms, following Orosius's old east–west trajectory.[12]

Postel's prime meridian – the one that indicated the world's midday – was the line that passed through the Holy Land and other neighbouring regions. It was near this line that Abraham had left Ur in Chaldaea, with his father Terah, to follow the course of the sun along a journey that, through Harran, Bethel, Egypt and the Negeb, eventually led him to Hebron in Judaea and to seal his covenant with God (Genesis 11.31–13.18). It was along the same line that Noah's Ark had grounded in Armenia after the Flood and the subsequent renewal of the earth. And it was along this line that the Incarnation of Christ had taken place. Thus linked in a dynamic east–west relationship, the four meridians not only marked the daily passage of the sun, but also the main stages of human history and destiny. Sunrise over the meridian of paradise in the east heralded the shining of the full light of noon over the Holy Land, and midday in the Holy Land coincided with sunrise over Europe. As the light of the day was extending slowly westwards, so the light of divine revelation was gradually dawning over the threshold of the New World, carried there by missionaries preaching the gospel.

Fig. 10.1 (opposite). Diagrams of the meridian of paradise and the terrestrial globe, and the movement of the sun from east to west, according to Guillaume Postel.

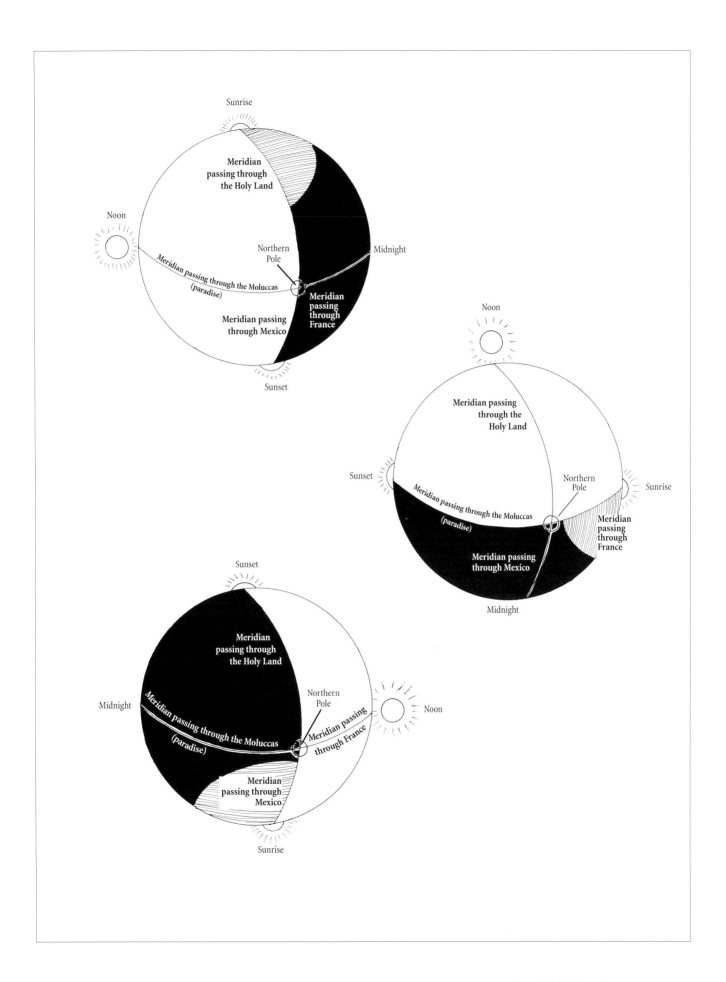

According to Postel's proclamation, the world was in its final age, and a great new age was at hand.

The process of the return of the world to God, however, was still incomplete. Postel's prime meridian in the Holy Land divided the world into two hemispheres. The eastern hemisphere stretched from the Pacific Ocean to the Holy Land. This was the solar half of the world, the most luminous and the most spiritual. Down its centre ran the Moluccan meridian of paradise. All the world's religions had originated in this geographically and morally superior half. The western hemisphere, in contrast, was the lunar, nocturnal, and material side of the globe. It extended from the eastern Mediterranean to the New World. The age that was yet to come would bring spiritual light to this benighted secular hemisphere until the two hemispheres – the luminous eastern and the nocturnal western – were fused into a single brilliantly illuminated globe.

Down the centre of the western hemisphere ran, according to Postel, the meridian of France. The positioning of the line was significant. Postel believed it was France, situated in the west of the ancient world, that had become the repository of the light of salvation and the place where God's people would gather to achieve universal harmony. The French held the rights and authority of primogeniture as they descended from Noah through Japhet and Japhet's eldest son, Gomer. It was up to them to reconquer Jerusalem, establish a universal empire and to promote the teachings of Christianity all over the world. Since France had been divinely chosen to rule the world for Christ, Postel repeatedly appealed to its rulers to carry out their mission and exercise their right to universal rule for the sake of the whole of Christendom. If French rulers were to neglect their duties, warned Postel, it was up to other peoples to realize the divine scheme on earth.[13]

The final triumph of Christendom, however, was to be achieved through both military action and a peaceful spiritual campaign.[14] Christian missionaries, such as the members of the Company of Jesus, which had been recently founded in the West by Ignatius of Loyola and of which Postel was a member, were spreading the Word both westwards and eastwards to convert the peoples of the New World and those of far eastern Asia. For Postel, the eastward expansion of Christianity was highly significant, for it signified the recovery of primordial bliss. The duality of the conversion process, with Christian missionaries going both west with the sun and east against the sun, was ensuring that the entire inhabited sphere would in due course be embraced by the light of salvation. That moment of divine restitution, when the West became one with the East, would mark the return of the world to its primordial state of intimacy with God.[15] Postel's argument that mankind's access to divine revelation and universal concord was ordered by the sun's revolution was a way of turning Ptolemy's mathematical description of the inhabited earth to Christian eschatological use. In Postel's geography, longitude ruled both space and history.

A New Genre: Sacred Geography

Postel's vision of history reveals a good deal about sixteenth-century attitudes and about the way Ptolemaic geography was being used to accommodate religious belief. After the decline of the *mappa mundi*, new ways of understanding the past in cartographical terms had to be found to express a new historical and sacred geography. The practice of accommodating isolated biblical features within the mathematical framework of measurable coordinates continued after Waldseemüller, Münster and Gastaldi. It was not pure chance, however, that led to episodes of salvation history appearing, as we have seen, on sixteenth-century regional Ptolemaic maps, where the larger scale of the map loosened the constraints of the grid. At the core of early modern encyclopedic culture lay the task of reconciling biblical, classical, and modern geography. The Reformation and the spread of printing gave a new impulse to the creation of maps of biblical history. Amongst these, maps showing the location of the earthly paradise were compiled in the context of both sacred geography (to illustrate the Bible and biblical commentary) and historical geography (as maps in historical atlases).

From the sixteenth century onwards, interest in sacred geography – the use of contemporary geographical knowledge to identify and describe lands and countries mentioned in the Bible – was reinforced by a renewed concern to explain the literal sense of the Bible, especially as a counter to the over-use of allegory and philosophical manipulation. Christian typology had encouraged the inclusion of the temporal dimension (and the earthly paradise) on medieval *mappae mundi*, and it did not lose ground in the Renaissance or as a result of the Reformation. A Christocentric reading of Scripture was still advocated, but now biblical scholars, especially those in Protestant circles, were insisting even more than had their predecessors on the primacy of the literal sense, and were laying yet more emphasis on the need to study the original text and its human authors. The biblical text was scrutinized, above all by Hebraicists, for clues to the geographical mystery of Eden that would allow it to be mapped. Traces of Eden were thought to be discernible in place names inherited from ancient civilizations, albeit in distorted form as a result of the confusion of languages at Babel. The Tower of Babel, alluding to the linguistic phenomenon of a single human language before the confusion of tongues, was often shown close to the earthly paradise on sixteenth- and seventeenth-century maps of the biblical Near East, possibly as a way of drawing attention to the need for rigorous philological study if the exact site of Eden were ever to be identified. Eden was no longer isolated by physical obstacles, as in medieval lore, but its discovery in early modern times was obstructed by changes in biblical nomenclature.

Biblical history had always been an important facet of Christian exegesis. The medieval emphasis on the wider tapestry of universal history, however, narrowed and sharpened in the sixteenth century to focus on the placing of each event described in the Bible into its geographical setting with as much detail as possible. Biblical historical geography now constituted a specialized branch of historical knowledge. Moreover, the renewed emphasis on the literal and historical interpretation of the Bible matched the widespread humanist concern for textual clarity, as new tools were adopted to meet the demands of a wider literate readership. Explanatory historical maps were drawn and printed to accompany editions of the Bible or biblical commentaries, and their commercialization was a significant aspect of the Renaissance culture of print, and of the Protestant effort to spread the knowledge of Scripture. Exegetical charts and plans, for example, were included among the reader's aids that characterized Geneva editions of the Bible to illustrate biblical features such as the Exodus from Egypt, the division of Canaan among the Twelve Tribes of Israel, the journeys of Saint Paul, as well as the location and events of the Garden of Eden.[16] Wall maps (and, later in the sixteenth century, maps in atlases) detailing major biblical episodes also contributed to the understanding of the biblical truth as history.[17]

The new cartographical genre of the historical atlas emerged in the final quarter of the sixteenth century as a platform on which the past, classical as well as biblical, could be displayed. The historical atlas was a collection of maps that focused on particular themes or eras in history.[18] The genre was rooted in the maps of Ptolemy's *Geography* and in an early modern perception that Ptolemy's maps had to be regarded as artefacts belonging to the ancient world. By the late fifteenth century, beginning with the edition printed by Holle in Ulm in 1482, then increasing steadily in the sixteenth century, 'modern' maps (*tabulae modernae*) were already supplementing the corpus of 'old' maps (*tabulae veteres*) in the *Geography*.[19] After 1513 and the publication of Martin

Waldseemüller's edition of Ptolemy's *Geography*, with its full complement of twenty *tabulae modernae*, almost no illustrated edition of the *Geography* was without both old and new maps.[20] The traditional corpus of Ptolemaic maps might stand as an atlas of the classical world, but it was not an atlas of the modern world. In the historical atlas, different ages or different historical themes ('epochal zones') were represented independently, map by map, quite unlike the all-embracing medieval *mappae mundi,* on which past and present were layered onto a single geographical landscape. In the historical atlas, topics such as the peregrinations of the Patriarchs, the voyages of Saint Paul, and the martial exploits of Aeneas and of Alexander the Great were presented as a series of selected stills instead of being interwoven and intermixed with scenes and places important for other reasons, for example of contemporary importance. The maps showed the whole area of the chosen travel and only places and events relevant to the specific mapped journey were included. By 1579, when Abraham Ortelius started work on historical maps for the *Parergon*, in which he described geography as 'the eye of history', the transition from the separate historical map to the historical atlas was complete.[21]

Ortelius had taken advantage of improvements in the calculation of latitude and (to lesser extent) longitude to plot on an increasingly accurate cartographical background information coming from historical and biblical scholarship.[22] The relegation of paradise to the historical past was the result in part of the exegetical problem of identifying its contemporary location and in part of the new critical approach to history. The new genre of sacred geography was also structured by combining biblical and secular learning, as shown by the case of various sixteenth-century maps that placed *Paradisus* in northern Syria. The ancient locality of *Paradisus*, described by classical authors such as Strabo as not far from the source of the Orontes, was thought to preserve in its name a memory of the nearby biblical *paradisus*.[23] The place name appeared on sixteenth-century maps, such as the map of the Holy Land engraved by Paolo Forlani in 1566, and the historical map of the journeys of Saint Paul in the first issue of the *Additamentum,* the set of historical maps accompanying Ortelius's *Theatrum orbis terrarum* (1579)

Fig. 10.2 Abraham Ortelius, *Peregrinatio divi Pauli. Typus corographicus* (1584). Ortelius's map of the travels of Saint Paul was one of the historical maps he added to his *Theatrum orbis terrarum* (Antwerp: Christophorus Plantinus, 1584), after fol. 114. 35 × 50 cm. London, British Library, Maps C2.d.10(2). Detail showing the ancient city of *Paradisus* in Syria. *Paradisus*, marked on the map south-west of Palmyra and not far from the source of the river Orontes, had been mentioned by classical authorities and its name was later taken as evidence that the biblical earthly paradise had once been in the vicinity and as confirmation in general of a middle eastern location for paradise. Sixteenth- and seventeenth-century European scholars thought that the memory of the Garden of Eden was preserved and handed down in the linguistic heritage of ancient civilizations.

290   Mapping Paradise: A History of Heaven on Earth

(Figure 10.2).²⁴ In cases such as this, historical geography appeared to corroborate the historical reality of the biblical account.

Biblical scholars took the paradise question seriously and invested much philological and exegetical learning in their attempt to make biblical text and geographical and cartographical science agree. To find the right location for paradise was essential if the historical truth of the biblical account was to be upheld. In the early seventeenth century, Walter Raleigh was one of those who realized that reliance on fanciful medieval theories endangered the credibility of Scripture:

> *... if the truth of the Storie be necessarie, then by the place proved, the same is also made more apparent. For if wee should conceive that Paradise were not on the Earth, but lifted up as high as the Moone; or that it were beyond all the Ocean, and in no part of the knowne World, from whence Adam was said to wade through the Sea, and thence to have come to Iudaea, (out of doubt) there would be few men in the World, that would give any credit unto it. For what could seeme more ridiculuous then the report of such a place?*²⁵

As Thomas Gale noted in the introduction to his translation of Pierre-Daniel Huet's *Traité de la situation du paradis terrestre* (1691) – cited in the epigraph to this chapter – faith itself was at stake. The discovery of the site of paradise depended on the disclosure of the real meaning of a Genesis text that had been corrupted over time by changes in the human and natural world since the Fall. After Calvin, a phalanx of scholars rose up to defend the authority of Scripture in the face of scientific progress as well as religious controversy.

Confluent Streams: Paradise in Mesopotamia

As discussed in Chapter 9, Calvin relaunched the cartography of paradise, locating it in Mesopotamia, the region first mentioned by Steuchus. Calvin's idea was that, although the pristine beauty of Eden was lost after the Fall, something had remained to suggest the region's former paradisiacal condition, and that that region could be located and represented cartographically. The historical character of Calvin's map of the location of Eden, which featured some of the biblical place names associated with paradise (Havilah, Cush, Assyria, Tigris and Euphrates) as well as postdiluvian cities (Babylon and Seleucia), was only implicit. The absence of a map sign for paradise itself, and the fact that the lowest reaches of the Tigris and the Euphrates were not called by their biblical names of Gihon and Pishon, implied that Calvin's map reflected only the traces that betrayed the former existence of the lost Garden and not its exact site. The Garden had vanished. In a subsequent process of cartographical polarization, maps began to carry a vignette or a place name specifying the precise site of the former Garden, and thus took on a more explicitly historical character. The paradise that was now made clearly visible on maps copied or derived from Calvin's was obviously a paradise past. This historical character was further emphasized in the case of those maps on which paradise was marked together with other places mentioned in the Old Testament. Now that space was precisely measured by astronomical time, and not any longer structured by the progress of history, the historical dimension was adopted to define the scope of the regional and historical maps that situated biblical events. The character of the Garden of Eden as being simultaneously separate from and connected to post-lapsarian mankind, continued to be expressed cartographically, but in a completely different way from its former representation on *mappae mundi*. On the medieval maps, the four rivers connected a remote and separate paradise to the inhabited earth. On the Renaissance historical map, the inaccessible paradise of the past was depicted on the same geographical stage as contemporary regional geography. The

Fig. 10.3. Variant of Calvin's map of the location of Eden, from *La Saincte Bible* (Lyons: Sebastien Honoré, 1566), p. 2r. 10 × 9 cm. London, British Library, C23.e.10. Whereas Calvin had indicated the general vicinity of paradise within Mesopotamia, later authors narrowed the location of Eden still further. For Honorati's edition of the Geneva Bible, a vignette of the Fall, showing Adam and Eve with the serpent entwining the Tree of Knowledge, has been added to the map, below the name *Terre de Hevilah* and to the east of the single river.

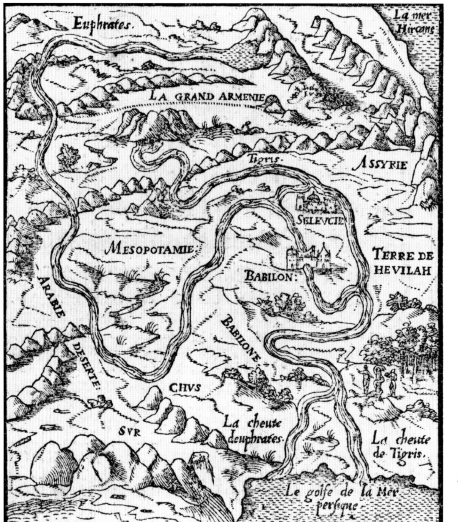

common geographical space made the link between two, extremely different, conditions that were separated by time but bound by sameness of site. The combination of biblical and secular place names on the Renaissance map reinforced the idea of a geographical continuum between past and present.[26]

Reflecting the new concern for historical precision, later authors amended Calvin's map in their attempts to narrow the search for the site of paradise in Mesopotamia. Crucial to all their arguments was the question of the single river (which Calvin had solved by creating a link between Tigris and Euphrates on his map) and of the courses of the Gihon and the Pishon. In Geneva in 1560, Calvin's map of Eden had been redrawn and added to the maps included in new editions of Geneva Bibles to help the reader visualize the place where human history had begun.[27] Almost at once, Calvin's portrayal of the four Mesopotamian rivers was found to be too vague and insufficiently informative, and his map was constantly being adapted for later editions of the Geneva Bible. The most important change was the addition of a tiny vignette of the Fall, as on the map the Lyons publisher Sebastien Honoré first used in 1565 for a French edition of the Geneva Bible (Figure 10.3).[28] The explicitness of Honoré's image of Adam and Eve with the serpent and the Tree of Knowledge provided a direct link back to the biblical account as well as an apparently exact indication of the site of the Garden of Eden. The creator of Honoré's map had diverged from Calvin in a second respect. He had placed the vignette for the Garden of Eden not at the confluence of the Tigris and the Euphrates near Babylon and Seleucia, which Calvin had said was the most likely place for paradise, but much further south, below the name *Terre de Hevilah* (Havilah was the most obvious reference on Calvin's map to Edenic geography). Yet another adaptation of Calvin's map would soon be the removal of the contrived link between the Tigris and the Euphrates to the north of Seleucia.[29] That link, which formed an island between Babylon and Seleucia, was the vital element in Calvin's argument that a single river of paradise flowed through Mesopotamia. Once the step had been taken, as on Honoré's map, to portray the scene of the Fall near Havilah, south of the confluence of the Tigris and Euphrates, Calvin's island was redundant.

The tendency was to simplify Calvin's map for the sake of clarity. Yet another way this was done is found on the map Antoine Regnault obtained for his *Discours du voyage d'outre mer au St Sépulcre de Jérusalem et autres lieux de la Terre Saincte* (1573) from one of the Lyons Bible printers (Figure 10.4).[30] On Regnault's map, the geography of paradise has been reduced to a striking simplicity. Two rivers, the Tigris and the Euphrates, first join together to form the single river that flows through the rectangular enclosure containing the garden of paradise, then divide again into the Gihon and the Pishon (so labelled on the map). Regnault shows Adam and Eve both within the Garden, picking the forbidden fruit, and also outside it, being expelled, a 'cartoon'

Fig. 10.4. Map of the location of Eden inserted into Antoine Regnault's *Discours du voyage d'outre mer au St Sépulcre de Jérusalem et autres lieux de la Terre Saincte* (Lyons: [n. pub.] 1573). 16 × 13 cm. London, British Library, G28 24. Regnault's account of his journey to the Holy Land includes a discussion, inspired by Calvin's views, of the location of paradise. His map shows the whole middle eastern region and incorporates the Holy Land and the shores of the Mediterranean, Caspian and Black seas. In the lengthy legend below the map, the region is praised as the most fertile and delightful on earth (and the most frequently represented by map makers). The text rehearses Calvin's arguments on the single river and the four heads, but a large rectangular Garden of Eden, surrounding the entire stretch of the single river, has been added to the map, on which 'A' marks the confluence, just outside paradise, of the Euphrates and Tigris rivers, and 'B' the division of the single river into two branches, again just outside Eden. On Regnault's map, as on other derivatives, Calvin's geography has been further simplified by the removal of the link north of Seleucia between the Tigris and the Euphrates.

The Afterlife of Paradise on Maps   293

Fig. 10.5. Peter Plancius's *Tabula geographica, in qua paradisus, nec non regiones, urbes, oppida, et loca describuntur* . . . for the beginning of Genesis, from a Protestant Dutch Bible (Amsterdam–Haarlem: Jacobszoon–Rooman, 1590). 23 × 17 cm. London, British Library, 3041.b.12. Plancius has given the Tigris and the Euphrates their Hebrew names, *Hidekel* and *Phrath*, and has identified the rivers Gihon and Pishon and the lands of Havilah and Cush. By showing the geography of the Holy Land in some detail on a map of Mesopotamia that extended as far west as Cyprus in the Mediterranean Sea, Plancius was able to present Eden both in its Mesopotamian context and within the wider context of salvation history. The post-Calvin sharpening of the focus on the exact site of Eden reflects the increasing emphasis on the historical aspect maps were now expected to show.

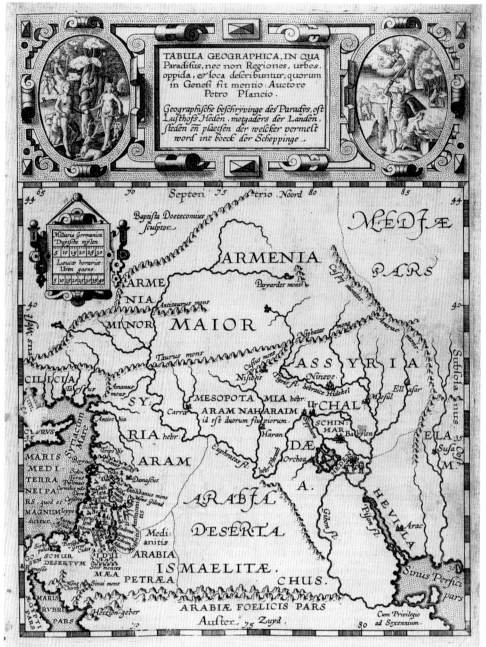

approach already displayed on the much earlier Hereford map (Figure 6.12). The lands surrounded by the streams of paradise, Havilah, Cush and Assyria are also marked.

Not every variant of Calvin's map of the location of Eden was a simplification; some were elaborated in the continuing search for the exact location of paradise. The Dutch clergyman and scholar, Peter Plancius, created a map of paradise for the set of maps he had engraved for Dutch Bibles from 1590 onwards (Figure 10.5).[31] Plancius placed the region of *Heden* and paradise (with Adam and Eve) west of the single river, but not strictly speaking within Mesopotamia (Eden here is placed in Chaldaea). Calvin's linking river is absent. An island is shown, though, with Babylon in the south of it, formed by the Euphrates and including a short section of the combined course of the Tigris and the Euphrates. Degrees of latitude and longitude are indicated along the borders of the map, and two scale bars are provided for the measurement of distance (in 'German miles' and hours). Paradise, it is clear, was to be measured in the mathematically accurate terms of Ptolemaic geography. Thus, according to Peter Plancius, Eden stretched from east to west for about 20 miles and was situated at approximately 34° N and 80° E.

The Problem of the Single River

Calvin's map established a new model for the cartography of the earthly paradise. While his solution focused attention on Mesopotamia, it also introduced a whole new range of problems. The most controversial aspect of the problem, as Plancius's map exemplifies, proved to be the identification of the single river of paradise out of which the four rivers were said to branch. Calvin's cartographical sleight of hand, the link between the Tigris and the Euphrates, would long continue to haunt the maps of his imitators. Towards the end of the sixteenth century, scholars accepted the fundamental elements of his brilliant exegetical solution – paradise in Mesopotamia, and the Gihon and Pishon as branches of the Tigris and the Euphrates – but tried to find ways of making his model hydrographically more convincing. The solution was to replace Calvin's notion of two streams joining into a single river before branching again towards their mouths by the idea of four branches separating from a single river and continuing individually in the natural direction of flow, that is, from north to south. In this refined model, the single river was identified with the Euphrates, and the branching streams with the dense network of tributaries in the Babylonian region. The Gihon and the Pishon, which Calvin had identified with the lowest reaches of the Euphrates and the Tigris just above where they empty into the Persian Gulf, were now incorporated into the Mesopotamian river network (Figure 10.6).

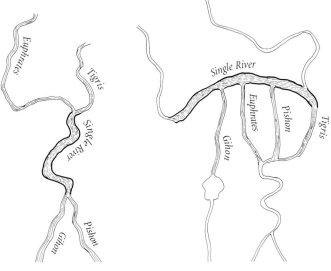

Fig. 10.6. Comparison of two models of the four rivers of paradise, one based on Calvin (see Fig. 9.4) and the other on Iunius (Fig. 10.8).

For the humanist and Protestant theologian Franciscus Iunius (François du Jon) and the Orientalist Iohannes Hopkinsonus (John Hopkinson), such a solution – an extension of Calvin's cartographical expedient – allowed paradise to be located within Mesopotamia, a single river (the Euphrates) to be shown flowing in the land of two rivers, and the four rivers of paradise to be portrayed separating from the single river and then flowing southward. Each attempt to meet the challenge of how to show a single-river Eden within Mesopotamia, however, introduced a sequence of yet other challenges, and each small change on the map either required correction on the next map or demanded an adjustment to the general interpretative framework. Bewildered by the often tiny variations from map to map, the modern researcher is tempted to dismiss them as confusingly insignificant. In fact, each change had its own logic.

Although he was a committed Calvinist, Matthaeus Beroaldus (Matthieu Beroalde or Bérould) sidestepped the issue of the rivers altogether by claiming that no single river had ever flowed in paradise. The map he drew of the region of Eden for his *Chronicum, Scripturae Sacrae autoritate constitutum* (*Chronicle Based on the Authority of Holy Scripture*, 1575), was later inserted into a Latin Bible published in Basle in 1578 and again in 1590 (Figure 10.7). In the words of its author, the map placed the courses of the four rivers before the eyes of his readers 'to make easier the understanding of paradise itself, and of the regions contiguous to Eden'.[32] The key to finding the location of paradise, Beroaldus explained, was to set aside the muddle of vain opinions and to scrutinize instead the sacred text, for this had been inspired by the Holy Spirit. It was also

Fig. 10.7 Map of the vast region of paradise (labelled *Paradisus*), from Matthaeus Beroaldus, *Chronicum, Scripturae Sacrae autoritate constitutum* (Geneva: A. Chuppinus, 1575), p. 88. 12.5 × 15 cm. London, British Library, C79.e.12 (1). Beroaldus shows four rivers flowing across the region of paradise (*Paradisus*). Places mentioned in Genesis – Assyria, Cush and Havilah – are denoted with inscriptions. The Tigris and the Euphrates come together in Mesopotamia, the Gihon flows between Egypt and Palestine, and the Pishon flows through Arabia into the Persian Gulf. The vignette of the Red Sea crossing is a direct reference to the Exodus of the Israelites from Egypt.

essential to appreciate that the Book of Genesis had been written in Hebrew and then translated into Latin, and that the two versions did not always agree. Where the Latin rendering of Genesis described a single river flowing out of Eden to water paradise, there should in fact be four rivers, for in Hebrew the singular and the plural forms are interchangeable, as, for example, in other passages in Genesis (1.12; 3.8) where the Hebrew refers to a single tree, but means several. Thus, the real sense of the statement 'and a river went out of Eden to water paradise' was that 'rivers went out of Eden to water paradise'. Accordingly, Beroaldus's map has four rivers flowing through paradise, which is shown as a huge region, populated by exotic animals, in the middle of which are represented Adam and Eve. The four rivers come from widely separated sources in different parts of this vast region, making paradise a large country, lying within Syria and extending from the eastern borders of the Holy Land (Beroaldus's map also portrays the Exodus) to the north-east of Mesopotamia. As it was watered by such great rivers, it followed that paradise, too, had to be a vast garden.[33]

Neither Iunius nor Hopkinson felt the need to go quite so far as Beroaldus and replace the single river with several on their historical maps of paradise. Nonetheless, their readings of Eden's geography also marked a departure from Calvin. In his

Praelectiones in Genesim (1582), Iunius referred to paradise as north of Chaldaea at the meeting point of Babylonia, Assyria and Mesopotamia. On his map, paradise is in the land that Ptolemy had called Auranitis, part of the region of Babylonia, which Iunius associated with the biblical Eden through its ancient names of Audanitis or Edenitis (Figure 10.8).[34] Quoting authorities such as Ptolemy, Herodotus, Strabo and Pliny, Iunius argued in his book that the single river that had watered the Garden of Eden was the main stream of the Euphrates, that the four rivers branching from it were (from west to east) the Gihon, the Prat or lower Euphrates, the Pishon, and the Hiddekel or Tigris, and that paradise was on the island enclosed by the Pishon and Tigris.[35]

Iunius's theories and map inspired the English Orientalist John Hopkinson to devote an entire pamphlet, the *Synopsis paradisi* (1593), to the paradise question. Like Iunius's, Hopkinson's Eden was a region in Babylonia, where the garden of Adam and Eve had been planted in a 'most delightful place' on the island formed by the Pishon

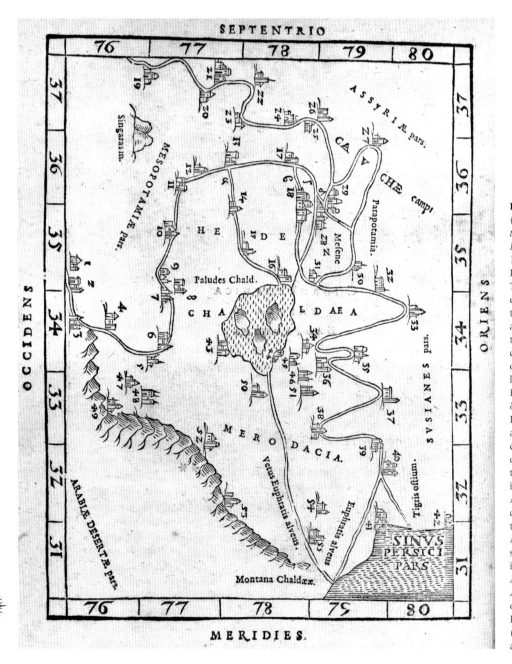

Fig. 10.8. Franciscus Iunius (François du Jon), *Chorographica tabula Babyloniae*, from his *Praelectiones in Genesim* (Heidelberg, 1582/9), before p. 1. 16 × 12 cm. London, British Library, 1015.b.8. The text that follows the map explains the keys to the signs: towns are identified by numbers and the various branches of the Euphrates, which Iunius considered to be the single river of paradise, are marked with Greek letters: α for the Gihon, β for the Prat or Euphrates, γ for the Pishon, δ for the Hiddekel or Tigris. The Hiddekel (δ) is shown on the map as a branch running into the main course of the Tigris, which comes down from the north and flows towards the sea. The land of Cush is described in the text as part of Arabia, and Havilah as corresponding to Susiana, the region mentioned in the classical sources. According to Iunius, *Heden* was situated to the north of Chaldaea, in a region called by Ptolemy *Auranitis*, a corruption from *Audanitis* or *Edenitis*. Paradise was on the island surrounded by the Pishon (γ) and the Hiddekel/Tigris (δ), where the place-name *Mesene* appears.

The Afterlife of Paradise on Maps 297

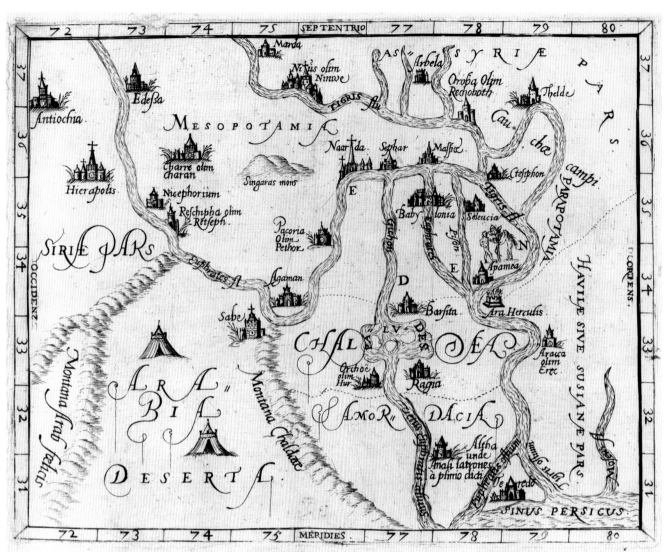

Fig. 10.9. Ioannes Hopkinsonus (John Hopkinson), map of paradise, from his *Synopsis paradisi, sive paradisi descriptio, ex variis … scriptoribus desumpta, cum chorographica … tabula* (Leiden: Plantinus, 1593). Final page. 15 × 19 cm. London, British Library, 570.f.1 (1). Copperplate. Paradise is placed between Seleucia and Apamea. Pecked boundary lines on the map indicate the threefold division of Babylonia described in the text: *Eden* in the north, *Chaldaea* in the middle, and *Amordacia* in the south. Just to the right (east) of centre – immediately below the letter 'D' of *EDEN*, the river Gihon runs into an extensive water body, labelled *paludes*, in an area identified in the text as the land of Cush, in southern Chaldaea.

and the Tigris.[36] On the map, which Hopkinson explained was based on the most excellent ancient and modern authors, the crucial part of the river network is closely modelled on Iunius's, except that Hopkinson did not reconnect the *paludes* formed by the Gihon to the lower Euphrates, which Iunius had marked by a single line in his elaboration of Calvin's model (Figure 10.9).[37] The labelling of the names of the Pishon and the Hiddekel/Tigris branch is also clearer, making the course of the single river more obvious. A vignette with the figures of Adam and Eve and the Tree of Knowledge marks the site of paradise in the eastern part of the region of Eden, between Apamea and Seleucia, matching the scriptural account of a paradise located east of Eden. Inasmuch as he placed paradise on an island between the rivers, Hopkinson (like Iunius) was agreeing with Calvin. However (again like Iunius) Hopkinson departed from Calvin's identification of the single river and its branches, arguing instead that the single river was the Euphrates, from which four streams flowed south, and labelling the rivers on his map. Hopkinson also followed Iunius in linking the Tigris and the Hiddekel branch of the single river to make the Hidekkel/Tigris branch part of a western arm of the main course of the Tigris in the vicinity of the city of Ctesiphon, the location of which he took from a map by Gasparus Vopelius (Caspar Vopel).[38] The name of the land of Havilah appears on the map just below this confluence, to the east of the Pishon/Tigris (known

298   Mapping Paradise: A History of Heaven on Earth

locally as the Pasitigris) as an alternative for part of Susiana. Hopkinson claimed that his map served to prove that paradise was not 'an obscure, airy and imaginary place, incorporeal, suspended in the air, located above or beyond the terrestrial globe', nor had it extended over the whole earth, as some had argued in the past.[39] Rather, it had been 'located on earth, ... confined to a determinable site ... [in the upper part of Babylonia, which] ... Scripture calls Eden and other reliable authors call Audanitis'.[40] After all, he concluded, it was surely from that ancient name, and its alternative Edenitis, that the name Eden was derived.[41]

Iunius's cartography of paradise was also adopted by the Flemish cartographer and engraver Iodocus Hondius. The distinguishing feature of Hondius's pocket edition of Mercator's *Atlas* of 1595, the *Atlas Minor* (1607), apart from its small size, was its complement of historical maps, which were intended to show not only modern places but also those of the ancient world according to biblical as well as classical authorities.[42] A page of explanatory text in Latin, in which Hondius relates that it had been left to him to produce Mercator's map of the earthly paradise, faces the first map (entitled simply *Paradisus*) in the historical section (Figure 10.10).[43]

The Mercator–Hondius *Atlas minor* presents paradise in Mesopotamia as the ultimate proof that previous theories about the location of Eden – those describing

Fig. 10.10. Iodocus Hondius, *Paradisus*, from Gerard Mercator and Iodocus Hondius, *Atlas Minor* (Amsterdam: Iodocus Hondius, 1607/10). 15 × 19 cm. London, British Library, Maps C3.a.3. The map shows the region from Mesopotamia in the east to the Holy Land and the Mediterranean Sea in the west. The river network around paradise (towards the right-hand side of the map) is much the same as Iunius's. An eye-catching scene in a circular cartouche portrays the moment of the Fall. The gravity of the sin just committed by Adam and Eve is underlined by the presence of a third figure, seemingly dark-skinned and probably representing Death. Three tiny figures in the background represent the fleeing pair and the Cherub.

The Afterlife of Paradise on Maps 299

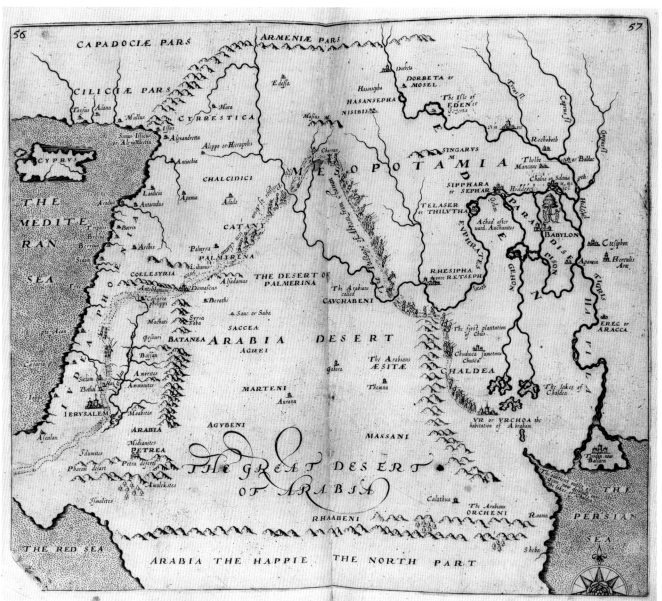

paradise as the whole earth or on the equator, for example – were foolish and nonsensical. The text opposite the map concludes with the comment that 'a more appropriate comment on these speculations would be to laugh at them, rather than to confute them'.[44] Greater tolerance and understanding of the errors of the past was shown by the English courtier, explorer and writer Sir Walter Raleigh, perhaps because he was writing on paradise as an adventurer and man of letters who had enjoyed royal favours but who was now (in 1603) disgraced and in prison in the Tower of London.[45] Raleigh's untitled map is one of several illustrating his *History of the World* (1614) (Figure 10.11). With Calvin, Iunius and Hopkinson, Raleigh believed that the Flood had wiped out a highly localized paradise within the region of Eden and that, since the rivers were still there, the former site of paradise could be indicated on a map. He advised his readers that 'In this leafe following, I have added a *Chorographicall* description of this terrestriall paradise, that the Reader may thereby the better conceive the preceding Discourse.'[46] The reason why it was difficult to find its true site was because middle eastern place names had been changed as different nationalities succeeded first the Jews and then each other throughout the region. Assyrians, Babylonians, Medes, Persians, Greeks, Romans, Turks, all had sought to extinguish the memory of their predecessors by altering the toponymy. Like Hondius, Raleigh took the broad geographical perspective, and his map also depicts the whole area from the Tigris to the Mediterranean Sea. Whereas Hondius had alluded but briefly to Abraham's connection with Ur, however, Raleigh portrayed the full length of Abraham's journey from Ur to the Holy Land as a zigzagging track. Less dramatically, he also marked part of the Exodus route. For such an experienced navigator as Raleigh, it is perhaps curious that he omitted to provide his map with any mathematical aid. There are neither graduations of latitude and longitude, nor a scale bar, only a north point.

Despite his awareness of Hopkinson's views on Eden, Raleigh's text and map show that for him the region of Eden – the legend naming it stretches obliquely from high up the Tigris to beyond the Tigris–Euphrates confluence – was not limited to the Babylonian region of Auranites, but extended well beyond it to include the whole of Mesopotamia and, in the north, parts of Assyria and Armenia, and in the south, Shinar (the land where the Tower of Babel was built, according to Genesis 11.2). Raleigh's paradise is similarly unconstrained and spreads, as the label *Paradise* on the map indicates, over the entire island formed by the confluence of the Tigris and the Euphrates. At the top of the map, in the north, the region of Eden starts at a small island in the upper reaches of the Tigris called 'The Isle of Eden'. This island, observed Raleigh, had been mentioned in letters written to the pope in 1553 by Syriac Nestorians and translated in 1569 by the Flemish diplomat and Orientalist, Andreas Masius.[47] For Raleigh, the island confirmed both the vastness of the land of Eden and its very existence in and around Mesopotamia. Paradise was in the southern part of Eden in a particularly favourable temperate latitude, 35 degrees north of the equator and 55 degrees south of the North Pole, where the climate was excellent and the vegetation luxuriant. There were, Raleigh admitted, other parts of the world, such as in the Far East or in the New World, where the vegetation was also wonderful and the climate no less perfect, but in those regions perfection was spoilt by some major imperfection, such as storms, earthquakes, or dangerous animals.[48]

Raleigh's text suggests that he had in mind both Hopkinson's model for the four rivers and Calvin's. Hopkinson's model would have suggested to him that branching of the four streams took place where the Euphrates formed the northern boundary of paradise. Calvin's model would have led Raleigh to envisage the entire river network around Babylon as the single river, in which case two of the four streams (the Tigris and the Euphrates) had to be thought of as parting from the single river on the north side of paradise, and the other two (the Gihon and the Pishon) on the south side. Raleigh was aware, following Beroaldus, that the Hebrew word for 'river' ('And a river went out of Eden to water the garden', Genesis 2.8) could refer either to a singular river or to a plurality of rivers, and he admitted to being uncertain whether there was one river (either the Tigris *or* the Euphrates) or two rivers (the Tigris *and* the Euphrates) 'going out of Eden to water paradise'. He noted, however, that the single river could be taken as the Euphrates, which included all the various branches flowing out of the vast region

Fig. 10.11a (opposite). Walter Raleigh, map of paradise, from his *History of the World*, 2nd edn (London: William Jaggard for Walter Burre, 1617), between pp. 56 and 57. 30 × 34.5 cm. London, British Library, C115.h.5. Raleigh's map contains aspects of both Calvin's and Hopkinson's models. The single river of paradise – the Euphrates – goes out of the vast region of Eden to water paradise through a complex network of stream from which the other three rivers emerge. The Tigris parts from the Euphrates on the north side of paradise in the sense that where the Hiddekel – the east–west continuation of the Euphrates – joins the main course of the Tigris, the latter, coming from the north, points towards Assyria. The other two rivers (the Gihon and the Pishon) branch off the network of the Euphrates on the south side of paradise and encircle the lands of Cush and Havilah, respectively. The figures of Adam and Eve and the Tree of Knowledge are depicted in the middle of the maze of streams at the confluence of the Tigris and the Euphrates near Babylon.

Fig. 10.11b. Diagram of Raleigh's idea of paradise and Eden.

Fig. 10.12. Walter Raleigh, *History of the World*, 2nd edn (London: William Jaggard for Walter Burre, 1617). Frontispiece. 30 × 18 cm. London, British Library, C115.h.5. The title page is that of the first edition, and bears the date 1614. The frontispiece of Raleigh's *History of the World* displays an allegory on the importance of history through the interplay of a number of personifications. History, the master of life, raises the world from Death and Oblivion to good or evil Fame, under the eye of Providence. Her work is helped by Truth and Experience, and supported by the pillars of the Witness of Time, the Herald of Antiquity, the Light of Truth, and the Life of Memory. The map representing the world is a rare example of a post-Renaissance world map, ruled by parallels and meridians and showing the Americas, that also features the Garden of Eden, vaguely marked in central Asia. The map is not intended, however, as a geographical display, but to symbolize the world as the stage of universal history.

302   Mapping Paradise: A History of Heaven on Earth

of Eden and the complex hydrological network watering paradise, and was divided into four branches, two to the north and two to the south. Raleigh insisted that this division took place within the region of Eden, but outside paradise itself, on its boundaries. He interpreted the reference in Genesis to the four 'heads' as indicating the specific point at which each of the rivers of paradise became independent of the single main stream, either as a branch flowing in a new direction or as a stream with its own name. In any case, each of the four 'heads' led towards an important region (Figure 10.11b). Nowhere in Genesis, argued Raleigh, was it stated that the branching of the rivers took place downstream. After its confluence with the Tigris at Apamea, the Pishon encompassed the land of Havilah (Susiana). The Gihon encircled the land of Cush, on the borders of Chaldaea and Arabia, before being lost in the lakes of Chaldaea (although, as Raleigh's map indicates, it had formerly flowed all the way to the sea). Merging with the main course of the Tigris just east of Seleucia, the Hiddekel showed the way to Assyria, upstream of the Tigris. The fourth river was the Euphrates itself, which ran through paradise.[49] In this way, Raleigh offered a vision of paradise's hydrography that modified Calvin's view with ideas taken from Hopkinson and, crucially, his own belief that Eden was a much larger region than Babylonia. The single river mentioned in Genesis was the Euphrates, as propounded by Hopkinson. This river received the waters from the vast region of Eden before watering paradise. The other three 'heads' mentioned in Genesis (the Pishon and the Gihon in the south and the Tigris in the north) branched off the river network around Babylon, as Calvin maintained. Raleigh's arrangement of the rivers served to anchor his choice of site for paradise. Convinced that the Garden of Eden had lain near the confluence of the Tigris and Euphrates, he placed a vignette showing Adam and Eve and the Tree of the Knowledge of Good and Evil next to the place sign for Babylon, within the large area labelled *Paradise*.

Variations on the Mesopotamian Theme
After Raleigh, the notion of the four rivers as four branches separating off from the Euphrates was abandoned, and the question of the rivers was resolved by a return to Calvin. For the rest of the seventeenth century the debate focused on Calvin's idea that the four 'heads' corresponded to the Tigris and Euphrates (to the north of a Mesopotamian paradise) and to the Gihon and the Pishon (to the south). Not all exegetes considered the matter clear enough to be shown on a map. In 1630, Cornelius a Lapide agreed that paradise had been placed in Mesopotamia, but insisted that the exact site could not be known. Convinced, nonetheless, that an earthly paradise had really existed, he accepted that Moses had described the rivers as he had seen them, even though they had changed their courses over time, making it difficult to reconstruct the original network. Instead of devising a map of his own, Cornelius advised his readers to study the confluence of the Tigris and Euphrates on maps by authorities such as Mercator and Ortelius.[50]

Evidence for the growing preference for Calvin's model rather than Iunius's is provided by the seventeenth-century Dutch map maker Claes Ianszonius Visscher. A map dated 1642, bearing Visscher's name, was engraved for Dutch folio editions of the Bible.[51] Visscher's paradise map seems to have been a commercial success and, together with his other Bible maps, was widely copied. Visscher's maps were translated and inserted into English Bibles by the polymath Joseph Moxon.[52] In 1671, Moxon published a small oblong treatise entitled *Sacred Geographie* and containing 'an explanatory discourse' on the maps, which were issued separately as a folio volume. Moxon explained that he had selected the book's format so that 'it might more conveniently ly [lie] open on the Maps while you are perusing them'.[53] The first map described in the *Sacred Geographie* was not one of Visscher's but a world map entitled *A Map of All the Earth and how after the Flood it was Divided among the Sons of Noah*, apparently created

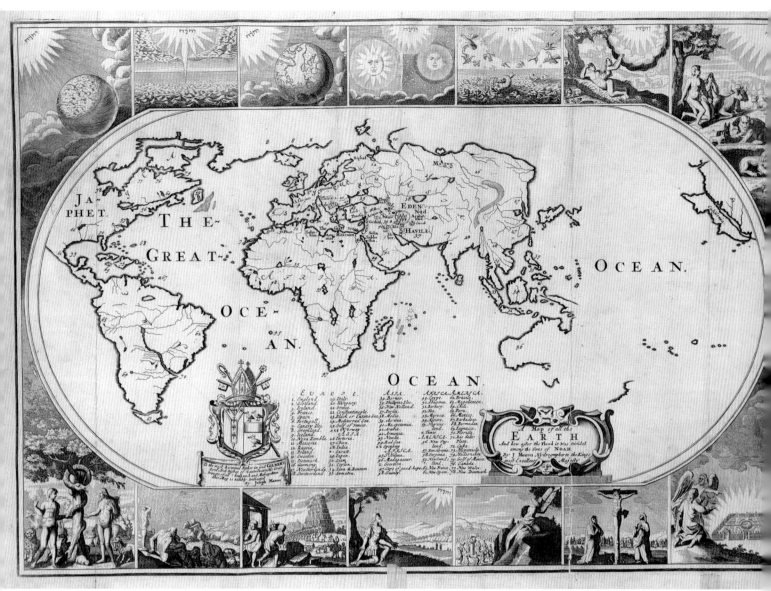

Fig. 10.13. Joseph Moxon, *A Map of All the Earth and how after the Flood it was Divided among the Sons of Noah* (1671), from *The Holy Bible* (London: the author, 1683). 32 × 46 cm. London, British Library, L18.d.5 (2). Note that North America, on this world map, drawn to show the division of the earth among the sons of Noah, was assigned to Japhet. The map is dedicated to Gilbert Sheldon, Archbishop of Canterbury from 1663 to 1677, and, like the other five that Moxon issued in his *Sacred Geographie* of 1671, was intended to be inserted in Bibles. Eden, Havilah and the land of Nod (Genesis 4.16) are marked not far from Mesopotamia and Persia. The sign used by Moxon for paradise (a fenced enclosure) was the standard sign on contemporary English regional maps for the private parks of the gentry. The map is surrounded by fourteen decorative insets of biblical scenes. Portrayed above is the week of creation, ending with God instructing Adam and Eve in paradise, and, below, the most important events in salvation history: the Fall, the Flood, the Tower of Babel, God's promise to Abraham, Moses receiving the tablets of the Law while the Israelites worship a golden calf, the Crucifixion. The last is the vision of the Heavenly Jerusalem.

Fig. 10.14 (opposite). Joseph Moxon, *Paradise, Or The Garden of Eden. With the Countries circumjacent Inhabited by the Patriarchs* (1671), English version of Claes Ianszonius Visscher's map of 1642, from *The Holy Bible* (London: the author, 1683), between pp. 2 and 3. 32 × 46 cm. The double-folio map shows the whole of the Middle East from the land of Nod (Genesis 4.16), east of the Tigris, to Cyprus in the Mediterranean Sea. The four rivers of paradise are given their biblical names. A single river passes through the Garden of Eden, and the vignettes of the Temptation and the Expulsion are found in the wooded area on its eastern bank. Other biblical events are shown on the map, such as the resting of Noah's Ark on a peak of Armenia, Jacob's journey back to Canaan, and Jonah's 'Flight from the Lord' into the whale's mouth in the Mediterranean Sea off Joppa. For the location of paradise, Visscher (and thus Moxon) had followed Calvin's model, except that he referred to the single river as the Pishon, from which he made the Gihon branch off to the west, south of paradise. On this map, the Pishon, which both Visscher and Moxon accepted was the Tigris, continues separately to the Persian Sea.

304   Mapping Paradise: A History of Heaven on Earth

by Moxon (Figure 10.13).⁵⁴ Despite the small scale of the map and the inevitable overcrowding of the geographical details of the Edenic region in Mesopotamia, the biblical names of the four rivers are clearly discernible, as are the names of Eden and Havilah, and the circular vignette of the Garden itself, albeit all slightly displaced eastwards. The second map, *Paradise, Or The Garden of Eden. With the Countries Circumjacent Inhabited by the Patriarchs* (Figure 10.14), was a faithful copy of Visscher's paradise map. The back of the map is filled with two texts in Dutch, one discussing the location of paradise and the other identifying the 26 features keyed by numbers to the map. Moxon's text is taken virtually word for word from the Dutch original on the verso of Visscher's map.

Visscher's Garden of Eden was situated in the land of Shinar, not far from the site of the Tower of Babel and the ancient city of Ur. The map shows the river running through paradise. A small part of the Garden is shown on the west bank, but much the greater part is east of the river. This arrangement reflected Visscher's conclusion that Adam must have dwelt on the eastern side of the river, for Genesis 3.24 describes how Cherubim were placed on guard only on the eastern side of the Garden where, by implication, there was no river to prevent Adam and Eve from re-entering paradise. In contrast, on the west was the wide and turbulent single river (which Visscher likened to the Rhine or the Danube) that kept the sinful pair out of paradise, for they lacked the means of crossing it. Visscher pointed out that Genesis makes no mention of any boat or ferry prior to Noah. For the same reason, Adam and Eve's son Cain was prevented from crossing another turbulent river, the Euleus, in his attempt to escape from his home in the land of Nod, to the east of Eden, yet further east in 'his flight from the

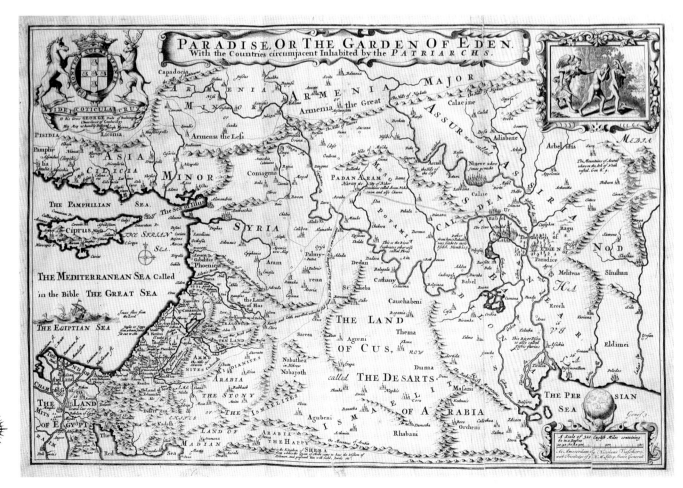

The Afterlife of Paradise on Maps 305

Fig. 10.15. Arthur Bedford, *The Place of Paradise, the Habitation of Adam and Seth, and the Land of Nod*, from his *The Scripture Chronology Demonstrated by Astronomical Calculations* (London: James and John Knapton, 1730), p. 117. 29.5 × 21.5 cm. London, British Library, L20.gg.70. Bedford's map, which situates several places of biblical history – such as Mount Ararat, Nineveh, Enoch and the land of Nod – locates paradise on the east side of the single river, opposite the place where the Tower of Babel and the city of Babylon were later built. Bedford agreed that a single river ran through Eden into paradise to water it, and that this river was the river formed by the junction of the Tigris and the Euphrates, which thereafter separated into the Pishon and the Gihon before emptying itself into the Persian Gulf. Bedford pointed out, however, that the Book of Genesis did not say that the river ran *through* the garden. In fact, he said, the river ran along the west side of paradise. Note the reversal of the two distributaries. Following Samuel Bochart and Pierre-Daniel Huet but contrary to tradition (see below), Bedford gives the name Pishon to the western branch and Gihon to the eastern branch.

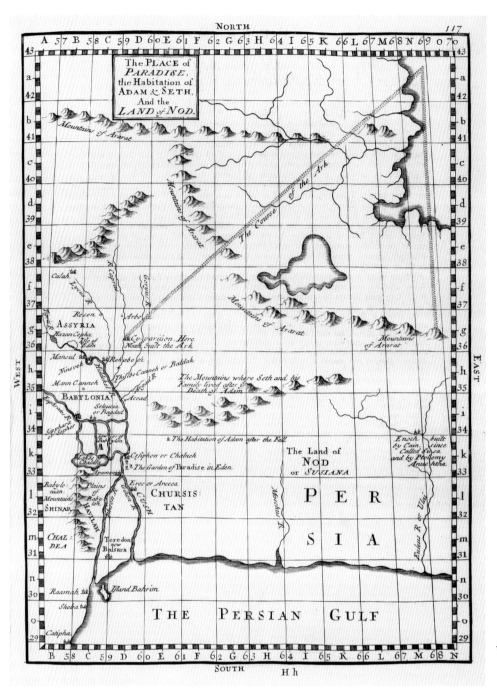

face of the Lord' (Genesis 4.16).[55] Visscher's (and so Moxon's) aim was to present the reader with a map on which correctly placed biblical features could be identified 'with very gret ease, speed, and delight', and he was critical of 'the eminent errour' of 'many Ancient Geographers' who had placed the Garden of Eden to the west side of the single river.[56] Seventy or so years later, the maps in Arthur Bedford's *Scripture Chronology Demonstrated by Astronomical Calculations* (1730) advanced Visscher's argument by showing the whole of paradise on the east side of the single river, an idea confirmed by Bedford's reference in his book to the single river as the western boundary of paradise (Figure 10.15).[57] Bedford added that part of Adam and Eve's work was to cut ditches from the river to water the Garden. Unlike Visscher and Moxon, but like Raleigh, Bedford included on his map the island of Eden in the river Tigris.[58]

306 Mapping Paradise: A History of Heaven on Earth

The seventeenth-century debate on the location of paradise in Mesopotamia reached its climax in the learned arguments of the French Protestant clergyman Samuel Bochart and his friend and disciple, Pierre-Daniel Huet, the Catholic Bishop of Avranches, France.[59] Both scholars developed and refined Calvin's ideas about the mapping of Eden. Both placed the Garden beside the single river formed from the confluence of the Tigris and Euphrates, and both reversed the Pishon and the Gihon, naming the Gihon as the easternmost of the two streams into which the single river eventually divided. This meant that Cush was now in the east and Havilah in the west. Bochart's outstanding linguistic ability – he was proficient in Hebrew, Syriac, Chaldaic and Arabic – equipped him particularly well to assemble philological evidence pointing to the site of the earthly paradise. He deployed his skills in the immensely learned two-volume work *Geographia sacra* (respectively, *Phaleg*, 1646, and *Chanaan*, 1651), in which Jewish and Christian learning, ancient history and geography, and linguistics all served to prove the primacy of the Bible and of the Christian faith over pagan cultures.[60] Bochart's hope was that, by unravelling the information hidden in the biblical names of the rivers, precious stones, and plants found in Eden, he could demonstrate the superiority of the Genesis account of the world's creation over any pagan cosmology. He pretended he had not had time to discuss the location of Eden in the *Geographia sacra* or to produce the detailed treatise on the question he had had in mind, which would have traced the location of paradise through a painstaking reconstruction of the original Hebrew place names.[61] In fact Bochart had explored the problem in the 1640s. He had filled two manuscript notebooks with his thoughts, outlining the opinions of earlier authors and discussing the various aspects of the question, in particular confuting the old identification of the Gihon with the Nile. He also sketched two maps that summarized Calvin's and Iunius's views and one map illustrating his own ideas.[62] He had also elaborated on the question in a sermon and two letters, one of which was in due course published in his collected works, the posthumous *Opera omnia*, edited and published in 1692 by Stephanus Morinus (Étienne Morin), his biographer and another of his disciples. The *Opera omnia* also included a treatise on the location of paradise assembled by Morinus 'according to Bochart's views' (*ad mentem Bocharti*), which, on the basis of the manuscript notebooks, can be taken as representative of Bochart's thinking. Likewise, it may be assumed that the map included by Morinus, which is an elaboration of the sketch in Bochard's notebook, also reflects Bochart's views (Figure 10.16).[63]

Bochart's map, *Edenis seu paradisi terrestris situs*, conflates the region of Eden and the Garden of Eden (the earthly paradise). More radical even than Raleigh, whose instinct was to avoid being trapped by the geographical minutiae relating to the various streams of the Euphrates, Bochart avoided portraying the streams altogether (although they are shown on another map in the *Opera*). What we see is the simple confluence of the rivers Tigris and Euphrates. Close by is Ctesiphon, and a vast wooded region labelled *Paradisus terrestris*. The encircling words *Edenis regio* define a region that includes the cities of Babylon, Seleucia and Apamea. The wooded area spreads across the rivers to include the confluence of the Tigris and Euphrates, the single river, and its division into the Pishon and the Gihon. The figures of Adam and Eve are seen disporting themselves within the Garden. The island of Eden in the Tigris is marked. A number of textual references scattered about the map relate to biblical history and classical geography.

The simplified delineation on Bochart's map of the main rivers of the Eden region hint at a return to Calvin and the dismissal of Iunius's views, a conclusion confirmed twice in the treatise. On the first occasion, Bochart declared his agreement with Calvin's location of paradise.[64] On the second, Bochart voiced his objection to Iunius's idea that the four streams branching from the Euphrates were the four rivers of paradise, pointing out that there was no foundation for such an interpretation. One of Bochart's arguments was that the ancient authorities described only three branches of

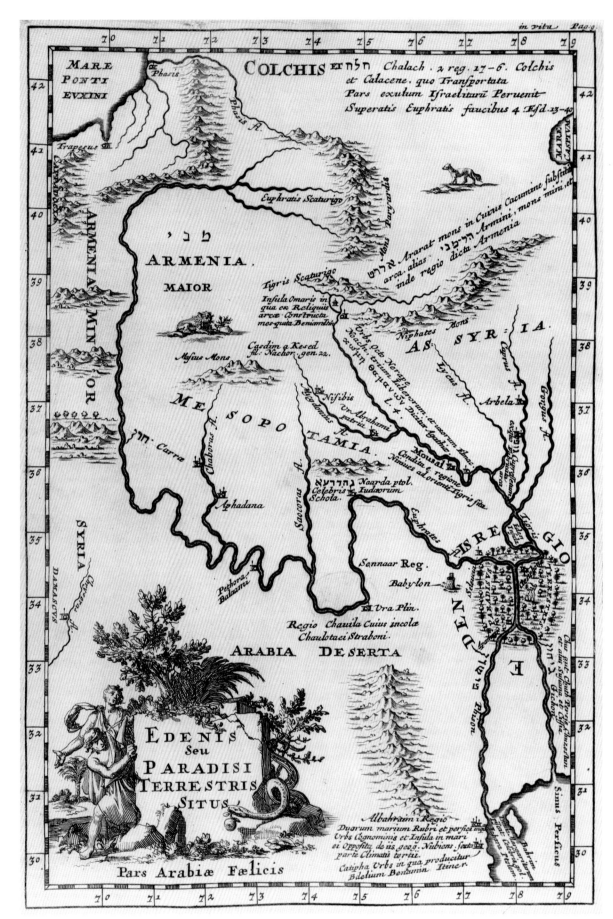

the Euphrates. Another was that in Genesis, the Pishon is mentioned first and the Euphrates last, implying that the author of Genesis (supposedly Moses, who would have been in Palestine) was describing the four rivers as he saw them, namely in geographical sequence starting with the one nearest to him (the Pishon) and ending with the one furthest away (the Euphrates). Iunius, in contrast, depicted the Gihon first and the Tigris last. Another of Iunius's errors, in Bochart's eyes, was to consider the Tigris (Hiddekel) as no more than a branch of the Euphrates, since the Bible qualified the Tigris as a great river (Daniel 10.4). Finally, Bochart pointed out, none of the streams associated with the Euphrates watered Assyria, Cush, or Havilah. Bochart did not hesitate to amend what he saw as Calvin's errors either. By switching, on his own map, the Gihon and the Pishon, Bochart was correcting what he considered to be Calvin's mistaken identification of Cush with Arabia and Havilah with Susiana and an erroneous labelling of the two rivers. The fact that Havilah was in the west, he said, was proved by the biblical allusion to the sons of Ishmael (the Arabs) living in the land that stretched 'from Havilah unto Shur, that is before Egypt' (Genesis 25.18). This had to mean that Shur, near Egypt, lay along the western boundary of Arabia, while Havilah, also in Arabia, lay along its eastern boundary, that is, on the western side of the Euphrates and the Tigris, which the Arabs never crossed. Other biblical passages (Isaiah 15.7 and Numbers 12.30), together with references in the works of Herodotus, Strabo, Diodorus Siculus and Pliny, were marshalled by Bochart to confirm Havilah's western location. Additional facts Bochart held to be relevant were that Arabia had been described by the ancients as a region rich in gold, pearls and precious stones, and that the name Havilah derived from the Hebrew word for sand (חול, *hol*).[65]

Although the Catholic Huet had been educated at the Jesuit school in Caen, he had also been Bochart's student, and he acknowledged that it was Bochart's *Phaleg* that converted him to the study of ancient languages. Huet was one of Bochart's most outstanding pupils, and his *Traité de la situation du paradis terrestre* (1691) can be seen as a continuation of his Protestant mentor's guiding principle of putting learning at the service of faith.[66] Huet's arguments differed little in essentials from Bochart's, except in one respect. Where Bochart was deliberately vague regarding the exact site of paradise, defining instead an extensive area centred on the confluence of the Tigris and Euphrates, Huet could not abandon the challenge of finding a precise location for paradise. Avoiding the whole issue of the river network, he looked farther downstream. As his English translator, Thomas Gale, pointed out, Huet's treatise was the last of a number of learned books aimed at corroborating the historical veracity of the Genesis account through the deployment of the rational arguments of secular history and geography.[67] Despite all the uncertainties, the range of opinions, and the fact that Bochart – his much-admired master – had failed to leave a definitive answer to the issue of the exact location of paradise, Huet pledged to shed light on it. At the same time, he allowed that the question of the location of the earthly paradise was not part of Christian dogma and that it was possible to countenance different opinions without danger of being considered heretical.[68] Instead of all those conflicting opinions, he urged, what was needed was a thorough re-examination of Moses's original words.

Huet found untenable the view that Eden was in the Babylonian region, above the confluence of the Tigris and the Euphrates, because that was not where the lands of Havilah and Cush were to be found, as he would demonstrate with his map (Figures 10.17, 10.18 and 10.19). He followed Bochart's reading of Calvin in situating Eden on both sides of the single river formed by the junction of the Tigris with the Euphrates, but he refuted Iunius and Hopkinson, pointing out that the area around paradise had changed since Moses's time and that of all the branches of the Euphrates discussed by them, the only natural stream was the river that flowed through Babylon, all others being man-made. The source of the river that had watered paradise lay well beyond Eden. The

Fig. 10.16 (opposite). Samuel Bochart, *Edenis seu paradisi terrestris situs*, in *Opera omnia*, ed. by Stephanus Morinus (Étienne Morin) (Leiden: Cornelius Boutesteyn et Iordanus Luchtmans, 1692), p. 9. 29 × 18.5 cm. London, British Library, 7.f.6. Bochart expounded his opinion on Eden's location in manuscript notebooks, a sermon and two letters, which were published after his death in his *Opera omnia*, together with this map. Bochart intended to locate paradise by carefully deciphering the original form of ancient Hebrew place names that, in his view, had been distorted over the centuries in various languages. He declared that he very much agreed with Calvin's location of paradise, but that he also intended to draw attention to the fact that Calvin located the Pishon in the east and the Gihon in the west and thus mistook Cush for Arabia and Havilah for Susiana. On the contrary, argued Bochart, Havilah was located in the west, and Cush in the east.

Fig. 10.17. Pierre-Daniel Huet, map of paradise, frontispiece of his *Traité de la situation du paradis terrestre* (Paris: Jean Anisson, 1691). 14.5 × 8 cm. London, British Library, 219.c.15. Huet, the Catholic Bishop of Avranches, had been educated at the Jesuit school of Caen, but he had also been taught by the Protestant clergyman Samuel Bochart, whom he accompanied to Stockholm in 1652 at the invitation of Queen Christina of Sweden, who was at this point in her mid-twenties and already a remarkable scholar. Huet acknowledged that reading Bochart's *Phaleg* had converted him to the study of ancient languages. The arguments he proposed were so similar to those of his friend and master Bochart that he was accused of plagiarism. Following Bochart's reading of Calvin, Huet situated the region of Eden along the canal formed by the the junction of the Tigris and the Euphrates, but located paradise further south, as shown in the maps that illustrate his celebrated treatise.

Fig. 10.18 (left). Pierre-Daniel Huet, *A Map of the Situation of the Terrestrial Paradise*, from his *A Treatise of the Situation of the Terrestrial Paradise*, English translation by Thomas Gale (London: James Knapton, 1694). Folded to face p. 1. 22 × 19 cm. London, British Library, 1017.e.21. The map in the English version of Huet's treatise is a close copy of the original French. According to Huet, the division of the single stream into the two distributaries at the head of the Persian Sea, denied by some of his contemporaries, was documented in the writings of reliable ancient authorities, such as Ptolemy, and the reports of modern travellers, such as the Portuguese Pedro Teixeira and the French Jean de Thevenot. Repeating Brochart's argument that Moses would have named first the river nearest to him (Pishon), Huet saw the westernmost of the two distributaries as the Pishon, Havilah as part of Arabia, the Gihon as the easternmost distributary, and Cush corresponding to the land of Susiana. The island formed by the Pishon and the Gihon was called Chader, formerly Messene.

Fig. 10.19 (above). Frontispiece to Pierre-Daniel Huet's *Traité de la situation du paradis terrestre* (Amsterdam: François Halma, 1701). 13 × 8 cm. London, British Library, G15918. With the help of a map, Moses explains to Geography personified the geographical meaning of his description of the Garden of Eden in Genesis. Huet's map of Paradise enjoyed a long life, being reprinted in subsequent editions of his treatise and in several eighteenth-century Bibles.

The Afterlife of Paradise on Maps   311

Fig. 10.20. Salomon van Til, *Tabula situm paradisi et regionum vicinarum referens*, from his *Dissertatio singularis geographico-theologica de situ paradisi terrestris*, in *Malachius illustratus* (Leiden: Iordanus Luchtmans, 1701). 26 × 21.5 cm. London, British Library, 854.h.23. 9303. Folded to face p. 1. Van Til located Eden and paradise north of the confluence of the Tigris and the Euphrates. He pointed out that, amongst the regions mentioned in the Bible, Eden was relatively small compared with such lands as Cush, Nod, Assyria, Havilah, and Sinar. Van Til explained that King Nebuchadnezzar had sought to solve the conflicting problems of flood and drought by constructing channels to divert all excessive flow above Babylon into a reservoir for irrigation. Two of Nebuchadnezzar's channels are shown on Van Til's map, the Naar Malca (between Massice and Seleucia), marked with dotted lines, and the Nahar Sares, shown flowing through a large rectangular reservoir. Also indicated as an artificial channel is the Pallacopa, which was constructed by Alexander the Great.

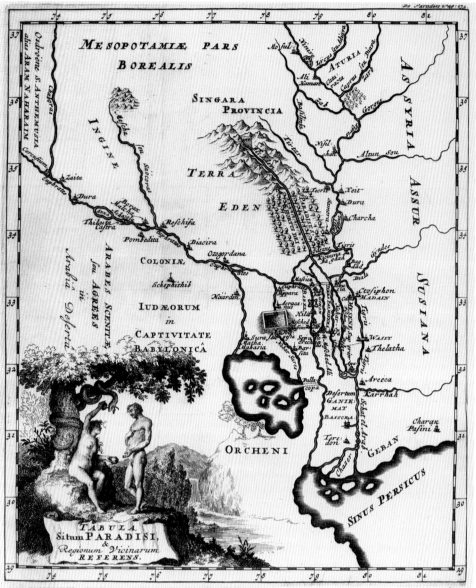

separation into the four streams (heads) also occurred outside the Garden: the Euphrates and Tigris upstream, above paradise, and the Pishon and Gihon below it. Finally, Huet agreed with Bochart that Calvin's identification of the Gihon and the Pishon was mistaken and that Calvin had displaced Havilah and Cush from their true positions.

Huet's map thus differs in certain details of Mesopotamian geography from Bochart's. The salient difference is Huet's placing of the Garden of Eden on the southern arm of the bend made by the single river south of Babylonia. Huet explained his reasoning: since the biblical text stated that the single river ran through the region of Eden and entered the Garden, which extended on both sides of the river in eastern Eden, the river must at that point have been flowing from west to east. Indeed, a striking characteristic of Ptolemy's Fourth Map of Asia was the large bend in the river formed by the combined Tigris and Euphrates (see, for example, Figures 9.5 and 9.6). This great bend was confirmed as taking place at 32° 39′ N and 80° 10′ E in Ptolemy's list of coordinates.[69] It was thus highly likely, concluded Huet, that paradise had been situated at the eastern end of the southern arm of the bend, close to the point where Ptolemy placed the city of Aracca (33° 45′ N and 80° 30′ E), the biblical Erech (Genesis

312 Mapping Paradise: A History of Heaven on Earth

10.10). Having in this way pinpointed the exact site of paradise, Huet was free to plot the streams of the Euphrates around Babylon as he liked, without further reference to paradise, now safely out of the way in the south. The validity of his rather southerly siting of paradise was confirmed, in his view, by reports of the area's outstanding beauty and natural wealth – a relict of its original state. The supernatural beauty of paradise had inspired the pagan poets as they described their *loci amoeni*, which included, amongst other places, the Fortunate Islands, the Elysian Fields and the Gardens of the Hesperides.[70]

In 1701, a few years after Huet had placed the Garden of Eden unusually far south in Mesopotamia, Salomon Van Til gave it an unusually northern location. Van Til focused again on the problem of the single river and offered yet another variation on the theme. Whereas Luther had claimed that the Flood had destroyed all of the rivers of paradise, and Calvin had objected that the four rivers still flowed in Mesopotamia and could be mapped, Van Til blamed the Babylonian King Nebuchadnezzar for altering the original river pattern of paradise through his hydraulic works.[71] Van Til's paradise was one of the few strictly Mesopotamian locations in the sense that his map shows the land of Eden (*Terra Eden*) as the land between the two rivers, and the Garden of Eden placed well north of the confluence of the Tigris and the Euphrates, between Tekrit and Baghdad and not far from the island of Eden (Figure 10.20).

Whereas Calvin and the others struggled to see the biblical single river in the confluence of the Tigris and the Euphrates, Van Til ignored both rivers and suggested that the single river (marked on his map as *flumen Eden*) was the completely separate stream that the locals called *Odeines* (a derivation of the name *Eden*).[72] The single river flowed in a straight course through paradise (*Hortus Dei*, 'Garden of God'), before sending out branches east and west to the Tigris and the Euphrates, and then parting again into the Gihon (the modern Cobnar) and the Pishon (the modern Diala). Cush, which bordered Arabia on the west, and Havilah, which bordered Persia on the east, are mentioned in Van Til's text, but are not indicated on his map. According to Van Til, whereas the Bible recorded that a single river flowed out of Eden, the Tigris and the Euphrates arose far outside Eden, in Armenia, and neither originated in the single river nor were in any way associated with it. As for the Gihon and the Pishon, Van Til followed Iunius, but rejected Iunius's view of the Euphrates as the single river by claiming that the branch connecting the Euphrates with the Tigris was artificial and the work of King Nebuchadnezzar. Van Til's point was that, irrespective of the Flood, it was Nebuchadnezzar's hydraulic improvements that led to the drying up and disappearance of a number of streams and to the alteration beyond recognition of the river network seen by Moses, but subsequently lost.

An Underground Solution

Athanasius Kircher was a German Jesuit scholar of wide-ranging erudition as well known for his knowledge of the natural sciences as for his remarkable ideas on biblical geography. He situated paradise in Mesopotamia, but looked underground to work out a consistent solution for the problem of its hydrography.[73] For his basic premises, Kircher followed the reformers Martin Luther and John Calvin. He adopted Luther's thesis that the Flood had devastated the whole earth, altering the courses of all rivers including those of paradise, and accepted Calvin's placement of the Garden of Eden in Mesopotamia. Kircher was one of the few Catholic writers on the subject of the earthly paradise to follow Calvin and to attempt to show its location by means of a map. His highly original ideas on world's hydrography, however, allowed him to go back to the essentially medieval notion of a single source for the four rivers that watered the Garden of Eden.

In his *Mundus subterraneus* (1664), Kircher presented a novel theory involving the underground circulation of all the world's waters. His theory was based on the

Fig. 10.21 (opposite). Athanasius Kircher, *Descriptio regionis Edeniae*, from his *Arca Noe* (Amsterdam, Ioannes Ianssonius, 1675), between pp. 22 and 23. 34 × 21 cm. London, British Library, 460.c.9. Kircher's work on Noah's Ark reveals his thirst for knowledge and his considerable interest in cartography. On this map, orientated to the west, the Tigris and the Euphrates are shown rising separately in the Armenian mountains and joining in Babylonia, well to the south, before dividing into the two distributaries that lead into the Persian Gulf. Brief texts on the map point out the abundance of timber in the cedar forests of Lebanon and the forests of Mesopotamia, the string of bitumen lakes south of Babylon, the pitch produced in the land of Shinar, and the sulphur deposits of Judaea. Also shown are the place of the sacrifice of Abel, the location of the city of Enoch (founded by Cain) and the Tower of Babel.

passage in Genesis in which God is said to have divided, on the second day of Creation, 'the waters from the waters' (Genesis 1.6), that is to say, the waters of the earth's surface from the waters remaining deep within the core of the earth. Kircher envisaged a complex system of interconnected subterranean channels and vents in which water and fire interacted before the water passed through a network of underground springs and reservoirs to reach the ocean. He discussed the location of the earthly paradise within the framework of his theory about the world's waters, devoting to the issue a chapter of his *Arca Noe* (1675). In this work, Kircher analysed all aspects relating to Noah's Ark: its size, weight and configuration of the Ark, how the various animals – two of every kind – were distributed within it, what they would have eaten, how long they spent sailing about, in which direction the Ark had drifted, and so on.[74] Noah, Kircher pointed out, was born 126 years after Adam's death, and was 55 years old when God commanded him to build the Ark, a task that was executed in Eden, which Kircher defined as a vast region including not only Babylonia and Mesopotamia, but also Armenia, Syria, Assyria, Lebanon, and the part of Judaea that lay east of the river Jordan.[75] His point was that the boat that saved mankind from complete destruction had been assembled in the same region as the one into which God had placed the first man and which in due course became the home of the Patriarchs. Kircher's west-orientated *Descriptio regionis Edeniae* is placed in the book as one of several drawings illustrating his discussion of the construction of the Ark, and particularly as a demonstration of the eminent suitability of the region of Eden for the ark building that is shown on the map in the central part of the region and west of the Euphrates (Figure 10.21).

Appropriately, for a map concerned with the construction of a boat within the region of Eden and not with the location of the Garden of Eden, paradise is not marked on the *Descriptio regionis Edeniae*. Kircher dealt with the location of paradise on another map in the *Arca Noe*, in which he devoted an entire chapter to the question 'Utrum paradisus terrestris diluvio destructus sit, et ubinam locorum fuerit' ('Whether the earthly paradise was destroyed by the Flood, and in what place it was'), concluding that it had been in Mesopotamia.[76] This second map is headed *Topographia paradisi terrestris iuxta mentem et coniecturas authoris* (*Topography of the Earthly Paradise according to the Judgement and the Conjectures of the Author*) and is orientated to the north (Figure 10.22). It is a distinctive map, less for the way the problem of the rivers was resolved than for the prominence given to the Garden of Eden and its internal details.

For the rivers, Kircher conformed to Calvin's naming of the western branch as the Gihon and the eastern as the Pishon, although he abandoned Calvin's model of the four heads.[77] His objective was to address the issue of the geographical change in Eden brought about by Adam's sin, and to show the pattern of the rivers both before and after the Fall. At first sight the map may appear to be a visual paradox that fails to match Kircher's text. It is unlikely, however, that Kircher should be reproached for inconsistency, since his map makes perfect sense if read as a chronology of human history, on which successive biblical episodes are shown in their appropriate geographical place. The medieval cartographical practice of displaying several layers of time on a single map seems to have been partially resumed on this regional and historical Renaissance map of paradise, another example of Kircher's original approach to the subject.[78] Thus, we are shown on the map both ante- and postdiluvian rivers, the event of the Fall itself, Adam and Eve living as farmers away from the Garden from which they had been banished, the sacrifice of Abel, his murder by Cain, the city of Enoch, and the landing of the Ark on Mount Ararat. In the text, Kircher explained that the antediluvian spring in the middle of paradise had drawn its abundant waters from a huge underground reservoir in the mountains of Armenia and that the four rivers had originally flowed out from it separately, one in each cardinal direction, watering first paradise itself and then, beyond the Garden, the lands of Havilah, Cush and

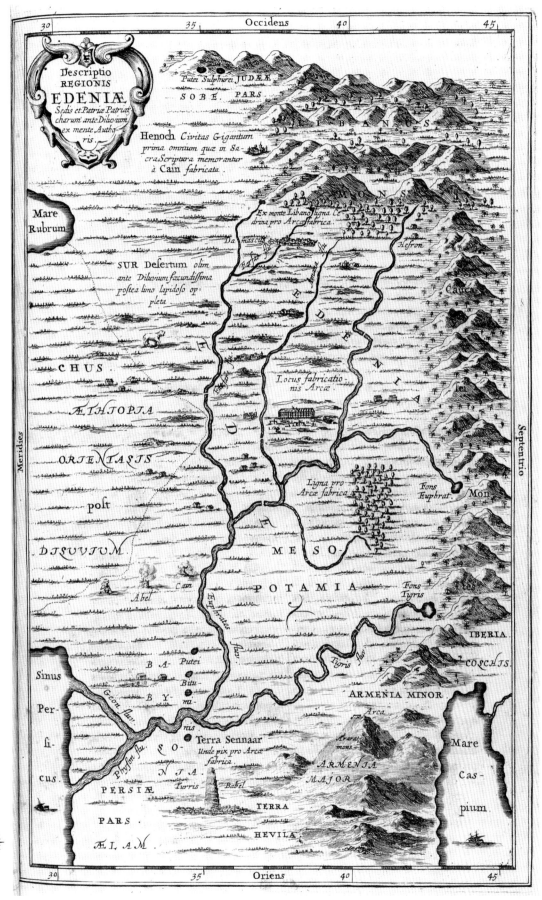

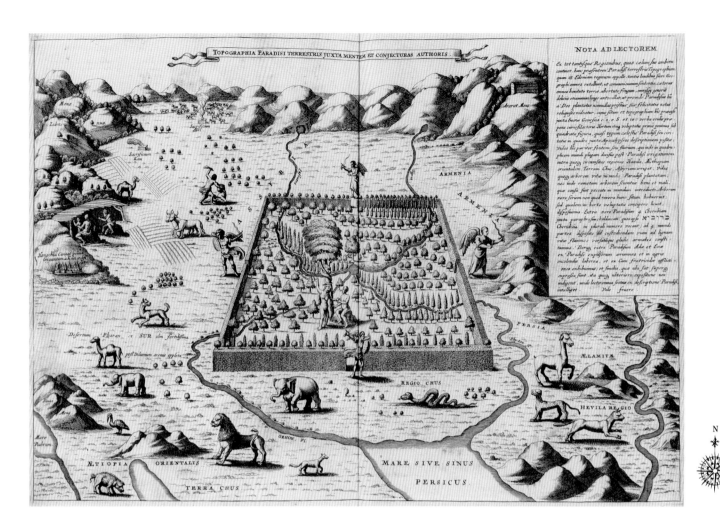

Fig. 10.22. Athanasius Kircher's map *Topographia paradisi terrestris iuxta mentem et coniecturas authoris*, from his *Arca Noe* (1675), between pp. 196 and 197. 28.5 × 43 cm. London, British Library, 460.c.9. Kircher's map of the topography of the earthly paradise shows both the original hydrography of Eden (with an antediluvian spring welling up in the middle of paradise), and the postdiluvian courses of the Tigris and the Euphrates. In a lengthy note in a cartouche Kircher praised the beauty and fertility of Eden. He also pointed out that the square shape of the Garden of Eden evoked the Heavenly Jerusalem as described in the Book of Revelation and served as a reminder that the earthly paradise had been crafted by God as a pre-figuration of the heavenly paradise. His observation points to the continuing importance of typological association in Christian theological thinking.

Assyria. After the Fall, however, God caused the spring to be blocked up so that its waters were forced back into the Armenian reservoir, from which they resurfaced in different parts of Armenia to give rise to the present river network. After this post-lapsarian change, Kircher explained, the single source in paradise disappeared and nobody had ever been able to see it.

The most striking feature of the map, however, is the prominence of a plan of the Garden of Eden. Apart from Regnault in 1573 (see above, Figure 10.4), paradise was normally marked on maps (if at all) with a pictorial place sign or vignette. On Kircher's map, by contrast, the Garden is prominently portrayed in bird's-eye perspective in the centre of the map. The square Garden is surrounded by what appears to be a thick and luxuriant hedge and is aligned to the cardinal points in such a way that an entrance on each side faces north, east, south and west respectively. Outside each entrance is a guardian angel brandishing a flaming sword.[79] The landscape within the Garden is shown in detail, its wooded nature contrasting with the openness of most of the surrounding region of Eden, even though Kircher admitted that such a visual rendering of the interior of the Garden could only be hypothetical.[80] The central point of the Garden is marked by a spring, and near it is the Tree of Life. In the south of the Garden is the Tree of the Knowledge of Good and Evil, with the serpent coiled around its trunk and Adam and Eve beside it.

Kircher's map of the topography of the terrestrial paradise offers an arresting comment on the fate of mankind. The Garden's original state of perfect harmony is implied in the orderly arrangement of the trees, the Fall is represented in the portrayal

316   Mapping Paradise: A History of Heaven on Earth

of Adam in the act of taking the forbidden fruit Eve has just plucked from the Tree of Knowledge, and the post-lapsarian situation is marked by the four Cherubim who are already in position in front of each gate to prevent mankind from re-entering the Garden. Outside the Garden are further indications of the consequences of the Fall. The Tigris and the Euphrates flow from their Armenian sources, and a serpent in the foreground crawls on its belly (Genesis 3.14). The rest of the map portrays the region already afflicted with the first miseries of human history: the struggle to provide shelter from the elements, the toiling at the plough, the drudgery of shepherding, the enmity of man to man, and, finally, the need – unknown in paradise – to earn God's friendship through sacrifice. Several historical moments are being portrayed in a single vignette. Whether the animals that also populate the landscape were intended by Kircher to represent those of the Ark, or those given names by Adam, Kircher's message of loss in this particular map is unambiguous: the earthly paradise has vanished from Mesopotamia.

Return to the Sources: Paradise in Armenia

Some exegetes found it easier to abandon Steuchus's and Calvin's contention of a paradisiacal single river in Mesopotamia and to search for paradise in Armenia, a region which in the sixteenth century included the area between the upper Euphrates and Lake Urmia, the Black Sea and the Syrian desert. The biblical Eden was located in a region defined with a classical place name. Sound reasons could be found for this change of focus, not the least of which was that the identity of two of the four rivers named in Genesis, the Tigris and the Euphrates, was uncontroversial, and both rivers were known to rise in Armenia. Once scholarly attention had shifted northwards, the reasoning went, the more problematical Gihon and Pishon were sure to be identified amongst the local rivers. Some of those who looked to Armenia wanted to avoid departing too far from Calvin's insistence on the survival of the original river network and tried to adhere to the notion of a single river. Others would have liked to return to the less complicated traditional idea of four rivers parting from a single source, but since no evidence had been found of such a source, they had to follow Luther's suggestion that it had been wiped out by the Flood.

Allusions to a possible Armenian location for paradise had already been made in the sixteenth century. In 1544, for example, Fausto da Longiano observed in his vernacular translation of Aeneas Sylvius Piccolomini's geographical work, the *Cosmographia*, that the Armenian rivers Araxes and Cyrus could be identified with the Pishon and the Gihon.[81] From 1601 on, editions of Ortelius's *Theatrum* included a map of the central part of the Old World, entitled *Geographia sacra* and dated 1598 (the year of his death), which Ortelius had created for the *Parergon*. In the accompanying text, Ortelius explained that the map was for the benefit of scholarly theologians who wanted to know in which part of the world sacred history had taken place and to locate the site of the earthly paradise (Plate 16).[82] Ortelius went on to claim he had done his best: 'we offer what we could do, since we could not do what we wished to do.'[83] Saying that he had followed the Septuagint Bible, Ortelius acknowledged that there were various opinions on the location of paradise, and referred to Hopkinson's map amongst others. Ortelius's map is as prudent and syncretic as his text. The name Eden appears in Armenia, written across the upper Euphrates, but paradise itself is not marked. In fact, the Gihon is identified with the Nile, an idea rejected by Aeneas Sylvius Piccolomini and most of the later Renaissance exegetes, and incompatible with an Armenian paradise. Moreover, on his map Ortelius managed to avoid taking a stand on the identity of the river Pishon, giving the name to three rivers: the Danube (Gregory Nazianzen's idea), the Hydaspis in

Fig. 10.23. Marmaduke Carver's map of paradise, from his *A Discourse of the Terrestrial Paradise Aiming at a More Probable Discovery of the Trye Situation of that Happy Place of our First Parents Habitation* (London: James Flesher, 1666), after p. 167. 17 × 10 cm. London, British Library, 4375.aa.12. The source of the single river is shown on the map as deep in the Armenian mountains, on the southern flank of Mount Taurus and within the region of Sophane, a place celebrated in ancient sources, such as Strabo, Tacitus and Pliny, as blessed in its temperate climate and exceptional fertility. Scholars might disagree as to the exact site of paradise, Carver said, but their debates did not mean that the earthly paradise had never existed.

Mesopotamia (Nicephorus Callistus's idea), and the Ganges, in connection with which he felt he had to apologize that the Ganges ran too far to the east to be included on the map.[84] The land of Havilah that was supposed, according to Genesis, to be encompassed by the Pishon is marked by a single name (*Evilath*) placed north of Mount Sinai.

Some seventeenth-century writers and map makers were more assertive in their identification of the earthly paradise in Armenia. Marmaduke Carver, a Yorkshire parson, devoted a short treatise to the 'not-unnecessary' question of the whereabouts of paradise.[85] In the prefatory 'Epistle' addressed to Gilbert, Lord Archbishop of Canterbury, Carver claimed that his objective was to defend the truth of the biblical description against the blasphemies of the 'Heathenish Infidels' and 'the more Heathenish Christians of these later times' who had the 'superlative insolence', as he put it, to 'scorn and deride' the paradise narrative 'as a mere *Utopia*, or Fiction of a place that never was, to the manifest and designed undermining of the Authority and Veracity of the Holy Text'.[86] Thomas Gale too, Huet's English translator, had already railed against the same intellectual development: to support the veracity of Scripture was the crucial motivation behind the post-Renaissance impulse to map paradise.[87] Dismissing previous allegorical fantasies and improbable locations, Carver praised the works of

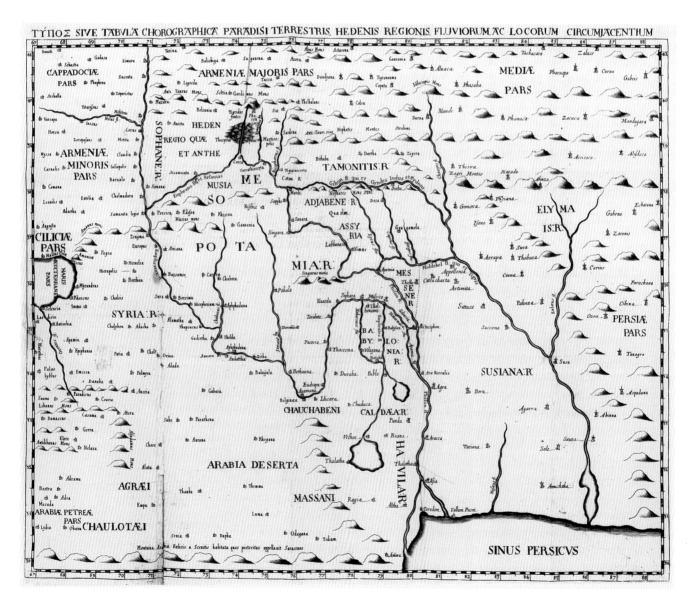

318   Mapping Paradise: A History of Heaven on Earth

Steuchus – the first, in his view, to locate paradise according to geographical truth – and of Iunius, Raleigh and Bochart, Carver reserving particular approval for the last for the way he had clarified and advanced the issue. Carver's own line of argument, though, was different. His idea was that the most likely place for paradise would have been in the region that included the sources of the Tigris and Euphrates. Carver described the wealth of such a region, which the ancients called *Anthemusia* and which he saw as corresponding to Eden, its precious stones and gold mines, and the many different species of trees, bearing rare and delicious fruits, and the medicinal herbs that flourished in the fertile soil. He noted, too, that relics of Eden and paradise survived in the place names recorded by the ancient writers. Ptolemy had mentioned *Bagrandavene*, a compound word that Carver saw as embedding the element *Adena*, i.e., Eden. Moreover, Pliny had called the place where the Tigris rises *Nymphaeum*, a term that Carver saw as the Latin equivalent for the biblical word 'paradise', and Strabo had written about *Syspereitis* and *Hysperatis*, names derived from *ipse paradisus*, 'that very paradise'.[88] Carver's map, *Τύπος sive tabula chorographica paradisi terrestris, Hedenis regionis, fluviorum, ac locorum circumiacentium* (*Type or Chorographical Map of the Earthly Paradise, the Region of Eden, the Rivers and the Surrounding Places*), was based on Ptolemy, but, although Ptolemy, like Strabo, had described both rivers as coming from separate springs, Carver preferred to lean on the authority of other classical writers, such as Sallust, Lucan, Boethius and Isidore, who had suggested a common source for the two rivers.[89] Accordingly, Carver's map shows what he explained in the text: that the single river of paradise rose in Greater Armenia, watered the Garden, and then divided into four streams (Figure 10.23). The rivers had retained their original courses despite the Flood, and had kept their biblical names, two issues on which Carver agreed with Calvin.

As the map shows, Carver's Euphrates and Tigris flow as a single stream before diverging westward and eastward respectively. The Euphrates borders on the regions of Mesopotamia, Syria and Babylonia before rejoining the Tigris. After branching eastward from the Euphrates, the Tigris passes through the Gordiaean mountains. As it enters Assyria, it divides to form a new branch, the Gihon. In the text, Carver tells his readers that the river Gihon flows through a region that corresponds to the biblical Cush.[90] A second bifurcation occurs near Apamea, where the Hiddekel branches off from the Tigris to flow into eastern Assyria. After this division, the Tigris, which is now called the Pishon, joins with the Euphrates and continues its course towards the land of Havilah, marked on the map west of Susiana. This arrangement of rivers allowed Carver to point to the location of the Garden of Eden with reasonable precision:

> *As for the Limits of Eden, I think it lies not in the wit of any man at this day to set us out punctually and exactly how large or narrow the compass of that Countrey was in Moses's Chorography: yet seeing himself hath told us that the Spring of this River was in Eden, he hath left us assured that it was either the same, or at least a part of that Countrey which Secular Geographers call Sophane, lying betwixt the Mountains Masius and Anti-Taurus.*[91]

The Garden of Eden, Carver concluded, lay between the source and the first forking of the single river of paradise in the neighbourhood of Corra or Charan. He had considered the possibility that God had turned the Garden of Eden into a sulphurous lake (as with the Dead Sea in the case of Sodom and Gomorrah) and that this lake was the nitrous Lake Thospites that Ptolemy showed on his map of the area, also marked on Carver's map.[92] In the end, however, Carver admitted that all these ideas about the location of Eden needed refinement and improvement by scholars more skilled than he was and who had more time at their disposition.[93]

Fig. 10.24. Augustin Calmet, *Carte du paradis terrestre*, from his *Commentaire littéral sur tous les livres de l'Ancien et du Nouveau Testament*, 2nd edn, 8 vols (Paris: [n. pub.] 1724–6), I/1 (1724). 42 × 55.5 cm. London, British Library, 9.h.l. Calmet's map of the earthly paradise was published as an inset on a double-folio map of part of the Old World that showed the lands inhabited by the sons of Noah. Calmet transferred Eden from Mesopotamia to Armenia by abandoning Calvin's idea that the course of the rivers remained intact after the Flood. In his view, the Flood and other natural disasters threw the single source of paradise into disarray, but the four rivers still rose not far from each other in Armenia. The map shows the original four rivers flowing out to the four cardinal points from the area around the lost single source: the Pishon (later called Phasis) northwards, the Tigris southwards, the Gihon (later called Araxes) eastwards, and the Euphrates westwards.

Fig. 10.25 (opposite). Pierre Moullart-Sanson, *Carte du paradis terrestre selon Moyse* (Paris: prope Aedes Regias, 1724). 18 × 28.5 cm. London, British Library, Maps K Top 3.47. In drawing his map of paradise 'according to Moses', the geographer Pierre Moullart-Sanson adopted Calmet's views and located the Garden of Eden, indicated by a tiny vignette of the Fall within a circular ring of trees, in Armenia. Moullart-Sanson's map was based on Calmet's (Fig. 10.24) and like Calmet's it does not show the common source of the four rivers (believed to have been destroyed by the Flood), even though the inscription on the right claims that the *tabula geographica* refers to the first age of the world, comprising the 1656 years before the Flood. On the left is an abridged version of the biblical account of the Garden of Eden, and a list of meanings and names for the river Tigris. The map also pinpoints the place where Moses wrote the Book of Genesis, in the Arabian desert, 2,551 years after the creation of the world.

Sixty years later, the distinguished French exegete Augustin Calmet was persuaded to persist with the search for the Garden of Eden despite all the different opinions and all those scholars who found paradise almost everywhere. He remarked that:

> *It would be better to imitate the restraint of the ancient Fathers on this issue. But it may also be that somebody could draw from this silence the wrong conclusions against the truth of Scripture. It is necessary to show more or less the location of the earthly paradise, by means of the remnants which survived for a long time, many of which still survive today, in the land where paradise was located.*[94]

Like his map, *Carte du paradis terrestre* (1724), Calmet's arguments for an Armenian paradise differed from Carver's (Figures 10.24 and 10.25).[95] Instead of worrying about an extant single stream, Calmet preferred the old and relatively straightforward notion of four existing rivers originally emanating from a single source subsequently obliterated by the Flood and other natural disasters, and looked for four rivers rising close together in Armenia. He justified seeking them in Armenia rather than in Mesopotamia by referring to Bochart and Huet, whom he admired for their learned and elegant theories, but who he considered had overstretched the meaning of the biblical text, which he saw as pointing to a single source within paradise for the four rivers. He insisted that as the Tigris and the Euphrates both rose in Armenia, there was no point looking for paradise in Mesopotamia, and that there was nothing in either the Bible or the ancient authorities to confirm that Havilah was near the Persian Gulf. Moreover, as Pliny had stated, the union of the Tigris and the Euphrates in Mesopotamia had occurred later in history.[96] In favour of Armenia, Calmet leant on the facts that Noah's Ark had landed there, that the region was well watered by rivers and had been praised for its fertility by the

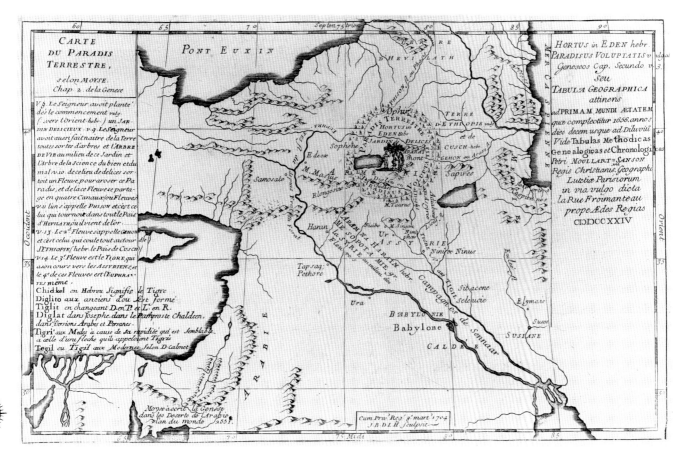

The Afterlife of Paradise on Maps 321

ancient authorities, and that many Armenian place names included an element derived from the name 'Eden', notably Aphadena, Atheanae and Adienum, and the ancient classical names for the region, Adonis or Madena.[97]

Calmet's commentary on the rivers was detailed. He abandoned Calvin's idea that their courses had remained intact after the Flood. True, the Flood had deprived paradise of its original beauty and most of its internal features, but its site survived.[98] Instead of a common source described by ancient writers, in Calmet's time the Tigris and the Euphrates sprang from different places within Armenia.[99] The sources of the other two rivers of paradise were also in Armenia: the Pishon corresponded to the river Phasis (which flowed into the Black Sea) and Gihon to the river Araxes (which flowed into the Caspian Sea). In support of his identification of the Armenian Phasis with the biblical Pishon, Calmet cited their circuitous courses. The biblical Pishon was described as having encircled the land of Havilah (*qui circuit omnem terram Evilat*), and the Phasis meandered so much that the ancients had had to construct a host of bridges over the river.[100] Both meanders and bridges are shown on Calmet's map. Calmet also cited Colchis, the part of Armenia that the ancients reported as having abundant gold and in which the Greek hero Jason had sought the Golden Fleece.[101] Yet again, Calmet thought he could discern the biblical name 'Havilah' in local place names such as those of the city of Cholva and the land of Haloen. There was also etymological evidence in Armenia for the Gihon. Calmet pointed to the Hebrew meaning of the term גִּיחוֹן (*Gihon*), which, he said, was suggestive of the river's impetuosity, and which suited perfectly the turbulence of the Araxes as it rushed down from its source on Mount Ararat. Along its course were further reminders of the biblical Gihon's watering of Cush, such as Cyta and the Cutheeans.[102]

Back to the Future: Paradise in the Holy Land

Convinced by his argument for an Armenian paradise, Calmet was critical of those who insisted that paradise had been in the Holy Land and dismissed their unlikely and audacious hypotheses as merely an excuse to show off their learning.[103] In fact, given the overall focus from the Renaissance onwards on the Middle East as the location of the earthly paradise, many commentators found the temptation of making a case for a Holy Land paradise irresistible, not least because such a location carried the power of Christian salvation history. Situating Eden in Palestine meant identifying Adam's paradise not only with the Promised Land of the Old Testament, but also with the place where the Son of God (the Second Adam) had redeemed mankind from the consequences of Adam's Fall. Linking the Garden of Eden and the land of the Crucifixion geographically, and cartographically, was a new way of expressing the old typological association between Adam and Christ; once again, everything had to change to ensure that nothing had changed.

In the history of the mapping of the earthly paradise, the Holy Land has always occupied a significant position, but this became even more apparent when some Renaissance map makers began to map the Garden of Eden in the land of Canaan. Whereas in the case of Mesopotamia or Armenia it was little more than a question of coaxing a fit between modern geography and the geography described in Holy Scripture, the identification of the earthly paradise with the Holy Land was an expression of a deeply felt spirituality that permitted a radical bending of geography in the interest of theology. Those who advocated a Holy Land paradise wanted to demonstrate that the names 'Eden', 'Promised Land' and 'Canaan' referred to one and the same region. Instead of tinkering with an established model such as Calvin's, the chief proponents of the paradise in the Holy Land paradise model – notably Jacques d'Auzoles

Lapeyre, Iohannes Herbinius and Jean Hardouin – created their own, highly personal routes towards the common objective. Their very different maps include some of the most ingenious cartographical layouts in connection with the location of the earthly paradise.

Although the maps representing the location of Eden in the Holy Land date from the seventeenth and eighteenth centuries, the underlying typological concept went back to the Middle Ages. The first to rearticulate the concept in terms of sixteenth-century geography was the Spanish humanist and theologian Michael Servetus. Servetus had produced the edition of Ptolemy's *Geography* that Calvin may have used for his map of Eden.[104] He also sent Calvin a preliminary draft of his *Christianismi restitutio*, a theological treatise in which Servetus claimed that God had planted paradise in Palestine, an idea that failed to meet with Calvin's approval. On the contrary, it is not too far-fetched to suggest that Calvin's urge to map Eden in Mesopotamia was a deliberate confutation of Servetus's ideas. Calvin certainly disliked much in Servetus's writings, not least Servetus's comment that contemporary Palestine was a desolate and desert land and his insistence on the spiritual rather than the material dimension of paradise.[105] To refute Servetus, Calvin needed to be able to point to a tangible testimony of the former earthly paradise.

Servetus's aim in the *Christianismi restitutio* was literally to bring about 'the restoration of Christianity' and the teachings of Christ as recorded in Holy Scripture. Calvin found the work heretical in its interpretation of the Trinity and its opposition to infant baptism. After several exchanges of letters, Calvin cut all communication with Servetus and become his implacable enemy. Servetus's *Christianismi restitutio* was secretely printed in 1553.[106] In the same year, at Calvin's prompting, Servetus was condemned by the Genevan authorities for heresy and burned at the stake. One of the charges levelled against Servetus was his blasphemous description of the Holy Land as territory *not* flowing in 'milk and honey', as described in the Bible.[107] But Calvin had misunderstood Servetus. What Servetus had wanted to emphasize – as had many Old Testament prophets – was that the current desolation and barrenness of the Holy Land was the direct result of human sin. He explained that God had planted Adam's paradise in the same land where his Son was later born, a land that was associated in the Bible with the Garden of Eden (Joel 2.3; Isaiah 51.3; Ezekiel 36), but a land that after the Fall had lost most of its beauty and was promised to Abraham not as a place of tangible well-being but as 'a shadow of paradise', meaning in prefiguration of the true spiritual paradise given by Christ.[108] Christ's redemption had transformed the squalid physical landscape into a wholly spiritual environment by virtue of the fact that it was from that place that access to heaven was granted. Likewise, the Tree of Life and the Tree of the Knowledge of Good and Evil growing in the middle of the Garden of Eden were prefigurations of Christ's Incarnation and his twofold gift to mankind of body and spirit. According to Servetus, the Holy Land was geographically unique. Situated in the middle of the inhabitable earth, it was accessible from all other lands, and all the seas of the world went towards it. The Fall and the Flood may have destroyed the single source, but the four rivers of paradise still flowed in the region. Like the destruction of the Temple later, the destruction of paradise was to encourage mankind to repentance. For Servetus, the new Christian paradise, a token of the eternal one, was present now, within the human heart, and Christ was the single source of the four rivers watering the whole world. Whereas for Adam, paradise had been a material garden, and for the Israelites a corporeal paradise ruled by Law, for Christians paradise was the spiritual reality of eternal life in Christ.[109]

By the end of the sixteenth century, the idea of an earthly paradise in the Holy Land seems to have gained wide currency. In Germany, Heinrich Bünting adopted Luther's approach to the location of Eden in his *Itinerarium Sacrae Scripturae* (1585),

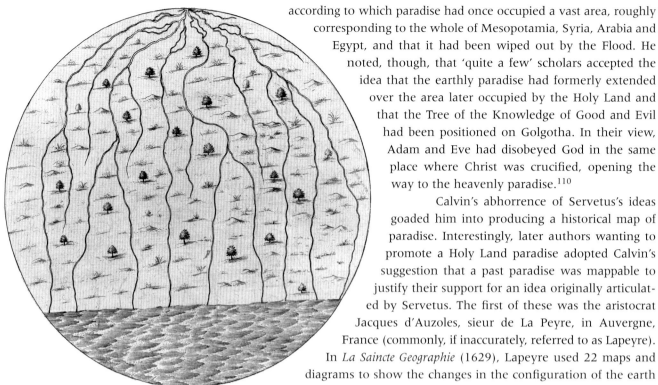

Fig. 10.26. Jacques d'Auzoles Lapeyre, *Premiere forme de la terre et des eaux après leur distinction*, from his *La Saincte Géographie, c'est-à-dire, exacte description de la terre et veritable demonstration du paradis terrestre* (Paris: Antoine Estiene, 1629), p. 60. 19 × 22 cm. London, British Library, 744.d.13. Lapeyre claimed that, although Moses was writing for a Jewish unsophisticated audience and accordingly abstained from expounding profound theological mysteries, what he narrated about Eden was to be taken as true history, and that to understand the situation of paradise, it was important to acknowledge that the configuration of the earth before the Flood was very different from its postdiluvian condition. A series of maps in his book illustrates the terrestrial globe at various stages of its creation. The entire earth was originally covered in water, but the map representing the situation on the third day shows a great change, for God had gathered the waters into a single place, a seventh of the globe's surface (an idea inspired by the apocryphal Book of Ezra). After the third day, the other six parts were dry and ready for cultivation.

according to which paradise had once occupied a vast area, roughly corresponding to the whole of Mesopotamia, Syria, Arabia and Egypt, and that it had been wiped out by the Flood. He noted, though, that 'quite a few' scholars accepted the idea that the earthly paradise had formerly extended over the area later occupied by the Holy Land and that the Tree of the Knowledge of Good and Evil had been positioned on Golgotha. In their view, Adam and Eve had disobeyed God in the same place where Christ was crucified, opening the way to the heavenly paradise.[110]

Calvin's abhorrence of Servetus's ideas goaded him into producing a historical map of paradise. Interestingly, later authors wanting to promote a Holy Land paradise adopted Calvin's suggestion that a past paradise was mappable to justify their support for an idea originally articulated by Servetus. The first of these was the aristocrat Jacques d'Auzoles, sieur de La Peyre, in Auvergne, France (commonly, if inaccurately, referred to as Lapeyre). In *La Saincte Geographie* (1629), Lapeyre used 22 maps and diagrams to show the changes in the configuration of the earth since the creation and to argue that the Garden of Eden was in Palestine. His unusual and remarkable theory allowed for both the disappearance of paradise through the profound changes caused by the Flood and its survival in an invisible form discernible only by divine grace.[111] His argument was that God had selected one particular place on earth as the site of the most important events of salvation history: Eden was the land of Adam's natural perfection and Fall, and was also the very same land inhabited by Noah after the Flood, the land promised to Abraham and celebrated by the prophets, and, finally, the land of Christ. After the Tower of Babel, and the confusion of tongues, Eden was called by different names – Canaan, Palestine, Judaea, Promised Land, Holy Land – but it remained the same place.

The identification of Eden with the land of Canaan highlighted for Lapeyre the typological association between Adam and Christ. Paradise, he explained, had been planted in the eastern part of Eden, in the area now corresponding to Galilee. Adam had been created in the Damascene Fields (the western part of Eden, outside paradise) and then put by God into the Garden of Eden, where he stayed with Eve in a state of innocence and perfection for nine months.[112] He had been created in the spring, the same season that Christ, the Second Adam, had been conceived by the Virgin Mary. He was expelled from paradise on the same day as Christ's birth, 25 December, and then lived with Eve in Judaea, outside paradise but still within the wider region of Eden. He was buried at the site of the future Crucifixion. In parallel, the Second Adam was miraculously conceived at Nazareth, in the eastern part of Eden (that is, within paradise), where he stayed for nine months in the Virgin's womb, his own earthly paradise. In accepting the entry of the Holy Spirit into her body, the Virgin had crushed the head of the serpent in the same place as Eve had succumbed to the serpent's temptation, thus fulfilling God's prophecy spoken at the moment of the Expulsion (Genesis 3.15). Christ was born in Bethlehem and sacrificed in Jerusalem, both places situated within Eden but outside paradise.[113]

Lapeyre called his work a 'sacred geography' in accordance with his claim that he had drawn his arguments from Scripture and the doctrine of the Church, which alone constituted 'the firm pillar of truth'.[114] The Bible, which, he said, contained the

324   Mapping Paradise: A History of Heaven on Earth

key to all sciences, was 'an endless treasury, not subject to any rust', a book that must be read with humility and reverence, while human geographical theories were destined to become obsolete and to be contradicted by subsequent ideas.[115] Ancient geographers did not know about the New World, and their theories about the inhabitability of the globe now sounded untenable.[116] Secular geographers taught nothing that was not already recorded in the Bible. The historian who wrote without the support of sacred geography and sacred history was like a blind man discussing colour. Likewise, mapping paradise required adherence to every letter in Scripture:

> *Once our geographical maps have been drawn in this way to show that it is not necessary to search for the earthly paradise outside the earth, it is now necessary to consider the sacred text exactly, and in this way to mark the true place of paradise on earth as far as we are able.*[117]

The key factor in Lapeyre's exposition of the location of paradise was the change wrought on the earth by the Flood. A series of maps in his treatise illustrated the terrestrial globe both at its creation and at its re-creation by the Flood. For the first two days the entire earth was covered in water. One map, the *Premiere forme de la terre et des eaux après leur distinction*, represented the earth on the third day of Creation and showed the first great change in its configuration, when God gathered the waters in a single place (Figure 10.26).[118] The single fountain mentioned in Genesis 2.6, which, in Lapeyre's view, irrigated the entire earth and removed the need for rain, is shown at the top of the map. Another map, *La Saincte Geographie depuis lorigine du monde iusques au deluge*, showed exactly the same antediluvian earth, with the single continent, the single sea and the single source, but this time projected as a rectangle over which is superimposed a grid of latitude and longitude (Figure 10.27a).[119] The tropics and the equator are marked, as are the four rivers and paradise. In his text, Lapeyre described the rivers as flowing over the surface of the land, underground, and across mountain chains. On his map, these distinctions are not made, and the courses of the rivers delineate, for part of their course, the continents as they appeared after the Flood. The coordinates allowed Lapeyre to locate Eden (the Holy Land) exactly, and its site is marked by a tiny rectangle at about 70° E and 35° N (no. 1 on the map).

Lapeyre was nothing if not systematic. From this global overview he proceeded to present, map by map, a series of increasingly focused views until he could show a close-up of the interior of paradise. Yet another of Lapeyre's maps portrayed the antediluvian region of Eden, with paradise within it (Figure 10.28a). The untitled map is described in the text as a map of the region of Eden;[120] it is in a square format with graduations of latitude and longitude around the sides. Eden is described as another square, containing a lozenge-shaped paradise. The western extremity of the Lebanon range falls within paradise and from this the single river springs before following a circuitous route through the Garden to divide into four just before emerging from paradise to flow out first over Eden and then over the rest of the world. The last map in this sequence, entitled *Le Paradis terrestre ou le iardin de Eden ou le iardin des delices*, showed paradise itself (Figure 10.29a).[121] The way in which Lapeyre described the internal configuration of Eden is reminiscent of Kircher's rendering (Figure 10.22), except that Lapeyre showed paradise as a lozenge-shaped garden rather than a square, and his conception of paradise's hydrography was different from Kircher's. Lapeyre's Garden of Eden extended between the cities of Enoch and Cain (the latter identified as near Mount Carmel) that had been marked on the previous map. The single river, rising from Mount Lebanon in the north-eastern corner, irrigates the Garden through a network of right-angled channels that produce a striking chequerboard pattern. The 24 squares, Lapeyre explained, symbolize the whole of mankind; twelve squares stand for the Twelve Tribes of Israel,

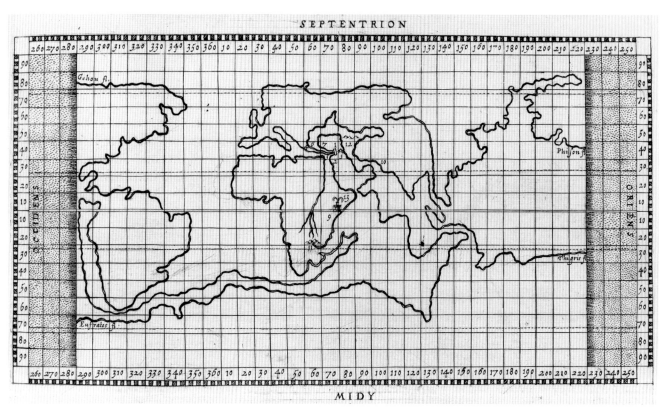

Fig. 10.27a. Jacques d'Auzoles Lapeyre, *La Saincte Geographie depuis lorigine du monde iusques au deluge,* from his *La Saincte Géographie, c'est-à-dire, exacte description de la terre et veritable demonstration du paradis terrestre* (Paris: Antoine Estiene, 1629), p. 89. 14 × 22 cm. London, British Library, 744.d.13. The earth is depicted here within a rectangular format as a single landmass surrounded by a single sea with latitudes and longitudes. Eden is marked by a tiny rectangle at about 70° E and 35° N (no. 1 on the map). The place of Adam's creation, the Damascene Fields, is marked at 32° N and 66° E (no. 2). The four rivers (nos 4, 5, 6, 7) have their sources in Eden and flow after long and devious courses into the single sea. Havilah is identified in the text as no. 8 and described as corresponding to the Moluccas, although the number is missing from the map. Cush (no. 9) is in Ethiopia, and Assyria is marked as no. 10. The map also features Mount Lebanon (no. 3), the Mountains of the Moon (crossed by the Gihon and marked as no. 11), and Mount Ararat (no. 12). The single source (no. 13) is shown at the centre of the map, at the equator.

and the other twelve for the twelve apostles of the Christian Church. The place where the river separates into four is labelled as the site of the well of living waters, mentioned in the Song of Solomon and described as originating in Mount Lebanon (Song of Songs 4.15). A scale bar for appreciating the extent and dimensions of the earthly paradise bears witness to Lapeyre's preoccupation with measurement.

Lapeyre provided his readership with a cartographical analysis of the earth before and after the Flood. He reckoned that the Flood lasted for one year and ten days, 1665 years after the Creation, and changed the face of the earth. He complemented his rectangular map of the antediluvian earth with a similarly structured map of the postdiluvian earth, *La Saincte Geographie depuis le deluge iusques à la confusion des Langues qui advint lan du monde deux mille quinse, qui à eu de duree 350 ans* (Figure 10.27b).[122] It is immediately clear that the Flood had had a huge impact on the configuration and landscape of the earth. Instead of a single continent, there are now four (those of the Old World and the Americas), together with the polar masses and numerous islands. Instead of a single sea surrounding the terrestrial landmass, there are now several seas. The Flood also affected the courses of the four rivers of paradise. The destruction of the fountain meant that the world has had henceforth to rely on rain for its surface water. For Lapeyre, the map of the postdiluvian world was an image of the corruption of nature as a consequence of human sin. The fragmenting of the single primordial continent brought about the differentiation of the world, creating distance and separation.[123] After the Flood, paradise disappeared from human sight. When Noah disembarked from the Ark and settled in Eden, he knew that paradise had once been there, but was unable to see it.[124] Finally, Lapeyre drew a parallel between the four rivers that watered the world and the four (Noah and his three sons) who had peopled it.

Lapeyre went on to consider the postdiluvian region of Eden and then the Garden of Eden itself. The territory covered in the *Carte ou table particuliere de la region d'Eden apres le deluge* is identical to that that existed before the Flood (Figure 10.28b).[125]

326   Mapping Paradise: A History of Heaven on Earth

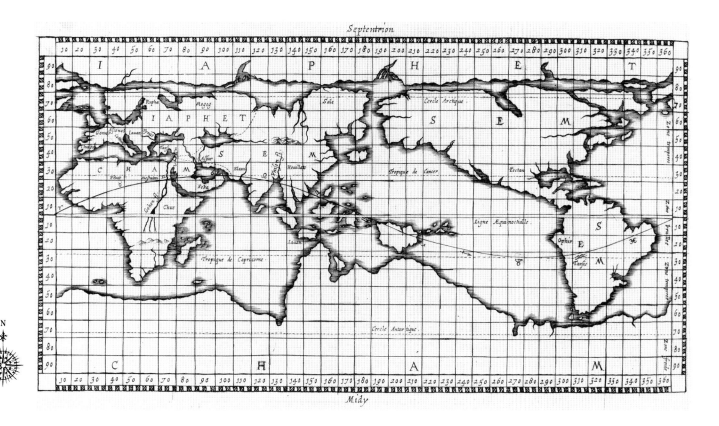

From southern Anatolia to the delta of the Nile, the outline of the Mediterranean coast ghosts the course of the antediluvian rivers Euphrates and Gihon. The Taurus and the Lebanon ranges now terminate in the sea. South of the Lebanon range, the area of a lozenge-shaped paradise is indicated by a broken line. As on the map describing the lands surrounding paradise before the Flood, the region of Eden is contained within a square, but half of this is now occupied by the sea. Taken together, the two maps (ante- and postdiluvian Eden) reveal the drama of the Flood. Once the north-western half of the region has been submerged, the postdiluvian paradise appears to be located along the Mediterranean shore of the land of Canaan. There is no sign of the four rivers of paradise that had featured so prominently on the earlier map. Instead of the single river that had flowed tranquilly through paradise, the map now shows a river in the north-eastern corner of Canaan. The Flood has covered the waters of the four rivers in many other places of the earth, now submerged. The original source of the Tigris, the Euphrates, the Gihon and the Pishon was in the land of Eden, but after the Flood the four rivers sprang from different places far from each other.[126] The Flood also changed the course of the single river that watered paradise, but in the north-east of Canaan, between the cities of Tyre and Sarepta, Lapeyre marked an abundant stream flowing in haste to the sea, 'as if', he said, 'it found it distasteful to see this holy place inhabited by unworthy sinners'.[127] Before the Flood, the single river had divided into four at the site of what is now the city of Tyre, shown with four springs therein, as if echoing the prophet Ezekiel's description of the king of Tyre being in Eden, in the garden of God (Ezekiel 28.13), and in accordance with the geographers' practice of putting the well of living waters, mentioned in the Song of Solomon (Song of Songs 4.15), near Tyre.[128]

For his cartographical close-up of the postdiluvian Garden of Eden, Lapeyre studied a number of maps of the Holy Land and other parts of the world, including those of Breydenbach, Brocardus and Ortelius. In the end, he decided that of these by

Fig. 10.27b. Jacques d'Auzoles Lapeyre, *La Saincte Geographie depuis le deluge iusques à la confusion des Langues ...* , from his *La Saincte Géographie, c'est-à-dire, exacte description de la terre et veritable demonstration du paradis terrestre* (Paris: Antoine Estiene, 1629), p. 134. 14 × 22 cm. London, British Library, 744.d.13. The Flood altered the earth's configuration. From the original one-sea and one-continent globe were formed many continents, seas and islands. Lapeyre's map shows how the postdiluvian earth was divided among Noah's sons: to Sem were assigned the eastern lands, old and new Indies, to Cam the southern lands, Africa and Magellanica, and to Japhet, the western and northern lands. Whereas the map of the world before the Flood was centred on the meridian passing through the fountain of all waters and close to paradise, that is to say, on the Middle East and on eastern Africa, the postdiluvian map is centred on the line of longitude 180° E – exactly opposite the prime meridian – so that it presents an unbroken view of the whole world from the western coasts of Africa and Europe to the eastern coasts of the New World.

The Afterlife of Paradise on Maps   327

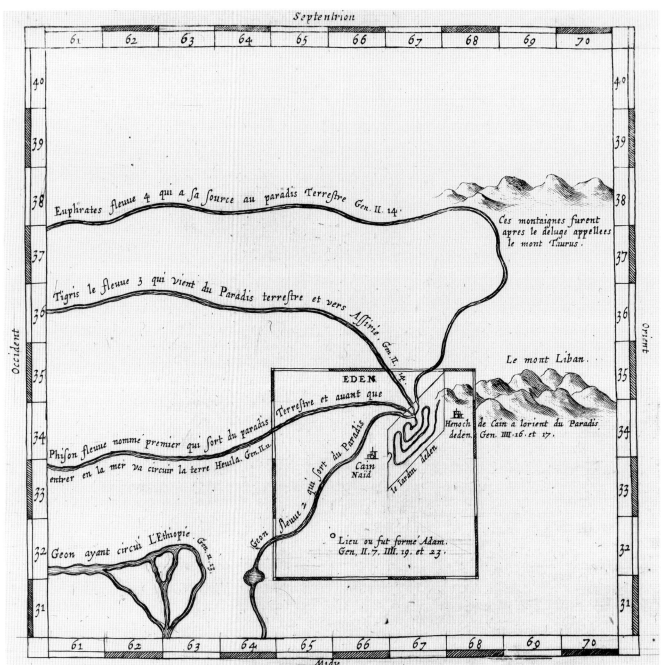

Fig. 10.28a. Jacques d'Auzoles Lapeyre, the antediluvian region of Eden, from his *La Saincte Géographie, c'est-à-dire, exacte description de la terre et veritable demonstration du paradis terrestre* (Paris: Antoine Estiene, 1629), p. 91. 18 × 18 cm. London, British Library, 744.d.13. The Ptolemaic grid allowed Lapeyre to position and measure paradise precisely and to translate his measurements into German, French and Italian leagues. The region of Eden is shown extending from 32° N to 35° N, and from 65° E to 68° E. Each side of Eden was said by Lapeyre to measure 3.5 degrees. Given that each degree corresponded to 15 German leagues, each side of the square accordingly represented 37.5 German leagues (one German league corresponding to 7.4 km). The two short sides of paradise (which occupied the best part of Eden) represented 15 German leagues (or 1 degree each) and the two long sides 22.5 German leagues (1.5 degrees each). The source of the single river was placed at 34° N and 67° E. The four rivers are seen flowing out of paradise and Eden to water the world. The mountains that would become the Taurus range are portrayed in the north. The two cities of Enoch and Cain (where Cain lived) flank paradise to east and west, respectively. The place of Adam's creation is shown in the southern part of Eden.

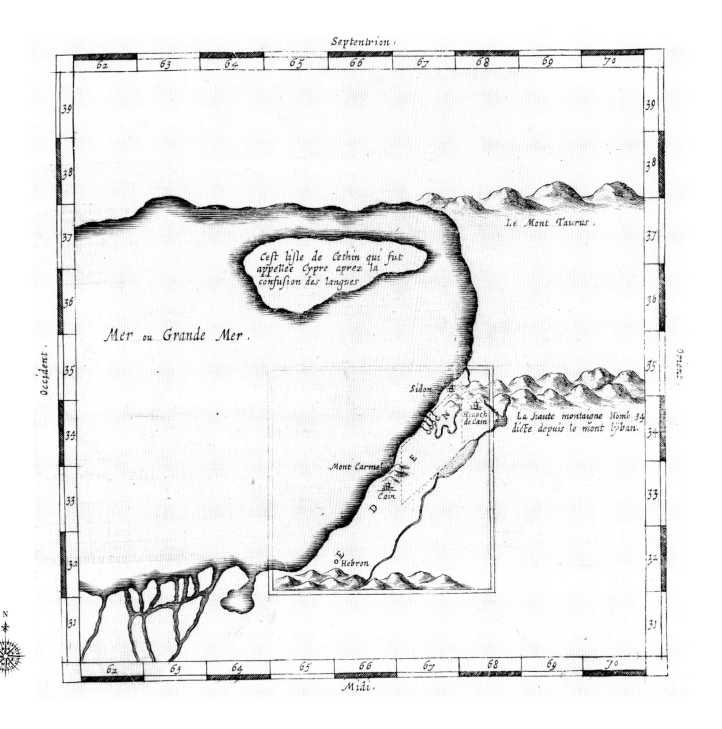

Fig. 10.28b. Jacques d'Auzoles Lapeyre, *Carte ou table particuliere de la region d'Eden apres le deluge*, from his *La Saincte Géographie, c'est-à-dire, exacte description de la terre et veritable demonstration du paradis terrestre* (Paris: Antoine Estiene, 1629), p. 135. 18 × 18 cm. London, British Library, 744.d.13. The map shows the postdiluvian condition of Eden. The waters of the Flood have inundated the north-western part of the region of Eden and the four rivers have disappeared into the sea. The eastern shore of the Mediterranean bisects the former garden of paradise, the remnant of which has become the land of Canaan stretching along the coast. Close to Tyre (unmarked but south of Sidon) where before the Flood the single river had divided into four rivers, Lapeyre placed four springs conforming to those referred to by Ezekiel in connection with the king of Tyre's sojourn in Eden (Ezekiel 28.1–13) and to the geographers' recognition of the site of the well of living waters near Tyre (Song of Songs 4.15).

The Afterlife of Paradise on Maps   329

far the best was Christiaan van Adrichem's printed topographical map of the Holy Land of 1590.[129] Using Adrichem's map as his base map, Lapeyre showed the position of paradise in the Holy Land of his day. By depicting the Holy Land and the Mediterranean coast from the Lebanon mountains to the mountains of Sur and the Nile delta, he managed to suggest the situation of paradise (Figure 10.29b).[130]

Lapeyre took pains to clarify the thinking behind his maps. He compared his geographical enquiry about paradise and the land of Canaan to the study of Hebrew. A great number of languages, he said, all now confused, entered human history with the building of the Tower of Babel, but the 'original' language of humanity – Hebrew – had remained alive so that scholars could learn this 'original' human tongue (however imperfectly because of their sins). In the same way, God changed the sources and courses of the rivers of paradise, leaving intact only the land of Canaan, so that the devoted scholar might eventually discover that the Holy Land (the Promised Land) is the same as the Garden of Eden. Lapeyre believed in a clear distinction between the consequences of the Fall and of the Flood. The Fall had not affected the splendour of the Garden of Eden, and it was only after the Flood that paradise vanished from human sight to the extent that Noah failed to recognize it. The disappearance of paradise had left a bundle of conflicting opinions, though, with some people believing in its destruction by the Flood and others in its survival somewhere. Lapeyre was reassuring: what an individual thought, he said, was not an issue of faith. At the same time, he was anxious to impress on those who read his book that his ideas did not contradict the Church's teaching.[131] Paradise was still there, where it had always been; what had changed was only its visibility. Holy Scripture is rich in examples of people looking but not seeing: Balaam failed to see the angel who prevented him from reaching his ass (Numbers 22.1–35), Mary Magdalene mistook the resurrected Christ for a gardener at

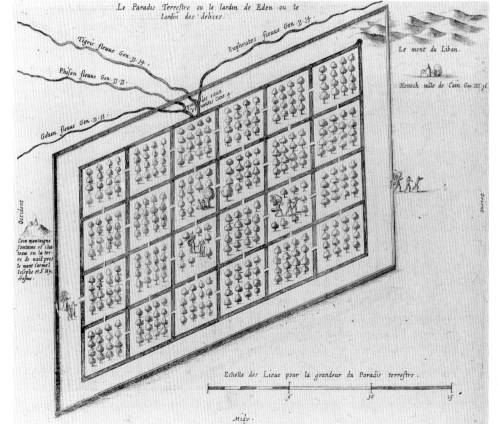

Fig. 10.29a. Jacques d'Auzoles Lapeyre, *Le Paradis terrestre ou le iardin de Eden ou le iardin de delices*, from his *La Saincte Géographie, c'est-à-dire, exacte description de la terre et veritable demonstration du paradis terrestre* (Paris: Antoine Estiene, 1629), p. 92. 19 × 22 cm. London, British Library, 744.d.13.
A detailed map of the lozenge of paradise shows how the single river rising in Mount Lebanon flows into the Garden of Eden in such a way as to form 24 squares separated by channels, over which there are bridges. Each square is filled with neat rows of trees. The Tree of Life and the Tree of the Knowledge of Good and Evil, slightly larger than all the others, are shown just above the central point of the garden. As it emerges from the garden, the river separates into four streams. The perimeter of paradise is defined by a wall. A series of vignettes portrays the main events: God escorting Adam into paradise from the west, his conversing with Adam and Eve, the Temptation, the Expulsion and, finally, Adam and Eve in flight from the Garden of Eden towards the east.

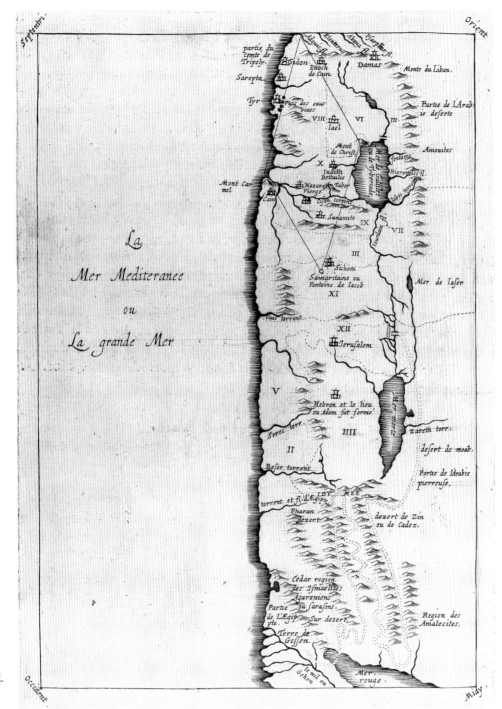

Fig. 10.29b. Jacques d'Auzoles Lapeyre, Map of the Holy Land according to Christiaan van Adrichem, from his *La Saincte Géographie, c'est-à-dire, exacte description de la terre et veritable demonstration du paradis terrestre* (Paris: Antoine Estiene, 1629), p. 164. 19 × 22 cm. London, British Library, 744.d.13. Lapeyre searched for a base map on which he could set paradise into the geographical context of the Holy Land. His preference was for the map drawn by Christiaan van Adrichem (for his *Theatrum Terrae Sanctae*, published posthumously in 1590), which he judged to be the most accurate among the maps of many 'excellent cosmographers and geographers'. On his own map, Lapeyre outlined the lozenge-shaped paradise in the land of Eden with lines joining Mount Lebanon, the Mount of the Beatitudes, Mount Carmel and the Fountain of Jacob or Mountain of Sichem (the dwelling-place of Abraham and Jacob, Genesis 12.6, 7; 34.5; 32. 18, 19; 34.2 etc.); and the place where Christ met the Samaritan woman, John 4.1–30). In the north-east of Canaan, near the city of Tyre, four springs mark the site where, before the Flood, the single river divided into four rivers. The division of the land of Canaan amongst the Twelve Tribes of Israel is indicated by a network of single lines and Roman numbers. The wandering route of the Jews in the Exodus towards the Promised Land is marked by a double dotted line.

the empty tomb, and the disciples on the road to Emmaus all failed to recognize the resurrected Christ (John 20.14–18; Luke 24.28–32). With divine grace, though, God may open human eyes to see the Garden of Eden in the same way as the son of a carpenter may be seen as the Son of God and the bread of the Eucharist may be seen as the body of Christ.[132] Lapeyre found confirmation of the invisible presence of paradise in the Holy Land in the biblical account of the Transfiguration, the occasion on which Elijah (who had been waiting in the Garden of Eden) was seen with Moses and the resurrected Christ on Mount Tabor, in the south of Galilee, but in the middle of Eden (Matthew 17.1–13; Mark 9.2–13; Luke 9.28–36).[133]

The Afterlife of Paradise on Maps 331

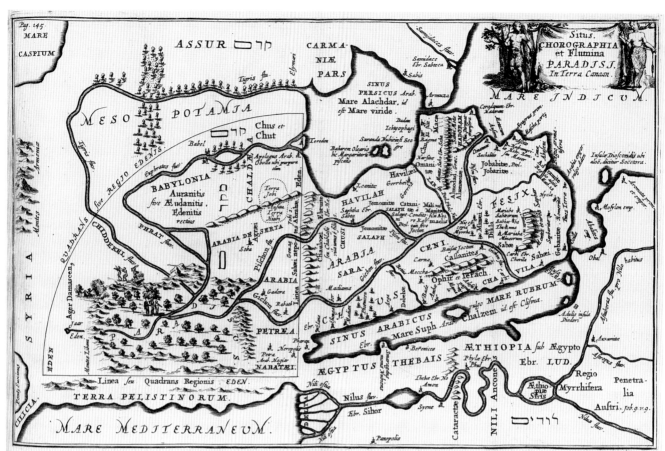

Fig. 10.30. Iohannes Herbinius, *Situs, chorographia et flumina paradisi in terra Canaan*, from his *Dissertationes de admirandis mundi cataractis* (Amsterdam: Ianssonius, 1678), between pp. 144 and 145. 17.5 × 27 cm. London, British Library, 233.i.27. Ioannes Herbinius's east-orientated map shows the whole of the Middle East, from the Nile to the Tigris, in its antediluvian state. The land of Eden is shaped like a quadrant. The straight western side is aligned with the Mediterranean coast of Palestine. On the south, Eden faces Arabia. The map features tiny vignettes in the wooden area of paradise with the Fall and the Expulsion. Herbinius also indicated the Damascene Fields (*Ager Damascenus*), where Adam was created, and the mountain chain through which he entered paradise and through which he later passed when expelled from paradise.

The paradoxes in Lapeyre's thinking must have been challenging for his contemporaries. For a start, he held that paradise was at one and the same time invisible and yet present in the Holy Land. Then, he combined concepts not normally associated with each other, blending, for example, the medieval belief that the Ganges and the Nile were the Pishon and the Gihon of paradise with the Renaissance idea (propagated by Luther) that the Flood destroyed paradise. Similarly, Lapeyre's literal reading of Genesis and his adherence to the letter of the biblical account of the Expulsion was combined with the idea of an antediluvian earth (one continent, one sea) different from the modern physical world. Unlike those who proposed a whole-earth paradise – and who also wanted to emphasize the difference between the pre- and postdiluvian world, but for whom the Expulsion was to be read allegorically as a change of state – Lapeyre held that Adam and Eve had been banished from one small but well-defined part of the earth to finish their lives in a different part. Not least, Lapeyre was unusual for the way in which he combined his interest in the cartographical form of visual expression and the tangibility of maps with his inclination towards a spiritual interpretation of the reality of paradise.

However unusual it may have been, the detailed cartography of paradise elaborated by Lapeyre served to demonstrate that the Garden of Eden existed on earth and offered a novel version of the traditional typological link between Adam and Christ. After Lapeyre, other people produced yet other arguments, and created different maps in support of the idea that Garden of Eden was in the Holy Land. Iohannes Herbinius was one of those confronting the issue of the location of paradise in the final quarter of the seventeenth century. He used his study of waterfalls, the *Dissertationes de admirandis mundi cataractis* (1678), as a vehicle for his rebuttal of all false opinions about the

332 Mapping Paradise: A History of Heaven on Earth

location of paradise and to urge that the Garden of Eden be 'restored' to Palestine.[134] Like Lapeyre, Herbinius embraced the notion that the Flood had altered the original geography of paradise, but, like the single map he produced, his thinking was much less radical. Herbinius was dismissive of both those who denied the material existence of paradise (whom he dubbed Cabalists) and those who advocated placing paradise in unlikely places. To see paradise in Mesopotamia or in Armenia was not entirely unreasonable, but in his view the true place of paradise was the Holy Land, the country of the prophets, the Patriarchs, and Christ. Indeed, the Son of God lived in Canaan and his feet never trod upon Mesopotamian or Armenian soil. Holy Scripture itself, he noted, refers in many places to the Holy Land as 'the garden of God'.[135]

The map in Herbinius's book describes 'the place, the chorography and the rivers of paradise in the land of Canaan' (*Situs, chorographia et flumina paradisi in terra Canaan*) (Figure 10.30).[136] The map was drawn to illustrate Herbinius's idea that the biblical mention of paradise as planted *ab oriente* ('from the east') meant that the garden was not actually in Mesopotamia but that it sloped down from the borders of Mesopotamia to the land of Canaan.[137] Herbinius had not allowed himself to become bogged down in the problem of how to map the four rivers. He had simply taken, he said, a modern map of Palestine and asked the Holy Spirit to guide him.[138] In this way he was led to conclude that the single river that had watered paradise was the Jordan, which, he erroneously claimed, in Hebrew meant 'river of Eden'.[139] On his map, the Jordan's single underground source is placed to the north of the Lebanon mountains, and the river flows across paradise before separating outside it into the four branches. The Tigris then flows towards Assyria, the Euphrates through Babylonia, the Pishon through *Arabia Deserta* and *Arabia Felix* (the land of Havilah) into the Persian Gulf, and the Gihon through *Arabia Petraea* (the land of Cush) into the Red Sea. The scene of the Fall is marked in the Genesareth valley, where, as Herbinus's text explains, the Lake of Genesareth (Sea of Galilee) and the Dead Sea had been formed after the Flood from the waters of the obstructed single river.[140] Herbinius, unlike Lapeyre, used his map to describe only the lost paradise and not a modern-day Palestine.

The third proponent of the Holy Land paradise thesis was the French Jesuit, Jean Hardouin. One of Hardouin's works was a lengthy commentary on Pliny the Elder's *Naturalis historia* (1685).[141] By the time the second edition was published, in 1723, Hardouin had prepared a treatise, accompanied by a double-folio map, on the location of the earthly paradise, *De situ paradisi terrestris*, which he inserted between Pliny's Books Six and Seven on the grounds that the rivers *Salsum* and *Achum* described by Pliny were to be understood as the Gihon and the Pishon of Genesis.[142] Hardouin wanted to show how the biblical description of the Garden of Eden perfectly fitted Pliny's description of the Middle East. The fact that Pliny's text had survived was already, in Hardouin's mind, the work of divine providence.[143] By sidestepping the knotty problem of trying to account for the four rivers and their single source, Hardouin found a way of disposing of the need to refer to the Flood as an explanation of their postdiluvian pattern. He played on the word *dividitur* (Genesis 2.10). The division into four, he explained, referred not to the physical division of the single river but to proportions. What the Genesis text meant was that when paradise was compared with the lands that were watered by the four rivers, however fertile and wonderful each of these may have been, none had even a fourth of the beauty and perfection of the land that was watered by the single river.[144] Even a quarter of the splendour of paradise far exceeded anything offered by Assyria (watered by the Tigris), Mesopotamia (watered by the Euphrates), or Cush and Havilah (watered by the Pishon and the Gihon, both in Arabia Felix). In Hardouin's ingenious reading of Genesis, the division of the rivers described in the text was no more than an expression of praise for the unimaginable quality of paradise.[145]

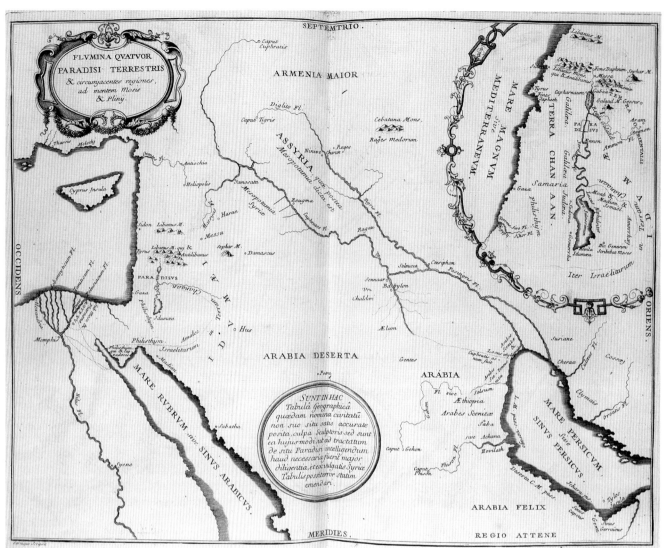

Fig, 10.31. Jean Hardouin's map of paradise, from his *De situ paradisi terrestris disquisitio* in *Caii Plinii Secundi Historiae naturalis libri XXXVII*, 2nd edn, 2 vols (Paris, Antonius-Urbanus Coustelier, 1723), I, between pp. 358 and 359. 33.5 × 44 cm. London, British Library, 685.k.2. Hardouin's map matches his text, in which he states that paradise had once stretched along the Jordan, in Galilee. Hardouin identified the Jordan as the river Paneas, described by Pliny as flowing into the Arabian Gulf, and showed it rising outside the Garden at the *Fons Daphium*. The Gihon (identified as Pliny's *flumen salsum*) forms the northern border of Cush, and to the south of it the Pishon (Pliny's *flumen Achanum*) traverses Arabia Felix and borders Havilah.

Hardouin's reading of Genesis resulted in a map of the location of paradise without the four rivers within paradise (Figure 10.31).[146] Instead, the four regions that were watered by the rivers are shown in one part of the Middle East and paradise in quite a different part. On his map, the Tigris and the Euphrates are seen to flow to the south from Armenia through Mesopotamia (noted on the map as the former Assyria) to the Persian Gulf. The Gihon and the Pishon also flow into the Persian Gulf, but from the west. Far away in the west, in the Holy Land, the word *Paradisus* and a minuscule rectangle outline the Garden of Eden. The Holy Land paradise is described in greater detail in an inset at the top right of the map. Paradise is situated between Lake Genesareth and the Dead Sea, with the single river (the Jordan) flowing through it. The only reason, Hardouin said, that Moses had not named the single river as the Jordan was because the river was not known by that name at the time of the Creation. Confident in his interpretation, Hardouin concluded his treatise with a prayer: 'May the Almighty God let me enter one day, with his mercy and the merits of Jesus Christ our Saviour, the paradise in heaven, in the same way as I had the opportunity of flattering myself to have found, with his help, paradise on earth.'[147]

334   Mapping Paradise: A History of Heaven on Earth

1. See Catherine Delano-Smith and Elizabeth M. Ingram, *Maps in Bibles, 1500–1600: An Illustrated Catalogue* (Geneva: Droz, 1991).
2. For examples of post-sixteenth-century printed maps relating to the Bible see Joost Augusteijn and Wilco C. Poortman, *Karten in Bijbels (16e–18e eeuw)* (Zoetmeer, Uitgeverij Boekencentrum, 1995), pp. 78–82; Eran Laor, *Maps of the Holy Land: Cartobibliography of Printed Maps, 1475–1900* (New York–Amsterdam: Alan R. Liss–Meridian Publishing Company, 1986); Kenneth Nebenzahl, *Maps of the Bible Lands: Images of Terra Sancta through Two Millennia* (London: Times Books; New York: Abbeville Press, 1986); Wilco C. Poortman, *Bijbel en Prent*, 2 vols (The Hague: Uitgeverij Boekencentrum, 1983–6).
3. Versions of the Sanudo–Vesconte map continued to be used as the modern map for the Holy Land in printed editions of the *Geography* until the middle of the sixteenth century. In 1540 Sebastian Münster supplied a new, south-orientated, map for his edition of the *Geography*, and in 1548 Giacomo Castaldi likewise a new map, this time with north at the top. The map in Bernardus Sylvanus Ebolensis's Venetian edition of 1511 may look 'modern', but the difference is no more than cosmetic; the underlying map is the traditional Ptolemaic map.
4. See Adolf E. Nordenskiöld, *Facsimile-Atlas to the Early History of Cartography with Reproductions of the Most Important Maps Printed in the XV and XVI Centuries* (1889; repr. New York: Dover, 1973), pp. 25–6 (Entry 28).
5. For Postel's vision of history see Marion L. Kuntz, *Guillaume Postel, Prophet of the Restitution of All Things: His Life and Thought* (The Hague–Boston–London: Nijhoff, 1981), and William J. Bouwsma, *Concordia mundi: The Career and Thought of Guillaume Postel (1510–1581)* (Cambridge, MA: Harvard University Press, 1957), pp. 251–92. For his cosmology see Frank Lestringant, 'Cosmologie et mirabilia à la Renaissance: L'Exemple de Guillaume Postel', *Journal of Medieval and Renaissance Studies*, 16/2 (Fall 1986), pp. 253–79, repr. 'Cosmographie pour une restitution: Note sur le traité "Des merveilles du monde" de Guillaume Postel (1553)', in Marion Leathers Kuntz, ed., *Postello, Venezia e il suo mondo* (Florence: Olschki, 1988), pp. 227–60.
6. Jean Oporin, the Basle printer who had published Postel's work in support of Servetus (*Apologia pro Serveto*, 1555) and had agreed to publish the work on paradise, was afraid of censorship. The manuscript is now in Basle, Universitätsbibliothek, MS A IX 99, fols 199–225. See Carlos Gilly, 'Guillaume Postel et Bâle', and Marcel Destombes, 'Guillaume Postel cartographe', in Guy Trédaniel, ed., *Guillaume Postel: 1581–1981* (Paris: Guy Trédaniel, 1985), pp. 41, 368 n. 12, and François Secret, *Bibliographie des manuscrits de Guillaume Postel* (Geneva: Droz, 1970), p. 48. Postel referred to the arctic location of paradise in the *Cosmographicae disciplinae compendium* published in 1561, and is cited for this by Thomas Malvenda, *De paradiso voluptatis: Quem Scriptura Sacra Genesis secundo et tertio capite describit commentarius* (Rome: [n.p.], 1605), p. 27 and Arturo Graf, *Miti, leggende e superstizioni del Medio Evo*, 2 vols (Turin: Loescher, 1892–3), I (1892), p. 13.
7. Postel's woodcut world map on a polar projection, entitled *Polo aptata nova charta universi*, was produced in 1581, but is known today only from a single exemplar, from a later reprinting (1621) with altered imprint and date, now in the Bibliothèque du Service Historique de la Marine, Vincennes; see Rodney W. Shirley, *The Mapping of the World: Early Printed World Maps, 1472–1700* (London: Early World Press, 2001), pp. 166–7; Destombes, 'Guillaume Postel cartographe' (1985), pp. 366–71; Destombes, 'An Antwerp *Unicum*: An Unpublished Terrestrial Globe of the 16th Century in the Bibliothèque Nationale, Paris', *Imago Mundi*, 24 (1970), p. 87. The northern Pole is occupied by four large islands (as on Mercator's map of 1569), but there is no map sign for paradise.
8. Guillaume Postel, *De paradisi terrestris loco*, Basle, Universitätsbibliothek, MS A IX 99, fol. 202r. Destombes, 'Guillaume Postel cartographe' (1985), pp. 361–71.
9. Guillaume Postel, *De paradisi terrestris loco*, Basle, Universitätsbibliothek, MS A IX 99, fols 200r–201v.
10. Ibid., fol. 200v.
11. Postel's theory about the meridian of paradise is expounded in his *Des merveilles du monde, et principalement des admirables choses des Indes et du Nouveau Monde* (Paris: J. Ruelle, 1553). See Lestringant, 'Cosmologie et mirabilia à la Renaissance' (1986; 1988).
12. Lestringant, op.cit., pp. 258–9.
13. By the time he compiled his manuscript describing a polar paradise, however, Postel seems to have abandondoned not only the idea of a Moluccan paradise, but also the idea that France was the favoured nation and had turned his attention instead to the Germans. On Postel's historical and political vision see Yvonne Petry, *Gender, Kabbalah and the Reformation: The Mystical Theology of Guillaume Postel (1510–1581)* (Leiden–Boston: Brill, 2004), pp. 51–69; and the various articles by Claude-Gilbert Dubois on the subject gathered in Dubois, *La Mythologie des origines chez Guillaume Postel* (Orléans: Paradigme, 1994); François Secret and Georges Weill, *Vie et caractère de Guillaume Postel* (Milan: Archè, 1987), pp. 199–203; Bouwsma, *Concordia mundi* (1957), pp. 213–30.
14. Bouwsma, *Concordia mundi* (1957), pp. 231–50.
15. Guillaume Postel, *Des merveilles du monde* (1953), fols 10r, 38v; see Lestringant, 'Cosmologie et mirabilia à la Renaissance' (1986; 1988); Bouwsma, *Concordia mundi* (1957), pp. 235–6, 270–81; Henri Berbard-Maitre, 'L'Apologie de la Compagnie de Jésus de Guillaume Postel à la fin de l'année 1552', *Recherches de Sciences Religieuses*, 38/2 (1952), pp. 209–33.
16. The reader's aids of the Geneva editions were piled on until the Bibles 'began to resemble those medieval Bibles laden down with glosses, which the Reformers had so vehemently rejected': Bettye Thomas Chambers, *Bibliography of French Bibles: Fifteenth- and Sixteenth-Century French-Language Editions of the Scriptures* (Geneva: Droz, 1983), p. XIII. See also Francis Higman, 'La Présentation typographique des Bibles genevois du XVIe siècle et pratique de la lecture', in Francis Higman, *Lire et découvrir: La Circulation des idées au temps de la Réforme* (Geneva: Droz, 1998), pp. 573–81.
17. Multi-sheet wall maps showing the Exodus, for example, were produced by Lucas Cranach (?1515/18), Gerard Mercator (1537), Wolfgang Wissenburg (1537/8) and Tilemann Stella (1557): see Catherine Delano-Smith,'Maps in Bibles in the Sixteenth Century', *The Map Collector*, 39 (1987), pp. 2–14, esp. Table 2.
18. The historical atlas has been defined as 'specialized collections of "maps for history" designed specifically to illustrate moments of time': Walter Goffart, *Historical Atlases: The First Three Hundred Years, 1570–1870* (Chicago–London: University of Chicago Press, 2003), pp. 1–2. On historical mapping as an aspect of the development of thematic mapping, and its relation to the emergence of history as an academic subject, see Jeremy Black, *Maps and History: Constructing Images of the Past* (New Haven–London: Yale University Press, 1997), p. 7.
19. For Ptolemy's *Geography* in the late fourteenth and early fifteenth century see Oswald A. W. Dilke, 'Cartography in the Byzantine Empire', in Brian Harley and David Woodward, eds, *History of Cartography*, I, *Cartography in Prehistoric, Ancient and Medieval Europe and the Mediterranean*:

20 The exception was Mercator's edition of the *Geography* (1578), which included only the 'old', Ptolemaic maps and was intended as an atlas of the classical world.
21 The phrase 'Historiae oculus geographia' appears on the title page of the *Parergon* accompanying Ortelius's *Theatrum orbis terrarum* (Antwerp: ex officina Plantiniana, apud Ioannem Moretum, 1601). For details of Ortelius's maps see Robert W. Karrow, *Mapmakers of the Sixteenth Century and their Maps* (Chicago: Speculum Orbis Press, 1993) and Peter Van der Krogt, *Koeman's Atlantes Neerlandici*, III/A (MS't Goy-Houten, The Netherlands: HES–De Graaf, 2003). For some recent interpretations of Ortelius's historical outlook, see Jean-Marc Besse, *Les Grandeurs de la terre: Aspects du savoir géographique à la Renaissance* (Lyons: ENS, 2003), pp. 261–375; Giorgio Mangani, *Il 'mondo' di Abraham Ortelio: Misticismo, geografia e collezionismo nel Rinascimento dei Paesi Bassi* (Modena: Franco Cosimo Panini, 1998); Mangani, 'Abraham Ortelius and the Hermetic Meaning of the Cordiform Projection', *Imago Mundi*, 50 (1998), pp. 59–83.
22 Goffart, *Historical Atlases* (2003), pp. 28–38; Black, *Maps and History* (1997), pp. 9–10. Ortelius was not the first to compile explicitly historical maps. Francesco Bertelli, for example, had already published a map of ancient Greece (1564).
23 Strabo, *Geography*, XVI.2.19. See *Paulys Real-Encyclopädie der classischen Altertumswissenschaft*, XVIII.2.2 (1949), col. 1134.
24 Like all the maps in the *Parergon*, the *Peregrinatio Divi Pauli: Typus corographicus* was entirely Ortelius's work. It bears the date 1579, and was included in the *Additamentum* together with two other historical maps, those of ancient Greece and Rome: Van der Krogt, *Koeman's Atlantes Neerlandici* (2003), pp. 76–84; Karrow, *Mapmakers of the Sixteenth Century* (1993), Maps 1/129–31, p. 16.
25 Walter Raleigh, *The History of the World* (London: Walter Burre, 1614), p. 33.
26 A treatment of sixteenth- and seventeenth-century regional maps of paradise is in Ute Kleinmann, 'Wo lag das Paradies? Beobachtungen zu den Paradieslandschaften des 16. und 17. Jahrhunderts', in *Die flämische Landschaft: 1520–1700* (Lingen: Luca Verlag, 2003), pp. 279–85 (illustrations, pp. 286–303), esp. 279–81; *Het aards paradijs: Dierenvoorstellingen in de Nederlanden van de 16de en 17de eeuw* (Antwerp: Koninklijke Maatschappij voor Dierkunde van Antwerpen, 1982), pp. 137–52.
27 The first edition of the Bible in which Calvin's map appeared was the one published by François Jaquy, Antoine Davodeau and Jacques Bourgeois (1560). A close copy of the French woodcut was included in the same year in the first English-language edition of the Geneva Bible: *The Bible* (Geneva: Rouland Hall, 1560).
28 Honoré used a woodblock that had been cut some three years earlier; see Delano-Smith and Ingram, *Maps in Bibles* (1991), pp. 5–6.
29 For reproduction, see Delano-Smith and Ingram, *Maps in Bibles* (1991), pp. 7–9.
30 The separately printed map of the 'situation du Iardin d'Eden', not in Delano-Smith and Ingram, *Maps in Bibles* (1991), has been inserted, like the other three maps. In the table of contents relating to the preliminary pages, Regnault mentions the map of Eden and the map of the Holy Land, but it is difficult to see why he should think that either map (or the map of Exodus, taken from a Geneva edition of the Bible, which is also included) was in any way relevant to his travelogue. Like the fourth map (signed by Antoine Pinaeus and dated 1564) that purports to show places mentioned in the text, Regnault seems to have taken the maps from stock held by a Lyonnais printer of Bible commentaries, as he also appears to have done in the case of the narrative woodcuts.
31 Plancius's map, *Tabula geographica, in qua Paradisus, nec non regiones, urbes, oppida, et loca describuntur, quorum in Genesio fit mentio*, was included in a number of Dutch Bibles printed in 1590; see Delano-Smith and Ingram, *Maps in Bibles* (1991), pp. 9–10 and Figure 12.
32 Matthaeus Beroaldus, *Chronicum, Scripturae Sacrae autoritate constitutum* (Geneva: Antonius Chuppinus, 1575). The map is at the end of Chapter 7, which is entitled *De paradiso, in quo varie est a nostris erratum, non assequutis quae in eius topographia proponuntur* (pp. 78–88). Letterpress text above the map reads 'Sciographia regionis Eden, quae longe lateque patuit per Syriam, et Arabiam Syriae affusam, item Mesopotamiam. In qua fluviorum Paradisi potissimum habita fuit ratio: ut iis ante oculos propositis, ipsius Paradisi, et vicinarum Edeni regionum, facilior esset comprehensio.'
33 Beroaldus, *Chronicum, Scripturae Sacrae autoritate constitutum* (1575), p. 79, held that, after sin, the Garden of Eden lost its beauty, but, given that Moses was able to describe it in detail, the Garden must have survived the Flood.
34 Franciscus Iunius, *Praelectiones in Genesim* (Heidelberg: [n.p.], 1589), pp. 99–101. The place name was also rendered in other forms, such as *Auchanitis* (Münster map, Figure 9.8) and *Auchanitis* (Servetus map, Figure 9.9).
35 Franciscus Iunius, *Praelectiones in Genesim* (1589), pp. 95–107. Similar arguments, but without the map, were repeated in the edition of the Geneva Bible Iunius edited with Immanuel Tremellius: *Testamenti Veteris Biblia Sacra sive, libri canonici priscae Iudaeorum Ecclesiae a Deo traditi*, ed. by Immanuel Tremellius and Franciscus Iunius (Geneva: Ioannes Tornaesius, 1590), pp. 5–8. Iunius placed the head of the Euphrates in Armenia maior, and noted that neither Arabia deserta nor Mesopotamia were fertile.
36 John Hopkinson, *Synopsis paradisi sive paradisi descriptio, ex variis diversarum tum linguarum, tum aetatum scriptoribus desumpta; cum chorographica eiusdem tabula* (Leiden: Plantinus, 1593), p. 18: 'Sin vero situm horti spectes, erat loco oportunissimo consitus sub coelo pulcherrimo, solo laetissimo et foecundissimo, fluminibus ad aspectum amaenissimis, ad gustum suavissimis undique irriguo.'
37 The map is printed on the final page. See Hopkinson, *Synopsis paradisi sive paradisi descriptio* (1593), p. 18: 'Sed ut haec de omnia quae exposuimus, clarius pateant, χωρογραφικῶ Babyloniae tabulam hinc subiicimus, quam ex probatis auctoribus tam vetustis quam novis, facta inter se sententiarum diversarum collatione, concinnavimus.'
38 No map specifically of Mesopotamia by Caspar Vopel is known. Hopkinson must have been referring to Vopel's multi-sheet woodcut map of the world (Cologne, 1545). Vopel's map has not survived, but Bernard van den Putte published a supposedly faithful reproduction in Antwerp, 1570: Karrow, *Mapmakers of the Sixteenth Century* (1993), pp. 560–2. For a reproduction of Van den Putte's version, see Shirley, *Mapping of the World* (2001), pp. 148–9. However, the city of Ctesiphon is not marked on the Van de Putte's map, from which it was presumably omitted in the process of engraving because of shortage of space. I am grateful to David Cobb, Robert Karrow, and Rodney Shirley for help in establishing these points.
39 Hopkinson, *Synopsis paradisi sive paradisi descriptio* (1593), p. 17: 'Ex quibus intelligimus Paradisum non esse locum imaginarium, aereum, insensilem, et incorporeum in aere suspensum; supra aut ultra terrarum orbem, et, ut ait ille, extra solis coelique vias; ut multi stulte commenti sunt: neque esse late diffusum per magna

40   terrarum spacia; ut alii non minus stulta persuasione crediderunt.'
40   Ibid.: 'sed esse in terris, et terrarum aliqua parte situm, certoque loco circumscriptum, nempe in Babyonia: sed in eius parte superiore, quae in scriptura Eden et apud alios, probatos auctores Audanitis appellatur, procul dubio sic dicta ab Eden, quasi Edenitis, ut antea observatum est'.
41   Ibid.
42   Gerard Mercator and Iodocus Hondius, *Atlas minor* (Amsterdam: Iodocus Hondius, 1610), p. 669.
43   Gerard Mercator, *Atlas sive cosmographicae meditationes de fabrica mundi et fabricati figura*, 3 vols (Düsseldorf: A. Busius, 1595), I, p. 32: 'De paradiso, ubi fuerit, et quae eius flumina, in veteri geographia restituta demonstrabo'; 'I will show the restored ancient geography where paradise was, and what were its rivers.' (translation taken from the CD-Rom, The Lessing J. Rosenwald Collection, Library of Congress, p. 100). The figure of Death appears to be closely modelled on an engraving by Heinrich Aldegrever (Bartsch, VIII.406.137/1541) where Death appears to be turning the handle of an instrument to produce the music that accompanies Adam and Eve's first steps outside the gate of paradise.
44   Gerard Mercator and Iodocus Hondius, *Atlas minor* (1610), p. 670: 'Quorum commentum ridere est aequius, quam refutare.'
45   Walter Raleigh, *The History of the World* (1614), p. 28: 'And it is true, that many of the Fathers were farre wide from the understanding of this place. I speake it not, that I my selfe dare presume to censure them, for I reverence both their Learning and their Pietie, and yet not bound to follow them any further, than they are guided by truth: for they were men; *Et humanum est errare.*' See Anthony Grafton, April Shelford, and Nancy Siraisi, *New Worlds, Ancient Texts: The Power of Tradition and the Shock of Discovery* (Cambridge, MA–London: Harvard University Press, 1992), pp. 207–12.
46   Walter Raleigh, *The History of the World* (1614), p. 56. The location of paradise is discussed at pp. 28–56.
47   Walter Raleigh, *The History of the World* (1614), p. 45. Andreas Masius, *De Paradiso commentarius … adiecta est etiam divi Basilii … ad haec duae epistulae populi nestoriani ad Pontificem rom.* (Antwerp: Christophorus Plantinus, 1569), p. 264. See Henry De Vocht, 'Andreas Masius (1514–1573)', *Miscellanea Giovanni Mercati*, IV (1946), p. 8.
48   Walter Raleigh, *The History of the World* (1614), p. 55.
49   Ibid., pp. 46–7, 49–56.
50   Cornelius a Lapide, *Commentaria in Pentateuchum Mosis* (Paris: E. Martin, 1630), pp. 56–8.
51   *De gelegentheyt van t'Paradys ende t'Landt Canaan, mitsgaders de erst bewoonde landen der Patriarchen* …. It is difficult to be certain for which edition of the Bible the map was initially intended, especially as such maps could be inserted at any time into any unbound volume. Of the two volumes now located in the British Library consulted for the present book, one is the first edition of the Bible approved by the Dutch States General (1636/7, Leiden, H. J. Van Wouw; pressmark L.11.e.2), and the other is a 1649 edition of the States General version (The Hague, H. J. Van Wouw; pressmark L.12.h.5, coloured).
52   Moxon had spent six years in Holland (1637–43). See Carey S. Bliss, *Some Aspects of Seventeenth-Century English Printing with Special Reference to Joseph Moxon* (Los Angeles: University of California Press, 1965), esp. pp. 14–24.
53   Joseph Moxon, *Sacred Geographie or Scriptural Maps* (London: the author, 1671), sig A4v.
54   The map survives in at least two states (printed from different plates), the (presumably) earlier of which lacks the key to places and countries numbered on the map, e.g. British Library, Maps CC.2.a.12, and London, Royal Geographical Society, 264.h.15; the latter is reproduced in Shirley, *Mapping of the World* (2001), p. 477, and Fred Plaut, 'Where is Paradise? The Mapping of a Myth', *The Map Collector* 29 (1984), pp. 2–7. The exemplar studied for this book comes from a 1683 edition of the King James Bible: British Library, pressmark L.18.d.5(2). The map is missing from the recently rebound separate folio volume of maps accompanying the British Library's text of Moxon's *Sacred Geographie* (pressmark C.48.e.1(2).
55   Moxon, *Sacred Geographie* (1671), p. 14. The Euleus, a Greek form of the Chaldaean Ulai (Daniel 8.2, 16), was the river mentioned in classical sources as being near Susa and flowing into the Tigris: William Smith, ed., *Dictionary of Greek and Roman Geography*, 2 vols (London: John Murray, 1856–7), I (1856), p. 598. On the same page Moxon also compared the width of the Pishon with the Rhine and the Danube, noting, as Visscher did, that the Bible makes no mention of a ferry or a boat before Noah's Ark, which meant Adam and Eve could not have crossed the river, and gives the biblical names of the four rivers. Not being able to flee further east, Cain had to establish his city of Enoch, later called Susa (*Shushan*, on the map), on the west side of the river (p. 16).
56   Moxon, *Sacred Geographie* (1671), p. 14.
57   Arthur Bedford, *The Scripture Chronology Demonstrated by Astronomical Calculations* (London: James and John Knapton, 1730). The chapter 'Of the Place of Paradise' is at pp. 101–10. In the smaller of the two folios in which Bedford's book was issued (such as British Library, pressmark L.20.gg.7), which seems to have been printed first, instructions are given on the penultimate page that all ten maps were to be placed at the end of the book, followed by the six architectural plans and elevations. In fact, three maps are given page numbers and placed, correctly, in the chapter to which they relate: *A Map of the Countries Directing to Paradise* (p. 114), *A Map of the Countries of Paradise* (p. 115), and *The Place of Paradise, the Habitation of Adam and Seth and the Land of Nod* (p. 117). For all three maps an alphabetical table of places, letter-keyed to a locational grid on the map, is provided. Two different versions of the second and third maps have been added to the nine at the end of the volume (i.e. after p. 774), making 12 maps in all.
58   The 'Isle of Eden' is marked on the maps on pp. 115 and 117.
59   Franco Motta, '"Geographia sacra": Il luogo del paradiso nella teologia francese del tardo Seicento', *Annali di Storia dell'Esegesi*, 14/2 (1997), pp. 477–506.
60   Here I have consulted Samuel Bochart, *Geographia sacra, seu Phaleg et Canaan*, ed. by Petrus de Villemandy (Leiden–Utrecht: Cornelius Boutesteyn, Samuel Luchtmans, Guilielmus Van de Water, 1707). Bochart began the *Geographia sacra* by explaining his arguments to some friends who were surprised by his opinions on the location of Eden. His intention was to show the errors of his predecessors and to prove that the geography of the Bible was insufficiently known. Bochart's manuscript notes are in Caen and Paris. See April G. Shelford, *Faith and Glory: Pierre-Daniel Huet and the Making of the Demonstratio Evangelica (1679)*, PhD Dissertation, Princeton University (1997), p. 251.
61   Bochart, *De serpente tentatore, paradiso terrestri, nonnullisque aliis*, in *Geographia sacra* (1707), p. 834.
62   Samuel Bochart, *Paradisus, sive de loco paradisi terrestris; Du lieu du paradis terrestre*, Caen, Bibliothèque Municipale, MS 11 and 14; the maps are in MS 14, fols 107v (illustrating Calvin's

opinion), 103r (illustrating Iunius's opinion), 108v (illustrating Bochart's own opinion). Bochart's arguments are produced in Latin and French and remained interrupted. See also Claudine Poulouin, *Le Temps des origines: L'Eden, le Déluge et 'les temps reculés': De Pascal à l'Encyclopédie* (Paris: Honoré Champion, 1998), pp. 235–6, 246–51, esp. p. 246 n. 63, with a reproduction of Bochart's drawings on Iunius's and Calvin's theories on Plates III and IV, between pp. 250 and 251.

63 Stephanus Morinus, *De clarissimo Bocharto et eius scriptis*; Bochart, *De serpente tentatore, paradiso terrestri, nonnullisque aliis*, in Bochart, *Geographia sacra* (1707), pp. 5 and 833. Stephanus Morinus, *Paradisi terrestris delineatio ad mentem Bocharti: Indices denique accurati et mappae geographicae suis locis insertae sunt*; Bochart, *Paradisus terrestris situs variae dissertationes, et epistolae, quae huic editioni accesserunt*, in Samuel Bochart, *Opera omnia*, ed. by Johannes Leusden, Petrus de Villemandy, 3 vols (Leiden–Utrecht: Cornelius Boutesteyn, Samuel Luchtmans, Guilielmus Van de Water, 1712), I, pp. 9–28 and 29–30 (the map does not appear in the 1692 edition from the same publisher). Bochart also wrote a sermon on Genesis in which he touched on the various meanings of the paradise narrative: *Du Pentateuque; c'est à dire, des cinq livres de Moise, et principalement du premier, dit la Genese*, in *Sermons sur divers textes*, 3 vols (Amsterdam: Jacques Desbordes, 1714), I, pp. 157–236. See also Motta, '"Geographia sacra"' (1997), p. 496.

64 Bochart, *Paradisus terrestris situs* (1712), p. 29: 'Inter eas omnes diversas explicandi rationes nulla mihi videtur probabilior, quam quae a Calvino affertur'. 'Of all the different explanations, none seems to be more likely than the one put forward by Calvin'; *Du Lieu du paradis terrestre*, Caen, Bibliothèque Municipale, MS 14, where Bochart praised Calvin's model and illustrated it with a sketch, noting that: 'This opinion is very ingenious. It is thought that Calvin was its first author'; 'Cette opinion est très ingenieuse. On tient que Calvin en est le premier auteur' (fol. 107v).

65 Morinus, *Paradisi terrestris delineatio ad mentem Bocharti*; Bochart, *Paradisus terrestris situs* (1712), I, pp. 9–30. The same arguments are expounded in Bochart's manuscript notebooks (see note 62) where Iunius's opinion is confuted in *Du Lieu du paradis terrestre*, Caen, Bibliothèque Municipale, MS 14, fols 102v–106v, and Calvin's ideas are discussed at fols 106v–109v. In the manuscript, Bochart also offers a detailed commentary on Genesis 2.8–14 (fols 109v–136r).

66 Pierre-Daniel Huet, *Traité de la situation du paradis terrestre* (Paris: Jean Anisson, 1691); the map is folded and inserted before the first chapter; another edition was published in the same year in Amsterdam with a different frontispiece and re-engraved map. A Latin version, *Tractatus de situ paradisi terrestris*, was published in *Thesaurus antiquitatum sacrarum*, 34 vols (Venice: Ioannis, Gabriel Herthz, 1744–69), VII, ed. by Blasius Ugolinus (1747), pp. 500–79. On Huet see Elena Rapetti, *Pierre-Daniel Huet: Erudizione, filosofia, apologetica* (Milan: Vita e Pensiero, 1999); on Huet and paradise Pouloin, *Le Temps des origines* (1998), pp. 223, 234, 246; Shelford, *Faith and Glory* (1997), pp. 250–9, 321–2; Jean-Robert Massimi, 'Montrer et démontrer: Autour du *Traité de la situation du paradis terrestre* de P. D. Huet (1691)', in Alain Desrumeaux and Francis Schmidt, eds, *Moïse géographe: Recherches sur les représentations juives et chrétiennes de l'espace* (Paris: Vrin, 1988), pp. 203–25. At the invitation of Queen Christina of Sweden (r. 1632–54), Huet accompanied Bochart to Stockholm in 1652. On the relationship between Bochart and Huet, see Shelford, *Faith and Glory* (1997), pp. 218–324; Shelford, 'Amitié et animosité dans la République des Lettres: La Querelle entre Bochart et Huet', in Suzanne Guellouz, ed., *Pierre-Daniel Huet (1630–1721)* (Paris–Seattle–Tübingen: Papers on French Seventeenth Century Literature, 1994), pp. 99–108.

67 Pierre-Daniel Huet, *A Treatise of the Situation of the Terrestrial Paradise* (London: James Knapton, 1694), translated into English by Thomas Gale, who added a note 'To the Reader', in which he commented: 'Our first Parents, Reader, were turned out of Paradise for their Disobedience: Many of their Posterity endeavour by their Disbelief of Moses's Writings, to turn Paradise out of the World. To stop and correct this Humour, several Learned Books have been put in Print; and this, the last of all' (sig. A2). In his Preface, Huet warned his readers not to expect 'an Elegancy of Speech, nor Fineness of Thoughts: You must, on the contrary, prepare your selves to a dry Reading, to a thorny Inquiry, to the Tediousness of Citations, and to hear some Greek and Hebrew Words' (p. 3); Huet, *Traité de la situation du paradis terrestre* (1691), p. 4: '… ne cherchez pas icy l'élegance du discours, ni l'agrément des pensées. Préparez-vous au contraire à une lecture seche, à une recherche épineuse, à l'enuy des citations, et à essuyer quelque Grec et quelque Ebreu.'

68 Huet, *Traité de la situation du paradis terrestre* (1691), pp. 239–40 (pp. 167–8 in the English edition, 1694).

69 Huet, *Traité de la situation du paradis terrestre* (1691), pp. 19–21 (p. 13 in the English edition, 1694). Here Huet attributes all the maps found in Ptolemy's *Geography* to Agathos Daimon, crediting the so-called Agathodaimon endorsement, the note added in some manuscripts to Book VIII of Ptolemy's *Geography*, which ascribed the drawing of the world map to an $\grave{\alpha}\gamma\alpha\vartheta\grave{o}\varsigma\ \delta\alpha\acute{\iota}\mu\omega\nu$ ('good spirit'). See Dilke, 'Cartography in the Byzantine Empire' (1987), p. 271.

70 Huet, *Traité de la situation du paradis terrestre* (1691), pp. 207–8 (p. 45 in the English edition, 1694).

71 Salomon Van Til, *Dissertatio singularis geographico-theologica de situ paradisi terrestris*, in *Malachius illustratus* (Leiden: Iordanus Luchtmans, 1701). See Charles W. J. Withers, 'Geography, Enlightenment, and the Paradise Question', in David N. Livingstone and Charles W. J. Withers, eds, *Geography and Enlightenment* (Chicago–London: University of Chicago Press, 1999), p. 75; Fred Plaut, 'General Gordon's Map of Paradise', *Encounter* (June/July 1982), p. 29. Most of Van Til's text is a systematic review of earlier theories on the location of paradise.

72 Salomon Van Til, *Dissertatio singularis geographico-theologica de situ paradisi terrestris* (1701), p. 101: 'Hoc, qui praestabit fluvius (in Edene oriundus) debuit necessario quaeri in terra interamnana, Ephratem inter et Tigrim intercepta: ac intra interamnam regionem adhuc infra se duos fluvios emittente, ad quos quoque liber pateat per divisionem ulteriorem aditus. Potuit autem id praestare fluvius aliquis, regionem Samorrae inter Tigrin et Saocoram perluens, qui ostium ad Tigris tipam exhibet spectabile, cuique incolae nomen Odeines imponunt; prout tabula nostra delineavit.'

73 On Kircher see Janet Brown, 'Noah's Flood, the Ark, and the Shaping of Early Modern Natural History', in David C. Lindberg and Ronald L. Numbers, eds, *When Science and Christianity Meet* (Chicago–London: University of Chicago Press, 2003) pp. 115–17; Denis Cosgrove, 'Global Illumination and Enlightenment in the Geographies of Vincenzo Coronelli and Athanasius Kircher', in Livingstone and Withers, eds, *Geography and Enlightenment* (1999), pp. 33–66. On Kircher's notion of nature in relation to paradise see Ivan Cantoni, '*Tempora labuntur irrequieta cyclis*: Tempo e cosmologia nella filosofia della natura di Athanasius Kircher', in Walter Tega, ed., *Le origini della*

74 Athanasius Kircher, *Arca Noe, in tres libro digesta* (Amsterdam: Ioannes Ianssonius, 1675), pp. 22–186. In addition to the two maps relating to paradise (inserted between pp. 22 and 23 and between pp. 196 and 197), and numerous diagrams and illustrations, Kircher's *Arca Noe* contains two other maps, both of the world. One shows the four continents before and after the Flood (inserted between pp. 192 and 193, reproduced in Shirley, *Mapping of the World* (2001), p. 485), and the other shows the distribution of the descendants of the sons of Noah and the languages they spoke (inserted between pp. 222 and 223).

75 Athanasius Kircher, *Arca Noe*, pp. 22–3.

76 Ibid., pp. 197–204.

77 On this map, Kircher gives the name Gihon to the western branch, and Pishon to the eastern branch. On a later map, *Chorographia qua Noe arcam egressus ex monte Ararat*, inserted between pp. 12 and 13 in his *Turris Babel, sive Archontologia ...* (Amsterdam: Ioannes Ianssonius, 1679), Kircher made the Gihon the eastern branch and the Pishon the western. When describing the delightful land of Sennar, or Babylon (p. 18), Kircher referred to his early map, but without commenting on the different identification of the two rivers. Huet, who looked at the later map and who had also switched the usual order of the two rivers, wondered what sources Kircher had used: Huet, *Traité de la situation du paradis terrestre* (1691), pp. 112–13 (pp. 78–9 in the English edition, 1694).

78 Athanasius Kircher, *Arca Noe*, pp. 201–2: 'Fontem igitur, qui per subterraneos Canales ex montium crateribus derivatus, in medio Paradisi, inde in quatuor capita divisus, oriebatur, illum postea mutatus rebus per exteriores alveos, terram edeniam irrigare voluisse, haud incongruis coniecturis asseveramus.' Kircher tries to put forward 'conjectures and guesses that are not inconsistent'.

79 The number of angels was justified by Kircher by the fact that the Hebrew word *Cherubim* was the plural of *Cherub*.

80 Note to the reader on the map: 'Arborum vero seriem non quod revera hunc situm habuerint; sed qualem in horto voluptatis concipere licuit, disposuimus.'

81 Fausto da Longiano, 'Del paradiso terrestre', in Aeneas Sylvius Piccolomini, *Discrittione de l'Asia et Europa*, transl. by Fausto da Longiano (Venice: Vincenzo Vaugris, 1544), fols 379r–v.

82 The map is the first in the *Parergon*, the historical section of Ortelius's *Theatrum orbis terrarum*. The text introducing the map is in the verso. On Ortelius's historical maps in general, see Goffart, *Historical Atlases* (2003), pp. 30–5, and Peter H. Meurer, 'Ortelius as the Father of Historical Cartography', in Marcel Van den Broecke, Peter Van der Krogt, and Peter H. Meurer, eds, *Abraham Ortelius and the First Atlas: Essays Commemorating the Quadricentennial of his Death, 1598–1998* (Houten, The Netherlands: HES, 1998), pp. 133–59.

83 'Dedimus quod potuimus, quando ut voluimus non licuit. Valentiam potius quam volentiam en offerimus.'

84 'Phison unum e Paradisi fluminibus, quem quidam gangem interpretantur, qui nimis versus Ortum fluit, haec tabula (ignosce benevole spectator) propter suam angustiam recipere non potuit: eius autem situm cognosces ex alia nostra "Aevi veteris geographiae" in hoc parergo nostro "Tabula."'

85 Marmaduke Carver, *A Discourse of the Terrestrial Paradise Aiming at a More Probable Discovery of The True Situation of that Happy Place of our First Parents Habitation* (London: James Flesher, 1666). The paradise issue is defined 'not-unnecessary' at p. 167. Carver's name does not appear on the title page, but the dedicatory epistle is signed with the initials C. M.. Carver was vicar at Harhill, Yorkshire. As he told his readers in the prefatory epistle (sig. A1), Carver published his treatise 26 years after having composed it, resisting meanwhile the temptation to add new material.

86 Carver, *A Discourse of the Terrestrial Paradise* (1666), sig. A3v.

87 See the epigraph to this chapter.

88 Carver, *A Discourse of the Terrestrial Paradise* (1666), pp. 154–62.

89 Ibid., pp. 31–9. Lucan, *Pharsalia*, III.256–66; Boethius, *De consolatione philosophiae*, II.1; Isidore, *Etymologiae*, XIII.21, ed. by Wallace Martin Lindsay, 2 vols (Oxford: Clarendon Press, 1911), who quoted Sallust, *De bello Jugurthino*, IV.77.

90 Carver, *A Discourse of the Terrestrial Paradise* (1666), pp. 85–108.

91 Ibid., p. 135.

92 Ibid., pp. 151–3.

93 Ibid., p. 167: 'Though all these evidences laid together (which surely are as great as well may be expected in a Subject of this nature) have not raised our confidence to such a height as some have attained to, (and, as we suppose, upon far weaker grounds); yet we verily believe that if they whom God hath blessed with abler parts, more skill in the Tongues, History, Geography, etc. a larger freedom from other imployments and distractions, with a more plentiful supply of Books, and other accommodations for such a study, (all which we want) would resume this Argument, and apply their pens to the farther search of this *not-unnecessary Question*, they might here (sooner then in any other place yet discovered) find out *the true place of the Situation of the Terrestrial Paradise*.'

94 Augustin Calmet, *Commentaire littéral sur tous les livres de l'Ancien et de Nouveau Testament*, 2nd edn, 8 vols (Paris: [n.p.], 1724–6), I/1 (1724), p. 20: 'Il seroit peut-être mieux d'imiter la modestie des anciens Peres sur cette matiere; mais peut-être aussi qu'il y auroit des personnes qui pourroient tirer de ce silence quelque mauvaise conséquence contre la verité de l'Ecriture. Il faut essayer de montrer à peu près la situation du paradis terrestre, par des vestiges qui ont subsisté long-tems, et dont plusieurs subsistent encore aujourd'hui, dans le pays où il étoit.'

95 The map entitled *Geographie sacrée ou terres inhabitées pas les fils de Noé*, with the *Carte du paradis terrestre* inset at the top right, is one of the seven maps and plans maps with which Calmet illustrated his commentary on the Old Testament. He comments on paradise in *Commentaire littéral*, I/1 (1724), pp. 20–9.

96 Pliny, *Naturalis historia*, VI.27.

97 Augustin Calmet, *Commentaire littéral*, I/1 (1724), p. 22.

98 Ibid., p. 29: 'Mais nous ne doutons point que le lieu où fut planté le Paradis ne subsiste encore, quoique dépouillé de ces beautez qui le rendoient si agréable, et de la plûpart des qualitez et des circonstances qui pourroient nous le faire distinguer aujourd'hui'; 'We do not doubt, however, that the place where paradise was planted still survives, though deprived of its beauties which made it so delightful, and of most of its features and circumstances, which could make us recognize it today.'

99 Calmet quotes various authorities: Strabo, Quintus Curtius, Xenophon, Procopius.

100 Pliny, *Naturalis historia*, VI.4: 'Pontibus centum viginti pervius.'

101 The reference to Jason's Fleece is found in Strabo, *Geography*, XI.2.19, and Pliny, *Naturalis historia*, XXXIII.3. On the fertility of the region, see: Song of Songs 5.11, Jeremiah 10.9, Daniel 10.5. Pliny observed that Ophaz or Uphaz was the gold of Phasis (*Naturalis historia*, XXXIII.4); that the best bdellium and onyx (mentioned in

102 Genesis as abundant in Eden) came from Bactriana (XII.9); and the best emeralds came from Tartaria and Scythia (XXXVII.5).
102 Augustin Calmet, *Commentaire littéral*, I/1 (1724), p. 26.
103 Ibid., p. 28.
104 See above, Chapter 9, p. 277 and Figure 9.76.
105 Michael Servetus, *Christianismi restitutio* ([n.p.], 1553; repr. Nuremberg: Rau, 1790), p. 376: 'Desolatum et desertum ibi docet propheta paradisi locum, quem Christus restituit.' The maps in Servetus's editions of the *Geography* were printed from woodblocks cut in 1522 for Lorenz Fries's Strasbourg edition of 1522 (also used by Iohannes Grüningerus, Strasbourg, 1523). The letterpress on Fries's *Terra Sancta* included a note to the reader observing the merchants's report that, far from being richly fertile, the Holy Land was 'uncultivated, infertile, and lacking in all pleasantness' ('incultam, sterilem, omni dulcedine carentem'). The text is repeated in Servetus's edition of Ptolemy of 1535 but not in the edition of 1541.
106 Virtually every copy of Servetus's *Christianismi restitutio* was destroyed and only three exemplars are known to exist today: see Charles Donald O'Malley, *Michael Servetus: A Translation of his Geographical, Medical and Astrological Writings* (London: Lloyd-Luke, 1953), pp. 195–6.
107 O'Malley, *Michael Servetus* (1953), pp. 195–6, who quotes John F. Fulton, *Michael Servetus: Humanist and Martyr* (New York: Reichner, 1953).
108 Michael Servetus, *Christianismi restitutio* (1553; 1790), p. 372: 'ut vestigium quoddam, et paradisi umbram illa contineat'.
109 Ibid., pp. 372–6.
110 Henricus Bünting, *Itinerarium Sacrae Scripturae, das ist ein Redebuch über die ganze Heilige Schrift* (Magdeburg: Paul Donat, 1585), p. 57.
111 Jacques d'Auzoles Lapeyre, *La Saincte Géographie, c'est-à-dire, exacte description de la terre et veritable demonstration du paradis terrestre* (Paris: Antoine Estiene, 1629). See Poulouin, *Le Temps des origines* (1998), pp. 237–44.
112 To make sense of how Lapeyre conceived of the position of paradise and the Damascene Fields within the region of Eden, see Figures 10.28a and 10.29a.
113 Lapeyre, *La Saincte Géographie* (1629), pp. 31–2, 110–11, 155, 174–82, 184–5. The burial place of Adam is discussed at pp. 31–2.
114 Ibid., p. 3: 'ferme colomne de la verité'.
115 Ibid.: 'Aux lecteurs sur le subiet de la saincte geographie'; 'De ce Thresor … infiny, qui n'est sujet à aucune rouilleure'.
116 Ibid.: 'Tous les iours, en ce qui n'est de la Foy, il nous est licite de décroire s'il nous plaist ce que nous avons autresfois creu: toutes les choses passent et ont leur temps, *mais la parolle de Dieu demeure eternellement*: de laquelle bien entendue conformément à la doctrine de l'Eglise, tout ce qui en provient, est solide, certain, et indubitable.'
117 Ibid., p. 93: 'Nos plans Geographiques ainsi dressez pour faire voir qu'il ne faut pas chercher le Paradis terrestre hors de la Terre, il faut maintenant exactement considerer le sainct texte, et par iceluy marquer tant qu'il nous sera possible le veritable lieu du Paradis terrestre.'
118 Ibid., p. 60.
119 Ibid., p. 89.
120 Ibid., p. 91.
121 Ibid., p. 92.
122 Ibid., p. 134.
123 Frank Lestringant, *Le Livre des îles: Atlas et récits insulaires, de la Genèse à Jules Verne* (Geneva: Droz, 2002), pp. 42–52 and, esp. 327, has pointed out that the archipelago was for Lapeyre an image of the Fall.
124 As the waters receded, the dove sent by Noah came back to him with the branch of an olive tree, very likely from one of the trees in paradise, according to Lapeyre, *La Saincte Géographie* (1629), pp. 130–1.
125 Ibid., p. 135.
126 Ibid., p. 171.
127 Ibid., p. 171: 'car au lieu qu'auparavant il ne faisoit que couler doucement dans le iardin d'Eden, l'arrousant par mille et mille tours et retours qu'il y pouvoit faire; aujourd'huy elle coule avec impetuosité, et se va soudainement rendre dans la Mer Mediterranée, entre les villes de Tyr et de Sarepta, comme si elle avoit quelque deplaisir de voir habiter ce sainct lieu par des hommes pecheurs, et qui en sont indignes.'
128 Ibid., p. 182.
129 Christian von Adrichom's map, *Situs Terrae Promissionis SS bibliorum intelligentiam exacte aperiens*, was created before his death in 1585, together with ten other maps. The manuscript of his *Theatrum Terrae Sanctae* was printed by George Braun in Cologne in 1592. For a reproduction of the two-sheet copperplate map, see Nebenzahl, *Maps of the Bible Lands* (1986), Plate 35, pp. 94–7.
130 Lapeyre, *La Saincte Géographie* (1629), p. 164.
131 Ibid.; a brief note at the end of the table of contents preceding the treatise from H. Bachelier and I. Bandel, theologians of the University of Paris, certifies that the text not only does not contain doctrinal errors, but also offers learned remarks worthy to be published: it contained 'plusieurs belles et doctes remarques qui meritent d'estre communiquées au public'.
132 Ibid., pp.128–9.
133 Ibid., pp. 183–9. Lapeyre confirms that Enoch and Elijah, whom medieval tradition held to be residents of paradise, are now in the earthly paradise, alive in soul and body. On Enoch and Elijah in paradise see above, Chapter 3, p. 55–6.
134 Iohannes Herbinius, *Dissertationes de admirandis mundi cataractis* (Amsterdam: Ioannes Ianssonius, 1678). The tract on paradise, pp. 136–88, bears the title *Dissertatio quinta: De cataractis paradisiacis ubi occasione cataractarum, situs paradisi verus ac genuinus contra vexatam hactenus communem utopiorum, Indianorum, Armeniorum, Mesopotamiorum et aliorum opinionem in Palaestina, Terra Sancta, asseritur ac restituitur.*
135 Genesis 13.10; Ezekiel 28.13; 31.16; Judges 9.15; 1Kings 5.6.
136 Iohannes Herbinius, *Dissertationes* (1678), inserted between pp. 144 and 145.
137 Ibid., p. 154: 'Quicunque constitutus in Deserto Arabiae Petraeae, sub eodem cum Mesopotamia climate, Paradisum ab Oriente Euphratis versus Terram Canaan ad Occidentem declinat, ille Paradisum non in Mesopotamia, sed extra Mesopotamiam versus Occidentem ponit.'
138 Ibid., pp. 161–2.
139 The Hebrew word for 'Jordan' is יַרְדֵן (*yarden*) and not *Jaar Eden*, as claimed by Herbinius, who wanted the words to mean 'river of Eden'. In Hebrew, 'the river of Eden' is נְהַר־עֵדֶן (*nehar-Eden*). Herbinius's conclusion was thus based on a misreading and a false etymology.
140 Genesis 8.2. The Flood had also destroyed the river's underground source.
141 See Anthony Grafton, 'Jean Hardouin: The Antiquary as Pariah', *Journal of the Warburg and Courtauld Institutes*, 62 (1999), pp. 241–67. Hardouin's five-volume edition of Pliny's *Naturalis historia* made him famous.
142 Jean Hardouin, *De situ paradisi terrestris disquisitio, sive de Plinii cum Mose convenientia in paradisi fluminibus indicandis*, in *Caii Plinii Secundi Historiae naturalis libri XXXVII*, 2nd edn, 2 vols (Paris: Antonius-Urbanus Coustelier, 1723), I, pp. 359–68, comment on p. 359. Jean Delumeau, *Une histoire du paradis*, I, *Le Jardin des délices* (Paris: Fayard, 1992), p. 189 (p. 144 in the English edition), and Massimi, 'Montrer et démontrer' (1988), give 1716 as the date of the first edition: see also Poulouin, *Le Temps des origines*, p. 244 n. 57; Motta, '"Geographia sacra"'

(1997), p. 502, indicates 1723. Hardouin's treatise on paradise was later translated into French and republished without the map in 1730 in a silloge of geographical and historical treatises edited by Antoine-Augustin Bruzen de la Martinière: Jean Hardouin, *Nouveau Traité sur la situation du paradis terrestre (ou Conformité de Pline avec Moïse, par rapport à la position des fleuves du paradis terrestre)*, in *Traitez geographiques et historiques pour faciliter l'intelligence de l'Écriture Sainte* (La Haye: Van der Poel, 1730), I, pp. 1–172.

143 Jean Hardouin, *Nouveau Traité sur la situation du paradis terrestre* (1730), I, pp. 4–5: 'J'espére, qu'après la lecture de ce traité, les personnes qui ont du goût pour la langue Latine, quelles qu'elles soient, regarderont comme un effet de la Providence divine, que les ouvrages de Pline soient parvenûs jusqu'à nous; car quoique cet Auteur n'eût pas la connoissance du vrai Dieu; il est le seul de tous les Anciens qui ait fait mention dans ses livres de Géographie des fleuves du Paradis Terrestre, et il se trouve qu'il les met au même endroit où Moïse les a placés.' He also defends the Vulgate as the most authoritative and accurate version of Scripture.

144 Ibid., pp. 74–5: 'Et la partie de ces Terres, ajoute Moïse, sçavoir celle qui se trouve la plus agréable et que le Tigre arrose, fut le partage d'Assur, fils de Sem: le Pays agréable qui est autour des sources de l'Euphrate échut aux enfans de Japhet; et celui qui est autour des sources du Gehon et du Phison, aux enfans de Chus, petit-fils de Cham et ce Païs porte le nom de Terre d'Hevilath et de Terre de Chus. Aucune de ces contrées, dis-je, n'a pas plus de la quatrième partie de la beauté de la Terre que le Seigneur vous réserve, où il planta au commencement le Paradis pour le premier homme.'

145 Hardouin implied that the Holy Land had retained some of its original paradisiacal glory at the time the Book of Genesis was being written, ibid. p. 99: 'du tems de Moyse en voyoit encore quelque chose de cette ancienne beauté du Paradis.'

146 Hardouin's map, *Flumina quatuor paradisi terrestris et circumiacentes regiones, ad mentem Mosis et Pliny*, is inserted between pp. 358 and 359. The map of paradise drawn by the prolific cartographer Guillaume Delisle in 1719 and published in 1764, after his death (1726) by his brother Joseph-Nicholas, referred to by Goffart, *Historical Atlases* (2003), p. 226, and reproduced in *Het aards paradijs* (1982), p. 150, is clearly a copy of Hardouin's map.

147 Jean Hardouin, *Nouveau Traité sur la situation du paradis terrestre* (1730), I, p. 172: 'Fasse le Dieu tout-Puissant que comme j'ai lieu de me flatter d'avoir par son secours trouvé le Paradis Terrestre, je puisse par sa misericorde et les merites de Jesus Christ notre Sauveur entrer un jour dans le Paradis Celeste.'

The Eclipse of the Theological Eden

The very greatest minds stood, and some of them still stand today,
under the puissant thrall of the mystery encompassing the First Book of Moses.
Friedrich Delitzsch, *Babel und Bibel*, transl. by Thomas J. McCormack (1902)

From Late Antiquity to the end of the Middle Ages paradise was portrayed on maps of the world as a remote and inaccessible place. During the Renaissance, its historical site was being exactly plotted on regional maps as if within human reach. By the eighteenth century, the consensus had fragmented, and an accumulation of new theories shifted paradise back into the vaguer realm of possibilities, as certainty in some respects was counteracted by doubt in others. After the Renaissance attempt to situate the Garden of Eden with mathematical and Ptolemaic precision, according to latitude and longitude, the search in the West for paradise drifted back to the intellectual situation that characterized Late Antiquity and the Middle Ages, when each proposed location for paradise fuelled as much doubt as certainty. As seen in the last chapter, a profusion of diverse and sometimes highly idiosyncratic maps showing the location of paradise on earth succeeded Calvin's mid-sixteenth-century model. That stream of creativity continued beyond the eighteenth century and the Enlightenment. Throughout the nineteenth and twentieth centuries – even at the beginning of the twenty-first century – philologists, scientists and historians, but increasingly rarely theologians, piled theory after theory onto the heap that had been accumulating since Augustine's time, through the Middle Ages, and after Luther and Calvin. By the turn into the nineteenth century, yet another shift in the mapping of the location of the earthly paradise had taken place. In the Renaissance and up to the Enlightenment, mapping paradise had been part of the exegesis of Holy Scripture as the authoritative text. After the eighteenth century, however, mainstream theologians seemed content to set aside any unresolved questions about paradise and to replace the heated debates of previous centuries about its location with debates that were no less heated but that concerned other matters, such as evolution, ecumenism, the Church's social doctrine, the reform of liturgy, modernism . . . They did not show much concern for the location of the lost Garden of Eden, nor did they invest their intellectual energies in trying to create cartographical evidence to corroborate the Scriptures. They had bigger fish to fry. They could not entirely ignore the problem of the location of the earthly paradise, though. Something had to be said about it in, for example, the reference books and encyclopedias that were being produced, edition after edition, throughout the nineteenth century. So the biblical commentators tended either to allude to one of the traditional Renaissance locations of paradise, such as Mesopotamia, Armenia, or Central Asia, or to refrain from expressing any opinion other than that the problem was 'insoluble' or, at best, just part of a 'mythical geography'.[1] Their lack of engagement created a vacuum. Where exegetes and theologians were silent, a wide range of secular scholars were not shy of stepping into the breach. After well over a millennium of sober exegetical mapping by clerics and pastors, the earthly paradise was abandoned to the zeal of the non-professional enthusiast. Today, too, those who set themselves the task of mapping the location of the earthly paradise usually do so with the unbridled energy of a beginner who has little if anything to do with mainstream theology.

The Theologian Gives up

The *New Map of the Garden and Land of Eden* compiled by an English country parson for the family Bible he published in London in 1782 offered a cartographical summary of the theories that had emerged since the sixteenth century together with a new theory (Figure 11.1).[2] On Paul Wright's map, broken lines hint at the presence of Eden in different parts of the Near East according to the various authorities: notably Calmet for Armenia, Calvin and Huet for Mesopotamia, and Hardouin for the Holy Land.[3] Since each location featured a specific river pattern, and all these rivers could not be accommodated on one map with any degree of legibility, Wright made no attempt to summarize the different hydrological patterns as he had for the site of the Garden of Eden. However, the river names on his map seem to imply that Wright favoured one theory in particular, that of a paradise submerged in the Persian Gulf by the waters of the Flood, with the Garden of Eden distinguished from the land of Eden and the four rivers flowing into a single river encircling paradise. Wright shows how the sea has covered most of the land of Havilah, defined by the Pishon, a distributary of the Euphrates. Likewise, the land of Cush (Chushestan) is contained by the Gihon (Tab), which rises in the Zagros range. A note on the map acknowledges the 'celebrated Divines' as the originators of this idea.[4] In a footnote to the text, Wright remarked that 'the various conjectures' about paradise were more 'curious than useful' and that 'its situation [was] difficult to be known'.[5] The issue of the location of the Garden of Eden was presented to Bible readers as an irresolvable and, in the final analysis, a not particularly important puzzle. A similar approach had been taken some twenty years earlier by the teacher of languages and literature Pierre Luneau de Boisjermain, who intended to map, in his *Atlas historique*, the major historical vicissitudes of humankind in the various regions of the earth. Luneau managed to produce only three of the twenty maps he had in mind, but on the map of Europe, Asia and the Americas intended to depict the two first centuries after creation, he gave alternative locations for the earthly paradise, one in Palestine and one in Armenia, mentioning also the theory that there may have been a Mesopotamian Eden. Luneau also made it clear, in an inscription in the lower left corner of his map, that he was describing the earth as it appeared at his time, but that he considered it likely that the Flood had changed its original configuration (Figure 11.2).[6]

The continuous multiplication of theories and maps – even in the nineteenth century – made it increasingly obvious that a single, authoritative solution would never be found for the paradise question. This negative conclusion was not, however, perceived as a disaster. After the eighteenth century, it was no longer a vitally urgent matter whether the Garden of Eden had been situated in Armenia, Mesopotamia or anywhere else. For centuries the Bible had been trusted implicitly as the word of God and as a document inspired in its entirety by the Holy Spirit. In the course of the eighteenth century the notion emerged that the text of Scripture was, like any other text, open to philological and historical enquiry, and that its production could be related to external circumstances. In the new historical-critical methodology (a logical if extreme development of the traditional literal and historical reading of Scripture) the Bible began to be analysed as separate units of text and as representing different traditions. Biblical scholars began to accept that – although they still considered it as having been inspired by God – the scriptural text they were reading might represent the work of different authors. The geographical references found in Genesis were now understood to reflect

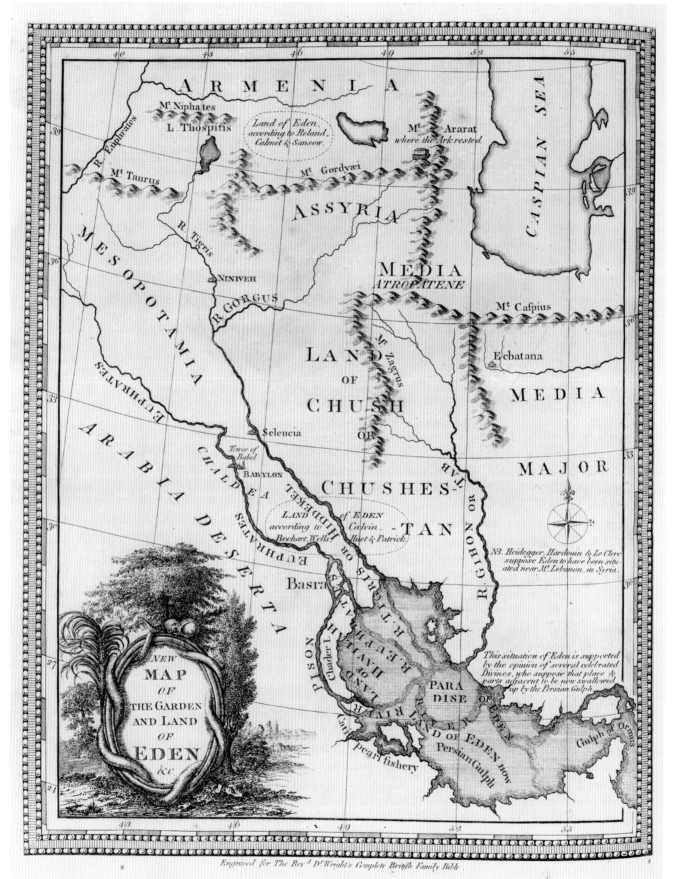

344　Mapping Paradise: A History of Heaven on Earth

the individual knowledge of the ancient writer who, while expressing ideas common in his day, had been inspired by God to convey principally moral teachings, not a geographical or even historical record. The deployment of cartography in the service of biblical exegesis was becoming unnecessary. The Bible was increasingly seen as a *religious* text and no longer as the infallible depository of scientific fact.

Together with the other books of the Pentateuch (Exodus, Leviticus, Numbers and Deuteronomy) Genesis had hitherto been attributed by both Jews and Christians to Moses. Now Moses's authorship was questioned and his books subjected, as Goethe put it, 'to probe and knife' by scholars such as Henning Bernhard Witter (1711) and Jean Astruc (1753).[7] The questioning of Moses's authorship had already begun in the seventeenth century, with such figures as Benedict Spinoza, Thomas Hobbes and Richard Simon. In the eighteenth century, Astruc in fact defended Moses's authorship, but he introduced an important critical method in which distinct sources (which he said that Moses himself had used) were identified by which name was used for the deity. The issue of authorship continued to be debated throughout the nineteenth century until, in 1878, Julius Wellhausen confronted the question with his synthesis of earlier exegetical efforts and his clarification of the so-called 'Documentary Hypothesis'.[8] Wellhausen noted that the Pentateuch had been recognized as deriving from a variety of sources that were eventually brought together in the post-exilic period. This meant that the repetition of the creation story in the first and second chapters of Genesis could be explained by the idea that the two versions came from two distinct documents, the Priestly and the Jahwist respectively. The abandonment of the traditional idea of a single, Mosaic authorship in the nineteenth century opened the way to a new sensitivity to the different literary forms found in the Bible. The legitimacy of the Bible's moral teaching continued to be accepted, however, even though the idea that the biblical stories comprised a single historical narrative had now been undermined.

As the end of the nineteenth century approached, the Catholic Church started to direct attention to the importance of achieving a more sophisticated and more critical understanding of Scripture, even though, for the most part, its early reaction to historical-critical scholarship was hostile. Given that the Bible had divine as well as human authorship, and that the words of the text were selected by the latter – so the argument went – it was essential to disclose the original human author's conscious intent and take his cultural conditioning into account. In a series of papal encyclicals, the Church admitted that the Bible was not about scientific truth but about faith and morals.[9] The new approach was confirmed in the course of the twentieth century. At the same time that it recommended prudence in the general dissemination of modern critical views, Catholic teaching began to encourage biblical scholars to take every advantage of modern scholarship and criticism, an injunction that applied to the paradise narrative no less than to any other part of the biblical text.[10] A Papal Biblical Commission, reporting on 30 June 1909, established that the faithful had to extract from the key concepts in the Book of Genesis those that were to be taken as historical truths: the creation of the world and of man by God; the happiness of the first man in the original state of grace; his transgression of the command given by God to test his obedience; his exclusion from that primordial state of innocence and the promise of future redemption.[11] A similar conclusion was reached by many of the Protestant churches, notwithstanding the range of their exegetical traditions. As Michael Ramsey, Archbishop of Canterbury and head of the Church of England, put it in 1962, 'the central fact of Christianity is not a Book but a Person – Jesus Christ, himself described as the Word of God'.[12] In the twentieth century, for Catholics and Protestants alike, the latitude and longitude of the Garden of Eden were no longer needed to validate the authority of the Scriptures in support of the Christian faith.

Most Christian theologians today prefer to stress the symbolic meaning of the

Fig. 11.1 (opposite). Paul Wright, *New Map of the Garden and Land of Eden*, in Wright, ed., *The Complete British Family Bible* (London: for Alex Hogg, 1782). 29 × 23 cm. London, British Library, L15.d.7. Wright's map, showing Mesopotamia and Armenia, provides a cartographical synthesis of the principal ideas about the location of paradise that had been expounded between the sixteenth and the eighteenth centuries. The map was engraved by Robert Pollard and is placed facing the first chapter of Genesis. Some features of biblical geography are confidently marked, such as Mount Ararat, 'where the Ark rested', and the Tower of Babel in Babylon on the Euphrates. Recognizing the uncertainty of the location of paradise, Wright presents his readers with various suggestions. Eden is hinted at with hatched oval-shaped circles both in Armenia (top of the map) and Mesopotamia (bottom), while a note (just below the compass, on the right) points out that another possible site for Eden is near the Holy Land. The map also illustrates the theory that paradise had once been located in an area now entirely covered by the Persian Gulf. At the end of the eighteenth century, the place which had been shown on maps for centuries as remote and unlocatable and then, after the Renaissance, as within reach and precisely mappable, was shifting to an ineffable realm of possibilities.

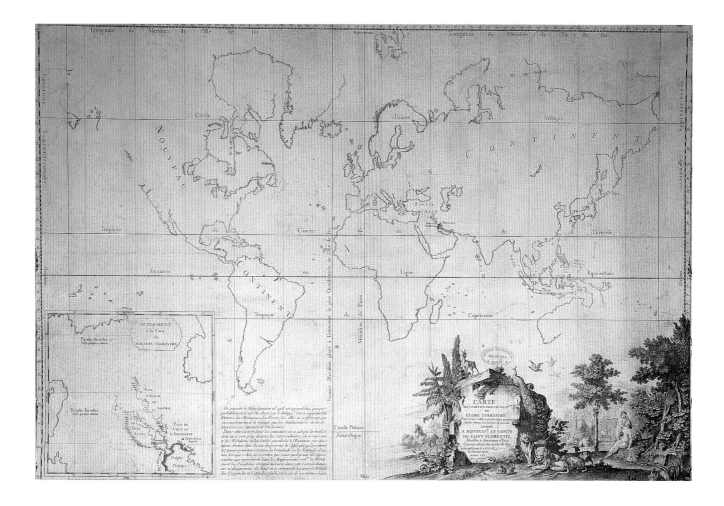

Fig. 11.2. Pierre Joseph François Luneau de Boisjermain, Carte des parties principales du globe terrestre pour servir à l'histoire des temps qui ont suivi la création (Map of the Principal Parts of the Terrestrial Globe to Illustrate the History of the Two First Centuries from the Creation of the World) (Paris: 1765). Philadelphia, Library of the American Philosophical Society, L965ter Large. The map was supposed to be part of an ambitious historical atlas of twenty maps, never published in full. As Walter Goffart has pointed out, Luneau's cartographical project intended to merge the biblical past into the wider picture of mankind's universal history.

paradise narrative even when they accept its historical validity. Rather than argue over the exact site of the Garden of Eden, they describe paradise as the state of human perfection that was lost through sin. To them, the paradise story is true in the sense that it presents, in the simple figurative style that accommodated the understanding of an ancient middle eastern audience, the two most important truths on which they base their striving for eternal salvation: the essential goodness of creation and the consequences of human disobedience. As can happen, the poet sometimes anticipates the theologian. In *Paradise Lost* (1667), the Protestant John Milton drew upon the abundant scholarly material of his day concerning the creation of the world and the Fall of Adam to expand the brief description in Genesis of the Garden of Eden in verse. As the stage for his epic about paradise lost, Milton located the Garden of Eden on the river Tigris, probably north of Seleucia, in Assyria.[13] The emphasis of Milton's poetry, however, was always on the individual's search for an inner paradise and the ultimate, celestial paradise. For Milton, the external beauty of the natural garden provided a way of revealing the inner paradise and of apprehending the mysteries of heaven. His poetry disclosed through the story of Adam and Eve the drama of the eternal struggle that takes place within the garden of the human soul. Through the depth and intensity he gave to his retelling of the biblical narrative, Milton had already moved towards presenting paradise in his epic as a symbol of a spiritual state.[14] A few years later, in a tract called *Disputatio geographica de paradiso terrestri* (1696), the theologian Balthasar Rhau wrote that if only ruins of the former Garden of Eden were to be found on earth (in his view, too, in Mesopotamia), it would be better to search for the spiritual paradise available to man in the soul, in the Church, in heaven.[15]

A Babylonian Paradise

A new development in the history of the mapping of paradise in the nineteenth century was the interest Assyriologists began to take in the question of the location of Eden. Many of these scholars had been theologically trained, but their area of professional expertise was not Christian theology as taught by the Church so much as the languages, history and antiquities of Assyria, and it was primarily their interest in the region that led them an interest in the mapping of the biblical Eden. Another factor encouraged the sudden entrance of 'lay' scholars into a field that had been traditionally reserved for biblical exegetes. The gap left by the fading interest of the professional theologians, who had ceased to question the place of paradise on the map of the world and whose minds had turned to other matters, happened to coincide with breathtaking archaeological discoveries coming from the excavation of ancient sites in the Near East and from the decipherment of cuneiform texts. The untiring efforts of field and armchair archaeologists, philologists and students of ancient texts and artefacts were shedding new light on the astonishing importance of Mesopotamian culture in the ancient world. The Old Testament could now be – indeed had to be – seen completely differently. With an enthusiasm for mapping paradise that professional biblical exegetes had long forgotten, scholars of near and middle eastern ethnography, literature and religion gave rise to a new cartography of a Babylonian Garden of Eden. Fresh life was being poured into a very old debate, which had seemed to have reached stalemate and to be on the verge of drying up.

Modern standard editions of the Holy Bible, even those Evangelical or Methodist editions most directly in the Calvinist line, such as Frank Charles Thompson's ongoing Chain-Reference Bible (1908–), do not usually offer a map of the location of Eden. They are more likely to contain an archaeological map of the Middle East. More or less at the same time as the location of paradise was being dropped from the complement of mainstream cartographical illustration of the Bible, a new subject, Assyriology, was emerging in universities and learned societies in the West. From the late nineteenth century onwards, a growing number of specialists in every aspect of the past (and sometimes the present) in Mesopotamia found they had something to say regarding the allusions in the Bible to the four rivers of paradise and the Tree of Life of the Garden of Eden. For them, the Bible was a guide to the mapping of a Babylonian Eden.

We may speculate about the motives underlying the new scholarly trend. It could be said that, after the revolution in the field of biblical studies, the Assyriologists' approach to the parallels they found between their new and exciting archaeological discoveries and the biblical account of the region was little more than a reincarnation of old ideas in disguise and the expression of a nostalgia for the reassuring, if old-fashioned, literal reading of the Bible. Assyriologists found themselves in a position to demonstrate the truth of the biblical text by means of the Mesopotamian evidence. Instead of looking as of old to the Bible for answers to their scientific questions, they turned to their scientific learning to corroborate the veracity of the biblical account. In firmly associating the new discoveries with Scripture, the potentially threatening findings of Assyrian and Babylonian antiquities could be made acceptable to Western Christian societies accustomed to a vision of their pre-Christian past as based uniquely on Graeco-Roman classical traditions. If the new discipline could be shown to be an important part of biblical history and a useful tool for the understanding of Holy Scripture, more room would be made in European museums for Mesopotamian art, and more sponsorship would be forthcoming for archaeological expeditions to the Middle East.

Assyriologists thus took their Bibles with them to excavations and metaphorically mounted the relevant extracts on their clipboards. In 1849, Austen Henry Layard suggested that there was a direct connection between mural bas-reliefs in the Assyrian palace at Nimrud depicting a 'sacred tree' and the Tree of Life of Genesis.[16] In 1881, Friedrich Delitzsch claimed openly that the reference in an Akkadian tablet to the Tree of Life described the Babylonian equivalent of the biblical paradise and interpreted a passage, in which he recognized the Sumerian word *edin*, as a direct reference to the Garden of Eden of Genesis.[17] A few years later, Archibald Henry Sayce thought he had located this *edin*, the Babylonian Garden of Eden, in Eridu, in southern Mesopotamia.[18] He also considered the Persian Gulf, described in cuneiform texts as 'the Salt River', to be the source (in the sense of mouth) of the four rivers of Genesis which flowed into it from their 'heads', or springs, in the north. In Babylonian seals depicting Ea, the god of *abzu* ('sweet water'), Sayce saw representations of the four rivers of paradise.[19] Layard, Delitzsch and Sayce were not alone in their interpretations; there were other historians and philologists who found representations of the Tree of Life of the Garden of Eden on Assyrian seals or on mural bas-reliefs.[20]

For Layard and Sayce (both good biblical scholars) the discovery of Mesopotamian culture proved that the Bible was historically trustworthy. Delitzsch, however, made different use of his knowledge of cuneiform texts and gave a new twist to the debate. He thought it was inappropriate to give the Bible so much importance. Unlike his colleagues, his aim was to stress the primacy of the ancient Mesopotamian civilization over the Jewish Bible, an approach that offended many sensitivities. In a series of lectures on *Babel und Bibel*, delivered to the prestigious German Oriental Society in Berlin in January 1902 and January 1903, and to literary societies in Barmen and Cologne in October 1904, Delitzsch showed how Hebrew and Eastern ideas, legends and myths were shared, and how the biblical world view depended on ancient Babylonian culture; 'Hebrew antiquity', he asserted, 'is linked together from beginning to end with Babylonia and Assyria.'[21] He pointed to a Babylonian story of the world's creation, a Babylonian account of the Flood, and an account of the Fall of man allegedly found on a Babylonian cylinder-seal. He also suggested that the ten antediluvian patriarchs recorded in the Bible were in fact Babylonian kings who reigned before the Flood. The thought that the Old Testament stories were of Babylonian origin dealt a fatal blow to the belief that everything in the Old Testament was original. More critically, it threatened belief in the text's divine inspiration. Not surprisingly, Delitzsch's thesis proved highly controversial. The German emperor Wilhelm II, who, with other dignitaries, had been present at Delitzsch's lecture, later dissociated himself from Delitzsch's theories, although he continued to support him as an Assyriologist.[22]

Delitzsch's persuasion of the importance of the ancient geography of Mesopotamia and neighbouring countries (Canaan, Egypt, and Elam) to the biblical text, and his insistence that the Genesis story was only a reflection of Babylonian myths led to his map of a Babylonian Garden of Eden. On this he showed Babylonia at the time of the Assyro-Babylonian Empire, identifying Cush with the land of the Kassites (Kassu) and Havilah with a region in southern Mesopotamia (Figure 11.3).[23] He placed the site of paradise on the Euphrates between Baghdad and Babylon. The four rivers were to be identified with the great canal to the west of the Euphrates that had been called the Pallacopas by the Greeks (which Delitzsch thought was the Pishon), the Shat-en-Nil (the Gihon), the lower Tigris (the biblical Hiddekel) and the Euphrates.[24]

A kind of cartographical *entente cordiale* was attempted in 1918 by William Willcocks. Willcocks was in Mesopotamia in the aftermath of the First World War as consultant to the Turkish government. He was an engineer, and his participation in the debate over the location of the earthly paradise grew out of his conviction that the problem could be solved by his knowledge and practical experience of the techniques of irrigation. In the end, he discussed the sites of the two paradises, the biblical and the

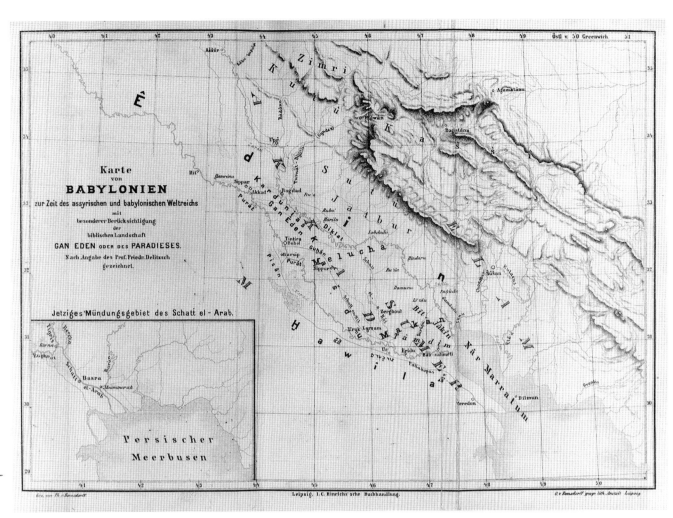

Sumerian, in his book *From the Garden of Eden to the Crossing of the Jordan*.[25] Sayce praised the pragmatic Willcocks for having transformed 300,000 acres of modern Babylonia into a Garden of Eden, but Willcocks himself claimed no expertise other than that of an engineer, describing himself as 'a novice in archaeology'.[26] For Willcocks, the key was, perhaps predictably, irrigation, which was for him the oldest applied science in the world. He traced the way, as he saw it, that ancient civilizations arose in irrigated valleys through the organization and discipline implied by a successful watering system. From this premise he went on to interpret the biblical description of the earthly paradise as indicating that the Old Testament Garden of Eden lay somewhere along the Euphrates where the Garden could have been watered by a natural flow all through the year. In his personal experience of the area, there were only two such localities: one was the region of the cataracts between Anah and Hitt, in Assyria, and the other was the marshland beginning at Nasiriyah, near the ruins of Ur of the Chaldaeans. Willcocks designated the first as the biblical Garden of Eden, the cradle of the Semitic race, and the second as the Garden of Eden of the Sumerians, their earliest settlement; he marked both on the first of the four maps in his book (Figure 11.4). He placed the biblical earthly paradise well north of the confluence of the Tigris and the Euphrates, and the Sumerian paradise in the marshlands of the Persian Gulf. The biblical Pishon corresponded to the flooded depressions of Habbania and Abu Dibis that lie between Ramadi and Kerbela and to the east of Havilah (between Egypt and Assyria), which can thus be described as 'encompassed' by the biblical river (Genesis

Fig. 11.3. Friedrich Delitzsch's map of Babylon, showing the Garden of Eden, from *Wo lag das Paradies?* (Leipzig: J. C. Hinrich, 1881). 16.5 × 23.5 cm. London, British Library, 2200.bb.13. Paradise was located by students of the ancient Near East through Mesopotamian images and texts. In this map of Babylon at the time of the Assyro-Babylonian Empire, Delitzsch situated paradise (*Gan Eden*) on the Euphrates, between Baghdad and Babylon, in the wider region of *Edin*. In the adjoining text, he described how the Hebrews shared many of their ideas, legends and myths with the Babylonians. In his view, the words *Babel* and the *Bible* were inseparable: 'the people of Israel, with its literature, appears as the youngest member only of a venerable and hoary group of nations …'

The Eclipse of the Theological Eden 349

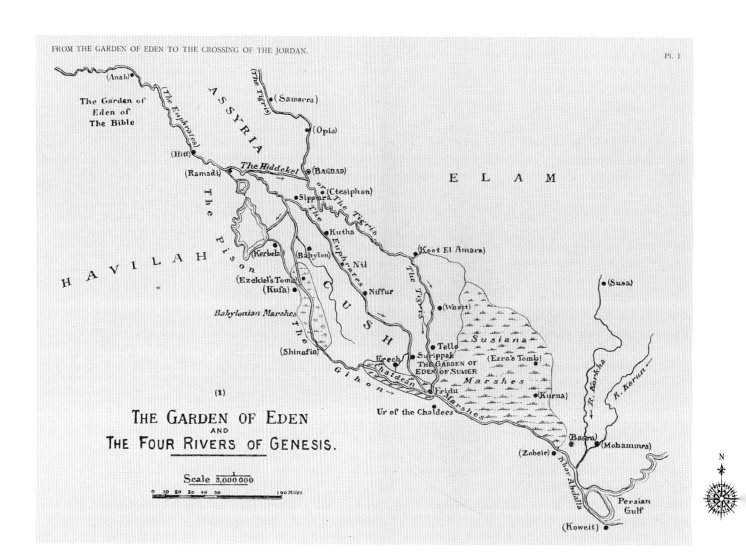

Fig. 11.4. William Willcocks, *The Garden of Eden and the Four Rivers of Genesis*, from his book *From the Garden of Eden to the Crossing of the Jordan* (Cairo: The French Institute of Oriental Archaeology, 1919). 24.5 × 29 cm. London, British Library, 03149.i.4. Scale given as 1:3,000,000. Willcocks located 'the Garden of Eden of the Bible' north of the confluence of the Tigris and the Euphrates, and 'the Garden of Eden of Sumer' in the marshlands of the Persian Gulf. He insisted that it was necessary to acquire a comprehensive understanding of the language of the Bible, noting that, for example, the independence of the United States of America would have been described biblically by the expression 'And George the Third begat George Washington, the beginning of whose dominion was Boston and New York and Philadelphia and Savannah.'

2.11). The Gihon corresponded to the modern river Hindia (or Pallacopas). The land of Cush was between Babylon and Erech. The river Hiddekel is shown on his map as a short stream flowing from the Euphrates to the Tigris at Baghdad, south of Assyria, and a branch of the Euphrates as flowing through Babylon.

In the text, Willcocks explained further details of the paradise narrative. The Tree of Life was a date palm and the Tree of the Knowledge of Good and Evil a vine, for example. He also drew attention to changes that had taken place in the region's physical geography over the last five or six thousand years, emphasizing the key point that the Euphrates had so deepened its bed in the cataract region, between Anah and Hitt, that river water could no longer overflow naturally into the gardens along the banks; waterwheels were thus needed to raise and direct the water out of the river and into irrigation channels. Without a man-made watering system, the region would be a desert. For Willcocks, the biblical Expulsion was to be explained as the eastward emigration of the local Semitic population, consequent on this process of desertification, while the flaming sword of the Cherubim was the heat haze over the lakes of bitumen and naphtha. For his explanation of the Sumerian paradise, Willcocks turned to Sayce's theory that the Babylonian plain had been formerly called Eden, although he rejected Sayce's notion of the sea entering the marshes to feed the four rivers. Willcocks's concluding hope – that paradise would be restored through the extension of irrigation in the marshland and the re-creation of the 'gardens of Eden, whose memory has lasted

350   Mapping Paradise: A History of Heaven on Earth

so long' – encapsulates the extent to which by his day the search for the earthly paradise had digressed from strictly theological realms.[27]

By the 1950s, however, the attempt to map a Babylonian paradise had become unfashionable among Assyriologists. There were just too many discrepancies between the biblical account and the picture that was emerging from the huge numbers of cuneiform texts that had been deciphered in the preceding half-century. Significantly, the years after the Second World War were the years when in European and American universities the departments of Assyriology began to be separated from the departments of biblical studies. After this divorce, what had been the mainstream approach in Assyriology became the occupation of a minority, who continued to make new claims for a southern Mesopotamian Eden. As recently as 2002, Manfred Dietrich proffered his idea, based on a comparison of the Genesis narrative with ancient texts, that the site of the Garden of Eden is at Eridu, in southern Mesopotamia, at the point where fresh river water meets the salt water of the sea (Figure 11.5). Dietrich pointed out that in Eridu there was once a temple dedicated to the Mesopotamian god who created mankind, Enki/Ea, traditionally represented in Mesopotamian art with four rivers. Like Sayce and others before him, he recognized these four rivers as precisely those mentioned in Genesis and whose mouths were in that area. He took his interpretation from a number of texts describing the creation of man in temple gardens and recording environmental changes at the head of the Persian Gulf. In the pairing of the Genesis rivers – the Pishon with the Gihon and the Tigris with the Euphrates – Dietrich also saw an allusion to the opposition of the two contemporary political powers, Elam and its implacable enemy, Babylon.[28]

Dietrich's single-minded preoccupation with the location of the earthly paradise is not shared by everybody in his field. Most mainstream Assyriologists today would not seek to map the Babylonian Eden. On the contrary, for many modern Assyriologists the naïveté of their predecessors, who took the Bible so literally and gave it such prominence in the study of the ancient Akkadian tablets, is something of an embarrassment, even though Assyriology owes its growth and many of its strengths to ideas established in the nineteenth century and often to the attempt to find archaeological evidence for biblical statements. There is still, however, an enduring 'mapping impulse' in the field. In the late twentieth century, some Assyriologists identified the Sumerian equivalent of the earthly paradise, the land of Dilmun, with Bahrain, and the effort to identify the ancient sites and match them to present-day geography is now deployed also in other directions, such as the location of the origin of monotheism.[29] Trends in Assyriology have parallelled its former academic partner, biblical studies, in the way intellectual positions have shifted between the literal and the symbolical at different times. Dietrich's theorizing is a reminder of how much of past practice and belief is carried forward into the present.

Fig. 11.5. Diagram of Manfred Dietrich's map of southern Mesopotamia showing the four rivers of paradise and the site of the Garden of Eden at Eridu (2002). According to Dietrich, Genesis lists the four rivers in geographical order from east to west. The Pishon would be the present-day Karun, which thus would have cut through the region of Elam and surrounded the land of Havilah, in south-east Iran. Havilah was an important mining district and source of gold in Antiquity, and it was also a key route centre for connections between the Indus valley, Arabia and the Mediterranean. The Gihon, for Dietrich, is the western stream of the region of Elam and is today known as the Kercha. The land of Cush is the land of the Kassiten (Kassu), the mountainous region north of Susiana, in the northern part of Elam. Three of the paradisiacal rivers flowed into a lagoon called Nar Marrati, 'the Salt river'; the Gihon, in fact, did not flow into the lagoon, but was close enough to be associated with the Pishon. Eden was thus in the area where the mouths of the four rivers flowed from Mesopotamia and Elam, and paradise was located at Eridu.

Ways of Thinking: Mainstream and Fringe

Looking back over two millennia of Christian exegesis, the degree to which mainstream thinking and so-called fringe thinking are but two sides of the same coin becomes clear. The two roles are interchangeable. Ideas that today are held by a minority, and considered at or close to the fringe, belonged yesterday to the mainstream. Nobody in the established Church today, either Catholic or Protestant, would ascribe major theological importance to the issue of the location of the earthly paradise, let alone attempt to indicate its site on a map. Yet, while theologians have, like the Assyriologists, turned their back firmly on the issue of the location of paradise, maps purporting to demonstrate the former whereabouts of the Garden of Eden are still being published in the name of religion. Their authors come from groups who, like the Jehovah's Witnesses, adhere to a highly literal interpretation of the biblical text (Figure 11.6).[30] It should not be overlooked, however, that isolated Assyriologists like Manfred Dietrich and religious literalists draw the substance of their arguments from notions once aired in mainstream scholarship. As we have seen throughout this book, from Augustine onward all those who placed the earthly paradise on a map of the world or (like Calvin) indicated its location on a regional map were at the forefront of Christian thought and were – in their time and, in many cases, for years afterwards – accepted as leading authorities in the interpretation of the Bible. The pendulum has been swinging from symbolic to literal interpretation and back again. Today, it happens to be more fashionable to read the Bible for a symbolic message than it was at other times in the past, but there were periods (notably between Augustine and Calvin) when the uncompromisingly symbolical stance set by Origen was disowned by leading exegetes. The tendency, then, was to pair symbolic interpretations with the fringe and literal interpretations with the mainstream. Today, the tendency in ecclesiastical and academic establishment thinking is to keep a guarded distance from those who insist on a literal interpretation of the narrative recorded in the Book of Genesis.

The long survey presented in this book on the mapping of paradise and the search for the place of heaven on earth enables us to appreciate how much the passionate impulse to map a garden of delight – a place for human perfection – has been shared by mainstream and fringe thinkers alike. Many, if not most, nineteenth- and twentieth-century maps of the earthly paradise have come from the intellectual fringe. Some of the places suggested in the last two centuries for the former Garden of Eden may take us by surprise, but most of the underlying ideas were not in fact all that original. The essence of each 'new' argument can often be found to be rooted in one or other of the earlier debates, with the notion that paradise was distant in time and exotic in nature sounding as a *basso continuo*. Of all the localities suggested over the last five centuries for paradise, the confluence of the Tigris and the Euphrates has proved the most enduring. However, some Renaissance and medieval ideas have also reappeared in modern guise.[31] There has been no waning in the association of the earthly paradise with Jerusalem. So we find, in 1801, Richard Brothers sketching a plan in which the Garden of Eden is shown in the centre of the built-up area of Jerusalem.[32] Self-assurance has never been lacking amongst those on the fringe. In 1796, Carl Josef Michaeler proposed a Mesopotamian paradise as the last word on the problem with his optimistic title of *Das Neueste über die geographische Lage des irdischen Paradieses* (*The Latest Evidence on the Geographical Location of the Terrestrial Paradise*), which he supplied with a map.[33]

Other places for the site of paradise have been found that echo earlier theories. In 1885, Moritz Engel placed paradise in an oasis in the Syrian desert, at Ruhbe, east of Damascus (Figure 11.7), and in the same year, William Fairfield Warren located paradise at the North Pole.[34] The germ of Warren's argument was that the Eden of

Fig. 11.6. Paradise in the Near East, according to the Jehovah's Witnesses' publication *From Paradise Lost to Paradise Regained* (Brooklyn, NY: Watchtower Bible and Tract Society of New York, 1958). 16×21.5 cm. The Jehovah's Witnesses take a literal view of the biblical text. In their opinion, the Genesis account of Creation and paradise was written by Adam himself, who wrote down what God told him about the creation of the universe and what had happened to himself and his wife Eve. The days of creation described in Genesis are taken not as consisting of twenty-four hours each, but as enduring for 7,000 years. Thus, near the end of the sixth creative day, nearly 42,000 years since he 'created the heavens and the earth', God gave life to man. Among the vignettes on the map, which places biblical history in its middle eastern setting, are scenes relating to paradise, which appears to be located along the Tigris, not far from Mount Ararat. Although the whole primitive earth is described as a beautiful paradise, the adherence to the letter of Scripture and the account of the Expulsion promotes the understanding amongst Jehovah's Witnesses of paradise as a secluded area. Jehovah's Witnesses also point out that the Bible, in which it is recounted how paradise was lost, also says how paradise will be restored to the earth.

primitive tradition, the cradle of the human race, had to have been at the natural centre of the northern hemisphere in a country submerged at the time of the Flood. He thought the Arctic Pole would have been the most likely place, in view of the extraordinary prevalence of light there: six months of daylight, four months of twilight/dawn, and two months lit up by the stars, the moon and the *Aurora Borealis*. Added to this, the unparalleled intensity of magnetic and electric forces at the Pole would have encouraged the development of higher forms of life. Traces of that polar paradise, Warren pointed out, can be found in world mythology. He also described the hydrography of paradise. The original single river came from the sky, originating in the rain that falls from heaven, which Warren envisaged as a celestial stream with celestial headwaters. The division of the single river took place on the heights at the Pole, and the four resulting rivers formed the chief streams of the circumpolar continent as they descended in different directions to the surrounding sea.[35]

As always, imaginative theories spawned imaginative maps. General Charles Gordon took a keen interest in sacred history, and in 1881 he claimed to have discovered the site of the lost Garden of Eden on Praslin, the largest island of the Seychelles, whose entire topography he found described in Genesis. The tree *Lodoicea Seychellarum*,

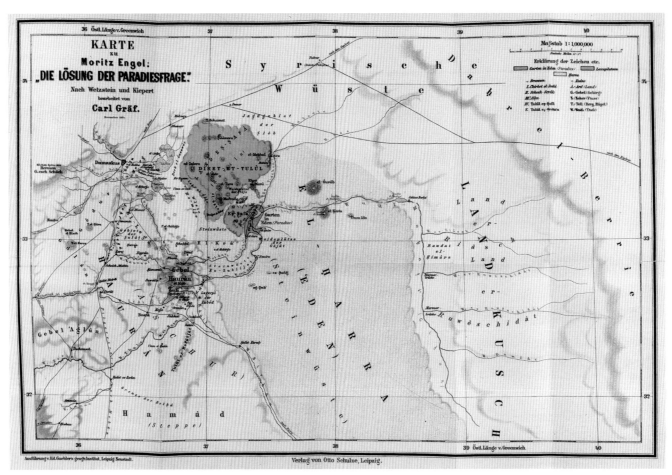

Fig. 11.7. Moritz Engel's map of paradise, from his book *Die Lösung der Paradiesfrage* (*The Solution to the Paradise Problem*)(Leipzig: Otto Schulze, 1885). 30 × 75.5 cm. London, British Library, 3155.cc.1. Engel, like several writers before and after him, claimed to have the final word on the subject. His map, drawn by Carl Gräf, illustrates Engel's thesis that Eden was in the Syrian desert. The Garden of Eden was in an oasis at Ruhbe, southeast of Damascus, where the four rivers flow out of their common source, not far from the meridian 37° E of Greenwich and the parallel 33° N of the equator.

whose strange fruit resembled the female pudenda, was formerly the Tree of Knowledge of Good and Evil, and the bread-fruit palm, *Artocarpus incisa*, was the former Tree of Life. The local climate matched the description of an agreeably temperate paradise, and the three-foot long snakes that abounded on the island were relicts of the Genesis narrative. Gordon illustrated his theory with a map that was posthumously published in the biography of him written by his brother Henry William together with two marginal diagrams showing how, according to Gordon, the four 'heads' of the stream described in Genesis ran into the river of paradise and not, as was usually thought, out of it (Figures 11.8 and 11.9).[36] Thus, instead of a single river flowing from Eden to water the Garden, which then branched into four, Gordon saw four 'heads', or headwaters – those of the river Gihon, a now dessicated course south of Jerusalem, the Nile (that Gordon identified as the Pishon), the Tigris and the Euphrates – flowing into one river to water the Garden. Before the Flood, Eden had spread over a huge region now covered by the Indian Ocean.

Gordon was not alone in looking to the Indian Ocean in a search for paradise. In 1905, A. P. Curtis printed a map called *The Land of Eden and Havilah*, showing the Indian Ocean and surrounding continents, Asia, Africa and Australia (Figure 11.10).[37] Printed boldly across the Indian Ocean, into which paradise had been submerged, are the words 'The Land of Eden'. The postdiluvian coastlines follow the courses of the four rivers. Pencilled notes in what appears to be Curtis's own hand on the three maps deposited at the British Library indicate that the single stream is the Jordan and that this river unites all the waters surrounding the land of Eden. The Pishon divides Eden from the Eastern Ocean and encircles the land of Havilah, which is Australia, with its wealth

354   Mapping Paradise: A History of Heaven on Earth

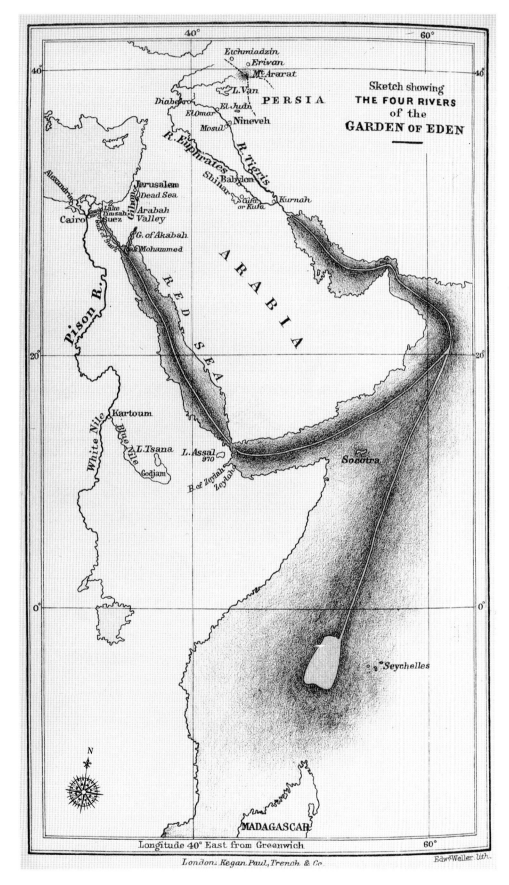

Fig. 11.8. Charles G. Gordon's *Sketch Showing the Four Rivers of the Garden of Eden*, from Henry William Gordon, *Events in the Life of Charles George Gordon, from its Beginning to its End* (London: Kegan Paul Trench, 1886), between pp. 262 and 263. 20.5 × 12 cm. London, British Library, 2406.f.11. Gordon's map shows that paradise was not far from the equator, in Praslin, the largest island of the Seychelles, where, according to Gordon, all the features mentioned in Genesis can be found. Gordon suggested that the Tigris and the Euphrates had come together in Mesopotamia to flow south to connect with the great central river of Eden. The Gihon and the Pishon joined each other at a point now covered by the Red Sea and then flowed as a single river to the north-east of what became after the Flood the Island of Socotra. He identified the Nile as the Pishon and the region of Godjam – 'rich in gold' – near the Blue Nile as the land of Havilah. He also saw a parallel between Babylon and Israel. Babylon was placed on the Euphrates and Nineveh, its enemy, on the Tigris. Israel was watered by the Gihon, and Egypt, its enemy, by the Pishon.

The Eclipse of the Theological Eden   355

Fig. 11.9a. Charles G. Gordon's marginal diagram of *The Four Rivers and the Great River*, from Henry William Gordon, *Events in the Life of Charles George Gordon, from its Beginning to its End* (London: Kegan Paul Trench, 1886), p. 262. 2.5 × 3 cm. London, British Library, 32406.f.11. Gordon held that before the Flood all the rivers of the world originated in water from melting ice at the two poles and that they all flowed towards a habitable zone around the equator. He acknowledged that a certain Reverend Bery (who may have read Calvin) had suggested the possibility of reading the four 'heads' (or springs) of Genesis as the 'origins' of the single river and not as being originated by it. The four river sources were therefore the upper part of the paradise river, running into rather than out of it.

Fig. 11.9b. The second of Charles G. Gordon's marginal diagrams, ibid., p. 263. 3 × 3 cm. Instead of one river flowing from Eden to water the Garden and then branching into four (as shown in this sketch), Gordon thought of four rivers, from four head streams, flowing into one to water the garden (as in Figures 11.8 and 11.9a).

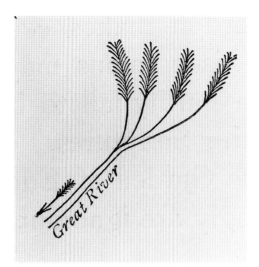 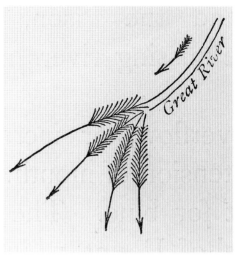

of pure alluvial gold; the Gihon surrounds the land of Cush, which is Africa; and the Tigris or Hiddekel flows from Assyria, and the Euphrates (*Perath* or *Prat*) defines the southern coast of Eden. Curtis clearly had in mind the old idea that before the Flood paradise existed on a very different-looking earth from that of modern times, so he showed on his map the antediluvian lands, later submerged by the Flood, just as Lapeyre had done in the seventeenth century.

Paradise was placed in more familiar regions by other twentieth-century writers: in the Arabian desert, by the Orientalist Albert Herrmann (1931), and, more recently, by Kamal Salibi (1996); in Mesopotamia, by Richard Hennig (1950); in Armenia, by Lars Ringbom, who even pointed to a specific plateau near the Caspian Sea (1958), and by David Rohl (1998) (see Prologue, Figures 0.4 and 0.5).[38] Again as in the past, identifying the former site of paradise could provide an opportunity for promoting nationalism, as in the case of Franz von Wendrin who claimed, not long after the First World War, that the earthly paradise was formerly in eastern Germany on the Mecklenburg-Pomeranian boundary at Demmin (Figure 11.12).[39] Others, giving a new twist to another old idea, claimed that the earthly paradise was above the earth, in the heavenly sphere. Thus, in 1935, James A. Nicholson dismissed the Garden of Eden as an invention elaborated by fanatical Jewish priests out of a Babylonian myth, insisting instead that it occupied the space within the circle of circumpolar stars when viewed from the latitude of Mesopotamia (Figure 11.11).[40] Nicholson explained that the Expulsion represented the driving out of the Man (the constellation Cepheus) and the Woman (the constellation Cassiopeia) by the precession of the equinoxes, although they would return in the future, in about 3,500 years' time.

The idea of a submerged Eden also had its twentieth-century advocates. In 1962, the Baptist Albert R. Terry claimed to have received confirmation directly from God, through dreams and signs, that the site of Eden lay below the Mediterranean Sea.[41] Terry's map shows the site of the former Garden to the south of Crete and Cyprus (once mountains, now islands), with the Pishon in what is now Turkey (either the Halys or the Sarus), and the Gihon as the Nile. It also shows how the four rivers of paradise had been diverted by the Flood – 'no doubt the greatest spectacle on earth' – which in wiping everything out had left the great bend of the Euphrates (Figure 11.13).[42] Then, right at the end of the twentieth century, in 1998, there was posted on the Internet the not entirely novel claim that paradise had been at the centre of the world (Figure 11.14).[43] The point of departure for this particular unveiling was the *Table of Nations* in the mid-second-century rewriting of Genesis and part of Exodus (the *Book of Jubilees*)

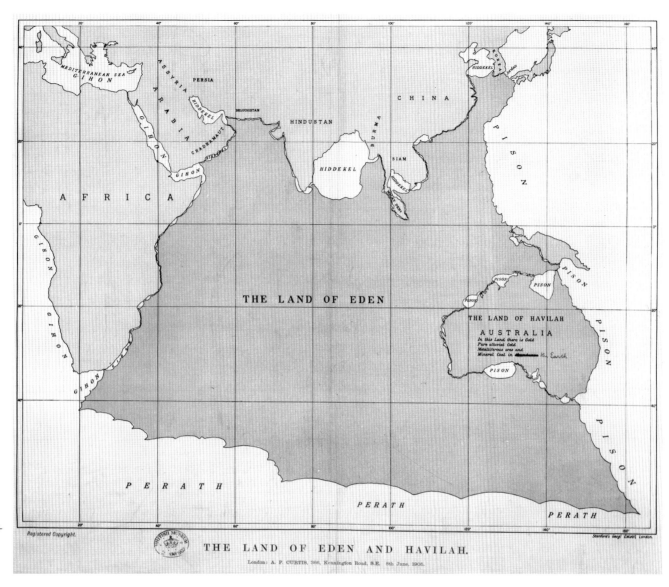

Fig. 11.10. A. P. Curtis, *The Land of Eden and Havilah* (London: the author, 1905). 42 × 33.5 cm. London, British Library, 2406.f.11. Curtis showed the land of Eden as a vast region now submerged by the Pacific and Indian oceans. He believed that the European search for a southern continent in the Pacific and Indian oceans, which ended with Captain James Cook's voyages and resulted in no such discovery, had originated in a chart dating from before the eighteenth century on which the continent was represented as if above sea level. All that remained of this now submerged land were a few islands, rocks and barrier reefs scattered throughout the sea.

that lists the 70 tribes or peoples named in Genesis 10 according to their distribution amongst the three sons of Noah.[44] From the list and from the statement in the *Book of Jubilees* that Eden, Mount Sinai and Sion are three holy places 'the one facing the other' (8.19), the Internet geographer was inspired to describe the location of the Garden of Eden as in the delta of the river Nile, where Moses had been born, and exactly at the centre of the terrestrial landmass (including the Eurasian, African, and American continents), or 32° E, 32° N. The Internet seems, at the time of writing, to be fast becoming the home of the current mapping of the location of paradise. Three papers produced between 2001 and 2005 have also been posted on the Internet, all by a mathematics lecturer from the University of Bergamo in northern Italy, Emilio Spedicato, who claims that the Garden of Eden was situated in the Hunza Valley in northern Pakistan, not far from the Pamir.[45] Like several of his predecessors, he supports his suggestion by emphasizing the effect of the Flood on the courses of the four rivers (which he sees as having been further altered by subsequent natural catastrophes) and by pointing to past misleading translations and present revealing etymologies in support of his thesis. The river 'going out of the place of delight, to water paradise', later divided into four (Genesis 2.10) was, according to Spedicato, a 'snow field'; the Gihon was the river presently

The Eclipse of the Theological Eden 357

named as Pandji-Amu Darya; the Pishon was the river now variously called Mastuj, Yarkhand, Konar, Kabul, joining the Indus after the city of Peshawar; the Hiddekel was the river Mintaka-Tarim; and the fourth river, called in Hebrew פְּרָת (*Prat*), was the Hunza river: its name meant 'the river of food/fertility'. The place called Assyria refers possibly to the city of Tashkurgan (whose name means 'the gate to the mountains of the Garden of Eden'); the land of Cush corresponds to present Hindu Kush; the land of Havilah (whose name probably means 'land of Abel') is the present region of Kabul

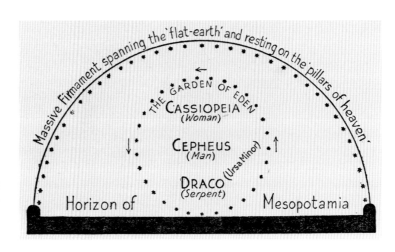

Fig. 11.11. James E. Nicholson, *The Garden of Eden*, from his book *The Probable 'Garden of Eden': A Long-Forgotten Star Myth* (London: Watts, 1935, for private circulation only), p. 10. 5.5 × 9.5 cm. London, British Library, 08560.f.13. Nicholson suggested that the Garden of Eden was just a portion of the sky, with Adam, Eve and the Serpent being only constellations: Cassiopeia, Cepheus, and Draco.

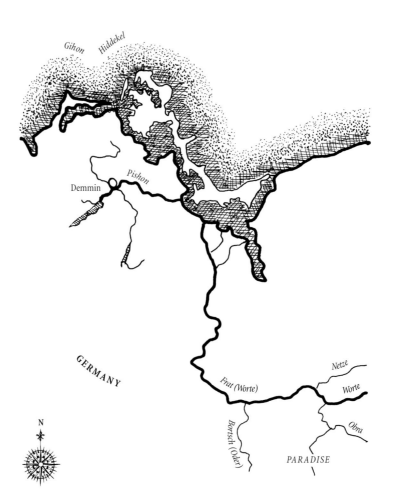

Fig. 11.12. Diagram of Friedrich von Wendrin's map of paradise, from his book *Die Entdeckung des Paradies* (Leipzig: Braunschweig, 1924). Von Wendrin argued that the Flood had altered the configuration of the earth and thought he recognized the four rivers of Genesis in the river system of eastern Germany.

358   Mapping Paradise: A History of Heaven on Earth

(which means 'soul of Abel'). The biblical word Cherubim referred to the Khunjerab pass. All these identifications, Spedicato argues, are confirmed by a range of Sumerian texts and by the apocryphal books of Enoch, which preserved a memory of real places and events. Finally, Spedicato provides a map of his Pakistani paradise, which he has compiled by marking the courses of the four rivers on a pre-existing map of the region.

The beginning of the twenty-first century, we see, has not brought an end to the theorizing about the location of the former earthly paradise. What was once a torrent of suggestions may have dwindled to a trickle, but the same old ideas keep resurfacing, for instance in publications based upon computer generated geographical information, just as they did in the nineteenth century.[46] In 1871 the conflation of the Gihon with the Nile was revived when David Livingstone expressed his conviction that if he could discover the sources of the Nile he would be led to discover the site of the original paradise.[47] Livingstone was followed by Gerald Mossey, for whom equatorial Africa was the most likely site of the Garden of Eden.[48] After the changes in the European cultural, intellectual and religious contexts during the eighteenth century under the influence of the Enlightenment, the problem of the location of the Garden of Eden was taken up in various secular disciplines.[49] The outstanding Swedish botanist Carl von Linné, for example, cited paradise as the source of the world's flora and fauna. Von Linné explained that the astonishing variety of the world's plants and animals was a result of their propagation from the primordial island of paradise and their dispersal, as the waters covering the rest of the globe gradually receded, to colonize the emerging landmasses.[50] Johann Gottfried Herder discussed the origin of languages in India and,

Fig. 11.13. Albert R. Terry's map of the Garden of Eden and the four rivers, from his book *The Flood and Garden of Eden: Astounding Facts and Prophecies* (Elms Court, [Ilfracombe, Devon]: Arthur H. Stockwell, 1962; repr., 1963), p. 12. 10 × 14 cm. London, British Library, 311.cc.44. The Garden of Eden, Terry insisted, was submerged by the Flood and its site is now under the Mediterranean. The four rivers, indicated on the maps by dotted lines, either disappeared or their courses were altered and their names changed. He claimed that his theory about the location of Eden had been directly inspired by God. In a text composed of numbered paragraphs and packed with scriptural quotations, he warned that the Apocalypse could be imminent because of the armaments race and technological progress.

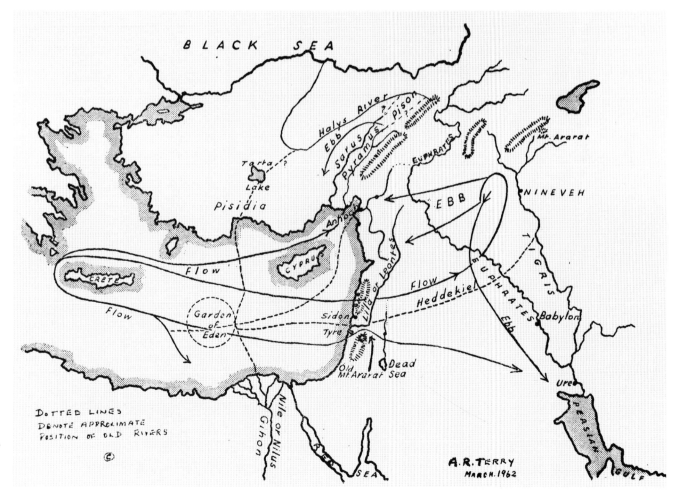

Fig. 11.14. Diagram of Chris Ward's map of Eden as the centre of the world (1998). Ward claimed to be the first to unveil the true and exact location of the Garden of Eden. Since the *Book of Jubilees* states that Eden, Sinai and Sion face each other, he drew a line to connect Sion and Mount Sinai; then, by creating an equilateral triangle from this line, he identified the site of the Garden of Eden at the opposite angle, in the vicinity of the city of al-Mansurah. From this Egyptian Eden, Adam and Eve were banished east out of the Garden, towards the region generally considered to be the cradle of civilization, Mesopotamia.

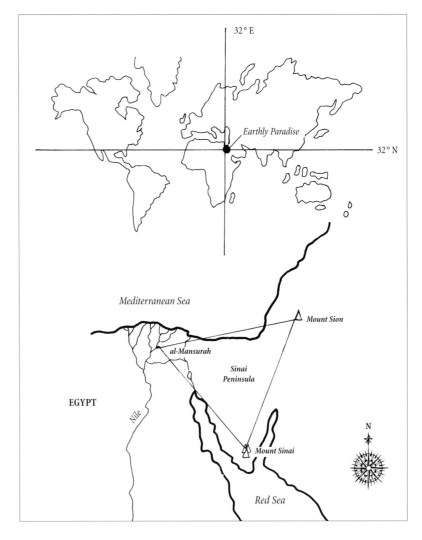

as had almost every scholar before the Renaissance, saw the Ganges as one of the rivers of paradise.[51]

More generally, the traditional theological way of thinking of the Garden of Eden as the home of the first man and woman, Adam and Eve, was transformed in academic circles into the scientific search for the origin of mankind. In this form, the question has involved mainstream scholars as well as those on the fringe. Of course, the concept of the original site of humankind is not the same as the idea of paradise. Comparative philologists, mythologists and archaeological ethnographers were not searching for the delightful place of the biblical paradise but for the the cradle of mankind. Their solutions – many including maps in support of their ideas – ranged from Greenland to Central Africa and from America to Central Asia.[52] The favoured hypothesis (adopted by, among others, the nineteenth-century pioneering scholars Graf, Durazzo and Coli, discussed in Chapter 1) was that the original birthplace of mankind was in Central Asia, either on the high plateau of the Pamirs or in a now submerged prehistoric continent below the northern Indian Ocean called Lemuria.[53] The problem of accounting for the arrival of human beings in the Americas led to the idea of a 'Lost Atlantis', a former land connection between the eastern and western hemispheres, of which the archipelagos of the Canaries, Madeira, and Azores may be relicts.[54] Maps were compiled to demonstrate the progressive dispersion of the human race over the globe according to these theories. As the nations of Europe began to colonize the New

World, especially in the eighteenth century, their encounters with indigenous peoples were liable to be coloured by their ideas about paradise. For Diderot, for example, the simplicity of the languages and the 'natural' way of life of native Tahitians were 'close to the origin of the world', and similar ideas were expressed about the islanders of Mauritius and St Helena.[55] The idea of paradise also inspired the creation of botanical gardens for didactic purposes and for the study of natural history; it stimulated nationalistic theories; and it was invoked in the context of palaeontological discoveries of early man.[56] After paradise was located in Africa in the fifteenth century, the claim was made 500 years later that the origin of mankind was in eastern Africa, in Kenya or Ethiopia, on the grounds of archaeological evidence in these regions of the earliest hominids.[57]

Enquiries about the original cradle of the human race, however, were no longer based on a sacred text, nor did they have recourse to mythology or religion, nor did they theorize a 'paradise' or garden of any kind. The failure to find a final answer, therefore, was not explained by the difficulties, inevitable uncertainty and mystery that surround a God-given account. The main reason given for the difficulty of plotting precisely the original site of mankind was that the empirical evidence needed to verify such a discovery was insufficient and unable to form the basis of a conclusive theory. The continuing elaboration of all these old themes, however, does bring to mind William Fairfield Warren's comment: 'Evidently the naturalists and the ethnologists, the comparative mythologists and *Kulturgeschichtsschreiber*, have not yet solved the problem. Their "mother-region" of the human race is as elusive and Protean as are any of the terrestrial Edens of theology, or of legend, or of poetry.'[58]

Attempts to locate paradise on a map are today viewed by mainstream theologians as the vain exercises of fringe and fundamentalist mavericks. As we have seen in this book, however, mainstream thinking in the past was very different. A literal approach to the Bible has nearly always been an integral part of theological history, and we should not forget that it is the theology of profound thinkers like Augustine and Calvin that lies at the origin of modern mainstream theological thinking. In biblical exegesis, the see-saw of opposing tendencies lasted for centuries rather than (as has been the case with Assyrian studies) just decades, but in biblical studies and Assyriology alike the superseded legacies of the past were quickly swept under the carpet. However much they may have been superseded, these forgotten ways of thinking cannot be simply dismissed out of mind, for they have contributed to present ways of thinking. They are an indissoluble of intellectual history. We set out wanting to understand the medieval and Renaissance mapping of the earthly paradise. We have found that nineteenth- and twentieth-century cartographers and historians have been far too quick to brand as simple-minded and naïve the products of a history of which they themselves are part. In the final analysis, we seem to have learnt from the enduring debate over the location of paradise more about the people involved than about the issue itself. Different situations and contexts offer different choices and approaches as the winds of history blow scholarly thought first this way and then that. Whether a medieval map maker, a Renaissance biblical scholar, or a Middle or Near East archaeologist is involved, we have come to appreciate that the dynamic tension between the symbolic and the literal approaches may represent an inherent and fundamental polarity in the human mind. Both approaches merit to be even-handedly credited with intellectual integrity and imaginative vigour.

1. According to John MacClintock and James Strong, eds, *The Cyclopedia of Biblical, Theological, and Ecclesiastical Litterature*, 12 vols (New York: Harper and Brothers, 1867—87), III (1873), 'Eden', pp. 52—5, the Garden of Eden was located in Armenia. The problem was abandoned as insoluble by William A. Wright, 'Eden', in William Smith, *Dictionary of the Bible: Comprising its Antiquities, Biogeography, Geography, and Natural History*, 3 vols (London: John Murray, Walton and Maberly, 1863), I/1, pp. 482—7 (see also the epigraph to the Epilogue). Christian Friedrich Dillmann, 'Eden', in Daniel Schenkel, ed., *Bibel-Lexikon: Realwörterbuch zum Handgebrauch für Geistliche und Gemeindeglieder*, 5 vols (Leipzig: Brockhaus, 1869—75), I (1869), pp. 42—50, placed paradise in the Himalayas, north of India. In the chief Roman Catholic encyclopedia, Heinrich Joseph Wetzel and Benedikt Welte, eds, *Kirchenlexikon oder Encyklopädie der katholischen Theologie und ihrer Hilfswissenschaften*, 12 vols (Freiburg im Breisgau: Herder, 1886—1901), IV (1886), 'Eden', p. 120, located the antediluvian paradise in Armenia. In the entry on *paradis* in Frédéric Lichtenberger, ed., *Encylopédie des sciences religieuses*, 13 vols (Paris: Sandoz et Fischbacher, 1877—82), the whole story of Genesis 2 is declared a 'philosophical myth'. Francis Brown, 'Eden', in the biblical encyclopedia edited by Schaff on the basis of Johann Jacob Herzog, *A Religious Encyclopaedia: or Dictionary of Biblical, Historical, and Practical Theology: Based on the Real-Encyklopädie of Herzog, Plitt, and Hauck*, ed. by Philip Schaff and others, 3 vols (New York: Funk & Wagnalls, 1882—4), I (1882), pp. 689—92), enumerated a variety of opinions advocated by others, but refrained from expressing any opinion of his own. Rudolf Rüetsch, 'Eden', in Herzog's encyclopedia on which Schaff based his work, Johannes Jacob Herzog, Gustav Leopold Plitt, Albert Hauck, eds, *Real-Encyklopädie für protestantische Theologie und Kirche ...*, 2nd edn, 18 vols (Leipzig: Hinrichs, 1877—88), IV (1879), pp. 34—8, declared it necessary to deny the story of Eden a strictly historical character, branding it as 'a bit of mythical geography': 'Alle diese versuche ... sind verfehlt und mussten fehlschlagen ... wegen der Beschaffenheit der Schilderung Genesis 2 selbst, welche nun einmal der Natur der Sache nach keinen streng-historischen Charakter hat, sondern das Gebilde der dichtenden Sache, ein Stück mythischer Geographie ist' (p. 35). Wilhelm Pressel, however, contributor to the same encyclopedia (entry 'Paradies'), made an elaborate argument in favour of the location of paradise at the junction of the Tigris and the Euphrates. See his map with a Mesopotamian Garden of Eden near Qurna, *Karte des Orbis Mosaicus*, folded at the end of his *Geschichte und Geographie der Urzeit von der Erschaffung der Welt bis auf Mose* (Nördingen: Beck, 1883).
2. Paul Wright, ed., *The Complete British Family Bible: Being a New Universal Exposition and Commentary on the Holy Scriptures ... Illustrated with Notes and Annotations, Theological, Critical, Moral, Historical, Chronological and Explanatory* (London: for Alex Hogg, 1782). Wright describes himself on the title page as 'Vicar of Oakley and Rector of Snoreham [sic] in Essex & late of Pembroke Hall, Cambridge'.
3. For paradise in the Holy Land, the map quotes Heidegger, Hardouin and Le Clerc; for paradise in Armenia, Reland, Calmet, and Sanson; for paradise in Mesopotamia, Calvin, Bochart, Wells, Huet, and Patrick.
4. The unidentified 'celebrated Divines' were presumably clerics at the University of Cambridge.
5. Wright, ed., *The Complete British Family Bible* (1782), sig. B2v (Genesis 2.8).
6. Pierre Joseph François Luneau de Boisjermain, *Carte des parties principales du globe terrestre pour servir à l'histoire des deux premiers siècles depuis la création du monde* (Paris: 1765). See Goffart, *Historical Atlases* (2003), pp. 156, 490–1. The inscription reads: 'On présente le globe terrestre tel qu'il est aujourd'hui, quoique probablement il a été alteré par le Déluge'.
7. Goethe was quoted by Friedrich Delitzsch, *Babel und Bibel: Ein Vortrag* (Leipzig: Hinrichs, 1902), p. 32: 'Jean Astruc, welcher im Jahre 1753, wie Goethe sich ausdrückt, zuerst "Messer und Sonde an den Pentateuch legte"'.
8. On the documentary hypothesis and Old Testament criticism see Hans-Joachim Kraus, *Geschichte der historisch-kritischen Erforschung des Alten Testaments*, 3rd edn (Neukirchener-Vluyn: Neukirchener Verlag, 1982), pp. 155–7; Hans W. Frei, *The Eclipse of Biblical Narrative: A Study in Eighteenth- and Nineteenth- Century Hermeneutics* (New Haven–London: Yale University Press, 1974). See also Cees Houtman, *Der Pentateuch: Die Geschichte seiner Erforschung neben einer Auswertung* (Kampen: Kok, 1994); Emil G. Kraeling, *The Old Testament since the Reformation* (London: Lutterworth, 1955). A summary of the issues involved is in Raymond E. Brown, Joseph A. Fitzmeyer, and Roland E. Murphy, eds, *The New Jerome Biblical Commentary* (Englewood Cliffs, NJ: Geoffrey Chapman, 1968; repr. 1990), pp. 3–13, 1146–65.
9. Since Leo XIII's encyclical *Providentissimus Deus* (1893), the Catholic Church has seen the advantage of scientific, linguistic and exegetical studies and taught that the views of the biblical authors in matters of science were not invested with scriptural infallibility. On the various pronouncements of the Catholic Church, see Brown, Fitzmeyer, and Murphy, eds, *The New Jerome Biblical Commentary* (1968; 1990), pp. 1166–74.
10. In the encyclical *Divino afflante Spiritu* (1943), Pius XII encouraged Catholic biblical scholars to take every advantage of modern scholarship and criticism. He also declared that the theory of evolution was a scientific hypothesis that did not contradict Christian faith (*Humani generis*, 1950). In 1961 came the first warning from the Holy Office of the dangers of scandalizing the faithful, and another was issued in 1964 by the Pontifical Biblical Commission. More recently, in a letter to the Pontifical Academy of Sciences (22 October 1996), John Paul II stressed that the Church's teaching relates to the wider horizon of man's position within the cosmos and that it objects only to reductive materialistic readings that do not refer to a transcendental dimension. See Saturnino Muratore, 'Magistero e Darwinismo', *La Civiltà Cattolica*, 3518 (1997/I), pp. 141–5; Brown, Fitzmeyer, and Murphy, eds, *The New Jerome Biblical Commentary* (1968; 1990), pp. 1164–5, 1173.
11. *Encyclopedia of Biblical Theology*, ed. by Johannes B. Bauer, 3 vols (London–Sydney: Sheed and Ward, 1970), II, p. 631.
12. Michael Ramsey, Archbishop of Canterbury, 'The Authority of the Bible', in *Peake's Commentary on the Bible*, ed. by Matthew Black and Harold H. Rowley (London: Nelson, 1962), pp. 1–7. See also Samuel H. Hooke, 'Genesis', ibid., pp. 175–207, esp. 175–81. On Genesis and the various Protestant theological traditions see, for example, Frederick Gregor, 'The Impact of Darwinian Evolution on Protestant Theology in the Nineteenth Century', in David C. Lindberg and Ronald L. Numbers, eds, *God and Nature: Historical Essays on the Encounter between Christianity and Science* (Berkeley–Los Angeles–London: University of California Press, 1986), pp. 369–90.
13. Joseph E. Duncan, *Milton's Earthly Paradise* (Minneapolis: University of Minnesota Press, 1972), p. 220.
14. See Helen Wilcox, 'Milton and Genesis: Interpretation as Persuasion', in Gerard P. Luttikhuizen, ed., *Paradise Interpreted: Interpretations of Biblical Paradise: Judaism and Christianity* (Leiden–Boston, MA: Brill, 1999),

pp. 197–208, and Duncan, *Milton's Earthly Paradise* (1972), esp. pp. 15–18, 29–37, 115–87, 220–68; Milton's use of geographical lore has been the object of various studies: see Robert R. Cawley, *Milton and the Literature of Travel* (New York: Gordian Press, 1970), and Alan H. Gilbert, *A Geographical Dictionary of Milton* (New York: Russell and Russell, 1968).

15 Balthasar Rhau, *Disputatio geographica de paradiso terrestri* (Frankfurt an der Oder: Iohannes Coepselius, 1696), pp. 47–9.

16 Austen Henry Layard, *Nineveh and its Remains*, 2 vols (London: John Murray, 1849), II, p. 472.

17 Friedrich Delitzsch, *Wo lag das Paradies?* (Leipzig: W. De Gruyter, 1881), pp. 79–80 and n. 41.

18 Archibald Henry Sayce, *The Religions of Ancient Egypt and Babylonia: The Gifford Lectures on the Ancient Egyptian and Babylonian Conception of the Divine: Delivered in Aberdeen* (Edinburgh: T. and T. Clark, 1902), p. 385; *The Hibbert Lectures, 1887: Lectures on the Origin and Growth of Religion as Illustrated by the Religion of the Ancient Babylonians* (London: Williams and Norgate, 1887), pp. 238–42, 238.

19 Sayce, *The Religions of Ancient Egypt and Babylonia* (1902), p. 385.

20 See the historiographic study in Mariana Giovino, *A History of Interpretations of the 'Assyrian Sacred Tree': 1849–2004*, PhD Dissertation, University of Michigan, Ann Arbor (2004).

21 Friedrich Delitzsch, *Babel and Bible: A Lecture on the Significance of Assyriological Research for Religion Delivered before the German Emperor*, transl. by Thomas J. McCormack (Chicago: Open Court Publishing Company, 1902), p. 3; *Babel und Bibel* (1902), p. 5: 'das Hebräische Altertum von Anfang bis zu Ende gerade mit Babylonien und Assyrien verkettet ist'. Delitzsch intended to reveal the contradictions, inconsistencies and violence in the Old Testament stories, which he believed Christians should not hold sacred. See Mogens Trolle Larsen, 'The "Babel/Bible" Controversy and its Aftermath', in Jack M. Sasson, ed., *Civilizations of the Ancient Near East*, 3 vols (New York: Charles Scribner's Sons, 1995), I, pp. 95–106, and Larsen, 'Seeing Mesopotamia', in Ann C. Gunter, ed., *The Construction of the Ancient Near East, Culture and History*, 11 (1992), pp. 126–8.

22 See Adolf Harnack, *Letter to the Preussische Jahrbücher on the German Emperor's Criticism of Prof. Delitzsch's Lectures on 'Babel und Bibel'*, transl. by Thomas Bailey Saunders (Oxford: Williams and Norgate, 1903). The Emperor wrote a letter to dissociate himself from the theological views of Delitzsch, even though his authority as an Assyriologist was not disputed.

23 *Karte von Babylonien zur Zeit des assyrischen und babylonischen Weltreichs mit besonderer Berücksichtigung der biblischen Landschaft GAN EDEN oder der PARADIESES*, in Delitzsch, *Wo lag das Paradies?* (1881). The map is referred to on the title page and is folded and bound in at the end of the volume.

24 Delitzsch, *Wo lag das Paradies?* (1881). See also Felice Finzi, *Ricerche per lo studio dell'antichità assira* (Turin: Loescher, 1872), p. 433.

25 William Willcocks, *From the Garden of Eden to the Crossing of the Jordan* (Cairo: French Institute of Oriental Archaeology, 1918; 2nd edn, London: E. and F. N. Spon, 1919, with imprint added in the margin). Plate I, the map of *The Garden of Eden and the Four Rivers of Genesis*, is inserted at the end of the first chapter, together with Plate II, a map showing the area between Egypt and Babylon.

26 Willcocks, *From the Garden of Eden to the Crossing of the Jordan* (1919), 'Preface', pp. V–VI.

27 Ibid., p. 27.

28 Manfred Dietrich, 'Das biblische Paradies und der babylonische Tempelgarten: Überlegungen zur Lage des Gartens Eden', in *Das biblische Weltbild und seine altorientalischen Kontexte*, ed. by Bernd Janowski, Beate Ego, and Annette Krüger (Tübingen: Mohr Siebeck, 2001), pp. 280–323; Dietrich, 'Der "Garten Eden" und die babylonischen Parkanlagen im Tempelbezirk', in *Religiöse Landschaften*, ed. by Johannes Hahn and Christian Ronning (Münster in Westphalia: Ugarit Verlag, 2002), pp. 1–29.

29 See Bendt Alster, 'Dilmun, Bahrain, and the Alleged Paradise in the Sumerian Myth and Literature', in Daniel T. Potts, ed., *Dilmun: New Studies in the Archaeology and Early History of Bahrain* (Berlin: Reimer, 1983), pp. 39–74. See also Ed Noort, 'Gan-Eden in the Context of the Mythology of the Hebrew Bible', in Luttikhuizen, ed., *Paradise Interpreted* (1999), pp. 21–36, esp. 32. I am grateful to Mariana Giovino for having shared with me her ideas on the developments of Assyrian studies in the nineteenth and twentieth centuries. For the background of these developments, see her PhD Dissertation, *A History of Interpretations of the 'Assyrian Sacred Tree': 1849–2004* (2004), Chapter 1.

30 *From Paradise Lost to Paradise Regained* (Brooklyn, NY: Watchtower Bible and Tract Society of New York, 1958).

31 For an example of a Mesopotamian location of paradise in nineteenth-century travel literature see John F. Newman, *A Thousand Miles on Horseback* (New York: Harper & Brothers, 1875), p. 69.

32 Brothers based his assertion on the text of Ezekiel 40–8: Richard Brothers, *A Description of Jerusalem: Its Houses and Streets, Squares, Colleges, Markets, and Cathedrals, the Royal and Private Palaces, with the Garden of Eden in the Centre, as Laid Down in the Last Chapters of Ezekiel …* (London: George Riebau, 1801). The plan of the Garden of Eden, is at p. 13: *The Garden of Eden with its Twelve Palaces, Gardens & Cathedrals*.

33 Carl Josef Michaeler, *Das Neueste über die geographische Lage des irdischen Paradieses* (Vienna: Rötzler, 1796). See Jan Mokre, 'Kartographie des Imaginären: Von Ländern, die es nie gab', in Hans Petschar, ed., *Alpha and Omega: Geschichten vom Ende und Anfang der Welt* (Vienna–New York: Österreichische Nationalbibliothek–Springer, 2000), pp. 30–1.

34 Moritz Engel, *Die Lösung der Paradiesfrage* (Leipzig: Otto Schulze, 1885); William F. Warren, *Paradise Found: The Cradle of the Human Race at the North Pole* (Boston: Houghton, Mifflin and Co., 1885).

35 Warren claimed to be the first to advocate this location, although he acknowledged in a footnote to have come across a certain 'Pastellus' ('Who Pastellus was and what he wrote upon the subject remain to be investigated'): *Paradise Found* (1885), pp. 47–8. On Postel's location of Eden see above, Chapter 10, pp. 285–8.

36 Henry William Gordon, *Events in the Life of Charles George Gordon, from its Beginning to its End* (London: Kegan Paul Trench, 1886), pp. 261–9. The map, *Sketch of the Four Rivers of the Garden of Eden*, is inserted between pp. 262 and 263. The two marginal diagrams are in the footnote that runs over from p. 262 to p. 263. On Gordon's map, see Fred Plaut, *Analysis Analysed: When the Map becomes the Territory* (London–New York: Routledge, 1993), pp. 145–72, where Gordon's arguments are analysed from a psychological point of view; Plaut, 'Where is Paradise? The Mapping of a Myth', *The Map Collector*, 29 (December 1984), p. 7; 'General Gordon's Map of Paradise', *Encounter*, (June/July 1982), pp. 20–32.

37 A. P. Curtis, *The Land of Eden and Havilah* (London: the author, 1905). The three annotated copies now in London, British Library, pressmark Maps 700.(4)1–3, are numbered and were evidently intended (identified by the initial A.P.C. on the second map) to be treated as a set, together with a fourth unmarked map and Curtis's note. I am grateful to Francis Herbert for his help in identifying the map.

38 Albert Herrmann, *Die Erdkarte der Urbibel*

38 (Braunschweig: Kommissionsverlag von Georg Westermann, 1931); Kamal Salibi, *The Bible Came from Arabia* (London: Jonathan Cape, 1996); Richard Hennig, *Wo lag das Paradies?* (Berlin: Ullstein Verlag, 1950); Lars Ivar Ringbom, *Paradisus terrestris: myt, bild och verklighet* (Copenhagen–Helsinki: Ejnar Munksgaards–Akademiska Bokhandeln–Nordiska Antikvariska Bokhandeln, 1958); and David Rohl, *Legend: The Genesis of Civilisation* (London: Century, 1998), pp. 46–68: see above, Prologue, pp. 13–14.

39 Friedrich von Wendrin, *Die Entdeckung des Paradies* (Leipzig: Braunschweig, 1924).

40 James E. Nicholson, *The Probable 'Garden of Eden': A Long-Forgotten Star Myth* (London: Watts, 1935, for private circulation only). The illustration with the Garden of Eden is Figure 3, p. 10; the other two illustrations are *The Earth in the centre of its Celestial Sphere* (Figure 1, p. 4) and *Track of the Celestial North Pole ...* (Figure 2, p. 9).

41 Albert R. Terry, *The Flood and Garden of Eden: Astounding Facts and Prophecies* (Elms Court, Ilfracombe, Devon: Arthur H. Stockwell, 1962; repr. 1963). Maps of paradise submerged in the Mediterranean are found at pp. 4 and 12: detail of the garden and rivers, east Mediterranean.

42 Terry, *The Flood and Garden of Eden* (1962), p. 15.

43 See http://www.logoschristian.org/eden.html. [last accessed: 15 February 2006]. The author is Chris Ward.

44 See James M. Scott, *Geography in Early Judaism and Christianity: The Book of Jubilees* (Cambridge, Cambridge University Press, 2002), esp. pp. 23–43. See also above, Chapter 5, p. 118 n. 36. For various attempts in the second half of the twentieth century to map the divisions in detail, see Francis Schmidt, 'Naissance d'une géographie juive', in Alain Desreumeaux and Francis Schmidt, eds, *Moïse géographe: Recherches sur les représentations juives et chrétiennes de l'espace* (Paris, Vrin, 1988), pp. 12–30.

45 Emilio Spedicato, 'Eden Revisited: Geography, Numerics and Other Tales', *Report DMSIA, Miscellanea* 1/01 (2001); version revised in 2003 and 2005, http://www.unibg.it/dati/persone/636/419.pdf, (last accessed 15 February 2006).

46 Diana Hamblin, 'Has the Garden of Eden Been Located at Last?', *Smithsonian*, 18/2 (1987), pp. 127–35, returns to the idea of a near eastern paradise submerged in the Persian Gulf. Her ideas have been reproposed on the Internet; see http://www.ldolphin.org/eden (last accessed : 15 February 2006); see also Charles W. J. Withers, 'Geography, Enlightenment, and the Paradise Question', in *Geography and Enlightenment*, ed. by David N. Livingstone and Charles W. J. Withers (Chicago–London: University of Chicago Press, 1999), p. 69.

47 David Livingstone, Letter to Sir Roderick Murchison published in the *Athenaeum*: see Withers, 'Geography, Enlightenment, and the Paradise Question' (1999), p. 86, and Warren, *Paradise Found* (1885), p. 21.

48 Gerald Massey, *The Natural Genesis*, 2 vols (London: Williams and Norgate, 1883), II, p. 162: 'If there be an earthly original for the heavenly Eden, it will be found in equatorial Africa ...'

49 For a general overview, see Withers, 'Geography, Enlightenment, and the Paradise Question' (1999), pp. 67–92.

50 Carl von Linné, *Oratio de telluris habitabilis incremento* (Leiden: Cornelius Haak, 1744). See Janet Brown, 'Noah's Flood, the Ark, and the Shaping of Early Modern Natural History', in David C. Lindberg and Ronald L. Numbers, eds, *When Science and Christianity Meet* (Chicago–London: University of Chicago Press, 2003), pp. 132–5; Withers, 'Geography, Enlightenment, and the Paradise Question' (1999), pp. 81–2. Von Linné also left room in his system of scientific classification for some of the *monstruosi populi*, like the Wild Man and the *Cynocephali*.

51 Johann Herder, *Abhandlung über den Ursprung der Sprache* (Berlin: Voss, 1772). See Withers, 'Geography, Enlightenment, and the Paradise Question' (1999), p. 79.

52 See, for example, the map folded at the end of vol. I of Otto Caspari, *Die Urgeschichte der Menscheit*, 2 vols (Leipzig: [n.p.], 1873; repr. 1877), and Alexander Winchell, *Preadamites* (Chicago: Griggs & Co., 1880), p. 1.

53 See above, Chapter 1, pp. 20–3. For example, the French anthropologist Armand de Quatrefages, *The Human Species* (London: Kegan Paul, 1881), pp. 175–7, thought of Central Asian origins of man, in particular in Pamir. A summary of the arguments on Lemuria is found in Oscar F. Peschel, *The Races of Man and their Geographical Distribution* (London: King & Co., 1876), pp. 26–34. Eugène Beauvois, 'L'Élysée transatlantique et l'Eden Occidental', *Révue de l'Histoire des Religions*, 4/7; 4/8 (1883), pp. 273–318, 672–727, as well as Klaproth, Gobineau, and others, believed that the cradle of mankind had been in America. An overview is found in Warren, *Paradise Found* (1885), pp. 33–43, whose opinion on the location of paradise at the North Pole is discussed above.

54 See, for example, Franz Unger, *Die versunkene Insel Atlantis* (Vienna: [n. pub.], 1860); Ignatius Donnelly, *Atlantis: The Antediluvian World* (New York: Harper & Brothers, 1882); Heinrich G. A. Engler, *Versuch einer Entwicklungsgeschichte der Pflanzenwelt*, 2 vols (Leipzig: [n.p.], 1879–82), I, p. 82.

55 Denis Diderot, *Supplément au Voyage de Bougainville*, ed. by Herbert Dieckmann (Geneva: Droz; Lille: Giard, 1955). On European reactions to Mauritius and St Helena see Withers, 'Geography, Enlightenment, and the Paradise Question' (1999), pp. 82–4.

56 Withers, 'Geography, Enlightenment, and the Paradise Question' (1999), pp. 68–9, 84–5.

57 Julian Ford, *The Story of Paradise* (New York: Harper & Row, 1981); see Withers, 'Geography, Enlightenment, and the Paradise Question' (1999), p. 69.

58 Warren, *Paradise Found* (1885), p. 43.

Epilogue: Paradise Then and Now

It would be difficult to find any subject in the whole history of opinion which has so invited and at the same time so completely baffled conjecture, as the Garden of Eden. ... Theory after theory has been advanced, but none has been found which satisfies the required conditions. ... The site of Eden will ever rank, with the quadrature of the circle and the interpretation of unfulfilled prophecy, among those unresolved and perhaps insoluble problems which possess so strange a fascination.

William A. Wright, 'Eden', in Smith's *Dictionary of the Bible* (1863)

We have made a long conceptual journey in this book. We have travelled from the beginnings of Christianity in the first century, when paradise on earth emerged from Judaism, through Late Antiquity, the Middle Ages, and the Renaissance to the present day. We have encountered the earthly paradise in a diversity of guises, fashioned from words on the page and lines on a map. It would be surprising were an issue related to some of the most deeply reflective thoughts contributing to almost two thousand years of Christian teaching about the ultimate destiny of all human life not to have been moulded and remoulded generation by generation. What is striking, however, is the way that since the Reformation each author of the numberless works on the question of paradise makes a point of ridiculing the theories of his predecessors, by listing them briefly one after the other in a dismissive survey, a topos that has endured to the present day.

In a way, the practice of listing untenable theories had been anticipated in Late Antiquity, when the Church Fathers would rehearse what they saw as Origen's faults, reproaching him for manipulating the word of God and turning it into allegory to fit his philosophical constructs. Similarly, in the Middle Ages, each scholastic philosopher would systematically point to an identical set of objections to the existence of an earthly paradise in order to confute one predecessor or another. The almost routine inventorizing of ridiculed theories of the 'simple-minded', however, and their weird speculations on the location of paradise was an invention of the Renaissance. It started in embryonic form with Augustinus Steuchus's sharp comment on the absurdity of imagining paradise in a purely mystical way or of thinking that it was on the equator or beyond the inhabited earth.[1] Steuchus was followed by Luther, who enumerated all the various opinions by which earlier exegetes had tortured themselves in amazing ways about the location of paradise, dismissing each as a waste of time.[2] It was the Lutheran scholar Sebastian Münster, though, who formulated what became the standard listing of irrational locations of paradise. He poured scorn on the way paradise had been popularly seen in a range of places such as the Far East, the tropics and the equator, or as a high mountain that reached the sphere of the moon.[3] Münster's ironic tone and dismissive list of medieval opinions about paradise was adopted first by Calvin and then, word for word, by a succession of authors.[4] With the passing of time, the list of preposterous and misguided views was expanded to include eventually not only medieval opinions but also ideas about the location of the earthly paradise favoured in the Renaissance, such as the suggestion that paradise had once comprised the entire earth.

Henceforth, after the sixteenth century, each writer – who always claimed to be offering the last word on the subject – has also added his piece to the list of ridiculed locations. In the early seventeenth century, Walter Raleigh dismissed allegorical readings and imaginative theories about the location of paradise as 'Castles in the Aire, and in mens fancies, vainely imagined'.[5] Later, Lapeyre listed the many opinions on the location of paradise put forward before him as 'ridiculous'.[6] Marmaduke Carver censured both those who had disregarded the letter of the text in order to plant a cabalistic or allegorical paradise of their own, and those who had sought for paradise in the most unlikely places: 'how much rubbish hath been digged up, and dust raised'![7] The most comprehensive list, and the model used by later writers, was Bishop Huet's:

> *They placed it in the third Heaven, in the fourth, in the Orb of the Moon, in the Moon it self, upon a Mount near the Orb of the Moon, in the middle Region of the Air, out of the Earth, upon the Earth, under the Earth, in a hidden place and far beyond the Knowledge of Men. They placed it under the artick Pole, in Tartaria, in the place where now is the Caspian Sea. Others have placed it as far as the extremity of the South, in the Land of Fire. Many will have it to be in the East, either along the sides of the River Ganges, or in the Isle of Ceilan, deriving also the name of Indies from the word Eden, which is the name of the Province where paradise stood. They have placed it in China, and beyond the East also, in a place uninhabited. Others in America, others in Africa under the Aequator, others in the Aequinoctial-East, others upon the Mountains of the Moon, from which they thought the Nile sprung. The greatest part in Asia, some in the great Armenia, others in Mesopotamia, or in Assyria, or in Persia, or in Babylonia, or in Arabia. Or in Syria, or in Palaestina. Some also would have honoured with it our Europe, and which is beyond the greatest Impertinency, placed it at Hedin a City in Artois, upon no other ground than the Affinity of that name with the word Eden. I do not despair, but some Adventurer, to have it nearer to us, will one day undertake to place it at Houdan.*[8]

Augustin Calmet differed from Huet in his choice of place, but this did not prevent him from reproducing Huet's list to scoff with him at the zeal of past paradise-locaters.[9] The habit of replicating the list continued, feeding the topos that through medieval ignorance paradise was chased all around the world and placed in any region that remained unknown. In this way the list of almost random and implausible locations was transmitted to nineteenth-century historians, such as Arturo Graf and Pompeo Durazzo, who, as we saw in the first chapter, dismissed the medieval mapping of paradise as superstitious and bizarre. Right at the end of the nineteenth century, Edoardo Coli admitted taking his list of the bizarre locations of paradise directly from Calmet (as well as from Cornelius a Lapide):

> *From the fourteenth century onwards, one wanted to find it in all the most bizarre sites that could be imagined: in India, in addition to Ceylon, in the Cachemir valley, in China, Armenia, Syria, Persia, Babylon, Tartaria, Arabia, at the sources of the Nile, in Palestine near the Jordan, on the Alps, in the Caspian Sea, in America, in the Land of Fire, at the South Pole, underground. We should also mention those who identified paradise with the whole earth, those who suspended it in the middle region of the air, those who located it in an unknown antictone, or on a mountain close to the moon, or on the moon itself or in its sphere, or in the third or fourth heaven, or those who again confused it with the empyrean, when they did not deny it as a material reality . . .*[10]

When it was his turn, Coli updated the list with a reference to the most recent theories, commenting that 'the ardour of this search invaded also the learned who masked it as the investigation on the ethnical original home of the human races'.[11] Twentieth-century historians followed Coli's example in deriving much of their material on the Middle Ages from seventeenth- and eighteenth-century biblical commentaries, especially those of Malvenda and Pererius as well as Cornelius a Lapide and Calmet. In the process, they absorbed, as had their nineteenth-century predecessors, the dismissive attitude of the Renaissance towards the allegedly irrational character of medieval theories. The old list of the bizarre locations of paradise still survives in modern historical literature.[12] Many aspects of the modern approach to the medieval mapping of paradise also have a post-Reformation origin. The over-reliance on secondary (that is, early modern) sources led to a lack of discrimination and to generalization about original medieval thinking. For instance, the late antique, exclusively allegorical reading of Genesis, and the early medieval idea that paradise was in the lunar orbit were bundled together under the label 'otherworldly', despite the fact that after Augustine such an interpretation of the paradise narrative had been abandoned, and that by the later Middle Ages nobody would have assumed that paradise reached up to the moon. Modern historians who look back at the Middle Ages have also been misled into giving undue emphasis to Moses Bar-Cepha's treatise on paradise by their over-reliance on seventeenth- and eighteenth-century accounts of medieval theories. Bar-Cepha, a ninth-century bishop of Mosul, originally wrote in Syriac; his treatise was rendered into Latin by Andreas Masius in 1569.[13] This work enjoyed great popularity after its translation but had played no role in medieval Latin exegesis.

The lesson that we learn from recognizing the persistence of the list of implausible locations is that the dismissive attitude of scholars since the nineteenth century towards medieval mapping of paradise noted in Chapter 1 has not been based on first principles, but is part of a tradition that dates back to Luther and Calvin. Now, however,

Fig. 12.1. The dead Tree of Adam, from *The Times* of 23 December, 1944. This photograph of the dead Tree of Knowledge near the confluence of the Tigris and the Euphrates was published in *The Times* towards the end of Second World War. Nostalgia for paradise always intensifies in times of trouble. The Garden of Eden has been located in a number of different places in Mesopotamia since the Renaissance and the Reformation.

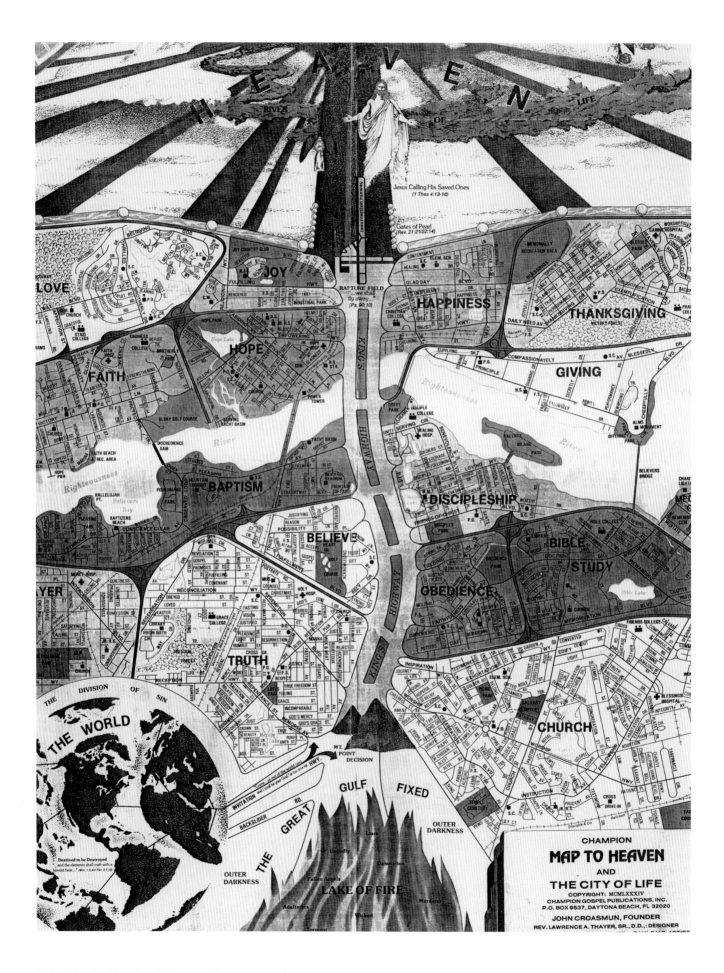

Fig. 12.2 (opposite). Lawrence A. Thayer, Mac McElwain, Paul Sale, *Map to Heaven and the City of Life* (Daytona Beach, FL: Champion Gospel Publications, 1984). This modern Christian map of paradise shows the way to heaven. The world, bottom left, is represented by a globe with its five continents clearly visible. Scriptural quotations remind the reader that the world is destined to be destroyed. If the individual is to reach the river of life in heaven, at the top of the map, Christ's invitation to walk along the streets and avenues of the city of life has to be accepted.

Fig. 12.3. Detail of Fig. 12.2. The city's streets include Godly Court, Glad Day Boulevard, Healing Road, God's Mercy Street, Angelic Drive, and much else.

the Renaissance inventory of past mistakes has been adopted for a new use. The aim of modern scholars has been to write history, not to find a rational geography for the earthly paradise. Unlike Luther, Calvin and all the others, the modern scholar does not have to argue against allegorical fantasies or risible locations in defence of Scripture. He has seen his role as emphasizing the triumph of modern science over medieval religion and superstition, which is why Renaissance ideas were also included in the parade of odd and imaginative theories, albeit with, significantly, a better press and a better understanding of the original context. What had been for post-Reformation writers a list of bizarre past opinions discrediting the Christian faith, was for modern scholars a display of how the advance of science and progress of geographical exploration forced paradise into countless different locations before it vanished altogether.

None of those 'last words' on the location of paradise was ever effectively final. Even as they rehearsed again and again the list of 'ridiculous' earlier opinions, Renaissance writers were generating a new flowering of 'imaginative' theories, those surveyed in Chapter 10, which differ from medieval theories mainly in searching for a lost paradise. The paradigm shift has not, however, put an end to the search for an earthly paradise. The idea with which this book started, David Rohl's suggestion in 1998 (now recognizable as a variant of Calmet's idea) that Eden was in north-west Iran has been taken up by *The Sunday Times*.[14] 'Hardy travellers' who accepted the invitation to sign up with a travel agency for the journey of a lifetime to the Garden of Eden would have enjoyed the wealth of archaeological features, but would not have found the biblical terrestrial paradise that they could re-enter. A similar disappointment must have been experienced by the nineteenth-century English explorer, William Heude, who, on 19 January 1817, reached the confluence between the Tigris and the Euphrates, as he journeyed from India to Arabia. Heude wrote:

> *The confluence of two such majestic streams forming an island gulf of great extent, is certainly an imposing sight: I could find nothing, however (except in the contrast with the surrounding desert) that could mark Korna as the fertile happy spot which had been assigned to man before his fall. The few trees, and the little cultivation it may boast, are certainly as a garden in the midst of a barren, black, desolated wilderness: without this wilderness, however, it would be only a marsh overgrown with rushes, a few palms, and fifty or sixty miserable huts.*[15]

Fig. 12.4. *Paradise under the Ceiling*, from Ilya and Émilia Kabakov, *The Palace of Projects* (London: Artangel, 1998). The Russian artists claim that the project shows how individuals can achieve perfection for themselves, instead of searching for remote paradises or succumbing to totalitarian dreams. They encourage all of us to create our own personal paradise simply by making a space under the ceiling of our own room and filling this with much-loved objects, such as small, toy-like figures of birds, animals, fish and plants and perpetually burning lamps.

The sense of disappointment lingers. On 23 December 1944, *The Times* of London reported that the Tree of Knowledge on the outskirts of Qurna, a town near Basra and at the junction of the Euphrates and the Tigris, had died and that the dead trunk was broken and uninspiring (Figure 12.1). It was suggested that, although the tree had not been particularly old and the authentic site of the garden of Adam was far from definitely agreed on, a garden of flowers, trees and fruits should be planted on this traditional site. In the 1970s, the spot was turned into a tourist attraction, but the war

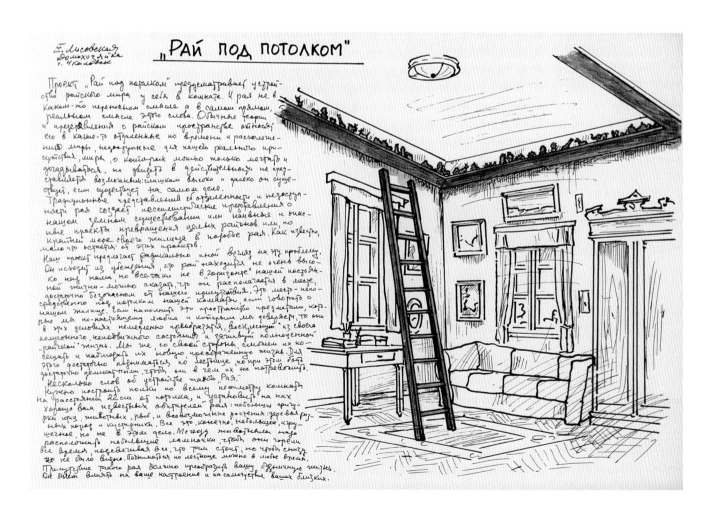

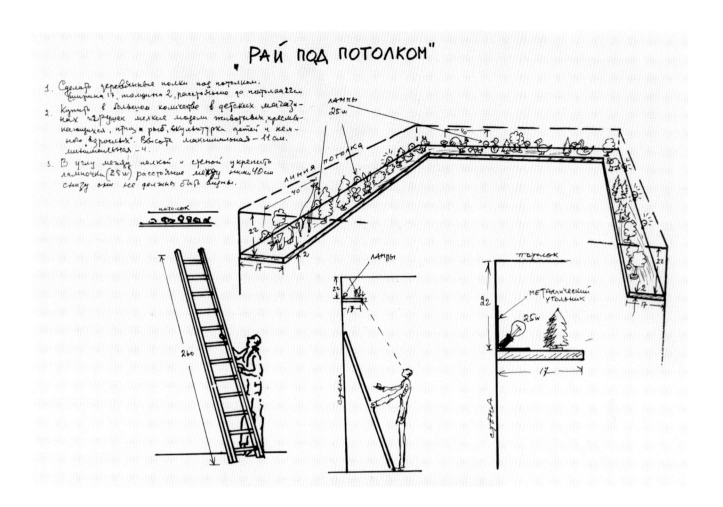

between Iran and Iraq from 1980 to 1988 led to its neglect and disrepair. In the course of another conflict, in 2003, the Western media reported that the site was a desolate wasteland of cracked paving stones and bullet holes, and that Adam's Tree was dead. Today, a plaque with an inscription in Arabic and English still directs attention to the sanctity of an ancient eucalyptus tree nearby: 'On this holy spot where Tigris meets Euphrates this holy tree of our father Adam grew, symbolising the Garden of Eden on earth. Abraham prayed here two thousand years before Christ.'[16]

It is clear that no surviving biblical earthly paradise will be found or mapped. A paradise in heaven, though, may still be mapped for the religious. Figures 12.2 and 12.3 reproduce a detailed allegorical route map in the style of a city atlas, entitled *Map to Heaven and the City of Life*, published in Florida in 1984. The map shows a path leading from the world along the King's Highway towards Heaven and passing between such districts as Truth, Prayer, Obedience, and Bible Study, and over the river of Righteousness to Love, Hope and Happiness and, eventually, past the Field of Rapture, to the waiting Christ.

Whether the approach is openly religious or not, mankind still longs for a paradise on earth. In 1998, the Russian artists Ilya and Émilia Kabakov exhibited in London one of their visionary plans. Transforming the venue into a *Palace of Projects*, they displayed, with a sometimes melancholic humour, a number of surreal models, designs and drawings for improving the quality of human life. One suggestion was that putting on a pair of white tulle wings for a few moments each day was a way of uplifting the individual's moral life. Another was to use energy rings to overcome the force of gravity in order to inhabit the air and thus solve the problem of overpopulation on the

Fig. 12.5. *Paradise under the Ceiling*, from Ilya and Émilia Kabakov, *The Palace of Projects* (London: Artangel, 1998). The artists provide detailed instructions for building an indoor paradise that can be approached whenever desired: visitors to their exhibition were invited to construct a couple of wooden shelves just below the ceiling in their rooms (17 cm wide, 2 cm thick, at a distance of 22 cm from the ceiling); then, from a toy shop, to buy large quantities of miniature models of animals, reptiles, birds and fish, sculptures of children, and a few 'adults' (maximum height 11 cm, minimum height 4 cm); then, finally, to install lamps (25 watt) in the corner between the shelf and the wall, at intervals of 40 cm in such a way that they are invisible from below. Paradise, the Russian artists were suggesting, can be created in one's own room.

ground.[17] Yet another project was called *Paradise under the Ceiling*. This provided a paradise in one's own room. The artists advised setting aside space below the ceiling and packing it with familiar, trusted and much-loved objects in order to create a paradise-like realm that would be always within reach (Figures 12.4 and 12.5). Their point was that paradise need be neither remote nor inaccessible:

> *The traditional notion about the remoteness and inaccessibility of paradise creates a pessimistic notion about our earthly existence or else naïve and dangerous projects for transforming entire regions, or at least one's own residence, into a likeness of paradise. As everyone knows, little remains from these projects. Our project proposes a radically new view of this problem. It starts with the conviction that paradise is located not very high over us, but nevertheless not at the 'horizon' of our constant life ... This place is directly below the ceiling of our room ... If we fill this space with objects that we love and trust, then in these conditions they will become transformed, 'resurrected' from their semi-sleep, immobile state and will begin to live a full 'paradisial' life. We, for our part, can visit them and observe their new, transformed life. For this all you have to do is climb up the ladder.*[18]

To the modern mind, true paradises are paradises lost (to use Proust's words). The Kabakovs, however, deny this pessimistic vision. Although they make no mention of a paradise awaiting at the end of time, they nevertheless join the ranks of medieval map makers in their view that a paradise lost carries the promise of a paradise regained, and that this paradise is accessible at any time. All that is needed is a ladder that can be climbed here and now.

1. Augustinus Steuchus, *Recognitio Veteris Testamenti ad hebraicam veritatem* (Venice: Aldus, 1529), pp. 23v–24r.
2. Martin Luther, *Lectures on Genesis Chapters 1–5*, in *Luther's Works*, ed. by Jaroslav Pelikan (St Louis, MO: Concordia, 1958), I, p. 88; *Genesisvorlesung*, ed. by Gustav Koffmane and Otto Reichert, in *D. Martin Luthers Werke*, XLII (Weimar: Hermann Böhlaus Nachfolger, 1911), p. 67 (see above, Chapter 9, pp. 266–7).
3. Sebastian Münster, *Cosmographia* (Basle: Henricus Petrus, 1545), p. 28.
4. On Calvin, see above, Chapter 9, pp. 273–4.
5. Walter Raleigh, *The History of the World* (London: Walter Burre, 1614), p. 56.
6. Jacques d'Auzoles Lapeyre, *La Saincte Géographie, c'est-à-dire, exacte description de la terre et veritable demonstration du paradis terrestre* (Paris: Antoine Estiene, 1629), pp. 14–18.
7. Marmaduke Carver, *A Discourse of the Terrestrial Paradise Aiming at a more probable Discovery of The True Situation of that Happy Place of our First Parents Habitation* (London: James Flesher, 1666), letter to 'The Judicious and Ingenuous Reader, especially the professed Divine'.
8. Pierre-Daniel Huet, *A Treatise of the Situation of the Terrestrial Paradise* (London: James Knapton, 1694), transl. by Thomas Gale, pp. 3–5; *Traité de la situation du paradis terrestre* (Paris: Jean Anisson, 1691), pp. 5–7: 'On l'a placé dans le troisiéme ciel, dans le quatriéme, dans le ciel de la Lune, dans la Lune mesme, sur une montagne voisine du ciel de la Lune, dans la moyenne region de l'air, hors de la terre, sur la terre, sous la terre, dans un lieu chaché et éloigné de la connoissance des hommes. On l'a mis sous le Pole Arctique, dans la Tartarie, à la place qu'occupe presentement la mer Caspie. D'autres l'ont reculé à l'extrémité du Midy, dans la Terre du feu. Plusieurs l'ont placé dans le Levant, ou sur les bords du Gange, ou dans l'Isle de Ceilan, faisant mesme venir le nom des Indes du mot d'Eden, nom de la Province où le Paradis estoit situé. On l'a mis dans la Chine, et mesme par delà le Levant, dans un lieu inhabité; d'autres dans l'Amerique, d'autres en Afrique sous l'Equateur, d'autres à l'Orient équinoctial, d'autres sur les montagnes de la Lune, d'où l'on a crû que sortoit le Nil; la pluspart dans l'Asie, les uns dans l'Armenie majeure, les autres dans la Mesopotamie, ou dans l'Assyrie, ou dans la Perse, ou dans la Babylonie, ou dans l'Arabie, ou dans la Syrie, ou dans la Palestine. Il s'en est mesme trouvé qui en ont voulu faire honneur à notre Europe, et ce qui passe toutes les bornes de l'impertinence, qui l'ont établi à Hédin, ville d'Artois, fondez sur la conformité de ce nom avec celuy d'Eden. Je ne desespere pas que quelque avanturier, pour l'approcher plus prés de nous, n'entreprenne quelque jour de le mettre à Houdan.' Other examples include Albertus Aloysius, *De operibus sex dierum et De terrestri paradisi lectiones* (Venice: Bartholomaeus Ginami, 1618), pp. 165–6, and Cherubino Pasolino, *Il paradiso terrestre aperto alli curiosi* (Forli: Colantonio Zampa, 1677), pp. 10–21.
9. Augustin Calmet, *Commentaire littéral sur tous les livres de l'Ancien et du Nouveau Testament*, 2nd edn, 8 vols (Paris: [n. pub.], 1724–6), I/1 (1724), p. 20: 'Il n'y a peut-être aucune question dans l'Ecriture qui ait tant partagé les sentiments des Ecrivains, que celle de la situation du Paradis terrestre; et rien ne fait mieux voir l'embarras de cette matière, que cette contrariété d'opinions. On l'a placé dans le troisième Ciel, dans le quatrième, dans le Ciel de la Lune, dans la Lune même, sur une montagne voisine du Ciel de la Lune, dans la moyenne region de l'air, hors de la terre, au dessus de la terre, sous la terre, dans un lieu caché et éloigné de la connaissance des hommes: On l'a mis sous le Pole arctique, dans la Tartarie, à la place qu'occupe à present la mer Caspie. D'autres l'ont reculé à l'extrémité du Midy, dans la Terre de feu. Plusieurs l'ont placé

dans le Levant, ou sur les bords du Ganges, ou dans l'Isle de Ceïlan, faisant même venir le nom des Indes du mot *Eden*. On l'a mis dans la Chine, ou même par delà le Levant, dans un lieu inhabité. D'autres dans l'Armenie; d'autres dans l'Afrique sous l'Equateur; d'autres à l'Orient équinoxial; d'autres sur les montagnes de la Lune, d'où l'on croyait que sortoit le Nil. La plûpart dans l'Asie; les uns dans l'Armenie majeure; les autres dans la Mésopotamie, ou dans la Syrie, ou dans la Perse, ou dans la Babylonie, ou dans l'Arabie, ou dans l'Assyrie, ou dans la Palestine. Il s'en est même trouvé qui en ont voulu faire honneur à notre Europe, Philon et Origenes ont cru que ce Paradis étoit purement spirituel. Les Seleuciens soûtenoient qu'il étoit invisible.'

10 Edoardo Coli, *Il paradiso terrestre dantesco* (Florence: Carnesecchi, 1897), pp. 120–1: 'dal Trecento in poi lo si volle trovare in tutti i più strani siti immaginabili: in India, oltre che a Seilan, nella valle del Cachemir, nella Cina, in Armenia, in Siria, in Persia, a Babilonia, in Tartaria, in Arabia, alle sorgenti del Nilo, in Palestina presso al Giordano, sulle Alpi, nel mar Caspio, in America, nella Terra del fuoco, al polo antartico, sotto terra. Si ricorderanno anche quelli che lo identificavano con tutta la terra; quelli che lo sospendevano in "media aeris regione" quelli che lo mettevano in un'anticone ignota, o sopra un monte alla Luna vicino, o nella Luna stessa o nel suo cielo, o nel terzo o nel quarto cielo o tornavano infine a confonderlo con l'Empireo, quando non lo negavano affatto come materiale.' Coli remarks that in drawing the list from his sources, Calmet and Cornelius a Lapide, he tried to put the various locations in some chronological order: 'ci siamo sforzati di riportar le varie ubicazioni secondo il più probabile ordine cronologico con cui furono escogitate.'

11 Ibid.: 'L'ardore di questa ricerca invase anche i dotti che la palliarono come investigazione dei focolari etnici delle stirpi umane …'

12 For example, Joseph E. Duncan *Milton's Earthly Paradise* (Minneapolis: University of Minnesota Press, 1972), pp. 79–81. Further examples are found in Ute Kleinmann, 'Wo lag das Paradies? Beobachtungen zu den Paradieslandschaften des 16. und 17. Jahrhunderts', in *Die flämische Landschaft: 1520–1700* (Lingen: Luca Verlag, 2003), p. 280: 'Im Laufe der Jahrhunderte wurde der Garten Eden so in Armenien, Mesopotamien, Arabien, Indien, China, Ceylon, auf den Kanarischen Inseln, in verschiedenen Gegenden von Amerika und Europa und sogar am Nordpol lokalisiert', and Charles W. J. Withers, 'Geography, Enlightenment, and the Paradise Question', in *Geography and Enlightenment*, ed. by David N. Livingstone and Charles W. J. Withers (Chicago–London: University of Chicago Press, 1999), p. 70: 'The precise location of Paradise varies in early accounts: from Ceylon, to China, Japan, the polar regions, the Canaries, an island or mountain in the equatorial regions or torrid zones, even on the moon.'

13 Andreas Masius, *De paradiso commentarius …* (Antwerp: C. Plantinus, 1569), pp. 12–226. See above, Chapter 10, p. 337 n. 47. See, for example, Corin Braga, *Le Paradis interdit au Moyen Âge: La Quête manquée de l'Eden oriental* (Paris: L'Harmattan, 2004), pp. 130, 378; Jean Delumeau, *Une Histoire du paradis*, I, *Le Jardin des délices* (Paris: Fayard, 1992), pp. 30, 64–5 (pp. 17, 43, in the English edition, 1994); Coli, *Il paradiso terrestre dantesco* (1897), pp. 69–75.

14 Peter Martin, 'The Secret Garden', *Sunday Times Magazine* (11 October 1998), pp. 44–50.

15 William Heude, *A Voyage up to the Persian Gulf and a Journey Overland from India to England in 1817* (London: Longman, Hurts, Rees, Orme and Brown, 1819), p. 56. An engraving signed T. Fielding with 'Korna in the Garden of Eden' is inserted between pp. 56 and 57.

16 See Manfred Dietrich, 'Das biblische Paradies und der babylonische Tempelgarten: Überlegungen zur Lage des Gartens Eden', in *Das biblische Weltbild und seine altorientalischen Kontexte*, eds Bernd Janowski, Beate Ego and Annette Krüger (Tübingen: Mohr Siebeck, 2001), pp. 306 and 307, Plate 8.

17 Ilya and Émilia Kabakov, *The Palace of Projects* (London–Madrid: Artangel–Centro de Arte Reina Sofia, 1998), nos. 1, 29, pp. 26–8, 110–11.

18 Ibid., no. 24, p. 88.

Bibliography

PRIMARY SOURCES

Manuscripts

Albi, Bibliothèque Municipale,
  MS 29, late eighth century (Albi map).
Basle, Universitätsbibliothek,
  MS A IX 99, 1561 (Guillaume Postel, *De paradisi terrestris loco*).
Berlin, Staatsbibliothek Preussischer Kulturbesitz,
  MS Theol. Lat. Fol. 561, twelfth century (Beatus of Liébana, *Commentarius in Apocalypsin*).
Bischofszell (Canton of Thurgovia, Switzerland), Ortsmuseum, Dr.-Albert-Knoepfli-Stiftung,
  'The Brendan Map', twelfth century.
Brussels, Bibliothèque Royale de Belgique/Koninklijke Bibliotheek van België,
  MS 3897–3919, 1119 (Guido of Pisa, *Liber historiarum*).
  MS 9260, (Jean Mansel, *La Fleur des histoires*).
Burgo de Osma, Archivo de la Catedral,
  MS 1, 1086 (Beatus of Liébana, *Commentarius in Apocalypsin*).
Caen, Bibliothèque Municipale,
  MS 11 and 14, c.1640 (Samuel Bochart, *Paradisus, sive de loco paradisi terrestris*; *Du lieu du paradis terrestre*)
Cambridge, Parker Library, Corpus Christi College,
  MS 66, late twelfth century (Honorius Augustodunensis, *Imago mundi*).
Copenhagen, Det Kongelige Bibliotek,
  G.K.S. 2020-4to, thirteenth century (Lucan, *Pharsalia*).
Deventer, Athenaeumbibliotheek,
  MS I, 81 (old 1791), thirteeenth century (Sallust, *De bello Iugurthino*).
Einsiedeln, Benediktinerabtei, Stiftsbibliothek,
  MS 263 (973), tenth century (miscellaneous).
Florence, Biblioteca Medicea Laurenziana,
  MS Med. Pal. 89, fifteenth century (Gregorio Dati, *La sfera*).
  MS Plut. 27 Sin., 8, c.620 (Isidore, *Etymologiae*).
  MS Plut. IX.28, eleventh century (Cosmas Indicopleustes, *Christian Topography*).
Florence, Biblioteca Nazionale Centrale,
  MS Portolano 1, 1457 (Genoese world map).
Florence, Biblioteca Riccardiana,
  MS 881, thirteenth century (Guido of Pisa, *Liber historiarum*).
Genoa, Biblioteca Durazzo Giustiniani,
  MS A IX 9, second half of the fifteenth century (Lambert of St-Omer, *Liber floridus*).
Girona, Museu de la Catedral,
  Num. Inv. 7 (11), tenth century (Beatus of Liébana, *Commentarius in Apocalypsin*).
Hamburg, Staats- und Universitätsbibliothek,
  MS Theol. 2029, fourteenth century (Peter of Poitiers, *Compendium historiae in genealogia Christi*).
Lisbon, Arquivo Nacional da Torre do Tombo,
  MS 160, 1189 (Beatus of Liébana, *Commentarius in Apocalypsin*).
London, British Library,
  Add. MS 10049, twelfth century (Jerome, *Liber locorum*).
  Add. MS 11695, 1106 (Beatus of Liébana, *Commentarius in Apocalypsin*).
  Add. MS 14788, 1148 (*Biblia sacra latina*).
  Add. MS 18850, c.1423 (Bedford Book of Hours)
  Add. MS 24070, c.1436–9 (Iohannes Gmunden, *Tabulae astronomiae*).
  Add. MS 27376, c.1325 (Marino Sanudo, *Liber secretorum fidelium crucis super Terrae Sanctae recuperatione et conservatione …*).
  Add. MS 28106, late twelfth century (Stavelot Bible).
  Add. MS 28681, c.1265 (Psalter) .
  Arundel MS 44, twelfth century (*Speculum virginum*).
  Burney MS 3, thirteenth Century (*Biblia*)
  Cotton MS Tiberius B. V., c.1025–50 (Dionysius of Alexandria, *Periegesis*, Latin transl. by Priscian).
  Harley MS 218, thirteenth century (Honorius Augustodunensis, *Imago mundi*).
  Harley MS 2799, 1172 (Arnstein Bible).
  Harley MS 3954, c.1430 (*The Book of Sir John Mandeville*).
  Royal MS 14.B.IX, early fourteenth century (Peter of Poitiers, *Compendium historiae in genealogia Christi*).
  Royal MS 14.C.IX, c.1350 (Ranulf Higden, *Polychronicon*).
  Royal MS 14.C.XII, c.1350 (Ranulf Higden, *Polychronicon*).
London, College of Arms,
  Muniment Room 18/19, c.1390–1415 (Evesham map).
London, Lambeth Palace Library,
  MS 371, c.1300 (text and map inserted in a copy of Nennius, *Historia Britonum*).
Macon, Bibliothèque Municipale,
  MS Franç. 2, c.1473–80 (Augustine, *De civitate Dei*).
Madrid, Biblioteca de la Real Academia de la Historia,
  MS 76, 946 (Isidore, *Etymologiae*).
Manchester, John Rylands University Library,
  MS Lat. 8, late twelfth century (Beatus of Liébana, *Commentarius in Apocalypsin*).
Milan, Biblioteca Ambrosiana,
  MS F. sup. 150, twelfth century (Beatus of Liébana, *Commentarius in Apocalypsin*).
Modena, Biblioteca Estense,
  C.G.A. 1, c.1450–60 (Catalan Estense world map).
Munich, Bayerische Staatsbibliothek,
  MS Cgm. 564, 1455 (Peter of Poitiers, *Compendium historiae in genealogia Christi*).
  MS Clm. 721, fifteenth century (John of Udine, *Compilatio librorum historialium totius bibliae ab Adam ad Christum*).
  MS Clm. 10058, late twelfth century (Isidore, *Etymologiae*).
New York, Pierpont Morgan Library,
  MS 644, c.940–5 (Beatus of Liébana, *Commentarius in Apocalypsin*).
Oxford, Bodleian Library,
  MS Laud. Misc. 674, 1438 (tables of latitude and longitude attr. to William Worcester).
Paris, Bibliothèque Nationale de France,
  Cartes et Planes, Rés. Ge AA 562, c.1492 (nautical chart attr. to Christopher Columbus).
  Cartes et Planes, Rés. Ge B 1118, c.1275–91 (Carte Pisane).
  MS Franç. 9140, late fifteenth century (Bartholomaeus Anglicus, *Livre des proprieté des choses*, French transl. by Jean Corbechon).
  MS Lat. 1366, late twelfth century (Beatus of Liébana, *Commentarius in Apocalypsin*).
  MS Lat. 4126, fourteenth century (Ranulf Higden, *Polychronicon*).
  MS Lat. 5510, early fourteenth century (William of Tripoli, *De statu Sarracenorum*).
  MS Lat. 8878, late eleventh century (Beatus of Liébana, *Commentarius in Apocalypsin*).
  MS Lat. 14435, thirteenth century (Peter of Poitiers, *Compendium historiae in genealogia Christi*).
  MS Lat. 16198, fourteenth century (Ptolemy, *De dispositione spherae*).
  MS Lat. 16679, twelfth century (Macrobius, *Commentarii in Somnium Scipionis*).
Paris, Bibliothèque Ste-Geneviève,
  MS 782, c.1370 (*Les Grandes Chroniques du temps de Charles V*).
Parma, Biblioteca Palatina,
  MS 1614, 1512 (Vesconte Maggiolo, *Atlante nautico*).
Mount Sinai, St Catherine,
  MS Gr. 1186, eleventh century (Cosmas Indicopleustes, *Christian Topography*).
Stuttgart, Württembergische Landesbibliothek,
  MS Theol. Fol. 100, fifteenth century (John of Udine, *Compilatio librorum historialium totius bibliae ab Adam ad Christum*).
Turin, Biblioteca Nazionale Universitaria,
  MS I.II.1, early twelfth century (Beatus of Liébana, *Commentarius in Apocalypsin*).
Utrecht, Rijksuniversiteitsbibliotheek,
  MS 737, fifteenth century (Girard of Antwerp, *Historia figuralis ab origine mundi usque ad … 1272*).
Vatican City, Biblioteca Apostolica Vaticana,
  Borgiano XVI, c.1430 (Borgia world map).
  MS Pal. Lat. 1362b, 1448 (world map by Andreas Walsperger).
  MS Vat. Gr. 699, ninth century (Cosmas Indicopleustes, *Christian Topography*).
  MS Vat. Lat. 5698, fifteenth century (Ptolemy, *Geography*).
  MS Vat. Lat. 6018, late eighth century (miscellaneous).
Venice, Biblioteca Nazionale Marciana,
  MS Fondo Ant. It. Z.76. (4783), 1436 (world map by Andrea Bianco).
Verona, Biblioteca Civica
  MS 3119, 1442 (Giovanni Leardo, *Mapa Mondi/Figura Mondi*).
Vercelli, Archivio e Biblioteca Capitolare,
  Vercelli Map, c. 1191–1218
Vicenza, Biblioteca Civica Bertoliana,
  MS 598a, 1448 (Giovanni Leardo, *Mapa mondi/Figura mundi*).
Vienna, Österreichische Akademie der Wissenschaften,
  Sammlung Woldan, K-V(Bl): WE 3, c.1485–90 (Wieder-Woldan map of the world).
Wolfenbüttel, Herzog August Bibliothek,
  MS Helmst. 442, fifteenth century (Bartholomaeus Anglicus, *De proprietatibus rerum*; *Chronica mundi*).

Printed Editions

Abu Ma'shar, *On Historical Astrology: The Book of Religions and Dynasties (On the Great Conjunctions)*, ed. and transl. by Charles Burnett and Keiji Yamamoto, 2 vols (Leiden: Brill: 2000).
——, *The Abbreviation of the Introduction to Astrology together with the Medieval Latin Translation of Adelard of Bath*, ed. and transl. by Charles

Burnett, Keiji Yamamoto and Michio Yano (Leiden–New York: Brill, 1994).
Acta Apostolorum Apocrypha, ed. by Constantinus Tischendorf, Ricardus Adelbertus Lipsius, and Maximilianus Bonnet, 2 vols (Hildesheim: Georg Olms, 1959).
Adrichom, Christian von, *Theatrum Terrae Sanctae* (Cologne: George Braun, 1592).
Ailly, Pierre de, *Ymago mundi*, ed. by Edmond J. P. Buron, 3 vols (Paris: Maisonneuve Frères, 1930).
Albert the Great, *Opera omnia*, ed. by Auguste C. Borgnet, 38 vols (Paris: Vivès, 1890–9).
Alcuin, *Super Genesim*, *PL* C, cols 515–66.
Alexander of Hales, *Summa theologica*, 4 vols (Ad Claras Aquas, Quaracchi, Florence: apud Collegium S. Bonaventurae, 1924–48; repr. 1979).
Alexandri Magni iter ad paradisum, ed. by Julius Zacher (Königsberg: Th. Theile, 1859).
Alighieri, Dante, *Divina Commedia*, ed. by Natalino Sapegno 3 vols (Florence: La Nuova Italia, 1979).
Aloysius, Albertus, *De operibus sex dierum et De terrestri paradisi lectiones* (Venice: Bartholomaeus Ginami, 1618).
Ambrose, *Expositio in Lucam*, ed. by Mark Adriaen, *CCSL* XIV (Turnhout: Brepols, 1957).
—, *Hexameron*, ed. by Karl Schenkl, *CSEL* XXXII/1 (Prague–Vienna–Leipzig: Tempsky–Freytag, 1897; repr. New York–London: Fathers of the Church, 1962).
Andrew of St Victor, *Opera*, I, *Expositio super Heptateuchum*, ed. by Charles H. Lohr and Rainer Berndt, *CCCM* LIII (Turnhout: Brepols, 1986).
Apocryphal Gospels, Acts and Revelations, transl. by Alexander Walker, *Ante-Nicene Christian Library*, ed. by Alexander Roberts and James Donaldson, 24 vols (Edinburgh: Clark, 1867–72), XVI (1870).
Aristotle, *Opera cum Averrois commentariis*, 11 vols (Venice: apud Iuntas, 1562–74; repr. Frankfurt am Main: Minerva, 1962).
Pseudo-Aristotle, *De inundatione Nili*, transl. and ed. by Danielle Bonneau, *Études de Papyrologie*, 9 (1971), pp. 1–33.
—, *De causis proprietatum et elementorum*, ed. by Stanley L. Vodraska, PhD Dissertation, Warburg Institute, University of London (1969).
Augustine, *De Genesi contra Manichaeos*, ed. by Dorothea Weber, *CSEL* XCI (Vienna: Österreichische Akademie der Wissenschaften, 1998).
—, *De doctrina christiana*, ed. by Roger P. H. Green (Oxford: Clarendon Press, 1995).
—, *Confessions*, ed. by James J. O'Donnell, transl. by Henry Chadwick, 3 vols (Oxford–New York: Oxford University Press, 1992).
—, *The Literal Meaning of Genesis*, transl. and annotated by John H. Taylor, 2 vols (New York–Ramsey, NJ: Newman Press, 1982).
—, *Confessiones*, ed. by Lucas Verheijen, *CCSL* XXVII (Turnhout: Brepols, 1981).
—, *De musica*, ed. by Giovanni Marzi (Florence: Sansoni, 1969).
—, *De Trinitate*, ed. by William J. Mountain and François Glorie, *CCSL* L and LA (Turnhout: Brepols, 1968).
—, *Sermones de Vetere Testamento*, ed. by Cyrillus Lambot, *CCSL* XLI (Turnhout: Brepols, 1961).
—, *De doctrina christiana*, ed. by Klaus D. Daur, *CCSL* XXXII (Turnhout: Brepols, 1962).
—, *Enarrationes in Psalmos*, ed. by Eliquis Dekkers and Johannes Fraipont, 3 vols, *CCSL* XXXVIII–XL (Turnhout: Brepols, 1956).
—, *De civitate Dei*, ed. by Bernard Dombart and Alfons Kalb, *CCSL* XLVIII (Turnhout: Brepols, 1955).
—, *In Iohannis Evangelium tractatus*, ed. by Radbodus Willems, *CCSL* XXXVI (Turnhout: Brepols, 1954).
—, *De Genesi ad litteram libri duodecim*, ed. by Joseph Zycha, *CSEL* XXVIII/1 (Prague–Vienna–Leipzig: F. Tempsky–G. Freytag, 1894).
—, *Contra Faustum*, ed. by Joseph Zycha, *CSEL* VI/1 (Prague–Vienna–Leipzig: F. Tempsky–G. Freytag, 1891).
—, *Contra Iulianum*, *PL* XLV, cols 1049–608.
Pseudo-Augustine, *Dialogus questionum LXV*, *PL* XL, cols 733–51.
Avicenna, *The Canon of Medicine*, ed. and transl. by Oskar Cameron Gruner (London: Luzac, 1930).
Bacon, Roger, *Communia naturalium, liber secundus: De celestibus*, ed. by Robert Steele, *Opera hactenus inedita*, fasc. IV (Oxford: Clarendon Press, 1913).
—, *Opus maius*, ed. by John H. Bridges, 2 vols (Oxford: Clarendon Press, 1897–1900).
Balbus, Ioannes, *Catholicon* (Mainz: [n. pub.], 1460; repr. Westmead, England: Gregg, 1971).
Barnabas: *Der Barnabasbrief*, ed. by Ferdinand R. Prostmeier (Göttingen: Vandenhoeck und Ruprecht, 1999).
—, *Épître de Barnabé*, ed. by Pierre Prigent and Robert A. Kraft, *SC* CLXXII (Paris: Cerf, 1971).
Bartholomaeus Anglicus, *De proprietatibus rerum* (Frankfurt am Main: Wolfgang Richter, 1601; repr. Frankfurt am Main: Minerva, 1964).
Al-Battani, *Opus astronomicum*, ed. by Carlo A. Nallino, 3 vols (Milan: Hoepli, 1899–1907).
Beatus of Liébana, *Commentarius in Apocalypsin*, ed. by Eugenio Romero Pose, 2 vols (Rome: Istituto Poligrafico e Zecca dello Stato, 1985).
—, *In Apocalipsin*, ed. by Henry A. Sanders, 2 vols (Rome: American Academy, 1930).
Beauvois, Eugène, 'L'Élysée transatlantique et l'Eden occidental', *Révue de l'Histoire des Religions*, 4/7; 4/8 (1883), pp. 273–318, 673–727.
Becanus, Goropius (Jan van Gorp), *Origines antwerpianae, sive Cimmeriorum becceselana novem libros complexa* (Antwerp: Christophorus Plantinus, 1569).
Bede, *De temporibus liber*, ed. by Charles W. Jones, *CCSL* CXXIIIC (Turnhout: Brepols, 1980).
—, *De temporum ratione*, ed. by Charles W. Jones, *CCSL* CXXIIIB (Turnhout: Brepols, 1977).
—, *Ecclesiastical History of the English People*, ed. by Bertram Colgrave and Roger A. B. Mynors (Oxford: Clarendon Press, 1969).
—, *Libri quatuor in principium Genesis usque ad nativitatem Isaac et electionem Ismahelis adnotationum*, ed. by Charles W. Jones, *CCSL* CXVIIIA (Turnhout: Brepols, 1967).
—, *In Lucam*, ed. by David Hurst, *CCSL* CXX (Turnhout: Brepols, 1960).
Bedford, Arthur, *The Scripture Chronology Demonstrated by Astronomical Calculations* (London: James and John Knapton, 1730).
Benjamin of Tudela, *The World of Benjamin of Tudela: A Medieval Mediterranean Travelogue*, ed. by Sandra Benjamin (Madison, WI–London: Fairleigh Dickinson University Press–Associated University Presses, 1995).
—, *The Itinerary*, ed. and transl. by Ezra H. Haddad (Baghdad: Eastern Press, 1945).
Beroaldus, Matthaeus, *Chronicum, Scripturae Sacrae autoritate constitutum* (Geneva: Antonius Chuppinus, 1575).
The Holy Bible (London: Joseph Moxon, 1683).
The Holy Bible (London: Robert Barker, 1611)
Biblia (Amsterdam–Haarlem: Jacobszoon–Rooman, 1590).
The Bible (Geneva: Rouland Hall, 1560).
La Saincte Bible (Lyon: Sebastien Honoré, 1566).
Biblia, das ist, die gantze Heilige Schrifft Deudsch (Wittenberg: Hans Lufft, 1534; 2nd edn, 1536).
Biblia latina cum Glossa ordinaria, facsimile of the *editio princeps...* (Strasbourg: Adolph Rusch, 1480/1), ed. by Margaret T. Gibson and Karlfried Fröhlich, 2 vols (Turnhout: Brepols, 1992).
Biel, Gabriel, *Collectorium circa quattuor libros Sententiarum*, ed. by Wilfridus Werbeck and Udo Hofmann, 2 vols (Tübingen: Mohr, 1973–92).
Bignami-Odier, Jeanne, 'Une lettre apocryphe de saint Damase à saint Jérôme sur la question de Melchisedech', *Mélanges d'Archéologie et d'Histoire de l'École Française de Rome*, 63 (1951), pp. 183–90.
Bochart, Samuel, *Sermons sur divers textes*, 3 vols (Amsterdam: Jacques Desbordes, 1714).
—, *Opera omnia*, ed. by Iohannes Leusden, Petrus de Villemandy, 3 vols (Leiden–Utrecht: Cornelius Boutesteyn, Samuel Luchtmans and Guiliemus Van de Water, 1712).
—, *Geographia sacra, seu Phaleg et Canaan*, ed. by Petrus de Villemandy (Leiden–Utrecht: Cornelius Boutesteyn, Samuel Luchtmans, Guiliemus Van de Water, 1707).
—, *Opera omnia*, ed. Stephanus Morinus (Étienne Morin) (Leiden: Cornelius Boutesteyn et Iordanus Luchtmans, 1692).
Bonaventure, *Opera omnia*, 10 vols (Ad Claras Aquas, Quaracchi, Florence: apud Collegium S. Bonaventurae, 1882–1902).
Book of Enoch, ed. by Matthew Black (Leiden: Brill, 1985).
Bosio, Giacomo, *Crux triumphans et gloriosa* (Antwerp: ex officina Plantiniana apud Balthasarem et Ioannem Moretos, 1617).
Brothers, Richard, *A Description of Jerusalem: Its Houses and Streets, Squares, Colleges, Markets, and Cathedrals, the Royal and Private Palaces, with the Garden of Eden in the Centre, as Laid Down in the Last Chapters of Ezekiel...* (London: George Riebau, 1801).
Bünting, Henricus, *Itinerarium Sacrae Scripturae, das ist ein Redebuch über die ganze Heilige Schrift* (Magdeburg: Paul Donat, 1585).
Calmet, Augustin, *Commentaire littéral sur tous les livres de l'Ancien et du Nouveau Testament*, 2nd edn, 8 vols (Paris: [n. publ.], 1724–6).
Calvin, John, *Genesis*, ed. by Alister McGrath and J. I. Packer (Wheaton, IL–Nottingham: Crossway Books, 2001).
Isaiah, ed. by Alister McGrath and J. I. Packer (Wheaton, IL–Nottingham: Crossway Books, 2000).
—, *Sermons sur la Genèse: Chapitres 1,1–11,4*, ed. by Max Engammare, *Supplementa Calviniana* XI/1 (Neukirchen-Vluyn: Neukirchener Verlag, 2000).
—, *A Commentary on Daniel*, 2 vols (Edinburgh: Calvin Translation Society, 1852–3; repr. Edinburgh, Banner of Truth Trust, 1966; 1986).
—, *A Commentarie ... upon the first booke of Moses called Genesis*, transl. by Thomas Tymme (London: John Harison and George Bishop, 1578).
—, *Institution de la religion chrestienne* (Geneva: Jean Martin, 1565).
—, *Commentarii in Isaiam prophetam* (Geneva: I. Crispinius, 1559).
—, *Commentaire ... sur le premier livre de Moyse, dit Genese* (Geneva: Jean Gerard, 1554).
—, *Commentarii in Isaiam Prophetam* (Geneva: Crispinius, 1551).
Caminha, Pêro Vaz de, 'Lettre au Roi Dom Manuel', in Pêro de Magalhães de Gândavo, *Histoire de la province de Santa Cruz que nous nommons le Brésil*, transl. by Henri Ternaux, ed. by Philippe Billé (Nantes: Le Passeur–Cecofop, 1995), pp. 125–49.
Carver, Marmaduke, *A Discourse of the Terrestrial Paradise Aiming at a More Probable Discovery of the True Situation of that Happy Place of our First Parents Habitation* (London: James Flesher, 1666).
Caspari, Otto, *Die Urgeschichte der Mensheit*, 2 vols (Leipzig: [n. pub.], 1873; repr. 1877).
Cecco d'Ascoli, *Commentary to the Sphere of Sacrobosco*, in Thorndike, *Sphere of Sacrobosco* (1949), pp. 343–411.
Celan, Paul, *Gesammelte Werke in sieben Bänden*, III (Frankfurt am Main: Suhrkamp, 2000).
Chronica minora saec. IV.V.VI.VII, ed. by Theodor Mommsen, *MGH AA*, XI (Berlin: Weidmann, 1893).
Chrysostom, John, *In Genesim*, *PG* LIII.
Christian of Stavelot, *Expositio in Matthaeum*, *PL* CVI, cols 1261–504.

Clement of Alexandria, *Stromata*, in *Les Stromates: Stromate V*, ed. by Alain Le Boullec, transl. by Pierre Voulet, 2 vols, *SC* CCLXXVIII, CCLXXIX (Paris: Cerf, 1981).

Columbus, Christopher: *Select Documents Illustrating the Four Voyages of Columbus*, transl. and ed. by L. Cecil Jane, 2 vols (London: Hakluyt Society, 1930–3).

Corpus antiphonalium officii, ed. by Réné Jean Hesbert, *RED SM*, Fontes IX (Rome: Herder, 1968).

Corpus glossariorum latinorum, ed. by Georg Goetz, 7 vols (Leipzig: Teubner, 1888–1923; repr. Amsterdam: Hakkert, 1965).

Cosmas Indicopleustes: *The Christian Topography of Cosmas, an Egyptian Monk*, ed. by John M. McCrindle (London: Hakluyt Society, 1987).

——, *Cosmas Indicopleustès: Topographie chrétienne*, ed. by Wanda Wolska-Conus, 3 vols, *SC* CXLI, CLIX, CXCVII (Paris: Cerf, 1968–73).

——, *The Christian Topography of Cosmas Indicopleustes*, ed. by Eric Otto Winstedt (Cambridge: Cambridge University Press, 1909).

Curtis, A. P., *The Land of Eden and Havilah* (London: the author, 1905).

Dawson, Christopher, ed., *The Mongol Mission: Narrative and Letters of the Franciscan Missionaries in Mongolia and China in the Thirteenth and Fourteenth Centuries* (London–New York: Sheed and Ward, 1955; repr. New York: AMS Press, 1980).

Delitzsch, Friedrich, *Babel und Bibel: Ein Vortrag* (Leipzig: Hinrichs, 1902).

——, *Babel and Bible: A Lecture on the Significance of Assyriological Research for Religion Delivered before the German Emperor*, transl. by Thomas J. McCormack (Chicago: Open Court Publishing Company, 1902).

——, *Wo lag das Paradies?* (Leipzig: W. De Gruyter, 1881).

Diderot, Denis, *Supplément au Voyage de Bougainville*, ed. by Herbert Dieckmann (Geneva: Droz, 1955).

Dietrich, Manfred, 'Der "Garten Eden" und die babylonischen Parkanlagen im Tempelbezirk', in *Religiöse Landschaften*, ed. by Johannes Hahn and Christian Ronning (Münster in Westphalia: Ugarit Verlag, 2002), pp. 1–29.

——, 'Das biblische Paradies und der babylonische Tempelgarten: Überlegungen zur Lage des Gartens Eden', in *Das biblische Weltbild und seine altorientalischen Kontexte*, ed. by Bernd Janowski, Beate Ego, and Annette Krüger (Tübingen: Mohr Siebeck, 2001), pp. 280–323.

Dionysius of Alexandria, *Epistolae*, *PG* X, cols 1271–344.

Dionysius the Carthusian, *Opera selecta*, ed. by Kent Emery, *CCCM* CXXI–CXXIA (Turnhout: Brepols, 1991–).

——, *Opera omnia*, 42 vols (Montreuil-sur-mer–Tournai–Parkminster: typis Carthusiae Sanctae Mariae de Pratis, 1896–1935).

Dobschütz, Ernst von, *Das Decretum Gelasianum de libris recipiendis et non recipiendis* (Leipzig: Hinrichs, 1912).

Donnelly, Ignatius, *Atlantis: The Antediluvian World* (New York: Harper & Brothers, 1882).

Duns Scotus, John, *Opera omnia*, ed. by Lucas Wadding, 12 vols (Leiden: [n. publ.], 1639).

Durandus of St Pourçain, *In Petri Lombardi Sententias theologicas commentariorum libri IIII*, 2 vols (Venice: ex typographis Guerraea, 1571; repr. Ridgewood, NJ: Gregg, 1964).

El libro del conoscimiento de todos los reinos (The Book of Knowledge of all Kingdoms), ed. and transl. by Nancy F. Marino (Tempe, AZ: Arizona Center for Medieval and Renaissance Studies, 1999).

Elliott, James K., *The Apocryphal New Testament: A Collection of Apocryphal Christian Literature in an English Translation* (Oxford: Clarendon Press, 1993).

Engel, Moritz, *Die Lösung der Paradiesfrage* (Leipzig: Otto Schulze, 1885).

Engler, Heinrich G. A., *Versuch einer Entwicklungsgeschichte der Pflanzenwelt*, 2 vols (Leipzig: [n. publ.], 1879–82).

Ephrem the Syrian, *Hymns on Paradise*, transl. by Sebastian Brock (Crestwood, NY: St Vladimir's Seminary Press, 1990).

——, *Hymnen über das Paradies*, ed. and transl. by Edmund Beck (Rome: Herder, 1951).

Epiphanius, 'Epistula … ad Iohannem episcopum a Sancto Hieronymo translata', in Jerome, *Epistulae*, ed. Hilberg (1910; 1996), pp. 395–412.

——, *Ancoratus*, *PG* XLIII, cols 11–236.

——, *Panarion*, *PG* XLI, cols 173–1200.

Eratosthenes, *Die geographischen Fragmente*, ed. by Hugo Berger (Leipzig: Teubner, 1898).

Eucherius, *Formulae spiritalis intelligentiae, Instructionum libri duo*, ed. by Carmela Mandolfo *CCSL* LXVI (Turnhout: Brepols, 2004).

——, *De situ hierusolimitane urbis atque ipsius Iudaeae epistola ad Faustum presbyterum*, in *Itinera hierosolymitana saeculi IIII–VIII*, ed. by Paul Geyer, *CSEL* XXXIX (Prague–Vienna–Leipzig: Tempsky, 1898), pp. 123–34.

Eusebius, *Life of Constantine*, transl. by Averil Cameron and Stuart G. Hall (Oxford: Clarendon Press, 1999).

——, ΕΙΣ ΤΟΝ ΒΙΟΝ ΤΟΥ ΜΑΚΑΡΙΟΥ ΚΩΝΣΤΑΝΤΙΝΟΥ ΒΑΣΙΛΕΩΣ, ed. by Ivar A. Heikel, in *Eusebius Werke*, I (Leipzig: Hinrichs, 1902).

——, *Chronicorum libri duo*, *PG* XIX, cols 99–598.

——, *Historia ecclesiastica*, *PG* XX, cols 9–906.

Expositio totius mundi et gentium, ed. by Jean Rougé, *SC* CXXIV (Paris: Cerf, 1966).

Ezra: *Die Esra-Apokalypse (IV. Esra)*, ed. by Albert Frederik J. Klijn (Berlin: Akademie Verlag, 1992).

——, *Der lateinische Text der Apokalypse des Esra*, ed. by Albert Frederik J. Klijn (Berlin: Akademie Verlag, 1983).

——, *Apocalypsis Esdrae, Apocalypsis Sedrach, Visio Beati Esdrae*, ed. by Otto Wahl, *Pseudepigrapha Veteris Testamenti*, IV (Leiden: Brill, 1977).

——, *Revelation of Ezra*, in *Apocryphal Gospels, Acts and Revelations*, transl. by Walker (1870), pp. 468–76.

Fidelis, Ludovicus, *De mundi structura* (Paris: [n. publ.], 1556).

Finzi, Felice, *Ricerche per lo studio dell'antichità assira* (Turin: Loescher, 1872).

Flavius Josephus, *Jewish Antiquities*, Loeb Classical Library, transl. by Henry St. John Thackeray, Ralph Marcus, and Louis H. Feldman, 9 vols (London: Heinemann; Cambridge, MA: Harvard University Press, 1926–65).

Ford, Julian, *The Story of Paradise* (New York: Harper and Row, 1981).

From Paradise Lost to Paradise Regained (Brooklyn, NY: Watchtower Bible and Tract Society of New York, 1958).

Geminus, *Introduction aux phénomènes*, ed. and transl. by Germaine Aujac (Paris: Belles Lettres, 1975).

Glaber, Rodulfus, *The Five Books of History*, ed. and transl. by John France (Oxford: Clarendon Press, 1989).

Godfrey of Viterbo, *Pantheon*, in *Illustres veteres scriptores qui rerum a Germanis per multas aetates gestarum historias vel annales posteris reliquerunt*, ed. by Iohann Pistorius, 2 vols (Frankfurt am Main: [n. publ.], 1584; repr. 1613), II.

Gordon, Henry William, *Events in the Life of Charles George Gordon, from its Beginning to its End* (London: Kegan Paul Trench, 1886).

Gregory of Nyssa, *De beatitudinibus*, in *Opera*, VII/2, ed. by Werner Jäger, Hermann Langerbeck, Heinrich Dörrie, John F. Callahan (Leiden: Brill, 1992).

——, *De hominis opificio*, *PG* XLIV, cols 123–256.

Gregory the Great, *Moralia in Iob*, ed. by Marcus Adriaen, *CCSL* CXLIIIA (Turnhout: Brepols, 1979); CXLIIIB (Turnhout: Brepols, 1985).

Grosseteste, Robert, *Die philosophischen Werke des Robert Grosseteste, Bischofs von Lincoln*, ed. by Ludwig Baur (Münster in Westphalia: Aschendorff, 1912).

Hamblin, Diana, 'Has the Garden of Eden Been Located at Last?' *Smithsonian*, 18/2 (1987), pp. 127–35.

Hardouin, Jean, *Nouveau Traité sur la situation du paradis terrestre (ou Conformité de Pline avec Moïse, par rapport à la position des fleuves du paradis terrestre)*, in *Traitez geographiques et historiques pour faciliter l'intelligence de l'Écriture Sainte*, ed. by Antoine-Augustin Bruzen de la Martinière (La Haye: Van der Poel, 1730), I, pp. 1–172.

——, *De situ paradisi terrestris disquisitio, sive de Plinii cum Mose convenientia in paradisi fluminibus indicandis*, in *Caii Plinii Secundi Historiae naturalis libri XXXVII*, 2nd edn, 2 vols (Paris: Antonius-Urbanus Coustelier, 1723), I, pp. 359–68.

Harff, Arnold von, *Pilgerfahrt*, in *The Pilgrimage of Arnold von Harff Knight from Cologne, through Italy, Syria, Egypt, Arabia, Ethiopia, Nubia, Palestine, Turkey, France and Spain, which He Accomplished in the Years 1496 to 1499*, ed. by Malcolm Henri Letts (London: Hakluyt Society, 1946).

Harnack, Adolf, *Letter to the Preussische Jahrbücher on the German Emperor's Criticism of Prof. Delitzsch's Lectures on 'Babel und Bibel'*, transl. by Thomas Bailey Saunders (Oxford: Williams and Norgate, 1903).

Hennig, Richard, *Wo lag das Paradies?* (Berlin: Ullstein Verlag, 1950).

Herbinius, Iohannes, *Dissertationes de admirandis mundi cataractis* (Amsterdam: Ioannes Ianssonius, 1678).

Herder, Johann, *Abhandlung über den Ursprung der Sprache* (Berlin: Voss, 1772).

Herrmann, Albert, *Die Erdkarte der Urbibel* (Braunschweig: Kommissionsverlag von Georg Westermann, 1931).

Herzog, Johann Jacob, Gustav Leopold Plitt and Albert Hauck, eds, *Real-Encyclopädie für protestantische Theologie und Kirche …*, 2nd edn, 18 vols (Leipzig: Hinrichs, 1877–88).

Heude, William, *A Voyage up to the Persian Gulf and a Journey Overland from India to England in 1817* (London: Longman, Hurts, Rees, Orme and Brown, 1819).

Higden, Ranulf, *Polychronicon Ranulphi Higden, monachi cestrensis, together with the English Translations of John Trevisa and of an Unknown Writer of the Fifteenth Century*, ed. by Churchill Babington (vols I, II) and Joseph R. Lumby (vols III–IX), *RBMAS* XLIV, 9 vols (London: Longmans, Green, and Co., 1865–86).

Hippolytus, *Commentary on Daniel*, ed. by Georg Nathanael Bonwetsch, *GCS*, Neue Folge, VII (Berlin: Akademie Verlag, 2000).

——, *De Antichristo*, ed. by Enrico Norelli (Florence: Nardini, 1987).

——, *Die Chronik*, ed. by Adolf Bauer and Rudolf Helm, *GCS* XLVI (Berlin: Akademie Verlag, 1955).

Honorius Augustodunensis, *Imago mundi*, ed. by Valerie I. J. Flint, *Archives d'Histoire Doctrinale et Littéraire du Moyen Âge*, 49 (1982), pp. 7–154.

Hopkinson, John, *Synopsis paradisi sive paradisi descriptio, ex variis diversarum tum linguarum, tum aetatum scriptoribus desumpta; cum chorographica eiusdem tabula* (Leiden: Plantinus, 1593).

Huet, Pierre-Daniel, *Tractatus de situ paradisi terrestris*, in *Thesaurus antiquitatum sacrarum*, 34 vols (Venice: Ioannis Gabriel Herthz, 1744–69), VII, ed. by Blasius Ugolinus (1747), pp. 500–79.

——, *Traité de la situation du paradis terrestre* (Amsterdam: François Halma, 1701).

——, *A Treatise of the Situation of the Terrestrial Paradise*, transl. by Thomas Gale (London: James Knapton, 1694).

——, *Traité de la situation du paradis terrestre* (Paris: Jean Anisson, 1691).

Hugh of St Victor, *De archa Noe*, and *Libellus de formatione arche*, ed. by Patrice Sicard, CCCM CLXXVI (Turnhout: Brepols, 2001).

—, *De tribus maximis circumstantiis gestorum*, English transl. in Mary J. Carruthers, *The Book of Memory: A Study of Memory in Medieval Culture* (New York: Cambridge University Press, 1990), pp. 261–6.

—, *De tribus maximis circumstantiis gestorum*, ed. by William M. Green, *Speculum*, 18 (1943), pp. 488–92.

—, *Didascalicon*, ed. by Charles H. Buttimer (Washington, DC: Catholic University Press, 1939).

—, *De vanitate mundi*, PL CLXXVI, cols 703–40.

—, *Didascalicon*, PL CLXXVI, cols 739–838.

—, *De sacramentis fidei christianae*, PL CLXXVI, cols 173–618.

—, *De sacramentis legis naturalis et scriptae dialogus*, PL CLXXVI, cols 17–42.

—, *In scripturam sacram*, PL CLXXV, cols 9–28.

—, *Adnotationes elucidatoriae in Pentateuchon*, PL CLXXV, cols 29–114.

—, *Commentarii in hierarchiam coelestem S. Dionysii Areopagitae*, PL CLXXV, cols 926–30.

Iacobus de Voragine, *The Golden Legend*, transl. by William Granger Ryan, 2 vols (Princeton, NJ: Princeton University Press, 1993).

Irenaeus, *Adversus haereses*, ed. by Adelin Rousseau, Louis Doutreleau, and others, 5 vols SC CCLXIII, CCLXIV, CCXCIII, CCX and CCXI (2nd edn), C, CLII, CLIII (Paris: Cerf, 1952–74).

Isidore, *De natura rerum*, ed. by Antonio Laborda (Madrid: Instituto Nacional de Estadística, 1996).

—, *Etymologiae*, ed. by Wallace Martin Lindsay, 2 vols (Oxford: Clarendon Press, 1911).

—, *Chronica*, in *Chronica minora*, ed. by Mommsen (1893), pp. 391–497.

—, *Mysticorum expositiones sacramentorum seu quaestiones in Vetus Testamentum*, PL LXXXIII, cols 207–424.

—, *Chronicon*, PL LXXXIII, cols 1017–58.

—, *Allegoriae quaedam Scripturae Sacrae*, PL LXXXIII, cols 97–130.

Pseudo-Isidore, *De ortu et obitu patrum*, ed. by César Chaparro Gómez (Paris: Belles Lettres, 1985).

—, *Liber de ordine creaturarum*, ed. by Manuel C. Díaz y Díaz (Santiago de Compostela: Universidad de Santiago de Compostela, 1972).

—, *In libros Veteris et Novi Testamentum*, PL LXXXIII, cols 155–208.

Itineraria et alia geographica, CCSL CLXXV (Turnhout: Brepols, 1965).

Iulius Honorius, *Cosmographia*, in *Geographi latini minores*, ed. by Alexander Riese (Heilbronn: Henninger, 1878), pp. 21–55.

Iunius, Franciscus (François de Jon), *Praelectiones in Genesim* (Heidelberg: [n. publ.], 1589).

Jerome, *Apologie contre Rufin*, ed. by Pierre Lardet, SC CCCIII (Paris: Cerf, 1983).

—, *In Matthaeum*, ed. by Émile Bonnard, 2 vols, SC CCXLII and CCLIX (Paris: Cerf, 1977–9).

—, *Commentarium in Hiezechielem*, ed. by François Glorie, CCSL LXXV (Turnhout: Brepols, 1964).

—, *Hebraicae quaestiones in libro Geneseos*, ed. by Paul de Lagarde, in *Opera*, I/1, CCSL LXXII (Turnhout: Brepols, 1959).

—, *Epistulae*, ed. by Isidorus Hilberg, CSEL LIV (Vienna–Leipzig: Tempsky, 1910; repr. Vienna: Österreichische Akademie der Wissenschaften, 1996).

—, *Liber locorum*, in *Onomastica sacra*, ed. by Paul de Lagarde (Göttingen: Horstmann, 1887; repr. Hildesheim: Olms, 1966).

Pseudo-Jerome, *Ex dictis sancti Hieronymi*, in *Scriptores Hiberniae minores*, ed. by Robert E. McNally, I, CCSL CVIIIB (Turnhout: Brepols, 1973), pp. 225–30.

John of Damascus, *De fide orthodoxa: Versions of Burgundio and Cerbanus*, ed. by Eligius M. Buytaert (St Bonaventure, NY: Franciscan Institute, 1955).

—, *De fide orthodoxa*, PG XCIV, cols 790–1228.

John of Monte Corvino, *Letters*, in Dawson, ed., *The Mongol Mission* (1955; 1980), pp. 224–31.

—, *Epistolae*, in *Sinica Francescana*, I, *Itinera et relationes* (1929), pp. 340–55.

John of Plano Carpini, *The Story of the Mongols*, transl. by Erik Hildinger (Boston: Branden, 1996).

—, *Ystoria Mongalorum*, in *Sinica Francescana*, I, *Itinera et relationes* (1929), pp. 27–130.

John of Udine, *Die Historienbibel des Johannes von Udine (Ms 1000 Vad)*, ed. by Renate Frohne (Bern: P. Lang, 1992).

Joinville, Jean, sire de, *The History of Saint Louis*, English translation of *Histoire de Saint Louis* (ed. by Natalis de Wailly) by Joan Evans (London–New York: Oxford University Press, 1938).

Jordan de Sévérac, *Mirabilia descripta*, in Wilhelm Baum and Raimund Senoner, eds, *Indien und Europa im Mittelalter* (Klagenfurt: Kitab, 2000), pp. 104–51.

—, *Les Merveilles de l'Asie*, ed. by Henri Cordier (Paris: Librairie Orientaliste Paul Geuthner, 1925).

Kabakov, Ilya and Émilia, *The Palace of Projects* (London–Madrid: Artangel–Centro de Arte Reina Sofia, 1998).

Kircher, Athanasius, *Turris Babel, sive Archontologia* ... (Amsterdam, Ioannes Ianssonius, 1679).

—, *Arca Noe, in tres libro digesta* (Amsterdam: Ioannes Ianssonius, 1675).

Al-Khwarizmi: *The Astronomical Tables of Al-Khwarizmi: Translation with Commentaries of the Latin Version*, ed. by Heinrich Suter and Otto Neugebauer (Copenhagen: Kommission hos Munksgaard, 1962).

Lamberti S. Audomari canonici liber floridus codex autographus Bibliothecae Universitatis Gandavensis, ed. by Albert Derolez (Ghent: Story-Scientia, 1968) (facsimile edition and transcription of Ghent, Rijksuniversiteit Bibliotheek, MS 92).

Lapeyre, Jacques d'Auzoles, *La Saincte Géographie, c'est-à-dire, exacte description de la terre et veritable demonstration du paradis terrestre* (Paris: Antoine Estiene, 1629).

Lapide, Cornelius a, *Commentaria in Pentateuchum Mosis* (Paris: E. Martin, 1630).

Layard, Austen Henry, *Nineveh and its Remains*, 2 vols (London: John Murray, 1849).

Latini, Brunetto, *Li Livres dou tresor*, ed. by Francis J. Carmody (Berkeley: University of California Press, 1948).

Lichtenberger, Frédéric, ed., *Encyclopédie des sciences religieuses*, 13 vols (Paris: Sandoz et Fischbacher, 1877–82).

Linné, Carl von, *Oratio de telluris habitabilis incremento* (Leiden: Cornelius Haak, 1744).

Lowe, Elias A., ed., *Codices latini antiquiores*, I (Oxford: Clarendon Press, 1934).

Luneau de Boisjermain, Pierre Joseph François, *Carte des parties principales du globe terrestre pour servir à l'histoire des temps qui ont suivi la création* (Paris: 1765).

Luther, Martin, *Table Talk*, transl. by William Hazlitt (London: Harper Collins, 1995).

—, *Lectures on Genesis Chapters 1–5*, in *Luther's Works*, ed. by Jaroslav Pelikan (St Louis, MO: Concordia, 1958), I.

—, *Tischreden*, ed. by Ernst Kroker, 6 vols (Weimar: Hermann Böhlaus Nachfolger, 1912–21).

—, *Genesisvorlesung*, ed. by Gustav Koffmane and Otto Reichert, in *D. Martin Luthers Werke*, XLII (Weimar: Hermann Böhlaus Nachfolger, 1911).

—, *Tischreden*, ed. Iohannes Aurifaber (Eisleben: Urban Gaubisch, 1566).

MacClintock. John, and James Strong, eds, *The Cyclopedia of Biblical, Theological, and Ecclesiastical Literature*, 12 vols (New York: Harper and Brothers, 1867-87).

Macrobius, *Commentarii in Somnium Scipionis: Commentary on the Dream of Scipio*, ed. and transl. by William H. Stahl (New York: Columbia University Press, 1952).

—, *In Somnium Scipionis expositio* (Brescia: Boninus de Boninis, 1483).

Malalas, John, *Chronographia*, transl. by Elizabeth Jeffreys, Michael Jeffreys, Roger Scott, and others (Melbourne: Australian Association for Byzantine Studies, 1986).

Malvenda, Thomas, *De paradiso voluptatis: Quem Scriptura Sacra Genesis secundo et tertio capite describit commentarius* (Rome: [n. publ.], 1605).

Mandeville, John: *The Book of Sir John Mandeville*, ed. by Michael C. Seymour (Oxford: Clarendon Press, 1967).

Marignolli, John of, 'Recollections of Eastern Travels', in Yule, ed. and trans., *Cathay and the Way Thither*, new ed. Cordier, III (1914), pp. 177–269.

—, *Relatio*, in *Sinica Francescana*, I, *Itinera et relationes* (1929), pp. 524–60.

Martianus Capella, *De nuptiis philologiae et Mercurii*, ed. by Adolfus Dick (Stuttgart: Teubner, 1978).

Martin, Peter, 'The Secret Garden', *Sunday Times Magazine* (11 October 1998), pp. 44-50.

Masius, Andreas, *De paradiso commentarius ... adiecta est etiam divi Basilii ... ad haec duae epistulae populi nestoriani ad Pontificem Rom.* (Antwerp: Christophorus Plantinus, 1569).

Massey, Gerald, *The Natural Genesis*, 2 vols (London: Williams and Norgate, 1883).

Mercator, Gerard, *Atlas sive cosmographicae meditationes de fabrica mundi et fabricati figura*, 3 vols (Düsseldorf: A. Busius, 1595).

Mercator, Gerard, and Iodocus Hondius, *Atlas minor* (Amsterdam: Iodocus Hondius, 1607; 1610).

Michaeler, Carl Josef, *Das neueste über die geographische Lage des irdischen Paradieses* (Vienna: Rötzler, 1796).

Moullart-Sanson, Pierre, *Carte du paradis terrestre selon Moyse* (Paris: prope Aedes Regias, 1724).

Moxon, Joseph, *Sacred Geographie or Scriptural Mapps* (London: the author, 1671).

Münster, Sebastian, *Cosmographia* (Basle: Henricus Petri, 1545).

Navigatio Sancti Brendani: De reis van Sint Brandaan: Een reisverhall uit de twaalfde eeuw, ed. by Willem Wilmink, intro. by Willem P. Gerritsen (Amsterdam: Uitgeverij Prometheus–Bert Bakker, 1994).

—, *The Voyage of Saint Brendan: Journey to the Promised Land*, transl. by John J. O'Meara (Portlaoise: Dolmen, 1985).

—, *The Anglo-Norman Voyage of St Brendan*, ed. by Ian Short and Brian Merrilees (Manchester: Manchester University Press, 1979).

—, *Navigatio Sancti Brendani Abbatis: From Early Latin Manuscripts*, ed. by Carl Selmer (Notre Dame, IN: University of Notre Dame Press, 1959).

Newman, John F., *A Thousand Miles on Horseback* (New York: Harper & Brothers, 1875).

Newton, Isaac, *Mathematical Principles of Natural Philosophy and his System of the World*, transl. Andrew Motte and Florian Cajori (Berkeley: University of California Press, 1934).

Nicholas of Lyre, *Postilla super totam Bibliam*, in *Biblia sacra cum glossa ordinaria*, ed. by Leander de Sancto Martino, Ioannes Gallemart, and others, 6 vols (Douai: Baltazar Bellerus; Antwerp: Ioannes Keerbergius, 1617).

Nicholson, James E., *The Probable "Garden of Eden": A Long-Forgotten Star Myth* (London: Watts, 1935, for private circulation only).

Nogarola, Ludovico, *Dialogus qui inscribitur Timotheus, sive de Nilo* (Venice: V. Valqrysius, 1552).

Odoric of Pordenone, *Relatio*, in *Sinica Francescana*, I, *Itinera et relationes* (1929), pp. 413–95.

—, *The Eastern Parts of the World Described*, in Yule, ed. and trans., *Cathay and the Way Thither*, new ed. Cordier, II (1913).

Origen, *Homiliae in Numeros*, ed. by Wilhelm A. Baehrens and Louis Doutreleau, 3 vols, *SC* CCCCXV, CCCCXLII, CCCCLXI (Paris: Cerf, 1996–2001).

——, *Commentaire sur Saint Jean*, ed. by Cécile Blanc, 5 vols, *SC*, CXX, CLVII, CCXXII, CCXC, CCCLXXXV (Paris: Cerf, 1966–92).

——, *Homélies sur Ézéchiel*, ed. by Marcel Borret, *SC* CCCLII (Paris: Cerf, 1989).

——, *In epistolam Pauli ad Romanos: Commento alla lettera ai Romani*, ed. and transl. by Francesca Cocchini, 2 vols (Casale Monferrato: Marietti, 1985–6).

——, *Traité des principes*, ed. by Henri Crouzel and Manlio Simonetti, 5 vols, *SC* CCLII, CCLIII, CCLXVIII, CCLXIX, CCCXII (Paris: Cerf, 1978–84)

——, *Homélies sur le Lévitique*, ed. by Marcel Borret, *SC* CCLXXXVII (Paris: Cerf, 1981).

——, *Homélies sur la Genèse*, ed. by Louis Doutreleau, *SC* VIIbis (Paris: Cerf, 1976).

——, *Contre Celse*, ed. by Marcel Borret, 5 vols, *SC* CXXXI, CXXXVI, CXLVII, CL, CCXXVII (Paris: Cerf, 1967–76).

——, *In Matthaeum*, ed. by Ernst Benz, and others, *Origines Werke*, X, *GCS* (Leipzig: Hinrichs, 1935).

——, *Homilien zum Hexateuch*, ed. by Wilhelm A. Baehrens, *Origenes Werke*, VII, *GCS* (Leipzig: Hinrichs, 1921).

——, *In Genesim*, *PG* XII, cols 45–146.

Orosius, *Historiae adversos paganos*, ed. by Marie-Pierre Arnaud-Lindet, 3 vols (Paris: Belles Lettres, 1990–1).

Ortelius, *Theatrum orbis terrarum* (Antwerp: apud Christophorum Plantinum, 1584; ex officina Plantiniana, apud Ioannem Moretum, 1601).

Ovid, *Metamorphoses*, ed. by William S. Anderson (Leipzig: Teubner, 1977).

——, *Metamorphoses*, transl. by Mary M. Innes (Harmondsworth–New York: Penguin, 1955).

Pasolino, Cherubino, *Il paradiso terrestro aperto alli curiosi* (Forli: Colantonio Zampa, 1677).

Pecham, John, *Tractatus de sphera*, ed. and transl. by Bruce R. MacLaren, PhD Dissertation, University of Wisconsin, Madison, WI (1978).

Pererius, Benedictus, *Commentariorum et disputationum in Genesim tomi quattuor*, 4 vols (Ingolstadt: David Sartorius, 1590).

Perez de Valencia, Jacob, *Commentaria in Psalmos* (Valencia: [n. publ.], 1484).

Peter Comestor, *Historia scholastica*, *PL* CXCVIII, cols 1045–722.

Peter Lombard, *Sententiae in IV libris distinctae*, ed. by Ignatius C. Brady, 3rd edn, 2 vols (Grottaferrata, Rome: Editiones Collegii S. Bonaventura Ad Claras Aquas, 1971).

Peter of Poitiers, *Sententiae*, ed. by Philip S. Moore and Marthe Dulong (Notre Dame, IN: University of Notre Dame Press, 1943).

——, *Sententiae*, *PL* CCXI, cols 789–1280.

Petrus Alphonsus, *Diálogo contra los Judíos*, introd. by John Tolan, Latin text ed. by Klaus-Peter Mieth, Spanish transl. by Esperanza Ducay, gen. ed. by Jesús Lacarra (Huesca: Instituto de Estudios Altoaragoneses, 1996).

Peschel, Oscar F., *The Races of Man and their Geographical Distribution* (London: King & Co., 1876).

Philo, [Works], Loeb Classical Library, transl. by Francis H. Colson and George H. Whitaker, 10 vols and 2 supplementary vols (London: Heinemann; New York: Putnam's Sons, 1929–62).

Philostorgius, *Historia ecclesiastica*, *PG* LXV, cols 459–628.

Piccolomini, Aeneas Sylvius, *Dialogo su un sogno: Dialogus de somnio quodam*, ed. by Alessandro Scafi (Turin: Aragno, 2004).

——, *Europa*, ed. by Adrianus van Heck (Vatican City: Biblioteca Apostolica Vaticana, 2001).

——, *Epistola ad Mahomatem II (Epistle to Mohammed II)*, ed. by Albert R. Baca (New York: P. Lang, 1990).

——, *Commentarii rerum memorabilium quae temporibus suis contigerunt*, ed. by Adrianus van Heck (Vatican City: Biblioteca Apostolica Vaticana, 1984).

——, *Selected Letters*, transl. and ed. by Albert R. Baca (Northridge, CA: San Fernando Valley State College, 1969).

——, *Opera inedita*, ed. by Giuseppe Cugnoni (Rome: Salviucci, 1883; repr. Farnnborough: Gregg, 1968).

——, *Der Briefwechsel*, ed. by Rudolf Wolkan, 3 vols (Vienna: Hölder, 1909–18).

——, *Opera geographica et historica (Cosmographia seu rerum ubique gestarum historia locorumque descriptio)* (Helmstadt: Sustermann, 1699).

——, *Opera omnia*, ed. by Marcus Hopperus (Basle: Henricus Petri, 1551).

——, *Discrittione de l'Asia et Europa di papa Pio II e l'historia de le cose memorabili fatte in quelle, con l'aggiunta de l'Africa, secondo diversi scrittori, con incredibile brevità e diligenza*, transl. by Fausto da Longiano (Venice: Vincenzo Vaugris, 1544).

——, *Dialogus* (Rome: Iohannes Schurener, 1475).

Pico della Mirandola, Giovanni, *Conclusiones nongentae: Le novecento tesi dell'anno 1486*, ed. by Albano Biondi (Florence: Olschki, 1995).

——, *Commento sopra una canzone d'amore*, ed. by Paolo De Angelis (Palermo: Novecento, 1994).

——, *Discorso sulla dignità dell'uomo*, ed. by Giuseppe Tognon and Eugenio Garin (Brescia: La Scuola, 1987).

——, *Heptaplus*, transl. by Douglas Carmichael (Indianapolis, IN: Bobbs; New York: Merrill, 1965).

Pineda, Juan de, *Los treynta libros de la monarchia ecclesiastica o historia universal del mundo*, 5 vols (Barcelona: Hieronymo Margarit, 1620).

Plato, *Statesman*, ed. and transl. by Julia Annas and Robin Waterfield (Cambridge–New York: Cambridge University Press, 1995).

——, *Statesman*, ed. by Christopher J. Rowe (Warminster: Aris and Phillips, 1995).

Polo, Marco, *The Description of the World*, ed. by A. C. Moule and Paul Pelliot (London: Routledge, 1938).

——, *Il Milione*, ed. by Luigi Foscolo Benedetto (Florence: Olschki, 1928).

Postel, Guillaume, *Cosmographicae disciplinae compendium* (Basle: Jean Oporin, 1561).

——, *Apologia pro Serveto* (Basle: Jean Oporin, 1555).

——, *Des merveilles du monde, et principalement des admirables choses des Indes et du Nouveau Monde* (Paris: J. Ruelle, 1553).

Pressel, Wilhelm, *Geschichte und Geographie der Urzeit von der Erschaffung der Welt bis auf Mose* (Nördlingen: Verlag der C. H. Beck'schen Buchandlung, 1883).

Ptolemy: Lennart Berggren, J., and Alexander Jones, *Ptolemy's Geography: An Annotated Translation of the Theoretical Chapters* (Princeton, NJ–Oxford: Princeton University Press, 2000).

——, *Almagest*, ed. and transl. by Gerald J. Toomer (London: Duckworth, 1984).

——, *Geographia*, ed. by Michael Servetus and Bilibaldus Pirckheimer (Lyons: Hugues La Porte, 1541).

——, *Geographia universalis*, ed. by Sebastian Münster and Bilibaldus Pirckheimer (Basle: Henricus Petri, 1540).

——, *Cosmographia* (Ulm: Lienhart Holle, 1486).

Purchas, Samuel, *Purchas his Pilgrimage; or, Relations of the World and the Religions Observed in all Ages …* (London: W. Stansby, for H. Fetherstone, 1613).

Quatrefages, Armand de, *The Human Species* (London: Kegan Paul, 1881).

Rabanus Maurus, *De universo*, *PL* CXI, cols 9–614.

——, *In Genesim*, *PL* CVII, cols 439–670.

Raleigh, Walter, *The History of the World* (London: Walter Burre, 1614; 2nd edn, William Jaggard for Walter Burre, 1617).

Ravennatis anonymi Cosmographia et Guidonis Geographica, ed. by Moritz E. Pinder and Gustav Parthey (Berlin: Fridericus Nicolaus, 1860; repr. Aalen: Zeller, 1962).

Regnault, Antoine, *Discours du voyage d'outre mer au St. Sépulcre de Jérusalem et autres lieux de la Terre Saincte* (Lyon: [n. pub.], 1573).

Reisch, Gregorius, *Margarita philosophica*, ed. by Adam Wernherus (Freiburg im Breisgau: Ioannes Schottus Argen., 1503).

——, *Margarita philosophica* (Strasbourg: Grüningerus, 1515).

Rhau, Balthasar, *Disputatio geographica de paradiso terrestri* (Frankfurt an Oder: Iohannes Coepselius, 1696).

Ringbom, Lars Ivar, *Paradisus terrestris: Myt, bild och verklighet* (Copenhagen–Helsinki: Ejnar Munksgaards–Akademiska Bokhandeln–Nordiska Antikvariska Bokhandeln, 1958).

Robertus Anglicus, *Commentary to the Sphere of Sacrobosco*, in Thorndike, *Sphere of Sacrobosco* (1949), pp. 143–246.

Rohl, David, *Legend: The Genesis of Civilisation* (London: Century, 1998).

Rüst, Hanns, *Mapa mundi*, separately-printed broadside, Augsburg, c.1480 (Munich: Ludwig Rosenthal's Antiquariat, 1924).

Rudimentum noviciorum (Lübeck: Lucas Brandis, 1475).

Sacrobosco, Iohannes de, *Tractatus de spera*, in Thorndike, *The Sphere of Sacrobosco* (1949), pp. 76–142.

Sale, Antoine de la, *La Salade*, in *Oeuvres complètes*, ed. by Fernand Desonay, I (Paris: Droz, 1935).

Salibi, Kamal, *The Bible Came from Arabia* (London: Cape, 1996).

Sanudo, Marino, *Liber secretorum fidelium crucis super Terrae Sanctae recuperatione et conservatione…*, in Jacques Bongars, *Gesta Dei per Francos*, II, *Liber secretorum fidelium crucis super Terrae Sanctae recuperatione etc.* (Hanover: typis Wechelianis apud heredes Ioannis Aubrii, 1611; repr. Jerusalem: Masada Press, 1972).

Sayce, Archibald Henry, *The Religions of Ancient Egypt and Babylonia: The Gifford Lectures on the Ancient Egyptian and Babylonian Conception of the Divine: Delivered in Aberdeen* (Edinburgh: Clark, 1902).

——, *The Hibbert Lectures, 1887: Lectures on the Origin and Growth of Religion as Illustrated by the Religion of the Ancient Babylonians* (London: Williams and Norgate, 1887).

Schaff, Philip, and others, eds, *A Religious Encyclopaedia: or Dictionary of Biblical, Historical, Doctrinal, and Practical Theology: Based on the Real-Encyklopädie of Herzog, Plitt, and Hauck*, ed. by Philip Schaff, 3 vols (New York: Funk & Wagnalls, 1882–4).

Schedel, Hartman, *Liber chronicarum* (Nuremberg: Anton Koberger, 1493).

Schenkel, Daniel, ed., *Bibel-Lexikon: Realwörter–buch zum Handgebrauch für Geistliche und Gemeindeglieder*, 5 vols (Leipzig, Brockhaus, 1869–75).

Schopenhauer, Arthur, *The World as Will and Idea*, transl. by Richard B. Haldane and John Kemp, 3 vols (London: Trübner, 1883–6).

——, *Die Welt als Wille und Vorstellung*, 3rd edn, 2 vols (Leipzig: Brochaus, 1859).

Scot, Michael (attr.), *Commentary to the Sphere of Sacrobosco*, in Thorndike, *Sphere of Sacrobosco* (1949), pp. 247–342.

Sedulius, *Opera omnia*, ed. by Johannes Huemer, *CSEL* X (Vienna: C. Geroldi filius, 1885; repr. New York: Johnson Reprint, 1967).

Servetus, Michael, *Christianismi restitutio* ([n.p.]: [n. pub.], 1553; repr. Nuremberg: Rau, [1790]).

Servius, *In Vergilii carmina commentarii*, *Aeneis*, ed. by Georg Thilo and Hermann Hagen, 3 vols (Leipzig: Teubner, 1881–1902).

Severian, *De mundi creatione*, *PG* LVI, cols 429–500.

Sextus Iulius Africanus, *Chronographia*, in Martin J. Routh, *Reliquiae sacrae*, 2nd edn, 3 vols (Oxford: Oxford University Press, 1846), II, pp. 238–309.

——, *Chronographia*, *PL* X, cols 65–93.

Simon of St-Quentin, *Histoire des Tartars*, ed. by Jean Richard (Paris: Geuthner, 1965).
Sinica Franciscana, 3 vols (Ad Claras Aquas, Quaracchi, Florence: apud Collegium S. Bonaventurae, 1929–36), I (1929), *Itinera et relationes fratrum minorum saeculi XIII et XIV*, ed. by Anastasius Van den Wyngaert.
Smith, William, *Dictionary of the Bible: Comprising its Antiquities, Biogeography, Geography, and Natural History*, 3 vols (London: John Murray, Walton and Maberly, 1863).
Solinus, *Collectanea rerum memorabilium*, ed. by Theodor Mommsen, 2nd edn (Berlin: Weidmann, 1958).
Speculum virginum, ed. by Jutta Seyfarth, CCCM V (Turnhout: Brepols, 1990).
Steuchus, Augustinus, *Recognitio Veteris Testamenti ad hebraicam veritatem* (Venice: Aldus, 1529).
Strabo, Walafrid (Remigius of Auxerre), *In Genesim*, PL CXXXI, cols 51–134.
Terry, Albert R., *The Flood and Garden of Eden: Astounding Facts and Prophecies* (Elms Court, Ilfracombe, Devon: Arthur H. Stockwell, 1962; repr. 1963).
Tertullian, *De baptismo*, ed. by Raymond F. Refoulé, SC XXXV (Paris: Cerf, 2002).
——, *De anima*, ed. by Jan Hendrik Waszink (Amsterdam: J. M. Meulenhoff, 1947).
Testamenti Veteris Biblia sacra sive, libri canonici priscae Iudaeorum Ecclesiae a Deo traditi, ed. by Immanuel Tremellius and Franciscus Iunius (Geneva: Ioannes Tornaesius, 1590).
Theodore of Mopsuesta, *In Genesim*, PG LXVI, cols 633–46.
Theodosius, *De situ Terrae Sanctae*, ed. by Paul Geyer, in *Itineraria et alia geographica* (1965), pp. 113–25.
Thomas Aquinas, *Summa theologiae*, Blackfriars edn, 61 vols (London–New York: Eyre and Spottiswoode–McGraw-Hill, 1964–81).
——, *Summa theologiae*, ed. by Pietro Caramello, Leonine edn, 3 vols (Turin–Rome: Marietti, 1948–50).
——, *Quaestiones quodlibetales*, ed. by Pierre Felix Mandonnet (Paris: Lethielleux, 1926).
Thorndike, Lynn, *The Sphere of Sacrobosco and its Commentators* (Chicago: University of Chicago Press, 1949).
Tyconius, *The Turin Fragments of Tyconius' Commentary on Revelation*, ed. by Francesco Lo Bue (Cambridge: Cambridge University Press, 1963).
Unger, Franz, *Die versunkene Insel Atlantis* (Vienna: [n. pub.], 1860).
Vadianus (Joachim Van Watt), *Pomponii Melae de orbis situ libri tres* (Paris: Christianus Wechelus, 1540)
——, *Epitome trium terrae partium, Asiae, Africae et Europae compendiarum locorum descriptionem continens* (Zurich: Christophorus Frosch, 1534).
Van Til, Salomon, *Dissertatio singularis geographico-theologica de situ paradisi terrestris*, in *Malachius illustratus* (Leiden: Iordanus Luchtmans, 1701).
Victorinus of Pettau, *De fabrica mundi*, in *Sur l'Apocalypse et autres écrits*, ed. by Martine Dulaey, SC CCCXXIII (Paris: Cerf, 1997), pp. 43–150.
Vincent of Beauvais, *Speculum maius*, 4 vols (Douai: Baltazar Bellerus, 1624; repr. Graz: Akademische Drück und Verlagsanstalt, 1965).
Visio Pauli, in Theodore Silverstein and Anthony Hilhorst, eds, *Apocalypse of Paul: A New Critical Edition of Three Long Latin Versions* (Geneva: Cramer, 1997).
Vita fabulosa sancti Macarii romani, servi Dei, qui inventus est iuxta paradisum, in *Acta sanctorum*, LVIII, entry for 23 October, pp. 566–71.
La Vie grecque d'Adam et Ève, ed. by Daniel A. Bertrand (Paris: Maisonneuve, 1987).
'The Books of Adam and Eve', by L. S. A. Wells, in *The Apocrypha and Pseudepigrapha of the Old Testament*, ed. by Robert H. Charles, 2 vols (Oxford: Clarendon Press, 1913), II, pp. 123–54.

Vollmer, Hans, ed., *Deutsche Bibelauszüge des Mittelalters zum Stammbaum Christi* (Potsdam: Akademische Verlagsgesellschaft Athenaion, 1931).
Walter of Châtillon, *Alexandreis*, in *The Alexandreis of Walter of Châtillon: A Twelfth-Century Epic*, transl. by David Townsend (Philadelphia: University of Pennsylvania Press, 1996).
——, *Alexandreis*, ed. by Marvin L. Colker (Padua: Antenore, 1978).
Warren, Willliam F., *Paradise Found: The Cradle of the Human Race at the North Pole* (Boston: Houghton, Mifflin and Co., 1885).
Wendrin, Friedrich von, *Die Entdeckung des Paradies* (Leipzig: Braunschweig, 1924).
Wetzel, Heinrich Joseph, and Benedikt Welte, eds, *Kirchenlexikon oder Encyklopädie der katholischen Theologie und ihrer Hilfswissenschaften*, 12 vols (Freiburg im Breisgau: Herder, 1886–1901).
Willcocks, William, *From the Garden of Eden to the Crossing of the Jordan* (Cairo: French Institute of Oriental Archaeology, 1918; 2nd edn, London: E. and F.N. Spon, 1919).
William of Auvergne, *The Trinity or the First Principle*, ed. and transl. by Roland J. Teske and Francis C. Wade (Milwaukee, WI: Marquette University Press, 1989).
——, *De universo*, in *Opera omnia*, 2 vols (Paris: E. Couterot, 1674), I, pp. 193–1074.
William of Conches, *Dragmaticon philosophiae*, ed. by Italo Ronca, in *Opera omnia*, ed. by Édouard Jeauneau, CCCM CLII (Turnhout: Brepols, 1997).
William of Rubruck, *The Mission of Friar William of Rubruck: His Journey to the Court of the Great Khan Möngke 1253–1255*, transl. by Peter Jackson, ed. by Peter Jackson and David Morgan (London: Hakluyt Society, 1990).
——, *Itinerarium*, in *Sinica Francescana*, I, *Itinera et relationes* (1929), pp. 164–332.
William of Tripoli, *De statu Sarracenorum*, in William of Tripoli, *Notitia de Machometo/De statu Sarracenorum*, ed. by Peter Engels, CISC SL IV (Würzburg–Altenberge: Echter–Oros, 1992).
William of Tyre, *Historia rerum in partibus transmarinis gestarum*, PL CCI, cols. 210–891.
Winchell, Alexander, *Preadamites* (Chicago: Griggs & Co., 1880).
Wright, Paul, ed., *The Complete British Family Bible: Being a New Universal Exposition and Commentary on the Holy Scriptures ... Illustrated with Notes and Annotations, Theological, Critical, Moral, Historical, Chronological and Explanatory* (London: for Alex Hogg, 1782).
Yule, Henry, ed. and trans., *Cathay and the Way Thither, being a Collection of Medieval Notices of China*, new edn by Henri Cordier, 4 vols (London: Cambridge University Press, 1913–6).
Zarncke, Friedrich, ed., 'Der Priester Johannes', *Abhandlungen der Philologisch-Historischen Classe der Königlich Sächsischen Gesellschaft der Wissenschaften*, 7 (1879), pp. 827–1030; 8 (1876), pp. 1–186.
Zurara, Gomes E. de, *Chronique de Guinée (1453)*, transl. by Léon Bourdon, and others (Paris: Chandeigne, 1994).

Websites:

http://www2.jpl.nasa.gov/srtm (last accessed: 15 February 2006): Shuttle Radar Topography Mission, SRTM.
http://www.logoschristian.org/eden.html (last accessed: 15 February 2006): Chris Ward's map of Eden.
http://www.unibg.it/dati/persone/636/419.pdf (last accessed 15 February 2006): Emilio Spedicato,'Eden Revisited: Geography, Numerics and Other Tales', *Report DMSIA, Miscellanea* 1/01 (2001); version revised in 2003 and 2005.
http://www.ldolphin.org/eden (last accessed 15 February 2006): Diana Hamblin, 'Has the Garden of Eden Been Located at Last?'

SECONDARY LITERATURE

Abeydeera, Ananda, 'In Search of the Garden of Eden: Florentine Friar Giovanni dei Marignolli's Travels in Ceylon', *Terrae Incognitae*, 25 (1993), pp. 1–24.
——, 'Aspects mythiques de la cartographie de Ceylan de l'Antiquité à la Renaissance', in François Moureau, ed., *L'Île, territoire mythique* (Paris: Aux Amateurs de Livres, 1989), pp. 9–15.
Actas del Simposio para el estudio de los codices del 'Comentario al Apocalipsis' de Beato de Liébana, 3 vols (Madrid: Joyas Bibliográficas, 1978–80).
Adler, William, *Time Immemorial: Archaic History and its Sources in Christian Chronography from Julius Africanus to George Syncellus* (Washington, DC: Dumbarton Oaks Research Library and Collection, 1989).
Alexander, Jonathan J. G., *Medieval Illuminators and their Methods of Work* (New Haven: Yale University Press, 1992).
Alexander, Philip S., 'The Fall into Knowledge: The Garden of Eden/Paradise in Gnostic Literature', in Morris and Sawyer, eds, *A Walk in the Garden* (1992), pp. 91–104.
Alexandre, Monique, *Le Commencement du livre: Genèse I–V: La Version grecque de la Septante et sa réception* (Paris: Beauchesne, 1988).
——, 'Entre ciel et terre: Les Premiers Débats sur le site du paradis (Gen. 2, 8–15 et ses réceptions)', in François Jouan and Bernard Deforge, eds, *Peuples et pays mythiques* (Paris: Belles Lettres, 1988), pp. 187–224.
Alfaric, Prosper, *L'Évolution intellectuelle de Saint Augustin*, I, *Du Manichéisme au Néoplatonisme* (Paris: E. Nourry, 1918).
Allen, Rosamund S., ed., *Eastward Bound: Medieval Travel and Travellers, 1050–1500* (Manchester: Manchester University Press, 2004).
Almagià, Roberto, *Monumenta cartographica vaticana*, I, *Planisferi carte nautiche e affini dal secolo XIV al XVII* (Vatican City: Biblioteca Apostolica Vaticana, 1944).
——, 'Il mappamondo di Albertin de Virga (1415)', *Rivista Geografica Italiana*, 21 (1914), pp. 92–6.
Alster, Bendt, 'Dilmun, Bahrain, and the Alleged Paradise in the Sumerian Myth and Literature', in Daniel T. Potts, ed., *Dilmun: New Studies in the Archaeology and Early History of Bahrain* (Berlin: Reimer, 1983), pp. 39–74.
Alverny, Marie-Therèse de, 'Le Cosmos symbolique du XIIIe siècle', *Archives d'Histoire Doctrinale et Littéraire du Moyen Âge*, 28 (1953), pp. 31–81.
Anderson, Andrew R., *Alexander's Gate, Gog and Magog and the Inclosed Nations* (Cambridge, MA: Mediaeval Academy of America, 1932).
Anderson, Gary A., Michael E. Stone and Johannes Tromp, eds, *Literature on Adam and Eve* (Leiden–Boston–Cologne: Brill, 2000).
——, and Michael E. Stone, eds, *A Synopsis of the Books of Adam and Eve*, 2nd edn (Atlanta, GA: Scholars Press, 1999).
Andrei, Osvalda, 'L'esamerone cosmico e le Cronografie di Giulio Africano', in *La narrativa cristiana antica: Codici narrativi, strutture formali, schemi retorici* (Rome: Institutum Patristicum Augustinianum, 1995), pp. 169–83.
——, 'La formazione di un modulo storiografico cristiano: Dall'esamerone cosmico alle *Chronographiae* di Giulio Africano', *Aevum*, 69 (1995), pp. 147–70.
Andrews, Michael C., 'An Early Printed Map in the Pierpont Morgan Library', *Geography Journal*, 65 (1925), pp. 469–70.
Andriani, Beniamino, *Aspetti della scienza in Dante* (Florence: Le Monnier, 1981).
Andrieu, Michel, *Les Ordines romani du haut moyen âge*, 5 vols (Louvain: Spicilegium Sacrum Lovaniense, 1931–48; 1961–74).

Arentzen, Jörg-Geerd, *Imago mundi cartographica: Studien zur Bildlichkeit mittelalterlicher Welt- und Ökumenekarten unter besonderer Berücksichtigung des Zusammenwirkens von Text und Bild* (Munich: Fink, 1984).

Armour, Peter, *Dante's Griffin and the History of the World: A Study of the Earthly Paradise (Purgatorio, cantos XXIX–XXXIII)* (Oxford–New York: Clarendon Press, 1989).

Armstrong, Arthur H., 'Man in the Cosmos: A Study of Some Differences between Pagan Neoplatonism and Christianity', in Willem den Boer, and others, eds, *Romanitas et Christianitas* (Amsterdam–London: North-Holland, 1973), pp. 5–14.

——, *Saint Augustine and Christian Platonism* (Villanova, PA: Villanova University Press, 1967).

Auffarth, Christoph, 'Paradise Now – But for the Wall Between: Some Remarks on Paradise in the Middle Ages', in Luttikhuizen, ed., *Paradise Interpreted* (1999), pp. 153–79.

Augusteijn, Joost, and Wilco C. Poortman, *Karten in Bijbels (16e–18e eeuw)* (Zoetmeer, Uitgeverij Boekencentrum, 1995).

Aujac, Germaine, 'Ptolemy', in Friedman, Mossler Figg, and others, eds, *Trade, Travel, and Exploration in the Middle Ages* (2000), pp. 507–11.

——, *Claude Ptolémée, astronome, astrologue, géographe* (Paris: Comité des Travaux Historiques et Scientifiques, 1993).

——, 'Les Images du monde entre rêve et réalité', *Revue de la Bibliothèque Nationale*, 45 (1992), pp. 2–13.

Avi-Yonah, Reuven S., *The Aristotelian Revolution: A Study of the Transformation of Medieval Cosmology, 1150–1250*, PhD Dissertation, Harvard University (1986).

Baert, Barbara, *Een erfenis van heilig hout: De neerslag van het teruggevonden kruis in tekst en beeld tijdens de Middeleeuwen* (Louvain: Universitaire Pers Leuven, 2001).

——, 'Seth of de terugkeer naar het paradijs: Bijdrage tot het kruishoutmotief in de middeleeuwen', *Bijdragen, tijdschrift voor filosofie en theologie*, 56 (1995), pp. 313–39.

Bagrow, Leo, 'Rust's and Sporer's World Map', and 'Essay of a Catalogue of Map-Incunabula', *Imago Mundi*, 7 (1950), pp. 32–6.

Baldelli Boni, Giovanni Battista, *Il Milione di Marco Polo*, 2 vols (Florence: Giuseppe Pagani, 1827).

Baloira Bértolo, Adolfo, 'El Prefacio del Comentario al Apocalipsis de Beato de Liébana', *Archivos Leoneses*, 71 (1982), pp. 7–25.

Balty-Fontaine, Janine, 'Pour une édition nouvelle du *Liber Aristotelis de inundatione Nili*', *Chronique d'Égypte*, 34 (1959), pp. 95–102.

Baránszky-Jób, László, 'The Problems and Meaning of Giovanni di Paolo's Expulsion from Paradise', *Marsyas*, 8 (1957–9), pp. 1–6.

Barber, Peter, 'Mito, religione e conoscenza: La mappa del mondo medievale', in *Segni e sogni della terra: Il disegno del mondo dal mito di Atlante alla geografia delle reti* (Novara: De Agostini, 2001), pp. 48–79.

——, 'The Evesham World Map: A Late Medieval English View of God and the World', *Imago Mundi*, 47 (1995), pp. 13–33.

——, 'Die Evesham-Weltkarte von 1392: Eine mittelalterliche Weltkarte im College of Arms in London: Von der Universalität zum Anglozentrismus', *Cartographica Helvetica*, 9 (January 1994), pp. 17–22.

——, and Michelle P. Brown, 'The Aslake World Map', *Imago Mundi*, 44 (1992), pp. 24–44.

——, 'Old Encounters New: The Aslake World Map', in Pelletier, ed., *Géographie du monde au Moyen Âge et à la Renaissance* (1989), pp. 69–88.

Barber, Richard, ed., *Bestiary: Being an English Version of the Bodleian Library, Oxford M. S. Bodley 764 with All the Original Miniatures Reproduced in Facsimile* (Woodbridge: Boydell Press, 1993).

Baron, Roger, *Études sur Hugues de Saint-Victor* (Paris: Desclée, 1963).

——, *Science et sagesse chez Hugues de Saint-Victor* (Paris: Lethielleux, 1957).

Bartelink, Gerhardus Johannes Marinus, 'De terugkeer naar het paradijs: Paradijsverhalen uit de Oudheid', *Hermeneus: Tijdschrift voor antieke cultuur*, 62 (1990), pp. 203–8.

Bartoli, Renata A., *La Navigatio Sancti Brendani e la sua fortuna nella cultura romanza dell'età di mezzo* (Fasano, Brindisi: Schena, 1993).

Bassett, Paul Merritt, 'The Use of History in the *Chronicon* of Isidore of Seville', *History and Theory: Studies in the Philosophy of History*, 15/3 (1976), pp. 278–92.

Baudet, Ernest H. P., *Paradise on Earth: Some Thoughts on European Images of Non-European Man* (New Haven: Yale University Press, 1965).

Bauer, Dieter, Klaus Herbers, and Nikolas Jaspert, eds, *Jerusalem im Hoch- und Spätmittelalter: Konflikte und Konfliktbewältigung-Vorstellungen und Vergegenwärtigungen* (Frankfurt am Main: Campus, 2001).

Baumgärtner, Ingrid, 'Die Wahrnehmung Jerusalems auf mittelalterlichen Weltkarten', in Bauer, and others, eds, *Jerusalem im Hoch- und Spätmittelalter* (2001), pp. 271–334.

Beazley, Charles Raymond, *The Dawn of Modern Geography*, 3 vols (London: John Murray, 1897–1901; repr. New York: Peter Smith, 1949).

Beck, Edmund, 'Symbolum-Mysterium bei Aphrahat und Ephräm', *Oriens Christianus*, 42 (1958), pp. 19–40.

Beckingham, Charles F., and Bernard Hamilton, eds, *Prester John, the Mongols, and the Ten Lost Tribes* (Brookfield, VT: Variorum, 1996).

Bejczy, István P., *La Lettre du prêtre Jean: Une utopie médiévale* (Paris: Imago, 2001).

Benjamins, H. S., 'Paradisiacal Life: The Story of Paradise in the Early Church', in Luttikhuizen, ed., *Paradise Interpreted* (1999), pp. 153–79.

Bennett Durand, Dana, *The Vienna–Klosterneuburg Map Corpus of the Fifteenth Century: A Study in the Transition from Medieval to Modern Science* (Leiden: Brill, 1952).

Beonio-Brocchieri Fumagalli, Maria Teresa, *Durando di S. Porziano: Elementi filosofici della terza redazione del 'Commento alle Sentenze'* (Florence: La Nuova Italia, 1969).

Berbard-Maitre, Henri, 'L'Apologie de la Compagnie de Jésus de Guillaume Postel à la fin de l'année 1552', *Recherches de Sciences Religieuses*, 38/2 (1952), pp. 209–33.

Bertola, Ermenegildo, 'La dottrina della creazione nel *Liber Sententiarum* di Pier Lombardo', *Pier Lombardo*, II (1957), pp. 35–40.

Bérubé, Camille, ed., *Homo et mundus* (Rome: Societas Internationalis Scotistica,1984).

Besse, Jean-Marc, *Les Grandeurs de la Terre: Aspects du savoir géographique à la Renaissance* (Lyons: ENS, 2003).

Bethmann, Ludwig, 'Nachrichten über die von ihm für die Monumenta Germaniae Historica benutzten Sammlungen von Handschriften und Urkunden Italiens aus dem Jahre 1854', *Archiv der Gesellschaft für Ältere Deutsche Geschichtskunde*, 12 (1872), pp. 253–5.

Bettoni, Efrem, 'The Originality of the Scotistic Synthesis', in Bonansea and Ryan, eds, *John Duns Scotus* (1965), pp. 28–44.

——, *Il problema della conoscibilità di Dio nella scuola francescana* (Padua: CEDAM, 1950).

Bini, Mauro, 'Dalla cosmografia classica alla cartografia del Quattrocento', in *Alla scoperta del mondo: L'arte della cartografia da Tolomeo a Mercatore* (Modena: Il Bulino, 2001), pp. 25–32.

Black, Jeremy, *Maps and History: Constructing Images of the Past* (New Haven–London: Yale University Press, 1997).

Blic, Jean de, 'L'Oeuvre exégétique de Walafrid Strabon et la *Glossa ordinaria*', *Recherches de Théologie Ancienne et Médiévale*, 16 (1949), pp. 5–28.

Bliss, Carey S., *Some Aspects of Seventeenth-Century English Printing with Special Reference to Joseph Moxon* (Los Angeles: University of California Press, 1965).

Bøe, Sverre, *Gog and Magog* (Tübingen: Mohr Siebeck, 2001).

Bonansea, Bernardine M., and John K. Ryan, eds, *John Duns Scotus: 1265–1965* (Washington, DC: Catholic University of America Press, 1965).

Bonneau, Danielle, *La Crue du Nil, divinité égyptienne* (Paris: C. Klincksieck, 1964).

Borst, Arno, *Der Turmbau von Babel: Geschichte der Meinungen über Ursprung und Vielfalt der Sprachen und Völker*, 4 vols (Stuttgart: Hiersemann, 1957–63).

Bouet, Pierre, *Le Fantastique dans la littérature latine du Moyen Âge: La Navigation de Saint Brendan* (Caen: Presses Universitaires de Caen, 1986).

Bouwsma, William J., *John Calvin: A Sixteenth-Century Portrait* (New York–Oxford: Oxford University Press, 1988).

——, *Concordia mundi: The Career and Thought of Guillaume Postel (1510-1581)* (Cambridge, MA: Harvard University Press, 1957).

Boyancé, Pierre, 'Études philoniennes', *Revue des Études Grecques*, 76 (1963), pp. 64–110.

Boyde, Patrick, *Dante Philomythes and Philosopher: Man in the Cosmos* (Cambridge: Cambridge University Press, 1981).

Braga, Corin, *Le Paradis interdit au Moyen Âge: La Quête manquée de l'Eden oriental* (Paris: L'Harmattan, 2004).

Bray, Dorothy A., 'Allegory in the *Navigatio Sancti Brendani*', *Viator*, 26 (1995), pp. 1–10.

Breisach, Ernst, *Historiography: Ancient, Medieval and Modern* (Chicago–London: University of Chicago Press, 1983).

Bremmer, Jan N., 'Paradise: From Persia, via Greece, into the Septuagint', in Luttikhuizen, ed., *Paradise Interpreted* (1999), pp. 1–20.

Brennan, Robert Edward, *Thomistic Psychology: A Philosophic Analysis of the Nature of Man* (New York: Macmillan, 1941).

Brincken, Anna-Dorothee von den, 'Mappamundi', in Friedman, Mossler Figg, and others, eds, *Trade, Travel, and Exploration in the Middle Ages* (2000), pp. 363–7.

——, 'Das Weltbild des irischen Seefahrer-Heiligen Brendan in der Sicht des 12. Jahrhunderts', *Cartographia Helvetica*, 21 (January 2000), pp. 17–21.

——, 'Mappe del Medio Evo: Mappe del cielo e della terra', in *Cieli e terre nei secoli XI–XII: Orizzonti, percezioni, rapporti* (Milan: Vita e Pensiero, 1998), pp. 31–50.

——, *Fines terrae: Die Enden der Erde und der vierte Kontinent auf mittelalterlichen Weltkarten* (Hanover: Hahnsche Buchandlung, 1992).

——, 'Die Ebstorfer Weltkarte im Verhältnis zur spanischen und angelsächsischen Weltkartentradition' in Kugler and Michael, eds, *Ein Weltbild vor Columbus* (1991), pp. 129–45.

——, 'Monumental Legends on Medieval Manuscript Maps: Notes on Designed Capital Letters on Maps of Large Size Demonstrated from the Problem of Dating the Vercelli Map (Thirteenth Century)', *Imago Mundi*, 42 (1990), pp. 9–23.

——, 'Gog und Magog', in Walther Heissig and Claudius C. Müller, eds, *Die Mongolen* (Innsbruck–Frankfurt am Main: Pinguin–Umschau, 1989), pp. 27–9.

——, *Cartographische Quellen: Welt-, See- und Regionalkarten* (Turnhout: Brepols, 1988).

——, 'Universalkartographie und geographische Schulkenntnisse im Inkunabelzeitalter (Unter besonderer Berücksichtigung des "Rudimentum Novicorium" und Hartmann Schedels)', in Bernd Moeller, Hans Patze, and Karl Stackmann, eds, *Studien zum städtischen Bildungswesen des späten Mittelalters und der frühen Neuzeit* (Göttingen: Vandenhoeck–Ruprecht, 1983), pp. 398–411.

—, 'Die Mongolen im Weltbild der Lateiner um die Mitte des 13. Jahrhunderts unter besonderer Berücksichtigung des "Speculum Historicale" des Vincenz von Beauvais OP', *Archiv für Kulturgeschichte*, 57 (1975), pp. 117–40.

—, '"Ut describeretur orbis universus": Zur Universalkartographie des Mittelalters', in Albert Zimmermann, ed., *Methoden in Wissenschaft und Kunst des Mittelalters* (Berlin: De Gruyter, 1970), pp. 249–78.

—, 'Mappa mundi und Chronographia: Studien zur *imago mundi* des abendländischen Mittelalters', *Deutsches Archiv für Erforschung des Mittelalters*, 24 (1968), pp. 118–86.

Bronder, Barbara, 'Das Bild der Schöpfung und Neuschöpfung der Welt als "orbis quadratus"', *Frühmittelalterliche Studien*, 6 (1972), pp. 188–210.

Brotton, Jerry, *Trading Territories: Mapping the Early Modern World* (Ithaca, NY: Cornell University Press, 1997).

Brown, Francis, and others, *A Hebrew and English Lexikon of the Old Testament* (Oxford: Clarendon Press, 1972).

Brown, Janet, 'Noah's Flood, the Ark, and the Shaping of Early Modern Natural History', in David C. Lindberg and Ronald L. Numbers, eds, *When Science and Christianity Meet* (Chicago–London: University of Chicago Press, 2003), pp. 111–38.

Brown, Lloyd Arnold, *The Story of Maps* (Boston: Little, Brown, 1949).

Brown, Raymond E., Joseph A. Fitzmeyer, and Roland E. Murphy, eds, *The New Jerome Biblical Commentary* (Englewood Cliffs, NJ: Geoffrey Chapman, 1968; repr. 1990).

Brown, Wesley A., *The World Image Expressed in the Rudimentum novitiorum* (Washington, DC: Geography and Map Division, Library of Congress, 2000).

Browning, Wilfrid R. F., *A Dictionary of the Bible* (Oxford: Oxford University Press, 1996).

Buarque de Holanda, Sérgio, *Visão do paraíso: Os motivos edênicos no descobrimento e colonização do Brasil*, 2nd edn (São Paulo: Brasiliense, 1969).

Burgess, Glyn S., and Clara Strijbosch, *The Legend of St Brendan: A Critical Bibliography* (Dublin: Royal Irish Academy, 2000).

Burnett, Charles, ed., *Adelard of Bath: An English Scientist and Arabist of the Early Twelfth Century* (London: Warburg Institute, 1987).

—, 'High Altitude Mountaineering 1600 Years Ago', *Alpine Journal*, 88/332 (1983), p. 127.

Buti, Giovanni, and Renzo Bertagni, *Commento astronomico della Divina Commedia* (Florence: Sandron, 1966).

Campbell, Mary B., *The Witness and the Other World: Exotic European Travel Writing, 400–1600* (Ithaca, NY: Cornell University Press, 1988).

Campbell, Tony, *The Earliest Printed Maps: 1472–1500* (London: British Library; Berkeley: University of California Press, 1987).

—, 'Portolan Charts from the Late Thirteenth Century to 1500', in Harley and Woodward, eds, *History of Cartography*, I (1987), pp. 371–463.

—, 'Census of Pre-Sixteenth-Century Portolan Charts', *Imago Mundi*, 38 (1986), pp. 67–94.

Cantoni, Ivan, '*Tempora labuntur irrequieta cyclis*: Tempo e cosmologia nella filosofia della natura di Athanasius Kircher', in Walter Tega, ed., *Le origini della modernità: Linguaggi e saperi nel XVII secolo*, 2 vols (Florence: Olschki, 1998–), II, pp. 361–90.

Capasso, Ideale, *L'astronomia nella Divina Commedia* (Pisa: Domus Galilaeana, 1967).

Capello, Carlo F., *Il mappamondo medievale di Vercelli (1191–1218?)* (Turin: C. Fanton, 1976).

Caquot, André, and others, eds, *In Principio: Interprétations des premiers versets de la Genèse* (Paris: Études Augustiniennes, 1973).

Caraci, Giuseppe, 'La produzione cartografica di Vesconte Maggiolo (1511–1549) ed il Nuovo Mondo', *Memorie Geografiche*, 4 (1958), pp. 223–90.

—, 'The Italian Cartographers of the Benincasa and Freducci Families and the So-Called Borgiana Map of the Vatican Library', *Imago Mundi*, 10 (1953), pp. 23–49.

—, 'Un'altra carta di Albertin da Virga', *Bollettino della Reale Società Geografica Italiana*, 63 (1926), pp. 781–6.

Cardini, Franco, 'Alla cerca del paradiso', in *Columbeis V* (Genoa: Università di Genova, 1993), pp. 67–88.

Carmody, Francis J., *Arabic Astronomical and Astrological Sciences in Latin Translation* (Berkeley: University of California Press, 1956).

Cary, George, *The Medieval Alexander* (Cambridge: Cambridge University Press, 1956).

—, 'Alexander the Great in Mediaeval Theology', *Journal of the Warburg and Courtauld Institutes*, 17 (1954), pp. 98–114.

Casella, Nicola, 'Pio II tra geografia e storia: La "Cosmographia"', *Archivio della Società Romana di Storia Patria*, 3rd series, 26 (1972), pp. 35–112.

Cattaneo, Angelo, *La mappamundi di Fra Mauro Camaldolese: Venezia 1450*, PhD Dissertation, Istituto Universitario Europeo (2005), to be published as *Fra Mauro's Mappamundi and Fifteenth-Century Venetian Culture* (Turnhout: Brepols).

—, Diego Ramado Curto, and André Ferrand Almeida, eds, *La cartografia europea tra primo Rinascimento e fine dell'Illuminismo* (Florence: Olschki, 2003).

—, 'Fra Mauro *Cosmographus Incomparabilis* and his *Mappamundi*: Documents, Sources, and Protocols for Mapping', ibid., pp. 19–48.

—, 'God in This World: The Earthly Paradise in Fra Mauro's Mappamundi Illuminated by Leonardo Bellini', *Imago Mundi*, 55 (2003), pp. 121–6.

Cavallo, Guglielmo, ed., *Due mondi a confronto 1492–1728: Cristoforo Colombo e l'apertura degli spazi*, 2 vols (Rome: Istituto Poligrafico e Zecca dello Stato–Libreria dello Stato, 1992).

Cawley, Robert R., *Milton and the Literature of Travel* (New York: Gordian Press, 1970).

Cazier, Pierre, *Isidore de Séville et la naissance de l'Espagne catholique* (Paris: Beauchesne, 1994).

Cerulli, Enrico, 'Fonti arabe del mappamondo di Fra Mauro', *Orientalia Commentarii Periodici Pontifici Instituti Biblici*, 4 (1935), pp. 335–8.

Ceserani, Remo, 'Note sull'attività di scrittore di Pio II', in Domenico Maffei, ed., *Enea Silvio Piccolomini Papa Pio II* (Siena: Accademia Senese degli Intronati, 1968), pp. 99–115.

Chabás, José, and Bernard R. Goldstein, *The Alfonsine Tables of Toledo* (Dordrecht–Boston–London: Kluwer, 2003).

Chambers, Bettye Thomas, *Bibliography of French Bibles: Fifteenth- and Sixteenth-Century French-Language Editions of the Scriptures* (Geneva: Droz, 1983).

Champeaux, Gérard de, and Sébastien Sterckx, *Introduction au monde des symboles* (Yonne: Atelier Monastique de l'Abbaye Ste-Marie de la Pierre-qui-vire, 1966).

Chekin, Leonid S., 'Easter Tables and the Pseudo-Isidorean Vatican Map', *Imago Mundi*, 51 (1999), pp. 13–23.

Clark, Elizabeth A., *The Origenist Controversy: The Cultural Construction of an Early Christian Debate* (Princeton, NJ: Princeton University Press, 1992).

Clarke, Katherine, *Between Geography and History: Hellenistic Constructions of the Roman World* (Oxford: Clarendon Press, 1999).

Coli, Edoardo, *Il paradiso terrestre dantesco* (Florence: Carnesecchi, 1897).

Colish, Marcia L., *Peter Lombard*, 2 vols (Leiden–New York–Cologne: Brill, 1994).

Collins, John J., *The Apocalyptic Imagination: An Introduction to Jewish Apocalyptic Literature*, 2nd edn (Grand Rapids, MI–Cambridge: Eerdmans, 1998).

Congar, Yves-M. J., 'Le Thème de *Dieu-Créateur* et les explications de l'hexameron dans la tradition chrétienne', in *L'Homme devant Dieu: Mélanges offerts au père Henri de Lubac*, 3 vols (Paris: Aubier, 1963–4), I (1963), pp. 189–222.

Connolly, Daniel, 'Imagined Pilgrimage in Itinerary Maps of Matthew Paris', *Art Bulletin*, 81 (1999), pp. 598–622.

Copleston, Frederick C., *Aquinas* (Harmondsworth: Penguin, 1955).

Cortesão, Armando, *History of Portuguese Cartography*, 2 vols (Coimbra: Junta de Investigações do Ultramar, 1969–71).

Cosgrove, Denise, 'Maps, Mapping, Modernity: Art and Cartography in the Twentieth Century', *Imago Mundi*, 57/1 (2005), pp. 35–54.

—, *Apollo's Eye: A Cartographic Genealogy of the Earth in the Western Imagination* (Baltimore–London: Johns Hopkins University Press, 2001).

—, ed., *Mappings* (London: Reaktion, 1999).

—, 'Global Illumination and Enlightenment in the Geographies of Vincenzo Coronelli and Athanasius Kircher', in Livingstone and Withers, eds, *Geography and Enlightenment* (1999), pp. 33–66.

—, and Stephen Daniels, eds, *The Iconography of Landscape: Essays on the Symbolic Representation, Design and Use of Past Environments* (Cambridge: Cambridge University Press, 1988).

Crinò, Sebastiano, 'La scoperta della carta originale di Paolo dal Pozzo Toscanelli che servì di guida a Cristoforo Colombo per il viaggio verso il Nuovo Mondo', *L'Universo*, 22 (1941), pp. 379–405.

Crivellari, Giuseppe, *Alcuni cimeli di cartografia medievale* (Florence: Seeber, 1903).

Crociata, Mariano, *Umanesimo e teologia in Agostino Steuco* (Rome: Città Nuova, 1987).

Crosby, Alfred W., *The Measure of Reality: Quantification and Western Society, 1200–1600* (Cambridge: Cambridge University Press, 1997).

Cross, Richard, *The Physics of Duns Scotus: The Scientific Context of a Theological Vision* (Oxford–New York: Oxford University Press, 1998).

Crouzel, Henri, *Origène* (Paris–Namur: Lethielleux–Culture et Vérité, 1985).

—, 'L'Anthropologie d'Origène: De l'arche au telos', in Ugo Bianchi, ed., *Arché e telos: L'antropologia di Origene e di Gregorio di Nissa: Analisi storico-religiosa* (Milan: Vita e Pensiero, 1981), pp. 37–42.

—, 'Le Thème platonicien du "véhicule de l'âme" chez Origène', *Didaskalia*, 7 (1977), pp. 225–37.

—, *Théologie de l'image de Dieu chez Origène* (Paris: Aubier, 1956).

Croydon, F. E., 'Notes on the Life of Hugh of St. Victor', *Journal of Theological Studies*, 40 (1939), pp. 232–53.

Dahan, Gilbert, *L'Exégèse chrétienne de la Bible en Occident médiévale* (Paris: Cerf, 1999).

Dainelli, Giotto, *Missionari e mercadanti rivelatori dell'Asia nel Medio Evo* (Turin: Unione Tipografico-Editrice Torinese, 1960).

Dal Lago, Giovanni, 'Giovanni Leardo, *Mapa Mondi*', in Cavallo, ed., *Due mondi a confronto* (1992), I, pp. 159–62.

Daniel, Norman, *The Arabs and Mediaeval Europe* (London: Longman; Beirut: Librairie du Liban, 1975).

—, *Islam and the West: The Making of an Image* (Edinburgh: Edinburgh University Press, 1966).

Daniélou, Jean, *Le Signe du temple ou de la présence de Dieu* (Paris: Desclée, 1942; repr. 1975, 1990).

—, *From Shadows to Reality: Studies in the Biblical Typology of the Fathers*, transl. by Wulstan Hibberd (London: Burns & Oates, 1960).

—, *Philon d'Alexandrie* (Paris: Fayard, 1958).

—, 'Terre et paradis chez les Pères de l'Église', *Eranos-Jahrbuch*, 23, (1953), pp. 433–72.

———, *Sacramentum futuri: Études sur les origines de la typologie biblique* (Paris: Études de Théologie Historique, 1950).

Danzer, Gerald A., *Images of the Earth on Three Early Italian Woodcuts: Candidates for the Earliest Printed Maps in the West* (Chicago: Newberry Library, 1991).

Dauphiné, James, *Le Cosmos de Dante* (Paris: Belles Lettres, 1984).

Davidse, Jan, 'On Bede as Christian Historian', in L. A. J. R. Houwen and Alasdair A. MacDonald, eds, *Beda Venerabilis: Historian, Monk and Northumbrian* (Groningen: Egbert Forsten, 1996), pp. 1–15.

De doctrina Ioannis Duns Scoti, 4 vols (Rome: Cura Commissionis Scotisticae, 1968).

De Jonge, Marinus, and Johannes Tromp, *The Life of Adam and Eve and Related Literature* (Sheffield: Sheffield Academic Press, 1997).

De Vocht, Henry, 'Andreas Masius (1514–1573)', *Miscellanea Giovanni Mercati*, 4 (1946), pp. 425–41.

Dechow, John F., *Dogma and Mysticism in Early Christianity: Epiphanius of Cyprus and the Legacy of Origen* (Louvain–Macon, GA: Mercer University Press, 1988).

Degenhart, Bernard, and Annnegrit Schmitt, 'Marino Sanudo und Paolino Veneto', *Römisches Jahrbuch für Kunstgeschichte*, 14 (1973), pp. 1–137.

Delano-Smith, Catherine, 'Milieus of Mobility: Itineraries, Route Maps and Road Maps', in James R. Akerman, ed., *Cartographies of Travel and Navigation* (Chicago: University of Chicago Press, in press).

———, 'The Intelligent Pilgrim: Maps and Medieval Pilgrimage to the Holy Land', in Allen, ed., *Eastward Bound* (2004), pp. 107–30.

———, 'Smoothed Lines and Empty Spaces: The Changing Face of the Exegetical Map before 1600', in Isabelle Laboulais-Lesage, ed., *Combler les blancs de la carte: Modalités et enjeux de la construction des savoirs géographiques (XVIIe– XXe siècle)* (Strasbourg, Presses Universitaires de Strasbourg, 2004), pp. 17–34.

———, 'Maps and Religion in Medieval and Early Modern Europe', in Woodward, Delano-Smith, and Yee, *Plantejaments i objectius /Approaches and Challenges* (2001), pp. 179–200.

———, and Roger J. P. Kain, *English Maps: A History* (London: British Library, 1999).

———, and Elizabeth M. Ingram, *Maps in Bibles, 1500–1600: An Illustrated Catalogue* (Geneva: Droz, 1991).

———, 'Geography or Christianity? Maps of the Holy Land Before AD 1000', *Journal of Theological Studies*, 42/1 (1991), pp. 143–52.

———, 'Cartography in the Prehistoric Period in the Old World: Europe, the Middle East, and North Africa', in Harley and Woodward, eds, *The History of Cartography*, I (1987), pp. 54–101.

———, 'Maps in Bibles in the Sixteenth Century', *The Map Collector*, 39 (1987), pp. 2–13.

———, 'Cartographic Signs and their Explanation', *Imago Mundi*, 37 (1985), pp. 9–27.

Delehaye, Hippolyte, 'Les Martyrs d'Egypte', *Analecta Bollandiana*, 40 (1922), pp. 5–154.

Delumeau, Jean, *History of Paradise: The Garden of Eden in Myth and Tradition*, transl. by Matthew O'Connell (New York: Continuum, 1995).

———, *Une histoire du paradis*, I, *Le Jardin des délices* (Paris: Fayard, 1992).

Deluz, Christiane, 'Le Paradis terrestre, image de l'Orient lointain dans quelques documents géographiques médiévaux', in *Images et signes de l'Orient lointain dans l'Occident medieval: Littérature et civilisation* (Aix-en-Provence: Université de Provence; Marseille: Laffitte, 1982), pp. 143–61.

Demus, Otto, *The Mosaics of San Marco*, 2 vols (Chicago: University of Chicago Press, 1984).

Derolez, Albert, 'Lambert van Sint-Omaars als kartograaf', *De Franse Nederlanden/Les Pays-Bas Français, Jaarboek Annales* (1976), pp. 15–30.

Desrumeaux, Alain, and Francis Schmidt, eds, *Moïse géographe: Recherches sur les représentations juives et chrétiennes de l'espace* (Paris: Vrin, 1988).

Destombes, Marcel, 'Guillaume Postel cartographe', in Trédaniel, ed., *Guillaume Postel*, (1985), pp. 361–71.

———, 'An Antwerp *Unicum*: An Unpublished Terrestrial Globe of the 16th Century in the Bibliothèque Nationale, Paris', *Imago Mundi*, 24 (1970), pp. 85–94.

———, *Mappemondes AD 1200–1500: Catalogue preparé par la Commission des Cartes Anciennes de l'Union Géographique Internationale* (Amsterdam: N. Israel, 1964).

Deus et homo ad mentem Ioannis Duns Scoti (Rome: Societas Internationalis Scotistica, 1972).

Dictionnaire de théologie catholique, ed. by Alfred Vacant, Eugène Mangenot, Émile Amann, and others, 17 vols (Paris: Letouzey et Ané: 1923–72).

Dilke, Oswald A. W., and Margaret Oswald, 'Mapping a Crusade', *History Today*, 39 (1989), pp. 31–5.

———, 'The Culmination of Greek Cartography in Ptolemy', in Harley and Woodward, eds, *History of Cartography*, I (1987), pp. 177–200.

———, 'Cartography in the Byzantine Empire', ibid., pp. 256–75.

Dixon, Laurinda S., 'Giovanni di Paolo's Cosmology', *Art Bulletin*, 67/4 (1985), pp. 604–13.

Dizionario biografico degli Italiani, gen. ed. Alberto M. Ghisalberti (Rome: Istituto dell'Enciclopedia Italiana, 1960–).

Donner, Herbert, *Pilgerfahrt ins Heilige Land: Die ältesten Berichte christlicher Palästinapilger (4.–7. Jahrhundert)* (Stuttgart: Katholisches Bibelwerk, 1979).

Drecoll, Carsten, *Idrísí aus Sizilien: Der Einfluss eines arabischen Wissenschaftlers auf die Entwicklung der europäischen Geographie* (Egelsbach–Frankfurt am Main–Munich–New York: Hänsel–Hohenhausen, 2000).

Dronke, Peter, *Imagination in the Late Pagan and Early Christian World: The First Nine Centuries AD* (Florence: SISMEL, 2003).

Dubois, Claude-Gilbert, *La Mythologie des origines chez Guillaume Postel* (Orléans: Paradigme, 1994).

Duhem, Pierre, *Le Système du monde: Histoire des doctrines cosmologiques de Platon à Copernic*, 10 vols (Paris: Hermann, 1913–59).

Dumville, David N., 'Two Approaches to the Dating of "Navigatio Sancti Brendani"', *Studi Medievali*, 3rd series, 29 (June 1988), I, pp. 87–102.

Duncan, Joseph E., *Milton's Earthly Paradise* (Minneapolis: University of Minnesota Press, 1972).

———, 'Paradise as the Whole Earth', *Journal of the History of Ideas*, 30 (1969), pp. 171–86.

Durazzo, Pompeo, *Il paradiso terrestre nelle carte medievali* (Mantua: Arnaldo Forni, 1886).

———, *Il planisfero di Giovanni Leardo* (Mantua: Eredi Segna, 1885).

Dürst, Arthur, 'Die Weltkarte von Albertin de Virga von 1411 oder 1415', *Cartographica Helvetica*, 13 (January 1996), pp. 18–21.

Eco, Umberto, *Semiotics and the Philosophy of Language* (London: Macmillan, 1984).

Edson, Evelyn, 'Reviving the Crusade in the Fourteenth Century: Sanudo's Schemes and Vesconte's Maps', in Allen, ed., *Eastward Bound* (2004), pp. 131–55.

———, and Emilie Savage-Smith, *Medieval Views of the Cosmos* (Oxford: Bodleian Library, 2004).

———, *Mapping Time and Space: How Medieval Mapmakers Viewed their World* (London: British Library, 1997).

———, 'World Maps and Easter Tables: Medieval Maps in Context', *Imago Mundi*, 48 (1996), pp. 25–42.

———, 'The Oldest World Maps: Classical Sources of Three Eighth-Century *Mappaemundi*', *Exploration and Colonization in the Ancient World*, 24 (1993), pp. 169–84.

Egry, Anne de, *Um estudo de o Apocalipse de Lorvão e a sua relação com as ilustrações medievais do Apocalipse* (Lisbon: Fundação Calouste Gulbenkian, 1972).

Ehlers, Joachim, *Hugo von St Viktor: Studien zum Geschichtsdenken und zur Geschichtsschreibung des 12. Jahrhunderts* (Wiesbaden: Steiner Verlag, 1973).

———, '*Arca significat ecclesiam*: Ein theologisches Weltmodell aus der ersten Hälfte des 12. Jahrhunderts', *Frühmittelalterliche Studien*, 6 (1972), pp. 171–87.

———, '*Historia, allegoria, tropologia*: Exegetische Voraussetzungen der Geschichtskonzeption Hugos von St. Viktor', *Mittellateinisches Jahrbuch*, 7 (1972), pp. 153–60.

Eichenberger, Walter, and Henning Wendland, *Deutsche Bibeln vor Luther* (Hamburg: Wittig, 1977).

Eldridge, Michael D., *Dying Adam with his Multiethnic Family* (Leiden–Boston–Cologne: Brill, 2001).

Encyclopaedia Judaica, ed. by Cecil Roth, 17 vols (Jerusalem: Encyclopaedia Judaica, 1972).

Encyclopedia of Biblical Theology, ed. by Johannes B. Bauer, 3 vols (London–Sydney: Sheed and Ward, 1970).

Engammare, Max, 'D'une forme l'autre: Commentaires et sermons de Calvin sur la Genèse', in *Calvinus Praeceptor Ecclesiae* (Geneva: Droz, 2004), pp. 107–37.

———, 'Portrait de l'exégète en géographe: La Carte du paradis comme instrument herméneutique chez Calvin et ses contemporains', *Annali di Storia dell'Esegesi*, 13/2 (1996), pp. 565–81.

Englisch, Brigitte, *Ordo orbis terrae* (Berlin: Akademie Verlag, 2002).

Erffa, Hans Martin von, *Ikonologie der Genesis: Die Christlichen Bildthemen aus dem Alten Testament und ihre Quellen*, 2 vols (Munich: Deutscher Kunstverlag, 1989–95).

Evans, Gillian R., ed., *Medieval Commentaries on the Sentences of Peter Lombard: Current Research*, 2 vols (Leiden–Boston: Brill, 2002–).

Faber Kolb, Arianne, *Jan Brueghel the Elder: The Entry of the Animals into Noah's Ark* (Los Angeles: J. Paul Getty Museum, 2005).

Falchetta, Piero, *The Fra Mauro World Map* (Turnhout: Brepols, forthcoming).

Farthing, John L., *Thomas Aquinas and Gabriel Biel: Interpretations of St Thomas Aquinas in German Nominalism on the Eve of the Reformation* (Durham, NC–London: Duke University Press, 1988).

Fasbender, Cristoph, and Reinhard Hahn, eds, *Brandan, die mitteldeutsche 'Reise'-Fassung* (Heidelberg: Universitätsverlag C. Winter, 2002).

Finzi, Felice, *Ricerche per lo studio dell'antichità assira* (Turin: Loescher, 1872).

Firneis, Maria G., Paul Uiblein and Hans Kaiser, 'Johannes von Gmunden um 1384–1442', in Hamann and Grössing, eds, *Der Weg der Naturwissenschaft von Gmunden zu Kepler* (1988), pp. 9–100.

Fischer, Theobald Friedrich, *Sammlung mittelalterlicher Welt und Seekarten italienischer Ursprungs und aus italienischen Bibliotheken und Archiven* (Venice: Ferdinand Omgania, 1886).

Flint, Valerie I. J., *The Imaginative Landscape of Christopher Columbus* (Princeton, NJ: Princeton University Press, 1992).

Fontaine, Jacques, *Isidore de Séville: Genèse et originalité de la culture hispanique au temps des Wisigoths* (Turnhout: Brepols, 2000).

Forest, Aimé, 'Guillaume d'Auvergne, critique d'Aristote', in *Études médiévales offertes à Augustin Fliche* (Paris: [n. pub.], 1952), pp. 67–79.

Frank, William A., and Allan B. Wolter, *Duns Scotus, Metaphysician* (West Lafayette, IN: Purdue University Press, 1995).

Frassetto, Michael, ed., *Western Views of Islam in Medieval and Early Modern Europe: Perception of Other* (New York: St. Martin's Press, 1999).

Frei, Hans W., *The Eclipse of Biblical Narrative: A Study in Eighteenth- and Nineteenth-Century Hermeneutics* (New Haven–London: Yale University Press, 1974).

Freudenberger, Theobald, *Augustinus Steuchus aus Gubbio: Augustinerchorherr und päpstlicher Bibliothekar (1497–1548) und sein literarisches Lebenswerk* (Münster in Westphalia: Aschendorff, 1935).

Fried, Johannes, 'Endzeiterwartung um die Jahrtausendwende', *Deutsches Archiv für Erforschung des Mittelalters*, 45/2 (1989), pp. 381–473.

Friedman, John B., Kristen Mossler Figg, and others, eds, *Trade, Travel, and Exploration in the Middle Ages: An Encyclopedia* (New York–London: Garland, 2000).

—, *The Monstrous Races in Medieval Art and Thought* (Cambridge, MA: Harvard University Press, 1981; repr. Syracuse, NY: Syracuse University Press, 2000).

Fulton, John F., *Michael Servetus: Humanist and Martyr* (New York: Reichner, 1953).

Galbraith, Vivian H., 'An Autograph Manuscript of Ranulf Higden's "Polychronicon"', *Huntington Library Quarterly*, 34 (1959), pp. 1–18.

Gallez, Paul, 'Walsperger and his Knowledge of the Patagonian Giants', *Imago Mundi*, 33 (1981), pp. 91–3.

Ganoczy, Alexandre, *La Bibliothèque de l'Académie de Calvin* (Geneva: Droz, 1969).

García-Aráez Ferrer, Hermenegildo, *La miniatura en los codices de Beato de Liébana (su tradición pictórica)* (Madrid: Enero, 1992).

Gardet, Clement, *L'Apocalypse figurée des ducs de Savoie* (Annecy: Gardet, 1969).

Gaspar, Camille, and Frédéric Lyna, *Les Principaux Manuscrits à peintures de la Bibliothèque Royale de Belgique*, 2 vols (Bruxelles: Bibliothèque Royale Albert 1er, 1937–45; repr. 1984–9).

Gasparrini Leporace, Tullia, *Il mappamondo di Fra Mauro* (Rome: Istituto Poligrafico dello Stato, 1956).

—, ed., *L'Asia nella cartografia degli Occidentali: Catalogo descrittivo della mostra* (Venice: Biblioteca Nazionale Marciana, 1954).

Gatti, Paolo, *Synonima Ciceronis: La raccolta Accusat, lacescit* (Trento: Università degli Studi, 1994).

Gatti Perer, Maria Luisa, ed., *'La dimora di Dio con gli uomini' (Ap 21,3): Immagini della Gerusalemme celeste dal III al XIV secolo* (Milan: Vita e Pensiero, 1983).

Gaullier-Bougassas, Catherine, *Les Romans d'Alexandre: Aux frontiers de l'épique et du Romanesque* (Paris: Honoré Champion, 1998).

Gautier Dalché, Patrick, 'The Reception of Ptolemy's Geography', in Woodward, ed., *The History of Cartography*, III (forthcoming).

—, 'Cartes de Terre Sainte, cartes de pèlerins', in Massimo Oldoni, ed., *Fra Roma e Gerusalemme nel Medioevo: Paesaggi umani ed ambientali del pellegrinaggio meridionale*, 3 vols (Salerno: Laveglia, 2005), I, pp. 573–612.

—, 'Le Sens de *mappa* (*mundi*): IVe–XIVe siècle', *Archivum Latinitatis Medii Aevii*, 62 (2004), pp. 187–202.

—, 'Le Paradis aux antipodes? Une *Distinctio divisionis terre et paradisi delitiarum* (XIVe siècle)', in Dominique Barthélemy and Jean-Marie Martin, eds, *Liber largitorius: Étude d'histoire médiévale offertes à Pierre Toubert par ses élèves* (Paris: Droz, 2003), pp. 615–37.

—, 'Les Diagrammes topographiques dans les manuscrits des classiques latins (Lucain, Solin, Salluste)', in Pierre Lardet, ed., *La Tradition vive: Mélanges d'histoire des textes en l'honneur de Louis Holtz* (Turnhout: Brepols, 2003), pp. 291–306.

—, 'Principes et modes de la représentation de l'espace géographique durant le haut Moyen Âge', in *Uomo e spazio nell'alto Medioevo* (Spoleto: Centro Italiano di Studi sull'Alto Medioevo, 2003), pp. 119–50.

—, 'Portolans and the Byzantine World', in Ruth Macrides, ed., *Travel in the Byzantine World* (Aldershot, Hants: Ashgate, 2002), pp. 59–71.

—, 'Un astronome, auteur d'un globe terrestre: Jean Fusoris à la découverte de la *Géographie* de Ptolémée', and 'L'Oeuvre géographique du cardinal Fillastre († 1428): Représentation du monde et perception de la carte à l'aube des découvertes', in Didier Marcotte, ed., *Humanisme et culture géographique à l'époque du Concile de Constance, autour de Guillaume Fillastre* (Turnhout: Brepols, 2002), pp. 161–75, and 293–355.

—, 'Décrire le monde et situer les lieux au XIIe siècle: L'*Expositio mappe mundi* et la généalogie de la mappemonde de Hereford', in *Mélanges de l'École Française de Rome*, 113 (2001), pp. 343–409.

—, 'Sur l'"originalité" de la "géographie" médiévale', in Michel Zimmermann, ed., *Auctor et Auctoritas: Invention et conformisme dans l'écriture médiévale* (Paris: École de Chartes, 2001), pp. 131–43.

—, 'Connaissance et usages géographiques des cordonnées dans le Moyen Âge latin (du Vénérable Bède à Roger Bacon)', in Louis Callebat and Olivier Desbordes, eds, *Science antique, science médiévale (autour d'Avranches 235)* (Hildesheim–Zurich–New York: Olms–Weidmann, 2000), pp. 401–36.

—, 'Le Souvenir de la *Géographie* de Ptolémée dans le monde latin médiéval (VIe–XIVe siècles)', *Euphrosyne*, 27 (1999), pp. 79–106.

—, '*Mappae mundi* antérieures au XIIIe siècle dans les manuscrits latins de la Bibliothèque Nationale de France', *Scriptorium*, 52/1 (1998), pp. 102–61.

—, *Carte marine et Portulan au XIIe siècle: Le 'Liber de existencia riveriarum et forma maris nostris Mediterranei' (Pise, circa 1200)* (Rome: École Française de Rome, 1995).

—, 'Nouvelles Lumières sur la *Descriptio mappe mundi* de Hugues de Saint-Victor', in *Géographie et culture: La Représentation de l'espace du VIe au XIIe siècle* (Aldershot, Hants: Ashgate, 1997), XII, pp. 1–27, edited version of 'La "Descriptio Mappe Mundi" de Hugues de Saint-Victor: Retractatio et additamenta', in Jean Longère, ed., *L'Abbaye parisienne de Saint-Victor au Moyen Âge* (Paris–Turnhout: Brepols, 1991), pp. 143–79.

—, 'De la glose à la contemplation: Place et fonction de la carte dans les manuscrits du haut Moyen Âge', in *Testo e immagine nell'alto Medioevo*, 2 vols (Spoleto: Centro Italiano di Studi sull'Alto Medioevo, 1994), II, pp. 693–771.

—, 'D'une technique à une culture: Carte nautique et portulan au XIIe et au XIIIe siècle', in *L'uomo e il mare nella civiltà occidentale: Da Ulisse a Cristoforo Colombo: Atti della Società Ligure di Storia Patria*, 32 (1992), pp. 284–312.

—, 'L'Espace de l'histoire: Le Rôle de la géographie dans les chroniques universelles', in Jean-Philippe Genet, ed., *L'Historiographie médiévale en Europe* (Paris: Éditions du Centre National de la Recherche Scientifique, 1991), pp. 287–300.

—, 'Entre le folklore et la science: La Légende des antipodes chez Giraud de Cambrie et Gervais de Tilbury', in *La leyenda, antropología, historia, literatura* (Madrid: Casa de Velázquez–Universidad Complutense, 1989), pp. 103–14.

—, *La 'Descriptio Mappe Mundi' de Hugues de Saint-Victor* (Paris: Études Augustiniennes, 1988).

Gelzer, Heinrich K. G., *Sextus Julius Africanus und die byzantinische Chronographie*, 2 vols (Leipzig: Hinrichs and Teubner, 1880–5, 1898; repr. Hildesheim: Gerstenberg, 1978).

Gentile, Sebastiano, 'Umanesimo e cartografia: Tolomeo nel secolo XV', in Cattaneo, Ramado Curto and Almeida, eds, *La cartografia europea* (2003), pp. 3–18.

Gibson, Margaret T., 'The Place of the *Glossa ordinaria* in Medieval Exegesis ', in Mark D. Jordan and Kent Emery Jr., eds, *Ad Litteram: Authoritative Texts and their Medieval Readers* (Notre Dame, IN: University of Notre Dame Press, 1992), pp. 5–27.

—, 'The Twelfth-Century Glossed Bible', *Studia Patristica*, 23 (1990), pp. 232–44.

Gil, Juan, 'Los terrores del año 800', in *Actas del Simposio … Beato de Liébana*, I (1978), pp. 215–47.

Gilbert, Alan H., *A Geographical Dictionary of Milton* (New York: Russell and Russell, 1968).

Gilhofer, H., and H. Ranschburg, 'Albertin de Virga: Weltkarte auf Pergament, mit Feder und in Farben gezeichnet, Venedig 141(5)', in *Versteigerungs-Katalog No. VIII zur Auktion am 14. und 15. Juni 1932 in Luzern* (Lucerne: H. Gilhofer and H. Ranschburg, 1932), pp. 17–8.

Gilly, Carlos, 'Guillaume Postel et Bâle', in Trédaniel, ed., *Guillaume Postel* (1985), pp. 41–78.

Gilmont, Jean-François, *Jean Calvin et le livre imprimé* (Geneva: Droz, 1997).

Gilson, Étienne, *History of Christian Philosophy in the Middle Ages* (London: Sheed and Ward, 1985).

—, *Les Métamorphoses de la Cité de Dieu* (Louvain: Publications Universitaires de Louvain, 1952).

—, *Jean Duns Scot: Introduction à ses positions fondamentales* (Paris: Vrin, 1952).

—, *La Philosophie de saint Bonaventure* (Paris: Vrin, 1924; repr. 1943).

Ginzberg, Louis, *The Legends of the Jews*, transl. by Henrietta Szold, 7 vols (Philadelphia: Jewish Publication Society of America, 1946–7).

Giovino, Mariana, *A History of Interpretations of the 'Assyrian Sacred Tree': 1849–2004*, PhD Dissertation, University of Michigan, Ann Arbor (2004).

Glacken, Clarence J., *Traces on the Rhodian Shore: Nature and Culture in Western Thought from Ancient Times to the End of the Eighteenth Century* (Berkeley: University of California Press, 1967).

Glorie, François, 'Mappa mundi; Indeculum quod maria vel venti sunt et (Pauli Orosii) Discriptio Terrarum e codice Albigensi 29', in *Itineraria et alia geographica* (1965), pp. 467–94.

Goetz, Hans-Werner, 'On the Universality of Universal History', in *L'Historiographie médiévale en Europe* (Paris: Éditions du Centre National de la Recherche Scientifique, 1991).

Goffart, Walter, *Historical Atlases: The First Three Hundred Years, 1570–1870* (Chicago–London: University of Chicago Press, 2003).

Goppelt, Leonhard, *Typos: Die typologische Deutung des Alten Testaments im Neuen* (Darmstadt: Wissenschaftliche Buchgesellschaft, 1969).

Gordon, Burton L., 'Sacred Directions, Orientation, and the Top of the Map', *History of Religions* 10 (1971), pp. 211–27.

Gössmann, Elisabeth, *Metaphysik und Heilsgeschichte: Eine theologische Untersuchung der Summa Halensis (Alexander von Hales)* (Munich: Hueber, 1964).

Gouguenheim, Sylvain, *Les Fausses Terreurs de l'an mil* (Paris: Picard, 1999).

Gow, Andrew, 'Gog and Magog on *Mappaemundi* and Early Printed World Maps: Orientalizing Ethnography in the Apocalyptic Tradition', *Journal of Early Modern History*, 2/1 (1998), pp. 61–88.

Grabar, André, *Christian Iconography: A Study of its Origin* (Princeton, NJ: Princeton University Press, 1968).

Graf, Arturo, *Il mito del paradiso terrestre* (Rome: Manilo Basaia, 1982).

—, *Il diavolo* (Rome: Salerno Editrice, 1980).

—, *Miti, leggende e superstizioni del Medio Evo*, 2 vols (Turin: Loescher, 1892–3).

—, *La leggenda del paradiso terrestre* (Turin: Loescher, 1878).

Grafton, Anthony, 'Jean Hardouin: The Antiquary as Pariah', *Journal of the Warburg and Courtauld Institutes*, 62 (1999), pp. 241–67.

—, April Shelford and Nancy Siraisi, *New Worlds, Ancient Texts: The Power of Tradition and the Shock of Discovery* (Cambridge, MA–London: Harvard University Press, 1992).

Greene, David, and Fergus Kelly, eds, *The Irish Adam and Eve Story from Saltair na Rann*, 2 vols (Dublin: Institute for Advanced Studies, 1976).

Greenhill, Eleonor S., *Die Stellung der Handschrift British Museum Arundel 44 in der Überlieferung des Speculum Virginum* (Munich: Hueber, 1966).

Greer, Rowan A., and James L. Kugel, *Early Biblical Interpretation* (Philadelphia: Westminster Press, 1986).

Gregor, Frederick, 'The Impact of Darwinian Evolution on Protestant Theology in the Nineteenth Century', in David C. Lindberg and Ronald L. Numbers, eds, *God and Nature: Historical Essays on the Encounter between Christianity and Science* (Berkeley–Los Angeles–London: University of California Press, 1986), pp. 369–90.

Grelot, Pierre, *Sens chrétien de l'Ancien Testament: Esquisse d'un traité dogmatique*, 3rd edn (Tournai: Desclée, 1962).

Gribaudi, Pietro, 'Il mito degli alberi del sole e della luna e dell'albero secco nella cartografia medievale', in *Atti del V. Congresso Geografico Italiano* (Napoli: Tocco e Salvietti, 1905), pp. 828–42.

Grimm, Reinhold R., *Paradisus coelestis, paradisus terrestris: Zur Auslegungsgeschichte des Paradieses im Abendland bis um 1200* (Munich: Fink, 1977).

Grosjean, Georges, ed., *Vesconte Maggiolo, 'Atlante nautico del 1512': Seeatlas vom Jahre 1512* (Dietikon-Zurich: Urs Graf, 1979).

Guelke, Leonard, 'The Relations between Geography and History Reconsidered', *History and Theory: Studies in the Philosophy of History*, 36/2 (1997), pp. 216–34.

Guenée, Bernard, *Histoire et culture historique dans l'Occident médiéval* (Paris: Aubier Montaigne, 1980; repr. 1991).

Guéret-Laferté, Michèle, *Sur les routes de l'empire mongol: Ordre et rhétorique des relations de voyage aux XIIIe et XIVe siècles* (Paris: Honoré Champion, 1994).

Guerrini, Luigi, 'Geografia e politica in Pio II', in Claudio Crescentini and Margherita Palumbo, eds, *Nymphilexis: Enea Silvio Piccolomini, l'umanesimo e la geografia* (Rome: Shakespeare and Company2, 2005), pp. 27–52.

Guillet, Jacques, 'Les Exégèses d'Alexandrie et d'Antioche, conflit ou malentendu?' *Recherches de Sciences Religieuses*, 34 (1947), pp. 257–302.

Guinot, Jean-Noël, 'La Typologie comme technique herméneutique', in *Figures de l'Ancien Testament chez les Pères* (Strasbourg: Centre d'Analyse et de Documentation Patristiques, 1989), pp. 1–34.

Gunter, Ann C., ed., *The Construction of the Ancient Near East*, *Culture and History*, 11 (1992).

Guzman, Gregory G., 'The Encyclopedist Vincent of Beauvais and his Mongol Extracts from John of Plano Carpini and Simon of St-Quentin', *Speculum*, 49 (1974), pp. 287–307.

Hackett, Jeremiah, ed., *Roger Bacon and the Sciences: Commemorative Essays* (Leiden–New York–Cologne: Brill, 1997).

Hahn-Woernle, Birgit, *Die Ebstorfer Weltkarte* (Ebstorf: Kloster Ebstorf, 1987).

Hailperin, Herman, *Rashi and the Christian Scholars* (Pittsburgh: University of Pittsburgh Press, 1963).

Hall, Thomas S., *Ideas of Life and Matter: Studies in the History of General Physiology, 600 BC–1900 AD*, 2 vols (Chicago–London: University of Chicago Press, 1969).

Hamann, Günther, and Helmut Grössing, eds, *Der Weg der Naturwissenschaft von Johannes von Gmunden zu Johannes Kepler* (Vienna: Verlag der Österreichischen Akademie der Wissenschaften, 1988).

Hamel, Christopher F. R. de, *Glossed Books of the Bible and the Beginnings of the Paris Booktrade* (Woodbridge: Brewer, 1984).

Hamman, Adalbert G., *L'Homme image de Dieu: Essai d'une anthropologie chrétienne dans l'Église des cinq premiers siècles* (Paris: Desclée, 1987).

Hans Holbein d.J.: Die Druckgraphik im Kupferstichkabinett Basel (Basle: Schwabe, 1997).

Hanson, Anthony T., *The Living Utterances of God: The New Testament Exegesis of the Old* (London: Darton, Longman and Todd, 1983).

Hanson, Richard P. C., *Allegory and Event: A Study of the Sources and Significance of Origen's Interpretation of Scripture* (London: SCM Press, 1959).

Haren, Michael, *Medieval Thought: The Western Intellectual Tradition from Antiquity to the Thirteenth Century*, 2nd edn (Toronto–Buffalo: University of Toronto Press, 1992).

Harley, Brian, *The New Nature of Maps: Essays in the History of Cartography*, ed. Paul Laxton (Baltimore and London: Johns Hopkins University Press, 2001).

—, and David Woodward, eds, *The History of Cartography*, II/1, *Cartography in the Traditional Islamic and South Asian Societies* (Chicago–London: University of Chicago Press, 1992) (for vol. III see under Woodward).

—, 'Deconstructing the Map', *Cartographica*, 26/2 (1989), pp. 1–20.

—, 'Maps, Knowledge and Power', in Cosgrove and Daniels, eds, *The Iconography of Landscape* (1988), pp. 277–312.

—, and David Woodward, eds, *The History of Cartography*, I, *Cartography in Prehistoric, Ancient and Medieval Europe and the Mediterranean* (Chicago: University of Chicago Press, 1987).

—, and David Wodward, with Germaine Aujac, 'Greek Cartography in the Early Roman World', ibid., pp. 161–76.

Harrison, Dick, *Medieval Space: The Extent of Microspatial Knowledge in Western Europe during the Middle Ages* (Lund: Lund University Press, 1996).

Harvey, Paul D. A., ed., *The Hereford World Map: Medieval World Maps and their Context* (London: British Library, 2006).

—, 'The Biblical Content of Medieval Maps of the Holy Land', in Dagmar Unverhau, ed., *Geschichtsdeutung auf alten Karten: Archäologie und Geschichte* (Wiesbaden: Harrassowitz, 2003), pp. 55–63.

—, 'The Sawley Map (Henry of Mainz) and Other World Maps in Twelfth-Century England', *Imago Mundi*, 49 (1997), pp. 33–42.

—, *Mappa Mundi: The Hereford World Map* (London: Hereford Cathedral–British Library; Toronto: University of Toronto Press, 1996).

—, and Raleigh A. Skelton, eds, *Local Maps and Plans from Medieval England* (Oxford: Clarendon Press, 1986).

—, *Medieval Maps* (London: British Library, 1991).

—, *The History of Topographical Map: Symbols, Pictures and Surveys* (London: Thames and Hudson, 1980).

Haslam, Graham, 'The Duchy of Cornwall Map Fragment', in Pelletier, ed., *Géographie du monde au Moyen Âge et à la Renaissance* (1989), pp. 33–44.

Hassinger, Hugo, 'Deutsche Weltkarten-Inkunabeln', *Zeitschrift der Gesellschaft für Erdkunde zu Berlin*, 9/10 (1927), pp. 455–82.

Hawkins, Peter S., 'Out upon Circumference: Discovery in Dante', in Westrem, ed., *Discovering New Worlds* (1991), pp. 193–220, repr. in Hawkins, ed., *Dante's Testaments* (1999), pp. 265–83.

—, ed., *Dante's Testaments: Essays in Scriptural Imagination* (Stanford, CA: Stanford University Press, 1999).

Hay, David M., ed., *Literal and Allegorical: Studies in Philo of Alexandria's Questions and Answers on Genesis and Exodus* (Atlanta, GA: Scholars Press, 1991).

Heine, Ronald E., 'Reading the Bible with Origen', in Paul M. Blowers, ed., *The Bible in Greek Christian Antiquity* (Notre Dame, IN: University of Notre Dame Press, 1997), pp. 131–48.

Hengevoss-Dürkop, Kerstin, 'Jerusalem – Das Zentrum der Ebstorf-Karte', in Kugler and Michael, eds, *Ein Weltbild vor Columbus* (1991), pp. 205–22.

Het aards paradijs: Dierenvoorstellingen in de Nederlanden van de 16de en 17de eeuw (Antwerp: Koninklijke Maatschappij voor Dierkunde van Antwerpen, 1982).

Higgins, Iain Macleod, *Writing East: The 'Travels' of Sir John Mandeville* (Philadelphia: University of Pennsylvania Press, 1997).

Higman, Francis, 'La Présentation typographique des Bibles genevois du XVIe siècle et pratique de la lecture', in Higman, *Lire et découvrir: La Circulation des idées au temps de la Réforme* (Geneva: Droz, 1998), pp. 573–81.

Hilhorst, Anthony, 'A Visit to Paradise: *Apocalypse of Paul* 45 and its Background', in Luttikhuizen, ed., *Paradise Interpreted*. (1999), pp. 128–39.

Hind, Arthur M., *Early Italian Engraving: A Critical Catalogue with Complete Reproduction of All the Prints Described*, 7 vols (London: Knoedler; New York: Quaritch, 1938–48).

Hoenen, Maarten J. F. M., and Alain de Libera, eds, *Albertus Magnus und der Albertismus: Deutsche philosopische Kultur des Mittelalters* (Leiden–New York–Cologne: Brill, 1995).

Hölscher, Ludger, *The Reality of the Mind: Augustine's Philosophical Argument for the Human Soul as a Spiritual Substance* (London–New York: Routledge & Kegan Paul, 1986).

Honigmann, Ernst, *Die sieben Klimata und die ΠΟΛΕΙΣ ΕΠΙΣΗΜΟΙ: Eine Untersuchung zur Geschichte der Geographie und Astrologie im Altertum und Mittelalter* (Heidelberg: Winter, 1929).

Houtman, Cees, *Der Pentateuch: Die Geschichte seiner Erforschung neben einer Auswertung* (Kampen: Kok, 1994).

Hübner, Wolfgang, *Zodiacus Christianus: Jüdisch-christliche Adaptationen des Tierkreises von der Antike bis zur Gegenwart* (Königstein: Anton Hain, 1983).

Hunter Blair, Peter, *The World of Bede* (London: Secker and Warburg, 1970).

Inglebert, Hervé, *Interpretatio christiana: Les Mutations des savoirs (cosmographie, géographie, ethnographie, histoire) dans l'antiquité chrétienne: 30–630 après J.-C.* (Paris: Institut d'Études Augustiniennes, 2001).

Ingram, Elizabeth M., 'Maps as Readers' Aids: Maps and Plans in Geneva Bibles', *Imago Mundi*, 45 (1993), pp. 29–44.

Iwańczak, Wojciech, 'Entre l'espace ptolémaïque et l'empirie: Les Cartes de Fra Mauro', *Médiévales* 18 (1990), pp. 53–68.

Jacob, Christian, 'Il faut qu'une carte soit ouverte ou fermée: Le Tracé conjectural', *Revue de la Bibliothèque Nationale*, 45 (1992), pp. 35–40.

Janvier, Yves, *La Géographie d'Orose* (Paris: Belles Lettres, 1982).

Johns, Jeremy, and Emilie Savage-Smith, 'The *Book of Curiosities*: A Newly Discovered Series of Islamic Maps', *Imago Mundi*, 55 (2003), pp. 7–24.

Jomard, Edmé-François *Les Monuments de la géographie; ou, Recueil d'anciennes cartes européennes et orientales* (Paris: Duprat, 1842–62).

Juste, David, *L'Astrologie latine au VIe au Xe siècle*, BA Dissertation, Université Libre de Bruxelles (1997).

Kaiser-Minn, Helga, *Die Erschaffung des Menschen auf den spätantiken Monumenten des 3. und 4. Jahrhunderts* (Münster in Westphalia: Aschendorff, 1981).

Kamal, Youssuf, *Monumenta cartographica Africae et Aegypti*, 5 vols (Cairo: 1926–51).

Kannengiesser, Charles, and William L. Petersen, eds, *Origen of Alexandria: His World and his Legacy* (Notre Dame, IN: University of Notre Dame Press, 1988).

——, ed., *Jean Chrysostome et Augustin* (Paris: Beauchesne, 1975).
Kappler, Claude, *Monstres, démons et merveilles à la fin du Moyen Âge* (Paris: Payot, 1980).
Karrow, Robert W., *Mapmakers of the Sixteenth Century and their Maps* (Chicago: Speculum Orbis Press, 1993).
Keates, John, *Understanding Maps* (New York: John Wiley, 1982).
Keil, Gundolf, 'Die verworfenen Tage', *Sudhoffs Archiv für Geschichte der Medizin und der Naturwissenschaften*, 41 (1957), pp. 27–58.
Kelly, John N. D., *Golden Mouth: The Story of John Chrysostom, Ascetic, Preacher, Bishop* (London: Duckworth, 1995).
Kennedy, Edward S., and Mary Helen Kennedy, *Geographical Coordinates of Localities from Islamic Sources* (Frankfurt am Main: Institut für Geschichte der Arabisch-Islamischen Wissenschaften, 1987).
Kervran, Louis, *Brandan: Le Grand Navigateur du Ve siècle* (Paris: Laffont, 1977).
Kimble, George H. T., *Geography in the Middle Ages* (London: Methuen, 1938).
——, 'The Laurentian World Map with Special Reference to its Portrayal of Africa', *Imago Mundi*, 1 (1935), pp. 29–33.
——, *The Catalan World Map of the R. Biblioteca Estense at Modena* (London: Royal Geographical Society, 1934).
Kitamura, Kunio, 'Cosmas Indicopleustès et la figure de la terre', in Desrumeaux and Schmidt, eds, *Moïse géographe* (1988), pp. 79–98.
Kitschelt, Lothar, *Die frühchristliche Basilika als Darstellung des himmlischen Jerusalem* (Munich: Neuer Filser-Verlag, 1938).
Klauck, Karl, 'Albertus Magnus und die Erdkunde', in Heinrich Ostlender, ed., *Studia Albertina* (Münster in Westphalia: Aschendorff, 1952), pp. 234–48.
Klein, Peter K., *Der ältere Beatus-Kodex Vitr. 14–1 der Biblioteca Nacional zu Madrid: Studien zur Beatus-Illustration und der spanischen Buchmalerie des 10. Jahrhunderts*, 2 vols (Hildesheim–New York: Georg Olms, 1976).
Kleinmann, Ute, 'Wo lag das Paradies? Beobachtungen zu den Paradieslandschaften des 16. und 17. Jahrhunderts', in *Die flämische Landschaft 1520–1700* (Lingen: Luca Verlag, 2003), pp. 279–303.
Klijn, Albert Frederick J., *Seth in Jewish, Christian and Gnostic Literature* (Leiden: Brill, 1977).
Kline, Naomi R., *Maps of Medieval Thought* (Woodbridge: Boydell Press, 2001).
Klug, Rudolf, *Der Astronom Johannes von Gmunden und sein Kalender* (Linz: Pirngruber, 1912).
Koch, Josef K., *Durandus de S. Porciano, O.P: Forschungen zum Streit um Thomas von Aquin zu Beginn des 14. Jahrhunderts* (Münster in Westphalia: Aschendorff, 1927).
Kraeling, Emil G., *The Old Testament since the Reformation* (London: Lutterworth, 1955).
Kraus, Hans-Joachim, *Geschichte der historisch-kritischen Erforschung des Alten Testaments*, 3rd edn (Neukirchener-Vluyn: Neukirchener Verlag, 1982).
Kretzman, Norman, and Eleonore Stump, eds, *The Cambridge Companion to Aquinas* (Cambridge: Cambridge University Press, 1993).
Kugler, Hartmut, and Eckhard Michael, eds, *Ein Weltbild vor Columbus: Die Ebstorfer Weltkarte: Interdisziplinäres Colloquium 1988* (Weinheim: VCH, Acta Humaniora, 1991).
——, 'Die Ebstorfer Weltkarte: Ein europäisches Weltbild im deutschen Mittelalter', *Zeitschrift für Deutsches Altertum und Deutsche Literatur*, 116/1 (1987), pp. 1–29.
Kühnel, Bianca, *The End of Time in the Order of Things: Science and Eschatology in Early Medieval Art* (Regensburg: Schnell-Steiner, 2003).
——, ed., *The Real and Ideal Jerusalem in Jewish, Christian and Islamic Art: Studies in Honour of Bezalel Narkiss on the Occasion of his Seventieth Birthday* (Jerusalem: Center for Jewish Art, Hebrew University of Jerusalem, 1998).
——, *From the Earthly to the Heavenly Jerusalem: Representations of the Holy City in Christian Art of the First Millennium* (Rome–Freiburg im Breisgau–Vienna: Herder, 1987).
Kuntz, Marion L., *Guillaume Postel, Prophet of the Restitution of All Things: His Life and Thought* (The Hague–Boston–London: Nijhoff, 1981).
Laborde, Alexandre de, *Les Manuscrits à peintures de la Cité de Dieu de Saint Augustin*, 2 vols (Paris: Societé de Bibliophiles François-Rahir, 1909).
Labrosse, Henri, 'Biographie de Nicolas de Lyre', *Études Franciscaines*, 17 (1907), pp. 489–505, 593–608.
——, 'Sources de la biographie de Nicolas de Lyra', *Études Franciscaines*, 16 (1906), pp. 383–404.
Lampe, Geoffrey W. H., and Kenneth J. Woolcombe, *Essays on Typology* (London: SCM Press, 1957).
Landes, Richard, Andrew Gow, and David C. Van Meter, eds, *The Apocalyptic Year 1000: Religious Expectation and Social Change, 950–1050* (Oxford: Oxford University Press, 2003).
——, 'Lest the Millennium Be Fulfilled: Apocalyptic Expectations and the Pattern of Western Chronography 100–800 CE', in *The Use and Abuse of Eschatology in the Middle Ages*, ed. by Werner Werbeke, Daniel Verhelst, and Andries Welkenhuysen (Louvain: Louvain University Press, 1988), pp. 137–211.
Lane, Anthony N. S., *John Calvin: Student of the Church Fathers* (Edinburgh: Clark, 1999).
Laor, Eran, *Maps of the Holy Land: Cartobibliography of Printed Maps, 1475–1900* (New York–Amsterdam: Alan R. Liss–Meridian Publishing Company, 1986).
Lascelles, Mary M., 'Alexander and the Earthly Paradise in Mediaeval English Writings', *Medium Aevum*, 5 (1936), pp. 31–104, 173–88.
Le Goff, Jacques, ed., *The Medieval World*, transl. by Lydia G. Cochrane (London: Collins and Brown, 1990).
——, *L'Imaginaire médiéval: Essai* (Paris: Gallimard, 1985).
Lecoq, Danielle, 'Saint Brandan, Christophe Colomb, et le paradis terrestre', *Revue de la Bibliothèque Nationale*, 45 (1992), pp. 14–21.
——, 'Le Temps et l'intemporel sur quelques représentations médiévales du monde au XIIe et au XIIIe siècles', in Bernard Ribémont, ed., *Le Temps, sa mesure et sa perception au Moyen Âge* (Caen: Paradigme, 1992), pp. 113–49.
——, 'La Mappemonde d'Henri de Mayence ou l'image du monde au XIIIe siècle', in Gaston Duchet-Suchaux, ed., *Iconographie médiévale: Image, texte, contexte* (Paris: Centre Nationale de la Recherche Scientifique, 1990), pp. 155–207.
——, 'La Mappemonde du *De Arca Noe Mystica* de Hugues de Saint-Victor (1128–1129)', in Pelletier, ed., *Géographie du monde au Moyen Âge et à la Renaissance* (1989), pp. 9–31.
——, 'La Mappemonde du *Liber Floridus* ou la vision de Lambert de Saint-Omer', *Imago Mundi*, 39 (1987), pp. 9–49.
Leloir, Louis, *Doctrines et méthodes de S. Éphrem après son commentaire de l'Évangile concordant*, *CSCO* CCXX (Louvain: CSCO, 1961).
Lemonnier, Arlette, ed., *Le Dessus des cartes: Art et cartographie* (Brussels: ISELP, 2004).
Lestringant, Frank, *Le Livre des îles: Atlas et récits insulaires, de la Genèse à Jules Verne* (Geneva: Droz, 2002).
——, 'Cosmologie et mirabilia à la Renaissance: L'Exemple de Guillaume Postel', *Journal of Medieval and Renaissance Studies*, 16/2 (Fall 1986), pp. 253–79, repr. 'Cosmographie pour une restitution: Note sur le traité "Des merveilles du monde" de Guillaume Postel (1553)', in Marion Leathers Kuntz, ed., *Postello, Venezia e il suo mondo* (Florence: Olschki, 1988), pp. 227–60.
Levison, John R., *Texts in Transition: The Greek Life of Adam and Eve* (Atlanta, GA: Society of Biblical Literature, 2000).
Lewis, G. Malcolm, 'The Origins of Cartography', in Harley and Woodward, eds, *The History of Cartography*, I (1987), pp. 50–3.
Lewis, Suzanne, *The Art of Matthew Paris in the 'Chronica majora'* (Berkeley: University of California Press; Aldershot, Hants: Scholar Press with Corpus Christi College, Cambridge, 1987).
Lexikon der christlichen Ikonographie, ed. by Egelbert Kirschbaum, 8 vols (Rome–Freiburg–Basle–Vienna: Herder, 1968).
Libera, Alain de, *Albert le Grand et la philosophie* (Paris: Vrin, 1990).
Liccaro, Vincenzo, *Studi sulla visione del mondo di Ugo di S. Vittore* (Trieste: Del Bianco, 1969).
Lieu, Samuel N. C., *Manichaeism in the Later Roman Empire and Medieval China*, 2nd edn (Tübingen: J. C. B. Mohr, 1992).
Ligota, Christopher, 'La Foi historienne: Histoire et connaissance de l'histoire chez S. Augustin', *Revue des Études Augustiniennes*, 43 (1997), pp. 111–71.
Lindberg, David C., 'The Medieval Church Encounters the Classical Tradition', in David C. Lindberg, Ronald L. Numbers, eds, *When Science and Christianity Meet* (Chicago–London: University of Chicago Press, 2003), pp. 7–32.
——, and Ronald L. Numbers, eds, *God and Nature: Historical Essays on the Encounter between Christianity and Science* (Berkeley–Los Angeles–London: University of California Press, 1986).
Lindemann, Rolf, 'A New Dating of the Ebstorf Mappamundi', in Pelletier, ed., *Géographie du monde au Moyen Âge et à la Renaissance* (1989), pp. 45–50.
Lindgren, Uta, 'Die Bedeutung Philipp Melanchthons (1497–1560) für die Entwicklung einer naturwissenschaftlichen Geographie', in *Gerard Mercator und seine Zeit: 7. Kartographisches Colloquium Duisburg 1994: Vorträge und Berichte*, ed. by Wolfgang Scharfe (Duisburg: Walter Braun, 1996), pp. 1–12.
Lippincott, Kristen, 'Giovanni di Paolo's "Creation of the World" and the Tradition of the "Thema Mundi" in Late Medieval and Renaissance Art', *Burlington Magazine*, 132/1048 (July 1990), pp. 460–8.
Livingstone, David N., and Charles W. J. Withers, eds, *Geography and Enlightenment* (Chicago–London: University of Chicago Press, 1999).
Lowrie, Walter, *Art in the Early Church* (New York: Pantheon Books, 1947).
Lubac, Henri de, *Exégèse médiévale: Les Quatre Sens de l'Écriture*, 4 vols (Paris: Aubier, 1959–64).
——, *Histoire et esprit: L'Intelligence de l'Écriture d'après Origène* (Paris: Aubier, 1950).
Luneau, Auguste, *L'Histoire du salut chez les Pères de l'Église: La Doctrine des âges du monde* (Paris: Beauchesne, 1964).
Luttikhuizen, Gerard P., ed., *Paradise Interpreted: Interpretations of Biblical Paradise: Judaism and Christianity* (Leiden–Boston, MA: Brill, 1999).
——, 'A Resistant Interpretation of the Paradise Story in the Gnostic *Testimony of Truth* (Nag Hamm. Cod. IX.3) 45–50', ibid., pp. 140–52.
MacEachren, Alan M., *How Maps Work: Representation, Visualization, and Design* (New York–London: Guilford Press, 1995).
Madathil, John O., *Kosmas der Indienfahrer: Kaufmann, Kosmologe und Exeget zwischen alexandrinischer und antiochenischer Theologie* (Thaur–Vienna: Kulturverlag, 1996).
Magné, Jean, *From Christianity to Gnosis and from Gnosis to Christianity: An Itinerary through the Texts to and from the Tree of Paradise* (Atlanta, GA: Scholars Press, 1993).
Malbon, Struthers Elizabeth, *The Iconography of the Sarcophagus of Junius Bassus* (Princeton, NJ: Princeton University Press, 1990).
Mangani, Giorgio, *Il 'mondo' di Abraham Ortelio: Misticismo, geografia e collezionismo nel Rinascimento dei Paesi Bassi* (Modena: Franco Cosimo Panini, 1998).

——, 'Abraham Ortelius and the Hermetic Meaning of the Cordiform Projection', *Imago Mundi*, 50 (1998), pp. 59–83.
Maraval, Pierre, *Lieux saints et pèlerinages d'Orient: Histoire et géographie des origins à la conquête arabe* (Paris: Cerf, 1985).
Marcolino, Venicio, *Das alte Testament in der Heilsgeschichte: Untersuchungen zum dogmatischen Verständnis des alten Testaments als Heilsgeschichtliche Periode nach Alexander von Hales* (Münster in Westphalia: Aschendorff, 1970).
Marcon, Susy, 'Il mappamondo di Fra Mauro e Leonardo Bellini', in Mario Piantoni and Laura de Rossi, eds, *Per l'arte da Venezia all'Europa: Studi in onore di Giuseppe Maria Pilo* (Venice: Edizioni della Laguna, 2001), pp. 103–8.
Markus, Robert A., *Signs and Meanings: World and Text in Ancient Christianity* (Liverpool: Liverpool University Press, 1996).
——, 'How on Earth Could Places Become Holy? Origins of the Christian Idea of Holy Places', *Journal of Early Christian Studies*, 2/3 (1994), pp. 257–71.
——, 'Augustine's *Confessions* and the Controversy with Julian of Eclanum: Manicheism Revisited', in Bernard Bruning, and others, eds, *Collectanea Augustiniana: Mélanges T. J. Van Bavel* (Louvain: Louvain University Press, 1990), pp. 913–25.
——, *The End of Ancient Christianity* (New York–Cambridge: Cambridge University Press, 1990).
——, *Conversion and Disenchantment in Augustine's Spiritual Development* (Villanova, PA: Villanova University Press, 1989).
——, *Saeculum: History and Society in the Theology of Saint Augustine*, 2nd edn (Cambridge–New York: Cambridge University Press, 1988).
Marrone, Steven P., *William of Auvergne and Robert Grosseteste: New Ideas of Truth in the Early Thirteenth Century* (Princeton, NJ: Princeton University Press, 1983).
Masnovo, Amato, *Da Guglielmo d'Auvergne a S. Tommaso d'Aquino*, 3 vols (Milan: Vita e Pensiero, 1945–6).
Massimi, Jean-Robert, 'Montrer et démontrer: Autour du *Traité de la situation du paradis terrestre* de P. D. Huet (1691)', in Desrumeaux and Schmidt, eds, *Moïse géographe* (1988), pp. 203–25.
Matos, Luís de, *L'Expansion portugaise dans la littérature latine de la Renaissance* (Lisbon: Fundação Calouste Gulbenkian, Serviço de Educação, 1991).
Maury, Alfred, *Essai sur les légendes pieuses du Moyen-Âge* (Paris: Ladrange, 1843).
Mayer, Anton, *Mittelalterliche Weltkarten aus Olmütz* (Prague: Geographisces Institut der Deutschen Universität in Prag, 1932).
Mayer, Cornelius P., *Die Zeichen in der geistigen Entwicklung und in der Theologie des jungen Augustinus* (Würzburg: Augustinus Verlag, 1969).
Mazzucco, Clementina, 'La Gerusalemme celeste dell'"Apocalisse" nei Padri', in Gatti Perer, ed., *La dimora di Dio con gli uomini (Ap. 21,3)* (1983), pp. 49–75.
McClung, William A., *The Architecture of Paradise: Survivals of Eden and Jerusalem* (Berkeley: University of California Press, 1983).
McCluskey, Stephen C., *Astronomies and Cultures in Early Medieval Europe* (Cambridge: Cambridge University Press, 1998).
McGinn, Bernard, *Apocalyptic Spirituality* (New York: Paulist Press, 1979).
McKenzie, Donald F., *Bibliography and the Sociology of Text*, The Panizzi Lectures 1985 (London: British Library, 1986).
McKim, Donald K., ed., *The Cambridge Companion to John Calvin* (Cambridge: Cambridge University Press, 2004).
Meine, Karl-Heinz, 'Zur Weltkarte des Andreas Walsperger, Konstanz 1448', in Wolfgang Scharfe, Hans Vollet and Erwin Herrmann, eds, *Kartenhistorisches Colloquium Bayreuth 1982* (Berlin: Dietrich Reimer, 1983), pp. 17–30.
Mendelsohn, Everett, *Heat and Life: The Development of the Theory of Animal Heat* (Cambridge, MA: Harvard University Press, 1964).
Mentré, Mireille, *Contribucion al estudio de la miniatura en Leon y Castilla en la alta Edad Media (problemas de la forma y del espacio en la ilustración de los Beatos)* (León: Institución 'Fray Bernardino de Sahagún', 1976).
Mercier, Raymond, 'Astronomical Tables in the Twelfth Century', in Burnett, ed., *Adelard of Bath* (1987), pp. 87–118.
Meurer, Peter H., 'Ortelius as the Father of Historical Cartography', in Marcel Van den Broecke, Peter Van der Krogt, and Meurer, eds, *Abraham Ortelius and the First Atlas: Essays Commemorating the Quadricentennial of his Death, 1598–1998* (Houten, The Netherlands: HES, 1998), pp. 133–59.
Meyer, Wilhelm, 'Die Geschichte des Kreuzholzes vor Christus', *Abhandlungen der Königlichen Bayerischen Akademie der Wissenschaften, Philosoph.-Philologische Klasse*, 16/2 (1881), pp. 103–60;
——, 'Vita Adae et Evae', ibid., 14/3 (1879), pp. 187–250.
Milano, Ernesto, *Il mappamondo catalano estense*, facsimile edition with commentary and transcription of the toponyms ed. by Annalisa Battini (Dietikon-Zurich: Urs Graf, 1995).
——, *La carta del Cantino e la rappresentazione della terra nei codici e nei libri a stampa della Biblioteca Estense e Universitaria* (Modena: Il Bulino, 1991).
Miller, Konrad, *Die Ebstorfkarte: Eine Weltkarte aus dem 13. Jahrhundert* (Stuttgart: J. Roth, 1900).
——, *Mappaemundi: Die ältesten Weltkarten*, 6 vols (Stuttgart: J. Roth, 1895–8).
Minnis, Alastair J., and A. Brian Scott, eds, *Medieval Literary Theory and Criticism, c.1100–c.1375: The Commentary Tradition* (Oxford–New York: Oxford University Press, 1988).
Mokre, Jan, 'Kartographie des Imaginären: Von Ländern, die es nie gab', in Hans Petschar, ed., *Alpha und Omega: Geschichten vom Ende und Anfang der Welt* (Vienna–New York: Österreichische Nationalbibliothek–Springer, 2000), pp. 21–42.
Monmonier, Mark S., *How to Lie with Maps* (Chicago–London: University of Chicago Press, 1991).
Moore, Edward, *Studies in Dante, Third Series: Miscellaneous Essays* (Oxford: Oxford University Press, 1903; repr. 1968).
Moore, Philip S., *The Works of Peter of Poitiers, Master in Theology and Chancellor of Paris (1193–1205)* (Notre Dame, IN: University of Notre Dame Press, 1936)
Moore, Rebecca, *Jews and Christians in the Life and Thought of Hugh of St Victor* (Atlanta, GA: Scholars Press, 1998).
Morgan, Nigel J., *The Lambeth Apocalypse: Manuscript 209 in Lambeth Palace Library* (London: Harvey Miller, 1990).
Morris, Paul, and Deborah Sawyer, eds, *A Walk in the Garden: Biblical, Iconographical and Literary Images of Eden* (Sheffield: Journal for the Study of the Old Testament Press, 1992).
Motta, Franco, '"Geographia sacra": Il luogo del paradiso nella teologia francese del tardo Seicento', *Annali di Storia dell'Esegesi*, 14/2 (1997), pp. 477–506.
Mundo, Anscario M., and Manuel Sanchez Mariana, *El Comentario de Beato al Apocalipsis: Catálogo de los códices* (Madrid: Biblioteca Nacional, 1976).
Munk Olsen, Birger, *L'Étude des auteurs classiques latins aux XIe et XIIe siècles*, 3 vols (Paris: Éditions du Centre National de la Recherche Scientifique, 1982–9).

Muratore, Saturnino, 'Magistero e Darwinismo', *La Civiltà Cattolica*, 3518 (1997/1), pp. 141–5.
Murdoch, John E., *Album of Science*, I, *Antiquity and The Middle Ages* (New York: Scribner, 1984).
Murray, Robert, 'The Theory of Symbolism in St. Ephrem's Theology', *Parole de l'Orient*, 6–7 (1975–6), pp. 1–20.
——, *Symbols of Church and Kingdom: A Study in Early Syriac Tradition* (London–New York–Cambridge: Cambridge University Press, 1975).
Nardi, Bruno, *Saggi di filosofia dantesca*, 2nd edn (Florence: La Nuova Italia, 1967).
Nebenzahl, Kenneth, *Maps of the Bible Lands: Images of Terra Sancta through Two Millennia* (London: Times Books; New York: Abbeville Press, 1986).
Neugebauer, Otto, *A History of Ancient Mathematical Astronomy*, 3 vols (Berlin–Heidelberg–New York: Springer, 1975).
Neuss, Wilhelm, *Die Apokalypse des Hl. Johannes in der altspanischen und altchristlichen Bibel-Illustration (das Problem der Beatus-Handschriften)* (Münster in Westphalia: Aschendorff, 1931).
Newton, Arthur Percival, ed., *Travel and Travellers of the Middle Ages* (London–New York: Routledge & Kegan Paul, 1926).
Noort, Ed, 'Gan-Eden in the Context of the Mythology of the Hebrew Bible', in Luttikhuizen, ed., *Paradise Interpreted* (1999), pp. 21–36.
Nordenskiöld, Adolf E., *Facsimile-Atlas to the Early History of Cartography with Reproductions of the Most Important Maps Printed in the XV and XVI Centuries* (1889; repr. New York: Dover, 1973).
North, John D., *Horoscopes and History* (London: Warburg Institute, 1986).
Ó Fearghail, Fearghus, 'Philo and the Fathers: The Letter and the Spirit', in Thomas Finan and Vincent Twomey, eds, *Scriptural Interpretations in the Fathers: Letter and Spirit* (Blackrock, Co. Dublin–Portland, OR: Four Courts Press, 1995), pp. 39–59.
O'Connor, Edward D., 'The Scientific Character of Theology according to Scotus', in *De doctrina Ioannis Duns Scoti* (1968), III, pp. 3–50.
O'Malley, Charles Donald, *Michael Servetus: A Translation of his Geographical, Medical and Astrological Writings* (London: Lloyd–Luke, 1953).
O'Reilly, Jennifer, 'The Trees of Eden in Medieval Iconography', in Morris and Sawyer, eds, *A Walk in the Garden* (1992), pp. 167–204.
Obermann, Heiko, *The Harvest of Medieval Theology: Gabriel Biel and Late Medieval Nominalism*, 2nd edn (Grand Rapids, MI: Eerdmans, 1967).
Olschki, Leonardo, *Storia letteraria delle scoperte geografiche: Studi e ricerche* (Florence: Olschki, 1937).
Orr, Mary A., *Dante and the Early Astronomers* (London–Edinburgh: Gall and Inglis, 1913).
Otten, Willemien, *From Paradise to Paradigm: A Study of Twelfth-Century Humanism* (Leiden: Brill, 2004).
The Oxford Companion to the Bible, ed. by Bruce M. Metzger and Michael D. Coogan (New York–Oxford: Oxford University Press, 1993).
Patch, Howard R., *The Other World, according to Descriptions in Medieval Literature* (Cambridge, MA: Harvard University Press, 1950).
Paulys Real-Encyclopädie der Classischen Altertumswissenschaft, ed. by Georg Wissowa, Wilhelm Kroll, and others, 47 vols (Stuttgart: Metzler, 1894–1963).
Peake's Commentary on the Bible, ed. by Matthew Black and Harold H. Rowley (London: Nelson, 1962).
Peckham, Lawton P. G., and Milan S. La Du, *La Prise de Defur et le voyage d'Alexandre au paradis terrestre* (Princeton: Princeton University Press, 1935; repr. New York: Kraus Reprint, 1965).
Pecoraro, Paolo, *Le stelle di Dante* (Rome: Bulzoni, 1987).
Pedersen, Olaf, *A Survey of the Almagest* (Odense: Odense Universitetsforlag, 1974).

Pelland, Gilles, *Cinq Études d'Augustin sur le début de la Genèse* (Tournai: Desclée, 1972).
Pelletier, Monique, 'Peut-on encore affirmer que la BN possède la carte de Christophe Colomb?' *Revue de la Bibliothèque Nationale*, 45 (1992), pp. 22–5.
—, ed., *Géographie du monde au Moyen Âge et à la Renaissance* (Paris: Éditions du Comité des Travaux Historiques et Scientifiques, 1989).
Pelster, Franz, *Kritische Studien zum Leben und zu den Schriften Alberts des Grossen* (Freiburg im Breisgau: Herder, 1920).
Pépin, Jean, *La Tradition de l'allégorie de Philon d'Alexandrie à Dante: Études historiques* (Paris: Études Augustiniennes, 1987).
Petchenik, Barbara, and Arthur Robinson, *The Nature of Maps* (Chicago–London: University of Chicago Press, 1976).
Petry, Yvonne, *Gender, Kabbalah and the Reformation: The Mystical Theology of Guillaume Postel (1510–1581)* (Leiden–Boston: Brill, 2004).
Picascia, Maria Luisa, *Un occamista quattrocentesco: Gabriel Biel* (Florence: La Nuova Italia, 1979).
Pietrzik, Dominik, *Die Brandan-Legende: Ausgewählte Motive in der frühneuhochdeutschen sogenannten 'Reise'-Version* (Frankfurt am Main: P. Lang, 1999).
Pirenne, Jacqueline, *La Légende du 'Prêtre Jean'* (Strasbourg: Presses Universitaires de Strasbourg, 1992).
Plaut, Fred, *Analysis Analysed: When the Map Becomes the Territory* (London–New York: Routledge, 1993).
—, 'Where Is Paradise? The Mapping of a Myth', *The Map Collector*, 29 (1984), pp. 2–7.
—, 'General Gordon's Map of Paradise', *Encounter* (June/July 1982), pp. 20–32.
Poirel, Dominique, *Livre de la nature et débat trinitaire au XIIe siècle: Le De tribus diebus de Hugues de Saint-Victor* (Turnhout: Brepols, 2002).
—, *Hugues de Saint-Victor* (Paris: Cerf, 1998).
Poirot, Éliane, *Les Prophètes Élie et Élisée dans la littérature chrétienne ancienne* (Turnhout: Brepols, 1997).
Poortman, Wilco C., *Bijbel en Prent*, 2 vols (The Hague: Uitgeverij Boekencentrum, 1983–6).
Pope, Hugh, *St. Thomas Aquinas as an Interpreter of Holy Scripture* (Oxford: Blackwell, 1924).
Pope-Hennessy, John, *Paradiso: The Illuminations to Dante's Divine Comedy by Giovanni di Paolo* (London: Thames and Hudson, 1993).
Pörck, Guy de, *Introduction à la Fleur des histoires de Jean Mansel (XVe siècle)* (Ghent: E. Claeys–Verheughe, 1936).
Postl, Brigitte, *Die Bedeutung des Nil in der römischen Literatur* (Vienna: Notring, 1970).
Poulouin, Claudine, *Le Temps des origines: L'Eden, le Déluge et 'les temps reculés': De Pascal à l'Encyclopédie* (Paris: Honoré Champion, 1998).
Prest, John M., *The Garden of Eden: The Botanic Garden and the Re-creation of Paradise* (New Haven: Yale University Press, 1981).
Quentin, Albrecht, *Naturkenntnisse und Naturanschauungen bei Wilhelm von Auvergne* (Hildesheim: Gerstenberg, 1976).
Quinn, Esther C., *The Quest of Seth for the Oil of Life* (Chicago: University of Chicago Press, 1962).
Quinn, John F., *The Historical Constitution of St Bonaventure's Philosophy* (Toronto: Pontifical Institute of Medieval Studies, 1973).
Rachewiltz, Igor de, *Papal Envoys to the Great Khans* (London: Faber and Faber, 1971).
Randles, William G. L., *De la terre plate au globe terrestre: Une mutation épistémologique rapide, 1480–1520* (Paris: Armand Colin, 1980).
—, 'Notes on the Genesis of the Discoveries', *Studia*, 5 (1960), pp. 20–46.
Rapetti, Elena, *Pierre-Daniel Huet: Erudizione, filosofia, apologetica* (Milan: Vita e Pensiero, 1999).
Ravaschietto, Giuliana, *Il viaggio dei tre monaci al paradiso terrestre* (Alessandria: Edizioni dell'Orso, 1997).
Relaño, Francesc, *The Shaping of Africa: Cosmographic Discourse and Cartographic Science in Late Medieval and Early Modern Europe* (Aldershot, Hants: Ashgate, 2002).
Rist, John M., *Augustine: Ancient Thought Baptized* (Cambridge: Cambridge University Press, 1994).
Roberts Lacy, Alan, *A Dictionary of Philosophy* (London: Routledge & Kegan Paul, 1976).
Robinson, Arthur H., *The Look of Maps* (Madison: University of Wisconsin Press, 1952).
Rohls, Jan, *Wilhelm von Auvergne und der mittelalterliche Aristotelismus* (Munich: Kaiser, 1980).
Roncière, Charles de la, 'Une carte de Christophe Colomb', *Revue des Questions Historiques*, 7 (1925), pp. 27–41
—, *La Carte de Christophe Colomb* (Paris: Champion, 1924).
Rosenthal, Erwin, 'Concerning the Dating of Rüst's and Sporer's World Maps', *Papers of the Bibliographical Society of America*, 47 (1953), pp. 156–8.
Rosien, Walter, *Die Ebstorfer Weltkarte* (Hanover: Niedersächsisches Amt für Landesplanung und Statistik, 1952).
Ross, David J. A., *Alexander Historiatus: A Guide to Medieval Illustrated Alexander Literature* (Frankfurt am Main: Athenäum, 1963; repr. 1988).
Rotsaert, Marie-Louise, *San Brandano: Un antitipo germanico* (Rome: Bulzoni, 1996).
Rowley, Harold H., *Darius the Mede and the Four World Empires in the Book of Daniel: A Historical Study of Contemporary Theories* (Cardiff: University of Wales Press, 1935).
Ruberg, Uwe, 'Die Tierwelt auf der Ebstorfer Weltkarte im Kontext mittelalterlicher Enzyklopädik', in Kugler and Michael, eds, *Ein Weltbild vor Columbus* (1991), pp. 319–46.
—, '*Mappae mundi* des Mittelalters in Zusammenwirken von Text und Bild', in Christel Meier and Uwe Ruberg, eds, *Text und Bild: Aspekte des Zusammenwirkens zweier Künste in Mittelalter und früher Neuzeit* (Wiesbaden: Reichert, 1980), pp. 550–92.
Runia, David T., *Philo of Alexandria and the Timaeus of Plato* (Leiden: Brill, 1986).
Rusch, William G., *The Later Latin Fathers* (London: Duckworth, 1977).
Russell, Jeffrey B., *Inventing the Flat Earth: Columbus and Modern Historians* (New York: Praeger, 1991).
Rüthing, Heinrich, 'Kritische Bemerkungen zu einer mittelalterlichen Biographie des Nikolaus von Lyra', *Archivum Franciscanum Historicum*, 60 (1967), pp. 42–54.
Salmon, Pierre, *Les Manuscrits liturgiques latins de la Bibliothèque Vaticane*, 5 vols (Vatican City: Biblioteca Apostolica Vaticana, 1968–72).
Salway, Benet, 'Travel, *Itineraria* and *Tabellaria*', in Colin Adams and Ray Lawrence, eds, *Travel and Geography in the Roman Empire* (London–New York: Routledge, 2001), pp. 22–66.
Sánchez Albornoz, Claudio, 'El Asturorum Regnum en los días de Beato de Liébana', in *Actas del Simposio … Beato de Liébana*, I (1978), pp. 19–32.
Sanford, Charles L., *The Quest for Paradise: Europe and the American Moral Imagination* (Urbana: University of Illinois Press, 1961).
Santarém, Manuel Francisco de Barros e Sousa, Viscount of, *Essai sur l'histoire de la cosmographie et de la cartographie pendant le Moyen-Âge et sur les progrès de la géographie après les grandes découvertes du XVe siècle*, 3 vols (Paris: Maulde et Renou, 1849–52).
Sanz, Carlos, *La Geographia de Ptolomeo* (Madrid: Victoriano Suarez, 1959).
Sarton, George, *Introduction to the History of Science*, 3 vols (Baltimore, MD: Williams & Wilkins, 1927).
Sasson, Jack M., ed. *Civilizations of the Ancient Near East*, 3 vols (New York: Charles Scribner's Sons, 1995).
Scafi, Alessandro, 'The Image of Persia in Western Medieval Cartography', in *Proceedings of the V European Conference of Iranian Studies* (Bologna: Societas Iranologica Europaea, Università di Bologna, 2006), pp. 223–34..
—, 'Defining *Mappaemundi*', in Harvey, ed, *The Hereford World Map: Medieval World Maps and their Context* (2006), pp. 345–54.
—, 'À la recherche du paradis perdu: Les Mappemondes du XIIIe siècle', in Jean Galard, ed., *Ruptures: De la discontinuité dans la vie artistique* (Paris: Musée du Louvre–École Nationale Supérieure des Beaux-Arts, 2002), pp. 17–57.
—, 'Paradise Found and Lost around 1500', in *Acts of the International Conference 'Vasco Da Gama: Men, Voyages and Cultures'*, ed. Joaquim R. Magalhães, 2 vols (Lisbon: Comissão Nacional para as Comemorações dos Descobrimentos Portugueses, 2002), II, pp. 435–52.
—, 'Les Colonnes d'Hercule dans la cartographie médiévale: Limite de la Méditerranée et porte du paradis', in Bertrand Westphal, ed., *Le Rivage des mythes: Une géocritique méditerranéenne*, I, *Le Lieu et son mythe* (Limoges: Presses Universitaires de Limoges, 2001), pp. 339–65.
—, 'Fra Mauro's World Map', in Friedman, Mossler Figg, and others, eds, *Trade, Travel, and Exploration in the Middle Ages* (2000), pp. 383–6.
—, 'Il paradiso terrestre di Fra Mauro', *Storia dell'Arte*, 93/4 (January–June 1999), pp. 219–27.
—, 'Mapping Eden: Cartographies of the Earthly Paradise', in Cosgrove, ed., *Mappings* (1999), pp. 50–70.
—, *The Notion of the Earthly Paradise from the Patristic Era to the Fifteenth Century*, PhD Dissertation, Warburg Institute, University of London (1999).
Scalise, Charles J., 'Origen and the *sensus literalis*', in Kannengiesser and Petersen, eds, *Origen of Alexandria* (1988), pp. 117–29.
Schäublin, Christoph, *Untersuchungen zu Methode und Herkunft der Antiochenischen Exegese* (Cologne–Bonn: [n. pub.], 1974).
Scheeben, Heribert Ch., *Albertus Magnus* (Bonn: Verlag der Buchgemeinde, 1932).
Schildgen, Brenda D., *Dante and the Orient* (Urbana: University of Illinois Press, 2002).
Schmidt, Francis, 'Naissance d'une géographie juive', in Desnrumaux and Schmidt, eds, *Moïse géographe* (1988), pp. 13–30.
Schramm, Albert, *Luther und die Bibel*, 2 vols (Leipzig: Hiersemann, 1923).
Schreiner, Susan E., *The Theater of His Glory: Nature and the Natural Order in the Thought of John Calvin* (Durham, NC: Labyrinth Press, 1991).
Schulz, Jürgen, 'Jacopo de' Barbari's View of Venice: Map Making, City Views, and Moralized Geography before the Year 1500', *Art Bulletin*, 60 (1978), pp. 425–74.
Scott, James M., *Geography in Early Judaism and Christianity: The Book of Jubilees* (Cambridge: Cambridge University Press, 2002).
Secret, François, and Georges Weill, *Vie et caractère de Guillaume Postel* (Milan: Archè, 1987).
—, *Bibliographie des manuscrits de Guillaume Postel* (Geneva: Droz, 1970).
Setton, Kenneth M., ed., *A History of the Crusades*, 6 vols (Madison: University of Wisconsin Press, 1969–89).
Shah, Mazhar H., *The General Principles of Avicenna's Canon of Medicine* (Karachi: Naveed Clinic, 1966).
Shelford, April G.. *Faith and Glory: Pierre-Daniel Huet and the Making of the Demonstratio Evangelica (1679)*, PhD Dissertation, Princeton University (1997).
—, 'Amitié et animosité dans la République des Lettres: La Querelle entre Bochart et Huet', in Suzanne Guellouz, ed., *Pierre-Daniel Huet (1630–1721)* (Paris–Seattle–Tübingen: Papers on French Seventeenth Century Literature, 1994), pp. 99–108.
Shirley, Rodney W., *The Mapping of the World: Early Printed World Maps, 1472–1700* (London: Early World Press, 2001).

Sicard, Patrice, *Diagrammes médiévaux et exégèse visuelle: Le Libellus de formatione arche de Hugues de Saint-Victor* (Paris–Turnhout: Brepols, 1993).

Sileo, Leonardo, ed., *Via Scoti: Methodologica ad mentem Ioannis Duns Scoti*, 2 vols (Rome: Antonianum, 1995).

Silva y Verástegui, Soledad de, 'Le "Beatus" navarrais de Paris (Bibl. Nat., Nouv. Acq. Lat. 1366)', *Cahiers de Civilisation Médiévale*, 40, (1997), pp. 215–32.

Silverberg, Robert, *The Realm of Prester John* (Athens, OH: Ohio University Press, 1996).

Simonetti, Manlio, 'Origene e i vignaioli perfidi', *Orphaeus*, 17 (1996), pp. 35–49.

——, 'Alcune osservazioni sull'interpretazione origeniana di *Genesi* 2,7 e 3,21', *Aevum*, 36 (1962), pp. 370–81.

Singleton, Charles S., *Dante's 'Commedia': Elements of Structure* (Baltimore: Johns Hopkins University Press, 1954).

Siniscalco, Paolo, 'Due opere a confronto sulla creazione dell'uomo: Il *De Genesi ad litteram libri XII* di Agostino e i *Libri IV in principium Genesis* di Beda', in *Miscellanea di studi agostiniani in onore di P. Agostino Trapè, OSA* (Rome: Istituto Patristico Augustinianum, 1985), pp. 435–52.

Sirinelli, Jean, *Les Vues historiques d'Eusèbe de Césarée durant la période prénicéenne* (Dakar: Université de Dakar, 1961).

Slessarev, Vsevolod, *Prester John: The Letter and the Legend* (Minneapolis: University of Minnesota Press, 1959).

Smalley, Beryl, *The Study of the Bible in the Middle Ages*, 3rd edn (Oxford: Blackwell, 1984).

Smith, John Clark, *The Ancient Wisdom of Origen* (Lewisburg–London–Cranbury, NJ: Bucknell University Press–Associated University Presses, 1992).

Smith, Jonathan Z., *Maps Is Not Territory* (Leiden: Brill, 1978).

Smith, William, ed., *Dictionary of Greek and Roman Geography*, 2 vols (London: John Murray, 1856–7)

Sorabji, Richard, *Time, Creation and the Continuum* (London: Duckworth, 1983).

Southern, Richard W., *Western Views of Islam in the Middle Ages* (Cambridge, MA: Harvard University Press, 1962).

Spicq, Ceslaus, *Esquisse d'une histoire de l'exégèse latine au Moyen Âge* (Paris: Vrin, 1944).

Stager, Lawrence E., 'Jerusalem ad the Garden of Eden', *Eretz-Israel: Archaeological, Historical and Geographical Studies*, 26 (1999), pp. 183–94.

Stange, Alfred, *Basiliken, Kuppelkirchen, Kathedralen: Das himmlische Jerusalem in der Sicht der Jahrhunderte* (Regensburg: F. Pustet, 1964).

——, *Das frühchristliche Kirchengebäude als Bild des Himmels* (Cologne: Comel, 1950).

Steele Hall, Thomas, *Ideas of Life and Matter: Studies in the History of General Physiology, 600 BC–1900 AD*, 2 vols (Chicago–London: University of Chicago Press, 1969).

Steer, Georg, 'Das *Compendium theologicae veritatis* des Hugo Ripelin von Strassburg', in Hoenen and Libera, eds, *Albertus Magnus und der Albertismus* (1995), pp. 133–54.

Steinhauser, Kenneth B., *The Apocalypse Commentary of Tyconius: A History of its Reception and Influence* (Frankfurt am Main–New York: P. Lang, 1987).

Stone, Michael E., *Armenian Apocrypha Relating to Adam and Eve* (Leiden–New York–Cologne: Brill, 1996).

——, *A History of the Literature of Adam and Eve* (Atlanta, GA: Scholars Press, 1992).

——, 'The Metamorphosis of Ezra: Jewish Apocalypse and Medieval Vision', *Journal of Theological Studies*, 33/1 (1982), pp. 1–18.

——, *Fourth Ezra: A Commentary on the Book of Fourth Ezra* (Minneapolis: Fortress, 1990).

Stookey, Lawrence H., 'The Gothic Cathedral as the Heavenly Jerusalem: Liturgical and Theological Sources', *Gesta*, 8 (1969), pp. 35–41.

Stopp, Klaus, 'Relation between the Circular Maps of the World of Hanns Rüst and Hans Sporer', *Imago Mundi*, 18 (1964), p. 81.

Stornajolo, Cosimo, *Le miniature della Topografia cristiana di Cosma Indicopleuste: Codice vaticano greco 699* (Milan: Hoepli, 1908).

Strijbosch, Clara, *The Seafaring Saint: Sources and Analogues of the Twelfth-Century Voyage of Saint Brendan*, transl. by Thea Summerfield (Dublin: Four Courts, 2000).

Stroll, Mary, 'The Twelfth-Century Apse Mosaic in San Clemente in Rome and its Enigmatic Inscription', *Storia e Civiltà*, 4/1–2 (1988), pp. 1–17.

Strube, Martha, *Die Illustrationen des Speculum Virginum* (Düsseldorf: Nolte, 1937).

Swain, Joseph Ward, 'The Theory of the Four Monarchies Opposition History under the Roman Empire', *Classical Philology*, 35/1 (1940), pp. 1–21.

Swanson, Jenny, 'The Glossa Ordinaria', in Gillian R. Evans, ed., *The Medieval Theologians* (Oxford: Blackwell, 2001), pp. 156–67.

Synan, Edward A., 'Albertus Magnus and the Sciences, in Weisheipl, ed., *Albertus Magnus and the Sciences* (1980), pp. 1–12.

Tardiola, Giuseppe, *Atlante fantastico del Medioevo* (Anzio: De Rubeis, 1990).

Taylor, Jerome, *The Origin and Early Life of Hugh of St. Victor: An Evaluation of the Tradition* (Notre Dame, IN: University of Notre Dame Press, 1957).

Taylor, John, 'Ranulf Higden', in Friedman, Mossler Figg, and others, eds, *Trade, Travel, and Exploration in the Middle Ages* (2000), pp. 252–4;

——, *The Universal Chronicle of Ranulf Higden* (Oxford: Oxford University Press, 1966).

Teixidor, Javier, 'Géographie du voyageur au Proche-Orient ancien', *Aula orientalis*, 7 (1989), pp. 105–15.

Terzoli, Riccardo, *Il tema della beatitudine nei padri siri: Presente e futuro della salvezza* (Brescia: Morcelliana, 1972).

Tesi, Mario, ed., *Monumenti di cartografia a Firenze (secc. X–XVII)* (Florence: Biblioteca Medicea Laurenziana, 1981).

Teske, Roland J., *Paradoxes of Time in Saint Augustine* (Milwaukee: Marquette University Press, 1996).

Thompson, Gunnar, *America's Oldest Map – 1414 AD: First Edition Draft of Technical Report* (Seattle: Misty Isles Press–The Argonauts, 1995).

Thorndike, Lynn, *A History of Magic and Experimental Science*, 8 vols (New York: Columbia University Press, 1934–58).

——, *University Records and Life in the Middle Ages* (New York: Columbia University Press, 1944).

Tobin, Thomas, *The Creation of Man: Philo and the History of Interpretation* (Washington, DC: Catholic Biblical Association of America, 1983).

Trapp, Joseph B., 'The Iconography of the Fall of Man', in C. A. Patrides, ed., *Approaches to Paradise Lost* (London: Edward Arnold, 1968), pp. 223–65.

Trédaniel, Guy, ed., *Guillaume Postel, 1581–1981* (Paris: Guy Trédaniel, 1985).

Trigg, Joseph W., 'Divine Deception and the Truthfulness of Scripture', in Kannengiesser and Petersen, eds, *Origen of Alexandria* (1988), pp. 147–64.

Tryggve, Kronholm, *Motifs from Genesis 1–11 in the Genuine Hymns of Ephrem the Syrian with Particular Reference to the Influence of Jewish Exegetical Tradition* (Lund: Gleerup, 1978).

Turner, Denys, *The Darkness of God: Negativity in Christian Mysticism* (Cambridge–New York: Cambridge University Press, 1995).

Uhden, Richard, 'Die Weltkarte des Isidorus von Sevilla', *Mnemosyne: Bibliotheca Classica Batavia*, 3rd series, 3 (1935–6), pp. 1–28.

——, 'Gervasius von Tilbury und die Ebstorfer Weltkarte', *Jahrbuch der Geographischen Gesellschaft zu Hannover* (1930), pp. 185–200.

——, 'Das Weltbild von Ebstorf', *Niedersachsen*, 33 (1928), pp. 179–83.

Vaccari, Alberto, 'La "teoria" esegetica antiochena', *Biblica*, 15 (1934), pp. 94–101.

Valois, Noel, *Guillaume d'Auvergne, évêque de Paris (1228–1249): Sa vie et ses ouvrages* (Paris: Irvington, 1980).

Van den Broeck, Roelof, *Studies in Gnosticism and Alexandrian Christianity* (Leiden–New York–Cologne: Brill, 1996).

Van der Horst, Pieter W., 'Philo and the Rabbis on Genesis: Similar Questions, Different Answers', in *Eratopokriseis: Early Christian Question-and-Answer Literature in Context* (Louvain: Peeters, 2004), pp. 55–70.

Van der Krogt, Peter, *Koeman's Atlantes Neerlandici*, III/A (MS 't Goy–Houten, The Netherlands: HES–De Graaf, 2003).

Van Meter, David C., 'Christian of Stavelot on Matthew 24:42, and the Tradition that the World will End on a March 25th', *Recherches de Théologie Ancienne et Médiévale*, 63 (1996), pp. 68–92.

Van Oort, Johannes, 'Augustinus und der Manichäismus', in Alois van Tongerloo, ed., *The Manichaean ΝΟΥΣ* (Louvain: International Association of Manichaean Studies, 1995), pp. 289–315.

Van Rompay, Lucas, 'Antiochene Biblical Interpretation: Greek and Syriac', in Judith Frishman and Van Rompay, eds, *The Book of Genesis in Jewish and Oriental Christian Interpretation: A Collection of Essays* (Louvain: Peeters, 1997), pp. 103–23.

Van Steenberghen, Fernand, *Aristotle in the West* (Louvain: Nauwelaerts, 1955).

VanderKam, James C., *Enoch and the Growth of an Apocalyptic Tradition* (Washington, DC: Catholic Biblical Association of America, 1984).

Vannier, Marie-Anne, *'Creatio', 'conversio', 'formatio', chez S. Augustin*, 2nd edn (Fribourg, Switzerland: Éditions Universitaires, 1997).

Vasoli, Cesare, 'Fonti albertine nel *Convivio* di Dante', in Hoenen and Libera, eds, *Albertus Magnus und der Albertismus* (1995), pp. 33–49.

Vaughan, Richard, *Matthew Paris* (Cambridge: Cambridge University Press, 1958).

Vázquez de Parga, Luis, 'Beato y el ambiente cultural de su época', in *Actas del Simposio … Beato de Liébana*, I (1978), pp. 33–46.

——, 'Un mapa desconocido de la serie de los "Beatos"', ibid., pp. 272–8.

Vezin, Gilberte, *L'Apocalypse et la fin des temps: Étude des influences égyptiennes et asiatiques sur les religions et les arts* (Paris: Éditions de la Revue Moderne, 1973).

Vignaud, Henry, *Americ Vespuce, 1451–1512: Sa biographie, sa vie, ses voyages, ses découvertes, l'attribution de son nom à l'Amérique, ses relations authentiques et contestées* (Paris: Leroux, 1917).

Von Löe, Paul, 'De vita et scriptis Beati Alberti Magni', *Analecta Bollandiana* 19 (1900), pp. 257–84.

Wallace, William A., 'The Scientific Methodology of St Albert the Great', in Gerbert Meyer and Albert Zimmermann, eds, *Albertus Magnus Doctor Universalis 1280/1980* (Mainz: Matthias Grünewald, 1980), pp. 385–407.

Wallace-Hadrill, David S., *Christian Antioch: A Study of Early Christian Thought* (Cambridge–New York: Cambridge University Press, 1982).

Wallis, Helen, ed., *Cartographical Innovations: An International Handbook of Mapping Terms to 1900* (Tring: Map Collector Publications–International Cartographic Association, 1987).

Wass, Meldon C., *The Infinite God and the Summa Fratris Alexandri* (Chicago: Franciscan Herald Press, 1964).

Wassermann, Dirk, *Dionysius der Kartäuser: Einführung in Werk und Gedankenwelt* (Salzburg: Institut für Anglistik und Amerikanistik, Universität Salzburg, 1996).

Watson al-Hamdani, Betty A., 'Beatus of Liébana versus Elipandus of Toledo and Beatus's Illuminated Commentary on the Apocalypse', in *Andalucia medieval* (Cordoba: Monte de Piedad y Caja de Ahorros de Cordoba, 1978), pp. 153–63.

Watson, Arthur, 'The *Speculum virginum* with Special Reference to the Tree of Jesse', *Speculum*, 3/4 (1928), pp. 447–9.

Watson, Gerard, 'Origen and the Literal Interpretation of Scripture', in Finan and Twomey, eds, *Scriptural Interpretation* (1995), pp. 75–84.

Wawrik, Franz, 'Österreichische kartographische Leistungen im 15. und 16. Jahrhundert', in Hamann and Grössing, eds, *Der Weg der Naturwissenschaft von Gmunden zu Kepler* (1988), pp. 103–18.

Weidlé, Wladimir, *The Baptism of Art: Notes on the Religion of the Catacomb Paintings* (Westminster: Dacre Press, 1950).

Weisheipl, James A., ed., *Albertus Magnus and the Sciences* (Toronto: Pontifical Institute of Medieval Studies, 1980).

—, 'The Life and Works of St. Albert the Great', ibid., pp. 13–51.

Westerdale Bowker, John, *The Targums and Rabbinic Literature: An Introduction to Jewish Interpretations of Scripture* (London: Cambridge University Press, 1969).

Westrem, Scott D., *The Hereford Map: A Transcription and Translation of the Legends with Commentary* (Turnhout: Brepols, 2001).

—, *Broader Horizons: A Study of Johannes Witte de Hese's Itinerarius and Medieval Travel Narratives* (Cambridge, MA: The Medieval Academy of America, 2001).

—, *Learning from Legends on the Bell Library Mappamundi* (Minneapolis: Associates of the James Ford Bell Library of the University of Minnesota, 2000).

—, 'Geography and Travel', in Peter Brown, ed., *A Companion to Chaucer* (Oxford: Blackwell, 2000), pp. 195–217.

—, 'Against Gog and Magog', in Sylvia Tomasch and Sealy Gilles, eds, *Text and Territory* (Philadelphia: University of Pennsylvania Press, 1998), pp. 54–75.

—, ed., *Discovering New Worlds: Essays on Medieval Exploration and Imagination* (New York–London: Garland, 1991).

White, Terence H., *The Book of Beasts: Being a Translation from a Latin Bestiary of the Twelfth Century* (London: Cape, 1954).

Whitfield, Peter, *The Image of the World: 20 Centuries of World Maps* (London: British Library, 1994).

Wieser, Franz R. von, *Die Weltkarte des Albertin de Virga aus dem Anfange des XV. Jahrhunderts in der Sammlung Figdor in Wien* (Innsbruck: Heinrich Schwick, 1912).

Wilcox, Helen, 'Milton and Genesis: Interpretation as Persuasion', in Luttikhuizen, ed., *Paradise Interpreted* (1999), pp. 197–208.

Wilke, Jürgen, ed., *Kloster und Bildung* (Göttingen: Vandenhöck und Ruprecht, forthcoming).

—, *Die Ebstorfer Weltkarte*, 2 vols (Bielefeld: Verlag für Regionalgeschichte, 2001).

Wilkinson, John, *Jerusalem Pilgrims before the Crusades* (Warminster: Aris and Phillips, 1977).

Willett, Tom W., *Eschatology in the Theodicies of 2 Baruch and 4 Ezra* (Sheffield: Sheffield Academic Press, 1989).

Williams, John, *The Illustrated Beatus: A Corpus of the Illustrations in the Commentary on the Apocalypse*, 5 vols (London: Harvey Miller, 1994–2003).

—, 'Isidore, Orosius and the Beatus Map', *Imago Mundi*, 49 (1997), pp. 7–32.

—, 'Purpose and Imagery in the Apocalypse Commentary of Beatus of Liébana', in Richard K. Emmerson and Bernard McGinn, eds, *The Apocalypse in the Middle Ages* (Ithaca–London: Cornell University Press, 1992), pp 217–33.

Wilms, Hieronymus, *Sant'Alberto Magno: Scienziato, filosofo e santo* (Bologna: Edizioni Studio Domenicano, 1992).

Winter, Heinrich, 'Notes on the World Map in the "Rudimentum novitiorum"', *Imago Mundi*, 9 (1952), p. 102.

Withers, Charles W. J., 'Geography, Enlightenment, and the Paradise Question', in Livingstone and Withers, eds, *Geography and Enlightenment* (1999), pp. 67–92.

Witte, Maria Magdalena, *Elias und Henoch als Exempel, typologische Figuren und apokalyptischen Zeugen: Zu Verbindungen von Literatur und Theologie im Mittelalter* (Frankfurt am Main–New York: P. Lang, 1987).

Woldan, Erich, 'A Circular, Copper-engraved, Medieval World Map', *Imago Mundi*, 11 (1954), pp. 13–6.

Wolf, Armin, 'Gervasius von Tilbury und die Welfen: Zugleich Bemerkungen zur Ebstorfer Weltkarte', in Bernd Schneidmüller, ed., *Die Welfen und ihr Braunschweiger Hof im hohen Mittelalter* (Wiesbaden: Harrassowitz, 1995), pp. 407–38.

—, 'Neues zur Ebstorfer Weltkarte: Entstehungszeit – Ursprungsort – Autorschaft', in Klaus Jaitner and Ingo Schwab, eds, *Das Benediktinerinnenkloster Ebstorf im Mittelalter* (Hildesheim: August Lax, 1988), pp. 75–109.

—, 'Die Ebstorfer Weltkarte als Denkmal eines Mittelalterlichen Welt- und Geschichtsbildes', *Geschichte in Wissenschaft und Unterricht*, 8 (1957), pp. 204–15.

Wood, Denis, *The Power of Maps* (New York: Guilford Press, 1992).

Woodward, David, ed., *The History of Cartography*, III, *Cartography in the European Renaissance* (Chicago, University of Chicago Press, forthcoming).

—, Catherine Delano-Smith, and Cordell D. K. Yee, *Plantejaments i objectius d'una història universal de la cartografia/Approaches and Challenges in a Worldwide History of Cartography* (Barcelona: Institut Cartogràfic de Catalunya, 2001).

—, with Herbert M. Howe, 'Roger Bacon on Geography and Cartography', in Hackett, ed., *Roger Bacon and the Sciences* (1997), pp. 199–222.

—, 'The Image of the Spherical Earth', *Yale Architectural Journal*, 25 (1991), pp. 4–15.

—, 'Roger Bacon's Terrestrial Coordinate System', *Annals of the Association of American Geographers*, 80 (1990), pp. 109–22.

—, 'Medieval *Mappaemundi*', in Harley and Woodward, eds, *The History of Cartography*, I (1987), pp. 286–370.

—, 'Reality, Symbolism, Time, and Space in Medieval World Maps', *Annals of the Association of American Geographers*, 75/4 (1985), pp. 510–21.

Wright, John K., *The Leardo Map of the World, 1452 or 1453, in the Collections of the American Geographical Society* (New York: American Geographical Society, 1928).

—, *The Geographical Lore of the Time of the Crusades: A Study in the History of Medieval Science and Tradition in Western Europe* (New York: American Geographical Society, 1925; repr. New York: Dover, 1965).

Wunderli, Peter, ed., *Reisen in reale und mythische Ferne: Reiseliteratur in Mittelalter und Renaissance* (Düsseldorf: Droste, 1993).

Wuttke, Heinrich, *Über Erdkunde und Karten des Mittelalters* (Leipzig: Mahler, 1853).

Zaganelli, Gioia, ed., *La lettera del Prete Gianni* (Milan: Luni, 2000).

Zaharopoulos, Dimitri Z., *Theodore of Mopsuestia on the Bible: A Study of his Old Testament Exegesis* (New York: Paulist Press, 1989).

Zahlten, Johannes, *Creatio mundi: Darstellungen des sechs Schöpfungstage und naturwissenschaftliches Weltbild im Mittelalter* (Stuttgart: Klett–Cotta, 1979).

Zinn, Grover A., '*Historia fundamentum est*: The Role of History in the Contemplative Life according to Hugh of St. Victor', in George H. Shriver, ed., *Contemporary Reflections on the Medieval Christian Tradition: Essays in Honor of Ray C. Petry* (Durham, NC: Duke University Press, 1974), pp. 135–58.

—, 'Hugh of Saint Victor and the Art of Memory', *Viator*, 5 (1974), pp. 211–34 .

Zurla, Placido, *Il mappamondo di Fra Mauro Camaldolese* (Venice: [n. pub.], 1806).

Index of Manuscripts

Page numbers in *italic* refer to illustrations; numbers in **bold** refer to colour plates.

Albi, Bibliothèque Municipale,
  MS 29 (Albi map) *138*
Basle, Universitätsbibliothek,
  MS A IX 99 (Guillaume Postel, *De paradisi terrestris loco*) 285-6, 335-6
Berlin, Staatsbibliothek Preussischer Kulturbesitz,
  MS Theol. Lat. Fol. 561 (Beatus of Liébana, *Commentarius in Apocalypsin*) 108, 122
Bischofszell (Canton of Thurgovia, Switzerland), Orstmuseum,
  'The Brendan map') *169*
Brussels, Bibliothèque Royale de Belgique/Koninklijke Bibliotheek van België,
  MS 3897–3919 (Guido of Pisa, *Liber historiarum*) *140*
  MS 9260 (Jean Mansel, *La Fleur des histoires*) *203*
Burgo de Osma, Archivo de la Catedral,
  MS 1 (Beatus of Liébana, *Commentarius in Apocalypsin*) **3a**, *108*
Caen, Bibliothèque Municipale,
  MS 11 and 14 (Samuel Bochart, *Paradisus, sive de loco paradisi terrestris; Du lieu du paradis terrestre*) 337
Cambridge, Parker Library, Corpus Christi College,
  MS 66 (Honorius Augustodunensis, *Imago mundi*) *143*
Copenhagen, Det Kongelige Bibliotek,
  G.K.S. 2020-4to (Lucan, *Pharsalia*) 86, 91, 116, 187
Deventer, Athenaeumbibliotheek,
  MS 81 (old 1791) (Sallust, *De bello Iugurthino*) *91*
Einsiedeln, Benediktinerabtei, Stiftsbibliothek,
  MS 263 (973) (miscellaneous) 90
Florence, Biblioteca Medicea Laurenziana,
  MS Med. Pal. 89 (Gregorio Dati, *La sfera*) 249
  Plut 27 Sin. 8 (Isidore, *Etymologiae*) 92
  MS Plut. IX.28 (Cosmas Indicopleustes, *Christian Topography*) 160, *160*, 184
Florence, Biblioteca Nazionale Centrale,
  MS Portolano 1 (Genoese map) *229*
Florence, Biblioteca Riccardiana,
  MS 881 (Guido of Pisa, *Liber historiarum*) 86, 140, *140*, 156
Genoa, Biblioteca Durazzo Giustiniani,
  MS A IX 9 (Lambert of St-Omer, *Liber floridus*) **5b**, *144*
Girona, Museu de la Catedral,
  Num. Inv. 7 (11) (Beatus of Liébana, *Commentarius in Apocalypsin*) 123, 124
Hamburg, Staats- und Universitätsbibliothek,
  MS Theol. 2029 (Peter of Poitiers, *Compendium historiae in genealogia Christi*) 83
Lincoln Cathedral,
  Hereford map **7**, *146*
Lisbon, Arquivo Nacional da Torre do Tombo,
  MS 160 (Beatus of Liébana, *Commentarius in Apocalypsin*) *111*
London, British Library,
  Add. MS 10049 (Jerome, *Liber locorum*) *141*
  Add. MS 11695 (Beatus of Liébana, *Commentarius in Apocalypsin*) **3b**, *113*
  Add. MS 14788 (*Biblia sacra latina*) 34
  Add. MS 18850 (Bedford Book of Hours) 73
  Add. MS 24070 (Gmunden, *Tabulae astronomiae*) *231*
  Add. MS 27376 (Marino Sanudo, *Liber secretorum fidelium crucis super Terrae Sanctae recuperatione et conservatione ...*) 200
  Add. MS 28106 (Stavelot Bible) 70
  Add. MS 28681 (Psalter) **8**, *148*
  Arundel MS 44 (*Speculum virginum*) 74
  Burney MS 3 (*Biblia*) **1**
  Cotton MS Tiberius B.V. (Dionysius of Alexandria, *Periegesis*, Latin transl. by Priscian) **6**, *139*
  Harley MS 218 (Honorius Augustodunensis, *Imago mundi*) 162
  Harley MS 2799 (Arnstein Bible) 167
  Harley MS 3954 (*The Book of Sir John Mandeville*) 54, 55
  Royal MS 14.B.IX (Peter of Poitier, *Compendium historiae in genealogia Christi*) 77, 79
  Royal MS 14.C.IX (Ranulf Higden, *Polychronicon*) **4**, *134*
  Royal MS 14.C.XII (Ranulf Higden, *Polychronicon*) *136*
London, College of Arms,
  Muniment Room 18/19 (Evesham map) **5**
London, Lambeth Palace Library,
  MS 371 (text and map inserted in a copy of Nennius, *Historia Britonum*) 131
Macon, Bibliothèque Municipale,
  MS Franç. 2 (Augustine, *De civitate Dei*) **10**
Madrid, Biblioteca de la Real Academia de la Historia,
  MS 76 (Isidore, *Etymologiae*) 92
Manchester, John Rylands University Library,
  MS Lat. 8 (Beatus of Liébana, *Commentarius in Apocalypsin*) 115, 124
Milan, Biblioteca Ambrosiana,
  MS F. sup. 150 (Beatus of Liébana, *Commentarius in Apocalypsin*) 123
Modena, Biblioteca Estense,
  C.G.A. (Catalan Estense map) 227
Munich, Bayerische Staatsbibliothek,
  MS Cgm. 564 (Peter of Poitiers, *Compendium historiae in genealogia Christi*) 83
  MS Clm. 721 (John of Udine, *Compilatio librorum historialium totius bibliae ab Adam ad Christum*) 214, 248
  MS Clm. 10058 (Isidore, *Etymologiae*) 154, 156
New York, Pierpont Morgan Library,
  MS 644 (Beatus of Liébana, *Commentarius in Apocalypsin*) *110*
Oxford, Bodleian Library,
  MS Laud. Misc. 674 (tables of latitude and longitude attr. to William Worcester) 250-51
Paris, Bibliothèque Nationale de France,
  Cartes et Planes, Rés Ge AA 562 (nautical chart attr. to Christopher Columbus) **14**, *216*
  Cartes et Planes, Rés. Ge B 1118 (Carte Pisane) 199, 245
  MS Franç. 9140 (Bartholomaeus Anglicus, *Livre des propriétés des choses*, French transl. by Jean Corbechon) 269-70, 281
  MS Lat. 1366 (Beatus of Liébana, *Commentarius in Apocalypsin*) 124
  MS Lat. 4126 (Ranulf Higden, *Polychronicon*) *171*
  MS Lat. 5510 (William of Tripoli, *De statu Sarracenorum*) 165
  MS Lat. 8878 (Beatus of Liébana, *Commentarius in Apocalypsin*) 22
  MS Lat. 14435 (Peter of Poitiers, *Compendium historiae in genealogia Christi*) 77, 83
  MS Lat. 16198 (Ptolemy, *De dispositione spherae*) 188
  MS Lat. 16679 (Macrobius, *Commentarii in Somnium Scipionis*) 90
Paris, Bibliothèque SteGeneviève,
  MS 782 (*Les Grandes Chroniques du temps de Charles V*) *164*
Parma, Biblioteca Palatina,
  MS 1614 (Vesconte Maggiolo, *Atlante nautico*) **12**
Mount Sinai, St Catherine,
  MS Gr. 1186 (Cosmas Indicopleustes, *Christian Topography*) 160, *160*, 184
Stuttgart, Württembergische Landesbibliothek,
  MS Theol. Fol. 100 (John of Udine, *Compilatio librorum historialium totius bibliae ab Adam ad Christum*) 213
Turin, Biblioteca Nazionale Universitaria,
  MS I.II.1 (Beatus of Liébana, *Commentarius in Apocalypsin*) **2b**, *114*
Utrecht, Rijksuniversiteit Bibliotheek,
  MS 737 (Girard of Antwerp, *Historia figuralis ab origine mundi usque ad ... 1272*) *168*
Vatican City, Biblioteca Apostolica Vaticana,
  Borgiano XVI (Borgia map) *210*
  MS Pal. Lat. 1362b (map by Andreas Walsperger) **13**, *234*
  MS Vat. Gr. 699 (Cosmas Indicopleustes, *Christian Topography*) *161*
  MS Vat. Lat. 5698, fifteenth century (Ptolemy, *Geography*) 199, 245
  MS Vat. Lat 6018 (miscellaneous) 98
Venice, Biblioteca Nazionale Marciana,
  Fra Mauro map **12b**, *236*
  MS Fondo Ant. Z.76 (4783) (map by Andrea Bianco) *208*
Vercelli, Archivio e Biblioteca Capitolare,
  (Vercelli map) *133*
Verona, Biblioteca Civica,
  MS 3119 (Giovanni Leardo, *Mapa mondi/Figura mondi*) map **2a**, *17*
Vicenza, Biblioteca Civica Bertoliana,
  MS 598a (Giovanni Leardo, *Mapa mondi/Figura mondi*) *209*
Wolfenbüttel, Herzog August Bibliothek,
  MS Helmst. 442 (Bartholomaeus Anglicus, *De proprietatibus rerum; Chronica mundi*) 248

General Index

Page numbers in *italic* refer to illustrations; numbers in **bold** refer to colour plates.

Abel 13, 40, 314
Abraham 68, 76, 79, 98, 134, 286, 301, 323, 331, 371
Adam 12, 13, 32, 36, 40, 41, 53, 55
    Augustine on 46, 56
    and Beatus of Liébana's map 111, 113–14, 115
    and Bedford's *Scripture Chronology* 306
    in Beroaldus's map 296
    and Bochart's map 307
    and Calvin's map 273, 274, 284
    and the climate of paradise 178
    creation of 83
    in Dante's *Divina Commedia* 183
    in the Ebstorf map 150
    in Ephrem the Syrian's *Hymns on Paradise* 162
    in the Evesham map 136
    Expulsion and the whole earth paradise doctrine 264–5
    in Giovanni di Paolo's panel painting 221, 222
    and the Hereford map 145, 149
    and the Holy Land paradise argument 322, 323, 324
    in Honoré's map 292
    in Hopkinson's map 298
    and Jehovah's Witnesses 353
    in Kircher's map 314, *316*, 316–17
    and the location of paradise 196
    in Fra Mauro's map 240
    Origen on 38–9
    and paradise on Christianized maps 91, 93
    and paradise in time 62, 63, 64, 65–6, 67, 68, 69, 70, 73, 75, 76, 77, 78, 79
    Peter Lombard on 50
    Philo on 37
    in the Psalter map 164
    and the Vatican (Pseudo-Isidorean) map 97, 98, 99
    in Raleigh's *History of the World* 301, 303
    in Regnault's map 293–4
    and Seth 55
    and Visscher's map 305
    and the waning of the idea of corporeal perfection 259–60, 261
    and Ward's *The Garden of Eden* 360
    in the Wittenberg woodcut 268–9
Adelard of Bath 171
Adrichem, Christiaan van 330, 331, 340
Africa
    and Adam 111
    in fifteenth-century *mappae mundi* 200, 201, 205, 207, 217
    in a map from a thirteenth-century manuscript of Honorius Augustodunensis's *Imago mundi* 162
    in the Lambeth Palace (Reading) map 131
    in Leardo's map 16, 17, 18
    as the location of paradise 218–30, 360
    in the Psalter (London) map 148
    and the origins of mankind 359
    in Virga's map 221
Ailly, Pierre de, 195, 244
the air, and the location of paradise 172, 174–5, *175*, 182, 192
Air Mauritius advertisement *10*, 11
Albert the Great 24, 179–81, 183, 188–9, 192, 196, 197, 218, 224, 239, 243, 252
Albi map *138*, 138–9
Alexander the Great 51, 57, 95, 128, 129, 131, 134, 136, 141, 152, 290
    in the Ebstorf map 150
    in the Hereford map *153*, 153

Alexander of Hales 174–5, *175*, 182, 183, 187, 192, 195, 259
Alphaeus of Cleophas 79
altitude of paradise 174–5, 194, 274
Ambrose 98, 105
Americas, and the origins of mankind 359, 360
Andrew, apostle 79
Angelo, Jacopo 199, 255, 278
angels
    four angels of the Sawley map 141–4
    in the Hereford map 149
Anglo-Saxon (Cotton) map **6**, *139*, 139–40, 156
Antioch, early Christian church of 39
the Antipodes, in Beatus of Liébana's map 111
the Apocalypse 105–7, 121–2, 144–6, 157, 359
    see also Beatus of Liébana; Second Coming of Christ
Apocalypse of Paul 56
the apostles, in Beatus of Liébana's map 109, 110, 113
Aquinas, Thomas *see* Thomas Aquinas
Arabic astronomy 246
    and the equatorial regions 173
    translations of 170–1
archaeological ethnographers 359
Arctic paradise 285–6, 352–3
Arin (mythical Indian city) 176, 228, 232, 233
Aristotle 170–1, 172, 173, 174, 175, 178, 180, 181–2, 186, 191, 193, 279
    and Fra Mauro's map 235
    and Peter of Poitiers 77
Armenia
    and Noah's Ark 285, 286, 314
    paradise in 317–22, 342, 343, 361
    and Raleigh's *History of the World* 301
Arnstein Bible maps *166*, *167*, 168, 169
Asia
    in Calvin's map 276–7
    and fifteenth-century *mappae mundi* 199, 201–2, 212
    in the Genoese map 230
    Jerome map of 140
    in the Lambeth Palace (Reading) map 131
    in Leardo's map 16, 17, 18
    in T-O maps 91
    in the Vatican (Pseudo-Isidorean) map 104
    in Virga's map 221
Assyria 126, 291
Assyriology 347–51, 352, 361
astrology, and the *Dies Aegyptiaci* 98
astronomical geography, and the location of paradise 170–9, 191
Astruc, Jean 345
Athens, in fifteenth-century *mappae mundi* 207
Atlantic coast
    and the African paradise concept 219, 230
    and fifteenth-century *mappae mundi* 207, 216, 217
    and nautical charts 199
Atlantis 226, 360
Atlas Mountains 211–12, 219
Augustine of Hippo, St 33, 36, 39, 44–7, 48, 49, 50, 51, 57–8, 80–1, 119, 126, 129, 174, 181, 259, 284, 342, 352, 367
    and Beatus of Liébana's map 105, 106, 111, 113, 114
    and the Western concept of an earthly paradise 160, 163
    on Enoch 56
    and fifteenth-century *mappae mundi* 202, 205
    in the Hereford map 131
    map from *De civitate Dei* **10**, 202
    and Fra Mauro's map 237, 239
    and the medieval mapping of paradise 88, 95, 102

and Neo Platonism 261
and Nicholas of Lyre 193
and paradise in time 62, 63, 64, 65, 66–7, 68–9, 70
Averroes 178
Avicenna 175–6, 177, 180, 181, 194
Azores 360

Babel, Tower of 127, 135, 136
    in the Ebstorf map 150
    in the Evesham map 136, 137
    in fifteenth-century *mappae mundi* 203, 207
    and the Holy Land paradise argument 324
    and sacred geography 288
    in Visscher's map 305
Babylon 128, 138, 179
    Babylonian paradise and Assyriology 347–51
    in the Estense map 228
    in fifteenth-century *mappae mundi* 202
    and Raleigh's *History of the World* 303
    in the Vatican (Pseudo-Isidorean) map 100
Bacon, Roger 24, 176–9, 183, 187–8, 192, 195, 218, 224, 255
Bahrain 351
Balaam 330
Balbus, Iohannes 192, 243
BarCepha, Moses 367
Barnabas, apostle 79
Bartholomaeus Anglicus 202, 205, 269–70
Bartholomew, apostle 79
AlBattani, *Astronomy* 171
Beatus of Liébana, *In Apocalypsin* **2b**, **3a**, **3b**, 22, 86, 104–16, *108*, *109*, *110*, 111, *112*, *113*, *114*, *115*, 117, 122–3, 124, 238
Beauvais, Vincent de 67
Beazley, Raymond, *The Dawn of Modern Geography* 23, 24
Beck, Julian 15
Bede 47, 48–9, 50, 51, 59, 65, 67, 69, 82, 88, 102, 170, 174, 179, 193, 195, 197
    and fifteenth-century *mappae mundi* 205
    and Fra Mauro's map 239
Bedford, Arthur, *The Place of Paradise 306*, 306
Bedford Book of Hours 72
Behn, Aphra 191
Benedictus Pererius 265
Benjamin of Tudela 180
Bennett Durand, Dana 250–1
Bernardino, Saint 196, 197
Beroaldus, Matthaeus, map of paradise 295–6, *296*, 301, 336
Bethlehem
    in the Evesham map 137
    in the Vatican (Pseudo-Isidorean) map 100
    in the Vercelli map 135
Bianco, Andrea, map (1463) 207–8, *208*, 219, 248
the Bible
    and Assyriology 347–51
    Authorized (King James) version 33–4, 41
    Book of Daniel 65, *269*, 269
    Book of Kings 56
    Book of Revelation 12, 56, 65, 69, 79, 95, 104–14
    Book of Wisdom 150
    Corinthians 56
    and early world maps 19, 94–104
    and fringe thinking 352
    Garden of Eden (or paradise) story *see* Genesis
    Geneva Bible 284, *292*, 292–3
    and the Holy Land paradise argument 323, 324–5
    and the location of paradise 176–7, 192–3, 194–5
    and mainstream theologians 361

392   Mapping Paradise: A History of Heaven on Earth

modern biblical scholars and the location of
Eden 343–6
Robertus de Bello's Bible **1**
and sacred geography 288
Song of Solomon 326, 327
Stavelot Bible 70, *71*
Vulgate 33, 47, 70, 193, 274
Wittenberg Bible **15**, 268, *269*, 269
see also Genesis
Biel, Gabriel 259, 260, 279
Black Sea, and nautical charts 199
Bochart, Samuel 25, 310, 319, 321, 337–8
Edenis seu paradisi terrestris situs 307–9, *308*
Boethius 319
Bonaventure 175, 187, 218
Book of Enoch 56
Book of Jubilees 88, 118, 356, 360
book production, and fifteenth-century *mappae mundi* 198
Book of Revelation 12, 56, 65, 69, 79, 95, 104–14
and the Ebstorf map 152
and the Sawley map 143–4
The Book of Sir John Mandeville 54, *55*, 241
Borgia, cardinal Stefano, map of *210*, 210, *211*, 211–12, *212*
Bosio, Giacomo 76
Braga, Corin 27
Brandis, Lucas 78, 205
Brendan map (*Navigatio Sancti Brendani*) 168, *169*
Brendan, Saint 52–3, 57, 226, 240
British Isles, in the Evesham map 137
Brothers, Richard 352, 363
Brown, Lloyd Arnold, *The Story of Maps* 24
Bünting, Heinrich, *Itinerarium Sacrae Scripturae* 323–4, 340

Cain 13, 40, 66, 127, 305–6, 314
Calais, in the Evesham map 137
Callistus, Nicephorus 318
Calmet, Augustin, *Carte du paradis terrestre 320*, 321–2, 322, 339–40, 343, 366, 367, 370, 372–3
Calvin, John 25, 254, 270–7, 282–3, 317, 319, 336, 342, 352, 365, 369
and Bochart's map 307
and the four rivers of paradise 272, 274–5, 276–7, 292–3, *295*, 295, 303
and the Holy Land paradise argument 322, 323
and Kircher 314
map of the location of Eden *272, 273*, 292
map to illustrate *Commentary on Genesis* 272, 273–7, 284, 291
and Raleigh's *History of the World* 301
and Regnault's map 293
and the single-river Eden 296, 298
Caminha, Pêro Vaz de 240
Campbell, Tony 29
Canaan 128, 322, 324, 327, 329, 330, *331*, 332, 333, 348
Canary Islands 232, 233, 360
Carpini, John of Plano *see* John of Plano Carpini
Carte Pisane 199
Carthage 126
in the Albi map 138
on Christianized maps 91, 94
in fifteenth-century *mappae mundi* 202
in the Vatican (Pseudo-Isidorean) map 104
Carver, Marmaduke 262, 280, 366, 372
map of paradise *318*, 318–19
Caspian Mountains 207
Caspian Sea 356, 366
and fifteenth-century *mappae mundi* 201–2, 215
and Noah's Ark 285
and William of Rubruck 246
Cassiodorus 89
Catholic Church
biblical scholarship 345, 362
views on the Garden of Eden 262, 265
see also Christian Church
Catholicon 192
Cecco d'Ascoli 176
Celan, Paul 153
Chaldaea 205, 286, 294, 297, 349

Cham, son of Noah 78, 79, 89, 131, 207, 258
Champeaux, Gérard de 24
Chekin, Leonard 99, 119–20
Cheremon 196
Cherubim, as guards of Eden 36, 40, 41, 75, 170, 176, 181, 192, 193, 196, 197, 254, 262, 265, 267, 270, 305, 316, 317, 350
China 12
in the Estense map 228
and fifteenth-century *mappae mundi* 201, 202, 212
in the Genoese map 230
Chosroes II, Persian king 55
Christ *see* Jesus Christ
Christian Church 12, 19, 82–3
in Beatus of Liébana's map 108–13, 115–16
concept of paradise in the Christian East 160–3
and fifteenth-century *mappae mundi* 201
fringe thinking 351–2
and the Holy Land paradise argument 323
and Hugh of St Victor's mental map 126, 127
missionaries 288
modern Christian map of paradise *368*, *369*, 371
and sacred geography 288–91
and Walsperger's map 233–4
see also Catholic Church
Christian of Stavelot 68
Christianized maps, paradise on 88–94, 118
Christina, Queen of Sweden 310
Chrysostom, John, Bishop of Constantinople 39, 76
the Church *see* Catholic Church; Christian Church
Cicero 86, 90, 91, 111
classical Antiquity, and fifteenth-century *mappae mundi* 201, 205
classical sites, in the Vatican (Pseudo-Isidorean) map 103, 104
climate in paradise 163–70, 194, 195
see also zonal maps
Colchis 322
Coli, Edoardo, *Il paradiso terrestre dantesco* 21–3, *22*, 24, 30, *182*, 189–90, 360, 366–7, 373
Cologne, in the Evesham map 137
Columbus, Christopher 11, 17, 21, 23, 25, 52, 217, 240–1, 242
nautical chart (attr.) *216*, **14**
Comestor, Peter 67, 69–70, 76–7, 195, 205
comparative philologists 359
Compendium theologicae veritatis 192
Constantine I, Emperor 55, 88, 196
Constantinople, fall of 197
Constantius III 75
Cook, Captain James 357
coordinates, paradise and tables of 230–40, 255
corporeal perfection, waning of the idea of 259–61
Cosmas Indicopleustes 26, 31
Christian Togography 160–2, *161*, 163, 184
Cotton (Anglo-Saxon) map **6**, *139*, 139–40, 156
Crates of Mellos 181, 250
quadripartite division of the globe *225*
Crete 128
Crivellari, Giuseppe 29
Crosby, Alfred W. 246
cuneiform texts 347, 348
Curtis, A.P., *The Land of Eden and Havilah* 353, *357*, 363
Cush, biblical land 13, *14*, 35, 37, 218, 272, 275, 313, 319, 322, 343, 350, 354, 357
Cyrus, Persian king 128

D'Ailly, Pierre *see* Ailly, Pierre de
Dalché, Patrick Gautier 31, 116–17, 121, 154, 245
Damascene Fields 78, 83, 196, 262, 332
Damascus 352
Daniel 65, 98, 126
biblical Book of 65, *269*, 269
Dante Alighieri 21–2, 23, 24, 25, 67, 189–90
Divina Commedia 182–3, 195, 223
Danube, river 40, 138, 212, 305, 317
David, King 68, 79, 103
Day of Judgement *see* Last Judgement

Dead Sea 319, 333, 334
in the Psalter (London) map *148*
Decius, Roman emperor 196
Decretum Gelasianum 99
Delano-Smith, Catherine 25, 31
Delitzsch, Friedrich 362
Babel und Bibel lectures 348
map of Babylon 348, *349*, 352
Delos island, in the Sawley map 143
Delumeau, Jean 26, 340–1
Diamond Mountains 228, 229–30
Dias, Bartholomeu 215
Diderot, Denis 361, 364
Dies Aegyptiaci 98, 98
Dietrich, Manfred 352, 363, 373
map of southern Mesopotamia *351*, 351
Dionysius of Alexandria 139
Dionysius the Carthusian 259, 279
Don, river, and fifteenth-century *mappae mundi* 205
Duchy of Cornwall map 152
Duncan, Joseph E., *Milton's Earthly Paradise* 24–5, 26, 279, 280, 281
Duns Scotus, John 193–5, 243, 259, 260–1, 279
Durandus of St-Pourçain 192, 243
Durazzo, Pompeo 21, 30, 360, 366

earthly paradise (*equivalent to* Garden of Eden) concept 191–2, 191–3
and astronomical geography 170–9
and changing cartography 257–8
and changing theology 258–61
and the Christian East 160–3
and climate 163–70
decline of belief in 254
and tables of coordinates 230–40
see also Garden of Eden and Genesis
the east
and fifteenth-century *mappae mundi* 199, 202, 207, 209, 212–14
as the location of paradise 125, 127, 144, 172, 174, 179, 181, 193–5
Easter, calculating the date of 29
Eastern Christian Church, concept of paradise 160–3
Ebstorf map **9**, 29–30, 63, 116, 128, 149–52, *150*, *151*, *152*, 158–9, 163
Eden *see* Garden of Eden and earthly paradise
Eden Project 15
Edson, Evelyn 29
Egypt
and Assyriology 148
and the location of paradise 172, 179–80
in the Vatican (Pseudo-Isidorean) map 100, 101, 104
Egyptians 12
Elijah, prophet 55, 56, 57, 61, 76, 146, 152, 196, 212, 239, 263, 331
Eliot, T.S. 62
Elysian Fields 164, 185, 196, 313
emanationism 44–5
Endeavour (space shuttle) 18, 19
Engel, Moritz, map of paradise 352, *354*, 363
Enlightenment 27, 28
Enoch, Old Testament patriarch 55–6, 57, 61, 76, 146, 152, 196, 212, 239, 263
Ephrem the Syrian 35, 271
Hymns on Paradise 162–3, 184
Epiphanius of Salamis 40–1, 43
epochal zones 129–31
paradise 131–7
sixth age 130–1, 144–5
equatorial zone
and the location of paradise 172, 173–6, 180–2, 192, 193, 194, 198, 224, 259
maps 166
and the mathematics of paradise *177*, 177–8
and Fra Mauro's map 239
paradise in equatorial Africa 226–30
and Thomas Aquinas 181–2
Eraclius, Emperor 55
Eratosthenes 172–3
Estense map 226–30, *227*, *228*
Ethiopia
and the African paradise concept 218–19

in the Estense map 230
in fifteenth-century *mappae mundi* 205, 207, 208
Etna, Mount 193
Eucherius, Bishop of Lyons 99–100, 120
Euphrates, river 13, 46, 47, 49, 102, 115, 126, 136, 139, 150, 180, 197, 198, 263
  and Armenian rivers 317, 319, 321, 322
  and Assyriology 348, 349, 350
  and Bochart's map 307–9
  confluence with the Tigris 367, 370–1
  and fifteenth-century *mappae mundi* 207
  and the Holy Land paradise argument 327, 334
  and Huet's map 309, 312
  and Mesopotamia 263, 264, 272, 275, 291, 292, 293, 295, 298
  and the Nile 179–80
  and paradise in the Bible 34, 36, 40, 41
  and Raleigh's *History of the World* 301–3
  and Van Til's map 313
Europe
  and fifteenth-century *mappae mundi* 207
  in in a map from a thirteenth-century manuscript of Honorius Augustodunensis's *Imago mundi* 162
  in the Lambeth Palace (Reading) map 131
  in Leardo's map 16, 17–18
  in Virga's map 221
Eusebius, Bishop of Caesarea 65, 88, 103
Eve 12, 13, 32, 36, 38, 40, 46, 53, 56
  and Beatus of Liébana's map 113–14
  and Bedford's *Scripture Chronology* 306
  in Beroaldus's map 296
  and Bochart's map 307
  and Calvin's map 284
  in the Ebstorf map 150
  in Ephrem the Syrian's *Hymns on Paradise* 162
  in the Estense map 228–30
  Expulsion and the whole earth paradise doctrine 265
  in Giovanni di Paolo's panel painting 221, 222
  and the Hereford map 145
  in Honoré's map 292
  in Hopkinson's map 298
  and Jehovah's Witnesses 353
  in Kircher's map 314, *316*, 316, 317
  in Fra Mauro's map 240
  and paradise in time 62, 63, 64, 65, 66, 67, 68, 69, 70, 73, 77, 79
  Peter Lombard on 50
  in the Psalter (London) map 164
  in Raleigh's *History of the World* 301, 303
  in Regnault's map 293–4
  and Seth 55
  and the Vatican (Pseudo-Isidorean) map 98
  and the waning of the idea of corporeal perfection 259–60
  and Ward's *The Garden of Eden* 360
  in the Wittenberg woodcut 268–9
Evesham map **5a**, 136–8, *137*, 156
Expositio [Descriptio] totius mundi 89
Ezekiel, prophet 29, 50, 69, 262
Ezra, prophet 56–7, 61

the Fall 183, 262–3, 267, 268, 269, 270, 271, 273, 274, 275, 316–17
  and the Holy Land paradise argument 323, 324, 330
Far East, and fifteenth-century *mappae mundi* 199, 202–9, 216
Al Farghani, *On the Elements of Astronomy* 171, 173, 175, 231
fifteenth-century *mappae mundi* 198–218
  Africa in 200, 201, 205, 207, 217
  Asia in 199, 201–2, 212
  disorientated maps 210–13
  and geographical discoveries 240–2
  mapping paradise in Africa 218–30
  see also Renaissance maps
Flavius Josephus 197
the Flood 179, 182, 191, 192, 194–5, 324
  and Assyriology 348
  and the four rivers of Eden 268, 313, 322

and the Garden of Eden 254, 262, 263, 264, 266–70, 271, 275, 359
and the Holy Land paradise argument 323, 326, 330, 333
and paradise in Armenia 317, 319, 322
and Raleigh's *History of the World* 300
see also Noah's Ark
Forlani, Paolo 290
Fortunate Islands 145, 226, 227, 238, 313
four rivers of paradise 33–4, 35, 36, 37, 40, 46, 49, 56, 75, 76, 89, 102, 114, 126
  and the African paradise concept 226
  in the Albi map 138, 139
  and Armenia 317–22
  and Assyriology 347, 348, 349–50, 351
  in the Brendan map 169
  and Calvin's map 272, 274–5, 276–7, 292–3, 295, 295, 303
  in the Ebstorf map 150
  in the Estense map 230
  in the Evesham map 136
  and fifteenth-century *mappae mundi* 214, 215
  and the Flood 268, 313, 322
  General Gordon's map of 353–4, *355*
  and geographical discoveries 242
  in Guido of Pisa's *Liber historiarum* 140
  and Herder 359
  in the Higden map 136
  and the Holy Land paradise argument 325, 326–7, 332, 333–4
  in Hopkinson's map 298
  and Huet's map 309–12
  and the invisible paradise on maps 138
  and Kircher's underground theory 313–16
  and the location of paradise 181, 195, 198
  in Fra Mauro's map 240
  and Mesopotamia 263, 268, 292–3, 295–303, 305
  and paradise in Mesopotamia 291
  and Raleigh's *History of the World* 301–3
  in Rohl's *Legend: The Genesis of Civilization* 13, *14*
  and the southern paradise concept 221, 222
  in Spedicato's map 357
  in Terry's map 256, *359*
  in Van Til's map 313
  in the Vercelli map 133
  in Walsperger's map 233
  and the whole-earth paradise doctrine 264, 267
Fourth Ezra 56
Fra Mauro *see* Mauro, Fra
France, Postel and the Moluccan meridian of paradise 286, 288
fringe thinking 351–61

Gabriel, Archangel 66
Gale, Thomas 284, 291, 318
Ganges, river 13, 36, 40, 46, 47, 49, 102, 126, 133, 136, 198, 318, 359
  in the Albi map 138
  in the Ebstorf map 150
  and fifteenth-century *mappae mundi* 207, 212
  in the Genoese map (1457) 230
  in the Jerome map of Palestine 140, 141
  in Fra Mauro's map 239
Garden of Eden (*equivalent to* earthly paradise)
  as an African paradise 218–20
  and Augustine's reading of Genesis 44–7
  in Beatus of Liébana's map 105, 113–16
  in Genesis 32–41
  Calvin's map of the location of 272, 273–7, 284
  and changing theology 258–61
  and Christianity 12, 27
  on Christianized maps 88–94
  climate 164, 165, 194, 195
  in Dante's *Divina Commedia* 183
  Dante's vision of the 21–2
  decline of belief in earthly 254, 261–4
  in the Ebstorf map 149–50, 163
  in Ephrem the Syrian's *Hymns on Paradise* 162–3
  and the equator 172, 173–6, 177–8, 180–2
  in the Estense map 228–30
  in the Evesham map 136

and fifteenth-century *mappae mundi* 207, 209, 212
and the Flood 254, 262, 263, 264, 266–70, 271, 275
and fringe thinking 351–61
and geographical discoveries 240–2
in the Hereford map 147, 149, 163
in the Higden map 136
and the Holy Land paradise argument 322–34
in Jerusalem 352
in Leardo's map 18
lists of implausible locations 365–71
in Fra Mauro's map 239–40
in medieval maps 24, 87, 160
medieval legends 51–7
and Mesopotamia 291–317
and modern biblical scholarship 343–6
naming the place 47–51
and paradise in time 62, 65–6, 69–80
in Piccolomini's *Dialogue* 197
and Renaissance maps 22, 25, 27, 257–8
in Rohl's *Legend: The Genesis of Civilization* 12–13, *14*
and tables of coordinates 230–40
in the Vatican (Pseudo-Isidorean) map 100, 102
in the Vercelli map 132, 133, 134
wall of fire surrounding 134, 144, 149, 150, 163, 203
in Walsperger's map 234
and the waning of the idea of corporeal perfection 259–61
and the whole-earth paradise doctrine 264–6
in Wright's map 343, *344*
and zonal maps 166
see also Cherubim; earthly paradise; the Fall; four rivers of paradise; Genesis; Tree of the Knowledge of Good and Evil; Tree of Life
Garden of Gethsemane 151
Garden of the Hesperides 97, 102, 313
Gastaldi, Giacomo 285, 288
Gautier (Gossouin) de Metz, *L'Image du monde* 91, 93
Gelasius, Pope 99
Genesis
  and the African paradise concept 218
  and Calvin's map of the location of Eden 272, 273, 275
  and the Holy Land paradise argument 325, 333–4
  and Jehovah's Witnesses 353
  and Kircher's underground theory 314
  and modern biblical scholarship 343–5
  paradise (or Garden of Eden) story in 32–41, 34, 69–76, 73
  Robertus de Bello's Bible **1**
  and the single-river Eden 295–6
  and Terry's map 356
  see also earthly paradise; Garden of Eden
Genoese map (1457) *229*, 230
geodetic measurement 19
geographical discoveries 240–2
Geographical Information Systems (GIS) 27–8, 358
Gerard of Cremona 171
Germanus, Nicholaus 255
Germany, paradise in 356, *358*
Gibraltar, Straits of 127
Gihon, biblical river 13, *14*, 46, 49, 133, 138, 140, 263, 275
  and Armenian rivers 317, 322
  and Assyriology 348, 349–50
  and Curtis's map 354
  in fifteenth-century *mappae mundi* 199, 215
  and the Holy Land paradise argument 327, 333
  and Huet's map *311*, 312
  and the location of paradise 197
  in Fra Mauro's map 239
  and Mesopotamia 295, 307, 314
  and paradise in the Bible 34, 35, 40, 41
  in Spedicato's map 357
  and Van Til's map 313
Gilson, Etienne 187, 189, 191

Giovanni di Paolo, *Creation of the World and Expulsion of Adam and Eve from Paradise 221*, 221, 222–3, 226, 249
Girard of Antwerp, *Tabulata Biblia* 166–8, *168*, 169
GIS *see* Geographical Information Systems
Glossa ordinaria 47, 49–50, 150, 170, 174, 193
Gmunden, Iohannes de, tables of latitude and longitude *231*, *232*, 232–3, 234, 235
gnostics 44, 57
Godfrey of Viterbo 52
Gog and Magog 95, 119, 129, 131, 140, 146, 152, 199, 248
　　and the African paradise concept 219, 226, 230
　　and Beatus of Liébana's map 113
　　in fifteenth-century *mappae mundi* 207, 208, 212, 217
　　and Fra Mauro's map 237, 238
　　in the Sawley map 141
　　in Walsperger's map 233
　　in the Wittenberg Bible 269
Golgotha 324
Gomorrah 135, 205, 319
Gordon, General Charles, map of the four rivers 26, 353–4, *355*, 363
Goropius Becanus (Jan van Gorp) 264, 265
Graf, Arturo 20–1, 23, 25, 30, 359–60, 366
the Great Khan 205, 208, 212, 230, 233
Greek Apocalypse of Ezra 56
Greek geographers, and medieval *mappae mundi* 86–7, 93–4
Greenland 359
Gregory the Great 50, 69, 82, 105
Gregory of Nyssa 43
Grosseteste, Robert 24, 175–6, 187
Guido of Pisa, map in *Liber historiarum* 86, *140*, 140, 156

Hardouin, Jean 323, 333–4, 341, 343
Harley, Brian 28
Havilah, biblical land 140, 180, 205, 263, 272, 275, 291, 293, 298–9, 303, 343
　　and Armenia 319, 321, 322
　　and Bochart's map 307, 309
　　and Curtis's map 354
　　in Rohl's *Legend: The Genesis of Civilization* 14
　　and Spedicato's map 357
　　in the Vatican (Pseudo-Isidorean) map 102
Heaven, in Dante's *Divina Commedia* 182, 183
Hebrew language 330
Hebron, in the Vatican (Pseudo-Isidorean) map 100, 101
Heimberg, Gregory 197
Helena, St 55
Hell, in Dante's *Divina Commedia* 182, 183
Hennig, Richard 356
Heraclius, Emperor 103
Herbinius, Iohannes 322–3, *323*, 340
　　eastorientated map *332*, 333
Herder, Johann Gottfried 359, 364
Hereford map **7**, 29–30, 128, 131, *145*, 146, 146–9, *147*, 152, *153*, 155, 157, 163, 164, 294
Hermann, Albert 354
Herodotus 94, 297, 309
Hese, Iohannes Witte de 241
Hesiod 164, 185
Heude, William 370, 373
Higden, Ranulf, *Polychronicon* **4**, 86, *134*, 134, *135*, 135–6, *136*, 171
Hippolytus of Rome 35, 65
history
　　Christian historians and the beginning of 64–9
　　and Hugh of St Victor's mental map 125–8
　　paradise as the beginning of 125
　　and temporal geography 128–31
Hobbes, Thomas 345
Holbein, Hans 268
Holle, Lienhart 255
Holy Land
　　medieval maps 120
　　paradise argument 322–34, 343, 361–2
Homer 185
Hondius, Iodocus 25, 301, 337

Mercator–Hondius *Atlas minor 299*, 299–300
Honoré, Sebastien, French edition of the Geneva Bible *292*, 292–3
Honorius Augustodunensis 67, 86, 91, 138, 141, 144, 150, 155, 158
　　Imago mundi 162, 163
Hopkinsonus, Iohannes (John Hopkinson), *Synopsis paradisi* 295, 296, 297–9, 309, 336–7
Horn of Africa 228
Huet, Pierre Daniel 25, 284, 291, 321, 338, 343, 366, 372
　　Map of the Situation of the Terrestrial Paradise 309–13, *310–11*
Hugh of St Victor 70, 125–8, 130, 141, 154–5, 166, 205, 280
humanist search for paradise 195–8
Hydapsis, river 317–18
Hyginus, monk 52

Ignatius of Loyola 288
India 12, 249
　　and the African paradise concept 218–19
　　in the Evesham map 136
　　and fifteenth-century *mappae mundi* 202, 205, 208, 209, 214
　　in the Higden map 136
　　in Honorius Augustodunensis's map 162
　　and the location of paradise 172, 179
　　in the Vatican (Pseudo-Isidorean) map 102
　　in the Vercelli map 133
Indian Ocean 215, 216, 224, 228, 230, 238, 257, 354, 357, 360
Ingram, Elizabeth M. 25
Iran, Eden in northwest 370
Iraq, site of Adam's tree in *367*, 370–1
Irenaeus, Bishop of Lyons 66
Isaac 103, 113
Isaiah, prophet 124, 152
Isidore of Seville 47–8, 69, 88, 95, 97, 170, 176, 195, 196, 319
　　and the African paradise concept 226
　　and the Albi map 138
　　and Beatus of Liébana's map 105, 107, 111
　　and Dante 183
　　De natura rerum 86, 163–4, 184
　　Etymologiae 58–9, 81–2, 90, 91, *92*, 103, 104, 121, 123, 185, 187
　　and fifteenth-century *mappae mundi* 205
　　and the Hereford map 145
Islam, and fifteenth-century *mappae mundi* 201
Iulius Honorius, *Cosmographia* 138, 156
Iunius Bassus 73
Iunius, Franciscus (François du Jon), map of paradise 295, 296–7, *297*, 298, 307, 309, 319

James the Great 79
James the Less 79
Japan 222
Japhet, son of Noah 79, 89, 95, 131, 207, 258, 288
Jason and the Golden Fleece legend 322, 339–40
Java 222
Jean, Sire de Joinville 52
Jehovah's Witnesses 352
　　publication *353*
Jericho
　　in the Vatican (Pseudo-Isidorean) map 100, 101, 105, 106
　　in the Vercelli map 135
Jerome 33, 35, 47, 65, 70, 82, 88, 99, 100, 103, 118, 119, 205, 274
Jerome maps of Asia and Palestine 140, *141*
Jeronimo de Girava 18
Jerusalem
　　in the Anglo-Saxon (Cotton) map 140
　　in Beatus of Liébana's map 105–7, 109, 116
　　on Christianized maps 89, 91, 93, 95
　　Christians and the Heavenly Jerusalem 12
　　in Dante's *Divina Commedia* 182, 183
　　in the Ebstorf map 151, 151–2, 159
　　in the Estense map 226
　　in the Evesham map 137
　　in fifteenth-century *mappae mundi* 203, 212
　　Garden of Eden in 352
　　in the Hereford map 145, 146, 147–8, 149

in the Hidgen map 136
in Honorius Augustodunensis's map 162
and Hugh of St Victor's mental map 127
in the Lambeth Palace (Reading) map 131
in Leardo's map 16
in Fra Mauro's map 237–8
and paradise in time 63–4
in Ptolemaic maps 257
Roman 88
and the sixth age 130
in the Vatican (Pseudo-Isidorean) map 100
in the Vercelli map 135
in Walsperger's map 233
Jesus Christ 55, 57
　　and the Apocalypse 105
　　Baptism of 76, 207
　　and Beatus of Liébana's map 109, 114, 115–16
　　Crucifixion 57, 63, 67, 68, 69, 75, 147, 149, 151
　　in Dante's *Divina Commedia* 183
　　in the Ebstorf map 149–52
　　in fifteenth-century *mappae mundi* 202, 207
　　genealogical tree of 76, 77–80, 78–9
　　in the Hereford map 147–9
　　and the Holy Land paradise argument 322, 323, 324, 330–1
　　in the Lambeth Palace (Reading) map 131
　　Nativity 207
　　and paradise on Christianized maps 93
　　and paradise in time 62, 63–4, 65, 66, 67, 68, 70, 73, 74
　　and the Psalter (London) map 148–9
　　Resurrection 135, 147, 149, 151–2, 242
　　in the *Speculum virginum* 74, 76
　　in the Stavelot Bible 70, 71
　　and the Vatican (Pseudo-Isidorean) map 97, 98, 99 101–2
　　see also Second Coming of Christ
John the Apostle 196–7
　　see also John the Evangelist
John the Baptist 98
John, Bishop of Jerusalem 40
John Cassian 70, 82
John of Damascus 164, 185, 195, 263
John the Evangelist 79, 272
　　see also John the Apostle
John of Marignolli *see* Marignolli, John of
John of Monte Corvino 202
John Paul II, Pope 362
John of Plano Carpini 202, 241
John of Seville 171
John of Udine, map by 77, 212, *213*, 213, 214
Jomard, Edmé-François 20
Jonah, prophet 98, 285
Jordan, river 101, 114, 115, 334, 354
Jordan of Sévérac 219
Josephus, Flavius 13, 35, 77
Judaea, and fifteenth-century *mappae mundi* 202, 205, 207
Judaism 12
　　and Christianized maps 89
Judas Iscariot 53, 79
Jude, apostle 79
Julius Caesar 86

Kabakov, Ilya and Émilia, 'Paradise under the Ceiling' 370, *371*, 371–2
Kappler, Claude 25
AlKhwarizmi (Khorazmian Tables) 171, 232
Kimble, George *Geography in the Middle Ages* 24
Kircher, Athanasius 325, 338–9
　　Arca Noe 314, *315*
　　Mundus subterraneus 313–17
Klein, Beat *see* Kühne, Hendrikje and Klein, Beat
Klosterneuburg coordinate tables *see* Vienna–Klosterneuburg coordinate tables
Kronos, Age of 164
Kühne, Hendrikje and Klein, Beat, *Map of Paradise* 11, *12*, *13*

La Sale, Antoine de, *La Salade 217*, 217–18, 248
Lambert of St Omer 86, 99
　　Liber floridus **5b**, *144*, 146, 157
Lambeth Palace (Reading) map *131*, 131–3
Lampedusa, Tomasi di, *The Leopard* 285

Index　395

Lapeyre, Jacques d'Auzoles 25, 340, 366
  La Sainte Geographie 322, 323, *324*, 324–32, *326*, *327*, *328*, *329*, *330*, *331*
Lapide, Cornelius a 303, 337, 366, 367
Last Judgement 80, 141, 152
  in Beatus of Liébana's map 105, 106, 107
  in the Evesham map 136
  and Hugh of St Victor's mental map 126
Latini, Brunetto 183
latitude
  and paradise in Mesopotamia 294–5
  and Ptolemaic maps 255, 257, 285
  tables of 230–40, *231*, *232*, 255
Layard, Austen Henry 347–8
Le Goff, Jacques, *L'Imaginaire médiéval* 25
Leardo, Giovanni, *Mapa mondi/Figura mondi* (or *mundi*) **2a**, *16*, 16–18, *17*, 19, 27, 208–9, *209*, 219, 228
Lebanon Mountains 325, 326, 330, 331, 333
Lelewel, Joachim 20
Leo XIII, Pope 362
Les Grandes Chroniques de Saint Denis du temps de Charles V 163, *164*
Liber secretorum fidelium 199, 200
Life of Adam and Eve 55–6
Linné, Carl von 359, 364
Living Theatre group 15
Livingstone, David 358–9, 363
London (Psalter) map *see* Psalter (London) map
Longiano, Fausto da 280, 317
longitude
  of Eden in Dante's *Divina Commedia* 183
  and paradise in Mesopotamia 294–5
  and Ptolemaic maps 255, 257, 285
  tables of 230–40, *231*, *232*, 255
Louis IX, king of France 180
Lucan 86, 319
Ludovicus Fidelis 264, 265
Lufft, Hans 268, 269
Luneau de Boisjermain, Pierre, *Atlas historique* 343, *346*, 362
Luther, Martin 266–70, 271, 281–2, 313, 317, 342, 365, 369, 372
Lutheran maps of paradise 25

Macarius of Rome, Saint 52, 86, 207
Macedonia 126, 128, 138
Macrobius 86, 90, 91, 111, 173, 174, 186–7
  commentary on Cicero's *Somnium Scipionis* 224, 224–5
Madeira 360
Maggiolo, Vesconte, map of Europe, Africa and part of Asia **12a**, 223, 250
Malina, Judith 15
Mandeville, Sir John *54*, 241
Manichaeans 44–5, 266
mankind, origins of 360–2
Mansel, Jean, *La Fleur des histoires* 202–3, *203*
mappae mundi, medieval world maps 84–116, 366
  and the African paradise concept 224
  Christianized 88–94
  Ebstorf map **9**, 63, 128, 149–52, *150*, *151*, *152*, 163
  Evesham map **5**, 136–8, *137*
  Higden's *Polychronicon* **4**, 86, *134*, 134, *135*, 135–6, *136*, *171*
  the invisible paradise on 138–40
  Lambeth Palace (Reading) map *131*, 131–3
  map in *Les Grandes Chroniques de Saint Denis du temps de Charles V* 163, *164*
  mapping God's creation and redemption 141–52
  and modern cartography 28–9
  nineteenth-century views on 19–23, 95
  origins of 84–7
  Psalter (London) map **8**, *148*, 149, 149, 163, *164*
  and Ptolemaic maps 255–7
  and Renaissance maps 277
  and sacred geography 288, 290
  Sawley map 141–4, *142*, *143*, 163
  space–time structure of 125–8, 130, 146–7, 152, 255
  and temporal geography 128–31
  twentieth-century views on 23–7, 95

Vatican (Pseudo-Isidorean) map 95–8
Vercelli map *132*, 132, *133*, 133–5
and Virga's map 219
William of Tripoli's map *165*, 165
winds in 163–4
and zonal maps 165, 191
  see also Beatus of Liébana; Hereford map; T-O maps
Marco Polo 18, 202, 208, 222, 230, 235, 246, 248
Margalho, Pedro 240
Marianus Scotus 65
Marignolli, John of 241
Martianus Capella 173
Mary of Cleophas 79
Mary Magdalene 330–1
Mary, the Virgin 63, 66, 73, 76, 98, 149
Masius, Andreas 367, 373
Massey, Gerald 364
Massimi, Jean Robert 25
mathematics of paradise 176–9
Matthew, apostle 79
Matthias, apostle 79
Mauritius 10, 11, 260
Mauro, Fra, map of **12b**, 235–40, *236*, 251
Maury, Alfred 20, 30
Mediterranean Sea
  and the African paradise concept 219
  and the east–west progression of history 127, 130
  in the Estense map 227
  in the Evesham map 137
  in fifteenth-century *mappae mundi* 199, 200, 205, 207, 208–9, 212, 215, 216, 217
  and the Flood 268
  in the Hereford map 146
  in the Jerome map of Palestine 141
  in Walsperger's map 233
Melanchthon, Philipp 269
Melchisedek, priestking 99
Melville, Herman 11
Mercator, Gerard 25, 337
Mercator–Hondius *Atlas minor* 299–300
Mesopotamia 202, 360
  and Assyriology 347, 348, 351
  as the location of paradise 243, 262–4, 270–7, 291–317, 322, 343, 363
Methuselah 103
Mexico 286
Michaeler, Carl Josef 352
Miller, Konrad 20, 29–30
Milton, John 24
  Paradise Lost 346
modern cartographical technology 27–8
Moluccan meridian, as the location of paradise 286–8
monasteries, and book production 199
Mongolian Empire 201–2, 212, 234
the Moon
  and the location of paradise 174, 181, *182*, 194, 196, 203, 367
  Mountains of the Moon 218, 223, 366
Morinus, Stephanus (Étienne) 307, 338
Moses 33, 39, 101, 172, 273, 274, 303, 319, 331
  and modern biblical scholars 345
Mossey, Gerald 359
Moullart Sanson, Pierre, *Carte du paradis terrestre selon Moyse* 320
mountain of paradise 172, 174–5, 176, 191–2, 203, 223
Mountains of the Moon *see* Moon
Moxon, Joseph 337
  maps in *Sacred Geography* 303–6, *304*, 306
Münster, Sebastian 365
  edition of Ptolemy's *Geography* 276, *277*, 285, 288
mythologists *see* nineteenth-century scholars

nautical (or navigation) charts 85, 191, 198, 199–200, 208–9, 215, *216*, 217, 230, 235
  and Virga's map 219
Navigatio Sancti Brendani 52–3
Nazianzen, Gregory 317
Nebuchadnezzar, king of Babylon 65, 126, 130, 313

Neo Platonism 260–1
Newton, Isaac 129
Niccolò de Conti 230, 235
Nicholas of Lyre 70, 192–3, 205, 237, 243
Nicholson, James, *The Garden of Eden* 356, *358*, 363
Nile, river 13, 36, 40, 47, 49, 52, 92, 121, 126, 136, 140, 248, 356, 359, 366
  and the African paradise concept 218, 223
  in the Albi map 138, *139*
  and fifteenth-century *mappae mundi* 205, 207, 214, 215–16, 217
  in Fra Mauro's map 239
  and the location of paradise 179–80, *181*, 195, 196, 197, 198
nineteenth-century scholars, and the origin of mankind 367–9
  Assyriologists 347–51, 352, 361
  and medieval maps 19–23, 95
  mythologists, and the origin of mankind 359, 361
Nineveh 226
Noah 68, 77, 78, 86, 89, 95, 103, 179
  and Eden 362
  three sons of 78–9, 89, 95, 125, 131, 133, 207, 258
  and Visscher's map 305
Noah's Ark 80, 126, 127, 135, 138, *139*, 262
  in Armenia 285, 286, 321
  in Bianco's map 207
  in Ephrem the Syrian's *Hymns on Paradise* 162
  in the Higden map 136
  Kircher's work on 314
  in Ptolemy's *Geography* 285
Nod, biblical land 13, 262, 305
Nogarola, Ludovico 264
North Pole paradise 285–6, 352–3
north-orientated world maps 217–18
northern hemisphere
  and the location of paradise 176, 196, 217–18, 234
  topological maps of the 166

Odoric of Pordenone 202, 241
Old Man of the Mountain 231, 238
Olomouc map 214–15, 248
Olschki, Leonardo 24, 25
Olympus, Mount 174
Oporin, Jean 335
Origen 33, 37, 38–9, 40, 42–3, 45, 58, 273, 352, 365
Orosius 48, 49, 86, 88, 97, 121, 126, 183, 195, 205, 286
  Historia adversus paganos 138
Ortelius, Abraham 41, 327
  map of the travels of Saint Paul *290*, 290–1
  and paradise in Armenia **16**, 317
Ottoman Empire 202
Ovid 164
  Metamorphoses 91

Pakistan, Garden of Eden in 357, *358*
Palestine
  and fifteenth-century *mappae mundi* 205
  Jerome map of 140, 141
  location of paradise in 322, 333
Paoline of Venice 199, 201
paradise *see* earthly paradise and Garden of Eden
Paris, in the Evesham map 137
Paris, Matthew 85
Patch, Howard, *The Other World* 24
Paul the Hermit 53
Paul, Saint 56, 63, 76, 78, 79, 140, 150, 152, 285, 289
  and Beatus of Liébana 109
  Ortelius's map of the travels of *290*, 290–1
Pecham, John 193
Persepolis, Persian city 128
Persia 138, 202, 205, 238
Persian Gulf 226, 268, 275, 321, 334, 343
Peter Lombard 47, 49, 50–1, 59–60, 76, 170, 180, 192, 193–4, 239, 244, 252, 279
Peter of Poitiers, *Compendium historiae in geneologia Christi* 76–80, *77*, 78–9, 205, 212, 214

Peter, Saint 76, 79
Petrus Alphonsus 173
Philip, apostle 79
Philip II, king of France 135
Philo 37–8, 39, 41–2, 45, 58, 266, 280
Philostorgius 173–4
Piccolomini, Aeneas Sylvius (later Pope Pius II) 195–8, 244–5, 262, 280, 317
Pico della Mirandola, Giovanni 260–1, 279–80
Pillars of Hercules 183, 205, 207, 226
Pineda, Juan de 264, 265
Pishon, biblical river 13, *14*, 46, 49, 102, 133, 138, 139, 140, 263, 275
    and Armenian rivers 317, 318, 322
    and Assyriology 348, 349, 351
    and Curtis's map 354
    and fifteenth-century *mappae mundi* 207, 214
    and the Holy Land paradise argument 327, 333
    and Huet's map 311, 312
    and the location of paradise 198
    and Mesopotamia 297–8, 307, 309
    and paradise in the Bible 33, 35, 37, 40–1
    in Spedicato's map 357
    and Van Til's map 313
Plancius, Peter 25
    Tabula geographica 294, 294–5
Plato 164, 185, 261
Plaut, Fred, 'General Gordon's Map of Paradise' 26, 363
Pliny the Elder 48, 49, 88, 102, 121, 131, 178, 197, 235, 297, 318, 319, 321, 333, 340
Poivre, Pierre 11
Polemius Sylvius 138
Polybius 94, 172–3, 186
Pompey 86
Pomponius Mela 235
Pontius Pilate 102
Portuguese voyages of discovery 238, 240
Postel, Guillaume 285–8, 335
    diagrams of the meridian of paradise 286, *287*
Pressel, Wilhelm, and paradise in Mesopotamia 362
Prester John 25, 199, 205, 207, 208, 209, 212, 246
    and the African paradise concept 219
    in the Genoese map 230
    and geographical discoveries 241
    and Fra Mauro's map 238
    and Walsperger's map 233
Priscian 139
Protestants
    on the Garden of Eden 262
    and modern biblical scholarship 345–6
    and sacred geography 288
Proust, Marcel 44, 372
Psalter (London) map **8**, *148*, 148, *149*, 149, 163, 164
Pseudo-Isidorean Vatican map *see* Vatican (Pseudo-Isidorean) map
psychoanalysis, fantasy maps in 26
Ptolemaic maps 130, 255–8, 285–6, 294–5
    and sacred geography 288–91
Ptolemy (Claudius Ptolemaeus) 178, 179, 180, 181, 194, 196, 197, 218, 278
    Almagest 171, 177–8, 231
    and Armenia 319
    Cosmographia 256
    and the equatorial zone 175–6, *177*
    Geography 15, 18, 86–7, 186, 199, 205, 215, 231, 233, 235, 236, 238, 254–5, 256, 257, 275–6, *276, 277*, 285, 290
    and the geography of Eden 297
    and Huet's map 311, 312–13
    and Fra Mauro's map 235, 236, 237–8
    and Servetus 323
Purchas, Samuel 265–6
Purgatory, in Dante's *Divina Commedia* 182, 183

quadripartite division of the globe 225–6

Raleigh, Walter 265–6, 291, 319, 366
    History of the World 278, 280, *300*, 300–3, *301*, 336, 337, 372

Ramsey, Michael, Archbishop of Canterbury 345, 362
Ravenna Cosmography 89, 241–2
Red Sea 129, 136
    and the African paradise concept 219, 224
    in the Albi map 139
    in the Anglo Saxon (Cotton) map 140
    in the Estense map 228
    in the Evesham map 137
    in fifteenth-century *mappae mundi* 203, 208
Regnault, Antoine, map of the location of Eden 293, 293–4
Reisch, Gregor, *Margarita philosophica* 222, 223, 224, *225*, 225–6, 242
Renaissance maps 342
    Garden of Eden in 22, 25, 27, 257–8
    and medieval *mappae mundi* 277
    and modern cartography 28–9
    and paradise in Armenia 317–18
    and paradise in the Middle East 270–7
    Ptolemaic maps 130, 255–8
    see also fifteenth-century *mappae mundi*
Renaissance writers, and the location of paradise 365–7, 369
Rhau, Balthasar 346
Rhine, river 305
Ringbom, Lars 356
Robertus Anglicus 176
Rohl, David 356, 369–70
    Legend: the Genesis of Civilisation 12–13, *14*
Roman Empire 84–5, 88, 89, 126, 128
    in the Albi map 138
    and fifteenth-century *mappae mundi* 207
    and the Vatican (Pseudo-Isidorean) map 103, 104
Rome
    on Christianized maps 91, 93
    in the Evesham map 137
    in fifteenth-century *mappae mundi* 203
    in the Higden map 136
    mosaics featuring paradise 73, *75*, 75
Romulus 196–7
Rudimentum noviciorum (*Elementary Book for Beginners*) **11a**, 77–8, *204*, 204–7, 247
Rüst, Hanns, woodcut map of the world **11b**, *206*, 206, 207, 247

sacred geography 288–91
Sacrobosco, Iohannes de, *Tractatus de sphera* 166, 166, 175, 187
St Helena 10, 11, 260
Salibi, Kamal 356
Sallust, Roman historian 86, 91, 94, 319
Salome 79
Santarém, Manuel Francisco de Barros e Sousa, 2nd Viscount of 20, 29, 161, 164
Sanudo, Marino 199, 200, 201, 245–6, 285
Sawley map 141–4, *142*, *143*, 157, 163
Sayce, Archibald Henry 348, 350
Schedel, Hartman, *Liber chronicarum 258*, 258
Schopenhauer, Arthur 11–12
science, and medieval maps 94–5
Scot, Michael 176, 187
Second Coming of Christ 57, 63, 65, 68, 76, 104, 106, 110, 130
    and Beatus of Liébana's map 116
    and the Ebstorf map 151
    and the Hereford map 145, 146, 148–9
Sem (Ham), son of Noah 78–9, 89, 95, 131, 207, 258
the Septuagint 33, 34–5, 41
Sergius, monk 52
Servetus, Michael
    Christianismi restitutio 323, 340
    edition of Ptolemy's *Geography* 277, *277*, 323
Seth, son of Adam 40, *55*, 57, 61, 241
Severian, Bishop of Gabala 35, 40
Sextus Iulius Africanus 64, 65, 103
the Seychelles 26, 353
Shakespeare, William 15
Sheba, Queen of 55, 226, 238
Shinar, Land of 305
Shuttle Radar Topography Mission (SRTM) 18, 19
Simon, apostle 79

Simon, Richard 345
Simon of St Quentin 202
Sisebut, Visigoth king 103
sixteenth-century maps 284–341
    and paradise in Armenia 317–18
    see also Ptolemaic maps
Smith, William, *Dictionary of the Bible* 362
Sodom 135, 205, 319
Solinus 48, 131, 197, 235, 244
Solomon, King 55
south-orientated world maps 211–14
southern hemisphere, and the location of paradise 172, 173, 176, 177, 178, 179, 180, 181, 183, 192, 193, 224–5
space–time structure, of medieval *mappae mundi* 125–8, 130, 146–7, 152, 255
Speculum virginum 74, 75–6
Spedicato, Emilio 357–8, 363
Spinoza, Benedict 345
Stavelot Bible 70, *71*
Sterckx, Sébastien 24
Steuchus, Augustinus 262–4, 272, 273, 274–5, 277, 280, 291, 317, 319, 365, 372
Strabo 49–50, 51, 94, 195, 197, 283, 290, 297, 309, 318, 319, 336
Sumerians 12, 349

T-O maps 89, 89–91, *90*, *91*, *92*, *93*, 95, 97, 116, 118, 131, 164, 204, 205
    and the African paradise concept 219
    hybrid T-O/zonal maps 165–6, 168, 169–70
Tacitus 318
Tahiti 260
Taittiriya Upanishad 84
Tardiola, Giuseppe 25–6, 121
Teixeira, Pedro 311
temperate zones, mapping 166, 168
temporal and spatial dimensions, of medieval *mappae mundi* 125–8, 130, 146–7, 152
Terry, Albert R., map of the Garden of Eden and the four rivers 256, *359*, 363
Tertullian 43, 123
Thaddaeus, apostle 79
Thayer, Lawrence A. et al., *Map to Heaven and the City of Life* 368, *369*, 371
Theodore of Mopsuestia 40
Theodosius 100
Theophilus, monk 52
Thevenot, Jean de 311
Thomas the Apostle (Saint Thomas) 79, 208, 209, 212, 230
Thomas Aquinas 32, 49, 70, 181–2, 188, 189, 194, 218, 259–60, 279
Thomas, Saint *see* Thomas the Apostle
Thompson, Frank Charles, Chain Reference Bible 347
Tiberius, Emperor 102
Tigris, river 13, *14*, 46, 47, 49, 102, 126, 136, 139, 150, 180, 197, 198
    and Armenian rivers 317, 319, 321, 322
    and Assyriology 348, 349, 350
    and Bedford's *Scripture Chronology* 306
    and the biblical story of paradise 36, 40, 41
    confluence with the Euphrates 367, 370–1
    and fifteenth-century *mappae mundi* 214
    and the Holy Land paradise argument 327, 334
    and Huet's map 309, 312
    and Mesopotamia 263, 264, 272, 275, 291, 292, 293, 295, 298–9, 307
    and Milton's *Paradise Lost* 346
    and the Nile 179–80
    and Raleigh's *History of the World* 301–3
    and Van Til's map 313
Toledo Tables 173, 231, 232
tourism, and paradise 11, 12, 13
Tree of the Knowledge of Good and Evil 147, 150, 162, 240, 292, 298, 316
    and Assyriology 350
    and General Gordon's map 353
    and the Holy Land paradise argument 324, 330
    report and photograph in *The Times* (1944) 367, 370–1

Tree of Life 69, 75, 125, 133, 136, 150, 162, 228, 260, 262, 265, 316
 and Assyriology 347, 348, 350
 and the Holy Land paradise argument 330
Trees of the Sun and the Moon 134, 140, 153, 203, 205, 207, 233
Tropic of Cancer 177, 178
Tropic of Capricorn 177, 178
tropical regions, and the location of paradise 172–3, 176
Troy 129
 on Christianized maps 91, 94
twentieth-century scholars 356–8, 361, 367
 and medieval maps 23–7, 95
Tyconius 105–6
Tyre 327, 329, *331*

Ulysses, in Dante's *Divina Commedia* 183, 191
Ur, city of 128, 286, 301, 305

Vadianus (Joachim van Watt) 264
Valencia, Jacob Perez de 218
Van Til, Salomon, *Tabula situm paradisi 312*, 313, 338
Vatican (Pseudo-Isidorean) map 95–104, *100, 101*, 116, 137, 185
Venice
 in fifteenth-century *mappae mundi* 207
 mosaic in San Marco church 75

Vercelli map *132*, 132, *133*, 133–5, 156
Vesconte, Pietro, map of 199–200, *200, 201*, 201, 208, 245, 285, 335
Vespucci, Amerigo 242
Vetus Latina 47, 49
Victorinus, Bishop of Pettau 66
Vienna–Klosterneuburg coordinate tables 231–3, 234
Vincent of Beauvais 202, 205
Virga, Albertin de, map of 219–22, *220*
Visscher, Claes Janzonius 303, 305–6
Vitry, Jacques de 205
Vopel, Gasparus 298, 336
Voragine, Jacobus de 79

Wagner, Richard 44
Waldseemüller, Martin 288, 289–90
Walker, Reginald Arthur 14
Walsperger, Andreas, map of (1448) **13**, 233–4, *234*
Walter of Châtillon, *Alexandreis* 51
Ward, Chris, map of Eden *360*
Warren, William Fairfield 352–3, 361
Weidlé, Wladimir 73
Wellhausen, Julius 345
Wendrin, Franz von, map of paradise 356, *358*
Westrem, Scott D. 119, 155–6
whole-earth paradise doctrine 264–6, 267, 277
Wieder-Woldan map *215*, 215–17, 248

Wilde, Oscar, *The Soul of Man under Socialism* 16
Wilhelm II, German emperor 348, 362
Willcocks, William 362–3
 The Garden of Eden and the Four Rivers of Genesis 348–50, *350*
William of Auvergne 171–2, 186
William of Conches 23–4, 173, 187
William of Rubruck 201–2, 241, 246
William of Tripoli's map *165*, 165
William of Tyre 180
Williams, John 109, 111
winds, in medieval maps 163–4
Wittenberg Bible **15**, 268, *269*, 269
Witter, Henning Bernhard 345
Wright, John Kirkland, *The Geographical Lore at the Time of the Crusades* 23–4, 188
Wright, Paul 361
 New Map of the Garden and Land of Eden 343, *344*
Wright, William A., 'Eden', 362, 365

Xenophon 35

Zebedee 79
zodiac signs, in Walsperger's map 233
zonal maps 165–70, *166, 167, 168, 169*, 185, 191, 199, *224*, 224–6
Zwingli, Ulrich 78

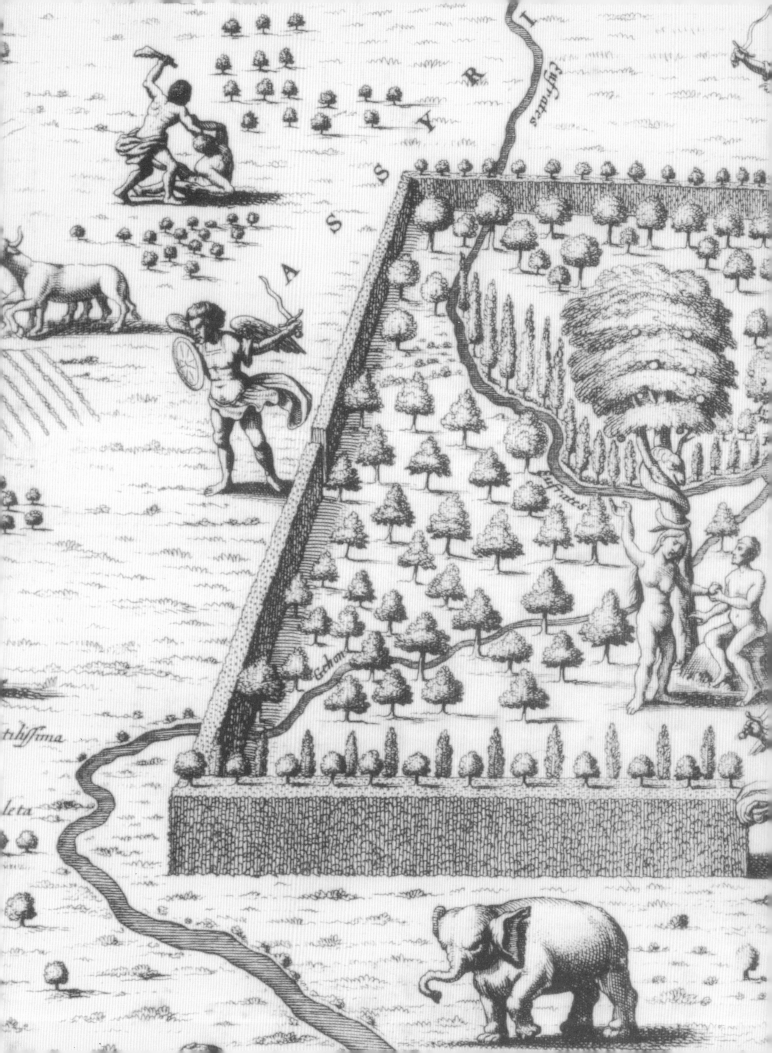